MILLETS

Ensuring Climate Resilience and Nutritional Security

The Editors

Dr. Vilas A. Tonapi has joined Indian Institute of Millets Research (IIMR) as Scientist in 1989. He was the first Officer Incharge of Centre on Rabi Sorghum, Solapur. From July 2011 to July 2013 he also headed Division of Seed Science & Technology, IARI, New Delhi. He has visited United States of America, Mexico, Malaysia, Singapore and Australia. He did his post-doctoral research at the University of Queensland, Australia and CRC for Tropical Plant Pathology, Brisbane, Australia. His specialization includes seed industry and seed systems development, IPRs, Plant Variety Protection and seed vigour and viability. He has received several awards, and some of these are Seedsmen Association medal and Award (2009) for service to seed industry, Consultancy to FAO (2008), Best Scientist award of Directorate of Sorghum Research (2006 & 2007), J.N.TATA Endowment award (2000-02), R.V. Swaminathan Gold Medal for highest OGPA & Best Thesis Award in Seed Technology at TNAU, Coimbatore (1988), ICAR JRF and SRF, National Merit Scholarship, Government of India, District Merit Scholarship, Government of Karnataka. He has 50 refereed research papers, 13 book chapters, 20 conference papers, 65 abstracts and 10 popular articles to his credit. He has authored 20 books, 10 technical bulletins, 10 folders, 10 posters, 23 reports, and edited 5 books.

Dr. J.V. Patil is the Director of Indian Institute of Millets Research (IIMR) Hyderabad. During his 28 years research career, he has made significant contributions towards crop improvement through the development of several varieties in sorghum (10), chickpea (5), safflower (1), mungbean (1), rajmabean (1) and pigeonpea (1). In sorghum, he introduced the concept of breeding for specific soil depths for rabi season. Concerted efforts towards this has led to the development of *Phule Vasudha* and *Phule Revathi* for heavy soils, *Phule Chitra* for medium soil and *Phule Anuradha* for shallow soils. He also conceptualized and developed several innovative technologies for sustainable production of *rabi* sorghum especially *in situ* moisture conservation strategies. He has also involved in development of sorghum varieties for specific end uses. He is elected fellow of National Academy of Agricultural Sciences -2012. He is a recipient of several international awards including CGIAR's Baudouin Team Award (2002) and Doreen Mashler Team Award (2002). For his contribution towards developing high yielding varieties in various crops he has received two state awards Baliraja's Late Annasaheb Shinde Smruti Krishi Sanshodhan Gauro Puraskar and Vasantrao Naik Krishi Puraskar awards. He is also recipient of Agricultural Literature award 2010 and 2012 from Govt. of Maharashtra.

MILLETS

Ensuring Climate Resilience and Nutritional Security

Editors

Vilas A. Tonapi

J.V. Patil

Indian Institute of Millets Research
Rajendranagar, Hyderabad – 500 030
Telangana, India

2015

Daya Publishing House®

A Division of

Astral International Pvt. Ltd.

New Delhi – 110 002

ISBN 978-93-5124-644-2 (Hardbound)
ISBN 978-93-5130-671-9 (International Edition)

Published by : **Daya Publishing House®**
A Division of
Astral International Pvt. Ltd.
– ISO 9001:2008 Certified Company –
4760-61/23, Ansari Road, Darya Ganj
New Delhi-110 002
Phone: 011-4354 9197, 2327 8134
E-mail: info@astralint.com
Website: www.astralint.com

Laser Typesetting : **Classic Computer Services**
Delhi - 110 035

Digitally Printed at : **Replika Press Pvt. Ltd.**

PRINTED IN INDIA

Dr. S. Ayyappan
Secretary (DARE) & Director General
Government of India,
Department of Agricultural Research & Education and
Indian Council of Agricultural Research
Ministry of Agriculture, Krishi Bhawan,
New Delhi – 110 011

Tel.(O): +91- 11-233822629/
23386711
Fax: +91-11-23384773
E-mail: dg.icar@nic.in

Foreword

सत्यमेव जयते

Sorghum, Pearl Millet and Small millets are important staple foods in the semi-arid tropics of Asia and Africa. Millets are highly nutritious than fine cereals, and are the principal sources of energy, protein, vitamins and minerals for millions of the poorest people in these regions. They grow well in harsh environments where other crops yield poorly. The area of millet cultivation in the country has declined very significantly over the past few decades. This could be primarily due to very low productivity, and susceptibility to biotic and abiotic factors. In this scenario, breaking yield plateau in millets is a major challenge to the millet breeders and scientists.

Promotion of value added products from millets in international markets could fetch higher foreign exchange. Indeed sorghum and millets are crucial to the world food economy because they contribute to household food security in many of the world's poorest, most food-insecure regions. In the main production regions in Africa and Asia, more than 70% of the sorghum crop and over 95% of the millet crop are consumed as food. A large proportion of farm households aim simply to produce enough grain to meet household requirements - and many often fail to meet even this limited goal. Only a small proportion of the harvest is traded, mostly on local food markets. In some countries like India, sorghum contributes to 1% of the total agricultural GDP while it goes up to 7% in Maharashtra and 5% in Karnataka the two major sorghum growing provinces in India. Hence it is required to overcome the issue of profitability of sorghum and millets production by research and extension on value addition, marketing research and liaison with the user industries. Simultaneously, seriously we need to pursue other avenues such as growing sorghum as a biofuel crop for which we have done significant work, in addition to promotion of millets as functional health foods.

Thus productivity improvement may be within the realms of reality in the near future. Different genotypes suited to different growing conditions may be essential to bring in all-round increase in productivity. The economic gains that may be augmented by addressing envisaged benchmarks will result in significant improvement in productivity, profitability and even export earnings. All these are expected to translate sorghum and millets farming into a healthy and prosperous proposition, justifying the public support for sorghum and millets research in the country. I am sure that this book on **"MILLETS: Ensuring Climate Resilience and Nutritional Security"** will be a valuable reference book for all millet workers, policy makers and industry. I congratulate all the authors, and the editors Drs. Vilas A Tonapi and JV Patil for bringing out this valuable publication on Millets in India.

(S. Ayyappan)

Dr. J.S. Sandhu
Deputy Director General (Crop Science)
Indian Council of Agricultural Research
Ministry of Agriculture, Krishi Bhawan,
New Delhi – 110 011

Tel.(O): +91- 11-23382545
Fax: +91-11-23097003
E-mail: ddgcs.icar@nic.in

Foreword

Crop improvement through plant breeding is "purposeful manipulation of plant species in order to create desired genotypes, traditionally pursued to ensure food security through quality seeds thus evolved. Millets continue to be important food and feed crops in developing world. Their versatility in multi-purpose use, stress adaptation and nutritive value makes them even more important among crops in the era of extreme climate variability and high incidence of dietary induced malnutrition.

Recent advances in sorghum and millets research and development in enhancing their yields, adaptation, stress resistance, nutritional value and processed products development discussed above contributes to increased economic value of these crops to the producers. The biofortified pearl millet and sorghum, sweet sorghum for bioenergy are some of the examples showcasing the potential of these crops in providing nutritional and energy security in developing world. Sorghum genome sequence is available and put to use for improving various traits. A consortium is sequencing the pearl millet genome and the draft genome sequence is expected soon. Efforts are underway to sequence the finger millet genome also. The challenge will be to make use of the genome information and developing customized research products and technologies to suit various climatic, foods, nutritional and product quality requirements

The progress of plant breeding in India has been phenomenal during the last five decades. The successful adoption of improved varieties along with improved cultivation practices helped to rapidly increase food grain production, from 50 million tonnes to more than 268.34 million tonnes during 2013-14. However, much of this increase has come from two major cereals, rice and wheat; the contribution of other crops is comparatively lower owing to their lesser area and marginal dry environments they are grown with minimum input use. The high yielding varieties have made more impact in high external input areas that are mostly irrigated. In low external input areas (dryland), the impact of improved cultivars on crop productivity has been limited. Sorghum and millets fall in the latter category. Genetic improvement of millets has progressed over the last several decades with impressive gains in productivity and adaptability of high-yielding cultivars. However, the traditional approaches have been slow in addressing the major loss causing or yield-destabilizing traits such as susceptibility to insects and diseases and drought stress.

Besides productivity enhancement the whole value chain should be looked into to make these crops more remunerative to farmers and processors. This calls for increased interest and investment from national governments and private sector for developing thriving integrated value chains for sorghum and millets. I congratulate both Drs. Vilas A. Tonapi and JV patil for putting together the valuable source of information in the form of book entitled "Millets: Ensuring Climate Resilience and Nutritional Security".

(JS Sandhu)

Preface

Millet crops (Sorghum, Pearl Millet & Small Millets) have been dominant components of rainfed agriculture on a regional / local basis. Their adaptation to harsher environments and diverse cultural and agro-climatic situations is well known. They are critical elements in the food, livelihood and nutrition security of the poor in particular. Some of the species are known for their specific attributes to withstand drought, heat, salinity and adaptability to poor soils. Nutritionally some of them are rich sources for micronutrients, dietary fibre and some attributes of therapeutic value besides industrial uses. Therefore, they provide sources for genes involved in many biochemical pathways and processes.

Millets used to be cultivated in an area of 35 – 37 m hectares in India. Past researches have largely been in the area of improving yields. The successful development and spread of hybrids and improved varieties of sorghum (jowar), pearl millet (bajra) and finger millet (ragi) made a significant impact on the productivity of these crops during the 70s and 80s. The private sector gained considerably from hybrid seed production and marketing of sorghum and pearl millet hybrids. In fact, the largest volume of hybrid seeds sold earlier was for sorghum and pearl millet hybrids. Later the area under millet crops reduced to 20 – 22 m hectares. The reduction has been more drastic in case of *kharif* sorghum and small millets. This has been largely due to change in food habits and their limited role in market economy.

The investments in R&D on these crops have also been limited. The private sector plays no attention to finger millet and small millets. The public distribution system supplies wheat and rice at subsidized prices. These are some of the reasons for the millets being pushed to periphery as orphan crops. Even though, the area under millets diminished, in view of their multifarious uses, the potential is very high. Millets crops now stand at a critical juncture. Keeping in view the present situation, there is a need to reorient millet R&D programmes both structurally and functionally as millets as a group stand distinctly as climate smart crops which have a significant role to play in the climate change scenario and also as crops which are highly drought tolerant and can usher in food, feed, fodder, nutritional and livelihood security to all in dryland ecosystems as "MIRACLE NUTRI-CEREALS" providing nutritional and health security to all, which in fact is the need of the hour. In addition, it is apt time to critically evaluate their role in the conservation of bio-diversity, human and animal nutrition, industrial uses and therapeutic diets in the form of functional foods.

The All India Coordinated Research Projects for various millet crops were set up during mid and late sixties. In view of the changes that have taken in millet areas and their geographies, their role in biodiversity, nutrition and health, and the current emphasis on alternate uses need more emphasis, it is necessary to redefine the goals and objectives for AICRPs under each crop. It is necessary to shift from the general improvement approaches, to projects with specified goals and inbuilt partnerships. This warrants both structural and functional changes for each of the projects and the respective centres. The tools may vary – conventional, biotechnological, biochemical or industry oriented. The establishment of the Indian Institute of Millets Research (IIMR) and the Society for Millets Research at this juncture is an important step and should provide the forum for both public and private sectors to come together, plan and support the multiple approaches and partnerships in millet improvement. In this regard, it shall be our endeavour to analyze the factors that prompted the rise and subsequent decline of millet crops, and focus on the areas of future importance to direct and promote R&D efforts through public-private and industry partnerships in focus areas.

This publication "Millets: Ensuring Climate Resilience and Nutritional Security" is a synthesis of information on crop improvement, development, utilization and value addition efforts made till date while identifying the areas that need real focus to make millets a profitable preposition. It is hoped that this publication will be a valuable source of information on millets for all the researchers, policy makers and the industry. We are highly thankful for the support from authors of various topics and secretarial assistance from Dr K V Raghavendra Rao, K sanathkumar and Harshal Gawali.

Vilas A.Tonapi

J.V. Patil

Contents

Part V: Nutritional and Health Benefits

Part VI: Utilization and Commercialization

Part VII: Seed Systems

Part I

Crop Improvement

Millets: Ensuring Climate Resilience and Nutritional Security (2015) *Pages* **3–43**
Editors: **Vilas A. Tonapi and J.V. Patil**
Published by: **DAYA PUBLISHING HOUSE, NEW DELHI**

Chapter 1

Millets Improvement and Utilization: Progress and Prospects

J.V. Patil, Vilas A. Tonapi and B. Dayakar Rao

ICAR-Indian Institute of Millets Research (IIMR)
Rajendranagar, Hyderabad – 500 030

Millets (sorghum, pearl millet and small millets) are the important food and fodder crops in semi-arid regions, and are predominantly gaining more importance in a world that is increasingly becoming populous, malnourished and facing large climatic uncertainties. These crops are adapted to wide range of temperatures, moisture-regimes and input conditions supplying food and feed to millions of dryland farmers, particularly in the developing world. Besides they also form important raw material for potable alcohol and starch production in industrialized countries. Among these crops, sorghum is the world's fourth most important cereal, in terms of both production and area planted. Millet, a general category for several species of small-grained cereal crops, is the world's seventh most important cereal grain. Roughly 90 per cent of the world's sorghum area and 95 per cent of the world's millet area lie in the developing countries, mainly in Africa and Asia. These crops are primarily grown in agro-ecologies subjected to low rainfall and drought. Most such areas are unsuitable for the production of other grains unless irrigation is available. Sorghum is widely grown both for food and as a feed grain, while millets are produced almost entirely for food.

The economic importance of the millets is increasing in terms of feed value, particularly that of sorghum though it is grown in contrasting situations in different parts of the world. Millets are grown in the harshest environments where there is limited scope for growing other crops. Millet production systems in Africa and Asia are generally characterized by extensive production practices and limited adoption of improved varieties, yields still average only 0.3 to 1.0 tonnes per hectare. While hybrids are being adopted in sorghum and pearl millet, most of the world's millet area remains under traditional varieties. Few farmers apply fertilizer or use improved moisture conservation practices. Therefore the yield levels remained low for long but increasing wherever improved hybrids and management practices are increasingly adopted like in India. After almost 67 years of Independence, malnutrition continues to plague India. Even while vast segments of resource-poor people suffer from undernutrition, particularly micronutrient deficiencies (hidden hunger), there is a growing incidence of obesity and chronic diseases like diabetes, cardiovascular diseases, cancer etc. Both the ends of this grim spectrum are at least partly due to changing food habits, loss of millets from the diet being one of the million millets offer unique advantage for health being rich in micronutrients, particularly minerals and B vitamins as well as nutraceuticals. These phytochemicals have been shown to mitigate above mentioned chronic diseases. Millets can also withstand environmental stress being resistant to drought and warming. Thus, these are grains for the future and "harbingers for evergreen revolution". Currently, they sustain the livelihood of over 60 per cent of small and marginal farmers. Time trends show marked reduction in area under millet cultivation due to variety of reasons. Thanks to technological breakthroughs, productivity and production have shown some increase despite erosion in farm area.

The sorghum and millets are crucial to the world food economy because they contribute to household food security in many of the world's poorest, most food-insecure regions. In the main production regions in Africa and Asia, more than 70 per cent of the sorghum crop and over 95 per cent of the millet crop are consumed as food. A large proportion of farm households aim simply to produce enough grain to meet household requirements - and many often fail to meet even this limited goal. Only a small proportion of the harvest is traded, mostly on local food markets. In some countries like India, sorghum contributes to 1 per cent of the total agricultural GDP while it goes up to 7 per cent in Maharashtra and 5 per cent in Karnataka the two major sorghum growing provinces in India. We hope to overcome the issue of profitability of sorghum and millets production by research and extension on value addition, marketing research and liaison with the user industries. Simultaneously, we will pursue other avenues such as growing sorghum as a biofuel crop for which we have done significant work, in addition to promotion of millets as functional health foods.

With our strategic planning based on this vision document, it is anticipated that overall, atleast 30 per cent increase in millet acreage over the current levels is expected to be attained by 2050 AD. Accordingly, the estimated production in sorghum is computed to be 30.06 million tonnes, of which *Rabi* production contribution alone will be 83 per cent of total production. Similarly 24.4 million tonnes in pearl millet, 4.40 million tonnes in finger millet, 1.8 million tonnes in other small millets with overall aggregate production of 60.60 million tonnes of Sorghum, pearl millet and small millets put together by the end of 2050 AD. Area gains in favour sorghum, pearl millet and small millets may come from other crops which are less remunerative (sorghum, pearl millet and small millet demand is going to higher due to value addition and due to their importance as functional health foods) and from areas where commercial crops like sugarcane cannot be continued due to lowering of water table making it difficult for them while it still favours sorghum, pearl millet and small millets which with minimal Irrigations will help to double the yields.

Sorghum

Sorghum is one of the main staple food for the world's poorest and most food insecure people across the semi-arid tropics. Globally, sorghum is cultivated on 41 million hectares to produce 64.20 million tonnes, with productivity hovering around 1.60 tonnes per hectare. With exceptions in some regions, it is mainly produced and consumed by poor farmers. India contributes about 16 per cent of the world's sorghum production. It is the fourth most important cereal crop in the country. In India, this crop was one of the major cereal staple during 1950's and occupied an area of more than 18 million hectares but has come down to 6.61 million hectares in 2013. The decline has serious concern on the cropping systems and the food security of these dry land regions of the country.

Thus, All India total sorghum production has registered a constant growth rate of 0.10 per cent per annum during the period 1969-70 to 2012-2013 which can be mainly attributed to negative production of *Kharif* sorghum rather than positive growth in *Rabi* sorghum production. Though, *Kharif* sorghum yield growth rates were relatively higher, it could not offset the declining growth rates in production, as the growth rates in *Kharif* sorghum area were negative and high. Just opposite is true in case of *Rabi* sorghum where the area decline was not sufficient to undermine the yield growth, thus resulting in positive production growth rates.

The overall increase in productivity of *Kharif* is far more than *Rabi* sorghum. However, the loss in both area and production is greater in *Kharif* sorghum than in *Rabi*. The coverage with high yielding varieties (HYVs) of sorghum is nearly 80 per cent in *Kharif* and potential under moderate input is also high (4-6 tonnes per hectare). The area loss may be due to the fact that the expansion in irrigation which has made other crops such as rice, sugarcane, cotton, etc., more attractive and remunerative thus rendering sorghum to be less competitive. The decline in consumption demand of sorghum grain was also a major factor for the decline in area. The increased productivity of sorghum has not been able to compensate the loss in area turning the production to be negative. However, the suitability of the improved varieties of sorghum to specific regions and local farming conditions remain unresolved. But sorghum is finding a niche area under rice fallows, under the zero tillage conditions, grain productivity as high as 7 to 8 tonnes per hectare has been realized, and now it occupies a prime area of more than 45000 hectares and is expanding further, where *Kharif* hybrids are prominently grown. The major thrust, in future, will be on enhancing area of sorghum in non-traditional areas and greater emphasis on alternate uses of sorghum and its utilization as a major food, feed, fodder, and fuel (bio-energy) for industrial utilization.

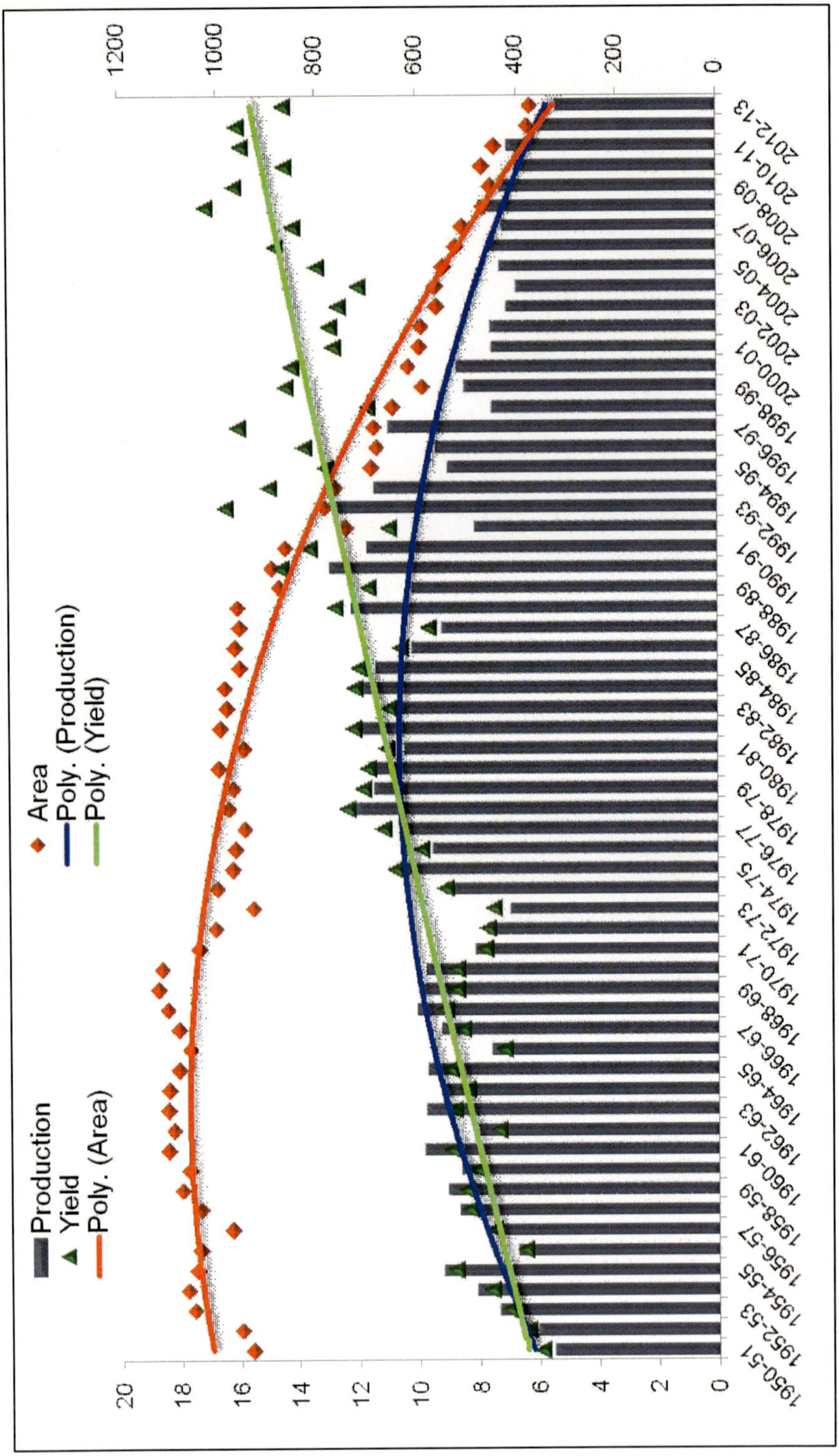

Figure 1: Area, Production and Yield of Sorghum in India.

(Area: in million hectares; Production: Million Tonnes and Yield kg per hectare)

Pearl Millet

Pearl millet is a major warm-season cereal grown on about 30 million hectares worldwide, largely in the arid and semi-arid tropical regions of Africa (18 million hectares) and Asia (10 million hectares) with India accounting for largest area (>9 million hectares). The diversification of cultivar base with mostly dual-purpose hybrids has led to 24 kg/hectare/year of grain yield increase during the last 15 years as compared to only 5.2 kg/hectare/year of yield increase during the pre-hybrid phase of 1950-1965. Development of improved crop cultivars is just one major component of technological interventions to enhance food and nutritional security. Improved crop management technologies with potential to substantially increase pearl millet grain yield have been developed.

Some of the agronomic interventions will also have significant effects on improving nutritional quality. Laboratory tested processing technologies and alternative food products have been developed, which have been shown to further enhance the bioavailability of these micronutrients. These technologies need to be tested for their commercial feasibility. Thus, an integrated approach of crop improvement backed with improved crop management, grain processing, and food products development; and appropriate policy support will enable pearl millet to play its rightful role in enhancing food and nutritional security.

Small Millets

There are six different small millets grown in India which include finger millet (*ragi*), foxtail millet (*kangni*), kodo millet (*kodo*), proso millet (*cheena*), barnyard millet (*sawan*) and little millet (*kutki*). The area under these crops during the last 6 decades has significantly shrunk from 8 million hectares in 1949-50 to around 2.3 million in 2012-13 This is also reflected in diminishing production, from around 4 million tonnes produced in late forties to around 2.5 million tonnes in 2011-12. The loss of area is very severe in all small millets other than finger millet. However, in the last 15 years, the finger millet also has lost ground and area has come down from 2.4 million to 1.2 million hectares and likely to lose further in the coming years.

Despite reduction in area, total production is not much affected. As regards productivity is conserved, finger millet has kept pace with most other major dry land crops in compound growth rates(CGR) for yield while the other small millets have shown little progress. By and large, the low productivity of these crops is largely due to meagre attention received in terms of inputs; water and technology backup which is further compounded by low value status of grains. The bulk of small millets production in India is of finger millet (80 per cent) and the remaining from kodo millet, little millet, foxtail millet, barnyard millet and proso millet in that order. Small millets in India are grown in Karnataka, Andhra Pradesh, Maharashtra, Tamil Nadu, Odisha, Jharkhand, Chhattisgarh, Madhya Pradesh and Uttarakhand. They being the components of traditional farming system one could see well evolved cropping systems. Agricultural progress in India has been technology driven. In the endeavour, rapid and efficient dissemination of technology in the target areas is very critical. Wheat, rice and cash crops because of their importance have got greater attention of development agencies.

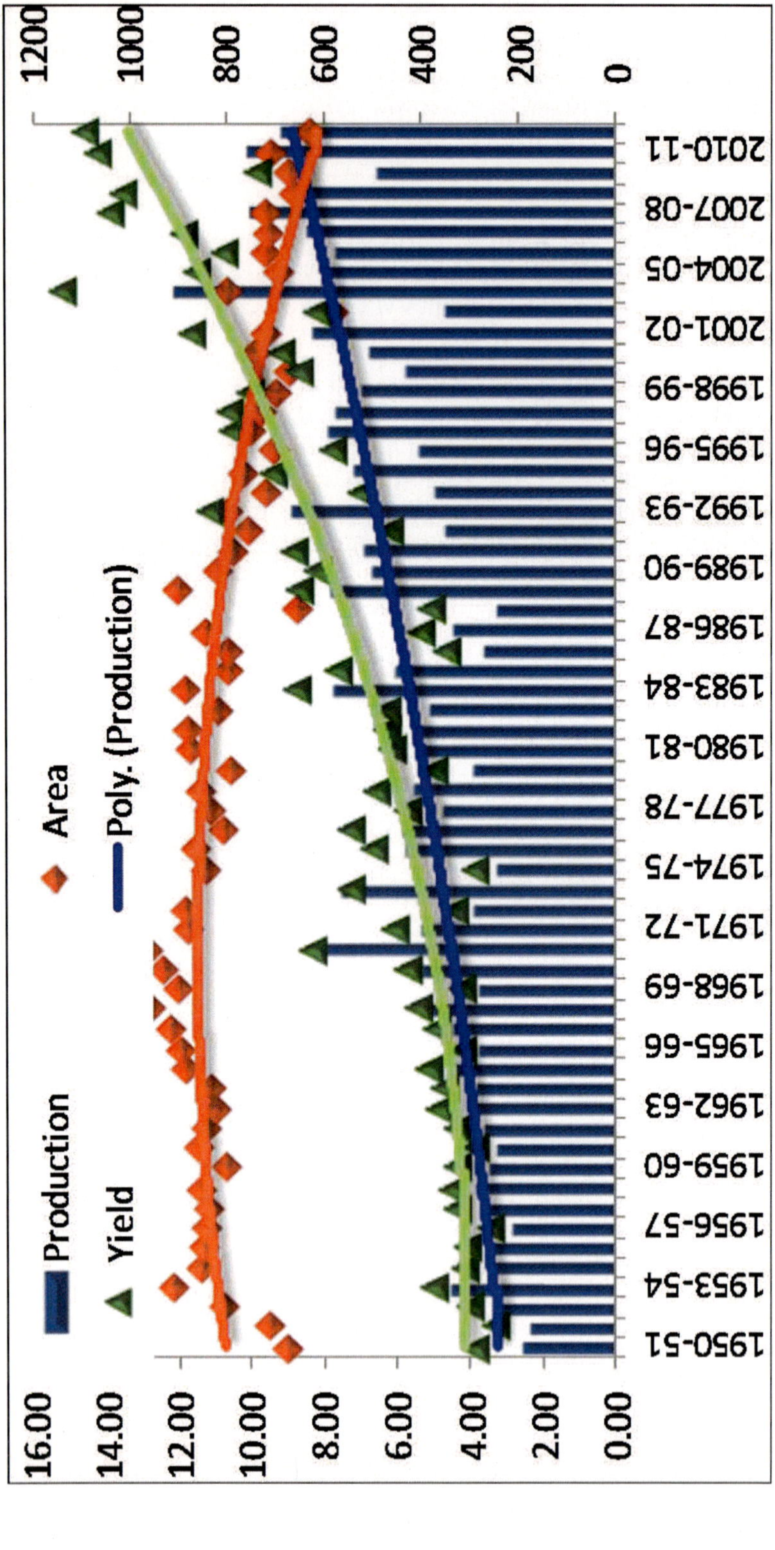

Figure 1.2: All India Trends in Area, Production and Yield of Pearl Millet.
(Area: in million hectares; Production: Million Tonnes and Yield kg per hectare).

Table 1: Area (Million hectares), Production (million tonnes) and Productivity (kg/ha) of Small Millets *vis-à-vis* other Millet Crops in India*

Crop/Year	*Category*	*1955-56*	*1965-66*	*1975-76*	*1985-86*	*1995-96*	*2005-06*	*2012-13*
Sorghum	Area	17.36	17.68	16.09	16.10	11.33	8.68	6.18
	Production	6.73	7.58	9.50	10.20	9.33	7.63	5.33
	Productivity	387	429	591	633	823	880	863
Pearl Millet	Area	11.34	11.97	11.57	10.65	9.32	9.58	7.20
	Production	3.43	3.75	5.74	3.66	5.38	7.68	8.74
	Productivity	302	314	496	344	577	802	1214
Finger Millet	Area	2.30	2.70	2.63	2.41	1.77	1.53	1.11
	Production	1.85	1.33	2.80	2.52	2.50	2.35	1.59
	Productivity	800	492	1064	1049	1410	1534	1428
Small Millets	Area	5.34	4.56	4.67	3.16	1.66	1.06	0.75
	Production	2.07	1.56	1.92	1.22	0.78	0.47	0.43
	Productivity	388	341	412	386	469	443	571

**Source*: Agricutural Cences, Directorate of Economics and Statistics, Department of Agriculture and Cooperation, Government of India.

Coarse cereals have slower dissemination of technology due to less importance accorded by the development agencies, cultivation in unfavourable environments and lower apparent superiority of the improved technology under stress environments. Thus, the exploitation of available technologies has been limited and there is a vast untapped potential or productivity potential in different coarse cereals crops. The potential available in different agroclimatic zones for the various small millets based cropping systems makes it evident that there is still large potential available in the technology recommended now and which could be bridged through the appropriate development strategies. Technological dissemination and development strategies should focus on specific technology package for different farming situations, and intensive extension education on the package. Rainfed farming production technologies, particularly water management and usage, greater coverage under improved cultivars, efficient water use, integrated nutrient and pest management and good crop husbandry can result in a rapid progress in small millets productivity, provided appropriate guidelines are formulated and implemented for comprehensive development of these crops in harmony with other components of cropping systems. On farm experiments, frontline and other demonstrations need to be conducted on an extensive scale. Seed production and supply is not satisfactory in general in coarse cereals and especially small millets.

The coverage under high yielding varieties is very poor in those areas where vagaries of monsoon are high resulting in uncertain production. In many states *ragi* and small millets seed production and supply is non-existent. Therefore, it is necessary to evolve a mechanism for assured production and supply of newly bred varieties. The evolution of such a mechanism is also necessary for the efficient conduct of central minikit programme as well as State/Centrally Sponsored Developmental activities. This can be further expanded and scope enlarged. Since seed is a critical input every effort should be made for efficient production and supply of seeds. Special seed programmes including Un- conventional methods of seed supply is also to be encouraged in coarse cereals. Involvement of Non-Governmental Organisations (NGOs) working in tribal/hill and remote dry land areas should be encouraged to undertake production and supply of truthfully labelled seeds. If necessary, seed subsidy has to be enhanced and its scope needs to be enlarged. On farm adaptive research conducted under NATP in Karnataka, Tamil Nadu, Andhra Pradesh, Odisha and Madhya Pradesh on intercropping of pigeon pea/field bean/black gram with small millets have shown enhanced grain yield as well as higher monetary returns to the farmers besides augmenting the legume production that also helped in minimizing the protein malnutrition. This technology can be further taken to other states as a developmental strategy to augment the pulse production in the country.

Evidently the growth rate in the area of total millets in India is on decline for all the possible periods during last four decades due to decline in the growth rates of area in minor millets, sorghum, and finger millet. The rate of decline is slow in case of pearl millet. In pearl millet the yield growth rates are high which sheltered the effect of rate of decline in area keeping growth in production on positive side. The situation in case of all other millets including production growth is negative due to inability of positive gains attained in yield to subside drastic decline in growth in area. The reason in the positive growth in yield is primarily due to the enhanced crop improvement R&D efforts by the IIMR and coordinating centers constantly striving to keep positive growth. The positive growth in yield during 2000-2013 is the highest for all millet crops except in finger millet.

Table 2: Compound Growth Rates of Area, Production and Yield of Millets for Various Periods

Crop	*Area (Million hectares)*	*Production (Million Tonnes)*	*Yield (kg/Hectare)*
SORGHUM			
1970-2000	–1.74	–0.07	1.70
1970-2013	–2.33	–1.11	1.27
1980-2013	–3.04	–2.13	0.96
1990-2013	–3.23	–2.66	0.62
2000-2013	–3.78	–1.74	2.11
BAJRA			
1970-2000	–0.91	0.93	1.85
1970-2013	–0.85	1.51	2.38
1980-2013	–0.90	1.99	2.91
1990-2013	–0.79	2.11	2.93
2000-2013	–1.12	2.45	3.60
RAGI			
1970-2000	–1.48	0.24	1.74
1970-2013	–1.92	–0.59	1.35
1980-2013	–2.47	–1.18	1.32
1990-2013	–2.43	–1.59	0.86
2000-2013	–2.90	–1.57	1.37
MINOR MILLETS			
1970-2000	–4.45	–3.64	0.85
1970-2013	–4.67	–3.87	0.85
1980-2013	–5.13	–4.14	1.04
1990-2013	–4.71	–3.72	1.04
2000-2013	–4.75	–1.26	3.65
TOTAL MILLETS			
1970-2000	–1.65	0.11	1.63
1970-2013	–1.91	–0.25	1.47
1980-2013	–2.29	–0.59	1.54
1990-2013	–2.22	–0.65	1.29
2000-2013	–2.56	0.15	2.37

Note: Above figures are in per cent per annum; *Source*: IIMR, Hyderabad, 2014.

(Area: in million hectares; Production: in million tonnes Yield: kg per hectare).

Research, Development and Utilization of Millets

Sorghum

With the release of CSH 1, the first commercial hybrid in 1964, sorghum became the second crop after maize in developing high yielding hybrids using cytoplasmic-genetic male sterility system. After CSH 1, 30 more hybrids at central level and a few at state levels were released as adapted to specific regions. These hybrids played a major role in pushing up productivity and production, particularly in the case of *Kharif* sorghum. Among the *Kharif* hybrids CSH 1, CSH 5, CSH 6, CSH 9, CSH 14 and CSH 16 need special mention as CSH 5 and CSH 6 had a yield potential of 34 q/ha which was raised to 40 q/ha in CSH 9 and further raised to 41.0 q/ha in CSH 16, CSH 23, CSH 25 and CSH 30 with distinct superiority in grain and fodder quality. High yielding varieties CSV 1 to CSV 30 at central level and many more at states level were released. Some of these varieties are dual-purpose type. By and large, varieties were less acceptable to farmers. Better preference was for dual-purpose varieties such as CSV 10, CSV 13, SPV 462, CSV 15, CSV 20, CSV 23 and CSV 27 in some pockets. The short duration variety CSV 17 is gaining popularity and is expanding in area. A major advantage of varieties over hybrids was their relative better grain quality and multiple resistance/tolerance to major pests and diseases. The reduced maturity duration of the HYVs led to its high vulnerability to grain mold damage during *Kharif*. Building reasonable resistance against grain mold is difficult. These biological limitations seriously restricted the economic advantage of their high yield. The government policies on production, pricing, procurement and distribution of cereals favoured fine cereals and placed coarse grains such as sorghum at a disadvantageous position.

Improvement of *Rabi* sorghum did not receive as much importance as *Kharif* sorghum until the nineties. Six hybrids and five varieties were hitherto centrally released for *Rabi*. In *Rabi* sorghum, the fodder yield is even more important than *Kharif* sorghum. From this point of view, a progressive success was achieved from the first *Rabi* hybrid CSH 7R to the latest hybrid CSH 19R. Unlike the *Kharif* cultivars, higher levels of resistance against major pest (shootfly) and disease (charcoal rot), stringent maturity duration requirements to suit different receding soil moisture regimes and adequate levels of thermo-insensitivity are essential in *Rabi* cultivars for better adaptability. Grain quality is also as much important as the grain yield, with quality benchmark being *Maldandi* (M 35-1). In adaptability criteria such as shoot fly resistance as well as the grain quality aspects, the varieties are superior to hybrids. The three *Rabi* varieties released, CSV 8R, CSV 14R, CSV 18 and *Swathi*, were better received than the *Rabi* hybrids such as CSH 7R and CSH 8R. However, the recently developed hybrids CSH 13 R, CSH 15R and CSH 19R are more productive, but the acceptability among farmers is not high as they do not want to invest on hybrid seeds every season during *Rabi* (dry season) without irrigation. However, the new range of soil depth specific varieties like CSV 26 and *Phule Anuradha* for shallow soil, *Phule vasudha* and *Maulee* for medium soils and CSV 22, CSV 29 and *Phule Revathi* for deep soils have achieved niche area with greater productivity and quality in relation to *Maldandi*.

Unlike the case of *Kharif* sorghum, biological and environment limitations posed difficult hurdles in *Rabi* sorghum productivity. Some of the newly developed varieties could gain acceptability in certain deep soil pockets with assured soil moisture. The yield potential of newly bred cultivars is only marginally higher than M 35-1, the widely grown local cultivar. Hence, viable management methods to narrow down the existing gap between the production potential of local and improved cultivars and their farm yield could become a short term approach to increase *Rabi* productivity. It appears that *Rabi* sorghum may continue to enjoy increasing demand for its food and fodder preference. Thus, its area, particularly in Maharashtra and Northern Karnataka, may not decline substantially and possibly stabilize between 4.5 and 5.0 million hectares. While sorghum is largely used as a feed grain world over, its domestic cost of production and quality limitations made it less competitive to maize. The current feed production in the country under organized sector is 2.7 million tonnes which is projected to go up to 3.9 million tonnes by 2015 AD. Domestic grain based industries also need 2.0 to 3.0 million tonnes grain. Generation of novel alternate uses requires promotional research in these areas, including cultivar development to suit specific end uses and economic competitiveness through high productivity and lower cost of production. The latter two considerations may also brighten up the chances for *Kharif* sorghum to emerge as an export feed grain.

Utilization of Sorghum in India

Unlike the case of *Kharif* sorghum, biological and environment limitations posed difficult hurdles in *Rabi* sorghum productivity. Some of the newly developed varieties could gain acceptability in certain deep soil pockets with assured soil moisture. The yield potential of newly bred cultivars is only marginally higher than M 35-1, the widely grown local cultivar. Hence, viable management methods to narrow down the existing gap between the production potential of local and improved cultivars and their farm yield could become a short term approach to increase *Rabi* productivity. It appears that *Rabi* sorghum may continue to enjoy increasing demand for its food and fodder preference. Thus, its area, particularly in Maharashtra and Northern Karnataka, may not decline substantially and possibly stabilize between 4.5 and 5.0 million hectares. The demand at the current growth appears to be around 6-8 million tonnes. This may, however, change in relation to the production and pricing of fine cereals. *Rabi* sorghum is highly valued as fodder during lean months. All these factors make *Rabi* sorghum production more profitable and economically sustainable despite its very low yield (grown under conserved moisture only). While sorghum is largely used as a feed grain world over, its domestic cost of production and quality limitations made it less competitive to maize. The current feed production in the country under organized sector is 2.7 million tonnes which is projected to go up to 3.9 million tonnes by 2015 AD. Domestic grain based industries also need 2.0 to 3.0 million tonnes grain. Generation of novel alternate uses requires promotional research in these areas, including cultivar development to suit specific end uses and economic competitiveness through high productivity and lower cost of production. The latter two considerations may also brighten up the chances for *Kharif* sorghum to emerge as an export feed grain. Global projection on future coarse grain demand for feed is at 2.4 per cent annual growth rate while its expected production growth rate is only 2.0

per cent. This demand-supply gap and expected low coarse grain reserve trends are favouring a long term increase in the international market price of coarse grains. Sorghum contributes to nearly 10 per cent of the current international trade volume of about 135 million tonnes coarse grains. Sorghum has a growing market in Asia. Its domestic price since last five years is steadily declining than that of maize. There is a growing opportunity for export oriented sorghum production as the international sorghum prices are increasing at 3.88 per cent per annum over past 5 years.

The utilization of *Kharif* sorghum grain as a raw material in various industries is increasing, given the limited prospects of rainy season (*Kharif*) sorghum for human consumption. Post-rainy season sorghum is a highly valued food grain, and expensive to be used as industrial raw material. The main industries currently using sorghum in India are the poultry feed, animal feed and alcohol distilleries. At present poultry feed sector is using approximately 2.0 million tonnes sorghum annually; animal feed sector uses about 0.60 million tonnes sorghum followed by alcohol distillers (about 0.49 million tonnes). But if government policy on allocating food grain to potable alcohol making is implemented, 4 million tonnes of sorghum may be needed for brewing industry alone. The likely scenario by 2050 AD for millets is depicted in Table 3.

Table 3: Pattern of Utilization of Sorghum and Pearl Millet Grain for various Alternate Uses

Utilization Patterns	*Type*	*Demand (million tonnes)*		
		Qty Utilized in 2008-09	*Projected Utilization 2050*	
			Scenario II	*Scenario III*
		Sorghum		
Direct use/	*Rabi*	4.0	4.83	10.76
human consumption	*Kharif* (partly)	0.1	0.97	2.15
Sub-total (*a + b*)		*4.1*	*5.79*	*12.92*
Other uses				
1. Poultry feed	*Kharif*	2.0	4.39	9.78
2. Animal feed	*Kharif*	0.6	1.79	4.00
3. Alcohol	*Kharif*	0.49	3.45	7.69
4. Seed +wastage	*Kharif+ Rabi*	0.15	0.28	0.62
Sub-total (*1 + 2 + 3 + 4*)		*3.20*	*9.91*	*22.08*
Total production		7.31	15.70	35.00
Utilization Patterns	*Type*	*Demand (million tonnes)*		
		Qty Utilized in 2004-05	*Projected Utilization 2050-51*	
			Scenario II	*Scenario III*
		Pearl millet		
Household consumption	*Kharif*	3.49 (44 per cent)	5.63 (25 per cent)	9.38 (25 per cent)
Sub-total (*a + b*)		3.49	5.63	9.38

Contd...

Table 3–*Contd...*

Utilization Patterns	*Type*	*Demand (million tonnes)*		
		Qty Utilized in 2004-05	*Projected Utilization 2050-51*	
			Scenario II	*Scenario III*
Other uses				
1. Poultry feed	*Kharif*	4.36	16.65	27.74
2. Alcohol	*Kharif*	(55 per cent)	(74 per cent)	(75 per cent)
3. Others (Animal feed, Seed +wastage)	*Kharif*	0.08 (1 per cent)	0.23 (1 per cent)	0.38 (1 per cent)
Sub-total (1 + 2 + 3)		*4.44*	*16.88*	*28.12*
Total production		**7.93**	**22.50**	**37.50**

Pearl Millet

Pearl millet is a multi-purpose cereal grown for grain, stover and green fodder on about 30 million hectares, primarily in the arid and semi-arid regions of Africa and Asia. Although its productivity in these environments is low, its ability to tolerate drought, heat and low soil fertility makes pearl millet attractive crop species. It responds well to improved moisture and soil fertility conditions. In terms of annual production, pearl millet is the sixth most important cereal crop in the world. India is a major pearl millet producing country with an area of more than 9.0 million hectares and production of about 8.74 million tonnes in 2012, where five states (Rajasthan, Maharashtra, Gujarat, Uttar Pradesh and Haryana) account for nearly 95 per cent of the pearl millet cultivated area.

It is quite interesting to compare the improvement in pearl millet productivity achieved during the last 28 years (1985-2013) than that achieved during 1960-85 period. During first 25 years (1960-85) of hybrid development, the crop productivity increased @ 6.3kg/hectare/year. This increase went up to over 20kg/hectare/year in next 25 years. The greater rate of improvement in pearl millet productivity during last 25 years is due to several reasons. Firstly, a far greater number of availability of pearl millet cultivars provided a wide range of choice for their cultivation in various agro-ecological regions. A total of 43 cultivars were released during 1960-85 in comparison to 112 hybrids/varieties released till date. As a result, there have been no major disease epidemics during last 25 years against quite a few prior to 1985. Secondly, involvement of private sector in seed production, distribution and marketing has helped provide quality seed of hybrids to farmers. Thirdly, greater adoption of production technology along with high-yielding hybrids proved synergistic in further augmenting the productivity gains. Pearl millet productivity has gone up from 0.54 tonnes per hectare during 1986-90 to 0.93 tonnes per hectare during 2006-13 registering a growth of 73 per cent improvement, which is highest among all food crops. This extent of improvement in pearl millet productivity has resulted in 40 per cent improvement in its grain production from 7.16 million tons in 1994-95 to 10.05 million tonnes in 2012-13, in spite of 13 per cent decline in crop area from 10.22 million hectares to 9.10 million hectares in the same period. Pearl millet showed more percent increase in production and productivity over the two major cereals rice and wheat in the 10th and 11th Five year plan period in spite of decrease in area. But when we compare the consumption pattern of pearl millet with rice and wheat during the period 1993-94 to 2009-10, it shows a steep decline in both rural and urban areas. Between 1972-73 and 2004-05, the annual per capita consumption of pearl millet at the all India level declined sharply by 67 per cent (4.1 to 1.4 kg) in urban areas and by 59 per cent (11.4 to 4.7 kg) in rural areas of India. It was also estimated that at the all India level during 1999-2000 about 54 per cent of pearl millet production was used as food and the remaining 46 per cent was used for other uses. By 2004-05, the utilization for other uses increased to 55 per cent while food use declined to 44 per cent (Basavaraj *et al.*, 2010). Another revelation from Parthasarathy *et al.*, also suggest that the trends in utilization of pearl millet for other uses (alcohol and poultry feed) have increased in the recent years to about 70 per cent of the production in Rajasthan and Haryana.

All India Coordinated Millet Improvement Project (AICMIP) was established in the year 1965 with its headquarters at the Indian Agricultural Research Institute, New Delhi. The headquarters of the project were shifted to Pune in 1977. Later on pearl millet was separated from the rest of the millet crops and the All India Coordinated Pearl Millet Improvement Project (AICPMIP) was established in 1985 with its headquarters at Pune as an independent coordinated project. The ICAR shifted, in July 1995, the headquarters of AICPMIP to Jodhpur in the state of Rajasthan, the state which occupies nearly half of pearl millet area of the country. AICPMIP has a network of fourteen centers in Rajasthan, Maharashtra, Uttar Pradesh, Karnataka, Andhra Pradesh, Madhya Pradesh, Punjab, Haryana, Tamil Nadu and Gujarat. The AICPMIP centers pursue mandated activities in pearl millet improvement, production and protection. The AICPMIP has played a pioneering role in developing a diverse range of improved breeding lines and parental lines of hybrids. These lines have been used extensively to develop and commercialize a large number of hybrids. These hybrids are currently cultivated on approx. 50 per cent of the total pearl millet area of 9-10 million hectares. Hybrids maturing in 80-85 days, when cultivated as an irrigated rains/summer season crop in parts of Rajasthan, Gujarat and Uttar Pradesh, have been reported to give as high as 4000-5000 kg per hectare of grain yield. With adaptive and nutritional features of pearl millet combined with high yield potential make it an important cereal crop that can effectively address the emerging challenges of global warming, water shortages, land degradation and food-related health issues. AICPMIP has also developed production-protection technologies specific to agro-eco regions of different states. The HYVs cover about 50 per cent of total pearl millet area, which is highest among coarse cereal crops. Area under HYVs is highest in Gujarat where almost whole area (>90 per cent) has come under hybrid coverage. Although Rajasthan has the highest area under pearl millet, adoption of HYVs in this state has been very low (25-30 per cent).

Small Millets

Research efforts in India have been in progress since the beginning of the 20 the century. But, the launching of coordinated crop improvement programs during late 1950s and 60s has contributed significantly by way of developing new superior varieties and concomitant production and protection technologies in all small millets.

The release of these varieties and production packages for general cultivation has helped in 5 fold increase in grain production from 50 to 250 million tonnes in the country. It is generally seen that this increase has largely come from two major crops –rice and wheat- and less from dry land crops such as millets and more so from small millets. The small millets have been the last priority crops in the agriculture developmental agenda in the country. Finger millet among small millets has received a little more attention than the rest especially in the southern states. Before the launching coordinated project in 1950s and 60s; the crop improvement in small millets was confined to a fewer states such as Tamil Nadu, Andhra Pradesh, Karnataka and Uttar Pradesh. The emphasis was on varietal improvement through selection of better types from local cultivars. In Tamil Nadu, Millet Research Station was established in 1923 at Coimbatore under the erstwhile Madras Presidency and the work later extended to Anakapalle (AP) and Hagari (Karnataka). Finger millet work in Karnataka dates back to 1900, initiated at Bangalore especially on finger millet and in Uttar Pradesh at Kanpur and Gorakhpur in 1944. As a pre-requisite to millet improvement, the inheritance of qualitative and quantitative characters, period of anthesis, pollination and crossing methods were studied at Coimbatore. The first finger millet variety released in the country was H 22 as early as 1918 in Karnataka. Interest in finger millet improvement got a fillip in Karnataka during 1950-60 and several new varieties such as Aruna, Udaya, K1, Purna, ROH 2 and Cauvery were released. Similarly, many varieties were released in other small millets also in many states. This included little millet variety Co 1 (1954); foxtail millet varieties Co1, Co2, Co 3 (1943), H1, H2 (1948), T 4 (1949); kodo millet varieties PLR 1(1942, T 2 (1949), Co 1 (1953), proso millet variety Co 1 (1954) and barnyard millet varieties T 46, T25 (1949). During fifties, with food production remaining stagnant and with raising population the importance of millet crops to Indian agriculture started gaining recognition, as they formed the important constituents of dry land agriculture. Project for intensification of research on cotton, oil seeds and millets was launched during this period with several centres working on millets. The first attempt to collect the germplasm of millets in the country was made in 1961 under the PL 480 project on "storage, Maintenance and distribution of millets germplasm resulted in the collection of nearly 3000 genetic stocks of various small millets, 718 in finger millet, 584 in kodo millet, 431 in little millet, 615 in foxtail millet, 250 in proso millet and 399 in barnyard millet.

Small millets have always been of local and regional importance and as a result have attracted little attention both at national and International level. Millets in general, started receiving with attention with the launching of All India Coordinated Millets Improvement Project (AICMIP) in 1969. Small millets improvement received the major boost during 1978-79 with the establishment of five crops specific lead research centres in the country under IDRC assistance. They were Almora in Uttarakhand (barn yard millet), Dholi in Bihar (proso millet), Dindori in Madhya Pradesh (Kodo millet), Semiliguda in Odisha (Little millet) and Nandyal in Andhra Pradesh (foxtail millet.) The IDRC project continued till 1985 and the "All India Coordinated Small Millets Improvement Project" (AICSMIP) was established in the year 1986 with head quarters at The University of Agricultural Sciences, Bangalore. The centres that were functioning under IDRC project became part of AICSMIP. With the inception of separate AICRP, research on small millets has been getting focused attention for developing varieties and other agro production and protection technologies suitable to different regions. Small millets are known for their suitability to dry land areas, hill and tribal agriculture and contribute to food and nutritional security of the disadvantaged regions. The work is multi-disciplinary and applied in nature in AICRP on Millets and IIMR will be taking up basic and strategic research apart from value addition and industrial utilization initiatives. In the past small millets scientists hardly had access to germplasm and worked with a handful of local collections which lacked diversity.

The germplasm conservation activities further gained momentum with NBPGR, New Delhi, playing a key role in augmenting the small millets collection. Recognizing the importance and conservation and easy access to germplasm, AICMIP established a separate germplasm unit at Bangalore in 1979. This is recognized as National Active Germplasm Site (NAGS) by ICAR/NBPGR and has the mandate to assist in collection, conservation, evaluation and documentation of small millets germplasm in the country. Presently the unit at Bangalore is maintaining one of the largest collections of more than 15000 accessions of 6 small millets (7122 in finger millet,2821 foxtail millet,1537 kodo millet, 939 proso millet,1657 little millet and 988 barnyard millet)in the country. It is envisioned that while single crops such as rice and wheat can succeed in producing food security for India, millets will do more. They will contribute immensely to securities of food, nutrition, fodder, fibre, health, livelihood and ecology, especially in changing climate scenario as they are very less water requiring drought tolerant nutri-cereals as potent functional foods.

Challenges in Millets Development

For centuries, sorghum and millets have been important staples in the semi-arid tropics of Asia and Africa. These crops are still the principal sources of energy, protein, vitamins and minerals for millions of the poorest people in these regions; these "coarse" cereals which in fact are "nutritious" cereals. They perform well in harsh environments where other crops grow or yield poorly. The growers in developing countries are a multitude of smallholder farmers including India. They are grown in traditional agricultural systems with limited water resources and usually without application of any fertilizers or other purchased inputs. Sorghum or great millet (*Sorghum bicolor* (L.) Moench), pearl millet (*Pennisetum glaucum* (L.) R. Br.) and finger millet (*Eleusine coracana* (L.) Gaertn.) are the three major millet crops cultivated in larger areas of India.

The progress of plant breeding in India has been phenomenal during the last five decades. The successful adoption of improved varieties along with improved cultivation practices helped to rapidly increase food grain production, from 50 million tonnes to more than 268.34 million tonnes during 2013-14. However, much of this increase has come from two major cereals, rice and wheat; the contribution of other crops is comparatively lower owing to their lesser area and marginal dry environments they are grown with minimum input use. The high yielding varieties have made more impact in high external input areas that are mostly irrigated. In low external input areas (dryland), the impact of improved cultivars on crop productivity has been limited. Sorghum and millets fall in the latter category. Uncontrolled population growth leading to conversion of agricultural land for other purposes, urbanization and government policies have pushed sorghum and millet cultivation into drier, more marginal lands. However, with demographic and agro-climatic transitions and the brighter prospects for alternate and new end-uses for sorghum and millets using new technology can perhaps help realize the importance of these crops for national economy where livelihood in dryland, nutrition, fodder and resource conservation (mainly water) are factors of immediate importance. Genetic improvement of millets has progressed over the last several decades with impressive gains in productivity and adaptability of high-yielding cultivars. However, the traditional approaches have been slow in addressing the major loss causing or yield-destabilizing traits such as susceptibility to insects and diseases and drought stress. This scenario was mainly due to the complex nature of these traits which are greatly influenced by environment and the absence of reliable genetic clues to manipulate them using the tools and know-how available with the erstwhile breeding programmes.

Issues Limiting Sorghum, Pearl Millet and Small Millets Development, Adoption and Utilization

Sorghum is a major food and fodder crop for dryland farmers in India and elsewhere. One of the major reasons for lower yields in this crop is due to the problem of insect pests such as stem borer and shoot fly, and diseases such as grainmold, ergot, and drought. Similarly, in pearl millet, the important losses of yields are mainly due to fungal diseases such as downy mildew, ergot and rust. One pearl millet hybrid, HB3, which was adopted widely in India during 1960s was devastated by downy mildew and had to be withdrawn from cultivation. Blast caused by the fungus *Magnoporthe grisea* is the most important disease of finger millet. In India, the disease appears regularly every year causing annual yield losses as high as 50 per cent in disease endemic areas and in favourable seasons. Classical breeding for blast resistance in small millets has not been rewarding, apparently due to the polygenic control of host resistance, significant environmental influence and the high level of variability present within the pathogen.

Sorghum

Grain Quality Deterioration of *Kharif* Sorghum

Designing of high yielding *Kharif* genotypes with reduced maturity duration enhanced their vulnerability to grain molds leading to serious problem of grain deterioration, decline in grain quality and market acceptance leading to low profitability to the producer despite high productivity. Protection of grain from mould damage through conventional breeding is a formidable challenge due to the multiplicity of fungal species involved, structure of grain, absence of high level of resistance, complex genetic basis of resistance, and association of available resistance with hard grain texture and coloured pericarp. Biotechnological approach using antifungal genes offers hope to achieve faster solution to conquer damage by grainmolds. Deterioration of grain quality due to molds became primary reason for the decreased preference of *Kharif* grain as food. Consequent non-expansion in demand for use as food, failure in creating diversified uses imposed restrictions on production and profitability to the farmers. These reasons led to partial diversion of sorghum area to more profitable alternate crops wherever such diversification and higher profitability is possible.

Vulnerability of HYVs to Biotic Stress

High yielding cultivars are more vulnerable to major pests such as shootfly, stemborer and earhead bug and major diseases like grainmold and charcoal rot. Leaf spot and rust that are important in forage sorghum. Immunity against these pests and diseases is not feasible. However, there is a need to build fair level of tolerance against most of these pests and diseases by conventional breeding and by biotechnological means through gene pyramiding.

New Dimensions in Nutritional and Quality Improvement of Grain and Forage

Research on grain processing, improving the limiting nutritional constituents, energy value and such other attributes which may impart value-addition to grain as food or feed or industrial raw material may improve sorghum demand.

Denial of Farmer-Friendly Public Policies

Farmer friendly public policies on production and procurement available to the producers of fine cereals have been denied to sorghum farmers. This lack of policy and public concern seriously affected its overall production, profitability to the producer and forced the farmers to gradually replace sorghum with economically attractive alternate crops which command higher MSP.

Seed Supply System

To fulfil the emerging needs for seeds, identification of appropriate new regions for large scale seed production through seed village concept, and development of self sustainable community seed systems to usher in seed security in village clusters requires earnest initiatives. Development of cost effective molecular tests for seed purity should also be the priority.

Pearl Millet

Pearl millet is the major cereal crop grown in the hottest, driest areas of the world where, rainfed agriculture is practiced. It is also very tolerant of acid soil conditions. Additional advantages of pearl millet include its excellent ability to ratoon following green fodder harvest, and the ease with which it can be transplanted compared to sorghum and maize. Finally, pearl millet, especially cultivars with medium or small grain size, stores better than either sorghum or maize. Downy mildew, smut, ergot, are often cited as the most important problems and issues to be addressed, in order of importance. Hence major research thrust need to be on:

- Downy mildew resistance; tolerance to drought and heat; and micronutrients (iron and zinc)
- Emphasis on dual-purpose hybrids, greater attention on hybrid development for the driest A_1 zone

- ✰ Development and promotion of green forage hybrids
- ✰ Application of biotechnology in breeding for downy mildew resistance; tolerance to drought and heat

Overcoming Vulnerability of HYVs to Abiotic and Biotic Stresses

Response of pearl millet to moisture stress at various growth stages has clearly established that yield losses are maximum when moisture stress coincides with grain filling stage, which is commonly referred to as terminal water stress. Various physiological and morphological traits have been examined as alternative selection criteria to further enhance tolerance to terminal drought. Conventional approaches to improve drought tolerance in pearl millet have a very short history and attempts have met' with some success. Various novel approaches have been attempted in pearl millet for enhancing yield under drought environments. These include use of adapted germplasm, genetic diversification of adapted 1andraces through introgression of suitable elite genetic material, and exploitation of heterosis to amalgamate drought tolerance and high yield. Genetic differences in tolerance to salinity and high temperature at both seedling and grain filling stages have been established and screening techniques standardized. The germplasm and breeding material with a higher degree of tolerance to high temperature and salinity have been identified in order to use them in breeding programs. When the rains end early, millet produces very little grain or leaf. Plus, up to a third of the crop is often lost to downy mildew. The genes that help plants withstand the three main threats - heat, drought and mildew - could vastly improve harvests. They are already being used in India by government and international research laboratories.

Improving and Enriching Quality of Pearl Millet Stover and Residues

Indeed it is the need of the hour to produce hybrid parent lines with enhanced stover quality suitable for use in commercial hybrid seed production through conventional and modern breeding techniques to determine the effects of individual stover quality Quantitative Trait Locus (QTLs) to improve on fodder quality of pearl millet stover for higher animal productivity and to promote the use of improved parental lines to public and private seed companies.

Nutritional and Quality Improvement of Grain and Forage

Research on grain processing, improving the limiting nutritional constituents, energy value and such other attributes which may impart value-addition to grain as food or feed or industrial raw material may further improve pearl millet demand.

Small Millets

The most urgent and the strong context for giving such a priority to millets, is provided by the coming decades of climate Change which confront us with three challenging scenarios. The first is rising global warming; the second, water scarcity that will acquire frightening proportions; and the third, the projected malnutrition that promises to engulf 70 per cent of the Indian population, particularly the poor and the vulnerable sections. Being hardy and robust crops, millets can withstand

and survive harsh climatic conditions. This makes them an ideal solution to the challenge of climate change. It is forecast that with the onset of climate change, wheat which is a thermal-sensitive crop would be hard-pressed to survive. At the same time, rice fields with 2"of standing water and heavy inputs of chemical fertilizers as required under the Green Revolution model of cultivation will release methane, a greenhouse gas into the atmosphere, thus rendering rice an utterly unviable crop. Many recent studies have pointed out that 20 per cent of diabetes suffered by the poor in India is caused by the rice distributed and consumed through PDS. So, if we are serious in confronting this problem, we should look at millets for a solution. They must not be seen just as a diabetic food or health food. They do not merely heal our wounds and illnesses, but have the ability to heal our planet. This is what makes them great food crops. Production of millets and small millets are subject to wide fluctuations and the area is declining. Excepting sorghum, pearl millet, and finger millet, no other millet have showed greater improvement in their cultivable area. The major challanges are as follows:

1. Improvement of yield potential in finger millet
   - ✰ Hybrid technology for higher productivity
   - ✰ Improving biomass production and harvest index
2. Development of genomic resources
   - ✰ Employing tools of comparative genomics
3. Gene discovery and allele mining from small millet genetic resource
   - ✰ Water and nutrients use efficiency
   - ✰ Nutritional quality
4. Improvement of nutritional quality (Grain and Fodder) and bio-fortification
   - ✰ Exploit with in species variability for nutrients
   - ✰ Improve bio-availability
5. Crop improvement for resistance to biotic and abiotic stresses
   - ✰ Blast disease in finger millet
   - ✰ Shoot fly in little, proso and foxtail millets
   - ✰ Tolerance to drought, temperature and salinity

Future Prospects

The emphasis on millet research has largely been on yield improvement through development of superior hybrids and varieties. Some attention was also paid to insect and disease resistance. With declining uses and areas under millet cultivation, it is necessary to have a relook at goals and objectives of future millet research both through genetic improvement and appropriate management and value addition to make millets as preferred industrial crops. In view of their greater role in bio-diversity, nutrition and health, livelihood enhancement and security the demand base for alternative uses needs attention, especially in relation to renewed millet improvement and development efforts at Indian Institute of Millets Research. The present operating environment needs to relook at various issues as analysed hereunder.

Millets: Biodiversity, Nutrition, Health and Sustainable Livelihoods

The shift to rice and wheat diets of the high input agriculture no doubt provides for more calories but not a balanced diet to the vulnerable groups. Added to this, the per capita availability and consumption of pulses has also come down. Due to marketization and influence of globalization, the consumption of refined/processed energy foods like carbohydrates, fats and fried foods, junk foods, etc., is on the increase. Hidden hunger due to lack of micronutrients, vitamins and other complements and several health and disease problems are attributed to poor nutrition which in turn is associated with a weak immune system and increases susceptibility to deficiency and infective diseases. Damage to the ecosystem through increasing chemicalization of agriculture aggravates the problem. Thus this lack of dietary diversification brings into focus the inter-related problems of biodiversity, nutrition and health in the overall issue of food security.

Demand Generation: Diversified Uses

Recognizing the decline in cultivated area and consumption of millets, attempts have been made to generate demand through multiple uses of millets in the areas of food, feed, forage, energy, industrial and other uses. The use of ragi malt has been in vogue since long. Researchers on sweet sorghum were initiated in late seventies. Now we have all the leads and best of the cultivars that will enable to take the cause of diversification, value addition and development of novel products and also functional foods, including second generation biofuel production to make millets as preferred industrial crops.

Dry and Wet Milling

Diversified uses for food, feed, industry, etc., require prior processing. Dry milling separates the kernel into its components, bran, germ and endosperm. Data is now available on varietal difference for various fractions. Wet milling involves hydration or steeping and milling carried out in water slurry in several steps to produce various kernel components. The components are used for various purposes like food, feed and industrial uses.

Food Uses

Whole meal flour is used in traditional foods like *roti*. There is data on the use of millet flours in various kinds of traditional preparations. Blended/fortified foods such as sorghum-soybean, sorghum gram, etc., have been studied. In the present days of marketing atta, there will be scope for such blended flours. Blends of flours with sorghum/millet with known therapeutic value further fortified with quality protein like soymeal can provide both for urban and rural markets. Weaning foods with malted millet and processed with pulses, sesame etc., have been formulated. Varietal differences for kernel attributes like corneous and sturdy endosperm, kernel hardness, water absorption, swelling properties, gel consistency, dough characters, biochemical attributes like amylase activity, proteolytic activity, have been studied. Depending on the product requirements, special varieties could be bred with product orientation either for domestic or processed foods. It is time to orient breeding programmes towards specific food products.

Animal Feed Shortages: Feed Grain Industry/Forage/Silage

The feed grain requirements are projected to be more severe than food grains. The World Bank has projected a deficit of 2.6 mt of concentrate feed and 251 mt of roughages for all categories of animals.

The Poultry Feed Industry

Commercial poultry feed manufacturers largely produce broiler feed. The layer feed is mixed by poultry producers. While maize is preferred in poultry feeds, sorghum has been the second candidate. The potential for replacement of maize at various levels has been shown in many studies. If maize prices are high, then they turn to sorghum. There is a need to consciously promote sorghum in poultry feeds. If productive yellow endosperm sorghum with high metabolic energy is bred, sorghum will find a more significant place in poultry feeds. Breeder of sorghum and millets should orient their studies towards meeting the needs of poultry feed industry. The high lysine and high betacarotene sorghums and the incorporation of the gene for tryptophan through genetic engineering methods would be important contributions in this direction. It is necessary to establish an interface with poultry feed industry and promote partnership studies and future efforts will be in this direction.

The Dairy Feed Industry

Besides feed grains, green fodder, dry fodder, oil cakes, etc., are important constituting dairy animal nutrition. According to the estimates approximately 10 m t of dairy feed is produced in India by the commercial sector (*i.e.* about 2 mt by Non-Coop. CLFMA members, 2 mt by the cooperative sector, and 6 mt by the unorganized commercial sector). The feed grain industry has an organized sector, Compound Livestock Feed Manufacturers Association of India (CLMFA) and a cooperative sector. Cooperative feed mills are looked after by NDDB. About 50 per cent of the feed rations are produced in the unorganized sector. Besides grain, green and dry fodders are also important in dairy nutrition. Sorghum can play a more useful role than maize. Dual purpose and sweet sorghum can satisfy the needs of both grain and forage. Hence there is a need to establish an inter-face with dairy nutrition researchers and organized feed manufacturers. The breeding programmes at AICRP- Sorghum Stations will also be reoriented towards feed, feed formulations, green forages and roughages.

Sorghum for Biomass and Bio-energy

Sorghum is recognized as one of the best crops for biomass-energy-feedstock. The initiation of research on sweet sorghums was in this direction. The introduced sweet sorghums were highly susceptible for stem borer. Over a period of time highly adaptable varieties like SSV 84, CSV 19SS, CSV 24 SS and sweet sorghum hybrid CSH 22SS have been developed. The sugars and starches in sweet sorghum could be fermented to ethanol and the liquid cellulose could be gasified to methane. The concept of high energy sorghums is now gaining ground. Some work has also been initiated on 'high energy sorghums'. The sorghum project has now commercialized the production of ethanol through cooperative efforts with some sugar factories. The collaborative efforts will be further strengthened to take ethanol production to commercial heights. Opportunities to produce biomass energy crops round the year would be explored by paying attention to the photo and thermo insensitivity, pest problems, the demarcation of areas of cultivation and related factors. The recent national bio-fuel policy of Govt. of India states that an indicated target of 20 per cent ethanol blending with petrol, viewed largely a measure of environmental sustainability and reduced dependence on fossil fuels will be implemented by 2017. The traditional route of producing ethanol as a by product of sugar industry will not meet this huge demand. Therefore, as demand for production of fuel ethanol, through renewable sources increases to unprecedented levels, feedstock for ethanol production will become more diverse. High biomass low lignin sorghum would be a best option owing to its wider adaptation, high water and input use efficiency apart from multiple uses. Therefore, sorghum being a water use efficient non-invasive crop having wider adaptability, if improved for biomass without significant trade-off on grain will meet the requirement of upcoming commercial lignocellulosic ethanol firms that has not only significant impact on sustaining the environment but also on livelihood opportunities of small farmers in semi-arid tropics.

Potable Alcohol/Alcoholic Beverages

Alcohol is largely produced from molasses, about 65 per cent for industrial use and 35 per cent for human consumption, mostly as Indian made foreign liquor (IMFL). Grain based alcohol is qualitatively superior and the demand is on the increase. The *Kharif* sorghums are ideal for this purpose, even if the grain is damaged due to rains. Of course, now there are methods to prevent damage due to rains and to obtain to clean grain. About 13 companies are licensed to produce grain based alcohol and the capacity is said to be 100 m tones. Varieties with high starch and low protein content and good sto*Rabi*lity are better suited for this purpose. The sorghum project has established linkages both with research laboratories and manufacturing companies and is working to realize genetic enhancement of *Kharif* sorghum for alchoholic beverages. The potential for *Kharif* sorghum is excellent in this direction and will be capitalized. Sorghums could also be produced in summer with high yields exclusively for potable alcohol. The by-product from alcohol production will be good animal feed. Sorghum has also the potential to be brewed into beer and this has been demonstrated. IIMR will establish links with breweries and explore extensive usage of sorghum in the beer industry.

Starch Industry

Maize is the major source for starch industry. Cassava and sorghum could be alternatives. Besides starch, a number of downstream products like modified starches, dextrin, liquid glucose, dextrin monohydrate, sorbitol, calcium gluconate, citric acid, etc., and by-products like steep water, oil, cake, fibre, etc., are produced in the industry. Starch is largely used in the textile, paper and food industries. If starch recovery from sorghum is improved, sorghum could be a potential source. The endosperm attributes for starch and the germ for oil need to be improved by genetic manipulations.

Exports

Despite the fact, we are the second largest producer of sorghum; we do not find recognition in the export market. It looks, we may have comparative advantage to export sorghum to neighboring countries in the middle east (Saudi A*Rabi*a, UAE, etc.), Sri Lanka, and Japan and near east. The major producer and exporter of sorghum, USA, produce brown/red sorghums with tannins. We need to educate that our sorghums are one of better quality devoid of tannins which have anti-nutritional factors. We will also study the market requirements of these countries and direct our programme to meet specific export needs, including development of red sorghum and colured sorghum hybrids for different markets. The demand for feed grain is increasing globally and we need to study the potential markets and their quality needs. We might even develop specific areas for cultivation of export oriented feed grains.

Future Improvement of Millets

The various factors leading to the decline of millets in Indian agriculture have been deliberated in detail. Millet based agricultural systems need to withstand biotic and abiotic stresses because of their cultivation mostly in unfavourable soil and climatic conditions. Further they also need to adjust to changing economic (prices and income) and policy induced stresses as has been the case in India where subsidized wheat and rice are supplied through the public distribution system. With this background in view, we will embark on future approaches for millet improvement. Promotion of genetic diversity, cropping system stability and economic advantage or parity will become the major criteria. The genetic approaches should promote genetic diversity and cropping system performance and stability.

Kharif Sorghum

The higher order yield of the currently available hybrids and varieties is around 4-5 tonnes/hectares. Average yields hover around one tonne per hectare only. The food use of *Kharif* sorghum is declining. Marginal yield improvements at the experimental level without perceptible changes in grain quality or a major advance in insect resistance etc., do not make much difference. Shootfly resistance is being incorporated in sorghum cultivars of economic worth. Quality standards of both grain and forage to meet international standards are being addressed. Alternate uses of *Kharif* sorghum in food products, feed grain, energy, nutritional and therapeutic foods, starch and other industrial products and exports, are being developed to have specific quality (chemical attributes) to render the processes and products more efficeint. Breeding goals and approaches will be product specific and will be developed at appropriate centres.

Rabi Sorghum

In view of the lack of significant progress in improving *Rabi* sorghum yields, more basic and fundamental genetic studies are warranted to understand the genetic processes involved in the improvement of yield and the adaptation processes. Studies on heterosis in diverse crosses, temperature implications, lodging, shootfly and shootbug resistances stability across early normal and delayed sowings in the respective regions need to be achieved. The non-senescence or 'stay green' trait has been exploited. However good progress has been made in evolving soil depth specific cultivars to enhance *Rabi* sorghum yields.

Nutritional Quality

In the past, the sorghum project carried out several genetic studies to transfer quality protein/high lysine trait from Ethiopian sorghums with shriveled seeds and also a purdue mutant. From the late, photosensitive Ethiopian sources, early dwarf high lysine types with shriveled seeds and normal plump seed types were developed. The transfer of high lysine to plump seeds was partially successful. The studies need to be continued using modern biotechnology tools. The β-carotene content in several yellow endosperm types was analyzed and the stability of carotene was also studied. The studies were not pursued. In bajra, there are beautiful golden coloured commercial types. The carotene content do not appear to have received attention. With the current interest in golden rice, it is time that the project initiates work on incorporating β-carotene in commercial types which will be valuable in both human and poultry nutrition. The minerals, micronutrient, vitamins, therapeutic attributes are important in millets, more so in *ragi* and small millets. It is time we start paying attention towards nutritional attributes.

Population Improvement

For incorporation of multiple traits and traits governed by quantitative genes, population improvement procedures are ideal. Using the Maldandi source of cytoplasm and allowing 4-5 years of inter-mating and crossing with diverse types, selfing was practiced and fertile types were isolated. They were designated FR lines. Some of them were agronomically good. When some of the FR lines were crossed to male steriles, surprisingly they did not restore fertility. We had a feeling whether the restoration was apomictic. Development of improved populations for incorporating multiple, traits is certainly a useful adjunct. In bajra, renewed attention is needed for systematic population improvement programmes.

Need for Critical Analysis and Characterization of Millet Germplasm Collection

Because of their evolution and adaptation to poorer soils and unfavourable agro-climatic conditions/environments, millets have developed unique attributed such as tolerance to drought, heat, adaptation to poorer soils and unfavourable conditions, nutritional and therapeutic values, processes of starch and vitamin bio-synthetics, etc. Some of these attributes have been discussed earlier. The germplasm collections of sorghum and *bajra* have been characterized for their morphological and physical attributes. Millet collections were, perhaps, not described. In the present context, the germplasm collections should be analyzed for such specific, physiological nutritional, therapeutic attributes and identify the genes controlling them and their bio-synthesis and gene markers characterizing such unique traits and processes. Such data bases will be of immense value to the improvement of millet crops by traditional as well as biotechnological methods and marker aided selection.

Public Private Partnerships (PPP)

Public-private partnerships are described as collaborative efforts between public and private sectors in which each sector contributes to the planning, resources, knowledge and capabilities needed to accomplish mutual objectives. Diversified use of millets for demand generation and improve incomes is market driven. They warrant collaborative research between agricultural research agencies, the concerned industry or even industrial research laboratories. Such collaborations, some of which have been initiated in the recent past, and more organized collaborative project could lead to the development of process efficiencies and marketable products. Some opportunities to be explored are in the areas of:

Dry and wet milling	–	Milling industry
Feed grains	–	Poultry feed and dairy feed manufacturing organizations with the private and cooperative sectors like NDDB
Ethanol	–	Sugar mills and ethanol manufacturing plants
Alcohol/beer	–	Breweries association
Starch and starch based production	–	Starch industry
Exports	–	Export organizations
Straw	–	Straw board manufacturers

Direct and Indirect Environmental Benefits

Millets are staple cereal grain and fodder crops grown by subsistence farmers in the hottest, driest regions of e Indian subcontinent where rainfed crop production is possible. Enhancing the reliability of millet grain and straw yields in such environments will contribute directly and sustainably to the alleviation of poverty and improvement of food security of farm households. Sorghum, pearl millet and small millets, are more tolerant of high temperatures than other cereals, are the ideal crops for such conditions. Efficient breeding for drought-prone environments requires selection strategies that are sufficiently independent of the variation inherent in the natural environment to permit identification and use of true genetic differences. Molecular markers linked closely to traits such as Sorghum, Pearl millet and Small millets grain and stover yield during drought will provide increased reliability for selecting and breeding of these traits. Developing new drought- resistant varieties however only reduces risk if resource-poor farmers have reliable access to high-quality seed at low cost.

Coping with the Effects of Climate Change, or Risk from Natural Disasters

Predicted climate change scenarios indicate that water shortages and shorter effective growing season lengths will be increasingly likely thus increasing the need for short-duration crops such as sorghum, pearl millet and small millets with enhanced drought tolerance. Development of millets will not only able to withstand such conditions but also will be able to provide a usable yield increase with resilience for poor people inhabiting drier regions.

Opportunities

The term millet includes a number of small-grained cereal grasses. Based on the grain size, millets have been classified as major millets which include sorghum and pearl millet and several small grain millets which include finger millet (*ragi*), foxtail millet (*kangni*), kodo millet (kodo), proso millet (*cheena*), barnyard millet (*sawan*) and little millet (*kutki*). The advantages of cultivation of these crops include drought tolerance, crop sturdiness, short to medium duration, low labour requirement, minimal purchased inputs, resistance to pests and diseases. Millets are C4 crops and hence are climate change compliant. There are varieties particularly in little millet and proso millet which mature in 60-70 days; yet providing reasonable and assured harvests even under most adverse conditions. India is a store-house of highly valuable genetic variability. Millets sequestrate carbon and thereby reduce the burden of green house gas. Millets have been called nutri-grains since they are rich in micronutrients like minerals and B-complex vitamins. Additionally millets are also rich in health promoting phytochemicals, and can be used as functional foods. Millet cultivation is the mainstay of rain-fed farming on which 60 per cent of Indian farmers depend. They provide food as well as fodder and can be mix-cultivated (polyculture) with pulses and vegetables. Despite these attributes, millets are losing their pride of place in production and consumption in India, but there has been a greater revival of these crops in recent years.

Sustained Production of Millets

Though India is the largest producer of millets in the world, between 1961 and 2013, there has been drastic reduction in the area under cultivation of millets but due to productivity gains in some varieties, total production of millets showed some increase despite shrinkage of area.

Reasons for Decline in Area

Almost 50 per cent area under millets has been diverted largely to soybean, maize, cotton, sugarcane and sunflower. A combination of factors like low remuneration as compared to other food crops, lack of input subsidies and price incentives, subsidised supply of fine cereals through Public Distribution System (PDS), and change in consumer preference (difficulty in processing, low shelf life of flour and low social status attached to millets), have led to shift from production of millets to other competing crops. There is a vast gap between productivity of millets in field demonstrations vs. farmers' field pointing to the need for robust extension activities, and transfer of state of the art farming practices, as well as access to good quality seeds to the farmers. Increasing marginalised cultivation is also an impediment to productivity. Demand for millets can be increased by: (i) Creating awareness regarding their environmental sustainability, nutritional and other health benefits, (ii) Making them available through PDS, (iii) Value addition, and (iv) Inclusion under feeding programmes like mid-day meal, Integrated Child Development Services (ICDS) feeding, and adolescent girls nutrition scheme (now under consideration of Ministry of Women and Child Welfare)

Estimates on Production Potential versus Demand

Estimated demand for millets by 2025 is 30-35 million tonnes. This has to be met by increasing productivity through choice of better varieties, good agronomic practices, effective extension activities and robust policy initiatives. A study by ICRISAT using

CERES-sorghum and CERES-pearl millet crop growth models and historical weather data, rain-fed potential yields and water balance of sorghum (*Kharif* and *Rabi*) in different locations, showed that the total yield gap (simulated rainfed potential yield - farmers' yield) in production zones ranged from 2130 to 2560 kg per hectare for *Kharif* sorghum, 280 to 830 kg per hectare for *Rabi* sorghum and 680 to 1040 kg per hectare for pearl millet. This indicates that productivity of *Kharif* sorghum can be increased by 3.0 to 4.0 times, *Rabi* sorghum by 1.4 to 2.7 times and pearl millet by 1.8 to 2.3 times from their current levels of productivity. Thus assuming an average of 2.5 to 3.0 times potential yield increase as a whole for millets, and current area coverage, the achievable production potential can be three times *i.e.*, 45-55 million tonnes. This production potential is achievable by bridging the yield gaps not only through supply side factors such as high yielding crop cultivars, nutrient management and integrated pest management but also addressing demand side issues - value addition by processing, nutritional labelling, alternate industrial utilization and policy measures so as to make millets cultivation more remunerative. Breeding strategies should be redesigned to orient product-specific cultivars. With increased demand and profitability, there will be incentive for farmers to shift to millet cultivation even on better lands. Such a shift will be environmentally and nutritionally beneficial.

Biofortification

Bio-fortification means enrichment/value addition in crop through genetic manipulations. This seed-based approach is farmer-empowering and can go a long way in reducing deficiency of micronutrients, particularly iron, zinc and vitamin A (Beta carotene) in Indian diets. Priority is needed for research in biofortification, preferably using conventional breeding and molecular breeding methodologies. A major thrust is given to bio-fortification research in pearl millet under CGIAR funded research programme by ICRISAT. The aim of bio-fortification in pearl millet is to develop varieties/hybrids with high iron and zinc content.

Seed Production

There should be an adequate support for seed production of hybrids/varieties of both public and private sector through public-private partnership (PPP). This should be initiated through signing of MoU between State Departments of agriculture, seed agencies (NSC/SFCI/SSCs) and private sector seed companies. The State Seed Committees may identify the varieties/hybrids as per State-specific needs. Participation of All India Coordinated Research Projects on pearl millet/sorghum/ small millets may be made mandatory in such committees. To promote cultivation of millets as industrial crops, seed delivery systems, community-based services, post-harvest processing, input supply and marketing, coupled with awareness regarding advantages of millet cultivation need to be strengthened.

Nutritional and Health Value of Millets

Millets in general are rich source of fibre, minerals and B- complex vitamins. Among the millets, pearl millet (*Bajra*) has the highest content of macronutrients, and micronutrients such as iron, zinc, Mg, P, folic acid and riboflavin. Finger millet (*ragi*) is an extraordinary source of calcium. Though low in fat content, it is high in PUFA (polyunsaturated fatty acids). It is also rich in essential amino acids, like lysine, threonine, valine, sulphur containing amino acids and the ratio of leucine to isoleucine is about 2. The chemical score (percentage of the most limiting amino acid compared to a standard protein like egg protein) of finger millet is about 50 which is relatively better than other millets, Sorghum (34) and pearl millet (43). Millets are a treasure-trove of micronutrients like B-complex vitamins and minerals whose deficiencies in India are rampant. They also contain fibre and health promoting phytochemicals which function as antioxidants, immune stimulants etc., and thus have potential to mitigate degenerative diseases such as diabetes, CVD, cancer etc. whose incidence is rising in India. This makes millets important candidates as functional foods.

Bioavailability of Nutrients from Millets

High fibre content and presence of some anti-nutritional factors like phytates and tannins in millets affect bioavailability of minerals. Few studies in humans have suggested that absorption of iron tends to be lower from millets than from rice or even wheat. Thus, the advantage of having higher content of micronutrients may to some extent be nullified by lower bioavailability. More studies using state of art techniques and different methods of cooking are needed to examine the bioavailability of micronutrients including minerals and B-vitamins to assess their nutritional advantage in vivo. Processing can improve the bioavailability of nutrients as well as functionality. Limited studies show that bioavailability as well as functionality differs with the type of processing and preparation. More work is needed to optimise both of these. Effects of physiological status (pregnancy, lactation), age, and nutritional status should also be investigated. Absorption of vitamins and minerals tends to respond to body's demand. Dietary fibre has health benefits like good bowel movement, and reduction in blood cholesterol and sugar. Besides fibre, millets are also rich in health-promoting phytochemicals like polyphenols, lignans, phytosterols, phyto-oestrogens, phytocyanins. These function as antioxidants, immune modulators, detoxifying agents etc. and hence protect against age-related degenerative diseases like cardiovascular diseases (CVD), diabetes, cancer etc.

Effect of Processing on Nutrients

Milling, roasting, soaking, malting, germination and fermentation have been found to reduce phytic acid and tannin contents of millets. The nutrient content of millet grain is relatively poor after milling but the bioavailability of certain nutrients, such as iron improves considerably. Weaning foods prepared by roasting of finger millet have higher iron bioavailability. Popping or puffing of finger millet enhances protein and carbohydrate digestibility. Soaking of millet flour prior to heating can activate phytases and thereby improve zinc availability. Malting, finger millet reduces tannin content (brown millet) and phytate and improves ionisable iron and soluble zinc significantly. Malting of pearl millet and finger millet reduces protein content, but improves protein efficiency ratio (PER), bioavailability and has pronounced effect in lowering anti-nutrients. Feeding isocaloric diets differing only in the type of cereal, *i.e.* pearl millet, sorghum, wheat and rice to rats showed significantly higher absorption and liver zinc and iron for the pearl millet and wheat-diet fed groups, than rice or sorghum fed groups. The weight gain was also the highest in the pearl millet group compared to sorghum, wheat and rice-fed groups.

Designer Functional Foods

In recent years, the term functional foods have been used for foods that promote health through prevention of specific degenerative diseases. This effect is due to the presence of health-promoting and bioactive phytochemicals in plant foods. Some of the known nutrients- vitamins, minerals, essential fatty acids also have benefits in terms of prevention of degenerative diseases, besides their known functions of preventing nutritional deficiency diseases. The term nutraceuticals (like pharmaceuticals) is used for such bioactive compounds having protective effect against degenerative diseases, in isolated form. It is now recognized that harmful oxygen species like free radicals and peroxides damage cells and initiate the process of diseases. Being non-glutinous, millets are safe for people suffering from gluten allergy and celiac disease. They are non-acid forming, and hence easy to digest. They are also non-allergenic.

Epidemiologically lower incidence of diabetes is reported in millet-consuming populations. The diabetes preventing effect of millets is primarily attributed to high fibre content. Some antioxidant phenols in millets also tend to have anti-diabetic effect. Reports from India are sporadic. Sorghum is rich in phenolic compounds and antioxidants. Among minor millets, fox tail and barnyard millet have low glycaemic index (40-50). Many processed traditional foods have been prepared from minor millets and pearl millets. However systematic studies to validate their glycemic index are needed. Millets being high in fibre, antioxidants and complex carbohydrates are potential candidates for having beneficial effects against diseases. A few *in vitro* and animal studies support this view but well controlled studies in human are needed. In conclusion, millets have potential for protection against age-onset degenerative diseases. This is an area where more work is needed since these diseases are increasing in India. As the largest producer of millets, India can capture world market with appropriate validated functional foods.

Post Harvest Processing Technologies for Sorghum and Millets

Commonly used processing technologies for millets are: milling including decortications or seed coat separation and size gradation into semolina and flour, popping, malting, fermentation and cold extrusion. In recent years, contemporary food processing technologies such as extrusion cooking for Ready to Eat (RTE) foods and breakfast cereals, pasta and vermicelli-noodles and bakery products, malting and brewing, wet milling for starch preparation are employed for these grains. Cost effective dehuller for SAM has not been developed though small prototype models are available. Development of simple technologies for preparation of semolina and flours; composite mini malting unit, Hot-air-based High Temperature Short Time (HTST) unit for sand less popping are some of the areas of future research. Central Institute of Agricultural Engineering (CIAE), Bhopal has developed a millet mill for small millets which can be popularized in millet catchment areas. Milling to separate seed coat or decortication reduces fibre and nutrients but improves consumer acceptability. Bran is a good source of dietary fibre and edible oil. Deoiled bran from millets has the advantage of having lesser silica than rice bran. Products involving heat treatment (popping, flaking, expanded grains (murmura type) par boiling have better shelf life, since heating deactivates lipase. They also have better aroma and texture, and can be converted in to a variety of products which can be used as snacks or made into complementary food for feeding programmes. Grains like barley and finger millet are amenable to malting (sprouting). Malting increases the bioavailability of micronutrients and amylase, and reduces bulk by breaking down starch. Amylaserich foods (ARF) like finger millet malt are used in industry. Incorporation of 5 per cent ARF is mandatory for energy-rich formulations used in states sponsored nutrition programmes. Texture and digestibility are also improved. SAM flours singly are not suitable for bakery products, but they can be mixed with wheat and converted to bakery products. Traditional foods like *roti* cannot be reheated but Mexican Tortillas can be. There is a tremendous scope for promoting SAM-based traditional (papad, murukku etc.) and contemporary foods (extrusion cooking-pasta, vermicelli, RTE foods etc.) as health foods because of their high nutrient and phytochemicals content. Several Home Science Colleges, industries, Directorates under ICAR, and NGOs are attempting to prepare and market traditional and contemporary millet-based foods. R&D work on parboiling of SAM is scanty even though the potential has been established.

Other Commercial Uses of Millets

Sorghum and millets are basically starchy grains and have potential for industrial level production of starches, dextrin and ethanol for food and allied applications. In most of the developed countries, Sorghum and Millets finds extensive usage as feed components for bird, cattle and pigs.

Recent Initiatives – A Value Chain Approach Initiative for Nutritional Security through Intensive Millets Promotion (INSIMP)

To promote cultivation and consumption of millets and millets-based products, the Government of India announced a programme under Rashtriya Krishi Vikas Yojna (RKVY), for INSIMP. The aim is to demonstrate improved production and post-harvest technologies in an integrated manner. Major millets such as sorghum, pearl millet and finger millet are covered. Apart from bringing more land under cultivation of millets, about 300 Post-Harvesting units have been established in the Andhra Pradesh, Gujarat, Madhya Pradesh, Mahrashtra, Tamil Nadu and UP. These units have increased the supply of raw material for value added products of Millets.

Employment Generation for Women through Production and Marketing of Millet-based Processed Foods

Some Home Science Colleges have involved Self Help Groups (SHGs) in preparing traditional millet-based products, teaching quality control measures, packaging and labeling and hand-holding for local marketing. A successful case study is from the University of Dharwad, where women of North Karnataka are making *roti* (salted pancakes) from sorghum, and pearl millet, and supplying to hotels "Khanavalis" retail and wholesale dealers, besides supplying to marriage and other social parties at the cost of Rs. 3-5 per *roti*. On an average each woman can prepare 50-60 *rotis*. A variety of other traditional and contemporary (noodles) millet-based products including small millets have been developed under projects like NAIP and IDRC.

SHG women have been trained in their production and help is being given for direct marketing as quality control and market are major challenges. Such a venture can be successful if either a government institution or NGO provides supervision and marketing.

Strategies for Creating Demand

1. The production and consumption of millets must be augmented with appropriate policy initiatives.
2. Consortium-mode research may be pursued for validating the advantages of millets as health and functional foods.
3. Traditional and non-traditional, ready to use, convenience foods and foods that can be used for complementary feeding programmes may be developed with proven nutrient content and bioavailability mapping.
4. Millet-based complementary foods such as *khichdi, upma, roti* etc. in feeding should be introduced in feeding programmes such as MDM, ICDS etc.
5. R&D on millets as fodder and forage for livestock feed security may be strengthened.
6. Commercialize and promote millet-based processed ready to eat snacks and convenience foods through public private partnership.
7. Awareness regarding nutritional, health and environmental advantages may be created thorough known communication strategies.

Strategies for Enhancing Supply

1. Development of varieties/hybrids of Sorghum and Millets with better recovery capacity on reversal of dry spell for harsh environment/drought prone areas.
2. Exploration of zero tillage for millets under rice fallows particularly for southern States.
3. Development of hybrids/varieties resistant/tolerant to salt/high temperature. Strengthening breeding programmes through conventional breeding, marker-assisted breeding as well as biotechnology for biofortification and other traits such as varieties with better root architecture.
4. Validation of high productive technology under real farming situations.
5. Effective deployment of trait-specific germplasm available in gene banks for genetic enhancement.
6. Evolving strategies for better seed production with public, private, NGO partnership and establishment of seed villages.
7. Research for better post-harvest management for enhancing the shelflife of millets and prevention of wastage.
8. R&D for integrated toolkit for farm mechanization. CIAE Bhopal has developed a millet mill suitable for small millets. This should be tested in

millet catchment area. Retro fitting of machinery used for rice/wheat/maize for millet foods processing.

9. Markets and entrepreneurship development through modern and innovative approaches.
10. Generation of scientific data to substantiate the claim of conservation of biological resources, low water consumption, agro-climatic limitations and high nutritious value of millets and their derivatives.
11. Promote production and consumption of millets through mixed/relay cropping with legumes and vegetables in homestead gardens.
12. R&D work to generate evidence-based information on the phytochemicals with nutraceutical characteristics and authenticate their health potential including anti-diabetes, anti-inflammatory and hypo-cholesterolemic properties, through clinical trials and nutritional studies. Functional foods for diabetes and obese populations based on Sorghum and Millets will have good market. Measurements of glycemic index should be done using specified WHO-FAO protocol. Such studies must be extended to the best preparations/recipes from millets with functional properties, through proper clinical trials.
13. Studies to examine the bioavailability of micronutrients from different preparations of millets.
14. Breeding to improve the levels of lysine and tryptophan and also screening the germplasms for specific end uses such as milling, popping, malting and vermicelli noodles etc.
15. Setting up of a training-cum-demonstration centre for integrated processing of sorghum and millets.
16. Undertaking surveys of sorghum and millets foods and allied industries, for bringing out a directory and share knowledge base for modernization of the Sorghum and Millets processing industries.

Scientific, technological and behavioural engineering involving convergence of efforts of agriculture scientists, food technologists, home scientists, policy makers, and media is needed to revalorise millets. Some recent initiatives to rejuvenate millets from production to Consumption, include: "Initiative for Nutritional Security through Intensive Millets Promotion" (INSIMP), under the Rashtriya Krishi Vikas Yojna of Government of India, "Revalorising Small Millets in the Rain-fed regions of South Asia (RESMISA) funded by International Development Research Centre (IDRC) and CIDA (Canadian funds), and IIMR-led value chain development approach for commercialisation of millets. Millets are an important component of the National Agriculture Innovation Projects of ICAR, and All India Coordinated Project in Home Science, Other policy initiatives include: price and procurement support for millets, inclusion of millets in the Mid day meal programme and, promotion of Nutrifarms. Demand growth for food, fodder and bio-energy triggered by expanding populations and dynamic patterns of their utilization has put Millets in an appropriate and opportune position. To produce maximum from limited available land, the onus for these target increases heavily rests on technologically driven yield improvements. For this purpose, there is a need to make our genetic improvement programmes more efficient, effective, rapid and precise by making use of cutting-edge technologies of genomics, proteomics and other tools and knowledge of new biology.

Way Forward

There are many opportunities for enhanced production and profitability of sorghum and millets. Despite a declining per capita consumption of sorghum and millets, they still remain cheapest grain accessible to the economically-deprived people especially in rural and remote areas. Resistance to abiotic stresses, diseases and pests, and introduction of processing technology to facilitate easy cooking and consumption and development of novel, functional and designer foods for specific health needs of children, diabetics and malnourished will greatly enhance the food value of millets. Therefore, we are developing strategies for resolving the constraints which inhibit the increased use of improved technologies in a cost-effective manner. The higher scope for industrial utilization, livestock development through quality forage and enhanced utilization of grain by poultry, and potable alcohol industries, and sweet stalk juice by biofuel industry will aid increased income to sorghum and millet cultivators, industries and exporters. Therefore, sorghum and millets will then be called as industrial crops rather than poor man's crops.

Another dimension deals with emerging class of health conscious urban people. ICAR is coming up with a mega research theme on health and value addition which will have dual benefits of increasing farm incomes besides offering consumers a range of health and nutritive products. This has bearing on our research agenda and the focus of IIMR's research priorities are to be re-assessed in this context so as to merge with ICAR goals. Sorghum and millets fits perfectly within this as they are nutri-cereals with high fibre and rich in minerals and vitamins. While many researchable issues such as processing for better shelf-life and bioavailability of minerals will be addressed, processing should become the means of eliminating inconveniences apart from reviving consumption. With findings that have emanated from NIN's clinical trials supporting the positive results of sorghum and millets consumption in the reduction of lifestyle diseases such as diabetes, the research on similar aspects will be continued to provide sorghum and millets their due niche position as a health food. Since IIMR is primarily a crop improvement institute there is a need to re-orient our breeding strategies by strengthening breeding for specific end-products which is a deviation from the past where the improvement of productivity levels were prime concern. Further, identification of genotypes suitable for end product characters and for higher recovery of processed products will also be taken up on priority. Also, shelf life enhancement of not only seed and grain, but also of flour, rava and other products will be on top agenda for value addition and sustainable market prospects. Though the promotion of sorghum through food processing may result in urban niche markets as health conscious urban class is opting for them, the mass consumption will thus get promoted through policy interventions such as INSIMP which again has all India coverage offering end to end solutions for demand creation efforts backed by the research efforts in the form of technology development on both processing machineries and products thereof meeting most of the rural producer's and consumer's needs alike and at much affordable cost. In wake of government's policy to include sorghum and millets in PDS, there is a need to ensure proper shelf life of grain/flour/rava the primary products which again a researchable issue to be addressed with regard to their placement in PDS and also in commercialization regime where the grain should not be lost in supply chain.

Limitations with Storage and Consumption of Millets

The awareness about the high nutritive values of millets is very low both among the producers and consumers. Besides, being small seeded and low priced commodity, the produce is not properly cleaned, graded and dried before they are brought to the markets, fetching low price to the farmers and poses storage problems. Besides, rains at the time of maturity/harvest also lead to poor quality of *Kharif* produce. The high fat, fibre, inert matters coupled with high moisture content/high humidity leads to infestation of insect/fungus during storage. Grains stored with high moisture content often develop Mycotoxins/Aflatoxin which is highly poisonous both for human and live stock. The shelf life of flour of millets is also low because of high fat content in the grains. The dark seed coat color of some millet such as ragi and bajra also possess problem in acceptance of food products of these commodities. Thus, enhancement of shelf life could be an important challenge to process grain to products and market them sustainability, which would be addressed on priority.

Commercialization

The market for health and wellness foods in India is estimated to be worth $7.5-10 billion by 2015 according to a study undertaken by the Tata Strategic Management group. The demand for fortified foods, whole grain, high fibre, low cholesterol, natural and organic foods will grow. The recent efforts of these public private partnerships under NAIP have led to commercialization of some millet products though scaling up is a challenge; however, they have proved the point of possibility of convenient millet foods on par with rice and wheat so as to offer choices to urban health conscious consumers. Similarly there are efforts from major private player such as Britannia industries Ltd through their nutria-choice segment and ITC's multigrain biscuits the millets appear to play a major role. In recent times Britannia Industries have entered a research MoU with IIMR (DSR) on sorghum biscuits and bakery products which is again a land mark. Apart from big players there are many small scale players who produce extruded snacks from sorghum, and pearl millet in Mumbai, Rajasthan, Tamil Nadu and Andhra Pradesh. Among the millets, finger millet based malt is more acceptable product both in traditional and non-traditional consuming areas. Some of these products have premium prices and are acceptable as up market products cater to high income groups. Gluten-free sorghum products are commercialized by an entrepreneur based in Kolkata catering to celiac patients. Processed Small millets are commercialized in a big way by entrepreneurs from Nasik, Maharashtra and are being export. Though substantial studies are not attempted on the bioavailability of minerals in millets the general awareness that is being created across the country and world over that millets are nutri-cereals and offer promise to health seeking population who desire to keep lifestyle diseases at bay. IIMR alone has entered more than 12 MoUs with private partners. However, advocacy and policy prescription would take the consumption to greater heights that would generate not only demand for millets but also raise standard of living of the poor dryland farmers growing millets.

Value Chain Approach: A Success Story

Value chain on sorghum and Millets: Under National Agricultural Innovation Project (NAIP) sub-project on "Creation of demand for millet food through Production to Consumption System (PCS) value-chain" with support from the World Bank IIMR led consortium comprising of, National Institute on Nutrition (NIN), ANGRAU and ITC Ltd. at Hyderabad from December, 2007. This was one of major effort in millet promotion where the pilot scale testing is done to create demand for millet foods through interventions at on farm level, processing technologies, nutritional labeling, marketing and entrepreneurship development. Thus end to end solution is provided linking from on farm production to consumers and will enable upscaling through entrepreneurship development.

Interventions through Product Specific on-Farm Production

The consortium with the help of ITC, its partner has established supply chain management in its place through it market assured e-choupal platform first time on pilot scale with sorghum farmers. The end product specific cultivation enabled the desired change in mindset of farmers that millets can be commercially grown under value chain approach. ITC provided backward integration by bulking and aggregation functions and linking farmers with rest of the value chain. Thus, the problem of supply of raw material (despite low marketed surpluses) for commercial production was resolved and value addition through these linkages made the success of all the stakeholders, especially our primary stakeholder, the farmer is biggest beneficiary through increased profitability per unit land they began to allocate better lands to sorghum cultivation which by all means an impact of the initiative, which may be replicated. The processing interventions at farm level have generated more awareness on commercialization of millets; it has high potential of generating employment in rural areas especially during the off season, more income from their produce and above all, positive impact in their overall farm practices changing their mindset from subsistence farming to commercialization. It will also provide them option to diversify from traditional typical farming.

Interventions on Nutritional Evaluation

NIN evaluated the effect of sorghum diet on growth of school children, among diabetes and studied that acceptability and growth in school children after consuming sorghum diet for more than 8 months in 'rice eating belt' and compared with controls. The growth of anthropometric indices was significantly better in sorghum group, the biochemical indices were also better for Hemoglobin, Calcium, Iron and Vitamin B12. Studies on GI show that low glycemic index for sorghum products ranged between 60 -70 while the glycemic load (~60-35) which were significantly lower than wheat control samples. These results were further corroborated with establishment of sorghum in reducing diabetes in terms of lowering of Gycosilated hemoglobin levels in subjects fed with sorghum processed products. Through these clinical trials, Glycosilated hemoglobin levels decreased from 7.9 to 7.3 g per cent; whereas Normal range <7.0 g per cent), corroborates the beneficial effects of sorghum among those ailing from diabetes. However, these results need to be validated further for large scale recommendations. However, in the pilot testing the findings are used for labeling of the commercialized products in the project. Similar studies and leads in millets will enhance their uptake and trigger initiatives in the direction of novel and nutritional designer foods.

Processing and Marketing Interventions

We developed and standardized 18 sorghum based processed food through employment of various processing technologies such as flaking, baking, extrusion, milling, popping etc through retrofitting machinery to suit sorghum processing (as there are no available machinery suitable for sorghum). The shelf life is enhanced. Short listed six products i.e sorghum based Multigrain *atta*, *Jowar suji*, flakes, pasta, vermicelli and biscuits which were labelled and packed for marketing to the consumers. Thus consumers were offered with choice of convenient and health foods. NIN data was used to label them. They found to be nutritionally superior to the existing on shelf products that are based on refined wheat and rice. IIMR has launched sorghum products on their brand name "*Eatrite*" on a pilot scale in Hyderabad. Eatrite products are now marketed through "Heritage Fresh" and ITC retail chain stores in Hyderabad. Simultaneously outsourced the event managers for popularization of sorghum products in urban markets. Massive awareness is created on sorghum as health and nutritious food through "Road shows" (100+) in public parks, malls, and institutes etc in Hyderabad and in exhibitions in creating awareness of sorghum to thousands of consumers through fabricated *Jowar Rath* in Pune, Bangalore, Jabalpur, Chennai, Coimbatore, New Delhi etc. We have also signed Memorandum of Understanding with 12 entrepreneurs It has also signed Research MoU with multi-national M/s Britannia Industries Ltd, Bengaluru for joint research to developed sorghum biscuits and bakery products of international standards. However, challenges loom around its success story such as sustainability, upscaling of the prototype technologies, Replication/Horizontal expansion and spread in the non-traditional consuming areas. Awareness on nutritional superiority is also a challenge. We are contemplating for horizontal expansion of marketing these products through private entrepreneurs either through technology licensing or brand franchisee MoUs. Innovative methods of popularization are being adopted to take the message across.

Entrepreneurship Development

In order to keep up the momentum and the sustainability of commercialization process, entrepreneurship development of the stakeholders is necessitated through interventions in food processing and product development and nutritional evaluation. Creating sustainable value chain has been one of the greatest challenges for the social scientists and research institutions at large. Therefore the ultimate goal of entrepreneurship development programmes is to disseminate thorough knowledge of post-harvest management which includes linkage of farmers with market, processing, nutritional importance of sorghum, and marketing. The stakeholders were trained who includes progressive farmers, rural entrepreneurs, NGOs, SHGs, small and medium scale processors, women group entrepreneurs on topics such as

1. Nutritional importance of sorghum and millets
2. Post Harvest Technologies of Sorghum/Millets
3. Branding, Packaging and labelling

Similarly, another two projects under NAIP value chain mode were attempted by UAS, Dharwad and Raipur to study on small millets. These projects successful stories may be available as a model for integrated approach for millets promotion across the country. The INSIMP, a Government of India's scheme on millets promotion is an offshoot of policy sensitization of the efforts of IIMR led consortium which is now operating across the country for the third year.

Millets- Long Term Perspective 2050

Possible Options for Working out Estimates for Long Term Planning of Millets Production

Three production scenarios for sorghum and other millets are listed below based on possible options/situations for working out estimates for long term planning. This would help in strategizing the research priorities in the context of changing scenarios as cultivation scenario is highly dynamic due to various issues listed and explained in the preceding chapters.

Scenario No.	*Description*
Scenario - I	Existing scenario
Scenario - II	Optimistic estimates based on anticipated improvement in yield levels from existing area
Scenario - III	Estimates based on anticipated favourable Policy push

Table 4: Millets Cultivation and Production: Projection up to 2050 AD

SORGHUM									
Ecology of Production area	*Scenario I*			*Scenario II*			*Scenario III*		
	Area (m ha)	*Yield (Kg/ha)*	*Prod. (m t)*	*Area (m ha)*	*Yield (Kg/ha)*	*Prod. (m t)*	*Area (m ha)*	*Yield (Kg/ha)*	*Prod. (m t)*
Grain Sorghum									
Kharif									
Assured Rainfall area	1.32	1378	1.82	1.32	3000	3.96	2.00	3000	7.00
Non-assured Rainfall area	1.99	718	1.43	1.99	1500	2.99	3.00	1500	4.50
***Kharif*-Total**	**3.31**	**1048**	**3.25**	**3.31**	**2750**	**6.95**	**5.00**	**2750**	**11.50**

Contd...

Table 1.4–*Contd...*

SORGHUM									
Ecology of Production area	*Scenario I*			*Scenario II*			*Scenario III*		
	Area (m ha)	*Yield (Kg/ha)*	*Prod. (m t)*	*Area (m ha)*	*Yield (Kg/ha)*	*Prod. (m t)*	*Area (m ha)*	*Yield (Kg/ha)*	*Prod. (m t)*
				Rabi					
Rainfed Area	3.51	616	2.65	3.51	1500	5.27	5.0	1500	7.50
Irrigated area	0.87	1597	1.39	0.87	4000	3.48	4.0	4000	16.00
***Rabi*-Total**	**4.38**	**1106**	**4.04**	**4.38**	**2750**	**8.75**	**9.0**	**2750**	**23.50**
Grain Sorghum-Total	**7.69**	**1077**	**7.29**	**7.69**	**2750**	**15.70**	**14.0**	**2750**	**35.00**
				Pearl Millet					
***Kharif*, *Rabi* and Summer**	**9.00**	**1214**	**10.92**	**9.00**	**2500**	**22.50**	**15.0**	**2500**	**37.50**
				Finger Millet					
***Kharif*, *Rabi* and Summer**	**1.11**	**1428**	**1.59**	**1.11**	**2500**	**2.78**	**4.0**	**2500**	**10.00**
				Other Millets					
***Kharif*, *Rabi* and Summer**	**0.75**	**750**	**0.43**	**0.75**	**1500**	**1.13**	**2.0**	**1500**	**3.00**
Total Millets-Grand Total	**18.55**	**1117**	**20.23**	**18.55**	**2400**	**42.11**	**35.0**	**2400**	**85.50**
				Multicut Fodder Sorghum					
Ecology of Production Area	*Area (m ha)*	*Green Fodder Yield (tonnes/ha)*	*Prod. (000 m t)*	*Area (m ha)*	*Green Fodder Yield (tonnes/ha)*	*Prod. (000 m t)*	*Area (m ha)*	*Green Fodder Yield (tonnes/ha)*	*Prod. (000 m t)*
***Kharif*, *Rabi* and Summer**	**18.70**	**92.0**	**17.20**	**18.70**	**100**	**18.70**	**27.0**	**110**	**29.70**
				SWEET STALK SORGHUM					
***Kharif/Rabi* Summer**	**–**	**–**	**–**	**–**	**–**	**–**	**1.0**	**150**	**6.67**

Source: Estimates worked out for Vision document 2050, IIMR, Hyderabad 2014.

Scenario – I: is existing situation with current levels of area, production and productivity which forms the bench mark for measuring likely improvement in respective parameters.

Scenario – II: deals with optimistic estimates with two assumptions. Firstly, current sorghum, pearl millet and small millets acreage will be retained. Secondly, there is going to be increase in productivity levels due to higher coverage of HYVs.

Scenario – III: assumes that the area under sorghum, pearl millet and small millets will impact positively due to favourable governmental policy (in the form of support for production, processing and value addition). Government policy is bound to be in favour of support for promotion of sorghum/millets due to increasing population growth rate and unmet demand for food consumption by the rice and wheat. Thus, it is expected that the Government policies are going to be strengthened for millet promotion during different plan periods. Further, it is expected that creation of awareness is being given more importance so as to generate consumption demand owing to nutritional merits of sorghum, pearl millet and small millets. The increasing incidence of lifestyle diseases which are linked with relatively poor nutritive composition of fine cereals especially rice, the promotion of nutri-cereals such as sorghum will be more pronounced owing to their superior composition of nutrients and minerals. It is also expected that millets as health food is being included in PDS which may gain some area under sorghum, pearl millet and small millets.

Overall, at least 88 per cent increase in millet acreage over the current levels is expected to be attained by 2050 AD if policy push and demand creation trend is going to be continued, even otherwise with current acreage if continued with more productivity enhancement through R&D is in place the increase in production will be doubled Accordingly, the estimated production in sorghum with policy push and R&D enhancement will be 35.0 million tonnes, of which *Rabi* production contribution alone will be 68 per cent of total production. Similarly 37.50 million tonnes in pearl millet continue to top, 10.0 tonnes in finger millet, 3.0 million tonnes in other small millets with overall aggregate production of 85.50 million tonnes of sorghum, pearl millet and small millets put together by the end of 2050 AD. Area gains in favour sorghum, pearl millet and small millet may come from other crops which are less remunerative (sorghum, pearl millet and small millet demand is going to higher due to value addition for food and non food uses and due to their importance as functional health foods) and from areas where commercial crops like sugarcane cannot be continued due to lowering of water table making it difficult for them while it still favours sorghum, pearl millet and small millets which with minimal Irrigations will help to double the yields. Even from non-traditional rice-fallow areas as second crop. Another area that has not been taken into account at all is area under multicut fodder sorghum. The present acreage under multicut fodder sorghum is one lakh hectares. The seed yield productivity is 15 quintals per acre. Thus a total of 37, 50,000 quintals of seed is produced, which @ 20 kg per hectare as seed rate will crop 18.7 million hectares area under multicut fodder sorghum. The 6-7 cuts from each hectares yield 92 tonnes of green fodder to the farmer. With the policy push and ever increasing demand for fodder required by domestic and dairy industry suggests that with improvement in yields and commensurate with demand, the fodder sorghum area by 2050 will be 27.0 million hectares. Sweet sorghum if policy push through blending of ethanol with petrol is implemented with remunerative pricing on par with petrol, the area under sweet sorghum will capture 1.0 million hectares by 2050 AD across the country and will supplement the sugarcane molasses based ethanol for blending with petrol. It can be inferred from above that the maximum production gains will be attained through policy interventions as in scenario III for millets to the extent of more than 400 per cent over the current situation. Such gains will enable meeting both food and nutritional security of millions. Thus productivity improvement may be within the realms of reality in the near future. Different genotypes suited to different growing conditions may be essential to bring in all-round increase in productivity. The economic gains that may be augmented by addressing envisaged benchmarks will result in significant improvement in productivity, profitability and even export earnings. All these are expected to translate sorghum and millets farming into a healthy and prosperous proposition, justifying the public support for sorghum and millets research in the country.

Millets: Ensuring Climate Resilience and Nutritional Security (2015) *Pages* **45–108**
Editors: **Vilas A. Tonapi and J.V. Patil**
Published by: **DAYA PUBLISHING HOUSE, NEW DELHI**

Chapter 2

Sorghum and Millet Genetic Resources

M. Elangovan and Vilas A. Tonapi
ICAR-Indian Institute of Millets Research (IIMR)
Rajendranagar, Hyderabad – 500 030

The term 'millets' refers to any of the small-seeded cereal and forage grasses used for food, feed or forage. Millets are small grained, annual, warm weather cereals of grass family that includes 8000 species within 600 genera, of which only 35 species comprising 20 genera have been domesticated (de Wet, 1986). The millets accounts for about 1 per cent of the food grains used in the world, and are important in areas, which are generally marginal for other agriculture. Millets are the sixth most important cereals of the world especially that of the semi-arid tropics of Asia and Africa. They play an important role in the diet of people in the semi-arid tropical areas. The protein content of most of these millets compares well with that of many pulses. Millets are mostly grown in poor soils as rainfed crops. Any serious pests and diseases do not affect them. The grains can be stored for a number of years under normal room temperature though many sorghums are poor in this trait Thus it acts to serve as famine reserves.

Genetic diversity provides farmers and plant breeders with options to develop, through selection and breeding, new, resilient and more productive crops. Collections of crop diversity held in *ex situ* genebanks appeared only in the 20th century. Altogether in national, regional and international institutes around the world holding, about 5.4 million samples of crop resources are held (Thomson *et al.*, 2004). It has been predicted that worldwide crop output must rise two-fold in the next couple of decades just to keep up with human population growth. This increase must occur – and be sustained – in the face of changing climates, diminishing land, stagnating yields, and degradation of other natural resources and global conflict (Imperial College Wye, 2003). FAO estimates that three fourths of the original varieties of agricultural crops have been lost from farmer's fields since 1900 (FAO, 1998) and an IUCN study found that an estimated 30,000 plant species disappear each year (Gardiner and Le Goulven, 2001).

Historical Background of Millets

Sorghum (*Sorghum bicolor* (L.) Moench) belongs to the family Poaceae and was domesticated in different areas of Africa. Seeds of sorghum, several millets, legumes and other edible plant remains were found at an ancient site, Nabta Playa, in southern Egypt close to the border with the Sudan. These remains were carbon-dated to be approximately 8,000 years old. Sorghum seeds were similar in size and structure to the modern wild types. This suggests the possibility of some cultivation for food and selection that may have contributed to domestication (*Wendorf et al.*, 1992). This is the earliest known archaeological record.

Pearl millet are absent from the early literature of the near-east like Egypt, Greece, and Rome. There is no biblical reference to the crop, nor it is mentioned in the works of Pliny, Dioscorides, or Theophrastus. Early literature tells us little beyond that pearl millet occurred in the semi-arid tropics of Africa before 1100 A.D. Pearl millet is not an ancient crop to India. Bulk of the evidence concerning the origin of pearl millet comes from the interpretation of present day distribution of the crop and its wild ancestor.

Origin and Domestication

Sorghum bicolor (L.) Moench origin and domestication taken place in Africa about 5000 – 8000 years ago. Indian subcontinent is the secondary origin of sorghum. The evidence for its cultivation was discovered at Saurastra dating back to about 4500 years. Sorghum was probably brought to India from eastern Africa during the first millennium BC. It is reported to have existed there around 1000 BC. The sorghums of India are related to those of northeastern Africa and the coast between Cape Guardafui and Mozambique. *Pennisetum glaucum* (L.) R. Br. is important millet in India, second to sorghum. It is originated in western Africa and was introduced to eastern Africa and the Indian sub continent some 2000 years ago. A large number of related species are found wild in Africa. The small millets indigenous to India are little millet (*Panicum sumentranse* Roth), kodo millet (*Paspalum scrobiculatum* L.) and Indian barnyard millet (*Echinochloa frumentacea* Link). On the contrary, three other small millets *viz.*, finger millet (*Eleusine coracana* (L.) Gaertn), foxtail millet (*Setaria italica* (L.) P. Beaurv) and common/proso millet (*Panicum miliaceum* L.) have originated elsewhere but reached India long ago. Finger millet originated in highlands of Africa and foxtail and common millet in Eurasia. The long history of cultivation of these three introduced crop species accompanied with human selection has resulted in generation of large variability especially in finger millet, giving India, a status of secondary centre of diversity.

Fuller *et al.* (2004) reported that, analysis of flotation samples from twelve sites in Karnataka and Andhra Pradesh in South India provides clear evidence for the predominant subsistence plants of the Neolithic period (2,800–1,200 cal B.C.). This evidence indicates that the likely staples were two millet-grasses (*Brachiaria ramosa* and *Setaria verticillata*) which were indigenous to the Indian peninsula. At some sites there is evidence for limited cultivation of few crops that originated in Africa, including hyacinth bean (*Lablab purpureus*), pearl millet (*Pennisetum glaucum*) and finger millet (*Eleusine coracana*).

A major step in the process of domestication of sorghum is the loss of the seed shattering characteristic (Mann *etal.*, 1983). Harlan (1975) asserted that domestication of sorghum occurred over time and in several areas where it was probably ennobled many times over several years. He envisions that early domestication occurred in an area extending from near the Ethiopian border, west through Sudan and up to Lake Chad. There is great diversity in this area as well as the presence of the primitive race *bicolor* (Harlan and de Wet, 1972).

Much of the agricultural history of sorghum has been for food, beverage, feed and building material. It has been used as an industrial crop during the last 100 years. Mechanization of its cultivation and harvesting has occurred primarily in the last 60-70 years. Industrial uses of this crop have been for feed, some for food, starch, the chemical industry and for fuel alcohol.

Pearl millet is most important among these millets and has the greatest potential of all millets. It is probably the most drought and heat tolerant of all the grain crops being associated with high temperature, light soils and semi-arid growing conditions. The important areas of production are hot, dry plains of southern Asia, and the southern periphery of the African Sahara.

Foxtail millets are high yielding and drought resistant; grows at higher elevation (180 m) and is frequently sown as an alternate crop with sorghum in black cotton soils when rainfall is deficient. China, India, Japan and Russia are important countries cultivating the crop.

Proso millets are relatively short-duration, quick-season, and quick-season irrigated crop with low moisture requirements. The grain after hulling makes nutritious flour for unleavened bread. This millet was grown in Russia, China, the Balkan countries and Northern India in old times.

Little millet appears related to proso millet but is generally shorter, has small panicles and seeds, and is grown on a limited scale with minimum care on poor lands. Little millet matures quickly and withstands both drought and cataloging. The crop is cultivated mostly in India.

Finger millet has a relatively wide range of adoption within moderate temperatures and moisture ranges. It is most widely cultivated on hilly, lateritic soils in the 51-104 cm rainfall belt of the tropics and sub-tropics. It has high yielding potential. The grain is very nutritious and has malting properties. Finger millet is highly self-pollinated. Intermediate elevations in South Central India and East Africa are the principal producing regions. The foothills of the Himalayas, Sri Lanka and Malaysia grow limited quantities of this millet.

Kodo millet is extremely grown on the poorest soils throughout India. It is reputed to be extremely hardy drought resistant and grows on stony or gravelly soils which would not support other crops. It is relatively long in duration requiring five or six months to mature.

Taxonomy

Sorghum bicolor (L.) Moench, is a member of Gramineae family, closely related to maize, predominantly self-pollinating crop, 2X, 2n=20. Pearl millet (*Pennisetum glaucum* (L.) R. Br.) taxonomically belongs to family *Poaceae* sub-family *Panicoideae* section *Penicillarium* and possesses 2n= 2x=14 chromosomes. It is a cross-pollinated annual C^4 crop species. Finger millet [*Eleusine coracana* (L.)] is a member of Gramineae family, self-pollinating with 4X, 2n=40. Foxtail millet's progenitor is thought to be common old world weed, *S. viridis* (2n=2x=20). The barnyard millet (2n=4x=36; 6n=56 also recorded) is very similar to wide-spread grass *E. crus-galli* (2n=4x, 6x=36, 54, 42 and 48 also recorded). The classification of millets of the poaceae family by Rachie (1975) is presented in Table 1.

Table 1: Classification of Millets of the Poaceae Adapted from Rachie (1975)

Tribe	*Species*	*Common names*
Paniceae	*Setaria italica*	Foxtail millet, Italian millet, German millet, Hay millet
	Panicum miliaceum	Proso millet, Common millet, Hershey millet, Hog millet, White millet, Broomcorn millet
	Brachiaria ramosum	Brown top millet
	Digitaria exilis	Fonio millet
	Echinochloa colonum	Shama millet
	Decompositum	Australian millet
	Frumentaceae	Japanese millet, Barnyard millet
	Paspalum scrobiculatum	Kodo millet
	Pennisetum americanum	Pearl millet
Chlorideae	*Eleusine coracana*	Finger millet

Adaptation

Sorghum is the fifth most important cereal crop in the world and grown on 44 million ha in 99 countries in Africa, Asia, Oceania, and the Americas. The Major sorghum producers are the USA, India, Nigeria, China, Mexico, Sudan and Argentina. Sorghum occupies 25 per cent or more of arable land in Mauritania, Gambia, Mali, Burkina Faso, Ghana, Niger, Somalia and Yemen, and >10 per cent of this area in Nigeria, Chad, Sudan, Tanzania and Mozambique. Most of it is grown for direct human use (>55 per cent). The grain is mostly consumed in the form of flat breads (Bhakri) and porridges (thick or thin); stover is an important source of dry season maintenance rations for livestock, especially in Asia; also an important feed grain (33 per cent), especially in the Americas.

Pearl millet is the sixth most important cereal crop, and an most important among millet other than great millet (sorghum) (>55 per cent of global millet production), grown in over 40 countries, predominantly in Africa and Asia, as a staple food grain and source of feed, fodder, fuel and construction material in the hottest, driest, semi-arid and arid regions where rainfed agriculture is practiced. It is cultivated in 29 m ha, supporting >100 m people—the poorest of the poor; most important to national food security in Namibia and Niger; the major producing countries are Senegal, Mali, Burkina Faso, Niger, Nigeria, Chad, Sudan, and India. It is also grown in Oceania and the Americas, predominantly as a forage and/or mulch component of minimum tillage-based cropping systems

Sorghum and pearl millet are the most important dry land summer cereals. Sorghum is generally cultivated in dry, hot areas (38-40° C) with an average annual rainfall of 400-750 mm, though it can be grown where rainfall is much higher. Sorghum also withstands water logging. Pearl millet is frequently cultivated on lightly textured, sandy soils with annual rainfall in the 300-500 mm range. Under comparable moisture conditions, sorghum will do better under cooler temperatures, while pearl millet does better under warmer conditions (House, 1995).

Most important "minor" small millet in the tropics (12 per cent of global millet area, similar to area of foxtail millet, which has declined rapidly over the past decade but was the main millet in China). It is grown in more than 25 countries in Africa (eastern and southern) and Asia (from Near East to Far East), predominantly as a staple food grain. Uganda, India, Nepal, and China are the major producing countries. It has high yield potential (>10 t/ha under optimum irrigated conditions) and grain stores very well.

Finger millet is a widely cultivated crop of the tropical and sub-tropical region of the world. In India it is grown from sea level to altitude upto 2100 m and in poor soils on hill slopes inhabited by tribal groups and marginal farmers. It is an annual, growing upto a metre in height tillers well and also branches freely.

Italian millet was domesticated in the highlands of Central China (Prasada Rao *et al.*, 1991). An annual with tillers, 1.0-1.5 m height, Inflorescence is a spike terminal borne on a thin peduncle. Many varieties of the crop with variation in height, duration colour of grain, length of bristles and in tillering habit are under cultivation.

Little millet is naturalised in India and Sri Lanka, an annual grass growing erect upto 0.75-1.0 m with tillering. Inflorescence is a panicle. Purple and medium purple pigmentation is noted besides green throughout the plant parts. Grain colour varies from light olive brown to olive brown.

Proso millet is originated at Mediterranean basin. This millet has a very short duration with low water requirement. It is an annual, normally about metre long, tillers freely.

Barnyard millet is domesticated in India, and is generally grown as rainfed crop. It can also be grown in partially waterlogged areas, such as low lands on banks of rivers that often get submerged. The crop requires little manuring. It is an annual, growing upto 0.75 m in height, tillering sparsely. Inflorescence is a panicle, contracted or pyramidal.

Kodo millet is domesticated in India. The crop is remarkably drought resistant. The coarse grains can easily be preserved as a good famine reserve. Kodo millet is an erect annual up to 0.75 m in height with tillering habit. Inflorescence is a panicle.

Plant Genetic Resources

Plant Genetic Resources (PGR) contributes significantly towards food security, environmental protection and sustainable development. They represent a critical component of crop improvement efforts aimed at increasing food security—both for short-term gain and for long-term increase in productivity. The plant genetic resources as components of agro-biodiversity cover not only the whole gamut of genetic resources (from advanced cultivars to primitive landraces, domesticates, semi-domesticates, wild and weedy relatives) but also the diversity of ecosystems and agro-ecosystems that continuously exploited for agriculture and forestry (Upadhyay *et al.*, 2006).

Millets Collections

Genetic diversity provides farmers and plant breeders with options to develop, through selection and breeding, new, resilient and more productive crops. Collections of crop diversity held in *ex situ* genebanks appeared only in the 20th century. Here are now approximately 1470 collections in national, regional and international institutes around the world holding, collectively, about 5.4 million samples of crop resources. These collections, at the moment mainly through inadvertent as well as safety duplications, are growing in size as well as increasing in number and are located in more than 100 countries around the world. Approximately 11 (equal to approximately 40 of unique accessions) are held in the 11 genebanks of the Future Harvest Centres of the Consultative Groups on International Agricultural Research (CGIAR) and more than a third are stores in 15 national genebanks (Thomson *et al.*, 2004).

Table 2: Millets Collections Conserved at National Genebank, NBPGR, New Delhi (As on 31st January, 2015)

Crop/Crop Group	*Number of Accessions Conserved*
Sorghum	20,376
Pearl millet	7,841
Minor millet	23,864
Millets	52,081

It has been predicted that, worldwide crop output must rise in two-fold in the next couple of decades just to keep up with human population growth. This increase must occur – and be sustained – in the face of changing climates, diminishing land, stagnating yields, and degradation of other natural resources and global conflict (Imperial College Wye, 2003).

FAO estimates that, three fourths of the original varieties of agricultural crops have been lost from farmers' fields since 1900 (FAO 1998) and an IUCN study found that an estimated 30,000 plant species disappear each year (Gardiner and Le Goulven, 2001).

Table 3: Sorghum and Millets Germplasm Augmented under Indo-International Programmes (1976-2008)

Year	*Collaborative Institute/Programme*	*Crops/Accessions (No.)***	*Areas Covered*
1965-72*	IARI/PL Rickefeller Foundation	Sorghum (2467), Pearl millet and others (1075)	India
1977, 1978	Indo-Soviet protocol	Millets (7300) and others	Russia
1980	FAO/IBPGR	Pearl millet, sorghum, finger millet and others (3922)	Malawi and Zambia
1984	Indo-Australia Mission	Foxtail millet, barnyard millet, wild Setaria and *Echinochloa* species and others (1000)	Deccan, central and peninsular India
1985, 1987 and 1989	Indo-Japanese	Minor millets including wild/weedy types (496)	Karnataka. Tamil Nadu. Andhra Pradesh, Coastal Maharashtra, Odisha. Bihar and Madhya Pradesh
1985	Indo-Japanese Mission	Finger millet, foxtail millet, kodo millet, barnyard millet, proso millet , little millet and wild types of *Setoria glauca* and *Paspalum* sp. (177)	Andhra Pradesh, Karnataka and Tamil Nadu
1986	ICRISAT	Sorghum (100)	Andhra Pradesh, Bihar and Madhya Pradesh
1987	Indo-Japanese Mission: ICRISAT	Finger millet, little millet, foxtail millet (82), pearl millet (176), sorghum (42)	Andhra Pradesh (Tirupati). Coimbatore (Tamil Nadu) and Karnataka
1987	ICAR/FAO (EBPGR)	Forage sorghum (331) and its wild relatives (15)	Kenya. Sudan, Ethiopia
1988	ICRISAT	Pearl millet (287), sorghum (97)	Rohtak. Hisar, Sirsa, Bhiwani (Haryana), Adilabad and adjoining region, East and north coastal region of Andhra Pradesh and south Tamil Nadu
1990-91	ICRISAT	Pearl millet (181). sorghum (62), foxtail millet (16), little millet (23), barnyard millet (5) and other (Odisha) millets (12)	Telangana region (Andhra Pradesh), Eastern Ghat region – Koraput, Kalahandi, Ganjam, Phulbani (Odisha)
1991-92	ICRISAT	Wild pearl millet (37)	Dehradun. Tehri. Chamoli and Almora (Uttarakhand)

Contd...

Table 3–*Contd...*

Year	*Collaborative Institute/Programme*	*Crops/Accessions (No.)***	*Areas Covered*
1992-93	ICRISAT (with CAZRI)	Sorghum (59) and pearl millet (81)	North Karnataka and adjoining regions (Andhra Pradesh), part of Rajasthan
1993-94	ICRISAT	Sorghum and wild sorghum (*S. roxburghi*) (319)	Bittur, Dharwad (Karnataka), Osmanabad, Mahboobnagar (Maharashtra) and Anantpur (Andhra Pradesh)
1999-2005	ICAR/NBPGR/World Bank	Sorghum and other millets	Parts of Andhra Pradesh, Tamil Nadu, Maharashtra. Uttar Pradesh and Rajasthan

Source: NBPGR Annual Reports 1975-2008 (12).

Table 4: Introductions made in Sorghum and Millets Crop (1976-2007)

Crop	*Accession (Nos.)*	*Countries*
Sorghum	35,470	Australia, Brazil, Canada, China, Colombia, Costa Rica. Egypt, France, Germany. Hungary. Indonesia, Italy, Japan, Japan, Kenya, Mali, Mexico, Myanmar, Niger, Nigeria, Pakistan, Philippines, Russia, Rwanda, Saudi A *Rabi*a, Singapore, Somalia, Sri Lanka, Sudan, Switzerland, Tanzania, Thailand, Uganda, USA, PDR Yemen, Zimbabwe
Pearl millet	13,246	Africa, Australia, Brazil, Burkina Faso, Burma, Colombia, Cameroon, Canada, East Africa, Central Africa, Eritrea, France, Germany, Ghana, Hungary, Iraq, Israel, Italy. Japan, Kenya, Maldives, Mali, Mexico, Morocco, Namibia, Netherlands., Niger, Nigeria, Pakistan, Philippines, Rhodesia, Senegal, Sierra Leone, South Africa, Sri Lanka, Sudan, Taiwan, Tanzania, Uganda, UK, Uruguay, USA, Zambia, Zimbabwe
Finger millet	4760	Australia, Belgium, Bhutan, Burundi, Cameroon, Canada. Costa Rica, East Africa, Eritrea. Ethiopia, France, Italy, Japan, Kenya, Malawi, Maldives, Mozambique, Nepal, Niger, Nigeria, Philippines, South Africa, Spain, Sri Lanka, Tanzania, Uganda, UK, USSR, Zambia, Zimbabwe
Little millet	2504	Argentina, Australia, Brazil, Bulgaria, Canada, China, East Germany, Ethiopia, France, Germany, Ghana, Greece, Hungary, Iran, Italy, Japan, Kenya, Malawi, Mexico, Nepal, New Zealand. Nigeria, Philippines. Poland. Saudi A *Rabi*a, South Africa, Sri Lanka, Syria. Uganda, UK, USA, Yugoslavia
Kodo millet	769	Argentina, Australia, Brazil, Israel, Italy, Japan, Kenya, Malawi, Nepal, South Africa, Sri Lanka, UK, Uruguay, USA
Barnyard millet	601	Australia, Canada, France, Germany, Ghana, Italy, Japan, Kenya, Malawi, Philippines, South Africa, Spain. USA, USSR. Germany. Yugoslavia
Foxtail millet	657	Argentina, Australia, Bhutan, Brazil, Burundi, Canada, China, East Africa, France, Italy, Japan, Kenya, Malawi, Maldives, Mongolia, Nepal, New Zealand, Nigeria. South Africa, Syria, UK. Uruguay, USA, USSR, Zambia

Source: Plain Germplasm Reporter (1975-2006).

Table 5: Trait Specific Introductions in Sorghum and Millets

Crop/EC No.	*Country*	*Trait*
		Sorghum
EC242786-91	Sudan	Drought hardy
ECl91789	Australia	High lysine content
EC315823-52	USA	Acid soil tolerant
EC 331138-48	USA	Resistance to Army worm and Anthracnose
EC 428874	Nigeria	Disease tolerant lines
EC 466525-526	USA	Tx2911. TAMBPK-59, varieties resistant to green mould, downy mildew and head smut
EC4826705-97	USA	Male sterile and fertility restorer lines Immediate application for basic research on effect on
EC496845-854	Canada	Multi cutting purpose type fodder
EC538941-46	USA	Maintainers of the Al cytoplasmic genetic male sterility system
EC558947-55	USA	Restores of the Al cytoplasm genetic male sterility system
EC587422-509	USA	Striga resistant
EC568885-86 EC562509-l3	USA	Male sterile line
EC568887	USA	Maintainer line
EC582502-508	USA	Isogenic lines for brown mid rib genes
EC 416988	Nigeria	Early open pollinated variety
Proso millet		
EC 552153 Var. Horizon	USA	Resistant to wheat aphid, head rot associated with skin maggot and European corn borer widely adapted, lodging tolerant, high-yielding
		Pearl millet
EC4l 6988	Nigeria	Early open pollinated variety
EC4 17004	Nigeria	Whitish yellow grain
EC473259 leaf spot	USA	Potential for increasing growth rate, resistant to rust, smut and
EC463922-961	USA	Male sterile and downy mildew resistant lines
EC469936-945 EC47039 1-397	Nigeria	Early maturing, drought tolerant and dual purpose types
EC473259	USA	Potential for increasing growth rate, resistant to rust, smut and leaf spot

Source: Plant Germplasm Reporter (1975-2006).

Table 6: Genebank Holdings of Millets at National Genebank, NBPGR (As on 31 st January 2015)

Crop/Common Name	Total
Sorghum	20,376
Pearl millet	7,841
Small millets	23,864
Finger millet	11,587
Kodo millet	2,362
Little millet	1,885
Foxtail millet	4,244
Barnyard millet	1,888
Proso millet	1,005
Total	**52,081**

Souce: www.nbpgr.ernet.in and Radhamani, J *et al.* (2011)

Table 7: Sorghum and Millets Germplasm Registered for Potential Valuable Traits at NBPGR

Crop	Trait	No of Accessions Registered
Pearl millet	Fertility restorer line	2
	MS line converted with high combining ability in different genetic background	9
	CMS line	1
	New cytotype	2
	Tetraploid male sterile line with A cytotype	1
Sorghum	Resistant to yellow sugarcane aphid, *Melanapsia sacehaui*	1
	CMS, thermo-sensitive with high yield and long panicle	1
	Roti and dough quality high protein (9.63 per cent), stover quality, shoot fly and anthracnose resistant	1
	Low HCN (Hydrocyanic acid) and Multiple resistance	1
	Drought tolerant	1
	Converted male and female parents of dual purpose sorghum hybrid, SPH 1148 with high yield	2
	Shoot fly resistance and high grain yield	2
	Sweet sorghum MS line	15
	Medium dwarf sweet sorghum male sterile line with yellow grain color	2
	Grain mould resistance	13
	Shoot fly tolerance with multiple resistance to leaf diseases and also agronomic superiority and yield	1

Contd...

Table 7–*Contd...*

Crop	*Trait*	*No of Accessions Registered*
	Medium early duration male sterile line	2
	Resistance to multiple foliar diseases	4
	Good flaking quality	1
Foxtail Millet	Genie dwarf	1
Barnyard millet	Easy Dc-hulling type	1

Source: Singh, AK (2006); Kak, A et al. (2009) MS-Male sterile line

Table 8: Potential Valuable Traits Identified in Millet Accessions Stored in NGB

Crop	*Sources of Resistances*	*Total No. of Acc.*
Sorghum	Shoot fly	40
	Stem borers	23
	Midge	47
	Head hug	2
	Coin plant hopper	1
	Sugarcane aphid	5
	Shoot bug	1
	Grain molds	13
	Downy mildew	11
	Sugary disease	6
	Shoot fly + Stem borer	4
	Shoot fly + sugarcane aphid	1
	Shoot fly + midge	1
	Grain molds + sugary disease	3
	Grain molds+ downy mildew	4
	Shoot fly + stern borer + downy mildew	1
	Downy mildew + sugary disease	1
	Grain molds+ downy mildew	1
Finger millet	Resistant to blast	85
	Low leaf area, low stomatal number, high photo-synthetic efficiency	4
	High leaf area, high dry matter, moderate to high photo-synthetic efficiency	3
	High water use efficiency under stress	1
	High biomass and high harvest index	8
	High test weight with high photosynthesis	3
	Total	**269**

Source: Radhamani, J *et al.* (2011).

The first major effort in the assembly of a World Collection of sorghum germplasm was in the 1960s by the ICAR-Rockefeller Foundation's Agricultural Research Programme in India. A total of 22,701 exotic germplasm have been introduced in India from different countries of the world for various sources of important traits. The major contribution is from the Ethiopia, Sudan, Nigeria, Uganda, Zimbabwe, Cameroon, and USA, Ethiopia, Uganda, and Sudan in East Africa, Nigeria, Mali, and Burkina Faso. Resistance to several diseases is found in the *conspicuum* of Nigeria. Alleles for high productivity with prospects for increased yield due to nodal tillering appear to be in combinations of *caudatum*, *durra*, and *caffrorums* from both West and East African regions. The Ethiopian *durras* are an excellent source for the stay- green (non-senescence) trait related to post-flowering drought tolerance trait. There are several examples of pearl millet introductions that have proved to be of direct commercial significance. The introduction of cytoplasmic-nuclear male-sterile line (Tift 23A) from the Coastal Plain Experiment Station, Tifton, Georgia, USA, which led to the development and release in India of the first commercial hybrid HB 1, is a landmark in direct commercial utilization of an introduction.

Finger millet is presumed to have been introduced first to peninsular India at least 5000 years ago. The antiquity of foxtail millet cultivation although not very clear, this might have come from China and cultivation gradually extended to different parts of Indian sub-continent. The proso millet, which originated in Manchuria was possibly introduced from China, first to Himalayas and gradually spread to other parts of the country.

Sorghum and pearl millet are the only two important millets crops included in mandate crops of ICG Centre (ICRISAT). The ICAR-RF assembled over 16,000 sorghum germplasm accessions from major sorghum areas, and ICRISAT acquired 11,961 accessions of this collection in 1974 that existed in India and USA. ICRISAT has assembled 21191 germplasm accessions of cultivated pearl millet and its wild species from 49 countries. It represents the largest collection of pearl millet assembled and conserved anywhere in the world. ICRISAT in collaboration with the All-India Coordinated Pearl Millet Improvement Project (AICPMIP) and National Bureau of Plant Genetic Resources (NBPGR) launched 11 missions in India and collected 913 accessions. One set of germplasm collected by the French organization, ORSTOM, from western Africa was contributed for conservation at ICRISAT. The International Plant Genetic Resources Institute (IPGRI) and ICRISAT in collaboration with NARS had launched collection missions to several millet growing areas in Africa and collected from Botswana, Cameroon, Central African Republic, Namibia, Ghana, Malawi, Nigeria, Tanzania, Togo, Uganda, Zambia, Zimbabwe and several other countries in western Africa. Wild relatives provide good sources of disease resistance for purpose, several *Pennisetum* species were collected from western Himalayas from high elevation which are likely to be good sources of cold tolerance. ICRISAT has assembled 671 accessions of 23 different species of *Pennisetum*. Besides the germplasm held at ICRISAT, which represents the largest collection of cultivated pearl millet and its wild relatives, several other institution also maintain some germplasm. A collection of over 2700 accessions, representing cultivated, weedy and wild relatives is maintained by the Services Scientifiques Centraus of ORSTOM at Bondy, France.

These accessions are from 10 western African countries and represent collections made during 1975-82 by ORSTOM in collaboration with IPGRI. The institute Senegalaise de Recherches Agricoles (ISRA) at Bambey, Senegal, has over 2400 accessions. It includes 1120 landraces and improved varieties from western Africa, and over 1300 introduced, or programme generated, advanced breeding lines. ICRISAT also deals with germplasm collection, characterization, conservation, and seed distribution of six small millets: finger millet, foxtail millet, kodo millet, little millet, proso millet and barnyard millet (Dar and Upadhyaya, 2004).

The germplasm collections maintained at ICRISAT Genebank include 44,822 accessions (Sorghum 14,647; pearl millet 21,191; chickpea 7,488; pigeon pea 5,988; groundnut 6,060; small millets 3,460) received from or jointly collected with the Indian National Programs. The NBPGR requested ICRISAT to restore this entire collection in 1998. As of July 2004, ICRISAT has already repatriated 43,913 accessions (98 per cent of the entire request). Thus the Indian NARS regained their precious heritage that was conserved at the ICRISAT Genebank (Dar and Upadhyaya, 2004). Attempts to recover remaining Indian accessions are going through NBPGR.

The National Genebank at National Bureau of Plant Genetic Resources holds a total of 14,831 millet and forage collections. Sorghum dominated among millets and finger millet in small millets group. A total of 3,603 sorghum (*Sorghum bicolor*) collections were made from central India and peninsular region, parts of Rajasthan and Indo-Gangetic plains. In pearl millet (*Pennisetum americanum*), 2,225 collections were made from parts of Rajasthan, Gujarat, Andhra Pradesh, Bihar, Jharkhand, Karnataka, Madhya Pradesh, Tamil Nadu and Haryana. A total of 12,786 collections in different small millets have so far been made. Four hundred and two indigenous forage collections were also made which include forage legumes from berseem and oat from parts of Haryana and Uttar Pradesh, Lucerne from parts of Gujarat, forage grasses and legumes from Lahaul Spiti and Kinnaur (Himachal Pradesh) and Medicago falcate from Ladakh Himalayas (Someswara Rao *et al.*, 2004).

At DSR, a total of 1720 accessions have been collected during 1999 – 2012. The list of sorghum germplasm collected at DSR is given in Table 9.

Diversity of Millets

There are five basic races of sorghum: *bicolor, guinea, caudatum, durra,* and *kafir* (Harlan and de Wet. 1972). *Bicolor* sorghum has spread over much of the old sorghum growing world, being found in India. It is the likely progenitor of the *kaoliongs* of China (Mann *et al.*, 1983). The race *guinea* arose from *bicolor* with the possibility of interaction with the wild race *arundinaceum* in the higher rainfall areas of West Africa. The *guineas are* now the dominant sorghum of West Africa, but have spread and are found in Tanzania and Malawi. The *guinea* race arose more than 2,000 years ago. The race *caudatum* also likely arose from *bicolo,* it is found in the same area. Today, the caudatums are most abundant from east Nigeria to eastern Sudan and southward into Uganda. The race *durra* was selected from early *bicolor* that had moved into India some 3,000 years ago. With Arab migration, the *durras* moved into Ethiopia around 615 A.D., and are today the dominant race in India, Ethiopia, the Nile Valley of Sudan and Egypt. Race *kafir* was probably derived from *bicolor* but there is also evidence of association with the wild race *verticilliflorum*. The kafirs are found primarily in eastern and southern Africa (Mann *et al.*, 1983). Sorghum found its way into the Americas after 1850.

Table 9: State-wise Sorghum Germplasm Collection in India by DSR during 1999–2012

Sl.No.	*State*	*No. of Accessions*
1	Andhra Pradesh	301
2	Maharashtra	300
3	Madhya Pradesh	214
4	Uttar Pradesh	172
5	Karnataka	156
6	Tamil Nadu	150
7	Gujarat	147
8	Rajasthan	135
9	Jharkhand	54
10	(blank)	47
11	Uttarakhand	30
12	Bihar	13
13	West Bengal	1
	Grand Total	**1720**

In pearl millet the genus *Pennisetum* includes approximately 140 species that are widely distributed in the tropics and sub-tropics. It is one of the largest genera in the tribe Paniceae, which is morphologically diverse and consists of the following five sections. *Gymnothrix, Brevivalvula, Pencillaria, Heterostachya* and Eu-*Pennisetum*. Pennsietum includes two reproductively isolated species. These are *Pennisetum purpureum* Schumach, a tetraploid perennial species which occurs throughout the wet tropics of the world and *Pennisetum americanum* (L.) Leeke, now called *P. gluvam* a diploid annual species, native to the semi-arid tropics of Africa and India. The latter contains three morphologically diverse subspecies. The subspecies *americanum* (*gluvam*) includes a wide array of cultivated pearl millet. Subspecies *monodii* from the Sahel in western Africa has been recognized as the wild progenitor of pearl millet. Subspecies *stenostachyum*, which is morphologically intermediate between subspecies *americanum* and *monodii*, is a weedy and mimetic form of and often associated with the cultivated pearl millet.

The species of finger millet (*Eleusine coracana*) is divided into two sub-species: *E. africana* and *coracana*. Each sub-species are divided into races and sub races. *Elongata, plana, compacta* and *vulgaris* are the main races available in India. There are two species in Italian millet *viz., Setaria pumila* and *Setaria italica*. *S. italica* is divided into sub species *viridis* and *italica*. *Italica* is further classified into races *moharia, maxima* and *indica* and further classified into sub races – *aristata, fusiformis, glabra, compacta, spongiosa, assamese, erecta, nana* and *profusa*. The following races and subraces are identified in little millet: Species – *Panicum sumatrense*, sub species – *psilopodium* and *sumatrense* race – *nana* and *robusta*, subrace – *laxa*, *erecta* and *compacta*. The inflorescence of proso millet is a lax panicle. Hairiness is a characteristic of the species. Various grades of hairiness are known (1) dense hairy plant parts, (2) less hairy and (3) hairless – all parts glabrous except lower leaf sheaths and nodes. Seven-grain colors have been recognized of which dark olive gray and buff yellows are met with in South Indian varieties. In barnyard millet, the species *Echinochloa colona* is divided into sub species *colona* and *frumentacea* and races *stolonifera, intermedia, robusta* and *laxa*. The kodo millet races are identified based on raceme morphology *viz., regulars, irregulars* and *variables*.

Landraces of Millets

At DSR, a total of 138 sorghum landraces are conserved. The maximum state-wise new landrace collections at NRCS are from Madhya Pradesh (39) followed by Maharastra (43) and Andhra Pradesh (16), Tamil Nadu (6), Rajasthan (5) and Uttar Pradesh (4). The important landraces are *moli jowar from* Madhya Pradesh, *irungu cholam* and *mathappu chola from Tamil Nadu, beed maldhandi, bidri bendri dagdi* and *khondya* from Maharashtra. Karnataka's *kodumurugu jola, allin jola* and *allur jola*. The *pachcha jonna* from Andhra Pradesh, *barmuda local, deshi chari* and *gudli local* from Rajasthan. *Valsangh maldandi local, Vadgaon dagdi maldandi, tongraligaon maldandi, tongraligaon dagdi, sultanpur local dagdi, sultanpur maldandi, harni jogdi* (*dagadi*), *harni jogdi, chungi maldandi, musti local* (*Maldandi*), *chungi kuch-kachi, baddi jowar, chakur maldandi* from Maharashtra, and *sai jonna* from Andhra Pradesh are considered by the farmers as drought tolerant landraces (Elangovan *et al.*, 2005).

Some of the millets landrace collections are reported in the East Coastal Zones by Someswara Rao *et al.* (2004). In the Improved varieties and hybrids have almost replaced the sorghums during the past 2 decades particularly in Ranga Reddy, Mahaboobnagar and Medak districts. Sorghum, a staple food in certain pockets of Eastern Ghats, lost its status due to the replacement of landraces having special taste relished by the people. In most of the traditional sorghum growing areas, the crop is replaced with remunerative crops such as sunflower. The important landraces of sorghum known traditionally are: *Alluhurijola, Annigiri, Billijola, Biyyapujonna, Chencholam, Dondajola, Ennigar, Gadajonna, Biyyapujonna, Geddajonna, Gidda jola, Gidduthinai, Haladikabar, Harinapulgar, Jahnara, Janhajonna, Kadabinijola, Kadejonna, Karigoni, Kempujola, Khedjonna, Kondajonna, Kondamuddajonna, Natujonna, Pachchajonna, Padijonna, Palepujonna, Podujonna Seethammavari jonna, Tella Jonna, Yerra Jonna.*

Pearl millet germplasm variability is known to occur in Visakhapatnam, Viziayanagaram, Srikakulam, Nalgonda, Mahaboobnagar, Kurnool and Prakasam districts. Pittaganti, a popular landrace of North coastal districts is an early maturing and highly tillering type. Variability occurs in plant height, stem thickness, tillering, spike length, size and shape and seed characters. The list of some of the pearl millet landraces collected from parts of Eastern Ghats are: *Gantlu, Peddaganti, Pittaganti, Poduganti, Punasaganti, Thellasodha, Bochchusodha, Edagaru, Javarisajje, Kambu, Kariganti, Kammisodha, Kattukambu, Kondaganti, Malakambu, Naatusodha, Palaganti, Peddaganti, Peddasodha, Pottukambu, Punasaganti.*

Important pearl millet landraces in Rajasthan are: *Desi, Dodasar, Jhakrana, Ardi, Bikaneri, Rajedu, Chadi, Bikaneri, Rajedu, Liladi, Ratanpur, Bishangarh, Rampur desi and Sulkaniya.*

In small millets, the important crops are finger millet, Italian millet proso millet, kodo millet, little millet and barnyard millet. Variability in plant height, tillering, finger compactness, number of fingers/ear and ears/plant exist in finger millet in the districts of Visakhapatnam, Viziayanagaram, Srikakulam, Prakasam, Anantapur, Chittoor, Ranga Reddy and Mahaboobnagar. Endemic local variability occurs in the other small millets. Some important landraces of small millets collected from parts of Eastern Ghats are:

- ✰ **Finger millet:** *Badamandia, Bainadia, Batimandia, Buradhachollu, Chhotamandia, Chinnachodi, Dasarachodi, Garuvuchodi, Gumpachodi, Karunsurattai, Korrachodi, Mettachodi, Muddachodi, Mushurimandia, Nadipichodi, Nallagunupachodi, Paruamandia, Peddachodi, Periyaragi, Punasachollu, Pyruchodi, Senduragi, Somamandiya, Tholakarichollu, Thoovalkevuru, Vangachollu, Vangchollu, Vellasurattai, Yerraragulu.*
- ✰ **Foxtail millet:** *Andukorralu, Arapandi Arjunakorra, Bandakorra, Bochukorra, Kukkatookakorralu, Nakkatokagaddi, Nallakorra, Palakorra, Pasupukorra, Punasakorra, Seethakorra, Senakorra, Senthinai, Tella Korra, Uppanthinai Vellanthinai Yerra Korralu.*
- ✰ **Pros millet:** *Barigalu, Panivaragu, Thirivaragu, Variga.*
- ✰ **Little millet:** Badasuan, Buradasamalu, Chinnasamalu, Dangarsuan, Ganjasama, Ghartia, Gurji, Kahadasuan, Kandahanki, Karumjhape, Karunsamai, Koslasuan, Kothusamai, Koyyasama, Maghisama, Maisama, Mallesama, Manchurisama, Matsuan, Nallasamalu, Odusamai, Pedda Sama, Pillusamai, Sindikolisama, Vellasama.
- ✰ **Kodo millet:** Aadulu, Allu, Arikalu, Jollu, Neetiarikalu
- ✰ **Barnyard millet:** Bonthalu, Burkasama, Chinnavoodalu, Kondavoodalu, Peddavoodalu, Punasavoodalu.

However, the area under minor millets has decreased very alarmingly because of the introduction of subsidized rice scheme in Andhra Pradesh and other Indian states. It has insidiously contributed in the replacement of small millets with cash crops such as chillies and other vegetables in North coastal districts and sunflower and castor in Telangana and Rayalaseema regions. This has resulted in loss of endemic diversity of small millets.

Sorghum Genetic Resources

Sorghum (*Sorghum bicolor* L.), an important crop in many parts of the world is grown for food, feed and industrial purposes. It is also a major crop in many parts of Africa and some Asian countries. Compared to other cereals, sorghum is more tolerant to many stresses, including heat, drought, salinity and flooding (Ejeta and Knoll, 2007). Early domestication of Sorghum occurred from Ethiopian borders extending west through Sudan and up to Lake Chad. In the Indian Subcontinent there is evidence for early cereal cultivation discovered at an archaeological site in western parts of Rojdi (Saurashtra) which dating back to about 4500 before present India is considered to be secondary center of origin of sorghum. The first major attempt toward assembling a world collection of sorghum was made by the Indian Agricultural Program of the Rockefeller Foundation, and 16, 138 accessions were assembled from the major sorghum growing areas. This collection was acquired by ICRISAT in 1974 through the All India Coordinated Sorghum Improvement Project (AICSIP), Rajendranagar. At present ICRISAT is holding 36, 774 acc. of World collection of sorghum.

The characterization of germplasm for the agro-morphological traits is useful and ready access of the material in the crop improvement programmes. More than 1100 accessions are characterized for use in the breeding programmes. The range of variation available in the sorghum is very high. The major variations are found in the traits *viz.*, time to flowering (49-124 days), plant height (100-481 cm), number of leaves (5-23), ear head length (8-55 cm), ear head width (3-18 cm), stem diameter (1-3 cm), stem fresh weight (50-1500 g/pt), stem dry weight (25-800 g/pt), brix (2-29 per cent), grain yield (20-126 g/pt), seed weight (1-3 g/100 seed) and time to maturity (89-165 days).

Sorghum is mainly grown in two seasons *viz.*, *Kharif* (rainy; June-October), and *Rabi* (post-rainy; November-March). It is also grown during late *Kharif* (maghi; August-December), summer (December-April). The fodder-purpose sorghum is cultivated irrespective of the season in most parts of the country. The evaluation of sorghum at DSR-Hyderabad, CRS-Solapur and 18 All-India Coordinated Sorghum Improvement Project (AICSIP) centres are being done to identify the stability of the trait in different agro-climatic zones. Sorghum is used for different purposes and mainly as food, fodder, feed and fuel. Different types of sorghum *viz.*, grain sorghum, fodder sorghum, dual-purpose sorghum, sweet sorghum, pop sorghum and hurda sorghum are used for different purposes.

Sorghum (*Sorghum bicolor* (L.) is an important cereal crop in India particularly under dryland conditions. It is one of the main nutritious coarse cereals, grown in about 8.45 m ha during both rainy (about 3.45 m ha) and post-rainy (about 5.00 m ha) seasons producing 7.15 m t of grain. Grain and fodder yields on farmers' fields are significantly lower (< > 1.0 t/ha) primarily due to the influence of various biotic and abiotic constrains.

There is great variability in sorghum, making selection possible for most traits of economic importance. Germplasm, in the form of breeding stock, collected accessions, and converted tropical cultivars, now moves readily worldwide and has contributed significantly to the crop's improvement in terms of yield, resistances and utilization/ quality traits (House, 1995). The utilization of sorghum germplasm in the ongoing research at various R&D institutes globally has released 30 varieties. Appa Rao *et al.* (1996) studied 2,343 Indian landraces from the sorghum ex-situ collection maintained at the ICRISAT and found higher phenotypic diversity among accessions from different states than those from within a state.

Sorghum (*Sorghum bicolor* (L.) is an important cereal crop in India particularly under dryland conditions. It is one of the main nutritious coarse cereals, grown in about 8.45 m ha during both rainy (about 3.45 m ha) and post-rainy (about 5.00 m ha) seasons producing 7.15 m t of grain. Grain and fodder yields on farmers' fields are significantly lower (< > 1.0 t/ha) primarily due to the influence of various biotic and abiotic constrains. In view of growing threat from climatic change, there is a need for the improvement of heat, drought and salinity tolerance in sorghum. Superior genetic resources are required to reduce the risk of loss from the adverse effects of abiotic stresses. Sources of abiotic stress tolerance are needed to stabilize the production/ productivity and quality. Sorghum germplasm collections from the hot spots will provide essential genetic variation to develop new varieties that can significantly impact agricultural systems. Germplasm utilization is increased if detailed characterization data is obtained for individual accessions, which should include qualitative and quantitative phenotypic traits, genotypic data, and responses to abiotic and biotic stresses. In addition, molecular characterization will help in the classification of germplasm as well as identification of most distinct groups. Molecular characterization together with phenotypic data helps in making proper decisions regarding the management and utilization of genetic resources.

The collection of sorghum germplasm from the semi-arid tropics and saline coastal areas of India will help to identify the potential sources for tolerance to major abiotic stresses like high temperature, drought and salinity. This activity will be vital to tackle the predicted climate change scenario. Directorate of Sorghum Research has important mandate on the exploration mission to collect germplasm from the hot spots of sorghum growing regions of the country especially in the states of Andhra Pradesh, Maharashtra, Karnataka, Gujarat, Rajasthan and Uttar Pradesh.

Sorghum Collections at National Active Germplasm Site

Landraces are the varieties nurtured and cultivated by the farmers through traditional method of selection over the decades. The "landrace" is a primitive cultivar grown by farmers and their successors since ancient times. The Directorate of Sorghum Research (DSR) has explored 9 major sorghum growing states and collected numerous accessions. Most of the popular landraces in these collections have specific traits and traditional utilities.

In these collections, the sorghum landraces are mostly named with the village name prefixed to the landrace (ex. *tongraligaon* maldandi, *sultanpur maldandi* etc), landraces are also named based on the grain colour (ex. *tella jonna, bili jola* etc.) and size (ex. *moti Rabi jowar*), ear head compactness (ex. *mudda jola*), use of the stem or grain (ex. *dagdi, gundu, allur etc.*) for varieties utilities. The landraces are being grown by farmers over the years for many reasons. The *moli jowar* grains are attractive and fetch higher price in the market compared to other released varieties in Madhya Pradesh. Some are used for specific food preparations. The *irungu cholam* flour is boiled with water to prepare food. The *mathappu cholam* is used to prepare a jelly like food known as *khali* in Tamil Nadu. The *beed maldandi* and *bidri* from Maharashtra are used to make quality jowar roti. Karnataka's *kodumurugu jola* and *allina jola* are used to make laddu's and pappads respectively. The *allur jola* from Karnataka is used to prepare allu dish during the nagapanchami (festival of snake god). The *pachcha jonna* from Andhra Pradesh, *barmuda local, deshi chari* and *gudli local* from

Rajasthan and *bendri dagdi* and *khondya* from Maharashtra are very good fodder types. *Valsangh maldandi local, Vadgaon dagdi maldandi, tongraligaon maldandi, tongraligaon dagdi, sultanpur local dagdi, sultanpur maldandi, harni jogdi* (*dagdi*), *harni jogdi, chungi maldandi, musti local* (*Maldandi*), *chungi kuch-kachi, baddi jowar, chakur maldandi* from Maharashtra, and *sai jonna* from Andhra Pradesh are considered by the farmers as drought tolerant landraces.

Andhra Pradesh

In Andhra Pradesh, majority of the landraces belongs to p*achcha jonna,* and *konda jola. Kakhi jonna, tella jonna* and *pili jola* are other important landraces collected. *Koya's* (tribe) grow *konda jonna.* The seedling is raised in the month of July-August and transplanted in the field during August-September. It matures in the month of December-January. The crop takes six months to maturity and exactly harvested during the "Sankranti" festival. *Konda Reddy's* does not prefer to grow released varieties because of plant height and body pain if they use the grains for food. *Lambada's* (tribe) used to grow the maghi sorghum landraces extensively, now the area has come down because of the loan provisions by the Government for cultivating commercial crop like cotton. Chintur mandal in Khammam district is the major maghi sorghum growing area.

The tribal community mainly uses the accessions as food and some are used for both food and fodder. The accessions collected from the higher elevation and remote forest area are source material for pest and disease resistance. The accession SEVS 7 (IC 347573) was with long awned fodder type. The accession SEVS 8 (IC 347574) is used as the tribal food. The accession SEVS 17 (IC 347583) is a *guinea* race and SEVS 19 (IC 347585) is a *durra-guinea* race sorghum collected at an elevation of 500 feet from the tribal village.

In the Rayalaseema regions, the accession EC 11 (IC 345713) was a sweet roti accessions. The accession EC 29 (IC 345731) was with very bold seed. The accessions EC 2 (IC 345704), EC 13 (IC 345715), and EC 18 (IC 345720) were with awns and EC 27 (IC 345729), EC 30 (IC 345732), EC 31 (IC 345733) with long awns belong to bicolour race, which is mostly the fodder sorghum. The accessions E 92 (IC 369123), E 94 (IC 369125), E 99 (IC 369130), and E 100 (IC 369131) are rated very good as standing crop in the field. The accessions E 88 (IC 369119), E 90 (IC 369121), E 91 (IC 369122), E 92 (IC 369123), E 94 (IC 369125), E 96 (IC 369127), E 97 (IC 369128), and E 100 (IC 369131) are yellow pericarp (*pacha jonna*) sorghum.

Gujarat

In Gujarat, the landrace *Utavali viz.*, EJN 14 (IC 0585184) and EJN 15 (IC 0585185) is cultivated extensively in the taluk of Kankraj. Farmers harvest these landraces along with the ear head at the time of maturity to feed the cattle. The landrace *Gundri viz.*, ERN 28 (IC 568543) and ERN 29 (IC 568544) is cultivated in one lakh acres in the Anjar taluk of Kutch. *Malwan viz.*, ERN 14 (IC 568529) and ERN 16 (IC 568231) is the famous dual-purpose landrace with great demand of seed from the farmers. Purification of this landrace from other landraces is important to undertake seed production of this landrace. This landrace is originated from the village Malwan,

Surendernagar district. *Gundri* and *Solapuri* [ERN 26 (IC 568541)] landraces are also the saline tolerant line. *Solapuri* landrace is imported from Solapur is a selection from M 35-1. The landraces colleted are *deshi, poru, sundia, char, rajka jowar* and *hundri*. *Gomati* (tribe) grow *deshi jonna* landraces. They cultivate sorghum with cotton, redgram, and vegetables in the Dong district of Gujarat

Thara village in the Kankraj taluk of Banaskantha district has big seed market. Palanpur is the famous fodder market in Gujarat. Kumbhasan region is called as *"Jowar ka Khan"* with 1500 ha areas of sorghum cultivation. Farmers purchase sorghum deshi seeds at the rate of Rs. 15 kg. Sri Babu Bhai Patel is the farmer in Gaglapur village, who makes selection of Malwan landrace since 5-7 years and produce seed to the farmers. He gets premium price for its selection and purification in comparison to the other deshi variety. The seed rate of Malwan is Rs 25/kg and other deshi costs Rs. 15/kg. The fodder price is Rs. 8/kg bundle as against Rs 5/kg for other deshi fodder. Farmers earn Rs. 30,000/ha as income from sorghum cultivations. Farmers are leasing the filed under the contract farming to the poor workers at the rate of 25:75 per cent to the worker: owner. Sorghum crop is rotated with blackgram, greengram, redgram, and groundnut to retain the soil fertility.

Trait specific germplasm were collected *viz.*, fodder (17), sweet stalk juice (2) and saline tolerant (6). The saline tolerant accessions were collected between the altitudes of 1 – 34 m MSL. Porbandar, Junagadh and Bharuch districts have sown late *Kharif* landraces. The accession EJN 4 is very early fodder deshi germplasm matures at 75 days, disease free, grain yield 2t/ha and fodder yield 5t ha, the stem and seeds are sweet and cattle eats fully.

Karnataka

In Karnataka, many a type of materials are of *bili jola. bili maldandi, farm maldandi and Maldandi bili* were of other name called for the local landraces. The local, *kanaggu jola, kodumurugu jola, allin jola, allur jola,* and *kadapu jola* were common. In Kannada language *bili* means white colour and *kanagu* means roughness. Every village has its own sorghum samples screened by them and being grown in that particular village. Many a type of materials are of Maldhandi with its own environmental attributes. The local, *joot, dukri, dagadi, Maldhandi, gundu jola* and *thella jola* are common. Majority of the sorghum grain is used for food as *roti*. Some accessions are used for preparing sweet chapattis. Most of the material collected was of drought tolerant material with stay green character. Dual-purpose sorghum is common. Pop type sorghum is also collected. Farmers are collecting their seed material from the selected earhead as a source for growing the next season.

The accessions EP 103 (IC 345192) were used for laddu preparation and EP 105 (IC 345194) for papped preparation. The accessions EP 108 (IC 345197) was very sweet in taste. The accession EP 17 (IC 305898) and EP 18 (IC 305899) are of good marketable quality with fine seed lustre. The accession EP 43 (IC 305924) and EP 44 (IC 305925) are used for preparing sweet chapattis. The accession EP 45 (IC 305926) is compact round shape head with very bold and bright white coloured seed, a good local sorghum of Maldhandi type.

Madhya Pradesh

In Madhya Pradesh, In the Bundelkhand region basmati (scented) jowar (E 228, E 254), basmati type jowar, *deshi jowar, khandva, baidara boria, hathikunta, doka, sabet baidara, peeli baidara, zunku, deshi jondi, chatkul, dugdugu, kathgav deshi, salegar, deola, and sabet deshi.* In Jhabua hills sorghum landraces are grown in the undulating terraces. The local, *sabet, dhawli* (means white in Maratee) are common.

Patai village in the Baigundpura tashil is the major deshi jowar cultivation region in Rewa district. About 32 – 40 per cent of the total *Kharif* area in the village is under sorghum cultivation. This area is commonly called *"Junariwala Shetra"*. Mehran, Swarguor, and Kushwaha are the main tribes cultivating the jowar in these regions. After 1983, soybean has replaced sorghum in many districts of Madhya Pradesh especially the sorghum hotspots like Sagar and Chattarpur.

The jowar is mainly used for preparing roti, laddu, meheri (dish with curd), popping (roasting the ear head by hanging in the fire), chat, ghadha, chadula (puri), kharka, samosa, kohri (ear heads boiled in water), dharia, and biscuits. The jowar is mainly used for preparing *roti, laddu, meheri* (dish with curd), popping (roasting the ear head by hanging in the fire), chat, ghadha, chadula (puri), kharka, samosa, kohri (ear heads boiled in water), dharia, and biscuits. Cloths or nylon bags are used to protect the ear head from bird's damage.

In the Malwa regions, the accession GGUB 19 (IC 319862) and GGUB 20 (IC 319863) were rare materials and the earlier accession was with attractive grains. The accession GGUB 14 (IC 319860) was with small ear head and coloured awn. The thick stemmed and juicy accession GGUB 34 (IC 319877) was a source of industrial sweet sorghum. The accession GGUB 44 (IC 319887) was a pop type, one of the alternate uses of the sorghum. Some of the accessions were 4 – 5 m height, big ear heads and stay green. The accession EB 1 (IC 332460) and EB 4 (IC 332463) are very bold seeds with big panicles. The accession EB 6 (IC 332465), EB 15 (IC 332474) and EB 20 (IC 332479) are very tall produces good biomass and yield. The accession EB 1 (IC 332460) and EB 2 (IC 332461) are belongs to the durra-bicolor race and EB 5 (IC 332464), EB 6 (IC 332465), EB 11 (IC 332470) and EB 12 (IC 332471) are belongs to rare bicolor race.

Maharashtra

In Maharashtra, the local landraces *jowri, khondya, local jowar, mutka, bidri, dagdi maldhandi bendri dagdi, detha, bhad* (add type in Marathi), *lahya phodi, ramgi, somgi,* and *nariyal jola* (shape like coconut) are prominent in this region. One *Sorghum halepense* collected from the Khoga river belt of Melghat Tiger Project region. Where it is grown in wild on the river banks and all along the road-side. *Korukoo's* (*tribe*) grow *somgi* and *deshi jonna* landraces. They cultivate sorghum with lablab, chillies, cotton, redgram, and tomato at the height of 650 – 870 m from MSL. The earhead shape of *bhad jonna* is like mango and the landrace grow very tall and the stalk sweetness also very good and it may be utilized in the sweet sorghum programme. The *ramgi jonna* is having very big ear head, very bold and flat seed may be utilized in the dual-purpose sorghum programme.

The *Rabi* sorghum landraces collection was based on some *Rabi* criteria *viz.*, drought tolerance, moderate yield, fodder and grain quality, good lustre, and thin stem without hollow in the middle. The collection was made from the most of the drought prevailing areas of Maharashtra. Most of the material collected was of drought tolerant material with stay green character and good stand in the field even after the drought.

The accessions EP 53 (IC 343552), EP 62 (IC 343561), and EP 63 (IC 343562) were with good lustre. The accessions EP 59 (IC 343558) were used for quality roti and EP 54 (IC 343553) for sweet roti preparation. The accession EP 60 (IC 343559) was with definite drought tolerance character. The accession PEC 22 (IC 392145) and PEC 23 (IC 392146) was used to prepare sweet roti. Most of the accessions are observed as *drought tolerant*, especially in parts of Bidar, Medak, Ranga Reddy, Lathur districts. Some accessions namely PEC 11 (IC 392134), PEC 12 (IC 392135), and PEC 16 (IC 392139) and PEC 31 (IC 392154), PEC 32 (IC 392155), PEC 33 (IC 392156), PEC 34 (IC 392157), PEC 35 (IC 392158) and PEC 36 (IC 392159) are *highly drought tolerant* and also in stay green condition. These are the great potential material for developing drought tolerant material. The accessions, PEC 13 (IC 392136), PEC 22 (IC 392145), PEC 24 (IC 392147), PEC 34 (IC 392157), PEC 35 (IC 392158) and PEC 36 (IC 392159) having big size panicles are potential material for high yield and drought tolerant. Most of the accessions are with stay green character, which are useful for fodder. The accession, PEC 7 (IC 392130), PEC 8 (IC 392131), PEC 9 (IC 392132), PEC 13 (IC 392136), PEC 15 (IC 392138), PEC 33 (IC 392156), PEC 34 (IC 392157) and PEC 36 (IC 392159) are with very bold seed. The accession, PEC 3 (IC 392126) with long awns and PEC 28 (IC 392151), PEC 29 (IC 392152) and PEC 30 (IC 392153) belong to bicolor race, which are used for fodder.

The accessions are mainly used as food and fodder. Some accessions were medium tall with very big panicles even in highly drought conditions are potential material for the sorghum improvement programme. The accessions PEC 10 (392133), PEC 13 (IC 392136), and PEC 15 (IC 392138) are tall height are used as a fodder. The accessions PEC 22 (IC 392145) and PEC 23 (IC 392146) are used for sweet roti. Sorghum roti is the common food and the excess will be exported to the market. The collection was made from the most of the drought prevailing areas of Maharashtra. The accessions EP 121 (IC 420936) was very tall (6ft) and sued for food and fodder. The fodder quality of EP 124 (IC 420939) was very good; it is a ratoon type and yield 10q/acre. The accession EP 137 (IC 420952) was 100 per cent stay-green type.

Rajasthan

In Rajasthan, the farmers were mainly using their traditional *deshi* sorghum. Most of the accessions collected were local landraces cultivated by the farmers over the decades. Farmers are maximum using sorghum as a fodder crop than for food. The use of sorghum as food depends on the availability of rain. Majority of the accession collected were fodder type. The accessions EJ 4 (IC 338984) and EJ 10 (IC 338990), EJ 11 (IC 338991) are sweet sorghum accessions. The accession E 2 (IC 338969), E 11 (IC 338976), E 12 (IC 338977), E 13 (IC 338978), E 14 (IC 338979), E 15 (IC 338980) and EJ 15 (IC 338995), EJ 16 (IC 338996), EJ 17 (IC 338997), EJ 24 (IC 339004),

EJ 25 (IC 339005), EJ 27 (IC 339007), EJ 30 (IC 339010, EJ 33 (IC 339013), EJ 41 (IC 339021), EJ 42 (IC 339022) are fodder type accessions. In western Rajasthan, the accession EJ 50 (IC 369250), and EJ 51 (IC 369251) are multicut fodder types. The accessions EJ 49 (IC 369249), EJ 53 (IC 369253), EJ 54 (IC 369254), EJ 55 (IC 369255), EJ 57 (IC 369257), EJ 59 (IC 369259), EJ 63 (IC 369263), EJ 67 (IC 369267), EJ 68 (IC 369268), EJ 71 (IC 369271, and EJ 74 (IC 369274) are fodder types. The accessions EJ 56 (IC 369256), EJ 70 (IC 369270), and EJ 72 (IC 369272) are dual purpose types.

Tamil Nadu

In Tamil Nadu, majority of the landraces belongs to *irungu cholam, senjolam* and *vellai cholam. periya manjal cholam, manjal cholam, pei cholam, nattu cholam, senkaton cholam,* and *karareddu cholam, thalaivirchchan cholam amd makkatai cholam* are other important landraces collected.

Farmers use their traditional sorghum landrace *viz., irungu cholam, vella cholam, sevappu cholam*. The accession EA 1 (IC 345243) and EA 6 (IC 345248) were used to prepare gelly like food *khali*. The accession EA 8 (IC 345250) is with pigmented seed. The accession EA 7 (IC 345249) is a fodder line. *Sivappu irunggu* and *Karuppu irunggu* are the most popular local landraces of these regions. *Karuppu irunggu* is the drought tolerant landrace and the only fodder in these dry areas. It has the sto*Rabi*lity capacity of upto three years. The accession with specific sorghum flour used for biscuit making in the Usilampatti areas of Madurai district is collected. Majority of the collections belongs to *bicolor* race which is a specific race for fodder. The accessions are only used for fodder and feed, rarely for food and seed for next generation. However only one farmer informed that some labours work in quarry use sorghum as food. They mixture sorghum and wheat flour to prepare *kali* (jelly like) to get more strength.

They are sold for Rs 20 and Rs 25/kg respectively in the market whereas the grain sorghum is being sold at Rs 10/kg. Another landrace *Tenkasi vellai* is sold in the market for Rs 1000/q whereas K8 and K11 is sold for Rs 800/q. Majority of the accessions were late *Kharif* types and were sown in the month of July-August and harvested during December-January. The crop was sown during August-September and matures in December – January. The crop takes six months to maturity and exactly harvested during the "Sankranti" festival. Sorghum is generally cultivated as a sole crop. Chillies, banana, sugarcane are grown under irrigated condition, besides the sorghum as rainfed crop.

Uttar Pradesh

In Uttar Pradesh, in the Ganga-basin regions the accession E 28 (IC 333386) was with red coloured anther. The accession E 50 (IC 33408), E 79 (IC 333437), E 86 (IC 333444) were very tall between 3.5 and 4.0m and with good biomass. The accession E 70 (IC 333428) and E 80 (IC 333438) were very tall with big heads. These accessions can be utilized for sweet sorghum breeding programme for high biomass and more yields. The accession E 77 (IC 333435) was potential material for any breeding programme with very big head. The twin seeded accessions can be utilized for breeding programme to increase the yield potential.

The accessions E 101 (IC 415792) and E 102 (IC 415793) are with bright white seeds. The accession E 108 (IC 415799), E 112 (IC 415803), E 140 (IC 415831), E 141 (IC 485132) is with very bold and twin seeded. The accession E 101 (IC 415792), E 102 (IC 415793), E 114 (IC 415805), E 116 (IC 415807), E 129 (IC 415820), and E 131 (IC 415822) are very tall with good biomass. The accession E 112 (IC 415803) is with big ear head.

Uttarakhand

A total of 30 accessions collected from central Uttarakhand. Accessions collected from 23 villages, 13 taluks and 5 districts. The maximum of 9 acc. each collected from Almora and Chamoli. The latitude range between 29.1825 and 30.3585 and the longitude range between 79.0078 and 79.4607 were covered for the exploration with the altitude variations between 813 and 1935 m from MSL. A total of 12 acc. collected above the 1500 m MSL. The local landraces *jonnel, jowari* and *bajra* are collected. Maximum frequency of guinea bicolor race (18 acc.), very tall (16), very loose (21), grayed orange seed colour (24), and grayed red glume colour. Maximum accessions collected are broom ear head shape and they are used as broom to sweep their home.

Some accessions are used for roti preparations mixed with other millets. The flours of some accessions are used as cattle feed. Farmers of these regions prefer only very tall types for forage purpose. All cultivations are only through organic farming. Sorghum landraces are being grown in these regions for more than 300 years. Sorghum is always grown with the other millets as a mixed-crop or border crop. Sorghum + Wheat flour used to prepare roti and cattle and poultry feed. Sorghum broom is used to avoid breakage in the floor crest.

Basmati (Scented) Jowar

Sorghum [*Sorghum bicolor* (L.) Moench] is originated from North-East quadrant of Africa. India is secondary centre of diversity (Singh, 2002). The first FAO-IBP report on the survey of crop genetic resources in their centers of diversity spotted out in India and important among is the river valley of Madhya Pradesh (Singh, 1973). The first major effort on the collection of sorghum genetic diversity was made under the auspices of the Rockefeller Foundation and the Indian Council of Agricultural Research (ICAR); and a total of 16,138 collections were assembled by 1970 (Rockefeller Foundation, 1970).

The present exploration was to collect the traditional cultivars of sorghum especially the unexploited basmati (scented) jowar in the Bundelkhand regions of Madhya Pradesh. Prasada Rao (1979) reported the collection of Basmati jowar in the Chattarpur district of Madhya Pradesh. Scented sorghum with the local name "Basmati" was collected for the first time from Karri and Sarwa villages near Bijawar (Prasada Rao and Murthy, 1979). Its grains are dimpled, chalky white in colour with a characteristics aroma resembling that of Basmati rice. Five scented sorghum germplasm *viz.*, IS18674, IS19907, IS19908, IS19910 and IS19912 has been reported in the ICRISAT list of promising sorghum germplasm, which are the collections from Madhya Pradesh (Elangovan *et al.*, 2006).

The regions were re-visited during *Kharif* 2008 to study the genetic erosion after thirty years of gap and explore the whole regions for the presence of basmati jowar. The exploration was headed by Dr M Elangovan, Senior Scientist (Genetic Resources), NRC for Sorghum, Hyderabad, Andhra Pradesh and was joined by Sri B.D. Narwaria, Assistant Director of Agriculture, Sagar district, Madhya Pradesh state. The very senior farmers (65-85 years old) were interviewed to get the information on the traditional cultivars of these regions especially the almost extinct aromatic cultivar basmati jowar. These farmers were well aware of the cultivation basmati jowar. They explained about the ear head compactness, shape, seed colour, seed size and aroma of the basmati jowar. They are prone to be eaten by birds as much as possible due to its aroma and taste of the grain. The scented jowar was collected from the same village where it has been collected thirty years back proves the existence of the traditional basmati jowar and proves the patronized by the local community. However, they are not cultivated as exclusive cultivar. Few aged farmers doubted the existence of basmati jowar because of its non-availability of the seed as an independent cultivar in the regions. Which some others had information on its availability along with other local landraces. They are mixed with other local popular landraces *viz., sabet baidara, peeli baidara, zunku, chatkul,* and *deshi sabet*.

Figure 1: Basmati (Scented) Jowar Ear Heads Collected from Katia Village.

Figure 2: Primary Branches of basmati (Scented) Jowar.

Figure 3: Caryopsis and Glume Colour and Coverage of (Scented) Jowar.

Figure 4: Caryopsis Colour and Size of Basmati (Scented) Jowar.

Based on the information available in the previous collection report and collected from the present farmers, the hunting for basmati jowar started in each and every field. Since basmati jowar was mixed with other landraces, only basmati jowar seeds (E 209, IC 568470) were identified by the aged farmer and collected from the Sadwa village in Chattarpur district. One-year-old panicles of basmati jowar (E 214, IC 568475) for which viability test to be done were collected from farmyard. The basmati jowar panicles (E 212, IC 568473) were also collected from the field in the Karri village in Chattarpur district. In the jowar field, smelling them individually helped to identify suitable panicles. Atleast 50-60 ear heads were smelled to collect one basmati jowar panicle. Finally, the panicles were identified has following characteristics: compactness, shape, seed colour, and aroma. The basmati jowar plant is very tall in height, semi compact panicle, white caryopsis, which are very bold, straw glume with dark red colour at the basal part of the glume and grains with 75 per cent coverage and belong to *durra* race. The whole panicle gives aroma like basmati. Even the flowering panicles smelled like basmati. The results of the search were encouraging. The basmati (scented) jowar was not only re-collected from the first ever collection location but explored in two more locations *viz.* Katia (E 228, IC 568489) and Kerwan (E 254, IC 568515) villages of Chattarpur district. In the process of natural crossing, basmati jowar also has been colleted in the name of *chatkul basmati* (E 213, IC 568474), *basmati sabet* (E 220, IC 568481), and *zunku basmati* (E 219, IC 568480) where the aroma was identified.

During next *Kharif*, we will explore the rest of the Bundelkhand regions for the existence of the traditional cultivar basmati jowar. Protection of these traditional Indian basmati (scented) jowar is the priority of the National Research Centre for Sorghum (NRCS), Hyderabad a unit under the Indian Council of Agricultural Research (ICAR), New Delhi joining with the Directorate of Agriculture, Madhya Pradesh state.

First Report on *Sorghum bicolor* ssp. *bicolor* Race *guinea* Sub-Race *conspicuum*

The most complete and definitive classification of cultivated *Sorghum bicolor* (Linn.) Moench is that of Snowden (1936). He has classified sorghum genus into 31 species, 158 varieties, and 523 forms, for a total of 712 taxa. Within the species *Sorghum bicolor* (Linn.) Moench were recognizing seven basic spikelet types: wild type, shattercane, *bicolor*, *guinea*, *caudatum*, *kafir*, and *durra*. Harlan and de Wet (1972) partitioned the primary gene pool, *Sorghum bicolor* (Linn.) Moench into five basic races, ten intermediate races, and six spontaneous races based on the glume coverage.

Guinea

In the Indian originated sorghum germplasm collections of 6201 accessions, *durra* race was observed in 3575 (58 per cent) accessions followed by *guinea* race with 773 (12 per cent). The *guinea* is the important sorghum race next to *durra*. It is basically a West African race, the dominant one of the Savannah sorghum belt. Some cultivars are remarkably tolerant of flooding (Harlan and Pasquereau, 1969). The morphological affinities and distribution indicate that the race *guinea* was probably derived from selection among wild members of the variety *arundinaceum*. The *guinea* race is further

Figure 5: *Sorghum bicolor ssp. bicolor* Race *guinea* Sub-Race *conspicuum*.

divided into three sub-races based on seed size namely, *margaritiferum* (small seeds), *guinea* (medium seed), and *conspicuum* (larger seeds). The spikelets of the *guineas* are derived and specialized and there is nothing resembling them in the wild forms. The long glumes and the open panicles are considered primitive, but otherwise the *guineas* must be considered specialized and highly evolved. Since the *guineas* are often grown in high rainfall areas, there has evidently been considerable selection for good weathering and storage qualities. Seeds are often hard and corneous, keep their colors well, and are rather resistant to insect damage under primitive storage conditions. The open pendulous panicles and gaping glumes probably help reduce mold damage under wet conditions. Grain flattened dorse-ventrally, sublenticular in outline, twisting at maturity 90 degrees between gaping involute glumes that are nearly as long to longer than the grain.

Conspicuum

Conspicuum is the sub-race of guinea was reported found in South Zambia, North Zimbabwe and in Bundelkhand area of India (Snowden, 1936). This sub-race was collected for the first time in India during the exploration undertaken by National Research Centre for Sorghum (NRCS) to Ganga basin and Bundelkhand area of Uttar Pradesh. This mission was undertaken under the auspicious of National Agricultural Technology Project (NATP) – Mission Mode - Plant Biodiversity). The material is having the important characteristics of reducing the mold damage due to the gaping glumes. This is to be utilized in sorghum improvement programme, as already the grain mold problem in this crop became the challenge to reduce the production of the crop. The accession was evaluated during *Kharif* 2002 and 2003 and the data is presented in Table 10.

Table 10: Characteristic Features of *Conspicuum*

Sl.No.	*Plant Characters*	*Snowden (1936)*	*Elangovan (2003)*
1.	Plant height	3 – 4.5m	2.5 – 3.5m
2.	Leaves	15 – 21	10 – 14
3.	Blades length (leaf length)	1m long	72 – 92cm
4.	Blades wide (leaf width)	10cm	5 - 9cm
5.	Panicle	Oblong to elliptic	Elliptic
6.	Panicle length	20 – 55cm	40 – 47cm
7.	Panicle width	8 – 20cm	5 – 7cm
8.	Panicle compactness	Loose and open, goose-necked	Loose
9.	Glumes	Equal	Equal
10.	Grains	Widely-gaping,	Widely-gaping
11.	Grain shape	Flattened	Flattened
12.	Grain colour	White, yellow or red	White

The list of sorghum landraces collected in India by the Directorate of Sorghum Research (DSR) from major sorghum states is presented in Table 11.

Table 11: List of Sorghum Landraces Collections

Sl.No.	Collector Number	Indigenous Collection Number	Vernacular Name	District	State
1.	EP 123	IC 420938	Jowri	Nashik	Maharashtra
2.	EP 124	IC 420939	Khondya	Ahmednagar	Maharashtra
3.	EP 127	IC 420942	Mutka	Pune	Maharashtra
4.	EP 128	IC 420943	Bidri	Pune	Maharashtra
5.	EP 132	IC 420947	Dagdi maldandi	Pune	Maharashtra
6.	EP 138	IC 420953	Bendri dagdi	Satara	Maharashtra
7.	E 101	IC 415792	Sabet deshi	Kanpur Dehat	Uttar Pradesh
8.	E 106	IC 415797	Jonnari	Unnao	Uttar Pradesh
9.	E 109	IC 415800	Jonndi	Rae Bareli	Uttar Pradesh
10.	E 143	IC 415834	Bani	Lucknow	Uttar Pradesh
11.	E 89	IC 369120	Tella Jonna	Mahaboobnagar	Andhra Pradesh
12.	PEC 2	IC 392125	Jungala Local	Lathur	Maharashtra
13.	PEC 5	IC 392128	Baddi Jowar	Lathur	Maharashtra
14.	PEC 7	IC 392130	Chakur Maldandi	Lathur	Maharashtra
15.	PEC 15	IC 392138	Maldandi	Bidar	Karnataka
16.	PEC 22	IC 392145	Sai Jonna	Medak	Andhra Pradesh
17.	PEC 26	IC 392149	Tandur Local	Ranga Reddy	Andhra Pradesh
18.	EC 1	IC 345703	Tella Jonna	Mahaboobnagar	Andhra Pradesh
19.	EC 11	IC 345713	Mudda Jonna	Kurnool	Andhra Pradesh
20.	EC 12	IC 345714	Raichur Jonna	Kurnool	Andhra Pradesh
21.	EC 20	IC 345722	Oola Jonna	Cuddapah	Andhra Pradesh
22.	EC 21	IC 345723	Eduakula Jonna	Cuddapah	Andhra Pradesh
23.	EC 33	IC 345735	Palar Jonna	Prakasam	Andhra Pradesh
24.	EC 34	IC 345736	Mudda Jonna	Prakasam	Andhra Pradesh
25.	SEVS 2	IC 347568	Pachcha Jonna	Khammam	Andhra Pradesh
26.	SEVS 3	IC 347569	Nattu Jonna	Khammam	Andhra Pradesh
27.	SEVS 7	IC 347573	Tella Jonna	Khammam	Andhra Pradesh
28.	SEVS 8	IC 347574	Konda Jonna	Khammam	Andhra Pradesh
29.	SEVS 17	IC 347583	Konda	Khammam	Andhra Pradesh
30.	SEVS 18	IC 347584	Kade Jonna	Khammam	Andhra Pradesh
31.	SEVS 20	IC 347586	Gompu Jonna	Khammam	Andhra Pradesh
32.	SEVS 26	IC 347592	Koda muddu jonna	Khammam	Andhra Pradesh
33.	EP 54	IC 343553	Gorani Maldandi	Osmanabad	Maharashtra
34.	EP 55	IC 343554	R - 8	Osmanabad	Maharashtra
35.	EP 57	IC 343556	Beedari	Beed	Maharashtra

Contd...

Table 11–*Contd...*

Sl.No.	*Collector Number*	*Indigenous Collection Number*	*Vernacular Name*	*District*	*State*
36.	EP 59	IC 343558	Beed Maldandi	Beed	Maharashtra
37.	EP 64	IC 343563	Sedam Maldandi Gurang	Aurangabad	Maharashtra
38.	EP 65	IC 343564	Gurang Maldandi	Aurangabad	Maharashtra
39.	EP 68	IC 343567	Dadar	Dhule	Maharashtra
40.	EP 78	IC 343577	Gurang	Ahmednagar	Maharashtra
41.	EP 79	IC 343578	Bandri	Ahmednagar	Maharashtra
42.	EP 80	IC 343579	Gurang Sabet	Solapur	Maharashtra
43.	EP 81	IC 343580	Gurang Dagdi	Solapur	Maharashtra
44.	EP 82	IC 343581	Sabet ganga gopargaon	Solapur	Maharashtra
45.	EP 84	IC 343583	Valsangh Maldandi local	Solapur	Maharashtra
46.	EP 86	IC 343585	Vadgaon Maldandi	Solapur	Maharashtra
47.	EP 87	IC 343586	Vadgaon Dagdi Maldandi	Solapur	Maharashtra
48.	EP 88	IC 343587	Tongraligaon Maldandi	Solapur	Maharashtra
49.	EP 89	IC 343588	Tongraligaon dagdi	Solapur	Maharashtra
50.	EP 90	IC 343589	Sultanpur local dagdi	Solapur	Maharashtra
51.	EP 91	IC 343590	Sultanpur Maldandi	Solapur	Maharashtra
52.	EP 92	IC 343591	Harni Jogdi (dagdi)	Solapur	Maharashtra
53.	EP 93	IC 343592	Harni Jogdi	Solapur	Maharashtra
54.	EP 94	IC 343593	Chungi Maldandi	Solapur	Maharashtra
55.	EP 95	IC 343594	Musti local (Maldandi)	Solapur	Maharashtra
56.	EP 96	IC 343595	Chungi Kuch-Kachi	Solapur	Maharashtra
57.	EP 97	IC 345186	Bile Maldandi	Raichur	Karnataka
58.	EP 102	IC 345191	Bili jola	Raichur	Karnataka
59.	EP 103	IC 345192	Kodumurugu jola	Raichur	Karnataka
60.	EP 104	IC 345193	Kanaggu jola	Raichur	Karnataka
61.	EP 105	IC 345194	Allin jola	Raichur	Karnataka
62.	EP 106	IC 345195	Bili jola	Raichur	Karnataka
63.	EP 107	IC 345196	Kanaggu jola	Raichur	Karnataka
64.	EP 112	IC 345201	Kadapu jola	Raichur	Karnataka
65.	EP 114	IC 345203	Maldandi Bili	Raichur	Karnataka
66.	EP 115	IC 345204	Allur jola	Raichur	Karnataka
67.	EP 117	IC 345206	Bili Maldandi	Kappal	Karnataka
68.	EP 120	IC 345209	Farm Maldandi	Bellary	Karnataka
69.	EA 1	IC 345243	Irungu cholam	Dindigul	Tamil Nadu

Contd...

Table 11–*Contd...*

Sl.No.	*Collector Number*	*Indigenous Collection Number*	*Vernacular Name*	*District*	*State*
70.	EA 2	IC 345244	Vella cholam	Dindigul	Tamil Nadu
71.	EA 4	IC 345246	Karum cholam	Dindigul	Tamil Nadu
72.	EA 6	IC 345248	Matthappu cholam	Dindigul	Tamil Nadu
73.	EA 10	IC 345252	Irungu cholam	Dindigul	Tamil Nadu
74.	EA 11	IC 345253	Sevappu cholam	Madurai	Tamil Nadu
75.	GGUB 19	IC 319862	Moli jowar	Jhabua	Madhya Pradesh
76.	GGUB 20	IC 319863	Chikni jowar	Jhabua	Madhya Pradesh
77.	GGUB 21	IC 319864	Pili local	Jhabua	Madhya Pradesh
78.	GGUB 22	IC 319865	Dhavali jowar	Jhabua	Madhya Pradesh
79.	GGUB 23	IC 319866	Dadar *Rabi* jowar	Jhabua	Madhya Pradesh
80.	GGUB 25	IC 319868	Syalu *Rabi*	Jhabua	Madhya Pradesh
81.	GGUB 27	IC 319870	Katrae	Jhabua	Madhya Pradesh
82.	GGUB 29	IC 319872	Local jowar	Jhabua	Madhya Pradesh
83.	GGUB 30	IC 319873	Lahi jowar	Jhabua	Madhya Pradesh
84.	GGUB 31	IC 319874	Barsati jowar	Jhabua	Madhya Pradesh
85.	GGUB 32	IC 319875	Badi jowar	Jhabua	Madhya Pradesh
86.	GGUB 34	IC 319877	Meethi jowar	Jhabua	Madhya Pradesh
87.	GGUB 35	IC 319878	Local Kajri jowar	Jhabua	Madhya Pradesh
88.	GGUB 36	IC 319879	Syalu *Rabi* jowar	Jhabua	Madhya Pradesh
89.	GGUB 37	IC 319880	Chikni pilia	Jhabua	Madhya Pradesh
90.	GGUB 38	IC 319881	Kale Tonsa ki Jowar	Jhabua	Madhya Pradesh
91.	GGUB 39	IC 319882	Bajri jowar	Jhabua	Madhya Pradesh
92.	GGUB 40	IC 319883	Moti *Rabi* jowar	Jhabua	Madhya Pradesh
93.	GGUB 43	IC 319886	Chikni jowar	Jhabua	Madhya Pradesh
94.	GGUB 44	IC 319887	Lahi jowar	Betul	Madhya Pradesh
95.	GGUB 46	IC 319889	Somji jowar	Betul	Madhya Pradesh
96.	GGUB 47	IC 319890	Local sehra	Betul	Madhya Pradesh
97.	GGUB 48	IC 319891	Ramtak local	Betul	Madhya Pradesh
98.	GGUB 50	IC 319892	Charsi local	Betul	Madhya Pradesh
99.	GGUB 51	IC 319893	Aathner Mohali	Betul	Madhya Pradesh
100.	GGUB 52	IC 319894	Badi Sahera jowar	Betul	Madhya Pradesh
101.	GGUB 54	IC 319895	Satpari	Khargaom	Madhya Pradesh
102.	GGUB 55	IC 319896	Kantolli	Khargaom	Madhya Pradesh
103.	GGUB 56	IC 319897	Safeda	Dewas	Madhya Pradesh
104.	GGUB 57	IC 319898	Gudagi	Vidisha	Madhya Pradesh

Contd...

Table 11–*Contd...*

Sl.No.	Collector Number	Indigenous Collection Number	Vernacular Name	District	State
105.	GGUB 58	IC 319899	Pili safed	Vidisha	Madhya Pradesh
106.	GGUB 59	IC 319900	Safed tinopal	Vidisha	Madhya Pradesh
107.	GGUB 60	IC 319901	Somkatch safeda	Vidisha	Madhya Pradesh
108.	GGUB 61	IC 319902	Mehara jowar	Vidisha	Madhya Pradesh
109.	GGUB 62	IC 319903	Pilimotia	Vidisha	Madhya Pradesh
110.	GGUB 63	IC 319904	Murga jowar	Vidisha	Madhya Pradesh
111.	GGUB 64	IC 319905	Somkath safea	Guna	Madhya Pradesh
112.	GGUB 65	IC 319906	Arom safed	Guna	Madhya Pradesh
113.	GGUB 67	IC 319907	Safed jowar	Vidisha	Madhya Pradesh
114.	GGUB 68	IC 319908	Mehara jowar	Vidisha	Madhya Pradesh
115.	E 1	IC 338968	Jowari local	Udaipur	Rajasthan
116.	E 3	IC 338970	Barmada local	Udaipur	Rajasthan
117.	E 4	IC 338971	Gudli local	Udaipur	Rajasthan
118.	EJ 3	IC 338983	Commedi	Pali	Rajasthan
119.	EJ 42	IC 339022	Deshi-Chari	Tonk	Rajasthan
120.	EB 1	IC 332460	Deshi sabet	Indore	Madhya Pradesh
121.	EB 2	IC 332461	Deshi Dhawli	Indore	Madhya Pradesh
122.	EP 1	IC 305882	Tamalwadi Dhagdi	Osmanabad	Maharashtra
123.	EP 9	IC 305890	Yermala joot	Osmanabad	Maharashtra
124.	EP 11	IC 305892	Yemala lakdi	Osmanabad	Maharashtra
125.	EP 12	IC 305893	Yermala dukri	Osmanabad	Maharashtra
126.	EP 13	IC 305894	Shirala dagdi	Osmanabad	Maharashtra
127.	EP 14	IC 305895	Shirala gota	Osmanabad	Maharashtra
128.	EP 16	IC 305897	Jamgaon dagdi	Solapur	Maharashtra
129.	EP 17	IC 305898	Jamgaon joot	Solapur	Maharashtra
130.	EP 22	IC 305903	Dagdi Maldandi	Solapur	Maharashtra
131.	EP 23	IC 305904	Dagdi local	Solapur	Maharashtra
132.	EP 24	IC 305905	Local Maldandi	Solapur	Maharashtra
133.	EP 37	IC 305918	Dharunyak Danda local	Gulberga	Karnataka
134.	EP 41	IC 305922	Hurdajola	Gulberga	Karnataka
135.	EP 42	IC 305923	Jevari local	Gulberga	Karnataka
136.	EP 45	IC 305926	Gundu jola	Gulberga	Karnataka
137.	EP 46	IC 305927	Bagadahali local	Gulberga	Karnataka
138.	EP 52	IC 305933	Hale jola	Bijapur	Karnataka
139.	ESRK 9	IC 597609	Broomcorn sorghum	Bageshwar	Uttarakhand
140.	EJN - 71	IC 597664	Dinelenia	Patan	Gujarat
141.	EJN - 73	IC 597666	Danchania	Patan	Gujarat
142.	EJN - 74	IC 597667	Chachadia	Patan	Gujarat

Sorghum Germplasm Characterization and Evaluation

The characterization of germplasm for the agro-morphological traits is useful and ready access of the material in the crop improvement programmes. More than 1100 accessions are characterized for use in the breeding programmes. The range of variation available in the sorghum is very high. The major variations are found in the traits *viz.*, time to flowering (49-124 days), plant height (100-481 cm), number of leaves (5-23), ear head length (8-55 cm), ear head width (3-18 cm), stem diameter (1-3 cm), stem fresh weight (50-1500 g/pt), stem dry weight (25-800 g/pt), brix (2-29 per cent), grain yield (20-126 g/pt), seed weight (1-3 g/100 seed) and time to maturity (89-165 days).

The preliminary characterization of 138 accessions of sorghum landraces was carried out at NRCS, Rajendranagar, Hyderabad during 2002-05. The centre is located at $17^{0}19'$ N latitude and $78^{0}24'$ E longitude and at an altitude of 538 m above MSL with temperature varying from 10^{0} C to a maximum of 30^{0} C during the cropping season. The accessions were raised in an augmented design in the sandy soil with check varieties (M35-1, Swati and CSV-15) in each block. The accessions were grown in 4 m rows with row spacing of 60 cm with plant to plant distance of 10 cm. Standard agronomic and plant protection practices were followed during the cropping season. The data on qualitative and quantitative descriptors were recorded using minimal descriptor developed by NBPGR (Mahajan *et al.*, 2000) and NRCS (Elangovan, *et., al.* 2004). Five representative plants in each accession were tagged for recording the qualitative and quantitative characters. Brix percentage of each accession was estimated by refractometer. The quantitative data was analysed statistically.

In these collections, the sorghum landraces are mostly named with the village name prefixed to the landrace (ex. *tongraligaon maldandi*, *sultanpur maldandi* etc), landraces are also named based on the grain colour (ex. *tella jonna, bili jola* etc.) and size (ex. *moti Rabi jowar*), ear head compactness (ex. *mudda jola*), use of the stem or grain (ex. *dagdi, gundu, allur etc.*) for varieties utilities. A similar observation was made by Teshome *et al.* (1997) because the midrib color, grain color, grain size, glume color, glume hairiness, and grain shape were the leading morphological characters used by the farmers naming the sorghum landraces. The list of 138 sorghum landraces collected is given in Table 10.

The landraces are being grown by farmers over the years for many reasons. The *moli jowar* grains are attractive and fetch higher price in the market compared to other released varieties in Madhya Pradesh. Some are used for specific food preparations. The *irungu cholam* flour is boiled with water to prepare food. The *mathappu cholam* is used to prepare a jelly like food known as *khali* in Tamil Nadu. The *beed maldandi* and *bidri* from Maharashtra are used to make quality jowar roti. Karnataka's *kodumurugu jola* and *allina jola* are used to make laddu's and pappads respectively. The *allur jola* from Karnataka is used to prepare allu dish during the nagapanchami (festival of snake god). The *pachcha jonna* from Andhra Pradesh, *barmuda local, deshi chari* and *gudli local* from Rajasthan and *bendri dagdi* and *khondya* from Maharashtra are very good fodder types.

Valsangh maldandi local, Vadgaon dagdi maldandi, tongraligaon maldandi, tongraligaon dagdi, sultanpur local dagdi, sultanpur maldandi, harni jogdi (dagdi), harni jogdi, chungi maldandi, musti local (Maldandi), chungi kuch-kachi, baddi jowar, chakur maldandi from Maharashtra, and *sai jonna* from Andhra Pradesh are considered by the farmers as drought tolerant landraces.

The accessions studied exhibited good variability in both qualitative and quantitative traits. In quantitative traits, the time to 50 per cent flowering (56 – 121), plant height (110 – 411 cm), earhead length (8 – 36.2 cm), stem fresh weight (110 – 955 g/plant), stem dry weight (30 – 557 g/plant) and grain yield (3.6 – 92 g/plant) showed good variability. The preliminary characterization of sorghum landraces revealed that stem fresh weight, stem dry weight, plant height, and grain yield are the most variable characters.

The qualitative characters have also shown variability. The higher seedling vigour (59 acc.), tan leaf pigmentation (128 acc.), dark green leaf colour (118 acc.), drooping leaf orientation (80 acc.), white midrib colour (55 acc.), elliptical earhead shape (80 acc.), semi compact earhead (91 acc.), straw glume colour (52 acc.), 50 per cent glume coverage (44 acc.), presence of awns (105 acc.), bold seed size (71 acc.), pearly white seed colour (81 acc.), non-senescence (stay green) plant type (133 acc.), durra race (78 acc.), and non-lustrus seed (52 acc.) recorded maximum frequency in qualitative characters.

The correlation coefficients of the quantitative traits revealed highly positive correlations for in time to 50 per cent flowering and time to maturity; number of leaves with leaf length, leaf width, plant height, stem thickness, stem fresh weight, and stem dry weight; leaf length with leaf width, plant height, stem thickness, stem fresh weight and stem dry weight; leaf width with stem fresh weight and stem dry weight; plant height with stem fresh weight and stem dry weight; stem thickness with stem fresh weight and stem dry weight; stem fresh weight with stem dry weight; and grain yield with 100-seed weight. Highly negative correlations were observed for grain yield with number of leaves, leaf length, plant height and earhead length; the 100-seed weight with leaf length, plant height, and earhead length; and the time to maturity with leaf width.

The variation in sorghum landraces for morphological characters has been reported by many workers in different countries. The evaluation of 152 accessions collected from Rwanda showed that the sorghum from Rwanda grows very tall (up to 500cm), and takes more days to flower during long days (rainy season) and less no. of days during short days indicating strong photoperiod sensitivity. Enormous variation was observed for panicle and grain characters (Appa Rao *et al.*, 1999).

Thirty four sorghum landraces collected from five agroecological sites (ecosites) in North Shewa and South Welo regions of Ethiopia have exhibited morphological variation for the 14 qualitative characters that showed two or more phenotypic classes. Panicle compactness and shape contributed relatively more to altitudinal and ecological differentiation. This differential distribution of landraces with panicle types with respect to compactness and shape revealed the adaptive significance of panicle compactness and shape that reflected the patterns of distribution of different races in north Shewa and south Welo. (Abdi *et al.*, 2002). The present study also observed the variation in panicle compactness.

The study to assess the phenotypic diversity and pattern of distribution among Sudanese sorghum landraces collected from different geographical regions revealed high phenotypic diversity among landraces with large range of variation for mean quantitative traits. Collections from Kassala showed a higher frequency of landraces with kernels that were more difficult to thresh. Landraces from Blue Nile tended to have greater agronomic eliteness with higher proportion of landraces with white kernels, poorly covered and were easy to thresh. Sorghums from the Upper Nile tended to have loose panicles with poorly covered kernels that may have resulted from the adaptation to high rainfall area in Southern region (Grenier *et al.*, 2004).

Field survey data of sorghum landrace populations grown by traditional farmers in four adjacent communities in the Ethiopian highlands were used to analyze and project the local risk of loss of individual landraces under three scenarios: wild population, traditional farming and agricultural modernization. The risk of loss in the wild population scenario is based on landscape ecology theory which evaluates total population size, spatial distribution and patch occupancy, with high values for each factor and the sum of the factors decreasing the risk. In the traditional farming scenario, the deliberate actions of farmers must be taken into account like farmers favour for various traits of the sorghum for specific reasons – yield, taste, sto*Rabi*lity, etc., which will balance plant numbers according to quantity and quality requirements. In the agricultural modernization scenario, introduction of high-yielding varieties tendency toward monocultures, use of agricultural chemical force the traits of the landraces that are most likely to be displaced and also bring in nature of the landraces that are likely to be retained on the basis of cultural or other factors (Tunstall *et al.*, 2001). Hence, the collection and protection of Indian sorghum landraces is very important.

Landraces are the sources of diversity for further use in crop improvement programmes. Unless landraces are collected, and conserved and protected, we can not have legal right on these diverse lines. Hence, they need to be registered first with NBPGR and then protect them under Plant Variety Protection Act 2001 in India, so that bio-piracy and extinction of these valuable resources, is prevented. It is the responsibility of all engaged in sorghum research to collect, conserve and protect them in a scheduled timeframe on mission mode basis.

Sorghum in India is grown in two seasons, namely rainy (*Kharif*) and post-rainy (*Rabi*) season in India. The requirements of varieties are different for two seasons. The emphasis is that the rainy season adapted varieties should be photosensitive, dwarf stature, and short duration to escape with infection by grain mold. On the other hand, varieties for the post-rainy season adaptation should be tall, high yielding with reasonable levels of terminal drought tolerance to sustain receding residual moisture. During the past four decades, remarkable progress has been made by diversifying the parental lines for yield (both grain and fodder), maturity, height, disease and insect resistance and quality improvement by utilization of indigenous and exotic germplasm.

Notable varieties developed during the early period through the landraces and still under cultivation are the Co-series in Tamil Nadu; the PJ *Kharif* and *Rabi* selections, Saoner, Ramkel, Aispuri, the Maldandi, Guntur and Anakapalle series of Andhra Pradesh; the bilichigan, fulgar white, fulgar yellow, kauvi, Nandyal, hagari, yanigar varieties of the erstwhile Mysore state (Rao, 1972). Indian germplasm line *karad local* was crossed with American material IS 3922 to develop the MS line 296A, which was the best combiner. By utilizing local variety Vidisha 60-1 at Indore centre, the ms line of CSH 18 (ms IMS 9A) was developed. The local variety, Vidisha 60-1 not only contributes for high stover yield but also for improved grain quality. The grain yield levels of rainy season hybrids have reached the plateau and there is a need to exploit unused germplasm and land races to diversify the genetic base (Elangovan, *et., al.* 2005).

Post-rainy season varieties were developed by crossing Indian locals, M 35-1 and IS 2644 with American germplasm lines. Marginal improvement was achieved for grain yield over the most popular local variety M 35-1. Recently released variety, CSV 216R (CSV 16) is a landrace selection from *Rabi* germplasm from Maharashtra. Another variety which became very popular, SPV 462, (from Coimbatore) was developed from multiple cross involving IS 2947 and IS 3687 from USA and IS 1151 and BP 53, locals of Maharashtra and Gujarat in India, respectively.

The need for further critical evaluation of germplasm and its utilization in grain and forage sorghum improvements is keenly felt. Sorghum collection missions may be further arranged on the specific targeted germplasm availability areas based on the usage like sweet sorghum, pop sorghum, dual-purpose sorghum and sorghums resistant to pests and diseases etc. Landraces are known to possess local adaptation and stability and may offer opportunities for direct utilization. On the other hand, improved varieties with diverse genetic base are likely to show wider adaptation and enhanced performance. Judicious combination of both in recombination breeding programmes can further lead to upgradation of yield potential. So far *caudatum* and *durra* types have been exploited. Inclusion of *guinea* germplasm may bring further increase in yield potential, tolerance to grain mold and lodging resistance.

Characterization and Evaluation of Sorghum from Karnataka, India

Sorghum [*Sorghum bicolor* (L.) Moench] also known as Jowar has originated and domesticated in Africa about 5000 – 8000 years ago (De Candolle 1890). Indian subcontinent is the secondary origin of this most important cereal. High diversity in sorghum is distributed throughout India. Elangovan (2004) reported that durra racial diversity is maximum in collection of Karnataka. Through number of surveys conducted earlier, about 1100 accessions were collected. In order to rescue the landrace diversity from unsurveyed and under surveyed areas, four explorations were undertaken during 2000-2003 in Karnataka under the National Agricultural Technological project (Mission Mode – Plant Biodiversity) and a total of 157 accessions were collected by National Research Centre for Sorghum (NRCS), Hyderabad. Characterization and evaluation of germplasm are the pre-requisite for the utilization of the available diversity in the crop improvement programme. Hence, the accessions were characterized to assess the variability and identify the promising accessions for different traits. Preliminary evaluation of 157 accessions of sorghum at the NRCS, Rajendranagar, Hyderabad and Solapur during *Rabi* 2002-03 and 2003-04 was conducted. The centres are located at $17^{0}19'$ N latitude and $78^{0}24'$ E longitude and at an altitude of 538 m above MSL and 17^{0} 04′ N latitude and 75^{0} 54′ E longitudes respectively with temperature varying with minimum of 10^{0}C to a maximum of 30^{0}C during the cropping season. The soil type is sandy loam. The accessions were raised in an Augmented Block Design (ABD) with check varieties (M35-1, Swati and CSV-15) in each block. The accessions were grown in 4 m rows with a row spacing of 60 cm and plant to plant spacing of 10 cm. Standard agronomic and plant protection practices were followed during the cropping season. The data on qualitative and quantitative characters were recorded using minimal descriptor developed by NBPGR (Mahajan *et al.*, 2000) and list of sorghum descriptors released by (Anon., 1993). Five representative plants in each accession were tagged for recording the qualitative and quantitative characters. Brix percentage of each accession was estimated by refractometer. Descriptive statistical analysis was done for the quantitative characters.

The 157 sorghum accessions were characterized and darker glumes contributed to grain mould resistance. The evaluated for 28 agro-morphological and one bio-chemical variability of glume colour available in the present study may be character. Wide range of variability was recorded in both utilized for screening for grain mold resistance in sorghum. The qualitative and quantitative characters. Majority of the range of variability and frequency observed in qualitative traits accessions showed tan type, dark green leaf colour, erect leaf are given in Table 1. The quantitative characters also showed orientation, and stay green types (non-senescence). However, wide variation in the evaluated sorghum germplasm. The results variation was observed in seedling vigour, midrib colour, earhead of descriptive statistical analysis are presented in Table 2. The shape, earhead compactness, glume colour, glume covering, seed characters of stem dry weight (70 - 665 g), stem fresh weight size, seed colour, and races. Sorghum landraces are consistent (110 - 1040 g), grain yield (3 – 75 g), brix (3 – 25 percent), and 100-for morphological characters, the importance of the midrib color, seed weight (0.6 – 4.5g) are the highly variable based on the CV grain color, grain size, glume color, glume hairiness, and grain percentage. The brix percentage ranged from 3 - 25. The chance shape were used by the farmers in naming the sorghum landraces of deriving sweet stalk sorghum may be possible from the high (Teshome *et al.*, 1997). Audilakshmi *et al.* (1999) evaluated brix accessions. The variability presents in the grain yield can twenty-two sorghum genotypes for grain mould response and be used to derive grain sorghum and stem fresh weight and observed that harder grain, higher levels of seed phenols, and stem dry weight can be used as fodder sorghum. Appa Rao, *et al.* (1999) evaluated over 4000 accessions from 11 major sorghum growing states in India for morphological and agronomical characters. He found that days to flowering, plant height, panicle length, erect and compact panicles are more frequent. The present study also observed the similar variability in plant height.

The correlation coefficients of grain yield/fodder yield showed that the grain yield and number of leaves, plant height and ear head width are showed positive correlation; and 100-seed weight also showed positive correlation with plant height, earhead width and grain yield. The dry fodder (stem dry weight) yield and some major fodder-yield attributing characters showed high positive correlation with stem fresh weight.

Characterization and Evaluation of Sorghum Germplasm from Andhra Pradesh, India

This paper discusses results of the preliminary characterization and evaluation of 95 accessions of sorghum germplasm collected from Andhra Pradesh under the National Agricultural Technology Project (Mission Mode – Plant Biodiversity). A total of 28 agro-morphological and one bio-chemical character were studied. The accessions studied exhibited good variability in both qualitative and quantitative traits. In qualitative traits, maximum diversity was observed in earhead compactness, glume colour, seed size, and seed colour. The days to 50 per cent flowering (60 – 121), plant height (103 – 270 cm), earhead length (8.96 – 39.65 cm), stem fresh weight (50 – 940 g), stem dry weight (30 – 290 g) and grain yield (14 – 83 g) showed good variability in quantitative traits. The correlation coefficients of the quantitative traits are also analysed.

Characterization and Evaluation of Sorghum Germplasm from Maharashtra, India

The results of preliminary characterization and evaluation of 99 accessions of sorghum germplasm collected from Maharashtra under the National Agricultural Technology Project (Mission Mode – Plant Biodiversity) is discussed in this paper. A total of 28 agro-morphological and one bio-chemical character were studied. The variability in both qualitative and quantitative traits is good. In qualitative traits, maximum variations were observed in midrib colour, earhead shape, glume colour, glume covering, seed size, and seed colour. The time to 50 per cent flowering (70 – 93 days), leaf length (50 – 75.7 cm), plant height (110 – 219 cm), stem fresh weight (50 – 510 g^{-1} plant), stem dry weight (25 – 400 g^{-1} plant), grain yield (10 – 92 g^{-1}plant) and time to maturity (102 – 134) showed good variability in quantitative traits. The correlation coefficients of the quantitative traits are also analysed. The potential donors for the grain yield attributing characters are being utilized in the *Rabi* sorghum improvement programme.

Characterization and Evaluation of Sorghum Germplasm from Madhya Pradesh, India

Sorghum germplasm exploration was undertaken in the Madhya Pradesh to collect, characterize and conserve the diversity of sorghum. Eighty-six accessions were collected and preliminary evaluation was conducted at National Research Centre for Sorghum, Hyderabad and All-India Coordinated Sorghum Improvement Project, Indore during *Kharif* 2003 and 2004. A total of 28 agro-morphological and one bio-chemical character were studied. The characters of grain yield (2 – 48 g^{-1}plant), stem dry weight (60 – 500 g^{-1} plant), stem fresh weight (170 – 940 g^{-1} plant), brix (3 – 23 per cent), 100-seed weight (0.9 – 3.4 g) and ear head length (11.3 – 33.9 cm) are highly variable based on the variance. The accession IC 319865, IC 319866, IC 319867, IC

319878 and IC 319880 were early flowering and early maturing lines. They are promising source materials for earliness. The accessions IC 319857 and IC 332464 are tall with high biomass. The accession IC 319861 is thick stemmed with high biomass and brix.

Evaluation

Sorghum is mainly grown in two seasons *viz.*, *Kharif* (rainy; June-October), and *Rabi* (post-rainy; November-March). It is also grown during late *Kharif* (maghi; August-December), summer (December-April). The fodder-purpose sorghum is cultivated irrespective of the season in most parts of the country. The evaluation of sorghum at DSR-Hyderabad, CRS-Solapur and 16 All-India Coordinated Sorghum Improvement Project (AICSIP) centres are being done to identify the stability of the trait in different agro-climatic zones. Sorghum is used for different purposes and mainly as food, fodder, feed and fuel. Different types of sorghum *viz.*, grain sorghum, fodder sorghum, dual-purpose sorghum, sweet sorghum, pop sorghum and hurda sorghum are used for different purposes.

Morphological and Molecular Diversity Studies Reveal wide Variability among Maldandi Landraces

Diversity among 82 Maldandi accessions was investigated using morphological and SSR markers to study the closeness and distinctness among the Maldandi landraces collected by the National and International programmes. Euclidian distance based on 17 quantitative traits put the accessions in two clusters with two out layers. Data on 19 qualitative traits put the accessions in one big cluster with six out groups. Many accessions could not be differentiated based on qualitative data. Sixteen out of 18 SSR markers detected polymorphism among the accessions with average PIC values of 0.36. Un-weighted neighbor joining clustering put the accessions into three clusters with 20, 16, 28 and 18 members respectively. The standard check, M 35-1 (a selection from the original Maldandi) could not be differentiated from EP 98 (Bile Maldandi; IC 345187), LG 2 (Angoli Maldandi), LG 10 (Karal Maldandi), IS 4501 (Maldandi devgaon) and IS 40791(Maldandi from Talegaon) based on qualitative data, and EP 64 (Sedam Maldandi Gurang; IC 343563) and IS 33839 (Maladandi from Jamgaon) using SSR markers. Based on quantitative data IS 4710 (Shallu Maldandi) was closest to M 35-1. Thirteen promising Maldandi accessions have been selected for further evaluation and utilization in the *Rabi* improvement programme for *Rabi* adaptation traits. The clustering of 82 Maldandi sorghum germplasm is presented in Figure 5.

Health and Nutritional Quality of Sorghum Genotypes

Sorghum and other millets are important staples in the semi-arid tropics of Asia and Africa. These crops are still the principal sources of energy, protein, vitamins and minerals for millions of the poor people in these regions. Considering the health benefits of sorghum, grain samples of 200 sorghum genotypes were analysed for the levels of iron and zinc, total phenolic compounds and antioxidant activity (TEAC), total phytate, fiber, trypsin inhibitor and cyanogens. Significant variation was observed among the genotypes for micronutrient contents and level of other nutritional

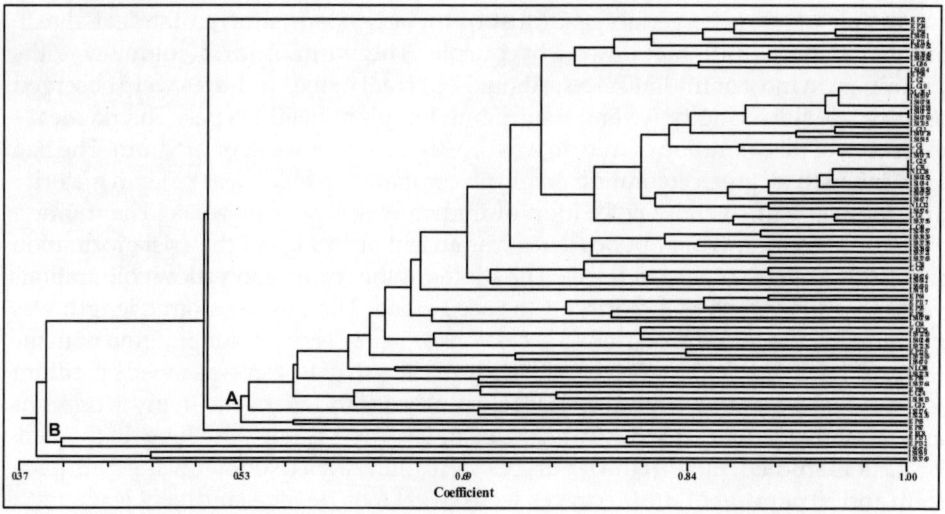

Figure 6: Sequential Agglomerative Hierarchical Nested clustering of the 82 Maldandi accessions based on Jaccard's similarity coefficients using 19 qualitative traits.

factors. The iron content in the grains varied from 6.3 to 168 ppm among the 200 genotypes with an overall mean of 41.7 ppm. Twenty genotypes recorded above 70 ppm, with four recording above 100 ppm, which is very high. The zinc content ranged from 3.6 to 87 ppm with an overall mean of 17.4 ppm. Fifteen genotypes had above 30 ppm zinc content. Germplasm accessions in general had higher values for iron and zinc compared to popular varieties and breeding lines. The highest polyphenols were present in SSG 59-3 (1271 gallic eq mg/100g), followed by Urja (1135 gallic eq mg/100g) and ELG 8 (IC 568344) (935 gallic eq mg/100g). The high content of polyphenols may be due to presence of glumes in these genotypes. Germplasm line IS 8525 contained maximum phytate (3909 mg/100g) followed by IS 14131 (3903 mg/100g) and DJ6514 (3415 mg/100g). The highest TEAC was shown by SSG 59-3 (2238 trolox eq mg/100g) followed by IS 8525 and URJA, all being the dark coloured varieties. The lower trend of cyanogens was observed in SPV 1762 (14.2 ppm), SEVS 20 (IC 347586) (14.7 ppm), EP 97 (IC 345186) (17 ppm), SPV1758 (18.2 ppm). The lower side of trypsin inhibitor was observed in EP 41 (IC 305922) (1.5 TIU), EP 127 (IC 415818) (1.8 TIU), SPV1759 (1.9 TIU). The ranking of sorghum types accreting to antioxidant quality indices is presented in Table 12. The selected accessions will be used in our program to improve mineral contents of food grain sorghums.

Preliminary Evaluation of *Rabi* Sorghum Landraces for DUS Characters

To observe the distinctness characters in the sorghum landraces collection at NRCS, a total of 86 acc. sorghum landraces collected under the National Agricultural Technology Project (NATP) Mission Mode–Plant Biodiversity were evaluated along with three checks (CSV 18, SWATHI and M35 - 1) at Hyderabad. Out of the 42 characters in the DUS guidelines prepared by the NRCS-2002, 40 characters were observed for the distinctness of the genotypes. Most of the accessions showed tan leaf sheath anthocyanin coloration followed by purple. The white midrib colour was the prominent in most of the landraces. About 77 percent of the landraces were observed as very late flowering types and medium to tall plant height types. The flag leaf - extension of discoloration of midrib was mostly absent or weak or medium. The flag leaf - intensity of green coloration of midrib compared to blade was paler in majority and the flag leaf - yellow coloration of midrib was absent or weak. The glume - anthocyanin coloration of pubescence was absent or weak and the arista formation was strong in most of the landraces. The stigma anthocyanin and yellow colourations were absent or weak in majority of the accessions. Though the stigma length was medium in most of the accessions 8 accessions were observed as long stigma with the range of 1.4–4.22 mm. The flower with pedicel - length of flower was mostly medium to long. The dry anther colour was largely pink grey and orange in many accessions and the anther length was mostly medium and in 6 accessions were long. In the stem characters studied, most of the landraces were high, stem diameter was small, juicy stem and 50 per cent of the landraces were sweet in taste. The landraces leaf length was medium and lead width was broad in majority of the accessions. The peduncle exertion was equ-distributed in all the scores except in ear covered by the leaf sheath. The panicle length and length of branches were short in most of the landraces. The panicles were mostly compact or semi compact and the shape of the panicles was symmetric or pyramidal. The neck of panicle visible length above sheath exertion was absent or weak in majority of the landraces. Majority of the glume color at physiological maturity of grain was green and the glume length was medium. The shattering of the plant was very low and majority of the accessions were freely threashable. In the grain characters, the grain form of most of the landraces was single, the caryopsis colour after threshing was pearly white, and the 1000-grain weight was high. Majority of the landrace grain dorsal and profile views were circular and the germ size was large. About 90 percent of the grains endosperm texture was ranged 25-50 per cent corneous. The grain - color of vitreous albumen was mostly white. The grain luster was equ-distributed in lustrous, medium and non-lustrous scores.

Correlation and Path Coefficient Analysis for Agronomical Traits in Forage Sorghum

Forage sorghum germplasm field experiment was conducted at Sorghum Research Station, Deesa (Gujarat) during *Kharif* 2008 by using 144 genotypes. The objective was to study the variability parameters and genotypic and phenotypic correlations and path analysis among agronomic traits. Significant variation among all the genotypes was recorded for all the characters. Most of the characters had higher genotypic and phenotypic coefficient of variation which is indicative that character expression in this sorghum population was genetic and can be exploited in breeding programs. The genotypes also exhibited varying degrees of heritability and genetic advance. Characters such as plant height, stem girth and dry and green fodder yield responded positively to selection because of high broad sense heritability and high genetic advance. Plant height, stem girth and leaf length were positively and significantly associated with green fodder and dry fodder yield per plant. Leaf

Table 12: Ranking of Sorghum Types According to Antioxidant Quality Indices

Rank	*Fe (ppm)*	*Zn (ppm)*	*Polyphenols (mg/100 g)*	*Phytate (mg/100 g)*	*TEAC (mg/100 g)*	*Cyanogen (-) (ppm)*	*Trypsin Inhibitor (-) (TIU)*	*Fibre (-) (per cent)*
1	EP 95	EP5	SSG 59-3	IS 8525	SSG 59-3	SPV 1762	EP 41	SLR 10
2	EP114	GGUB 40	URJA	IS 14131	IS 8525	SEVS 20	EP 38	EP6
3	SSG 59-3	GGUB 39	ELG 8	DJ 6514	ELG 8	EP 97	EP 127	IS 10284
4	GGUB 39	EP 18	IS 8525	E 36-1	URJA	SPV 1758	SPV 1759	EP 93
5	SPV 1794	PB 15881	EA 10	GGUB 34	EA 10	AKR 354	EP 24	SPV 1799
6	SPV 1801	EA 10	EP 108	EP 115	EA 11	SPV 1768	EP 10	SPV 1809
7	SPV 1616	EP 41	EA 11	EP 108	GGUB 34	PEC 7	EP 48	EP5
8	35B	CS3541	GGUB 34	IS 10284	N 13	2219B	EP 39	IS 8525
9	SPV 1411	GGUB 32	CSV 15	EP 68	EP 33	PEC 15	EP 22	SPV 1761
10	EA 6	EP 40	EP 107	CSV 19SS	GGUB 38	IS 26866	EP 21	EP 138
11	EP 115	GGUB 37	N 13	EA 10	GGUB 31	2077B	CSV 14R	SPV 1764
12	EP 41	IS 2122	CSV 17	IS 18551	B 58586	EC 1	EP 99	IMS 9B
13	CS3541	SURAT 1	GGUB 39	IS 2312	EP 95	SPV 1802	EP 64	IS 2312
14	SPV 1766	GGUB 31	B 58586	IS 9830	EP 107	CSV 22	AKR 150	EL 80
15	GGUB 41	SPV 1794	EP 95	EL 80	GGUB 37	EP 16	EP 57	EP 48
16	SPV 1805	SSG 59-3	SPV 1805	SSG 59-3	EC 34	EP 48	EC 21	EP 37
17	296B	EA 6	GGUB 38	GGUB 1	SEVS 3	IS 18551	EP 107	SPV 1757
18	GGUB 33	35B	CSV 18	SPV 1792	EP 108	EP 115	EP14	104B
19	EP 127	SPV 1755	CSV 14R	RS 673	AKMS 14B	SPV 1775	EP 37	SEVS 3
20	CSV 14R	EP 97	35B	GGUB 33	AKR 150	CSV 19SS	EP 97	EP 11
21	SPV 1760	SPV 1808	EP 115	EC 22	EC12	PEC 5	SPV 1411	EP 28
22	EA 11	EP 115	EP 42	SLR 10	GGUB 39	296B	EP 93	SPV 1762
23	SURAT 1	GGUB 34	EC 22	EP 65	27B	EP 91	EP 40	EP 16
24	CSV 19SS	SSV 84	EP 103	EA 6	IS 2312	EL 80	EC 13	SPV 1758
25	SPV 1759	EP 19	AKR 150	GGUB 31	SPV 1802	EP 120	EP9	EP 57

breath was positively and significantly associated with plant height, number of leaves per plant and leaf length. The path coefficient analysis indicated positive and significant correlation as well as high direct effect of dry fodder yield per plant, plant height, number of leaves per plant and stem girth on green fodder yield per plant.

Molecular Diversity of Lines likely to be Used in Breeding, and Fingerprinting Special Resources

Rapid purification of high quality genomic DNA was isolated from leaf tissue samples of 96 germplasm accessions using Mixer mill, minimum of 100 ng of DNA isolated from 100 mg leaf tissue and can be used for further assess the genetic variation using sorghum SSR markers. The 20 sorghum SSR primer pairs generated. A total of 104 alleles detected by silver-stained non-denaturing PAGE gels and the average number of alleles amplified per locus were 5.2 ranged from 2 to 8 alleles. The genetic diversity between genotypes was assessed on the basis of similarity coefficients and complemented with UPGMA cluster analysis. The correlation coefficients between the co-phenitic matrix computed from the dendrogram and the original similarity matrices were $r = 0.76$ ($t = 18.94$, $p = 1.00$) for the PAGE-generated marker data set. The coefficients of similarity ranged from 0.15 to 0.90 and these results suggested a good fit of the dendrogram generated from SSR data set. The dendrogram grouped the 96 sorghum genotypes into 21 major clusters at approximately the 30 per cent similarity level (Figure 6). Among the 96 germplasm lines tested for genetic diversity analysis two line shoot fly resistant germplasm line E 77 (IC 333435), stay green lines SEVS 3 (IC 347569) and SEVS 26 (IC 347592) were promising among the local germplasm.

Utilization

Sweet Stalk Sorghum

A total of 119 sweet sorghum germplasm lines were evaluated and used for making hand-crosses. The selected sweet-stalk sorghum germplasm lines originated from 13 countries and belong to 11 races. Hand crosses were made between these lines, and with exotic collections (15 accessions from USA), with SSV – 84 (sweet sorghum cultivar released at national level) and with local Indian germplasm with high brix (12 accessions). Out of 15 USA sweet sorghum germplasm evaluated, 4 lines showed the brix range between 18 and 24.

Dual-Purpose, Single-Cut, and Multi-Cut Forage Progeny Development

A total of 828 accessions Sudan sorghum germplasm and 22 released single, multi-cut forage and dual-purpose genotypes were used. 154 crosses were made among the single, multi-cut forage and dual-purpose lines, sweet sorghum lines with Sudan germplasm.

Exploitation of Indigenous and Exotic Sorghum Germplasm for Parental line Development

The *Kharif* sorghum germplasm *viz., zera zera, feteria* and *hegari* are better heterotic response than the *Rabi* germplasm *viz.,* Indian *durra* for parental line development.

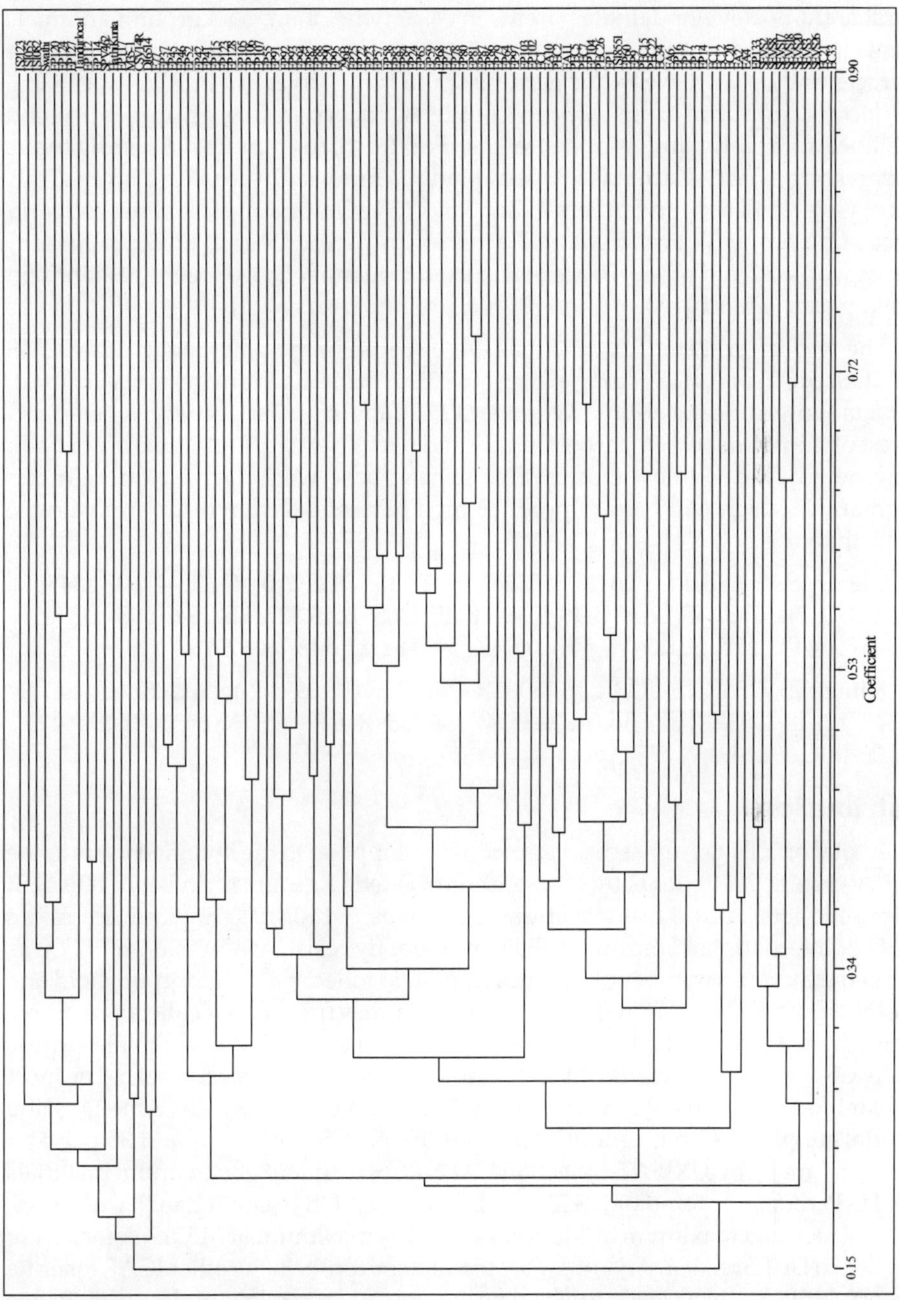

Figure 7

At Directorate of Sorghum Research (DSR), a new programme was initiated to create diversified parental lines by using Indigenous and Exotic sorghum germplasm. The main objective of the programme was to develop diversified genetic material with desirable traits, develop female parents in *caudatum/guinea* background and male parents in *durra* background. It was also envisaged to effect crosses among popular landraces and or with the exotic germplasm to bring diversity in R lines utilizing non-*durra* germplasm from African countries, incorporating grain quality and resistance to abiotic and biotic stresses. For this programme, 161 indigenous (Maharashtra 47 acc., Karnataka 33 acc., Madhya Pradesh 29 acc., Andhra Pradesh 17 acc., Gujarat 14 acc., and others 21 acc.) and 159 exotic germplasm lines (Nigeria 53 acc., Cameroon 37 acc., Sudan 17 acc., South Africa 15 acc., Ethiopia 11 acc., Zimbabwe 6 acc., and others 20 acc.) were used to identify B and R gene reactions on A1 cytoplasm (CMS 104A).

The fertility and sterility reactions were scored in the pairs made using CMS 104A. Out of 525 pairs (174 suing exotic and 351 using indigenous germplasm lines), 204 pairs showed clear sterility (B gene) and 175 pairs exhibited fertility (R gene) and the rest of the pairs showed partial sterility and partial fertility and a mixture of both. Many new B and R lines with diverse genetic base were derived through this programme, which will help to produce high heterotic *Rabi* hybrids and increase productivity.

B lines: IS 18068B, PEC 2B (IC 392125), IS 15870B, PEC 21B (IC 392144), IS 18452B, EP 106B (IC 345195), IS 22461B, IS 4540-1B, IS 33855B, IS 33733B, IS 6368B, EC 8-2B (IC 345710), PU 25B, SLK 3-1B, PEC 1B (IC 392124) and IS 18362-1B.

R lines: IS 5256R, IS 4413R, IS 18061R, PEC 32R (IC 392155), IS 15058-2R, EC 8R (IC 345710), IS 33832R, EP 55R (IC 343554), IS 15689R, EP 109R (IC 345198), IS 4411R, PEC 26R (IC 392149),

Distribution

A total of 3059 germplasm accessions were supplied to the bonafied users in the country during 2000 – 2007 by the Sorghum Genetic Resources division, NRCS. It comprised a total of 1093 *Rabi* germplasm accessions, 591 *Kharif* germplasm accessions and 1149 International Sorghum (IS) lines. Shoot fly resistant accession EP 117 was supplied maximum number of 12 times from *Rabi* collections while, grain mold and shoot fly resistant line E 77 has been supplied 8 times from *Kharif* collections. Shoot fly and grain mold resistant IS lines were supplied 5 times. The *Kharif* dual-purpose variety CSV 15 and *Rabi* variety M 35-1 were supplied maximum of 19 each. The most popular lines 296B and C 43 were supplied 5 and 7 times respectively. Among all the material supplied, 42 per cent of accessions are NRCS collections and 38 per cent accessions are from ICRISAT collections. At NRCS, recipient of maximum materials is Dr PG Padmaja, Entomologist (338 acc.). At AICSIP, Dr. Jaganmohan Rao of RARS, Madhira received maximum of 145 accessions. The maximum of 315 accessions were supplied to Dr T Sasikala, ANGRAU for research purpose. In the other ICAR institute category, NBPGR – New Delhi received maximum of 27 accessions under the exchange programme. Under other category, Mr. Narayana received 8 accessions of pop sorghum for multiplication and Tamil Nadu Agriculture Department received 15 accessions. A total of 570 accessions supplied to the sorghum varietal improvement programmes. The Agricultural Entomology division received maximum of 345 accessions to screen for insect resistance.

Conservation

Sorghum seeds show orthodox storage behavior (*i.e.* desiccation tolerant); hence, the germplasm is conserved *ex situ* as seeds in the genebank under controlled conditions. *Ex situ* conservation is cost efficient, and also makes it easier for scientists to access, study, distribute and use plant genetic resources. At DSR, all the germplasm is stored as active collection under medium-term storage conditions (4-10°C and 20-35 per cent RH). For each accession, about 300-400 g of seed is harvested at optimum maturity, usually 6-7 weeks after anthesis, from post-rainy season multiplication plots. The seeds are dried in the shade to about 8 per cent moisture content, threshed by hand, cleaned and stored in screw-capped airtight plastic containers. The germplasm is also conserved as base collection at -20°C for long-term storage at National Genebank, National Bureau of Plant Genetic Resources (NBPGR), New Delhi. Seeds stored under medium-term conditions (active collection) are used for distribution and research, while those stored under long-term conditions (base collection) are used only for regenerating active collections. Viability is monitored by germination tests at 5-year intervals in medium-term storage and 10-year intervals in long-term storage. Any sample having less than 75 per cent viability is identified for rejuvenation. The medium- and long-term storage rooms are constructed on a modular

Figure 8: Medium-term Storage at DSR.

principle with fabricated panels. The MTS room has a capacity of 110 m^3, capable of accommodating over 25,000 germplasm accessions (Figure 8). The storage rooms have both mobile and immobile shelving systems. The temperature and relative humidity inside the storage rooms are monitored regularly with thermo-hygrographs. Audible and visual electronic alarm systems have been installed to safeguard the seeds from any rise in temperature and relative humidity and to help maintain the desired conditions.

Documentation

Passport Data

A total of 1720 accessions collected under the project are documented with passport data for 48 field characters. For 979 accessions, Indigenous Collection number (IC No.) is received.

Conservation Data

A database on Status of Sorghum Genetic Resources of DSR along with passport data and evaluation data is prepared and connected to the LAN through DSR server for the easy retrieval of information on the availability of the required germplasm line from the desktop computer by the scientists of NRCS.

Diversity Mapping

With the help of DIVA GIS software developed by the CIP, Peru, sorghum diversity map is prepared for the NATP collections to identify the unexplored area for future exploration. Under the sorghum germplasm collection programme, 26 collection missions in Andhra Pradesh (66 acc.), Karnataka (151 acc.), Maharashtra (154 acc.), Madhya Pradesh (185 acc.), Rajasthan (85 acc.), Uttar Pradesh (154 acc.), Gujarat (107 acc.) and Tamil Nadu (139 acc.) have been undertaken and 1086 accessions were collected. The non-spatial information *viz.*, passport data, evaluation data and status of germplasm at medium-term storage, DSR were linked with the spatial information *viz.*, latitude and longitude for multi-query and preparation of maps. The diversity of sorghum races and landraces mapping have been done through this package. The hotspots of sorghum pests and diseases map, and district-wise sorghum released varietal adoption map are also in progress at DSR with the GIS software.

Indian Sorghum Landrace Mapping through GIS

India is the secondary origin of sorghum. The cultivation of sorghum reported 5000 years ago. Landrace is a farmer's variety developed and cultivated over the decades by the farmers. The richness of five basic sorghum races *viz., durra, caudatum, guinea, kafir* and *bicolor* and its ten intermediate races are also an indicator of the evolution of sorghum in the Indian regions. Geographical Information System (GIS) is an important tool to analyse the diversity of the races and landraces of sorghum its spread and distribution. The National Research Centre for Sorghum (NRCS) has been exploring the sorghum germplasm and maintaining the active collections from sorghum growing states of the country. Based on the GIS mapping (Figure 9) we found that *durra* is the dominant race and most of the *Rabi* (post-rainy) germplasm

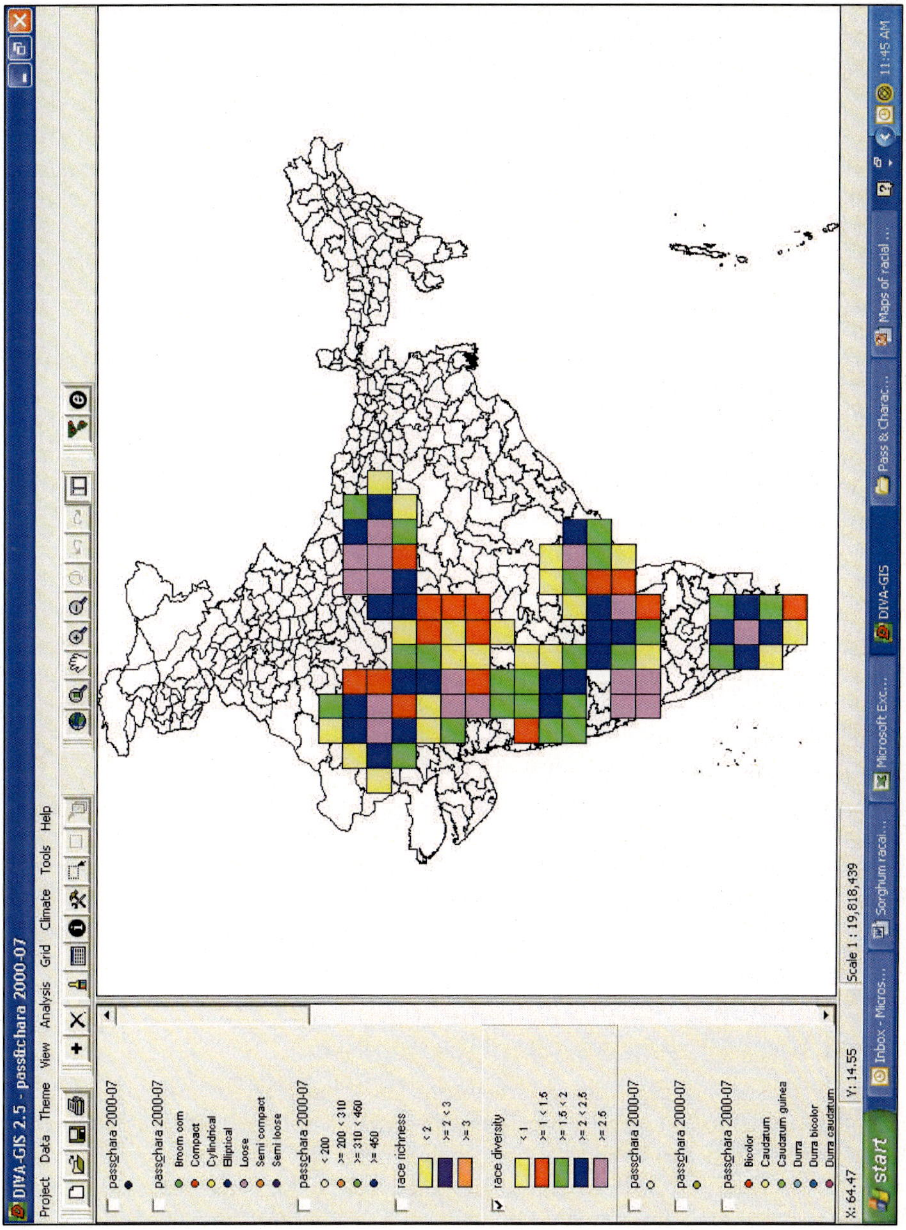

Figure 9: GIS Mapping of Sorghum Collections at DSR.

belong to this race. Durra races are prevalent in *Rabi* sorghum cultivating states *viz.*, Maharashtra, Karnataka, and Andhra Pradesh. *Caudatum* is the next dominant race and *Kharif* (rainy) germplasm lines belong to *caudatum*. The Madhya Pradesh, Uttar Pradesh, Rajasthan and Gujarat are *Kharif* sorghum growing states. The fodder sorghum germplasm basically belong to *bicolor* race and are prevalent in Tamil Nadu and Rajasthan. The *guinea* race belongs to the open glume type has the resistance to grain mold. It is found in the high rainfall areas of Uttar Pradesh. About 80 per cent of the breeding material belongs to *caudatum* race. We need to introduce breeding material for high yielding, drought tolerant, cold tolerant, shoot fly resistant, grain mold resistant and sweet sorghum materials from other countries. The existing Indigenous collections are having lustrous grain, high yielding, high biomass, high brix percentage, high protein and shoot fly resistant. These promising entries are being used in the sorghum improvement programme.

Potential High Popping Sorghum Germplasm

Sorghum can be used for diverse purposes. Pop sorghum and sweet sorghum are the plant types which promote the utilization of the crop in a commercial mode. In some parts of South India and in Ethiopia, sorghum grains are popped and used on special occasions and festivals. In spite of being superior to pop corn in tenderness and less hull, pop sorghum suffers commercial attention. Sorghum grain digestibility is limited and this decreases when the grain is cooked. The process of popping, flaking and extrusion increase the digestibility. Since pop sorghum has economical as well as nutritional values, there is a need to exploit it commercially while conserving the rich native diversity. Central India being the major sorghum growing area has large and diverse plant types of pop sorghum. Many of these cultivars have been collected, multiplied and studied for their popping qualities, similar to that of rice and maize. Some of these cultivars have also been grown on saline and alkaline soil situations. A replicated set of 50 pop sorghum germplasm collected from Karnataka was evaluated for the popping ability during 2005. The number of seeds used for popping, popped seed, un-popped seed and its popping percentage were analysed. The popped seed was observed as 73 per cent and un-popped as 17 per cent. A total of 13 pop sorghum germplasm *viz.*, POP 16 (IC 308651), POP 31(IC 308660), POP 32 (IC 308661), POP 37 (IC 308664), POP 38 (IC 308665), POP 47(IC 308671), POP 49 (IC 308673), POP 50 (IC 308674), POP 54 (IC 308678), POP 58 (IC 308682), POP 59 (IC 308683), POP 60 (IC 308684), and POP 62 (IC 308685) were identified with more than 70 percent popping ability (Table 22).

Potential High Protein Sorghum Germplasm

India is the secondary origin for sorghum. The variability of the crop exists in taxonomical, morphological and biochemical traits. A total of 150 *Rabi* sorghum germplasm collected from Andhra Pradesh, Karnataka and Maharashtra were subjected to proximate analysis during 2006 and 2007. The percent moisture ranged between 9.41 and 12.66, the percent fat 2.20 and 3.41, the percent starch 61.20 and 67.50 and the percent protein ranged between 9.95 and 17.60. The CV per cent of the protein per cent was 12.71. Nine germplasm lines were observed with >15 per cent protein. They are IS 36221 (15.52 per cent), IS 36223 (17.60 per cent), EP 26 (IC 305907)

Promising Germplasm Identified

Table 13: Promising Entries Identified from Germplasm Collections of Andhra Pradesh

Trait	*Promising Accessions*	*No. of Accessions*
Days to 50 per cent flowering (<55 days)	NSJB 6683 (IC 249106)	1
Number of leaves (>17)	SEVS 17 (IC 347583)	1
Leaf length (>82 cm)	SEB 12005 (IC 541863), SEB 12006 (IC 541864), SEB 12010 (IC 541868), SEVS 8 (IC 347574), SEVS 10 (IC 347576), SEVS 17 (IC 347583), SEVS 19 (IC 347585), SEVS 25 (IC 347591), NSJB 6602 (IC 249059), NSJB 6605 (IC 249062), NSJB 6606 (IC 249063), NSJB 6609 (IC 249066)	12
Leaf width (>8 cm)	SEB 11992 (IC 541850), SEB 12000 (IC 541858), SEB 12025(IC 541880), SEVS 26 (IC 347592), NSJB 6673 (IC 249096)	5
Plant height (>240 cm)	SEB 11999 (IC 541857), SEB 12005 (IC 541863), SEB 12006 (IC 541864), SEB 12020 (IC 541875), SEB 12021 (IC 541876), SEB 12023 (IC 541878)	6
Ear head length (>30 cm)	SEB 11988 (IC 541847), SEB 11992 (IC 541850), SEB 11993 (IC 541851), SEB 11996 (IC 541854), SEB 11997 (IC 541855)	5
Ear head width (>9 cm)	SEB 12000 (IC 541858), SEB 12005 (IC 541863), SEB 12025 (IC 541880)	3
Stem thickness (>2 cm)	SEB 12006 (IC 541864), SEB 12020 (IC 541875)	2
Brix (>18 per cent)	SEB 12004 (IC 541862)	1
Stem fresh weight (>700g/plant)	SEB 12006 (IC 541864)	1
Stem dry weight (>350g/plant)	SEB 12006 (IC 541864), SEB 12010 (IC 541868)	2
Grain yield (>70 g)	SEB 11995 (IC 541853), SEB 12003 (IC 541861), SEB 12009 (IC 541867), SEB 12017 (IC 541873), SEB 12025 (IC 541880)	5
100-seed weight (>3.5g)	SEB 12004 (IC 541862)	1
Days to maturity (<107days)	E 89 (IC 369120), E 90 (IC 369121), E 92 (IC 369123)E 100 (IC 369131), PEC 28 (IC 392151), SEVS 13(IC 347579)	6
	Total	**51**

Table 14: Promising Entries Identified from Germplasm Collections of Maharashtra

Trait	*Promising Accessions*	*No. of Accessions*
Days to 50 per cent flowering (<60 days)	ELG 1 (IC 568337), ELG 8 (IC 568344), ELG 10 (IC 568346)	3
Number of leaves (>17)	ELG 5 (IC 568341), ELG 13 (IC 568349), ELG 14 (IC 568350), ELG 17 (IC 568353), ELG 18 (IC 568354), ELG 20 (IC 568356), ELG 21 (IC 568357), ELG 25 (IC 568361)	8
Leaf length (>90 cm)	ELG 2 (IC 568338), ELG 3 (IC 568339), ELG 13 (IC 568349), ELG 16 (IC 568352), ELG 17 (IC 568353), ELG 18 (IC 568354), ELG 20 (IC 568356), ELG 21 (IC 568357), ELG 24 (IC 568360), ELG 25 (IC 568361)	10
Leaf width (>9cm)	ELG 2 (IC 568338), ELG 11 (IC 568347), ELG 13 (IC 568349), ELG 14 (IC 568350), ELG 20 (IC 568356), ELG 21 (IC 568357), ELG 25 (IC 568361), PEC 8 (IC 392131), PU 9, EP 20 (IC 305901)	10
Plant height (>350 cm)	ELG 8 (IC 568344), ELG 12 (IC 568348), ELG 13 (IC 568349), ELG 14 (IC 568350), ELG 15 (IC 568351), ELG 16 (IC 568352), ELG 17 (IC 568353), ELG 18 (IC 568354), ELG 20(IC 568356), ELG 21 (IC 568357), ELG 25 (IC 568361)	11
Ear head length (>30 cm)	ELG 1 (IC 568337), ELG 7 (IC 568343), ELG 8 (IC 568344), ELG 9 (IC 568345), ELG 10 (IC 568346), ELG 17 (IC 568353), ELG 19 (IC 568355)	7
Ear head width (>9 cm)	ELG 6 (IC 568342), ELG 7 (IC 568343), ELG 8 (IC 568344), ELG 9 (IC 568345), ELG 10 (IC 568346), PEC 2 (IC 392125), EP 76 (IC 343575)	7
Stem thickness (>3 cm)	PU 21	1
Brix (>19 per cent)	EP 61 (IC 343560), PU 22	2
Stem fresh weight (1000g/plant)	ELG 5 (IC 568341), ELG 18 (IC 56835), ELG 20 (IC 568356)	3
Stem dry weight (650g/plant)	ELG 5 (IC 568341), ELG 18 (IC 568354)	2
Grain yield (>90 g^{-1}plant)	ELG 9 (IC 568345), EP 89 (IC 343588)	2
100-seed weight (>4.5 g)	EP 74 (IC 343573), EP 80 (IC 343579)	2
Days to maturity (<108days)	EP 5 (IC 305886), EP 71 (IC 343570), EP 124 (IC 420939), PEC 7 (IC 392130), PEC 8 (IC 392131), PU 12, PU 19, PU 24	8
	Total	**76**

Table 15: Promising Entries Identified from Germplasm Collections of Karnataka

Trait	*Promising Accessions*	*No. of Accessions*
Days to 50 per cent flowering (<67 days)	POP 32 (IC 308661), POP 35 (IC 308663),CJV 34 (IC308628),CJV 35 (IC 308629)	4
Number of leaves (>20)	POP 8 (IC 308643)	1
Leaf length (>90 cm)	POP 5 (IC 308641)	1
Leaf width (>9.5 cm)	POP 8 (IC 308643), POP 27 (IC 308657), POP 121 (IC 308689), PEC16 (IC 392139), EP 42 (IC 305923), EP 49 (IC 305930)	6
Plant height (>300 cm)	POP 3 (IC 308640), POP 5 (IC 308641), POP 8 (IC 308643), POP 9 (IC 308644), POP 30 (IC 308659), POP 37 (IC 308664), POP 40 (IC 308667), POP 56 (IC 308680), POP 58 (IC 308682)	10
Ear head length (>30 cm)	POP 5 (IC 308641)	1
Ear head width (>9 cm)	POP 5 (IC 308641), POP 30 (IC 308659)	2
Stem thickness (>2 cm)	POP 5 (IC 308641), POP 8 (IC 308643), POP 30 (IC 308659), POP 37 (IC 308664), POP 45 (IC 308669), EP 28 (IC 305909), EP 45 (IC 305926)	7
Brix (>20 per cent)	POP 15 (IC 308650), EP 32 (IC 305913), CJV 11 (IC 308607), CJV 16 (IC 308611), CJV 17 (IC 308612), CJV 20 (IC 308615), CJV 25 (IC 308619)	7
Stem fresh weight (700g/plant)	POP 5 (IC 308641), POP 8 (IC 308643), POP 30 (IC 308659)	3
Stem dry weight (500g/plant)	POP 5 (IC 308641), POP 8 (IC 308643)	2
Grain yield (>70 g)	EP 102 (IC 345191), EP 106 (IC 345195), EP 117 (IC 345206)	3
100-seed weight (>4 g)	EP 38 (IC 305919), EP 97 (IC 345186), EP 99 (IC 345188), EP 100 (IC 345189), EP 102 (IC 345191), EP 106 (IC 345195), EP 109 (IC 345198), EP 110 (IC 345199), EP 113 (IC 345202), EP 117 (IC 345206)	10
Days to maturity (<108days)	EP 25 (IC 305906), EP 32 (IC 305913), EP 37 (IC 305918), EP 41 (IC 305922), EP 51 (IC 305932), EP 52 (IC 305933)	6
	Total	**63**

Table 16: Promising Entries Identified from Germplasm Collections of Madhya Pradesh

Trait	*Promising Accessions*	*No. of Accessions*
Days to 50 per cent flowering (<60 days)	ER 3 (IC 0585137), GGUB 23 (IC 319866), GGUB 24 (IC 319867), GGUB 25 (IC 319868), GGUB 26 (IC 319869), GGUB 27 (IC 319870), GGUB 28 (IC 319871), GGUB 35 (IC 319878)	8
Number of leaves (>20)	E 208 (IC 568469), E 209 (IC 568470), E 210 (IC 568471),E 215 (IC 568476), E 217 (IC 68478), E 229 (IC 568490)	6
Leaf length (>100 cm)	ER 21 (IC 0585155), GGUB 2 (IC 319848), GGUB 4 (IC 319850), GGUB 7 (IC 319853), GGUB 8 (IC 319854), GGUB 9 (IC 319855), GGUB 11 (IC 319857), GGUB 13 (IC 319859), GGUB 57 (IC 319898), GGUB 60 (IC 319901), GGUB 61 (IC 319902), GGUB 67 (IC 319907)	12
Leaf width (>11 cm)	E 201 (IC 568462), E 208 (IC 568469), E 211 (IC 568472), E 218 (IC 568479), E 222 (IC 568483), E 226 (IC 568487)	6
Plant height (>450 cm)	MP 3, MP 30, MP 39, MP 41, MP 48, E 215 (IC 568476)	6
Ear head length (>35 cm)	ER 4 (IC 0585138), ER 23 (IC 0585157), ER 26 (IC 0585160)	3
Ear head width (>12 cm)	ER 13 (IC 0585147), ER 23 (IC 0585157), ER 24 (IC 0585158), ER 26 (IC 0585160), MP 7, MP 17, MP 43	7
Stem thickness (>3 cm)	E 203 (IC 568464), E 208 (IC 568469), E 215 (IC 568476), E 216 (IC 568477), E 217 (IC 568478), E 222 (IC 568483), E 223 (IC 568484)	7
Brix (>20 per cent)	GGUB 19 (IC 319862), GGUB 29 (IC 319872), GGUB 33 (IC 319876), GGUB 43 (IC 319886), GGUB 49 (IC 319898), GGUB 57 (IC 319898), GGUB 60 (IC 319901), GGUB 61 (IC 319902), GGUB 63 (IC 319904), GGUB 64 (IC 319905), GGUB 68 (IC 319908), GGUB 69 (IC 319909)	12
Stem fresh weight (1500g/plant)	E 187 (IC 568448), E 210 (IC 568471), E 213 (IC 568474), E 214 (IC 568475), E 216 (IC 568477), E 223 (IC 568484), E 224 (IC 568485), E 227 (IC 568488), E 228 (IC 568489), E 229 (IC 68490), E 230 (IC 568491), E 248 (IC 568509)	12
Stem dry weight (1000g/plant)	E 187 (IC 568448), E 203 (IC 568464), E 210 (IC 568471), E 213 (IC 568474), E 215 (IC 568476), E 216 (IC 568477), E 227 (IC 568488), E 228 (IC 568489), E 229 (IC 568490), E 230 (IC 568491), E 248 (IC 568509)	11
Grain yield (>120 g)	ER 20 (IC 0585154), ER 25 (IC 0585159), ER 27 (IC 0585161), ER 33 (IC 0585167)	4
100-seed weight (>4 g)	E 233 (IC 568494), E 234 (IC 568495)	2
Days to maturity (<100days)	GGUB 23 (IC 319866), GGUB 24 (IC 319867), GGUB 25 (IC 319868), GGUB 36 (IC 319879), GGUB 38 (IC 319881), ER 13 (IC 0585147), MP 32	7
	Total	**103**

Table 17: Promising Entries Identified from Germplasm Collections of Uttar Pradesh

Trait	*Promising Accessions*	*No. of Accessions*
Days to 50 per cent flowering (<70 days)	ES 5 (IC 333350), ES 26 (IC 333371)	2
Number of leaves (>17)	E 125 (IC 415816), E 130 (IC 415821), E 137 (IC 415828)	3
Leaf length (>95 cm)	E 21(IC 333379), E 30 (IC 333388), E 40 (IC 333398), E 84 (IC 333442), ES 19 (IC 333364)	5
Leaf width (>8 cm)	E 101 (IC 415792), E 102 (IC 415793), E 105 (IC 415796), E 107 (IC 415798), E 108 (IC 415799), E 111 (IC 415802), E 143 (IC 415834), E 41 (IC 333399), E 44 (IC 333402), E 47 (IC 333405), E 56 (IC 333414), E 63 (IC 333421), E 82 (IC 333440)	13
Plant height (>360 cm)	E 18 (IC 333376), E 33 (IC 333391), E 80 (IC 333438), E 81 (IC 333439), E 83 (IC 333441), E 85 (IC 333443), E 86 (IC 333444), ES 9 (IC 333354), ES 22 (IC 333367)	9
Ear head length (>30 cm)	E 115 (IC 415806), E 116 (IC 415807), E 124 (IC 415815), E 125 (IC 415816), E 136 (IC 415827), E 138 (IC 415829), ES 5 (IC 333350)	7
Ear head width (>11 cm)	E 103 (IC 415794), E 104 (IC 415795), E 117 (IC 415808), E 118 (IC 415809), E 121 (IC 415812), E 128 (IC 415819), E 134 (IC 415825), E 144 (IC 415835)	8
Stem thickness (>2.1 cm)	E 105 (IC 415796), E 108 (IC 415799), ES 3 (IC 333348), ES 8 (IC 333353), ES 9 (IC 333354)	5
Brix (>20 per cent)	E 38 (IC 333396), E 39 (IC 333397)	2
Stem fresh weight (700g/plant)	E 20 (IC 333378), E 46 (IC 333404)	2
Stem dry weight (350g/plant)	E 20 (IC 333378), E 23 (IC 333381), ES 9 (IC 333354)	3
Grain yield (>50 g)	E 101 (IC 415792), E 104 (IC 415795), E 105 (IC 415796), E 107 (IC 415798), E 124 (IC 415815), E 126 (IC 415817), E 130 (IC 415821), E 138 (IC 415829), E 139 (IC 415830)	9
100-seed weight (>3.5 g)	E 51 (IC 333409), E 63 (IC 333421)	2
Days to maturity (<108 days)	E 82 (IC 333440), E 83 (IC 333441), ES 4 (IC 333349), ES 5 (IC 333350)	4
	Total	**74**

Table 18: Promising Entries Identified from Germplasm Collections of Rajasthan

Trait	*Promising Accessions*	*No. of Accessions*
Days to 50 per cent flowering (<50days)	EJ 9 (IC 338989), EJ 58 (IC 369258), EJ 61 (IC 369261)	3
Number of leaves (>17)	EJ 55 (IC 369255), EJ 76 (IC 369276), RAJ 4, RAJ 5, RAJ 9, RAJ 12, RAJ 14, RAJ 15, RAJ 16, RAJ 19, RAJ 21	11
Leaf length (>85cm)	EJ 75 (IC 369275), EJ 76 (IC 369276), E 2 (IC 338969), RAJ 4, RAJ 5, RAJ 6, RAJ 8, RAJ 9, RAJ 14, RAJ 16, RAJ 19, RAJ 21	12
Leaf width (>8 cm)	RAJ 6, RAJ 9, RAJ 15, RAJ 17, RAJ 34, RAJ 35, RAJ 36	7
Plant height (>360 cm)	EJ 76 (IC 369276), RAJ 6, RAJ 7, RAJ 8, RAJ 9, RAJ 10, RAJ 12, RAJ 13, RAJ 21, RAJ 25	10
Ear head length (>30 cm)	EJ 68 (IC 369268)	1
Ear head width (>12cm)	EJ 63 (IC 369263), EJ 64 (IC 369264), EJ 66(IC 369266), EJ 70 (IC 369270), EJ 71 (IC 369271)	5
Stem thickness (>2 cm)	EJ 65 (IC 369265), EJ 68 (IC 369268), RAJ 19	3
Brix (>20 per cent)	RAJ 9, RAJ 24	2
Stem fresh weight (700g/plant)	E 1 (IC 338968), E 2 (IC 338969), E 7 (IC 338973), EJ 2 (IC 338982)	4
Stem dry weight (350g/plant)	E 1 (IC 338968), E 2 (IC 338969), E 7 (IC 338973), EJ 2 (IC 338982), RAJ 14, RAJ 16, RAJ 19	7
Grain yield (>50 g)	EJ 65 (IC 369265), E 7 (IC 338973), EJ 25 (IC 339005), RAJ 6, RAJ 18, RAJ 35, RAJ 36	7
100-seed weight (>3.5 g)	E 1 (IC 338968), EJ 40 (IC 339020), EJ 41 (IC 339021)	3
Days to maturity (<100days)	RAJ 23, RAJ 25, RAJ 31, RAJ 33, RAJ 34	5
	Total	**80**

Table 19: Promising Entries Identified from Germplasm Collections of Gujarat

Trait	*Promising Accessions*	*No. of Accessions*
Days to 50 per cent flowering (< 50days)	EJN 4 (IC 0585174), EJN 7 (IC 0585177), EJN 9 (IC 0585179), EJN 11 (IC 0585181), EJN 15 (IC 0585185)	5
Number of leaves (>19)	EJN 16 (IC 0585186), EJN 20 (IC 0585190), ERN 5 (IC 568520), ERN 16 (IC 568531), E 149 (IC 568366), E 152 (IC 568369), E 153 (IC 568370), E 154 (IC 568371), E 167 (IC 568384)	9
Leaf length (>85 cm)	ERN 5 (IC 568520), ERN 6 (IC 568521), ERN 7 (IC 568522), ERN 17 (IC 568532), ERN 19 (IC 568534), E 147 (IC 568364), E 154 (IC 568371)	7
Leaf width (>8 cm)	ERN 5 (IC 568520), E 163 (IC 568380), E 164 (IC 568381)	3
Plant height (>400 cm)	EJN 1(IC 0585171), EJN 2 (IC 0585172), ERN 2 (IC 568517), ERN 6 (IC 568521), ERN 7 (IC 568522), ERN 8 (IC 568523), ERN 10 (IC 5685250), ERN 11 (IC 568526), ERN 15 (IC 568530), E 152 (IC 568369), E 153 (IC 568370), E 154 (IC 568371)	12
Ear head length (>35 cm)	E 146 (IC 568363), E 153 (IC 568370), E 154 (IC 568371), E 155 (IC 568372), E 159 (IC 568376)	5
Ear head width (>9 cm)	EJN 8 (IC 0585178), EJN 9 (IC 0585179)	2
Stem thickness (>2.5 cm)	ERN 2 (IC 568517), ERN 3 (IC 568518), ERN 5 (IC 568520), ERN 7 (IC 568522), ERN 26 (IC 568541), E 163 (IC 568380), E 169 (IC 568386), E 181 (IC 568398)	8
Brix (>18 per cent)	E 147 (IC 568364), E 151 (IC 568368)	2
Stem fresh weight (1200g/plant)	ERN 2 (IC 568517), ERN 3 (IC 568518), ERN 7 (IC 568522), ERN 8 (IC 568523), ERN 9 (IC 568524), ERN 11 (IC 568526), ERN 12 (IC 568527), ERN 14 (IC 568529), ERN 15 (IC 568530), ERN 16 (IC 568531), ERN 17 (IC 568532), ERN 19 (IC 568534)12	
Stem dry weight (800g/plant)	ERN 3 (IC 568518), ERN 7 (IC 568522), ERN 9 (IC 568524), ERN 11 (IC 568526), ERN 15 (IC 568530), ERN 17 (IC 568532), ERN 19 (IC 568534)	7
Grain yield (>70 g)	E 145 (IC 568362), E 171 (IC 568388)	2
100-seed weight (>3g)	E 145 (IC 568362)	1
Days to maturity (< 90days)	EJN 7 (IC 0585177), EJN 9 (IC 0585179), EJN 11(IC 0585181), EJN 15 (IC 0585185), EJN 24 (IC 0585194), E 172 (IC 568389)	6
	Total	**81**

Table 20: Promising Entries Identified from Germplasm Collections of Tamil Nadu

Trait	*Promising Accessions*	*No. of Accessions*
Days to 50 per cent flowering (< 58 days)	EG 66 (IC 568408), EG 67 (IC 568409) EG 68 (IC 568410)	3
Number of leaves (>17)	ERS 6 (IC 0585211), ERS 11 (IC 0585216), ERS 15 (IC 0585220), ERS 22 (IC 0585227), ERS 29 (IC 0585234), EG 92 (IC 568434), EG 93 (IC 568435), EG 95 (IC 568437), EG 96 (IC 568438), EG 100 (IC 568442)	10
Leaf length (>82 cm)	ERS 8 (IC 0585213), EG 95 (IC 568437), EG 96 (IC 568438), EG 51 (IC 541358)	4
Leaf width (>9cm)	ERS 15 (IC 0585220), ERS 24 (IC 0585229), ERS 25 (IC 0585230), ERS 29 (IC 0585234), ERS 32 (IC 0585237),EG 74 (IC 568416), EG 42 (IC 541349)	7
Plant height (>450 cm)	ERS 8 (IC 0585213), ERS 12 (IC 0585217), ERS 16 (IC 0585221), ERS 17 (IC 0585222), ERS 18 (IC 0585223), ERS 19 (IC 0585224), ERS 20 (IC 0585225), ERS 21 (IC 0585226), ERS 22 (IC 0585227), ERS 23 (IC 0585228), ERS 26 (IC 0585231), ERS 27(IC 0585232), ERS 28 (IC 0585233), ERS 29 (IC 0585234), ERS 33 (IC 0585238), ERS 35 (IC 0585240), ERS 36 (IC 0585241), EG 92(IC 568434), EG 95(IC 568437), EG 100 (IC 568442)	20
Ear head length (>35cm)	ERS 12 (IC 0585217), EG 33 (IC 541340), EG 41 (IC 541348), EG 43 (IC 541350)	4
Ear head width (>10cm)	ERS 1 (IC 0585206), ERS 10 (IC 0585215), ERS 12 (IC 0585217), ERS 19 (IC 0585224) ERS 26 (IC 0585231), ERS 28 (IC 0585233), ERS 33 (IC 0585238), ERS 35 (IC 0585240), EG 41 (IC 541348), EG 52 (IC 541359), EG 54 (IC 541361)	11
Stem thickness (>2.3cm)	ERS 8 (IC 0585213), ERS 14 (IC 0585219), ERS 25 (IC 0585230), ERS 28 (IC 0585233), EG 101(IC 568443)	5
Brix (>19 per cent)	EG 39 (IC 541346), EG 3 (IC 541310), EG 11 (IC 541318), EG 19 (IC 541326), EG 22 (IC 541329), EG 23 (IC 541330), EG 24 (IC 541331), EG 25 (IC 541332)	8
Stem fresh weight (700g/plant)	EG 92 (IC 568434), EG 93 (IC 568435), EG 95 (IC 568437)	3
Stem dry weight (400g/plant)	EG 92 (IC 568434), EG 93 (IC 568435), EG 95 (IC 568437), EG 96 (IC 568438), EG 98 (IC 568440), EG 100 (IC 568442)	6
Grain yield (>70 g)	EG 27 (IC 541334), EG 34 (IC 541341), EG 38 (IC 541345), EG 48 (IC 541355), EG 52 (IC 541359), EG 55 (IC 541362), EA 4 (IC 345246)	7
100-seed weight (>3g)	EG 1 (IC 541308), EG 2 (IC 541309), EG 4 (IC 541311), EG 7 (IC 541314), EG 13 (IC 541320), EG 23 (IC 541330), EG 34 (IC 541341), EG 55 (IC 541362)	8
Days to maturity (< 94days)	ERS 13 (IC 0585218), ERS 14 (IC 0585219), ERS 15 (IC 0585220), ERS 24 (IC 0585229), ERS 25 (IC 0585230), ERS 32 (IC 0585237), ERS 34 (IC 0585239)	7
	Total	**103**

Table 21: Promising Entries Identified from Germplasm Collections of Uttarakhand

Trait	*Promising Accessions Indigenous Collection No. (Collector No.)*	*No. of Accessions*
Ear head length (>50 cm)	IC 597603 (ESRK 3), IC 597604 (ESRK 4), IC 597610 (ESRK 10), ESRK 14, IC 597614 (ESRK 15), IC 597615 (ESRK 16), IC 597618 (ESRK 19), IC 597619 (ESRK 20), IC 597620 (ESRK 21), IC 597622 (ESRK 23), IC 597623 (ESRK 24), IC 597626 (ESRK 27), IC 597603 (ESRK 30)	13
Ear head width (>8 cm)	IC 597628 (ESRK 29), IC 597629 (ESRK 30)	2
Stem thickness (>2 cm)	IC 597629 (ESRK 30)	1
Stem fresh weight (>2000g/plant)	IC 597629 (ESRK 30)	1
Grain yield (>90 g)	IC 597601 (ESRK 1), IC 597609 (ESRK 9), IC 597621 (ESRK 22), IC 597629 (ESRK 30)	4
	Total	**21**

Table 22: Potential Pop Germplasm Identified

Sample No.	IC No.	Seed Taken (gm)		Popped Seed (wt)		Un Popped Seed (wt)		100 per cent Popped		100 per cent Unpopped	
		R1	*R2*	*R1*	*R2*	*R1*	*R2*	*R1*	*R2*	*R1*	*R2*
POP – 16	IC 308651	20	20	14.21	13.09	3.42	3.54	71.1	65.45	17.1	17.7
POP – 31	IC 308660	25	25	17.87	16.98	3.47	4.92	71.48	67.92	13.88	19.68
POP – 32	IC 308661	25	25	17.79	18.09	4.18	3.92	71.16	72.36	16.72	15.68
POP – 37	IC 308664	25	25	17.99	16.15	4.39	5.88	71.96	64.6	17.56	23.52
POP – 38	IC 308665	25	25	18.11	17.89	3.91	4.2	72.44	71.56	15.64	16.8
POP – 47	IC 308671	25	25	18.2	18.4	3.31	2.06	72.8	73.6	13.24	8.24
POP – 49	IC 308673	25	25	18.08	18.79	4.28	4.27	72.32	75.16	17.12	17.08
POP – 50	IC 308674	25	25	18.99	18.37	4.19	4.17	75.96	73.48	16.76	16.68
POP – 54	IC 308678	25	25	18.15	17.92	4.87	5.63	72.6	71.68	27.48	30.52
POP – 58	IC 308682	20	20	14.68	14.99	3	2.8	73.4	74.95	15	14
POP – 59	IC 308683	25	25	17.66	18.27	5.09	4.09	70.64	73.08	20.36	16.36
POP – 60	IC 308684	24	24	18.88	18.87	2.76	3.01	78.67	78.62	11.5	12.54
POP – 62	IC 308685	25	25	18.6	18.49	4.65	4.61	70.28	73.96	17.57	17.77
	Mean	24.15	24.15	17.63	17.41	3.96	4.08	72.68	72.03	16.92	17.43
	SD	1.79	1.79	1.41	1.64	0.70	1.03	2.22	3.79	3.74	5.11
	CV (per cent)	7.41	7.41	8.02	9.44	17.55	25.29	3.05	5.27	22.10	29.30

with 16.36 per cent, EP 62 (IC 343561) with 15.31 per cent, EP 104 (IC 345193) with 16.04 per cent, EP 105 (IC 345194) with 15.52 per cent, EP 115 (IC 345204) with 15.89 per cent and EP 118 (IC 345207) with 16.62 per cent. The sorghum landrace *danda local* (EP 26), *kanaggu jola* (EP 104), *allin jola* (EP 105) and *allur jola* (EP 115) were collected from Karnataka (Table 23). These high protein sorghum germplasm lines may be utilized as the source material in the sorghum nutritional quality improvement programmes.

Table 23: Proximate Analysis of Sorghum Hermplasm

Sl.No.	*Acc. No.*	*Moisture (Per cent)*	*Fat (Per cent)*	*Starch (Per cent)*	*Protein (Per cent)*
1.	IS 36210	10.49	2.80	66.60	9.95
2.	IS 36211	11.35	2.20	67.50	12.16
3.	IS 36212	10.09	3.05	65.90	12.45
4.	IS 36213	9.74	3.17	65.20	12.04
5.	IS 36221	10.99	3.05	64.50	15.52
6.	IS 36223	9.77	3.03	66.30	17.60
7.	EP 26	10.52	3.02	67.50	16.36
8.	EP 29	10.17	2.80	64.70	13.01
9.	EP 41	10.52	3.08	67.50	14.87
10.	EP 47	9.99	2.42	61.20	13.91
11.	EP 56	10.20	2.86	65.10	14.30
12.	EP 58	9.96	3.02	66.90	13.52
13.	EP 62	11.41	3.00	64.20	15.31
14.	EP 104	10.21	2.39	61.20	16.04
15.	EP 105	10.09	2.89	62.10	15.52
16.	EP 111	10.16	2.69	64.50	15.37
17.	EP 115	10.32	3.04	63.90	15.89
18.	EP 118	9.41	2.46	67.30	16.62
19.	CSV 14R	10.45	2.87	63.00	14.91
20.	CSV 20	12.66	3.04	63.80	12.56
21.	CSV 23	9.94	2.70	63.90	13.94
	Minimum	9.41	2.20	61.20	9.95
	Maximum	12.66	3.17	67.50	17.60
	Range	3.25	0.97	6.30	7.65
	Mean	10.38	2.84	64.88	14.34
	SE	0.15	0.06	0.41	0.39
	SD	0.71	0.26	1.94	1.82
	Confidence Level (95.0 per cent)	*0.31*	*0.12*	*0.86*	*0.81*
	CV (per cent)	*6.81*	*9.28*	*2.99*	*12.71*

Germplasm Management in Future

The genetic resources management in future will be of bargaining in nature. The country with wealth of genetic resources tries to use these materials as bargaining chips for exchange of other resources. Though we used sorghum germplasm extensively in the crop improvement programme, we find narrow genetic base in the released sorghum cultivars. The reason for such problem is that the focus was only on the grain yield traits. Now the trend of using sorghum has varied as food, fodder, feed, fuel and more. We need to exploit these un-exploited sorghum diversity to the present requirement and uses especially for health and nutritional quality traits.

References

Abdi, A., Bekele, E., Asfaw, Z. and A. Teshome, 2002. Patterns of morphological variation of sorghum (*Sorghum bicolor* (L.) Moench) landraces in qualitative characters in North Shewa and South Welo, Ethiopia. *Hereditas* 137: 161-172.

Annual Report National Bureau of Plant Genetic Resources (ICAR), Pusa Campus, New Delhi 110012 (1975-2006).

Appa Rao S, Prasada Rao, Mengesha MH and Gopal Reddy V (1999). Morphological diversity in sorghum germplasm from India. *Genet. Resour. Crop Evol.*, 43: 559-567.

Appa Rao, S., K.E. Prasada Rao, M.H. Mengesha, and V. Gopal Reddy, 1999. Morphological diversity in sorghum germplasm from India. *Genetic Resources and Crop Evolution* 43 (6):559-567.

Elangovan, M. Vilas A Tonapi and N. Seetharama, 2004. Collection, Characterization and Conservation of Sorghum Genetic Resources. ISBN 81-09335-04-9, pp 43.

Elangovan, M., Belum V. S. Reddy, S. Audilakshmi, S. Indira, B. U. Singh, V. Gopal Reddy, N. Kameshwararao, I P Singh and N. Seetharama, 2006. Sorghum *In*: Plant Genetic Resources: Foodgrain Crops, Ed. BS Dhillon, S. Saxena, A. Agrawal, and RK Tyagi, Narosa Publications, New Delhi. pp. 159 – 184.

Elangovan, M., Tonapi, VA., Prabhakar, Varanavasiappan, S. and Seetharama, N. 2005. Protection and Conservation of Indian Sorghum Landraces. National Research Centre for Sorghum, Rajendranagar, Hyderabad 500 030, Andhra Pradesh, India. 88pp. (ISBN 81-89335-06-05).

Grenier, C., Bramel, P. J., Dahlberg, J. A., El-Ahmadi, A., Mahmoud, M., Peterson, G. C., Rosenow, D. T., and G. Ejeta, 2004. Sorghums of the Sudan: analysis of regional diversity and distribution. *Genetic Resources and Crop Evolution*, 51 (5): 489-500.

Harlan, J. R., and de Wet, J. M. J. 1972. A simplified classification of cultivated sorghum. *Crop. Sci.* 12: 172-176.

Harlan, J. R., and Jean Pasquereau. 1969. Decrue agriculture in Mali. *Econ. Bot.*, 23: 70-74.

House LR (1995). Sorghum. One of the world's greate cereals. African Crop Science of Journal. 3:135-142.

Kak A. Srinivasan K and Sharma SK Information Bulletin No 2. Plant Germplasm Registration (2005-2008) National Bureau of Plant Genetic Resources, ICAR, New Delhi 110012, India (2009).

Mahajan RK, RL Sapra, U Srivastava, M singh and GD Sharma (2000). Minimal Descriptors (for Characterization and Evaluation) of Agri-Horticultural Crops (Part I). National Bureau of Plant Genetic Resources, New Delhi, 230p.

Plant Gertnpla.vm Reporter. Compiled by Germplasrn Exchange Division, National Bureau of Plant Genetic Resources (ICAR), Pusa Campus, New Delhi (1975-2006).

Prasada Rao, K. E. (1979). Sorghum and Millets germplasm collection in Madhya Pradesh. Progress report Genetic Resources – 8, Genetic Resources unit, International Crops Research Institute for Semi-Arid Tropics, Patancheru, Hyderabad, India.

Prasada Rao, K. E. and D. S. Murthy (1979). A Basmati (scented) sorghum from Madhya Pradesh. *Curr. Sci.* 48: 824-825.

Radhamani, J., Anjula Pandey, Kalyani Srinivasan and Vandana Tyagi, 2011. Conserving millet genetic resources in India. *Proc. Indian Natn. Sci. Acad* 77(4): 295-304.

Rao NGP. 1972. *Sorghum breeding in India: Recent developments*. Pages 101–142 in Sorghum in seventies (Rao NGP and House LR, eds.), Oxford and IBH Publishing Co.New Delhi, India.

Rockefeller Foundation, 1970. World Collection of Sorghum. List of pedigrees and origins. Rockefeller Foundation, Indian Agricultural Program, New Delhi, India.

Singh AK Information Bulletin No I. Plant Germplasm Registration (1996-2004) National Bureau of Plant Genetic Resources, ICAR, New Delhi 110012, India (2006).

Singh, A.K., 2002. Role of Indian agricultural heritage in conservation and enhancement of plant genetic resources. In: Y.L. Nene and S.L. Choudhary (eds). *Agricultural Heritage of India*. Asian Agri-History Foundation, Secunderanad, Andhra Pradesh, India. pp. 22-40.

Singh, H.B., 1973. Survey of Crop Genetic Resources in their center of diversity (Edited by O.H. Frankel) FAO-IBP First Report.

Snowden, J. D. 1936. The cultivated races of sorghum. Adlard and Son, London. 274 p.

Teshome, A., Baum, B. R., Fahrig,L., Torrance, J. K., Arnason, T. J., and J. D. Lambert, 1997. Sorghum [*Sorghum bicolor* (L.) Moench] landrace variation and classification in North Shewa and South Welo, Ethiopia. *Euphytica*, 97 (3): 255-263.

Tunstall, V., Teshome, A., and J.K. Torrance, 2001. Distribution, abundance and risk of loss of sorghum landraces in four communities in North Shewa and South Welo, Ethiopia. *Genetic Resources and Crop Evolution*, 48 (2): 131-142.

Upadhyaya, H.D., C.L.L. Gowda, H.K. Buhariwalla and J.H. Crouch. 2006a. Efficient use of crop germplasm resources: identifying useful germplasm for crop improvement through core and mini core collections and molecular marker approaches. *Plant Genet. Resour*. 4:25-35.

Millets: Ensuring Climate Resilience and Nutritional Security (2015) *Pages* **109–196**
Editors: **Vilas A. Tonapi and J.V. Patil**
Published by: **DAYA PUBLISHING HOUSE, NEW DELHI**

Chapter 3

Genetic Improvement in Sorghum

Prabhakar, C. Aruna, B. Venkatesh Bhat and A.V. Umakanth

ICAR-Indian Institute of Millets Research (IIMR)
Rajendranagar, Hyderabad – 500 030

Kharif Sorghum Improvement

Sorghum with its ability to survive in water limiting conditions provides an option for marginal farmers and fits very well in a sustainable agricultural model. It is a major dry land cereal grown as a dual purpose crop for food, feed and fodder over million hectares, primarily in Maharashtra, Karnataka, M.P., A.P., Rajasthan, Tamil Nadu, U.P. and Gujarat. Sorghum research in the country can be illustrated as a glaring example which has significantly contributed towards the green revolution in drier areas. Sorghum has originated in East Africa with its secondary centre of origin in India. The amount of genetic variability available in sorghum is very high because of its wide range of adaptation in tropical and temperate climates and free gene exchange among various races. In India, sorghum is grown in rainy and postrainy seasons, and the requirement of varieties is different for two seasons. Of late, we have reached the yield plateau in *Kharif* sorghum and there is an urgent need for diversification of the genetic base of our elite material to break the yield plateau.

Early Efforts for Improvement

Efforts have been made to improve the sorghum cultivars in India since 1930s. Most of the varieties till 1960s were the result of pure line selection practiced in local land races. The locals were tall, photosensitive, late maturing, flowering after the rainfall seized, and characterized by localized adoption and low harvest index. However, hybridization among the local cultivars reflected about 5 per cent yield improvement at experimental level which could not make any impact on production till the 1960s. Systematic effort for the improvement of rainy season sorghum has started during 1960s by manipulating the genes controlling height and maturity by introducing American germplasm. Remarkable progress has been achieved during the next four decades by diversifying the parental lines for yield, maturity, height, disease and insect tolerance, and quality by utilizing indigenous and exotic germplasm. The hybridization and selection up to a limited extent among improved land races could not bring genetic improvement to a perceptible level but basic traits for local adaptation were preserved. Notable among the varieties developed during this period and cultivated till recently are the Co series in Tamil Nadu, The Nandyal (NT), Guntur (G) and Anakapalle series of A.P., the PJ *Kharif* and *Rabi* selections, Saonar, Ramkel, Aispuri, the Maldandi and Dagadi selections of Maharashtra, Bilichigan, Fulgar white, Fulgar yellow, Kanvi, Nandhyal, Hagari, Yanigar varieties of Karnataka, Budh Perio (BP 53), Sundhia and Chasatio of Gujarath, selections of Gwalior and Indore from M.P., RS selections of Rajasthan (Rao, 1972). Since most of these locals are tall, photosensitive and late maturing with low yields, production of relatively short duration photoperiod insensitive sorghums had become the primary objective of almost all *Kharif* sorghum improvement programmes.

In rainy season genetic improvement of varieties was achieved during 60's by production of relatively short duration photoperiod-insensitive sorghums with short height by manipulating the gene for height and maturity by introducing American germplasm and adopting temperate x temperate and temperate x tropical crosses (Rao and Rana 1982). The major genotypic changes brought about during the 1960s triggered cultivar-input-management interaction and resulted in quantum jumps in productivity imparting stability to production. Over last 50 years several Indian sorghum breeders have used numerous germplasm lines belonging to different races. Sorghum yields have been increased by above 50 per cent within the last 30 years and much of this gain can be attributed to genetic diversity found within the species. Useful traits such as increased seed number, larger panicles, greater total plant weight, drought tolerance, disease resistance, greater leaf area indices, increased green leaf retention, and greater partitioning of dry matter have contributed to increased yields. This resulted in quantum jump in the productivity from 560 kg/ha in 1970 to 1000kg/ha in 2000.

The *Zera-zera* (an intermediate race between caudatum and guinea) landraces from Ethiopia and Sudan have proved to be useful sources of many traits such as excellent grain quality, high grain yield potential, tan plant, straw glume colour, resistance to leaf diseases, tolerance to grain weathering and desirable plant type. They were extensively used in sorghum breeding programs. Other groups of germplasm have also been used as sources of specific traits. Indian landraces (*Durra*) have provided sources of resistance to drought shoot fly and stem borer. The *kafir* race of sorghum in combination with *Durra* sorghums from eastern Africa provided the basis of the nuclear cytoplasmic male-sterility system that is the basis of all commercial hybrid sorghum breeding (Audilakshmi *et al.*, 2003). The *Guinea* race from west Africa has provided resistance to grain mould and *Bicolor* sorghums, which have contributed to the breeding of forage sorghums. In some cases, germplasm accessions were directly released as improved cultivars.

Hybrid Sorghum Development

In 1962, ICAR launched the "Accelerated Hybrid Sorghum Project" with an objective to initiate the hybrid breeding in sorghum. Through the efforts of Rockefellar Foundation, a wide range of germplasm was made available in India which involved male steriles, several converted lines and tropical varieties collected from Indian sub continent and several African countries. During the next five decades, remarkable progress has been achieved by diversifying the parental lines for yield, maturity, height, disease and insect tolerance and quality by utilizing indigenous and exotic germplasm.

Improvement during 1960s

The early (during 1962-69) efforts made to identify heterotic combinations among male sterile and converted dwarf lines, resulted in the development of the hybrids, CSH 1, CSH 2 and CSH 3 out of temperate x temperate and temperate x tropical crosses (Table 1). The first hybrid, CSH 1 was developed by crossing the MS line CK 60 with the germplasm line, IS 84. It was an early duration dwarf hybrid which was released in 1964 for cultivation all over India and resulted in quantum jump in the productivity. This hybrid became most popular with the farmers as it had high yield potential, suited to light soils and low rainfall areas. The second hybrid CSH 2 (released in 1965), was also based on same male sterile, MS CK 60 and a new R line IS 3691, which was also a yellow endosperm selection of Hegari from USA. Later on, a new male-sterile line 2219 B was identified from germplasm line kafir shallu and a hybrid CSH 3, was developed in 1970 by using 2219 A and IS 3691 (R line of CSH 2) (Table 2). CSH 2 and CSH 3 could not become popular due to seed production problem in both these hybrids due to their common male parent IS 3691 being shorter than respective male sterile lines.

Improvement during 1970s

The use of dwarfing, earliness and photo insensitivity traits was helpful to realize higher grain yield in these hybrids which ultimately proved to be a disadvantage for fodder production. Further genetic improvement was thus oriented to rectify the problems through genetic enhancement and use of improved derivatives. As new male steriles and dwarf derivatives became available, further hybrid breeding was based on these genetically enhanced indigenously bred parental lines. During 1970-1979, three hybrids, CSH 4, CSH 5 and CSH 6 were released. Hybrid CSH 4, based on MS 1036 A had better fodder yield. The MS line of this hybrid, MS 1036 A was developed from a cross of CK 60B and PJ 8K (a local variety from Maharashtra) and the R line was a selection from exotic IS 3924. Though this hybrid was good for grain yield, the hybrid was not acceptable from the point of grain quality as that of CSH1, due to shorter duration grain deterioration remained to be the major impediment for the spread of this hybrid.

Though grain yield was tripled by utilization of exotic breeding material in hybrid programme, this also resulted in consumers' non-preference, price differences between locals and hybrids and increased susceptibility to major diseases and pests which were the important bottlenecks for popularizing hybrids. The grain qualities of the hybrids were inferior to that of local varieties as selection pressure among temperate x tropical crosses was more on heterotic lines rather than on better quality of the grain. Keeping these challenges in mind, new male-sterile lines were developed in different genetic backgrounds. The MS lines and R lines were developed from derivatives of crosses between temperate x tropical lines. 2077A (the MS line of CSH 5) was developed from IS 2046, a germplasm line from Senegal. Similarly converting IS 3541, from Ethiopia into a photo-insensitive dwarf line, resulted in the R line, CS 3541. The hybrid CSH 5 (released in 1975), based on these converted lines MS 2077A and CS 3541, contributed not only to substantial yield improvement but also in increased tolerance for grain moulds and leafy diseases as compared to CSH 1. The hybrid was tolerant to mites and aphids also. This improvement was achieved by introduction of genes from tropical material, IS 3541. Another achievement was towards development of early maturing hybrid CSH 6 in 1976, developed from early MS line 2219A and CS 3541 taking only 95 days for maturing. This hybrid became very popular in inter-cultivation with pigeon pea due to its geometry and earliness. Both these hybrids showed less grain deterioration and good quality of seed along with remarkable improvement in yield.

Improvement during 1980s

Further increase in yield levels were achieved by the development of superior male sterile line 296 B. Indian germplasm line, Karad local was crossed with American material IS 3922 to develop the MS line 296A, which had the best combining ability. The MS line has a very compact panicle with more number of primary branches. During 1980-89, the hybrids CSH 9, CSH 10 and CSH 11 were released. CSH 9, a medium duration hybrid yields about 39 q/ha. This hybrid is widely adopted and is extensively grown. Later two hybrids, CSH 10 and CSH 11 based on the same male sterile line, 296 B were released. CSH 10, a tall hybrid though produces 30-40 per cent more fodder than CSH 9 could not be promoted due to seed production limitation. CSH 11, another hybrid of CSH 9 range, has marginal advantage in grain yield productivity. Due to smaller seed size, this hybrid is not popular among farmers. Among medium maturity hybrids of 110-115 days, CSH 5 and CSH 9 are most popular, with an average of 35 and 38 q/ha grain yield respectively and a matching fodder yield of 95-100q/ha.

Improvement during 1990s

In the subsequent decade (1990-99) most of the hybrids tested in AICSIP trials were based on 296 B with different restorers but could not make any remarkable dent for grain yield except for the hybrid CSH 13. Though the grain yield levels of this hybrid are marginally improved, the fodder yields are 40 per cent higher than that of CSH 9. It is an ideal dual purpose hybrid and is recommended for both grain sorghum and fodder sorghum growing areas of the country. The R line of this hybrid, RS 29 that contributes to heterosis for fodder yield is developed from SC 108, an American elite line and SPV 126 (a tall mutant of CS 3541). Another hybrid CSH 14 based on male sterile line AKMS 14A is an early duration hybrid with a maturity comparable to CSH 1 and has yield on par with CSH 9. Its female parent is a derivative of texas line, BTx 623 and male is derived from CS 3541. It provides a better choice for the areas of CSH 1 adaptation, light soil and deficient rainfall areas. The hybrid is popular with farmers of Karnataka and Maharashtra.

The introduction of the rainy season sorghum hybrids like CSH 1, CSH 5 and CSH 9 brought a remarkable increase in the sorghum production in India. After the release of the hybrid CSH 9 in 1981, the yield level of experimental hybrids has not advanced over this hybrid to any significant level and thus yield plateau is reached. However, useful diversification for early maturity and higher fodder yield has been achieved with the release of CSH 14 and CSH 13 respectively. Seed growers are facing problem in seed production of the hybrids particularly based on 296 A due to low temperature at flowering time in major seed production areas in Andhra Pradesh during October-November sowings. The need for diversification of female parent of hybrids was felt in view of above seed production problems and stagnating yield levels. As a result of intensive breeding programme to develop improved male sterile line and restorer a productive medium maturing hybrid CSH 16 was developed. It is based on a new male sterile line 27 A and a new restorer C 43. This hybrid showed further improvement in grain mould tolerance as the new genes from Ethiopian germplasm line IS 23549 were introduced in its R line. The R line has also a very compact panicle. In contrast to 296B, the MS line 27A of this hybrid has long and loose panicle with bold and round seed resembling postrainy season genotypes that is preferred by consumers. 27B has been developed from multiple crosses using germplasm lines like IS 3687, IS 3922 and 2219B. CSH 16 yields about 42 q/ha of grain yield and 97 q/ha of fodder yield. The hybrid has bold seed and is preferred by the farmers. It has better level of tolerance to grain moulds and downy mildew and its fodder quality is good. This was found to be good for ethanol production.

Another early maturing hybrid CSH 17 based on the male sterile line AKMS 14 A has been released for the states of Tamil Nadu, Gujarat, Madhya Pradesh and Rajasthan. It yields about 40q/ha of grain and 90q/ha of fodder. Another medium maturing hybrid CSH 18 developed at Indore centre yields about 41 q/ha of grain and 130 q/ha of fodder. The grain and fodder quality of this hybrid are superior. By utilizing local variety Vidisha 60-1, the MS line of CSH 18 was developed. The local variety, Vidisha 60-1 not only contributes for high stover yield but also for improved grain quality. The R line of this hybrid, Indore 12 is developed from SSV 53 and SPV 475, a variety developed from multiple crosses involving germplasm lines. By utilizing local variety Vidisha 60-1, the MS line of CSH 18 was developed.

Improvement since 2000 Onwards

CSH 23 is another early maturing hybrid released in 2005 for the states of Maharashtra, Karnataka, AP, MP, Gujarath, Rajasthan and UP. It takes 101-103 days to mature and yields about 43 q/ha of grain. One more medium maturing hybrid, CSH 25 was released for zone II in 2008. It is developed from the parents PMS 28A and C 43. It yields 43 q/ha of grain and 120 q/ha of fodder and was found to have good tolerance to shoot fly and grain moulds.

Another medium maturing hybrid, CSH 27 was released recently in 2012 for zone I involving the states of Rajasthan, N.Gujarat, UP, AP and Tamil Nadu. It is a dual purpose hybrid with 39 q/ha of grain and 136 q/ha of fodder yield. It is developed based on the parents 279A x CB 11. It has better level of tolerance to grain moulds. The latest hybrid which has been recommended for release in zone II comprising of the states of Maharashtra, Karnataka and MP, South Gujarat, North AP is CSH 30 which is an early maturing hybrid with good level of tolerance to grain moulds.

The varietal improvement programme has also been taken up simultaneously and till date CSV 1 to CSV 27 were released through AICSIP. However, popular recent *Kharif* varieties are SPV 462, CSV 15, CSV 20, CSV 23 and CSV 27.

Varietal Improvement for *Kharif*

In the first phase of varietal improvement, besides release of swarna (CSV 1), a pure line selection from IS 3924 in 1968, six more varieties derived from temperate x tropical crosses were released. CSV 1, CSV 2 and CSV 3 are early maturing (100-105 days) tend to yield 138-207 per cent higher than local. CSV 4 and CSV 5 are relatively dwarf varieties maturing in 110-115 days and combine good grain quality and resistance to grain deterioration even when caught in rains. CSV 6 is a relatively tall variety. The sources of resistance to grain mould and leaf diseases were located in zera-zera germplasm from Ethiopia, mechanism of resistance studied and resistance was consciously transferred in high yielding back ground. The resistance to these diseases was located in CS 3541 which is a dwarf derivative of IS 3541, a zera-zera from Ethiopia. Tan plant pigment conferring resistance to most of leaf diseases was discovered and it was augment in various breeding programmes of temperate x tropical crosses. Improved temperate and tropical germplasm from Ehiopia in various breeding programmes enabled to evolve many improved varieties eg. CSV 10, CSV 11, SPV 462, CSV 13, CSV15 etc. By now all the breeding programmes incorporated tan plant pigment in their *Kharif* nurseries.

The high yielding *Kharif* variety, CSV 15 was developed from SPV 462 and CSV 13. It is a dual purpose variety with grain yield as high as the hybrid CSH 5 and fodder yield equal to that of CSH 10. It yields about 36q/ha of grain yield and 121 q/ha of dry fodder yield. An early maturing variety, CSV 17 was developed by crossing CSV 15 with SPV 772. It is suitable to light soils and drought prone areas.

Another variety, CSV 20 was released with grain and fodder yields of 31q/ha and 133 q/ha respectively. The latest dual purpose varieties, CSV 23 and CSV 27 have high yield potentials of 22 and 28q/ha of grain and 155 and 193q/ha of fodder yields respectively.

Besides these there are a number of varieties and hybrids released for different states (Tonapi *et al.*, 2009). Private sector is also playing an important role in the development of *Kharif* sorghum hybrids. The centrally released *Kharif* sorghum hybrids, CSH 21, CSH 26, CSH 28 and CSH 29 were developed by different private seed companies.

In *Kharif* yield stagnation and the grain quality are the major problems that needs research attention. To break the yield plateau it is important to utilize the unutilised germplasm from the world collection and bring about the useful genetic diversity in the material (Aruna and Audilakshmi 2008; Reddy *et al.*, 2008). The *Kharif* grain is not as remunerative as that of *Rabi* produce because the seed is not round and lustrous. Now efforts are being made to make *Kharif* seed bold and lustrous by using guinea-caudatum and durra-caudatum lines. Now-a-days the utilization of

Table 1: Characteristics of Popular Sorghum Varieties and Hybrids

Varieties/Hybrids	*Grain Yield (q/ha)*	*Dry-Fodder Yield (q/ha)*	*Plant Height (cm)*	*Maturity (Duration) (Days)*	*Salient Features*
					Varieties
SPV 462 (CO 26/DSV 2)	33	97	208	110-115	Tall, dual purpose, bold round attractive seed, sweet stalk, tolerant to insect pests, grain deterioration and several leaf diseases.
CSV 13 (SPV 475)	35	97	181	110-114	Medium tall, medium bold seed semi-compact head, thin stem, tolerant to insect pests, grain deterioration and several diseases.
CSV 15 (SPV 946)	36	121	232	107-112	Tall, Dual purpose, medium bold round seed, glume small and straw colour, large ear head, oblong shape, semi compact. Resistant to all leaf spot diseases.
CSV 17	25	68	150	97	Early maturing, tan plant type with dark green leaves, white dull midrib colour, well exerted, cylindrical, semi compact panicle, creamy seed colour, free threshing, moderately resistant to shoot fly and stem borer.
CSV 20 (SPV 1616)	31	133	240	109	Tall, semi-compact panicle with dense clustering of grain in panicle branches, panicle shape oblong and pearly white bold seed, for All India cultivation.
CSV 23 (SPV 1714)	22	155	215	115	Dual-purpose variety, Normal rainfall sorghum growing regions of the nation The variety was found to be tolerant for shoot fly and stem borer
CSV 27 (SPV 1870)	28	193	235	115	Tall, Dual purpose variety, yellow green colour midrib, Well exerted semi-compact panicle, Greyed yellow seed, resistance to grain moulds, Non-lodging, non-shattering,
					Hybrids
CSH 5	34	93	174	110-115	Medium tall, thick juicy stem, medium bold seed, large long head, resistant to grain deterioration and leaf spot diseases.
CSH 6	34	81	161	95-100	Early maturing, thin stem, medium bold and hard seed resistant to grain deterioration, tolerant to leaf spots and other diseases.
CSH 9	39	98	182	110-115	Medium tall, thick juicy stem, vigorous growth, bold round seed, large semi compact head, tolerant to insect pests and diseases.

Contd...

Table 1–*Contd...*

Varieties/Hybrids	*Grain Yield (q/ha)*	*Dry-Fodder Yield (q/ha)*	*Plant Height (cm)*	*Maturity (Duration) (Days)*	*Salient Features*
CSH 10	36	120	233	105-110	Tall, dual purpose, vigorous plant, bold seed, tolerant to insect pests and diseases.
CSH 11	41	92	194	105-110	Medium tall, loose and large panicle, small seed, thick stem tolerant to insect and leaf spot diseases, some lodging under severe drought.
CSH 13	39	144	261	110-115	Tan, Tall, dual purpose, panicle broad, semi loose, spindle shape, seed round medium bold, light creamy in colour and free threshing. Tolerant to grain moulds and leaf diseases
CSH 14	38	88	181	102	Medium tall, semi loose panicle, bold seed, tolerant to grain mould and leaf spots.
CSH 16	42	95	210	110	Medium tall, long loose panicle with open apex, bold seed, highly tolerant to grain moulds and resistant to leaf diseases.
CSH 17	41	90	185	103	Tall, early maturing, panicle semi loose, white round seed, tolerant to shoot fly and stem borer.
CSH 18	41	131	212	112	Tall, midrib dull green, thick and juicy stem, white medium bold seed, medium maturity, resistant to grain moulds.
CSH 23	41	87	178	96-101	Tall (180cm), dull green midrib, medium bold, white seed. Early maturity (103 days) avoid terminal drought.For Zones – I and II:Maharashtra, Karnataka, AP, MP, Gujarat, Rajasthan, UP
CSH 25	43	128	205	110	*Kharif* hybrid- Rainfed *Kharif* sorghum areas under normal time of sowing especially for Maharashtra and Karnataka and central zone II areas.Tan, tall, white midrib, semi-compact ear- head, pearly white seed, medium almond shape, tolerant to grain-mold and shoot fly.
CSH 27	39	136	200	106	Tan, cylindrical semi compact earhead, white bold elliptical seed, medium maturity, tolerant to grainmold and resistant to lodging. Recommended for rainfed *Kharif* cultivation in Rajasthan, N.Gujarat, UP, AP and TN.
CSH 30	44	141	216	105	Tan plant color with symmetric semi compact panicle and white bold elliptical seed. It is early maturing with tolerance to grain moulds. Recommended for rainfed *Kharif* cultivation in Maharashtra, Karnataka, MP, South Gujarat, North AP

SPV 462 is released as CO 26 in Tamil Nadu and as DSV 2 in Karnataka.

Table 2: Germplasm Lines Used in the Development of *Kharif* Hybrids and Varieties

Sl.No.	Hybrid	Pedigree	Pedigree of the Parental Lines			
			MS Line	*Race*	*R Line*	*Race*
1.1	*Kharif Hybrids Released at National level*					
1.	CSH 1	CK 60 A x IS 84	CK 60 A'	Kafir (USA)	IS 84	Durra Caudatum
2.	CSH 2	CK 60 A x IS 3691	CK 60 A	Kafir (USA)	IS 3691	Durra Caudatum
3.	CSH 3	2219 A x IS 3691	2219 A	Sel from *Kharif* Shallu	IS 3691	Durra Caudatum
4.	CSH 4	1036 A x Swarna	1036 A	Kafir (USA)	Swarna Sel from IS 3924	Kafir Durra
5.	CSH 5	2077 A x CS 3541	IS 2046 x 3677 B	Kafir Not known	IS 3675 x IS 3541	Kafir Durra Zera-zera
6.	CSH 6	2219 A x CS 3541	2219	Sel from *Kharif* Shallu	IS 3675 x IS 3541	Kafir Durra Zera-zera
7.	CSH 9	296 A x CS 3541	IS 3922 x Karad Local	Kafir Durra *Kharif* local	IS 3675 x IS 3541	Kafir Durra Zera-zera
8.	CSH 10	296 A x SB 1085	IS 3922 x Karad Local	Kafir Durra *Kharif* local	SB 1085*	
9.	CSH 11	296 A x MR 750	IS 3922 x Karad Local	Kafir Durra *Kharif* local	MR 750 *	
10.	CSH 14	AKMS 14 Ax AKR 150	(MR 760 x BTx623) x AKMS 2B		CS 3541x 900* CS 3541 = IS 3675 x IS 3541 900*	Kafir Durra Zera-zera
11.	CSH 13 (K and R)	296 A x RS 29	IS 3922 x Karad Local	Kafir Durra *Kharif* local	SC 108 x SPV126 SC108SPV 126	Purdue Univ USA Tall Mutant of CS3541
12.	CSH 16	27 A x C 43	83 B x19983 B = CS3687xCS3922 CS 3687 is sel from IS 3687CS3922 Sel from IS 3922199B= 2219BxCS3922 2219BCS3922=IS3922	Durra Caudatum Durra Caudatum Sel.from *Kharif* Shallu Durra Caudatum	C43= CS 3541 xIS23549 CS 3541=IS 3675 xIS 3541	Guinea -Caudatum Kafir Durra Zera-zera

Contd...

Table 2–*Contd...*

Sl.No.	Hybrid	Pedigree	Pedigree of the Parental Lines			
			MS Line	*Race*	*R Line*	*Race*
13.	CSH 17	AKMS 14 A x RS 673	(MR 760 x BT623) x AKMS 2B		SPV 544 x K24-1 SPV544= (CS 3541x CO18) K 24-1= (CO27x1022)CS3541= IS 3675x IS 3541	Kafir Durra Zera-zera
14.	CSH 18	IMS 9A x Indore 12	IMS9A=2077A x (MA9BxVidisha60) 2077A= IS 2046 x 3677 B Vidisha 60	Kafir Not known Local *Kharif* Selection	Indore12= (SSV53x SPV475) SPV475= [(IS12622 x 555) x (IS3612 x 2219B) x E35-1]IS12622IS36122219B E35-1	Durra Bicolor Caudatum(Nigeria) Sel from *Kharif* Shallu Ethiopian early line
15.	CSH 23	MS 7A x RS 627	MS 7A	Selection from AKMS 14B		RS 71x M 35-1
16.	CSH 25	PMS 28A x C 43	PMS 28A	296B x 94040B	C 43	
17.	CSH 27	279A x CB 11	422B x IS 30469422B= SPV 422 x 2219B SPV422= GG 1483; 2219B= Kafir shallu	Grain mould tolerant germplasm line from Ethiopis belonging to GC race	SPV 1023 x C 43; SPV 1023= SPV 544 x K 24-1; SPV 544= CS 3541 x CO 18; K24-1= CO27 x 1022	
18.	CSH 30	415A x CB 33	AKMS 14B x IS 25017	Grain mould tolerant germplasm line from Sudan belonging to Caudatum race	GMRP 13 x CSV 15	Grain mould tolerant line from Parbhani

Contd...

Table 2–*Contd...*

			***Kharif* varieties**			
Sl.No.	*Variety*	*Pedigree*	*Pedigree of the Parental Lines*			
			Line 1	*Race*	*Line 2*	*Race*
1.	CSV 1	Sel from IS3924	IS 3924	Kafir Durra		
2.	CSV 2	IS 3922 xKarad local	IS 3922	Kafir Durra	Karad local	Sel from *Kharif* local
3.	CSV 3	IS 2954x BP 53	IS2954	Durra Caudatum	BP 53-IS18432	Durra
4.	CSV 4	IS3675x IS 3541	IS3675	Kafir Durra	IS 3541	Zerazera
5.	CSV 5	IS3687xAispuri	IS 3687	Durra Caudatum	Aispuri	Durra
6.	CSV 6	IS 3922xAispuri	IS3922	Kafir Durra	Aispuri	Durra
7.	CSV 9	CS3541(tall mutant)	IS 3675	Kafir Durra	IS 3541	Zerazera
8.	CSV 10	SB1066xCS3541	SB 1066	Sel from SC108 Purdue, USA	IS 3675 IS3541	Kafir Durra Zerazera
9.	CSV 11	(SC108-3 x CS3541)-11-1	SC 108 -3	Sel from SC108 Purdue, USA	IS 3675 IS3541	Kafir Durra Zerazera
10.	Co 26	(IS 2947xSPV232) x 1022	IS2947SPV232= 148x512148=IS3687x Aispuri IS1151512	Caudatum Durra Caudatum Durra India Derivative of BP53 *Kharif* local	1022	Not known
11.	CSV 13	(IS12622x555) x (IS3612xE35-1-52)	IS12622555=IS3687 x Aipuri	Durra Bicolor Ethiopia Durra Caudatum Durra India	IS 3612E35-1-52	Caudatum(Nigeria) Ethiopian early line

Contd...

Table 2–*Contd...*

			***Kharif* varieties**			
Sl.No.	*Variety*	*Pedigree*	*Pedigree of the Parental Lines*			
			Line 1	*Race*	*Line 2*	*Race*
12.	CSV 15	SPV 475xSPV462	SPV 475= (IS12622x555) x (IS3612xE35-1-52) IS12622555=IS3687 x Aispuri IS 3612 E35-1-52	Durra Bicolar Etiopia Durra Caudatum Durra India Caudatum(Nigeria) Ethiopian early line	SPV462= (IS 2947x SPV232) x1022 IS2947 SPV232= 148 x 512 148=IS3687x Aispuri IS1151 512	Caudatum Durra Caudatum Durra Derivative of BP53 *Kharif* local
13.	CSV 17	SPV 946 x SPV 772	SPV 946= SPV 475 x SPV 462 SPV 475= (IS12622x555) x (IS3612xE35-1-52) IS12622555=IS3687 x Aispuri IS 3612 E35-1-52 SPV462= (IS 2947x SPV232) x1022 IS2947 SPV232= 148 x 512 148=IS3687x Aispuri IS1151 512	Durra Bicolar DC DC (Nigeria) Ethiopian early line Caudatum Durra Caudatum Durra Derivative of BP53 *Kharif* local	SPV 772= SPV 233 x SPV 350 SPV 233= IC 419-1(2077 A x CS 3541) x Merta local SPV 350= (SC 108-3 x CS 3541)-3-1 CS 3541= IS3675x IS 3541	Zera-zera

Contd...

Table 2–*Contd...*

Sl.No.	Variety	Pedigree	*Kharif* varieties: Pedigree of the Parental Lines			
			Line 1	*Race*	*Line 2*	*Race*
14.	CSV 20	SPV 946 x Kh 89- 246	SPV 946= CSV 15		Kh 89-246= SPV 544 x SPV 526 SPV 544= CS3541 x Co18, CS3541= IS3675 x IS3541, SPV 526= CS3541 x MR-SGIRL	IS3675= *Kharif* Durra, IS3541= NYITHIN; MR-SGIRL= Midge resistant line
15.	CSV 23	SPV 861 x SU 248 SPV 861 x SU 248	SPV 861 Selection from – 35 (M) 12218	Sel from 35 (M) 12218	SU 248= SB 1066 x CS 3541	SB 1066= Purdev selection base No. 954
16.	CSV 27	(GJ 38 x Indore 12)-2-1-2-1	GJ 38 = GJ 35 x E 35-1	E35-1-Ehtiopian early line	Indore12= (SSV53xSPV475) SPV475= [(IS12622 x 555) x (IS3612 x 2219B) x E35-1] IS12622 IS361 22219B E35-1	Durra Bicolor Caudatum(Nigeria) Sel from *Kharif* Shallu Ethiopian early line

Kharif sorghum grain as a raw material in various industries is increasing, given the limited prospects of rainy season (*Kharif*) sorghum for human consumption. The main industries currently using sorghum in India are the poultry feed, animal feed and alcohol distilleries. The immediate need is to develop genotypes to suit different end products.

Breeding for grain mould resistance is a high priority area so that the market profitability and food value can be enhanced. Evolving grain mould resistant male steriles is the immediate requirement to strengthen the resistance in present day hybrids. Another problem during *Kharif* is shootfly. It is not infrequent to experience aberrant rainfall where onset of monsoon gets delayed. This builds up unsurmountable pressure of shoot fly causing economic losses in late plantings. Therefore, in order to stabilize the sorghum production, incorporation of grain mould and shoot fly resistance in times to come would provide an insurance against the crop losses.

Rabi Sorghum Improvement

Post-rainy (*Rabi*) sorghum is an important dry land crop grown in the Deccan Plateau on 5.6 m ha area in the states of Maharashtra (3.5 m ha), Karnataka (1.6 m ha) and Andhra Pradesh (0.48 m ha). It is the major source of food and fodder and as such it greatly influences the economic well being of the population in the region. It is mostly grown in medium (45 to 90 cm) and deep (>90 cm) black soils of dry semi-arid region under receding moisture conditions. The productivity of the crop in general is low (850 kg/ha). Moisture stress in GS-2 due to a unique situation of growing the crop on receding soil moisture in medium to shallow soils, susceptibility to shoot fly, charcoal rot and low temperature affecting crop growth as well as fertility restoration in hybrids are the major factors responsible for low yield.

Post-Rainy (*Rabi*) Sorghum in different Agroclimatic Zones Across States in India

- **Maharashtra**: Western Maharashtra plain zone (Zone V), scarce rainfall zone (Zone VI) and central Maharastra plateau zone (assured rainfall zone) (zone VIII), Eastern Vidarbha zone (Zone IX). The major *Rabi* sorghum districts include Solapur, Sangli, Satara-east, Pune east, Ahmadanagar, parts of Nasik, Dhule, Beed, Aurangabad, Parbhani and Osmanabad.
- **Karnataka**: Northern and central parts of the state consisting of entire Bijapur, Bagalkota, Bellary, Koppal districts, parts of Raichur, Gadag, Dharwad, Belgaum, Bidar and Gulbarga districts.
- **Andhra Pradesh**: Parts of Prakasham, Kurnool, Anantapur, Cuddapah, Ranga Reddy, Medak, Nizamabad, Mahabubnagar, and Adilabad encompassing agro climatic zones of North Telangana,(Zone IV) South Telangana (ZoneV) and scarce rainfall zone (Zone VI).
- **Tamil Nadu**: The major sorghum growing districts include Coimbatore, Dindigal, Thriuchirapalli, Salem, Perambalur, Karur, Dharmapuri, Erode, Vellore, Theni and Namakkal.
- **Gujarat**: Districts of Surat, Bharuch, and Baroda.

Varietal Improvement in Post-Rainy Sorghum in India

Historical Aspect

Focused breeding on *Rabi* sorghum was initiated in the early seventies. These breeding efforts led to the central release of varieties like Muguti. Heterosis breeding led to the central release of the hybrids like CSH-7R, CSH-8R and CSH-12R. These cultivars were notable in two respects. First, that none of these cultivars succeeded in gaining consistent acceptability from farmers at a scale to effect a discernible impact. Second, that the breeders came to realise the complexities in the genetic up-gradation of yield in *Rabi* sorghum. Following were the lessons learnt out of these programmes,

- ✰ The traditional and widely preferred land races are ineffective as donors of favourable gene complexes and as parents in heterosis breeding,
- ✰ The *Kharif* genetic diversity though very effective in achieving quick yield increase is bedeviled with low adaptability and grain quality, and
- ✰ Critical elements of *Rabi* sorghum adaptability are resistance to shoot fly, charcoal rot, moisture stress and low temperature.

The second phase of *Rabi* sorghum breeding with emphasis on hybrid cultivars was initiated in the late eighties. During the first eight years of this phase, 199 entries were evaluated in the AICSIP Initial Varietal Trials, out of which 43 were proposed to the Advanced Varietal Trials. This resulted in the identification of SPV 839 and its central release as CSV-14R. During this period 250 experimental hybrids were evaluated in the AICSIP Initial Hybrid Trials. These trials resulted in the identification of two hybrids, SPH 504 and SPH 677 for central release as CSH-13R and CSH-15R.Variety CSV-14R has no significant yield advantage over M35-1 and is comparable to the latter in most of the adaptability traits. CSH-13R has significant yield superiority over M35-1 but is highly vulnerable to shoot fly and low temperature and has inferior grain quality. The case of CSH-15R having a *Rabi* based MS line (104A) developed at Mohol Centre is different from CSH-13R has a marginal yield advantage over M35-1. *Rabi* sorghum hybrids will have a tangible impact only when the MS and R lines having the season adaptability and desired combining ability are developed (Rao, 1986).

Post-rainy season-adapted sorghums are characterized by response to shorter day length (photoperiod sensitivity), flowering and maturity (more or the less same time), irrespective of temperature fluctuations and sowing dates (thermo-insensitivity within the post-rainy season varieties).They are tolerant to terminal moisture stress and resistant to stalk rot/charcoal rot. They usually produce high biomass (grain and stover) and have high lustrous grain with semi-corneous endosperm. Tolerance to shoot fly, lodging (mechanical) and rust is also required. All these characters are exemplified best in M 35-1, a variety selected from a local landrace nearly 75 years ago at Mohol, Maharashtra, which produces high stable yields of grain and stover across different sowing dates. As a result, M 35-1, a landrace selection developed in 1937 still dominates the post-rainy season sorghum areas in India. Developing high-yielding post-rainy season-adapted varieties/hybrids is the main objective in almost all the crop improvement programmes. Though great efforts have been made to develop hybrids with wider adaptability to varied production environments, the results are not encouraging. Productivity depends not only on the moisture availability, but also on the soil types under which it is grown and the genotypes. Also, improvement of post-rainy sorghum did not receive as much emphasis and effort as the rainy sorghum until the nineties. Among the factors influencing adoption of sorghum varieties, farmers rated grain and fodder quality attributes as their first priority.

Several strategies were suggested to improve the yields of post-rainy sorghum. A multidisciplinary approach was one of them. However, conventional crop breeding techniques emphasized selection for yield per se or improvement of crop cultivars through incorporation of resistance to biotic and abiotic stresses. Donald (1968) proposed the ideotype approach to plant breeding in contrast to the empirical breeding approach of defect elimination and selection for yield per se. Ideotype breeding involves defining a crop production environment, designing a plant model, from morphological and physiological traits known to influence performance in that environment, and combining the traits into one plant type.

Much of the post-rainy season sorghum is grown on residual and receding soil moisture on shallow and medium-deep soils. Germplasm lines suitable to both the soil conditions and also to specific soils were identified by Pawar *et al.(2005)*. Under shallow soils, the hybrids were shorter, flowered and matured early and in medium-deep soils, mean leaf area, grain number and 1000 grains mass, grain and fodder yields were higher. Therefore, there is a need for the development of varieties adapted to specific soil situation in post-rainy season to enhance production and productivity levels. Hence, there is a need to evaluate the relative performance of post-rainy sorghum varieties for their productivity and identify ideal plant types suited to shallow, medium and deep black soils.

For the past 25 years, several programs in India attempted to improve varieties and/or hybrid parents through pedigree breeding approach. Though the area under *Rabi* sorghum is fairly consistent over many years, the progress in productivity is much slower compared to *Kharif* season (Kumar *et al.*, 2010). This very low rate of productivity calls for a change in production strategy including breeding. Post-rainy season-adapted sorghums grow well in short day length (photoperiod sensitivity), flower and mature irrespective of temperature fluctuations and sowing dates (thermo-insensitivity within the post-rainy season varieties), are tolerant to terminal moisture stress and resistant to stalk rot/charcoal rot, produce high biomass (grain and stover) and have large lustrous grain with semi corneous endosperm. Tolerance to shoot fly, lodging (mechanical) and rust are also required (Rao 1982, Rana *et al.*, 1985, Seetharama *et al.*, 1990).

Germplasm lines suitable to both the soil conditions and also to specific soils were identified by Pawar *et al.* (2005). Under shallow soils, the genotypes were shorter, flowered and matured early while in medium-deep soils, mean leaf area, grain number and 1000-grain mass, grain and fodder yields were higher (Rafiq *et al.*, 2003). The varieties like Phule Maulee and Phule Chitra were released for regions of Maharashtra with medium soil depth, Phule Anuradha for shallow soils and Phule Vasudha for deep soils of Maharashtra. These have been performing well in specific soil situations.

Table 3: *Rabi* Sorghum Varieties and Hybrids Recommended: State-wise

State	*Area of Adptation*	*Hybrids*	*Varieties*
Maharashtra	Shallow soils	Nil	Sel. 3, CSV 26, Mauli, Phule Anuradha, Mauli, Phule Chitra, Suchitra, Parbhani Moti, CSV14R, DSV 4 and CSH 15R
	Medium soils	CSH-15R	
	Deep soils	CSH-15R, CSH-19R, SPH-1666	CSV 18, CSV22, CSV 216R(Phule Yashoda), PKV Kranti, CSV 14R, DSV 5, SPV 2033, DSV 4
	Irrigated conditions	CSH-15R, CSH-19R, SPH-1666	Phule Vasudha, Phule Revati, CSV 216R (Phule Yashoda), DSV 4, CSV-29R
Karnataka	Dry zones	CSH-12R,CSH 15R	M 35-1, DSV 5
	Transitional zones	CSH-13R,CSH 15R	CSV 8R, CSV 14R
	Irrigated zone	CSH-12R,CSH-13R,CSH-15R	DSV 4, CSV-29R
			DSV5, CSV-29R
Andhra Pradesh	Early *Rabi* of Rayalaseema	CSH 13R, CSH 15R	CSV 8R, CSV 14R
	Normal *Rabi* areas	CSH-13R,CSH 15R	M 35-1, CSV 8R, CSV 14R
Tamil Nadu	Entire *Rabi* area	CSH 10, K.tall	
	Summer irrigated areas	CSH 13R	CO 24, CO 25
		CSH 13R, CSH 15R	CSV 14R
		CSH 3, CSH 5	CO 26, CO 24, CO 25
			CSV 8R, CSV 14R
Gujarat	Entire *Rabi* zone	CSH 13R, CSH 15R	CSV 14R,CSV 8R

Table 4: Salient Features of Popular *Rabi* Sorghum Varieties

Sl.No.	*Cultivar*	*Grain Yield kg/acre*	*Dry Fodder Yield tones/acre*	*Plant Height (cm)*	*Duration (Days)*	*Salient Features*
1.	Phule Revati	1250 kg/acre (3000-3500 kg/ha)	3.0-3.2 t/acre (7.5-8.2 t/ha)	220-240 cm	115-125 Days	Suitable for deep soils and under irrigated conditions. Pearly white round and very bold grain, tolerant to shoot fly and charcoal rot
2.	Phule Vasudha	1200-1400 kg/acre (3000-3500 kg/ha)	2.8-3.0 t/acre (7-7.5 t/ha)	180-210 cm	116-120 Days	Suitable for irrigated areas. Pearly white lustrous and Medium bold grain, tolerant to shoot fly and charcoal rot
3.	Phule Chitra	800-1000 kg/acre (2000-2500 kg/ha)	2.2-2.6 t/acre (5.5-6.5 t/ha)	200-220 cm	112-115 Days	Suitable for medium soils. Plumpy, lustrous and medium bold grain, good grain and fodder quality, tolerant to shootfly, charcoal rot, drought and non lodging
4.	CSV 216R (Phule Yashoda)	800-1000 kg/acre (2000-2500 kg/ha)	2.8-3.2 t/acre (7-8 t/ha)	220-250 cm	120-125 Days	Suitable for medium to deep soils in rainfed and irrigated areas. Pearly white round, lustrous and medium bold grain. Tolerant to shoot fly and drought.
5.	M35-1	832 kg/acre (2080 kg/ha)	2.44 t/acre (6.11 t/ha)	170-210 cm	115-118 Days	Suitable for medium to deep soils in rainfed areas. Oval shaped semi compact earheads. Pearly white round, lustrous and medium bold grain, tolerant to shootfly, charcoal rot and drought
6.	CSV 18	1400-1520 kg/acre (3500-3800 kg/ha)	3.4-3.6 t/acre (8.7-9.0 t/ha)	220-240 cm	120-126 Days	Suitable for irrigated areas. Yellow white and bold grain, tolerant to aphid and non lodging
7.	CSV 22	920 kg/acre (2300 kg/ha)	2.84 t/acre (7.1 t/ha)	180-200 cm	116-120 Days	Suitable for medium to deep soils in rainfed and irrigated areas. Pearly white round, lustrous and medium bold grain, tolerant to charcoal rot and shootfly,
8.	CSV 26	650-720 kg/acre (1300-1600 kg/ha)	1.8-2.4 t/acre (4.5-6.0 t/ha)	180-200 cm	112-115 Days	Suitable for shallow soils, tolerant to shoot fly, stem borer, charcoal rot and tolerant to drought. Medium bold grain with good fodder quality
9.	CSV 14R	908 kg/acre (2270 kg/ha)	2.22 t/acre (5.54 t/ha)	180-200 cm	115-120 Days	Suitable for medium to deep soils in rainfed areas. Pearly white round, bold grain, resistant to shootfly charcoal rot and drought equal to M35-1

Contd...

Table 4–*Contd...*

Sl.No.	*Cultivar*	*Grain Yield kg/acre*	*Dry Fodder Yield tones/acre*	*Plant Height (cm)*	*Duration (Days)*	*Salient Features*
10.	CSV 8R	1000-1200 kg/acre (2500-3000 kg/ha)	2.2-2.6 t/acre (5.0-6.0 t/ha)	200-230 cm	115-120 Days	Suitable for medium to deep soils. Creamy coloured bold gran tolerant to charcoal rot, shoot fly and drought
11.	Parbhani Moti	1280 kg/acre (3200 kg/ha)	2.44 t/acre (6.1 t/ha)	190-200 cm	125-130 Days	Suitable for medium soils in rainfed areas. Pearly white lustrous and bold round grain. Tolerant to shoot fly and charcoal rot. Responsive to fertilizers
12.	PKV Kranti	900-1000 kg/acre (3000-3500 kg/ha)	2.5-3.0 t/acre (5.0-7.5 t/ha)	180-200 cm	115-120 Days	Suitable for medium to deep soils and irrigated. Pearly white round and very bold grain, tolerant to shoot fly.
13.	SPV 2033	800-1000 kg/acre (2500-3000 kg/ha)	2.6-3.2 t/acre (5.0-7.5 t/ha)	180-200 cm	115-120 Days	Suitable for deep soil. Pearly white round and very bold grain, tolerant to shoot fly and charcoal rot
14.	Phule Anuradha	600-700 kg/acre (1500-1800 kg/ha)	1.2-1.6 t/acre (3.0-3.5 t/ha)	160-180 cm	100-112 Days	Suitable for shallow soils in rainfed areas. Early Pearly white round and medium grain, tolerant to shoot fly and charcoal rot. Early maturing.
15.	Phule Maulee	600-800 kg/acre (1500-2000 kg/ha)	1.8-2.0 t/acre (4.5-5.0 t/ha)	180-210 cm	110-115 Days	Suitable for shallow, medium soils in rainfed areas. Yellow white, lustrous and medium bold grain, tolerant to shoot fly and charcoal rot and drought
16.	DSV 5	900-1000 kg/acre (3000-3500 kg/ha)	2.6-3.2 t/acre (3.5-4.5 t/ha)	220-250 cm	120-125 Days	Suitable for deep soils, transitional and irrigated zones in Karnataka. Pearly white round and very bold grain, tolerant to shoot fly.
17.	DSV 4	1600-1680 kg/acre (4000-4200 kg/ha)	1.8-2.0 t/acre (4.5-5.0 t/ha)	180-200 cm	115-120 Days	Suitable for dry zones in karnataka. Creamy bold, round grain flat at one side and tolerant to diseases
18.	Phule Suchitra	1000-1200 kg/acre (2500-3000 kg/ha)	2.2-2.6 t/acre (5.5-6.5 t/ha)	200-220 cm	115 Days	Suitable for medium Pearly white, lustrous and medium bold grain, tolerant to shoot fly and charcoal rot and drought soil.

Therefore, there is a need for the development of varieties adapted to specific soil situation in *Rabi* season to enhance production and productivity levels. Broad adaptation in post-rainy sorghum varieties may not fetch the desired results, as the post-rainy environment is highly variable in terms of soil and climate. Breeding for specific soil depths has been undertaken in various public-sector programmes and significant progress has been achieved, one being the release of Phule Chitra for medium soils. As an example, on comparison of yield attributes and physiological traits of Phule Chitra with M 35-1, a widely cultivated postrainy season variety with broad adaptation and Phule Maulee, a variety suitable for medium soils, the ideotype was reflected in the higher per day grain and fodder productivity, higher harvest index, greater ear head exertion, higher relative water content and slow leaf senescence. Hence, while breeding for new varieties adaptable to medium soils, these traits have to be considered. However, there is further need to study the root characteristics of these varieties in controlled conditions to identify if any specific root trait contributes to the soil-specific adaptation of these varieties.

Table 5: Details on the Pedigree and Origin of Varieties

Sl.No.	*Variety*	*Pedigree*	*Origin*
1.	M35-1	Selection from Maldandi bulk	Mohol
2.	CSV-8R	R24(IS-2954 x M35-1)	NRCS
3.	SWATI	SPV-86 x M35-1	Rahuri
4.	SEL-3	Selection from Bidar local	Rahuri
5.	CSV-14R	M35-1 x (CS2947 x CS2644) x M35-1)	NRCS
6.	DSV-4(9-13)	E36-1 x SPV-86	Dharwad
7.	DSV-5(GRS-1)	Selection from Natte maldandi of Gulbarga	Dharwad
8.	CSV-216R (Phule Yashoda)	Selection from Dhulia germplasm	Rahuri
9.	RSLG-262 (Mauli)	Land race selection	Rahuri
10.	Parbhani Moti	Selection from GD 31-4-2-3	Parbhani
11.	CSV-18 (Parbhani Jyoti)	PVR 579	Parbhani
12.	PKV Kranti(AKSV-13R)	SPV-1201 x Ringni	Akola
13.	Phule Chitra	SPV-655 x RSLG-112	Rahuri
14.	Phule Vasudha	RSLG-206 x SPV-1047	Rahuri
15.	Phule Revati	CSV 216 x SPV 1501-1	Rahuri
16.	Phule Suchitra		Rahuri
17.	Phule Anuradha		Rahuri
18.	CSV-22	CSV 216 x RSP 2	Rahuri
19.	CSV-26	SPV 655 x SPV 1538	Solapur
20.	SPV-2033	[GRS 1 x CSV 216R] x CSV 216R	Bijapur

As a result, now, there are many better varieties for *Rabi* season, besides 5-decade old local variety M35-1, like the recently released national varieties/hybrids (CSV 216R, CSV 18, CSV 22, CSH 15R, CSH 19R, CSV 26, SPH 1666) and state released

(Shallow soils-Sel.3, Mauli, Phule Anuradha; medium soils- Mauli, Phule Chitra, Parbhani Moti, Phule Suchitra; deep soils/irrigated conditions-Phule Vasudha, Phule Revati, DSV 4, DSV 5, PKV Kranti and CSV 29R). In India, the following cultivars were released for cultivation under specific situations across states.

Hybrid Improvement in Post-Rainy Sorghum in India

Historical Aspect

Hybrid vigour and its commercial exploitation have paid rich dividends in *Kharif* sorghum leading to quantum jump in sorghum production (Rana *et al.*, 1997). However, the progress in *Rabi* sorghum is limited. Although several hybrids have been developed and released for *Rabi* season cultivation, the area covered with hybrids is almost negligible. Lack of appropriate hybrids with acceptable grain quality adapted to different agro-ecological situations of *Rabi* season characterized by terminal drought, low temperatures and biotic stresses like shoot fly infestation is a major constraint for higher productivity (Nimbalkar and Bapat, 1991, Rana *et al.*, 1997, Gite *et al.*, 2006).

Heterosis breeding led to the central release of the hybrids like CSH-7R, CSH-8R and CSH-12R. These cultivars were notable in two respects. First, that none of these cultivars succeeded in gaining consistent acceptability from farmers at a scale to effect a discernible impact. Second, that the breeders came to realise the complexities in the genetic up-gradation of yield in *Rabi* sorghum. Following were the lessons learnt out of these programmes:

- ✰ The traditional and widely preferred land races are ineffective as donors of favourable gene complexes and as parents in heterosis breeding,
- ✰ The *Kharif* genetic diversity though very effective in achieving quick yield increase is bedeviled with low adaptability and grain quality, and
- ✰ Critical elements of *Rabi* sorghum adaptability are resistance to shoot fly, charcoal rot, moisture stress and low temperature.

The second phase of *Rabi* sorghum breeding with emphasis on hybrid cultivars was initiated in the late eighties. During this period 250 experimental hybrids were evaluated in the AICSIP Initial Hybrid Trials. These trials resulted in the identification of two hybrids, SPH 504 and SPH 677 for central release as CSH 13R and CSH 15R. Variety CSV 14R has no significant yield advantage over M35-1 and is comparable to the latter in most of the adaptability traits. CSH 13R has significant yield superiority over M35-1 but is highly vulnerable to shoot fly and low temperature and has inferior grain quality. The case of CSH 15R having a *Rabi* based MS line (104A) developed at Mohol Centre is different from CSH 13R has a marginal yield advantage over M35-1. *Rabi* sorghum hybrids will have a tangible impact only when the MS and R lines having the season adaptability and desired combining ability are developed (Rao, 1986).

Hybrid Breeding at NRCS and AICSIP centres during 1999 to 2005

A Mission Mode Project on Development of Hybrid crops under National Agricultural Technology Project funded by World Bank was in operation specifically to develop *Rabi* sorghum hybrids at NRCS and 7 AICSIP centres working on *Rabi* sorghum at National level from 1999 to 2005 (Rana *et al.*, 2005). The objectives of the projects were to develop high yielding, shoot fly resistant *Rabi* sorghum hybrids of early to medium duration suitable for variable soil depths and develop suitable parental lines with resistance to shoot fly, charcoal rot, drought and low temperature by utilizing milo and non-milo cytoplasms to diversify the hybrid base. The highlights of research findings were as follows.

i) Development of Parental Lines/Hybrids

- ✰ More than 4900 new hybrids were developed and tested during 1998 to 2005 across seven centers.
- ✰ SPH 1010 bred at Akola centre was released and notified as CSH 19R. The grain yield increase was 33.3 per cent over CSH 15R (hybrid check) and 60.5 per cent over M 35-1(varietal checks). This hybrid is comparable for shoot fly tolerance to other checks, however, it is more tolerant to stemborer, head bugs and aphids than the earlier releases. Similarly it is tolerant to charcoal rot and leaf blight.
- ✰ New drought tolerant, early *Rabi* sorghum hybrid DSH 4R was developed and released for Karnataka during 2001. It yields 25-30 q/ha grains and 4-5 t/ha fodder. This hybrid is resistant to drought and moderately tolerant to charcoal rot. There is a great demand for this hybrid due to successive drought for three years in Karnataka.
- ✰ A new hybrid BRJH 111 which is high yielding and also resistant to shoot fly is in the final stages of release.
- ✰ New dual purpose high grain and fodder yielding hybrid DSH 6R is accepted for farm trial for Karnataka state for *Rabi* season 2004-05.
- ✰ New CMS lines developed across Rahuri, Dharwad, Bijapur, Parbhani, NRCS and Akola were 104A, 116A, 117A, 1409A early, 1409A medium, PMS 20A (tan type), 41A, 49A, 59A, 67A, 89A, 95A, 109A, 127A, 133A, 147A, 163A, 169A, 185A, 187A, 203A, 215A, 237A, DNA1, DNA2, DNA4, DNA5, BJMS 1A, BJMS 2A, BJMS 3A, BRJ 204 A, RS 20 A, RS 70A, RS 132A, RS 137A, RS 159A, RS 412A, AKRMS nos.– 14A, 43A, 46A, 47A, 63A, 67A, 68A, 72A and 82A.

ii) Utilising Non-Milo (A_2/M 31-2A/M 35-1 A) Cytoplasm for Breeding *Rabi* Hybrids

- ✰ Until now, all male-sterile lines in sorghum world-over (except in China) are based on *milo* cytoplasm. Therefore, diversification of male-sterile lines is an urgent need for future hybrids. Non-*milo* cytoplasm opens up new avenues to sterilize many of restorers on *milo* cytoplasm. If such high yielding and resistant lines are converted into CMS lines using non-*milo* cytoplasms, the hybrids with higher shoot fly resistance can be bred. Efforts in this direction have led to the development of RS 530 A2, ICSV 8603 A2, SPV 932 A2, RS 71 A2 at Rahuri, and PMS 12 A2, PMS 13 A2, PMS 15 A2, PMS 23 A2, PMS 14 A (M 35-1), PMS 18 A (M 35-1) at Parbhani. To develop new male-sterile lines with A2 cytoplasm, 8 paired crosses were made on RS 530 A2 and SPV 932 A2 by using desirable segregants from the F4 generation of following crosses at Rahuri: 104 B x 20 B A2, 104 B x 22 B A2 and 20 B A2 x 1409 B. At NRCS Hyderabad, 19 pairs of A2 based CMS lines are under BC 3 and BC 4 stages of conversion. Lines with bold-seed (MBMS 7B and SB 101 B), charcoal rot resistance (DCCR 1) and rust resistance (IS 3443) were utilized on M 31-2B at Dharwad to develop new B lines. *Rabi* locals *viz.*, Barshi Prakash, Muddihalijola, Honnutagi Local, RS 585 and SPV 570 which are maintainers on M 31-2A were advanced to BC 2 at Bijapur.

☆ For diversifying the base of restoration on A2/Maldandi cytoplasms, crosses involving following parents were implemented at various centres: SPV 655, SPV 1047, SPV 492, and PVK 801. At Hyderabad, 11 A2 restorers have been identified. At Parbhani MR 11, MR 13, MR 15, MR 27 and MR 29 were identified as capable of restoring on Maldandi cytoplasm and KR 198 and PVR 350 to be restoring on A2 cytoplasm. At Bijapur, the lines *viz.*, BRJ 62, BRJ 67, SPV 1491 and SPV 1556, CR 9, Jewaragi Local and SPV 1452 have been identified as restorers on Maldandi cytoplasm. At Dharwad, Barshizoot local, Dagdi local, Hagari local, SPV 570 and IS 4582 were identified as restorers on M 31-2A. During 2001-02, around 33 restorers on non-*milo* cytoplasm have been identified at NRCS, Rahuri and Bijapur. During 2003, restoration of SPV 1491, BRJ 67, BRJ 62 and CR 9 has been confirmed across both *Kharif* and *Rabi* seasons. Apart from this, a large number of experimental hybrids on non-*milo* cytoplasms were developed and evaluated at various centres. During 2003-04 period five stable restorers on Maldandi cytoplasm were identified *viz.*, IS 29411, 29406, 4587, Raichur Local and Ramke at Bijapur centre. The hybrids developed at Bijapur using M31-2A and BRJ62 is in advanced stage of testing in AICSIP trails. In Rahuri centre fifty two new experimental hybrids were developed on A_2 based CMS lines. Twelve conversion pairs on A_2 cytoplasm were evaluated. New MS lines *viz.*, PMS 47 A (A_2) and PMS 49 A (A_2) were developed at Parbhani centre.

iii) Evaluation of Parental Lines to Develop Hybrid Seed Production Technology

All the parental lines of already released and promising hybrids were examined for suitability for seed production by studying flowering behavior and synchronization. Flowering behaviour of parental lines indicated that restorer lines of most of the hybrids in general flowered 2-6 days late than respective A lines. The yield levels of B and R lines were higher than their respective A lines. In addition to this, nucleus and breeder seed production of *Rabi* sorghum hybrid CSH 15R is being produced every year and enough quantities of parental lines and hybrid seed were regularly produced.

iv) Studies on Cold Tolerance and Synchrony of Sorghum Parental Lines

- A total of 76 A, B and R lines were evaluated for synchrony at actual seed production sites at Nandyal in AP and Bellary in Karnataka and it has been observed that photoperiod and agro-ecology influence the synchrony of parental lines and are in much variation to flowering behaviour reported from various test locations and in the release proposals.
- A total of 142 lines consisting of hybrids/Varieties, B and R lines were evaluated for cold tolerance at Mahabaleshwar in Maharshtra and Indore in Madhyapradesh. The genotypes took on an average 171-185 days to flower at Mahabaleshwar and 97-150 days at Indore due to impact of cold. The minimum temperature at flowering was in the range of 5-18.9° C at Mahabaleshwar and 9.1- 26.6 °C during reproductive phase. The seed set varied between 10-100 per cent at Mahabaleshwar and between 1-100 per cent at Indore.
- A total of 57 promising B, R lines and hybrids/varieties are being screened for cold tolerance at wheat rust research station Mahabaleshwar and IGRI, Jhanshi. The data on crop growth and flowering in relation to cold have been recorded. The seed set data is awaited to identify the best cold tolerant lines.

v) Identification, Incorporation of Resistance to Biotic Stress in Parental Lines

- Following shoot fly resistant R lines were identified and shared for further breeding *viz.*, RSFR 9530, RSFR 9517, RSFR 9509, RSE 3, RSE 9715, RSV 207 and RSV 145. The lines *viz.*, RR 9822 (CS 3541 x CSV 15), RR 9823 (RSLG1770), RR 9825 (CSV 15 x IS2312), RR 9826 (SPV655 x SPV1090), RR 9827 (SPV1316), RR 9828 (Selection from Tandoor local), RSLG 871 (Local land race selection) and RSLG 971 (Local land race selection) were high yielding restorers, and can also be directly used for cultivation.
- Following lines were found to be promising for yield and foliar diseases *viz.*, SVDR 9678, SVDR 9662, SVDR 9672, SVDR 9693, SVDR 9749, SVDR 9750, SVDR 9740, SVDR 9646-1, SVDR 9751, SVDR 9752 and SVDR 9746. SPV 570 was found to be a good combiner. Superior R lines like RS 585, RS 586, RS 615, RS 645, RS 647, RS 653 and RS 654 which were involved in developing number of promising hybrids.
- Efforts in diversification of cytoplasm have led to the development of RS 530 A_2, ICSV 8603 A_2, SPV 932 A_2, RS 71 A_2 at Rahuri, and PMS 12 A_2, PMS 13 A_2, PMS 15 A_2, PMS 23 A_2, PMS 14A (M 35-1), PMS 18A (M 35-1) at Parbhani. At NRCS, two CMS lines *viz.*, RS 2214 A_2 and RS 2216 A_2 were developed with the shoot fly resistance comparable to the check 104B.

- ☆ Hybrid SPH 1452 bred on *Maldandi* source of male sterility is in the second year of testing under advanced hybrid trial at all India level. The lines *viz.*, BRJ 62, Barsizoot local, Dagadi local, Hagari local, SPV 570 and IS 4582 have been identified as restorers on M 31-2A cytoplasm while MR 15 and MR 29 were found to be restoring on M 35-1A cytoplasm.
- ☆ All the parental lines of already released and promising hybrids were examined for suitability for seed production by studying flowering behavior and synchronization. Flowering behaviour of parental lines indicated that restorer lines of most of the hybrids in general flowered 2-6 days late than respective A lines. The yield levels of B and R lines were higher than their respective A lines.

Recent Breeding Research Efforts for Hybrids at DSR

Attempts have been made in the past to breed for long ear head like 296B in *Rabi* sorghum. The CMS line 104A has been already exploited in the development of two commercial *Rabi* hybrids CSH 15R and CSH 19R. However, this and other new CMS lines, though have got good grain and genetic potential in *Rabi* background, they are still weak in combining ability for yield and also in resistance to shoot fly and charcoal rot. Also, they lack grain luster like M35-1. Besides, they are sensitive to low temperatures and cannot withstand terminal moisture stress. Hence, the strategies adopted for developing parental lines were as follows.

Female

- ☆ Improving the existing *Rabi* CMS lines (104A, 1409A, PMS 20A, AKR-45A, 401A, 185A)
- ☆ Initiate a new programme based on *caudatum/guinea* CMS lines and strengthen with characters of *Rabi* adaptation and grain quality

Male

- ☆ Popular land races have variability for grain characters and fertility restorations and identify them
- ☆ Cross among themselves and or with exotic germplasm to bring diversity in R lines utilizing non-durra germplasm from African countries.
- ☆ Convert them by incorporating grain quality/resistance to abiotic/biotic stresses.

Hence, in order to develop new parental lines with diversity and broaden the genetic base in *Rabi*, 161 indigenous and 159 exotic lines were initially used for introgression into *Rabi* background. Sterility and fertility reactions were scored in 525 pairs made using CMS 104A. Out of them, 204 pairs showed clear sterility (B gene) and 175 pairs exhibited fertile reaction(R gene). New B lines derived in the study were converted into A/B lines through back crossing and R lines were stabilized through pedigree breeding.

A total of 43 parental lines (18 B lines and 25 R lines) were evaluated for high yield, resistance and grain quality parameters. Performance of 18 B lines indicated that all of them were early in flowering (64 to 74 days) compared to check 104 B (76 days). However, these B lines were medium in height (172 to 230 cm). Panicle length was more in case of SLB 9 and SLB 19 and panicle width was higher in 10 lines. Twelve B lines were significantly superior in grain yield (64 to 81 g/plant) than the check 104B (58 g/plant). Superior B lines gave 10.3 to 39.6 per cent more grain yield than the check 104 B. All the B lines were having superior grain quality (shape, color and luster) than 104B. For charcoal rot, all the B lines were significantly superior to the check. Charcoal rot ranged from 14.3 to 28.3 per cent compared to check 104 B (34.2 per cent). Out of 25 R lines, 14 lines were early in flowering (68 to 70 days) compared to check RS-585 (73 days). The plant height, panicle length and width were on par with the check RS 585.Eighteen R lines (63 to 79 g/plant) were significantly superior to the check RS 585 (56 g/plant) and showed 12.5 to 41.0 per cent increase over the check. For charcoal rot, eleven R lines were significantly superior to the check. Charcoal rot ranged from 12.3 to 22.6 per cent compared to check RS-585(29.2 per cent). Thus, the above parental lines are diverse, early and superior in grain yield. They offer better resistance to charcoal rot and possess better grain quality. These diverse parental lines can be directly used for the development of commercial *Rabi* hybrids.

Genotype x environment interactions in 36 new *Rabi* sorghum hybrids were studied over 3 environments for flowering, maturity, grain and fodder yields. Significant differences were observed among environments and hybrids for the characters. The G x E interactions for the characters was significant and the significant mean squares due to environment (linear) indicated the existence of real differences among hybrids for characters. *Rabi* sorghum hybrids, namely 104A x SLR 31, 185A x SLR 13 and 104A x SLR 72" were found stable for earliness and for grain yield, the hybrids '185A x SLR 59, 185A x SLR 47, 104A x SLR 67, SL9A x SLR 57 and 104A x SLR 79 were the most stable under rainfed situation. For both grain and fodder yields, the stable hybrids were "185A x SLR 59, 104A x SLR 47, 104A x SLR 67, SL9A x SLR 57, 104A x SLR 79 and 104A x SLR 57. The identified hybrids can be commercially exploited for increasing the production and productivity of *Rabi* sorghum (Prabhakar *et al.,* 2013).

Assessment of heterosis for grain and fodder yields in 52 *Rabi* sorghum hybrids were evaluated across environments. Significant positive heterosis for grain yield was observed for 32 hybrids over checks CSH 15R and CSV 22, which ranged from 10.1 to 54.9 per cent. Superior hybrids namely BJMS 2A x SLR 13, AKR 45A x SPV 570, AKR 45A x SLR 28, AKR 45A x SLR 10, 1409A x RS 585, 104A x SLR 28, 104A x BRJ 358, 1409A x JP 1 1 5, 1409A x SLR 13 and 1409A x SLR 27 that exhibited more than 15 per cent heterosis and increase in grain yield were highly promising. The new hybrids identified in this study are required to be further tested in large scale trials for their performance, so that it would help to increase production and productivity of *Rabi* sorghum (Prabhakar *et al.,* 2013).

Promising *Rabi* sorghum parental lines comprising of male sterile A-lines (11), maintainer B-lines (11) and fertility restorer R-lines (14) were evaluated to understand the variability for seed setting potential. Results indicated that among the 11 male sterile (A-) lines, the seed set (per cent) ranged from 31 (BJMS 2A) to 77 (SL 12 A) with a mean of 62 per cent. All pair comparisons test indicated two homogenous groups for seed setting among the A-lines. Among the 11 maintainer (B-) lines, the seed set (per cent) ranged from 62 (185B) to 89 (SL 29B) with a mean of 79 per cent. Four homogenous groups for seed setting were noted among the B- lines. Among the 14 restorer (R-) lines, the seed set (per cent) ranged from 75 (SLR 72) to 89 (SLR 91) with a mean of 82 per cent. All pair comparisons test showed two homogenous groups for seed setting among the R-lines studied (Kannababu *et al.*, 2013).

Performance of Hybrids and Varieties

In recent years, very few hybrids are being tested in AICSIP trials under Advanced Varietal and Hybrid Trials. The hybrids SPV 1689, SPV 1690 and SPH 1663 stood 1^{st,} 3rd and 4th rank, respectively in AVHT Deep soil for grain yield during 2011 12. The hybrid SPH 1690 was the top fodder yield in AVHT Deep soil giving 14.5 per cent more yield the check. The hybrid SPH 1663 tested across locations for 3 years from 2009 10 to 2011 12 exhibited yield superiority of 6.39 per cent and 3.46 per cent over CSH 15R and CSV 22 for grain yield. For fodder yield it gave 11.11 per cent more yield than CSH 15R. The hybrids SPH 1689 was superior by 5.98 per cent and 14.39 per cent over CSH 15R for grain yield over 2 years. For fodder yield, it gave more than 10.07 per cent and 2.64 per cent over CSH 15R and CSV 22. Whereas the hybrid SPH 1690 gave 4.93 per cent and 13.26 per cent over CSH 15R and CSV 22 and for fodder yield, it gave 17.23 per cent and 9.32 per cent over CSH 15R and CSV 22 over 2 years.

Constraints in *Rabi* Sorghum Hybrid Breeding

Major constraints restricting the yield improvement in *Rabi* sorghum are:

1. Growing crop over large area in medium to shallow soils where the occurrence of drought is much faster than the deep soils;
3. Susceptibility to shoot fly and charcoal rot;
3. Low temperature affecting growth;
4. Lack of appropriate hybrids with required traits of *Rabi* adaptation, and limited response to applied nutrients under moisture stress situation.

Low Heterosis because of using Maldandi for the Improvement of both Parents

Difficulties in recombining grain and fodder yield, grain quality and resistance to biotic and abiotic stresses equivalent to popular variety, maldandi restricted genetic improvement to few varieties and hybrids only. Multiple trait selection is coupled with low genetic advance in *Rabi* sorghum. Both grain and fodder being of higher value than that of *Kharif*, building high grain yield and increased harvest index at the expense of fodder yield could not be favoured. This is one of the reasons why there is low magnitude of heterosis in *Rabi* compared to *Kharif*.

Contrary to *Kharif* hybrids, the heterosis in *Rabi* hybrids is insignificant because the landraces (which are low community performers) are used (mainly to maintain the consumer preferred grain size and lustre) in the development of both parents. It is envisaged that introduction of larger grain size and lustre in the female parents of *Kharif* hybrids by novel methods and hybridizing such female parents with *Rabi*

Table 6: *Rabi* Sorghum Hybrids Released at National Level

Sl.No.	*Hybrid*	*Year of Release*	*Pedigree of the Hybrid*	*Pedigree of the Parental Lines of the Hybrid*	*Centre which Developed*	*Area for which Recommended*
1.	CSH 7R	1977	36Ax168	36A = CK60B x PJ8K CK60B – Kafir (USA) PJ8K – Parbhani jowar (*Kharif*) 168=IS 3687-Durra Caudatum (USA) xAispuri (a *Kharif* local from Maharashtra)	AICSIPCU	Maharashtra, Karnataka, Andhra Pradesh
2.	CSH 8R	1977	36A x PD3-1-11	36A - See S.no. 1 PD3-1-11 = Temperate US dwarf x BP53 (Gujarat *Kharif* local)	Parbhani	Maharashtra, Karnataka, Andhra Pradesh, Gujarat
3.	CSH 12R	1986	296A x M148-138	296A - See *Kharif* hybrids S.no.7 M148 - 138 = Mutant of maldandi	Dharwad	Maharashtra, Karnataka, Andhra Pradesh, Gujarat
4.	CSH 13R	1991	296A x RS29	296A - See *Kharif* hybrids S.no.7 RS29 - SC108 x SPV 126 SC108 – Purdue (USA) SPV126 – tall mutant of CS 3541	NRCS	Maharashtra, Karnataka, Andhra Pradesh, Tamil Nadu, Gujarat
5.	CSH 15R	1995	104A x RS 585	104A = 296B x Swati 296B = See *Kharif* hybrids S.no.7 Swati = SPV 86 x M 35-1 RS 585 (CS 3541 x M35-1) x Nandyal *Rabi* local	NRCS	Maharashtra, Eastern parts of Karnataka, Andhra Pradesh, Tamil Nadu, Gujarat
6.	CSH 19R	2000	104A x R354	104A - See S.no.5 R354 - [(SPV-504(20KR)x (SPV 504 xR263)] x R-67-4	Akola	Maharashtra, Karnataka, Andhra Pradesh, Gujarat

Table 7: Salient Features of Popular *Rabi* Sorghum Hybrids under Cultivation at Present

Sl.No.	*Hybrid*	*Grain Yield kg/acre*	*Dry Fodder Yield tones/acre*	*Plant Height (cm)*	*Duration (Days)*	*Salient Features*
1.	CSH 15R	1278 kg/acre (3194 kg/ha)	2.23 t/acre (5.57t/ha)	180-200	112-115	Suitable for medium to deep soils for rainfed and irrigated areas. Tall, large semi compact panicle. Pearly white round and very bold grain, tolerant to shoot fly and charcoal rot and free threshing
2.	CSH 19R	1181 kg/acre (2953 kg/ha)	2.34 t/acre (5.84 t/ha)	225	117-120	Suitable for medium to deep soils for rainfed and irrigated areas. Pearly white,slightly flat, medium sized grain, tolerant to shoot fly and charcoal rot

based R lines would increase the yield levels of *Rabi* hybrids to fetch better farm incomes to the farmers.

Seed availability is not a constraint in rainy season sorghum but it is a major issue in postrainy season sorghum. In sorghum the seed (both hybrids and OPVs) for rainy season cultivation is multiplied in the postrainy season and hybrids are marketed just before next rainy season; so the holding time is short and companies realize returns on their investments quickly. In case of postrainy season adapted varieties (or hybrids), the seed produced in postrainy season has to be marketed in next postrainy season which means they need to wait for 5-6 months to market them and hence the returns are realized late. There are not many improved hybrids with all the farmers preferred traits. So it is primarily the public sector seed agencies and state agricultural universities catering the needs of farmers for sorghum seed supply in postrainy season.

Temperature Sensitivity in *Rabi* Hybrids

When minimum temperatures go below 10°C for several days during flowering, hybrids that are otherwise male fertile show male sterility as evidenced by partial to complete absence of seed set under bagging. Seed setting ability (under bagging) in hybrids at low temperatures is critical to the post rainy season hybrids and requires greater attention to ascertain the differences among the landraces for their ability to restore fertility in hybrids, especially under low temperatures, normally observed in post rainy season.

Fertility Restoration

Restoration frequency was higher on the A_1 (65 per cent) than on A_2 (56 per cent) CMS lines. Most of the landraces, including M 35-1, showed segregation for fertility restoration/sterility maintenance ability indicating the need to select for restoration ability within the landraces. This also explains the partial restoration observed when bulk pollen of M 35-1 was used by many workers (Vidyabhushanam, personal communication). lower levels of fertility restoration on *durra*-derived A_1 CMS lines than *caudatum*-derived A_1 CMS lines. This finding has a bearing on developing CMS lines involving *caudatum*-based germplasm lines adapted to postrainy season and testing for fertility restoration in hybrids.

The hybrids involving *caudatum*-based female parents and *durra*-based landraces showed high heterosis for grain yield. The *caudatum*-derived female lines were adapted to rainy season. Rao *et al.* (1986) argued that by eliminating temperature sensitivity (in relation to development) in both male and female parents, greater success could be achieved in breeding hybrids for postrainy season. However, male sterile lines bred from landraces adapted to postrainy season, as indicated earlier, showed temperature-induced restorer inefficiency. Also, the hybrids developed from landraces crossed to female lines derived from M 35-1 did not show high heterosis.

For *Rabi* hybrid development, CMS-based seed parents and restorers need to be diversified by creating separate gene pools through crossing between guinea-based B-lines and durra-based B-lines and between caudatum-based R-lines and guinea/

durra-based R-lines for various selected traits. However, banking on the limited variability by crossing among already developed B-lines leads to narrow genetic base of the resulting hybrids, although better parents could be quickly developed through this approach. However, for sustained and continuous improvement, fresh variability should be created by introgressing new sources of desired genes from landraces into elite agronomic background. Rana and Murty (1978) have reported that increase in number of seeds per panicle branch in short compact headed varieties (tropical) and increase in the panicle branches in the long panicle type (temperate) by introgression of genes from African germplasm result in yield heterosis. F_1's made on *caudatum*-based seed parents with durra-based pollinators resulted in high heterosis under post rainy season condition.

In order to develop superior high yielding *Rabi* hybrids, it is important to remove the existing weaknesses of parental lines eg. Low level of heterosis and inadequate level of resistance by way of genetic enhancement through incorporating high grain number, better harvest index and strengthening resistance to above biotic and abiotic stresses. Such genotypes may respond better to applied nutrients under receding moisture condition. Thus development of improved production technologies for economically cultivating the improved genotypes in dry and irrigated conditions by optimizing the use of critical inputs and to maximize the yield will overcome the *Rabi* sorghum production constraints. Simultaneous research on hybrid seed production technique and demonstration of production potential of hybrids in on-farm trials will facilitate expeditious transfer of technology, based on new genotypes and ultimate breakthrough in sorghum productivity. It will ensure food and fodder security in large sorghum growing drought prone and irrigated areas.

Conclusions

Sorghum improvement efforts, though had significant impact on increasing rainy season productivity, it did not have much impact on the productivity of postrainy season sorghums in India. The landrace pollinator-based hybrids had the required characters for postrainy season adaptation with farmers' desired grain quality traits, except that they lacked resistance to shoot fly and had intermediate grain luster. It is therefore expected that if improved male sterile lines with resistance to shoot fly and grain luster become available, they can contribute to enhanced acceptability of hybrids by farmers for postrainy season cultivation.

In sorghum, breeding hybrids with *Rabi* adaptation is a top priority. Male sterile and restorer parents with *Rabi* adaptation and high yielding hybrids with *Rabi* adaptation are also required. Research should also be initiated on the identification of environment sensitive male sterility and to facilitate two-line hybrid breeding. A long term goal is the fixation of heterosis through apomixis.

Genetic improvement strategy for increasing productivity of *Rabi* sorghum hybrids should offer flexibility to develop parental lines with following points in mind.

1. *Rabi* adapted parental lines capable of breeding high heterotic hybrids significantly superior to varieties under rained management conditions

2. Adoption of ideotype breeding by defining a crop production environment with reference to specific soil depths.
3. Identification and release of hybrids (early or medium duration) suitable to specific soil depths.
4. Genetic enhancement for resistance to shoot fly, charcoal rot, cold and drought. The benchmarks for resistance to biotic and abiotic stresses should be that of M 35-1
5. By eliminating temperature sensitivity in both male and female parents, greater success could be achieved in breeding hybrids for post-rainy season.
6. Improve the grain and fodder quality in the recently released varieties and the quality should be at least on par with M35-1
7. *Rabi* adapted and high input responsive hybrids for irrigated conditions.

Rabi Sorghum Improvement for Terminal Drought Tolerance

Rabi sorghum is cultivated by the poor farmers and grown under subsistence conditions. Lower yields are realized due to exposure to various biotic and abiotic stresses apart from poor management practices. Towards the end of *Kharif* season, the soil is completely charged with moisture to support the *Rabi* crop. The *Rabi* sorghum is grown on residual moisture disposing the crop to drought towards maturity termed as terminal drought (Murty *et al.*, 2007, Sajjanar *et al.*, 2011) apart from other factors such as heat and low soil fertility, and to pests and diseases. Terminal drought is the most significant cause of crop yield loss in *Rabi* sorghum (Srivastava *et al.*, 2010). The most severe droughts occurred across the area producing the majority of the *Rabi* sorghum and were predicted to devastate crop grain yield in about 25 per cent of seasons (Kholova *et al.*, 2013). Although, sorghum possesses excellent drought resistance compared to most other crops, improving its drought resistance would increase and stabilize grain and fodder production. It is recognized that genetic improvement is an important component of an integrated approach for stabilizing and improving crop production in drought environments.

Traits Governing Drought Tolerance

Symptoms of post-flowering drought stress susceptibility include pre-mature plant (leaf and stem) death or plant senescence, stalk collapse and lodging, and stalk rot or charcoal rot (*Macrophomina phaseolina*), along with a significant reduction in seed size, particularly at the base of the panicle (Garud *et al.*, 2002). Tolerance is indicated when plants remain green and fill grain normally. Such green stalks also have good resistance to stalk lodging and to charcoal rot. Such cultivars are referred to as having 'stay green' trait. In terminal stress environments, yield is greatly influenced by the dynamics of canopy development, which is regulated via maturity (Ravi Kumar *et al.*, 2009). Though transpiration efficiency is also known to vary considerably among diverse sorghum lines (Talwar and Vadez, 2011), it remains unutilized in breeding programs due to complexity in measuring the variation. Post rainy season cultivars such as M 35–1 and CSH 8 have a greater capacity to decrease their osmotic potential under stress than the rainy season cultivar CSH 6 (Seetharama *et al.*, 1982). Two independent major genes (OA I and OA 2), with some minor effects have been reported to control the inheritance of OA in sorghum (Basnayake *et al.*, 1995). Little if any progress in breeding for drought tolerance using either OA or any of the other physiological traits, has been documented in sorghum, partly because of poor understanding of these traits conferring drought tolerance and lack of procedures to impose reproducible stresses and rapid methods to measure these traits.

Characterization of the Drought Environments

Characterization of the target production region is the first and crucial step in undertaking a genetic improvement program aimed at improving yield and yield stability in drought prone environments. Production environments can be grouped into subsets of environments using canonical variate analysis (cluster analysis), by considering such factors as soil water balance, temperature regimes, the potential evapotranspiration of the growing environment (using long term climatic data), or based on GxE interactions. Thus, mean and variance of moisture deficits likely to occur during the growing season and length of the growing season can be calculated. This will assist in estimating the intensity and duration of stress the crop is likely to experience, and at which crop growth stage, thus guiding development of relevant screening methodology for evaluating germplasm lines. This strategy also will help in identifying the environments with similar drought patterns, to guide multilocational testing of genotypes developed for specific drought patterns.

Screening Methodologies

In most cases, screening for drought resistance involves evaluating the genotypes under field conditions with and without irrigation. Moisture response curves are generated by regressing the yield of individual entries against water applied or environmental yield, to calculate the stability of a genotype across a range of moisture availabilities using the stability analysis procedures. However, this approach does not consider the confounding effects of phenology on yield or the effects of yield potential on the slope of regression and hence on the intercept and drought resistance. To account for time to flowering, associated drought escape, and yield potential, Bidinger *et al.* (1987) proposed a drought resistance index (DRI), in which DRI is based on the residual variation in grain yield adjusted for experimental error. The index is positively correlated with yield under drought, and independent of yield potential and time to flowering.

Selection of breeding lines for stay-green trait was useful for screening for terminal drought tolerance. The stay-green trait is expressed only when the materials at post-flowering stage are exposed to severe moisture-stress. Comparison of yields on shallow Vertisols or on partially saturated deep Vertisols with yields with an irrigated control has been advocated to screen genotypes for terminal drought tolerance in receding moisture conditions. Line source (LS) sprinkler irrigation technique developed at Utah State University of USA was followed at ICRISAT, India for screening sorghum genotypes for terminal drought tolerance. Each side of the LS formed one replication. The field was uniformly irrigated until the crop reached boot stage and the LS were used at 50, 61 and 77 days after sowing to create a gradient of soil moisture (stress).

The amount of water received across the plot was measured in catch cans placed at crop height. The LS was also used to study the effect of soil moisture stress on charcoal rot incidence (Seetharama *et al.*, 1987). The rows of plants farthest from the LS showed disease earlier than those nearest. This was apparent for each of the 3 parameters of disease spread: percentage of soft stalks; number of nodes crossed and the length of fungal spread (Seetharama *et al.*, 1987).

Field screening at more than one location with different water regimes can be used to insure stress at different stages of growth. Rainout shelters and irrigation gradient systems can be used to supplement field screening. Two main strategies are employed by breeders for improving drought resistance in sorghum (i) direct selection for yield in drought environments and (ii) indirect selection for traits known to be associated with drought.

Indirect Selection Traits

1. Using 25 per cent Polyethylene Glycol (PEG Method). This method is used in seedling stage. It can reduce water potential equivalent to natural drought condition so that water absorption by roots is affected (Singh and Chaudhary, 1998).
2. Using 0.3 per cent solution of potassium iodide. This method is used in seedling stage. It can simulate leaf senescence and abscission equivalent to natural drought condition (Singh and Chaudhary, 1998).
3. *In vitro Screening method:* The immediate response by the different organs of a plant to water stress is decrease in turgor. This may be partially or fully adjusted by accumulation of solutes. The change in size and shape of the pollen grains when subjected to in vitro osmotic stress using PEG was considered as a measure of osmotic adjustment (OA). PEG (discriminative osmotic stress) at 40 per cent discriminated the pollen grains of *Rabi* genotypes (retained their size) compared to the *Kharif* genotypes (showed shrinkage). There is a close correspondence between intrinsic OA in pollen grains and high leaf OA. In addition the study indicates that OA is a drought-adaptive trait and could have evolved in the *Rabi* genotypes by virtue of their regular exposure to moisture stress (Patil and Ravikumar, 2011).

Trait Associations

The grain yield was positively correlated with relative water content of leaves (PLWC), leaf area index (LAI), biomass at harvest, grain number per panicle, test weight, earhead exertion, stay green at physiological maturity, per day production of grain and fodder yield and negatively correlated with leaf temperature, chlorophyll stability index (CSI) and stomatal frequency under drought stress (Nirmal and Patil, 2008) and also with long roots, coleoptile length, higher root/shoot ratio, leaf area, leaf dry matter (Ali *et al.*, 2009). There was a positive association between relative water contents and cell membrane stability but both of these traits were negatively correlated with residual transpiration and excised leaf weight loss. Head width, head weight, grain yield per plant, fresh and dry shoot weight were the most important characters explaining 77.65 per cent of the total variation (Ali *et al.*, 2011). Stay-green and grain yield were positively associated in a range of studies in India (Al-Naggar *et al.*, 2007), highlighting the value of retaining green leaf area under terminal drought. Grain number is generally the main determinant of differences in grain yield while grain size is a secondary yield determinant and is often negatively associated with grain number as observed in sorghum grown under terminal drought stress in southern India (Borrell *et al.*, 1999). The hybrids obtained through selective fertilization produced significantly higher mean grain yield compared to hybrids obtained through non-selective fertilization (PEG at 36 per cent was applied to stigma and stylar tissue one hour before pollination for pollen selection) (Ravi Kumar *et al.*, 2003). Pollen selection for osmotic stress tolerance in sorghum results in superior progeny in moisture stress environment and the study demonstrates the transmission of the selected trait from pollen generation to progeny.

Twenty land-races from India, Mali and Sudan were evaluated under hydroponics conditions in a growth chamber where a water stress treatment was imposed by adding polyethylene glycol-8000 to the nutrient solution, giving a solute water potential of -0.5 MPa, compared with a control solution at 003 MPa. Drought resistance, in terms of relatively less growth inhibition under stress, was higher in races from dry regions than in races from humid regions. Of all the physiological variables measured [carbon exchange rate, (CER), transpiration, transpiration ratio (CER/transpiration), leaf diffusive resistance, leaf water potential and osmotic adjustment], only osmotic adjustment under stress was generally correlated with average rainfall at each race's origin, indicating greater osmotic adjustment in land-races from drier regions. Races with a greater capacity for osmotic adjustment were characterized by smaller plants with high rates of transpiration and low rates of leaf senescence under stress. The carbon exchange rate per unit leaf area increased as live leaf area decreased under stress due to leaf senescence. Thus, drought resistant races under stress tended to have lower CER per unit live leaf area (but not per plant) than susceptible races (Blum and Sullivan 1986).

The SPAD chlorophyll meter reading was significantly correlated with total dry matter and grain yield but not associated with number of days to flowering. The highly significant relationship between drought sensitivity index and percent change in SPAD chlorophyll meter reading under water deficit conditions indicated that higher chlorophyll concentration is vital for adaptation to water deficit conditions during the post-flowering period (Talwar *et al.*, 2009).

The genotypes with LAI around 3.0, higher dry matter accumulation at maturity with higher partitioning efficiency and higher photosynthetic rate at flowering gave higher grain yield in the receding soil moisture conditions (Channappagoudar *et al.*, 2007).

Gene Action

Various morphological, physiological and biochemical characters also confer drought resistance and showed different types of inheritance pattern (monogenic and polygenic) and gene action (additive and non additive). Preponderance of non-additive gene action under moisture stress conditions was observed for harvest index

(Palanisamy and Subramanian 1984), grain yield per plant, panicle length, panicle weight, and plant height (Umakanth *et al.*, 2002, Rajguru *et al.*, 2004, Murumkar *et al.*, 2005, Jhansi Rani 2007) and 100-grain weight (Chinna and Phul 1988). However predominance of additive gene action was reported for harvest index (Swarnalatha and Rana (1988), for grain yield (Dhoble and Kale 1988), for stomatal number (Nandanwankar 1990) for panicle length, plant height and crop growth rate (Poor and Rezai 1996). Importance of both additive as well as non-additive gene action in the inheritance of plant height, number of leaves per plant, panicle length and yield per plant was also reported (Patel *et al.*, 1993, Dhole 2004). The preponderance of additive genetic variation and the potential for improving drought resistance was verified by the high estimates of heritability. Sankarpandian *et al.* (1993) reported high heritability for the characters green leaf area, stomatal count, number of closed stomata and dry matter production for all the stages in an association of physiological characters with drought resistance. Sankarpandian and Bangarusamy (1996) revealed the significant interactions between genotype and stress treatment for stomatal count, root length, number of closed stomata, dry matter production and grain yield per plant. Drought tolerant in gadambalia, a drought tolerant sorghum cultivar is associated with higher water extraction efficiency, fewer nodel roots per plant, and fewer metaxylem vessels per nodel root, smaller leaf area and well developed sclerenchyma (Salih *et al.*, 1999).

Bichkar (2005) reported predominant of additive gene action for adaxial stomatal density, whereas non-additive gene action was found to be important in the inheritance of relative leaf water content, chlorophyll stability index, above ground dry matter biomass, 100-grain weight, stomatal density and grain yield per plant. The correspondence between per se performance and GCA effects in desired direction was not consistent. Appreciable amount of heterobeltiosis and standard heterosis among hybrid was observed for all the traits.

Padhye (2006) reported the positive association of grain yield with transpiration rate and transpiration rate with photosynthesis rate, because photosynthesis rate is the major cause and grain yield is the effects. Since higher transpiration rate itself decides low WUE, negative correlation was observed between grain yield and WUE, though non-significant.

Thul (2007) reported preponderance of dominance gene effects along with dominance x dominance non allelic gene interactions effects were found important in the inheritance of relative leaf water content, chlorophyll stability index (CSI), stomatal density on upper surface, harvet index, total above ground dry matter and grain yield per plant. Additive as well as dominant gene effects were found important along with additive x additive gene interactiona in governing stomatal density (lower surface). Complementary gene action played important role in the inheritance of all the traits imparting to drought resistance. Exploitation of heterosis or development of composite variety would be effective in the improvement of these traits.

In the study of genetics of some physiological parameters related to drought tolerance in *Rabi* sorghum, Kandekar (2008) reported predominance of additive as well as dominance gene effects for the traits relative water content, chlorophyll stability index, adaxial and abaxial stomatal density, plant height, total above ground dry matter, harvest index and grain yield per plant. The dominance x dominance gene interaction effects was also found predominant in the inheritance of traits relative leaf water content, stomatal count, plant height, total above ground dry matter and grain yield per plant with duplicate epistasis indicated that hetirosis breeding would be effective for improvement of this traits.

Khot (2008) reported the preponderance of additive and dominant gene action in governing the characters days to 50 per cent flowering, chlorophyll stability index, number of leaves, panicle length, harvet index, dry matter per plant and grain yield per plant, whereas the magnitude of dominant gene action was higher for most of the traits. Presence of both additive × additive and dominance × dominance gene interactions were predominant for the traits panicle length, harvest index, dry matter per plant, and chlorophyll stability index. Duplicate epistasis was observed for all the characters in most of the crosses.

The characters *viz.*, stay green, root volume, leaf area index, plant height and harvest index in sorghum showed high value for phenotypic and genotypic co-efficient of variation (PCV and GCV), showed higher estimates of heritability and expected genetic gain indicating the presence of additive gene effect (Vinodhana *et al.*, 2009). The GCV was highest for excised leaf weight loss followed by grain yield suggesting considerable scope for selection of these traits (Ali *et al.*, 2009). Dry root weight exhibited the highest GCV among seedling traits while excise leaf weight loss among the flag leaf related characters (Ali *et al.*, 2009).

SCA was more important than GCA variance for most of the traits under drought stress and non-stress, except for LA and HI under non-stress. Dominance was appreciably greater and more affected by environment than additive variance for all cases, except for LA and RWC under drought where additive was larger than dominance. The degree of dominance "a" was overdominance (a>1.0) in all cases, except for RWC (no dominance, *i.e.* a=0) under drought. Narrow sense heritability estimates were generally higher under non-stress than under water stress and ranged from 12.5 per cent (1000-grain weight) to 50.5 per cent (LA) under non-stress and from 6.4 per cent (RWC) to 35.4 per cent (LA) under water stress (Al-Naggar *et al.*, 2007).

Breeding Approaches for Drought Tolerance

The development and use of crop cultivars adapted to water-stressed conditions is a long-term solution for improving and stabilizing crop productivity. Sorghum will likely become much more important in arid and semi-arid regions of the world as the demand for limited fresh water and global warming trends increases. Drought resistance in sorghum is a complex trait affected by several interacting plant and environmental factors. The growth stage (GS) at which moisture stress occurs is very important in determining the response or reaction of sorghum to water stress. Strategies for genetic enhancement of crop plants for drought tolerance have been widely discussed (Reddy *et al.*, 1980). It has been postulated that the genes for yield and tolerance to stresses are different, at least at some of the loci, and therefore, drought tolerance can be improved without sacrificing substantial yield.

Four basic approaches to the breeding for drought tolerance/resistance have been proposed. The first is to breed for high yields under optimal conditions *i.e.*, to breed for yield potential and then to assume that this will provide a yield advantage under sub optimal conditions. The second is to breed for maximum yield by empirical selection in the field in the target drought-prone environment. The success of this approach depends entirely on how variable the target environment is. It works well in the Indian postrainy season environment, which is very predictable, but not in the rainy season environment, which is highly unpredictable. The third approach is to incorporate the selected physiological and/or morphological mechanisms conferring drought tolerance into traditional breeding programs. The fourth breeding approach involves identifying key trait that confers drought tolerance at specific growth stages and its introgression into the high yielding background. This method involved selecting (through pedigree selection) breeding materials for specific traits such as (1) longer mesocotyl length for emergence under crust, and grain yield under drought prone and yield potential areas for early seedling stage drought, (2) and for grain yield under drought-prone and yield potential areas alternatively for midseason drought; and (3) for stay green and non-lodging and grain yield under drought prone and yield potential areas alternatively for terminal drought. Crosses were made between high yielding adapted lines and lines were selected for high yields under drought and/or with one or more drought related traits. Selections from F_2 onwards were made by evaluating the segregating material in alternate generations under specified drought (early, mid-season and terminal stage) and in yield potential environments. The F_5/F_6 pure lines are evaluated for drought yields, potential yields, and for specific drought related traits. Testing for yield under mild stress was adequate as the rankings of genotypes for potential and drought yields were similar, since the drought tolerant lines selected under mild stress had high yield potential in non-stress environments. However, breeding for drought tolerance has been slow due to G×E interaction effects (Sajjanar *et al.*, 2011).

Breeding for Drought Escape

In regions where end-of-season drought stress is common such as those in Indian peninsula in rainy and postrainy season environments, the most effective way to reduce losses due to drought is through the development of early maturing genotypes to enable them escape end-of season drought. The early maturing improved sorghum cultivars such as CSH 1 (100 days and 4 t ha^{-1}), CSH 6 (95 days and 3.2 t ha^{-1}) and NK 300 (88 days and 4 t ha^{-1}) produced better grain yields than long duration cultivars such as M 35-1 (105 days 1.9 t ha^{-1}) and SPV 86 (108 days and 3 t ha^{-1}) (Seetharama *et al.*, 1982). Selection for improved productivity under conditions of moisture-stress resulted in a genetic shift towards early flowering. Most of these studies have also confirmed the positive association between the long growth duration and yield potential, under potential conditions. The reduced yield potential in early genotypes may be compensated to some extent by increasing plant density. Under terminal water stress during the post-rainy season, short duration sorghum genotypes produce equal grain but less dry matter than late cultivars. Moisture stress is known to cause differing degrees of reduction in grain yield depending on the stage of the crop and frequency, duration and severity of moisture stress. While, sandy soil or shallow soil sites are best suited for pre-flowering field evaluation of stress response, heavier and deeper soils are best for evaluating post-flowering drought stress.

Heterosis Breeding

Rao and Khanna (1999) reported the superiority of sorghum hybrids over their parents for leaf area and dry matter production under both pre-flowering and post-flowering drought stress. Potential for hybrid breeding for drought tolerance in *Rabi* sorghum was reported by Patil *et al.* (2013). The increased performance of hybrids than their parents is due to greater growth rates and greater total biomass production and higher harvest index with or without an apparent increase in leaf photosynthetic rates (Sinha and Khanna, 1975). Bhale *et al.* (1982) found that some sorghum hybrids showed heterosis for proline accumulation (known to confer drought tolerance) under moisture stress. Evidences suggest that wider adaptability of hybrids is due to their relative tolerance to wide range of abiotic stresses including soil moisture-stress and related factors than varieties. The improvement of per se performance and combining ability of parents for agronomic traits and grain yield under drought stress should be given strategic importance, considering that parental per se performance and general combining ability in sorghum is strongly correlated with hybrid performance (Bhavsar and Borikar 2002).

Drought Tolerant Sources

At Muzaffarnagar, India, the lines B35, CO21, CO22, AS5078, K3, Murungapatti local, VS1564, VS1560, AS6616, AS8038, Tenkasil, MS7819, AS2059, AS8021, AS4289, CO24, AS2752, CO1 were found to be promising for drought (Vinodhana *et al.*, 2009). Phule Chitra (SPV 1546), a *Rabi* sorghum variety, identified as drought tolerant on medium type of soil performed well by maintaining leaf area, leaf relative water content, biomass at harvest, test weight, earhead exertion, leaf temperature, chlorophyll stability index and stomatal frequency (Nirmal and Patil, 2008). With regard to grain yield, the genotypes SPV 913, SPV 504, Nijamabad local, M 35-1 and IS 4565 performed well under receding soil moisture conditions (Channappagoudar *et al.*, 2007). The lines PEC 17, EP 57, EP 87, SLB 9 and RSLG 262 were identified as tolerant to drought in terms of higher grain yield, biomass and green leaf area retention (Talwar *et al.*, 2010). The genotypes, B 35, IS 22212, IS22335, IS 22697, IS 29323, IS 22243, IS 23418, IS 22794, IS 21756 and IS 22339 exhibited drought tolerance phenomenon and were grouped as drought tolerant under molecular level of genetic diversity (Rajarajan and Ganesamurthy, 2011).

Parental Line Development

In India, most of the sorghum is grown as a rain fed crop with limited inputs during both *Kharif* and *Rabi* seasons. In the Indo-Gangetic plains, there is some area under multi-cut forage sorghum that is irrigated. Sorghum is one of the crops we have been able to exploit heterosis. Release of 19 *Kharif*, 6 *Rabi*, two multi-cut forage and one sweet sorghum hybrids at national level and several at state level are a standing testimony to the success of Indian sorghum improvement programme. Much is achieved for *Kharif*, but a great deal is yet to be done to exploit heterosis in breeding hybrids for *Rabi* cultivation as well as for forage and sweet sorghum. Though outstanding and significant progress has been achieved in case of hybrid breeding, much needs to be done. Developing superior hybrids endowed with tolerance of different yield limiting biotic factors and wide adaptability is important to make sorghum farming a profitable enterprise.

Parental line development also has resulted in the development and release of improved genotypes as open pollinated varieties (OPVs). Many of the agronomically superior derivatives isolated from segregating populations of restorer × restorer crosses often fail to act as perfect restorers, but can yield satisfactorily paving the way for release as OPVs. Different AICSIP centres have been assigned the task of parental line and hybrid development with different objectives like earliness, shoot fly tolerance (Akola), grain mold, shoot fly and aphid tolerance (Parbhani), shoot fly and cold tolerance (Rahuri), charcoal rot and terminal drought tolerance (Bijapur), etc. Elite cultures which could not be promoted for All India release often gets released at State level by the respective State Agricultural Universities (SAUs). Table 8 summarizes the releases of improved cultivars by different stakeholders between 1964 and 2010. DSR has released 50 improved cultivars nationally for growing in major sorghum states in India between 1964 and 2010. The major sorghum states including their respective SAUs have released around 208 state notified cultivars which have location specific importance during the same period. A total of 258 improved cultivars have been released through NARS in India for sorghum crop improvement. The releases were at their peak during 1990s (76) followed by 1980s (59). In general, the no. of releases in the last decade has declined (51) when compared to earlier two decades (Kumara Charyulu *et al.*, 2011).

Table 8: Improved Cultivars Released by different Stakeholders (Kumara Charyulu *et al.*, 2011)

Released Period	*ICAR**	*Other Notified Varieties***	*Total*
1961-70	5	7	12
1971-80	12	48	60
1981-90	10	49	59
1991-00	12	64	76
2001-05	7	22	29
2006-10	4	18	22
Total	**50**	**208**	**258**

*: Only hybrids and varieties

**: Includes state releases and SAUs

In spite of release of a number of hybrids and location specific varieties over the period, the amount of genetic diversity in the available material under different types of sorghum like grain, forage and sweet sorghum has been reviewed and it was felt necessary to fast forward the breeding efforts in development of hybrid parents. It was also felt necessary that there should be more sharing of genetic material between DSR and AICSIP. With this background the activity of parental line development has been taken up with the main objective of diversifying the genetic base of parental lines for different types of sorghum. The priorities for diversification vary under different sorghum farming systems. For example, under *Rabi* sorghum, apart from grain and stover yields, quality of both grain and stover are important in addition to shoot fly resistance and tolerance to terminal drought. Under *Kharif* sorghum improving the biotic stress tolerance, and genotypes for alternate uses like poultry and animal feed, ethanol and food industry are the priorities. Similarly, priorities under sweet stalk sorghum are high brix and stalk yields apart from stem borer resistance, photo and thermo insensitivity. In case of forages, the emphasis is to enhance the green fodder yield with superior quality, low HCN and higher digestibility, and also develop female parents with forage traits.

New Parental Lines Developed in the Last 5 Years

At DSR several male sterile and restorer lines with superior grain yield, high brix, earliness, etc. adapted to *Kharif* and *Rabi* seasons have been identified. Nine CMS lines with superior grain yield, five superior for grain quality and five for brix content were registered with the NBPGR. The B lines like 27B, 422B and 356B were identified as best combiners for starch content. At CRS, Solapur 25 B lines and 24 R lines have been developed. Breeding for grain mold resistance has resulted in identification of superior lines with considerable resistance levels. Over the years, F 86 from B population was the best fertile for grain mold score with 3.9 GMS as compared to 4.0 of B58586 (resistant check) and 8.6 of 296B (susceptible check). F 23 was the best line from R population for grain mold score with 3.0 GMS as compared to 3.4 of B58586 and 9.0 of 296B. Two shoot fly resistant genotypes, NRCSFR 06-1 and NRCSFR 06-2 were developed with resistance on a par with the resistant check, IS 2312.

Among the AICSIP centres, at Akola, seven grain mold tolerant genetic stocks have been registered with NBPGR. A total of 22 CMS lines and 59 R lines were developed and at various stages of evaluation. At Bijapur, 10 new B lines on milo cytoplasm with near Maldandi grain features have been developed and in advanced stages of conversion. Ten restorers on milo source were developed. Five new A/B lines on Maldandi source of cytoplasm were developed and two new restorers on Maldandi have been developed. At Indore, four promising restorers, one A1 CMS line and two A2 CMS lines were developed and identified. At Rahuri, three *Rabi* sorghum CMS lines (185A, RMS 2010A, RMS 2016A) have been developed and 14 genetic stocks were registered with NBPGR.

Kharif Parental Line Development

Parents of released hybrids, elite parental lines available with DSR and AICSIP centres, selected parental lines from ICRISAT, released varieties, and germplasm lines from world collection formed the base material for this activity of hybridizing the traits of interest (Table 9). Nearly 900 B × B and R × R crosses were made at DSR during 2007-10 and the F_1s were raised. The F_2 segregating generations were shared with different AICSIP centres for advancement and location and trait-specific selection. A part of the selected progenies were received back at DSR for advancement and further selection, and the selected genotypes in the advanced generation were screened by the Entomologists/Plant Pathologists in the respective nurseries. The selections were also sent back to the respective AICSIP centres for further selection. The identified superior breeding lines are being evaluated for phenotypic superiority and male fertility maintenance/restoration capacity and subsequently will be used in the development of hybrids based on A-B-R system.

Table 9: Genotypes Ysed in the Hybridization Programme

Sl.No.	*Target Trait*	*No. of Lines Used*	
		B	*R*
1.	Shoot fly resistance	10	18
2.	Stem borer resistance	7	8
3.	Head bug resistance	2	–
4.	Grain mold tolerance	1	12
5.	Ergot resistance	3	–
6.	Leaf diseases (anthracnose, rust, ZLS)	11	6
7.	Sweet stalk	22	7
8.	Improving grain quality	–	14
9.	Agronomic superiority	91	23
10.	Others (forage, BMR, bold seed, earliness etc.)	–	19

Table 10: Centre-wise Distribution of F_4 Generation

Sl.No.	*Centre*	*F_3 Selections Received*	*Traits*	*Selections made and Sent Back (F_4)*
1.	Udaipur	127	SF, LD, Agro	437
2.	Indore	77	SF, GM, Agro	291
3.	Deesa	64	SF, Agro	184
4.	Coimbatore	105	SF, GM, LD, Agro	382
5.	Hisar	58	SF, SB	190
6.	Parbhani	67	SF, GM, Agro	193
7.	Palem	19	SF, GM, LD, Agro	90
8.	Pantnagar	188	LD	475
9.	Dharwad	173	SF, SB, GM, Agro	165
10.	Akola	52	SF, GM, Agro	65
11.	Ludhiana	80	LD	54
	Total	**1010**		**2526**

EAR: Earliness; SF: Shoot fly resistance; SB: Stem borer resistance; GM: Grain mold resistance; ERG: Ergot resistance; LD: Leaf diseases resistance; AGR: Agronomic superiority; GQ: Grain quality; SS: Sweet stalk.

From the F_1s raised during *Rabi* 2007-08 at DSR, 94 F_2s were distributed to 12 AICSIP centres for advancement and selection. A total of 1010 F_2 selections were received back from 11 centres, which were raised (F_3) at Hyderabad and Warangal, and TNAU, Coimbatore during *Rabi* and summer seasons of 2008-09, respectively. More than 2500 single plants were selected based on phenotypic superiority from these progenies and a part of the selections (F_4 seeds) were sent back to the concerned centres (Table 10) and another part is shared with Entomologists and Plant Pathologists for screening in respective nurseries.

The F_2 seeds of more than 500 restorer × restorer and maintainer × maintainer F1 crosses raised during *Kharif* and *Rabi* seasons of 2008-09 were distributed to 15 AICSIP centres for advancement during *Kharif* 2009 and location specific selection (Table 11). A part of the selections was also planted at DSR for advancement and selection based on phenotypic superiority.

Table 11: Distribution of F2 Generations to AICSIP Centres (*Kharif* 2009)

Sl.No.	*Centre*	*Trait-wise No. of Crosses*									*Total*
		EAR	*SF*	*SB*	*GM*	*ERG*	*LD*	*AGR*	*GQ*	*SS*	
1.	Udaipur	5	15				8	12			**40**
2.	Akola		12	4	8			11	5	5	**45**
3.	Dharwad		10		9		5	11	2		**37**
4.	Coimbatore	5	15		9			11		5	**45**
5.	Palem	2	10	4	9			11		6	**42**
6.	Surat		15		4			11		10	**40**
7.	Indore		15	4	5	2		11		3	**40**
8.	Parbhani		9	5	4			10	3	10	**41**
9.	Mauranipur		10		6			8			**24**
10.	Deesa	5	11		4			11		5	**36**
11.	Hisar		9	5				11		9	**34**
12.	Pantnagar						8	11		11	**30**
13.	Ludhiana						8	11		9	**28**
14.	Kovilpatti	3	15					11	3		**32**
15.	Rahuri									20	**20**
	Total	20	146	22	58	2	29	151	13	93	**534**

During *Rabi* 2008-09 a fresh set of about 350 restorer × restorer and maintainer × maintainer crosses were made and these were planted during *Kharif* 2009 and *Rabi* 2009-10. Out of these 283 fully fertile hybrids were selected and harvested for advancement. These F2s were distributed to the AICSIP centres during *Kharif* 2010 for further advancement (Table 12) following the same scheme as mentioned above. A part of the segregating generation was raised at DSR and advanced and phenotypic selections were effected at F5 and later generations. Some of these advanced breeding lines are being evaluated for grain yield superiority and shoot fly tolerance, and simultaneously for fertility restoration capacity and combining ability. The performance of selected restorer × restorer cross derivatives during *Kharif* seasons of 2011 and 2012 are given in Table 13. The superior maintainer × maintainer cross derivatives (Table 14) are being used in conversion programme for development of new CMS lines.

Table 12: Distribution of F_2 Generations to AICSIP Centres (*Kharif* 2010)

Sl.No.	*Centre*	*Trait-wise No. of Crosses*											*Total*
		EAR	*SF*	*SB*	*GM*	*LD*	*AGR*	*AGR GQ*	*GQ*	*Rabi*	*FOR*	*SS*	
1.	Udaipur	2	8				4				5		**19**
2.	Akola		8				15	4					**27**
3.	Dharwad				1		4	4	2				**11**
4.	Coimbatore		8				4					4	**16**
5.	Palem	5	14				14	4	2				**39**
6.	Surat						7	4	2				**13**
7.	Indore	1		2			9	4	2				**18**
8.	Parbhani						15	4	2				**21**
9.	Mauranipur						4	4	3				**11**
10.	Deesa			3			4				6		**13**
11.	Hisar			3					4		3		**10**
12.	Pantnagar					1					5	4	**10**
13.	Ludhiana										3	4	**7**
14.	Kovilpatti						4	4					**8**
15.	Phaltan						7					3	**10**
16.	Solapur		7	2	1	2	21			17			**50**
	Total	8	45	10	2	3	112	32	**17**	**17**	22	**15**	**283**

EAR: Earliness; SF: Shoot fly resistance; SB: Stem borer resistance; GM: Grain mold resistance; LD: Leaf diseases resistance; AGR: Agronomic superiority; GQ: Grain quality; SS: Sweet stalk; FOR: Forage; *Rabi*: *Rabi* adaptation.

Rabi Parental Line Development at CRS, Solapur

Combining Ability Studies of Landraces to Identify the Best Combining Landraces for Traits of *Rabi* Adaptation

Twenty hybrids and parents (1 A line and 20 testers- land races) were distributed to Rahuri and Solapur to identify the best combing landraces for the various traits of *Rabi* adaptation. The experiment was planted in RBD with 2 replications. The plot size consisted two rows each of 4.0 m length. Data was recorded on cold tolerance (per cent seed setting), grain yield, stover yield, and panicle length and grain quality. Data were collected for 10 plants from each row for the traits.

Table 13: Performance of Slected Advance Breeding Lines (R × R) for Grain Yield and Related Traits

Genotype	*Pedigree*	*Days to Flower*	*Plant Height (cm)*	*Panicle Length (cm)*	*Grain Yield (kg/ha)*	*TW (g)*
DSR 904	RS 673 × ICSR 7	70.8	197.5	27.8	4223	2.31
DSR 906	RS 673 × ICSR 91027	71.0	200.0	28.3	5308	2.17
DSR 907	RS 673 × ICSR 91027	70.0	213.8	26.1	4913	2.71
DSR 908	RS 673 × ICSR 93001	75.3	313.8	28.5	5220	2.82
DSR 911	NR 37 × ICSR 31	72.0	226.3	29.3	5703	2.93
DSR 926	RS 29 × ICSR 9	72.0	231.3	25.8	4988	2.61
DSR 928	RS 29 × ICSR 31	72.5	191.3	24.9	5220	2.58
DSR 929	RS 29 × ICSR 165	76.0	236.3	25.4	4332	2.65
DSR 942	C43 × ICSR 31	70.8	242.5	25.0	4858	2.75
DSR 944	Indore 12 × ICSR 16	73.0	267.5	24.8	5279	3.22
DSR 951	NR 34 × ICSR 13	80.3	233.8	27.3	5002	2.60
DSR 952	NR 34 × ICSR 13	72.5	236.3	22.8	5056	3.10
DSR 954	NR 34 × ICSR 13	73.5	197.5	27.5	4539	2.87
DSR 957	NR 184 × ICSR 13	73.0	238.8	29.4	4644	2.43
DSR 960	NR 184 × ICSR 13	76.0	191.3	28.9	4494	2.89
DSR 971	CS 3541 × CB 87	74.3	233.8	29.4	6080	2.26
DSR 979	RS 29 × NR 486	68.5	257.5	25.5	4377	2.71
DSR 982	RS 627 × AKR 354	76.3	296.3	24.5	5112	2.96
DSR 988	NR 486 × Indore 12	74.0	213.8	30.6	3674	2.48
DSR 989	NR 486 × Indore 12	76.3	230.0	28.8	3502	2.60
CSV 20	*(Check)*	*69.8*	*243.8*	*27.1*	*3958*	*2.79*
	Mean	*73.2*	*233.0*	*27.0*	*4785*	*2.69*
	C.D. (5 per cent)	*5.1*	*40.9*	–	*1254*	–

Over the locations (Rahuri and Solapur), 4 testers (RSLG-462, 425, 74 and 73) out of 20 were good combiners for grain yield, panicle length and per cent seed setting, showing significant high gca effects for these characters. These 4 lines, if used in crossing programme will give productive segregates. Two land races, RSLG-342 and RSLG -76 were showing significant gca effects for panicle length and seed setting per cent (Table 15). The best specific combiners for grain yield over two locations are given in Table 2. The crosses, 185 A ´RSLG-342, 185 × RSLG-347, 185 × RSLG-16 and 185 A ´ RSLG-76 were showing significant sca effects over two locations.

Table 14: Performance of Selected Advance Breeding Lines (B × B) for Grain Yield and Related Traits

Genotype	*Pedigree*	*Days to Flower*	*Plant Height (cm)*	*Panicle Length (cm)*	*Panicle Weight (g/plant)*	*Grain Yield (g/plant)*	*TW (g)*
DSR 961	219B × 416B	69.0	215.0	29.5	103.0	84.0	2.3
DSR 962	422B × 348B	62.0	150.0	24.0	96.9	77.2	2.6
DSR 963	463B × 418B	72.5	245.0	29.8	116.7	91.2	2.5
DSR 964	463B × 418B	71.5	230.0	28.3	126.9	103.2	2.2
DSR 966	7B × 422B	70.5	265.0	29.0	118.4	94.5	2.6
DSR 967	7B × 422B	77.5	337.5	31.5	115.9	94.7	2.6
DSR 968	296B × 264B	69.0	145.0	31.8	100.2	80.5	1.6
DSR 1180	ICSB 30 × 422B	67.0	117.5	30.5	106.5	80.2	3.1
DSR 1182	ICSB 31 × CK 60B	64.5	167.5	27.5	104.1	74.6	3.0
DSR 1185	ICSB 38 × 7B	71.0	150.0	28.5	97.2	78.8	3.5
DSR 1186	ICSB 38 × NSSB 26	73.0	172.5	32.0	105.0	72.8	2.8
DSR 1187	ICSB 38 × NSSB 26	73.0	162.5	33.5	98.2	73.5	3.3
DSR 1188	ICSB 82 × ICSB 95	69.5	140.0	26.0	98.8	74.3	3.4
DSR 1191	27B × 153B	70.0	160.0	33.0	83.9	74.8	3.0
DSR 1198	463B × 2077B	70.0	182.5	24.0	105.7	77.4	3.0
DSR 1200	IMS 9B × 2077B	74.0	190.0	28.0	107.1	82.8	3.4
DSR 1201	IMS 9B × 2077B	74.5	142.5	39.5	107.7	76.4	2.5
DSR 1207	261B × NSSB 13	73.5	172.5	26.5	105.2	80.5	2.9
DSR 1208	261B × NSSB 13	74.0	182.5	24.0	101.2	71.2	2.4
DSR 1213	1409B × 27B	64.5	165.0	32.5	118.9	86.9	2.9
27B	*(Check)*	*68.5*	*142.5*	*32.5*	*79.2*	*57.3*	*2.5*
	Mean	*72.2*	*197.2*	*27.2*	*98.9*	*72.0*	*2.7*
	C.D. (5 per cent)	*5.1*	*28.4*	*6.7*	*38.4*	*31.8*	*0.7*

Table 15: GCA Effects of *Rabi* Land Races for Various Traits Over Locations

	Grain Yield/Plant	*Panicle Length*	*Per cent Seed Setting*
		Testers	
RSLG -462	16.21**	2.32**	6.72**
RSLG -425	8.23**	1.29*	3.87*
RSLG -74	5.98*	0.99*	2.61*
RSLG -73	13.84**	1.58*	3.86*
		Line	
185 B	1.27	0.46	0.19

Contd...

Table 15–*Contd...*

	SCA effects for grain yield in crosses over locations	
Sl.No	*Cross*	*SCA Effects*
1.	185 A × RSLG-342	23.77**
2.	185A × RSLG-347	34.28**
3.	185A × RSLG-16	19.57**
4.	185 A × RSLG-76	12.32*

Parental Lines Trial Under Team Efforts

A parental line trial consisting of 32 parents contributed from cooperating centres was conducted at Rahuri, Solapur and Bijapur. The following parents were contributed from different centers.

Table 16

Sl.No.	*Centre*	*A/B Lines*	*R Lines*
1	Rahuri	1409 A and B, 104 A and B, RHRB-2, 8, 11 and 22	RR 9826, RR 2212, RR 2145, RR 06-3, RR 06-6, RR 06-1, RSV-1135, 1136, 1137, 1138
2	Bijapur	Milo based-BJMS 1A/B, BJMS 2A/B, BJMS 3A/B, BRJ 356A/B, M312A/B	BRJ 358, JP-1-15, SPV-570, BRJ 62 BRJ 67, BRL 112, BRL 127, BRL148, BRL154
3	Solapur	SL 8A/B, SL 10A/B, SL 11A/B, SL-51A/B, SL-56A/B	SLR-10, SLR-13, SLR-28, SLR-59, SLR-27, RS-585
4	Parbhani	PMS 20A/B	–
5	Akola	AKR-45A/B, AKR 63A/B, AKR 82A/B, 27A/B	AKR 354
6	Hyderabad	49A/B	CSV-14R

Observations were recorded on days to flowering, maturity, plant height, grain yield, synchronization of parents, sterility/fertility under bag, seed set percentage, seed size, seed shape, seed color, luster of grains. The results are presented in Table. Results indicated that there was good synchronization in all the *Rabi* parental lines. Hence, there is no need for staggered sowing of parental lines. The seed set was very good in *Rabi* parents. The yield potential of parental lines ranged from 556 to 2223 kg/ha. The performance of B lines (M31-2B, BRJ 356B, 1409B, SLA 11B) and R lines (RR 06-6, RR 2145, RR 9826, SLR 62 and RS-585) was very good.

Sweet Sorghum Improvement

Sweet sorghum, similar to grain sorghum except for its juice-rich sweet stalk is considered to be a potential bioethanol feedstock, and is expected to meet food, feed, fodder, fuel and fibre demands. Some sweet sorghum lines attain juice yields of about 78 per cent of total plant biomass, containing 15 to 23 per cent soluble fermentable sugar (by comparison, sugarcane has 14–16 per cent) (Reddy *et al.*, 2008). The sugar is composed mainly of sucrose (70–80 per cent), fructose and glucose. Most of the sugars are distributed in the stalk, with about 2 per cent in the leaves and inflorescences (Vietor and Miller, 1990), making the crop particularly amenable to direct fermentable sugar extraction. Besides high sugar accumulation (up to 22 per cent in stalks) and biomass production potential, sweet sorghum is tolerant to abiotic stresses that include drought, water logging, soil salinity and acid toxicity. Further, the stillage from sweet sorghum after the extraction of juice has higher biological value than the bagasse from sugarcane when used as fodder for animals, as it is rich in micronutrients and minerals. Apart from these, stillage contains similar levels of cellulose as sugarcane bagasse, therefore has a good prospect as a raw material for pulp product. The bagasse can also be used to cogenerate power of about 2.5 MW for every hectare of crop. Besides these uses, the whole plant biomass can also be used as a substrate for production of lignocellulosic ethanol, known as second generation biofuel.

R&D Efforts on Sweet Sorghum Breeding

Early R&D efforts were made at Nimbkar Agricultural Research Institute (NARI), Phaltan, Maharashtra to develop improved sweet sorghum cultivars by crossing locals with exotic lines that resulted in the identification of superior types with high cane yield, high brix content and moderate grain yield (Rajavanshi and Nimbkar 1996). At International Crops Research Institute for the Semi-Arid Tropics (ICRISAT), Patancheru, major attempts were made to evaluate and identify useful sweet sorghum high biomass germplasm from world collections (Seetharama *et al.*, 1987). Sweet sorghum research in India is being carried out at Directorate of Sorghum Research (DSR), Hyderabad, and at other All India Coordinated Sorghum Improvement Project (AICSIP) centers in Rahuri, Phaltan, Parbhani, Akola, Coimbatore and other places. DSR has been engaged in sweet sorghum research since 1989 and the sweet sorghum varieties and hybrids bred have the ability to produce extremely high stalk yields of up to 50 t ha^{-1}, with juice Brix reading between 18 per cent and 22 per cent and 1.5 to 2.5 t ha^{-1} grain.

Concerted research efforts during last two decades at DSR and its cooperating centres in different State Agricultural Universities and at ICRISAT have resulted in excellent sweet sorghum varieties for use in ethanol production by the sugar industries/alcohol distilleries and for use as green/dry fodder. Sweet sorghum improvement programs at AICSIP centers have resulted in the identification of several promising sweet sorghum varieties such as SSV 96, GSSV 148, SR 350-3, SSV 74, HES 13, HES 4, SSV 119 and SSV 12611 for TSS per cent and juice yield, GSSV 148 for cane sugar, NSS 104 and HES 4 for green cane yield, juice yield, juice extraction and total sugar content, and RSSV 48 for better alcohol yield.

Promising nationally released varieties/hybrids were SSV 84 (high brix), CSV 19SS (High stalk yield, shoot fly tolerance), CSV 24SS (High stalk and sugar yields) and hybrid CSH 22SS (High stalk and sugar yields). The productivity ranged between 40 and 50 t ha^{-1} (AICSIP 2004, 2005, 2006, 2007 and 2008). The performance of nationally released cultivars (Figure 1) during *Kharif* season across AICSIP locations is given in Table 17. The yields may vary according to the location, date of planting,

Table 17: Performance of Nationally Released Cultivars

Entry	*Year of Release*	*Pedigree*	*Days to Flower*	*Days to Maturity*	*Fresh Stalk Yield (tha^{-1})*	*Brix*	*Juice Yield (L ha^{-1})*	*Cal. Ethanol Yield (L/ha)*
SSV 84	1992	Selection from IS 23568	84	124	35-40	17-18	12000-14000	1000-1100
CSV 19SS	2005	RSSV 2 x SPV 462	78	120	35-40	17-18	12000-14000	1000
CSH 22SS	2005	ICSA 38 x SSV 84	82	119	44-52	17-18	14000-18000	1100-1300
CSV24SS	2011	NSS 1005B x (SSV 84 x 401B)	81	119	35-40	17-18	14000-15000	1100-1200

soil type, season (*Kharif* or *Rabi*), rainfall distribution etc. Stalk yields obtainable during *Rabi* will be 30-35 per cent less with reduced sugar content than *Kharif* and summer grown crops because of prevailing low night temperatures and shorter day lengths and their interaction with planting time.

DSR, ICRISAT and several AICSIP centres have developed a number of sweet sorghum breeding materials, experimental varieties and hybrids having high stalk sugar content and grain yields. In several multi-environment trials in India, cv SSV 84 has consistently yielded an average 37.5 t ha-1 of stalk yield with a stable per cent Brix of 18.6 per cent (Balaravi *et al.*, 1997).

Candidate Traits of Sweet Sorghum

The characteristics which make sweet sorghum a candidate crop for biofuel production are

- ✰ High biomass productivity (45–80 t ha^{-1}).
- ✰ High Brix (soluble sugars) per cent (16–20 per cent).
- ✰ Thick stems and juicy internodes with maintenance of stem juiciness until maturity.
- ✰ Photo- and thermo-insensitivity so that it can be grown throughout the year and fit into diversified cropping systems.
- ✰ Tolerance to shoot-pests and diseases.
- ✰ Good digestibility of residues when used as forage or for lignocellulosic ethanol production.
- ✰ Tolerance to mid-season and terminal drought.
- ✰ High water and nitrogen-use efficiencies.
- ✰ Suitability for specific conversion technologies (*bmr*).
- ✰ Grain yield (3.0 – 5.0 t ha^{-1}).

Sweet Sorghum Breeding

Sorghum is a self-pollinating crop basically and cross pollination is also found (0.6 to 6 per cent). Breeding methods that are used for self pollinated and cross pollinated crops can be used in sweet sorghum breeding to develop pure line varieties and hybrids. The most commonly used programmes in sweet sorghum improvement are short term programmes (pedigree method and backcross) and long term programmes (population improvement methods). Sweet sorghum breeding and improvement leads to the development of hybrids and parents that are useful as first and second generation (lignocellulosic) biofuel feed stock. Efforts are underway to breed and develop sweet sorghum hybrids and varieties with high biomass, rich in soluble sugars, which are highly advantageous and promising for the sugar and biofuel industry. The evaluation of four promising sweet sorghum lines ('Keller', 'BJ 248', 'Wray' and 'NSSH 104' ('CSH 22SS') along with the check 'SSV 84' indicated substantial genotypic differences for extractable juice, total sugar content, fermentation efficiency and alcohol production (Ratnavathi *et al.*, 2003). Genetic improvement should focus on stalk sugar, biomass quantity and quality and general agronomic traits (such as water and nutrient use efficiency) and, in particular, adaptation of sweet sorghum to colder, arid saline, and alkaline conditions. Further improvement in soluble sugar per cent, stalk yield and juice volume should be targeted in sweet sorghum to help improve the benefits to the industry and farmers without effecting grain yield. Improvement of population offers better opportunities for breaking the desired and undesired traits, which consequently provides scope for increased utilization of stress resistance (both biotic and abiotic). Population improvement of sorghum provides long term breeding strategy to develop superior varieties and hybrid parents. While population improvement programs are not the most common in sorghum breeding, they are an important source of genetic variation and improved traits (Rooney and Smith 2000). Brown mid rib (bmr) sorghum mutants have been identified, which give high biomass and have lower lignin in their cell walls compared to normal sorghum. Introgression of *bmr* into high biomass and stay green lines can be done as different bmr mutants are characterized (Vermerris *et al.*, 2007), and most of the bmr mutants resulted in increased yields of fermentable sugars followed enzymatic saccharification (Saballos *et al.*, 2008).

Studies were conducted at DSR, ICRISAT and elsewhere to understand the inheritance of extractable juice, total sugar content, fermentation efficiency and alcohol production in sweet sorghum. The evaluation of four promising sweet sorghum lines [Keller, BJ 248, Wray (varieties) and NSSH 104 (hybrid)] along with the check SSV 84 indicated substantial genotypic differences for extractable juice, total sugar content, fermentation efficiency and alcohol production (Ratnavathi *et al.*, 2003). An analysis of 53 ICRISAT-bred elite hybrids in both the rainy and postrainy seasons showed that the correlation and regression coefficients are significantly high for all the component traits of sugar yield (per cent Brix, stalk yield, juice weight and juice volume). The heritability for plant height, flowering time and one thousand seed weight was high, while calculated traits like stem juiciness, sugar concentration in stems, total sugar, juice glucose, juice fructose and juice sucrose had low heritability. The seasonal variations for correlation, regression and heritability are considerable in magnitude but in the same direction. At DSR, Hyderabad, India, Generation mean analysis of two crosses has showed predominant additive gene action for traits like sucrose and per cent Brix in stalk juice. However, for cane and juice yield, dominance gene action and dominance x dominance gene interaction were of higher magnitude in both the crosses. Since the traits important for high sugar content are having dominance and over-dominance inheritance, utilization of hybrid vigor by developing sweet sorghum hybrids is an attractive option. Also one of the parents with high sucrose content will suffice in getting good hybrids with high sugar and juice yield (AICSIP, 2007).

Current Progress in Breeding Improved Cultivars of Sweet Sorghum

At national level, 23 trial entries (IASSVHT) comprising 16 varieties, 4 hybrids along with 3 checks (CSV 24SS, CSV 19SS and CSH 22 SS) were evaluated at 12 locations during *Kharif* 2012. SPH 1711 with a flowering of 73 days was significantly early and had 8 per cent superiority over the check hybrid (79 days)For total biomass,

SPV 2196 was 11 per cent more yielding than CSV 24SS and 18 per cent significantly superior to CSV 19SS. It was also promising for fresh stalk yield too. For grain yield, none of the test hybrids were significantly superior to the check CSH 22SS. Among the test varieties, SPV 2199 recorded a superiority of 12 per cent over CSV 24SS. With respect to brix content, SPV 2198, SPV 2197, SPV 2135, SPV 2199 and SPV 2206 were promising. For juice yield, the hybrid SPH 1739 and varieties SPV 2196 and SPV 2200 exhibited significant superiority over respective checks. The hybrids SPH 1739 and SPH 1738 and varieties (SPV No's) 2196, 2200, 2202, 2197, 2074, 2241, 2201, 2205, 2195 and 2206 were promising for sugar yields and calculated ethanol yields (Figure 1).

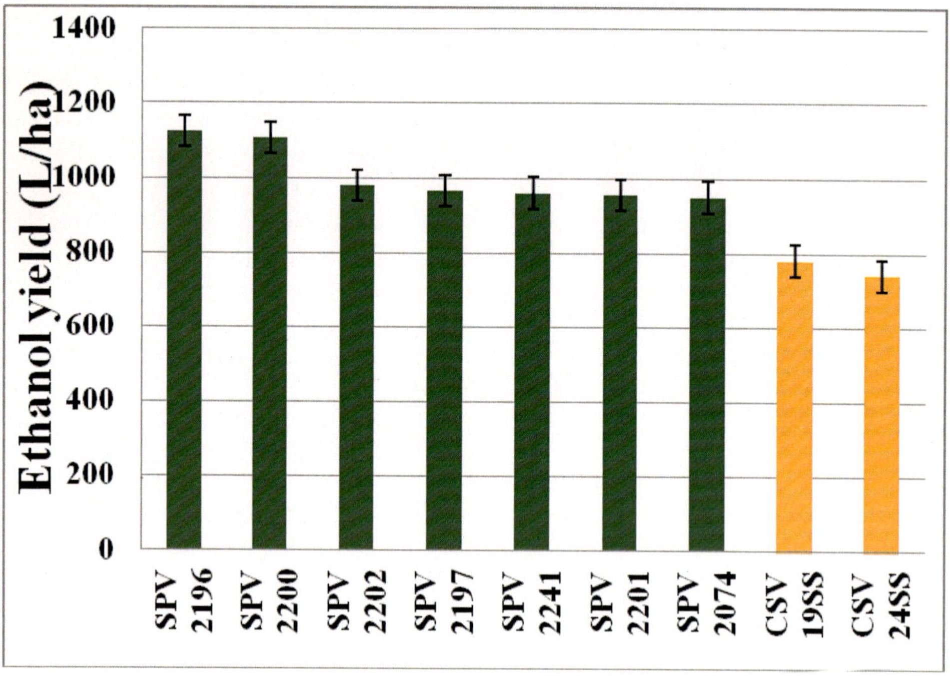

Figure 1: Current Levels of Ethanol Yields in Promising Varieties.

At DSR Hyderabad and its cooperating centres, several hundred experimental hybrids are produced and evaluated every year to identify the promising hybrids with high stalk and sugar yields. Majority of these experimental hybrids had all the desirable features such as optimum maturity period, plant height, green cane yield in the range of 35-50 t/ha with brix between 17-20 per cent and juice extractability ranging from 35-50 per cent.

Hybrid Parents

Female parents are critical for hybrid technology and as many as 35 A/B lines in sweet stalk background are developed in the national program. General combing ability (GCA) studies were made using both the seed parents and pollen parents and promising lines identified for high GCA. The B lines NSSB 20, NSSB 25 and NSSB

Table 18: Promising Initial and Advanced Sweet Sorghum Varieties and Hybrids for Stalk Yield, Biomass, Sugar Content and Bioethanol Yields (*Kharif*, 2012)

Sl.No	*Trait*	*Promising Hybrids and Varieties Superior to Checks*
1.	Time to 50 per cent flowering (d)	Hybrids: SPH 1711 Varieties: SPV 2198, 2199
2.	Time to maturity (d)	Hybrids: SPH 1711 Varieties: SPV 2198, 2199
3.	Plant height (cm)	Hybrids: None Varieties: SPV 2195, 2203
4.	Fresh biomass (t ha^{-1})	Hybrids: None Varieties: SPV 2196, 2074
5.	Fresh stalk yield (t ha^{-1})	Hybrids: None Varieties: SPV 2196, 2195
6.	Grain yield (Kg ha^{-1})	Hybrids: None Varieties: SPV 2199
7.	Juice brix per cent	Hybrids: None Varieties: SPV 2198, 2197
8.	Juice yield (L ha^{-1})	Hybrids: SPH 1739 Varieties: SPV 2196, 2200
9.	Juice extraction (per cent)	Hybrids: None Varieties: SPV 2200, 2201
10.	Total soluble sugars (per cent)	Hybrids: SPH 1738 Varieties: SPV 2135, 2197
11.	Sucrose (per cent)	Hybrids: SPH 1738,1713 Varieties: SPV 2198, 2199
12.	Sugar yield (t ha^{-1})	Hybrids: SPH 1739,1738 Varieties: SPV 2196,2200
13.	Bioethanol yield (L ha^{-1})	Hybrids: SPH 1739,1738 Varieties: SPV 2196,2200

1008 were superior for brix content (>18.5 per cent) while B lines NSSB 7, NSSB 11 and NSSB 1001 recorded highest fresh stalk yield. The parental line improvement programs emphasize on improving/developing sorghum seed parents for sugar content, brown mid-rib trait, resistance to stem borer, downy mildew and other leaf diseases. On the restorer front, the emphasis is on tan plant background (which is supposed to confer leaf disease resistance) apart from high sugar and stalk yields and resistance to stem borer. SSV 74, CSV 19SS were promising restorers with high fresh stalk yield while SSV 84 along with the former two were promising for total sugar content and computed ethanol yields and with good general combining ability. NSSV 258, NSSV 352 and RSCN 4006 were other promising restorers identified at DSR.

ICRISAT, Patancheru, India has also developed a number of sweet sorghum breeding materials, experimental varieties and hybrids having high stalk sugar content and grain yields (Reddy *et al.*, 2005). High sugar yielding varieties/restorers SPV 913 (NTJ2), SPSSV30, SPV 422, Serado, ICSR 93034, S35, ICSV 700, ICSV 93046, ICSV 25263, SP 4487-3, SP 4484-1, SP 4484-3, SP 4482-1, SP 4482-2 and SP 4481-1, and promising hybrid parents ICSB 264, ICSB 293, ICSB 321, ICSB 401, ICSB 405, ICSB 472, ICSB 474, ICSB 722 and ICSB 729 were identified (Srinivasa Rao eta al 2009).

Genetic Studies

At DSR, the inheritance of sugar concentration in stalk (brix), sucrose content, stalk and juice yield in sorghum were investigated and sugar concentration in stalk was found to be controlled by dominant genes and the traits associated with high ethanol production *i.e.* stalk yield, plant height and juice yield were governed by over-dominance. The traits *viz.*, brix, sucrose per cent, stalk yield, juice yield and plant height were found to be governed by dominance gene action (Audilakshmi *et al.*, 2010). In another study, plant height, brix per cent, total soluble solids, total sugar index, total biomass, fresh stalk yield, and juice yield manifested positive significant correlation with ethanol yield and selection could be done for these characters. Path coefficient analysis showed that fresh stalk yield (r=0.83), juice yield (0.57) and brix per cent (0.55) were the main contributors for ethanol yield (Rani *et al.*, 2012).

The predominant role of non-additive gene action for plant height, stem girth, total soluble solids, millable stalk yield and extractable juice yield, substantial magnitude of standard heterosis for all the traits related to ethanol production (stem girth: up to 5.3 per cent, total soluble solids (per cent): up to 7.4 per cent, millable stalk yield: up to 1.5 per cent and extractable juice yield: up to 122.6 per cent) indicates the importance of heterosis breeding for improving these traits (Sankarapandian *et al.*, 1994). The significant positive correlation of general combining ability effects with per se performance of parents in sweet sorghum facilitates quicker identification and development of sugar rich, high biomass yielding hybrid parents (Selvi and Palanisamy, 1990).

Heterosis was significant for plant height, green stalk yield, grain yield and commercial stalk sugar (Meshram *et al.*, 2005). Gene controlling the low sugar content in the stalk is partially dominant (Ma *et al.*, 1992). Magnitude of specific combining ability (SCA) was higher than the general combining ability (GCA) for dry fodder yields and plant height, indicating the predominance of the dominance gene action for these traits (Kadam *et al.*, 2001). G x E influence was present for plant height, stalk weight, reducing sugars and percent of total sugars (Bala Ravi *et al.*, 1997). Inheritance of stalk biomass, brix and stalk weight in sugar stalk was subject to both additive gene effect and non-additive gene effect, but mainly controlled by non-additive genes (Zhou *et al.*, 2005).

The wide range of variability for per cent Brix (from 13 to 24), sucrose per cent (from 7.2"15.5), stalk yield (from 24 to 120 t ha^{-1}), biomass yield (from 36 to 140 t ha^{-1}) and grain yield (from 1.5 to 7.5 t ha^{-1}) in sweet sorghum (Almodares *et al.*, 1997) indicates good scope for selection of sweet sorghum lines with high sweet-stalk yield coupled with high sucrose per cent. The best hybrids for total biomass, fresh stalk yield, juice yield, juice extraction, total sugar content and computed bioethanol yields after adequate testing across many locations were DMS 13A x Rio and DMS 23A x RS 647 and, these hybrids are recommended for commercial exploitation for ethanol production (Umakanth *et al.*, 2012).

In a study on combining ability over 3 locations involving five male sterile lines (NSSA No'S: 8, 10, 1007 and 1016 and RS 1220A) and six testers (SSV 84, M 11, SSV 74, RSSV 76 and RSSV 120) and their resultant 30 hybrids, the ANOVA indicated significant mean sum of squares due to lines, testers and line x tester interaction for almost all the characters under study except brix, juice extraction and total sugar index in case of lines. Greater diversity was found among the testers than the lines for traits of interest like total biomass, fresh stalk yield, juice yield, brix content, total sugar content, computed ethanol yields and grain yield. Gca effects indicated that the line RS 1220A with significant and desirable gca effects was the best general combiner for total biomass, fresh stalk yield, juice yield, brix and computed ethanol yields. Similarly, NSS 8A was good general combiner for fresh stalk yield, brix content, total sugar index and computed ethanol yields. NSS 1007A was promising for Brix and TSS (per cent). Among testers RSSV 9 was the best general combiner with significant and desirable effects for total biomass, fresh stalk yield, juice yield, total sugar index and computed ethanol yields. SSV 84 was best combiner for brix, TSS (per cent) and grain yield while SSV 74 was promising for total biomass, fresh stalk yield, juice yield, juice extraction, computed ethanol and grain yields (AICSIP 2010-11).

From these studies, it is quite evident that the traits are governed by multiple genes and both additive and dominance components of gene action have to be exploited while breeding for high stalk sugar and juice yielding genotypes. Under these conditions, mapping of QTL contributing to the traits will be very useful in selecting the desired genotypes from huge populations and enhance the pace of development of stabilized improved lines.

Breeding Targets

1. High biomass productivity (45-80 t ha^{-1})
2. High Brix per cent (18-20 per cent)
3. Thick stems and juicy internodes with maintenance of stem juiciness until maturity
4. Photo and thermo-insensitivity so that it can be grown throughout the year and fit into diversified cropping systems
5. Tolerance to shoot pests and diseases
6. Good digestibility of residues when used as forage
7. Tolerance to mid-season and terminal drought
8. High water and nitrogen-use efficiencies
9. Suitability for specific conversion technologies (*bmr*)
10. Grain yield (2.0-4.0 t ha^{-1}).

Biomass Maximization Studies in Sweet Sorghum

Sweet sorghum as a feed stock can supplement sugarcane for the supply of raw material to the industry. This necessitates the harvest of sweet sorghum either before or after the harvest of sugarcane. Ratnavathi *et al.* (2005) conducted studies at DSR and its 6 cooperating centres with quarterly plantings round the year for identifying suitable dates for maximum biomass production and juice yield. It was found that the planting date had a significant influence on green cane yield and juice extractability and highest values for these two characters were observed in June planting for *Kharif* crop. In a similar study to identify suitable date of planting for *Rabi* crop, it was found out that fresh stalk yields and juice yields were higher in November sowings compared to December sowings (Annual Report, NRCS 2006-07). With respect to the stage of harvesting for realising higher yields, harvesting at physiological maturity stage was found to be optimum in terms of juice volume, percent extractability and percent total sugars. The cane yield was significantly influenced by the stage of harvest and most of the genotypes tested have shown higher green cane yields at harvest maturity compared to milk stage and physiological maturity (Ratnavathi *et al.*, 2005).

Pilot Studies on Sweet Sorghum for Bioethanol

DSR took the lead in organizing a number of pilot scale studies with many distilleries and stakeholders such as M/s Sri Renuka Sugars, Belgaum, Karnataka, M/s Sagar Sugar, Nelavoy, Chittoor, Andhra Pradesh for ethanol production (Dayakar Rao *et al.*, 2004; Ratnavathi *et al.*, 2005,). The alcohol yield realized in these pilot studies was from 25 to 40 litres/tonne of stalks crushed. DSR also organized pilot studies successfully in collaboration with National Sugar Institute, Kanpur (Rao 2005; Shukla *et al.*, 2006), and several sugar and alcohol industries such as India Glycols, Ltd, Kashipur; KCP Sugars, Lakshmipuram; Navbharat Ventures, Samalkot etc. Currently, DSR and ICRISAT are collaborating with M/S Rusni Distilleries, Sangareddy, Andhra Pradesh and Tata Chemicals Limited, Nanded, Maharashtra for promoting sweet sorghum as a biofuel and bioenergy crop.

Looking Ahead

The road blocks for large scale cultivation of sweet sorghum are the ethanol pricing, the limited availability of genotypes suited to different agro-climatic conditions with all resistances, photoperiod sensitivity and non-availability of required quantity of feedstock suited to off-season crushing in sugar industries. There is a need to develop sweet sorghum cultivars that produce high stalk yield per unit time, input, energy and land area in different agro-climatic areas of the country. These cultivars should also be photo-and thermo-insensitive with desired levels of resistance/tolerance to various stresses and should be of different maturities to widen the harvest window which thereby ensures a continuous supply of feed stock to the industry. The predominant role of non-additive gene action for most of the sweet sorghum productivity traits like plant height, stem girth, total soluble solids, millable sweet-stalk yield and extractable juice yield indicates the importance of heterosis breeding for genetic enhancement of sweet sorghum. The wide variability in germplasm and hybrid parents offers bright scope for the development of high stalk yielding sugar-rich varieties and hybrids. The introduction of the *bmr* trait in sweet sorghums would result in a dual-purpose bioenergy crop that address both 1st and 2nd generation biofuel production issues.

Forage Sorghum Improvement

The genus *Sorghum* includes two economically important species *bicolor* and *sudanense*; while *bicolor* is used for food, feed, and forage in India, the other is specifically for forage purpose. Sorghum, sudangrass and sorghum-sudangrass hybrids are very versatile, multipurpose crops. Sorghum is the major supplier of green and dry fodder in India and its role becomes important during the lean period of winter and summer months. About 20 to 60 per cent of dry fodder supply in semi-arid area is dependent on sorghum. It is estimated that sorghum fodder constitutes 20-45 per cent of the total dry weight of feed of dairy animals during normal seasons and up to 60 per cent during the lean summer and winter seasons (Anonymous, 2007). The importance of sorghum is more pronounced in areas where livestock enterprise is one of the important livelihood strategies of people.

Forage sorghum is cultivated in about 3.0 million ha area in India each year (Anonymous, 2007). Forage sorghums are principally cultivated in Punjab, Haryana, Delhi, western and central Uttar Pradesh and adjoining areas of Madhya Pradesh. In these states, it is grown during *Kharif* and summer seasons, either as single-cut (mostly in *Kharif*, as rainfed) or as a multi-cut (summer and *Kharif*) forage crop. Intensive cropping, short growing season, poor growth of perennial grasses during winter, nutritional quality and the need for continuous supply of green fodder created demand for forage sorghum in northern India.

Sorghum is an Ideal a Forage Crop

Sorghum is an ideal forage crop due to its quick growth, high yielding ability (350-400 q/ha and 700-750 q/ha of green fodder from single cut and multi-cut varieties/hybrids, respectively), high dry mater content (25-30 per cent), better quality and its suitability for various forms of utilization like green chop, silage and hay. The stage of maturity is the most important factor influencing the quality and quantity of forage produced. In view of its importance research on improvement of sorghum for forage was initiated in the country during fourth 'Five year plan' under the aegis of AICSIP in order to develop both single and multi-cut varieties.

The major objectives in forage sorghum breeding are to develop varieties both for single cut and multi-cut with high tonnage, better quality, good seed yield and resistance to insect pests and diseases. The cultivar improvement before the seventies was mostly through the straight selection, applied to the available varieties, for local adaptation and a number of varieties were released. T-8B, 5-Tall, No. 9, 30-C and T-4 varieties of sorghum were released in Uttar Pradesh for grain as well as fodder production (Pathak and Singh 1963). In erstwhile state of Punjab, varieties like JS-20, JS 29/1 and JS 263 were developed and released exclusively for fodder purposes. Out of these JS 263 was sweet and juicier. Some of the old forage sorghum varieties like JS 20 had the potential to regenerate and provide two cuts. These varieties were early in flowering, possessed maximum early vigour and high growth rate. But such varieties were highly susceptible to foliar diseases and single tillered with poor seed potential.

History of Forage Sorghum Cultivar Development

Concerted breeding efforts for the improvement of forage sorghum were initiated in 1970 under the All India Coordinated Sorghum Improvement Project (AICSIP) and subsequently many improved varieties were developed. During seventies single cut varieties *viz.* Pusa Chari –1, Haryana Chari 73/53 and SL–44 were bred at IARI, New Delhi, HAU, Hisar and PAU, Ludhiana respectively. In 1977, multi-cut forage sorghum variety SSG 59-3 developed at HAU, was released at national level through pedigree selection of a cross between JS 263 (a sweet forage sorghum variety) x (sudan grass (with all the required multi-cut traits). This variety had all the desirable multi-cut traits such as early vigour, 6-12 synchronous tillers, faster growth, potential to give 4-5 cuts without significant reduction of forage yield in subsequent cuts, very high and best quality fodder with desirable level of resistance against foliar diseases and insect pests. Only drawback of this variety is that it has poor seed yield potential due to which the seed production of this variety is not economical and availability of seed of this variety to the farmers has always been scarce. Subsequently, in the 80s, variety PC 23 (Pusa Chari-23) was released as multi-cut variety of forage sorghum but it was highly susceptible to foliar diseases.

In early eighties three single-cut forage sorghum varieties PC 6 developed at IARI, HC-136 at HAU and UP Chari-1 developed at GBPAU and T, Pantnagar were released for general cultivation on all India basis. The important features of these varieties are high fodder yield, better quality and wide adaptability. UP Chari –1 has low HCN content while HC 136 has low tannin content. The variety HC 136 is also grown as a variety for two cuts as it provides equal quantity of quality fodder up to two cuts. Other single cut varieties, PC 9, RC 1, 2 and UP Chari 2 were also released in that decade. Later on two single cut varieties, *i.e.*, HC-171 and HC 260 developed at HAU, Hisar, were released for the whole country.

In the 1990s, development of multi-cut sorghum varieties and hybrid received lot of emphasis. At this juncture, private sector also came forward to join the efforts. Since then many hybrids like Punjab Sudex Chari, Pro-Agro Chari, MSFH-3, Hara Sona, Safed Moti,) and variety like UPMCH 503 have been released. Two sorghum varieties from Pantnagar *viz.* Pant Chari 3 (in 1989) and Pant Chari 4 (in 1995) are released for cultivation in Uttar Pradesh. MKV Chari-1 was released in Maharashtra in 1994 for Marathwada region (Shinde *et al.*, 1994). In addition to high green and dry fodder yields and wide adaptability, these varieties have high seed production, resistance to foliar diseases and better quality fodder. Co(FS)-29, a multi-cut variety derived from an inter-specific cross, was released by TNAU, Coimbatore during 2001, which has high biomass, tillering and regeneration potential. The multi-cut hybrid developed at GB Pant University of Agriculture and Technology, CSH 20MF has improved fodder yield and quality. Another multi-cut hybrid, CSH 24MF released recently had improvement in *in vitro* dry matter digestibility (IVDMD). The latest varieties possess improvement in terms of resistance to leaf spot diseases, stem borer and seed yield. Potential seed yield improved up to 14 to 15 q/ha under north Indian conditions. These varieties also exhibited higher per day productivity, dry matter digestibility (DMD) and total soluble sugars (TSS) with comparable protein content. Release of dual purpose variety CSV 15 for green forage production in North West

India was aimed at ensuring better seed supply and nutrition to livestock (Anonymous 2007). Several private sector hybrids are also popular for multi-cut forage. The latest list of forage sorghum cultivars released and their important characteristics have been compiled by Aruna *et al.* (2011).

Forage sorghum research has been undertaken several state agricultural universities and ICAR institutes during the past five decades. Haryana Agricultural University, Hisar; Punjab Agricultural University, Ludhiana; GB Pant University of Agriculture and Technology, Pantnagar; Directorate of Sorghum Research, Hyderabad; Indian Agricultural Research Institute, New Delhi; Gujarat Agricultural University, Anand; Navsari Agricultural University, Dantewada; Mahatma Phule Krishi Vishwavidyalay, Rahuri; Dr. Panjabrao Deshmukh Krishi Vidyapeeth, Akola; Indian Grassland and Fodder Research Institute, Jhansi; JNKVV, Jabalpur; Tamil Nadu Agricultural University, Coimbatore; Marathwada Agricultural University, Parbhani and Maharana Pratap University of Agriculture and Technology, Udaipur, among others, have made significant contributions for the advances in the yield potential and quality of forage sorghum in the country. Development and cultivation of multi-cut forage sorghum varieties/hybrids is the only alternative to meet the fodder requirement of ever increasing livestock population in India (Grewal *et al.*, 2005). Among private companies, Mahyco, Ankur seeds, Kanchan, Nuziveedu, Advanta, Krishi Dhan, JK Seeds, etc. are the major players while a large number of other seed companies together account for another major chunk. Private sector exclusively sells hybrids most of which are multi-cuts or yield at least two cuts. These cultivars possess traits introgressed from sudan grass genotypes. Some of the hybrids sold are also the result of a three-way cross made using suitable hybrid parents, mainly to increase the seed yield, without losing the hybrid vigour and multi-cut traits.

AICSIP - Enhancing Forage Sorghum Yield and Quality

AICSIP with the head quarters at Rajendranagar, Hyderabad and research centres located in agricultural universities and other institutions across the country in the major sorghum growing areas, has been constantly addressing the challenges in forage sorghum improvement since its beginning. Facilitating germplasm procurement, distribution and utilization, diversification of genetic base, laying down stringent standards of forage quality and resistance to stresses and collaborating with animal scientists have been the major milestones in augmenting the forage sorghum productivity in the country. Recognizing the major role played by private sector in catering to the huge demand for forage sorghum, licensing and IPR sharing with private entrepreneurs has promoted the synergy which has helped the public hybrids to reach the forage farmers at an accelerated pace. The present bench marks in the AICSIP system for forage sorghum are CSV 21F and CSV 30F (both are identified for release, to be notified) varieties for single-cut, SSG 59-3 variety for multi-cut and CSH 24MF hybrid for multi-cut.

Genetic Variability for Forage Yield, Quality and Associated Traits

Success in crop improvement depends largely on the extent of desirable genetic variability available for selection in crop improvement programs. Exhaustive studies in forage sorghum germplasm were undertaken by Mathur *et al.* (1991, 1992), Grewal *et al.* (1996), Yadav *et al.* (2002) and Vedansh *et al.* (2010). Mathur *et al.* (1991, 1992) evaluated 1500 accessions of ICRISAT gene bank for forage yield and its components in three locations (Delhi, Jhansi and Akola) during *Kharif* season of the year 1986. They noticed wide range of variability for traits related to forage yield resistance to pests and diseases. They grouped the germplasm into discrete groups based on the pattern of variability and scope for utilization in breeding. The results of these studies were useful for evolving breeding programmes depending on their specific objectives. Genetic variations detected using RAPD and ISSR markers among the sorghum accessions could be of much use for the introgression of new characters from wild counterparts to the cultivars, isolation of stable segregating markers and selection of improved varieties and conservation of germplasm resources (Chakraborty *et al.*, 2011). Recently, Jain and Patel (2012) evaluated 102 land races of forage sorghum and identified certain land races which may be included in the forage sorghum breeding to enhance forage-contributing traits, besides adaptability. High heritability coupled with genetic advance, in diverse material indicated that selection criteria based on leaf: stem ratio, stem girth, and total soluble solids may be useful for further developing forage sorghum cultivars with superior quality (Vedansh *et al.*, 2010).

Several morphological and biochemical traits have been determined for selection to improve adaptability, yield and quality of fodder as well as grain, and resistance to stresses, and their genetic behaviour is largely worked out. The effective sources of the desired characters for full expression in the suitable genetic background are also being identified. A considerable amount of genetic variability and heterosis exists for various forage yield and quality characters. Promising sorghum lines with good combining ability for different agronomic characters have been identified (Paroda and Lodhi 1981). Plant height, leaf number, and leaf size are the component traits of dry-matter yield. Protein content has a high direct effect on the digestibility of fodder, while tannin content is negatively associated with digestibility. The genetics of forage yield, its quality characters, and resistance to major insect pests and foliar diseases have been investigated (Lodhi and Dangi, 1981; Grewal *et al.*, 1987; Het Ram and Lodhi, 1992). Several findings of practical importance to forage sorghum management have been obtained, and can be used to increase forage sorghum production. These relate to sowing time, seed rate, fertilizer requirement, weed management, insect pests, and foliar diseases, management of hydrocyanic acid (HCN), harvesting, mixed cropping, and crop rotations (Lodhi and Grewal, 1988). Pahuja *et al.* (2013) tried to improve SSG 59-3 through mutagenesis and identified mutant lines with improved fodder quality (low HCN, high protein and IVDMD) which can be used in the breeding programme for the improvement in multi-cut forage sorghum varieties.

Since genetic improvement by selection is essentially based on selection for yield-contributing traits, several studies were conducted to identify such related traits and the magnitude of such association. Bangarwa *et al.* (1989) reported that plant height and leaf breadth had positive association with dry fodder yield. Selection based on higher number of broad leaves with taller plants is suggested for selection for higher dry matter yielding plants in forage sorghum (Bangarwa *et al.*, 1989). Warkad *et al.* (2008) concluded that dry fodder as economic yield could be increased by improving plant height, days to flowering and maturity. Plant height had a direct effect on dry fodder yield (Jain *et al.*, 2009).

Genetic Control and Inheritance of Traits Associated with Forage Yield and Quality

Enhancement of forage yield potential and quality of the forage for livestock consumption may be made by improving the traits associated with yield and quality. This needs a prior knowledge of genetic control of these traits and therefore several researchers have attempted to understand the genetic basis of such traits.

Seedling traits such as early vigour, resistance to insects (shoot fly) and diseases, growth related traits such as plant height, days to flower, number of tillers, tolerance to stem borer, shoot bug, foliar and viral diseases, regeneration potential, response to fertilizers, tolerance to erratic weather conditions and ability to recover from stress, etc. contribute to realization of forage yield potential. Total leaf area (a function of number of leaves, leaf length and leaf width) and stem form the actual quantifiable fodder yield. Ratios of leaf and stem weight, protein and fibre content, anti-nutritional factors, make up the quality part.

Easily measurable phenotypic traits such as plant height, days to flowering, stem girth, number of tillers, etc. have a direct bearing on fodder yield as they all form the part of biomass that forms fodder. Most of the traits mentioned above have a sizable environmental influence besides the genetic effects. Therefore, genetic control of the constituent traits have been studied by generation mean analysis (Grewal 1988; Khatri *et al.*, 2001a, 2001b; Pahuja *et al.*, 2003) and also derived from diallel (Ahluwalia and Rao 1980) and line x tester (Pandey and Shrotria, 2009) analyses.

Pahuja *et al.* (2003) reported preponderance of additive gene effects for most of the yield-contributing traits. Additive component was significant for plant height, number of tillers, leaf length and leaf breadth, whereas additive and dominance effects were significant for number of leaves and fodder yield. The type and magnitude of epistatic gene effects were not consistent across crosses. For further improvement in these traits, the authors suggested that simple pedigree selection based on progeny performance may be followed for additive effect controlled ones, and S_2 and reciprocal recurrent selection breeding to be adopted for traits controlled by additive as well as dominance gene effects.

Genetics of the traits that make up leaf yield were studied by Khatri *et al.* (2001b). Both additive (d) and dominance (h) gene effects were observed to be significant with preponderance of former for leaf breadth, whereas for leaf length and number of leaves non-additive gene effects were found to be important. Duplicate type of epistasis was found to be responsible for the inheritance of leaf number and leaf length in some crosses.

Both additive and dominance component of variation were important for regrowth and high heritability for narrow sense was also observed for regrowth in sorghum (Khatri *et al.*, 2001a). Owing to the non-additive gene effects as evinced by higher specific combining ability variance for dhurrin (HCN) content, Pandey and Shrotria (2009) suggested heterosis breeding, recurrent selection and other population improvement procedures to develop cultivars with lower dhurrin toxicity.

Grewal (1988) carried out generation mean analysis in ten crosses between two resistant and two susceptible parents to find the genetic basic of resistance to zonate leaf spot disease in forage sorghum. While both additive and dominance gene effects were significant, duplicate type epistasis was observed for the inheritance of this disease. Overdominance was demonstrated for resistance. Therefore, pedigree and backcross breeding were suggested to exploit the gene effects for disease resistance.

Protein content and digestibility of fodder sorghums exhibited preponderant additive gene action in their inheritance in addition to presence of non-additive genetic variance (Ahluwalia and Rao, 1980). Reciprocal recurrent selection after an initial straight selection for the quality characters was recommended for improving fodder quality in sorghum (Ahluwalia and Rao, 1980).

Boora (2003) identified SSR and RAPD markers that segregated with the gene for resistance to anthracnose, leaf blight and oval leaf spot in sorghum, using bulk segregant analysis. Mittal and Boora (2005) found resistance to leaf blight to be segregating as a single dominant trait. They found an SSR marker Xtxp 309 to be closely linked to the loci governing response to leaf blight and suggested Xtxp 309 to be useful marker for MAS and gene pyramiding.

Pandey (2010) developed low HCN transgenic sorghum by antisense approach with reduced HCN content. Stable T_3 plants exhibited HCN as low as 7 ppm with an overall mean of 70 ppm compared to mean HCN of 221 ppm in the control. The transgene can be introgressed into any superior forage genotype by backcross breeding to reduce HCN toxicity.

Development and Identification Parents for Hybrid Development

For successful and economic hybrid production, presence of male sterility, knowledge of genetic architecture, heterosis for various characters, good combining restorers for different economic traits are necessary. Commercial exploitation of heterosis in sorghum became possible after the discovery of cytoplasmic-genetic male sterility. Milo cytoplasm based grain sorghum parents were used as female parents while sorghum genotypes with forage traits and sudan grass introgression lines were used as male parents. The wide range of heterosis, the good combining and stable parents for forage yield and quality traits led to development of many forage sorghum hybrids (Grewal *et al.*, 2005).

In order to identify useful hybrid parents, series of studies on combining ability of forage sorghum breeding stocks were conducted by Indian breeders (Sanghi 1982; Parmar *et al.*, 2004; Parmar and Tikka 2005; Sumalini *et al.*, 2005; Agarwal *et al.*, 2005; Chand *et al.*, 2005; Iyanar and Khan 2005; Yadav and Pahuja 2007; Mohan *et al.*, 2007a and b; Singh and Shrotria 2008; Singh *et al.*, 2010; Bhatt and Baskheti 2011) through line x tester and diallel analyses. Traits such as days to 50 per cent flowering, plant height, leaf breadth and were observed to be more influenced additive gene effects while number of leaves, number of tillers and fodder yield were found to be controlled by both additive and non-additive gene effects (Parmar *et al.*, 2004; Sumalini *et al.*, 2005). This has lots of implications for trait based breeding. Pedigree selection based breeding can improve the traits governed by additive effects. Presence of large non-additive effect reassures the possibility of exploiting heterotic potential by the selection and combining of right parents for hybrid development. Singh and Shrotria (2008) reported that even traits such as leaf area, total soluble solids, crude protein content, dry matter digestibility and hydrocyanic acid content were predominantly influenced by non-additive genetic effects.

The genotypes that have gone into developing forage sorghum hybrids include (i) improved breeding lines developed and/or shared among the AICSIP partners (ii) parental lines developed and shared by from ICRISAT, especially female lines, and (iii) good germplasm lines or selections from them. Some of the good general combiners identified for fodder yield were Pusa Chari-6, and Pusa Chari-1 (Sanghi 1982), 2077A, 36A and M31-2A, C10-2, Vidisha, Rio, PC6, IS685, IS3287 and IS4776 (Pathak and Sanghi 1992), TNAU M 1A (Manickam and Das 1995), 18-3A and Ruchira (Sumalini *et al.*, 2005), S-250 (Chand *et al.*, 2005), 2077A and 111A (Iyanar and Khan 2005), ICS 13A, ICS 10286A, S 241 (Yadav and Pahuja 2007), Pant Chari-5 and UTMC- 532 (Singh and Shrotria 2008), HB 94004 A, SPA2 94012, 993100A, UTMC 532, Pant Chari 6 and PMC 03 1212 R (Singh *et al.*, 2010), 2219A, ICSA 363, M-20, Pant Chari-5 and UPMC-504 (Bhatt and Baskheti 2011) and SSG 59-3 (Sanghi 1982; Manickam and Das 1995; Yadav and Pahuja 2007). ICS 4A, ICS 79A and IS 3289 were good combiners for multi-cut traits (Mohan *et al.*, 2007a and b). Several promising heterotic combination with higher specific combining effects were also identified in the research mentioned above. To explore the possibility of using alternative CMS (non-milo cytoplasm) sources, Aruna *et al.* (2012) studied the effect of cytoplasm on forage yield and quality in sorghum. The CMS lines possessing A_3 cytoplasm were good combiners for important fodder yield and quality traits, indicating that A_3 cytoplasm based female lines can also be used as hybrid parents, diversifying the male sterility source of forage sorghum hybrids.

Way Forward

Forage sorghum improvement programs are beset with lack of information on variability and useful genetic stocks for various traits. Therefore, concerted and planned efforts are needed to collect, evaluate, catalog, and maintain germplasm exclusively for forage sorghum.

Major areas of improvement required for forage sorghum production in India include drought tolerant high biomass single-cut types and high seed yielding multi-cut hybrid parents. Another important area is to augment the protein content in all types of forages, especially in the cultivars from public sector. The future emphasis should be to increase per day dry matter production in single-cut cultivars. Drought tolerant types that can recover faster are preferred for both single and multi-cut types.

To get a regular supply of green fodder for a longer period of time, multi-cut varieties having profuse tillering, quick regeneration, faster growth and capability of giving a minimum of 4-5 cuttings should be developed. The high yield and multi-cut potential of Sudan grass should be further exploited in breeding programme to develop highly adapted and high yielding stable multi-cut variety of forage sorghum. Attention should be paid to stability in production of biomass and nutrient content through resistance breeding. This can be achieved through tile incorporation of resistance to biotic factors, improved tillering capacity, and quick growth.

The identification and development of male-sterile lines and restorers suitable for forage sorghum hybrid production are also important, as is a detailed study of the relationship between midrib color and fodder quality. Serious efforts are required to develop heterotic hybrid parents for higher forage yields and quality. New trait specific male sterile lines with diverse cytoplasms are required to be developed (Pahuja and Yadav 2008; Aruna *et al.*, 2012). At the same time, reasonable seed yields should be assured to benefit the seed farmers. The goal should be surpass the 100 t/ha green fodder yield by the end of the decade and achieve at least 20 per cent more protein and 5 per cent higher digestibility by 2025.

Breeding for Insect-Pest and Disease Resistance in Sorghum

In early 1960s, the first organized sorghum improvement programme was started in India when the Indian Council of Agricultural Research (ICAR) initiated an 'Accelerated hybrid sorghum project' in 1962. The intensive research efforts of this project resulted in the development of the first commercial hybrid, CSH 1, in 1964. Thereafter, the development and release of new sorghum hybrids and varieties marked a genetic breakthrough in the yield levels of sorghum in India. During the 80s and 90s focus was shifted to insect and disease resistance breeding. Initially it was mainly concentrated on the screening of germplasm for insect and disease resistance, identifying traits associated with resistance, inheritance of resistance and development and evaluation of segregating populations. As of now, significant improvements noticed particularly in developing screening techniques for major pest and diseases. New resistance germplasm lines have been identified for shoot fly (*Atherigona soccata*), stem borer (*Chilo partellus*), midge (*Stenodiplosis sorghicola*), aphids (*Schizaphis graminum* and *Melanaphis sacchari*), shoot bug (*Perigrinus maidis*) and for grain mold, charcoal rot and foliar diseases.

Breeding for Insect-Pest Resistance

Shoot Fly

The genetic differences among resistance genotypes were first reported by Ponnaiya (1951). The major task of screening 10000 germplasm and varieties world collections was systematically done at different locations (Singh *et al.*, 1968). Initially the "deadheart" was taken as the parameter for evaluating resistance. This work resulted in identification of number of varieties with comparatively less shoot fly damage. The identified lines were from the *Rabi* growing areas. (Rana *et al.*, 1984). The genetic studies revealed that the shoot fly resistance is very complex trait and quantitative in nature with various mechanisms of resistance (Rana *et al.*, 1975; Agarwal and Abraham, 1985; Singh *et al.*, 2004, Dhillon *et al.*, 2006). Resistance attributing traits like leaf glossiness, trichome density on lower surface of leaves, seedling vigour deters shoot fly from egg laying and confers resistance (Maiti and Gibson, 1983; Sharma *et al.*, 1997; Deshpande *et al.*, 2003; Kamatar and Salimath, 2003; Dhillon *et al.*, 2005; Gomashe *et al.*, 2012). The inheritance studies also revealed that the alleles for shoot fly resistance are contributed by both resistant and susceptible parents and thus requires methodologies to bring together the favourable alleles present in the resistant and susceptible genotypes. Resistance exhibits partial dominance under low infestation but appears to be partially recessive under high infestation (Rana *et al.*, 1981; Borikar and Chopde 1980). Insect populations vary from location to location and season to season, causing varying degrees of damage. In the absence of high levels of resistance, persistence or stability of even a low level of resistance is of considerable value. This can be handled by the application of advanced breeding technologies like marker assisted breeding (Aruna *et al.*, 2011). Identification of varieties with such inherent genetic characteristics is useful for a resistance breeding program. Subsequent screening in the All India Coordinated Sorghum Improvement Project (AICSIP) has identified number of resistant lines. During *Kharif* 2009, 1321 experimental lines (1250 lines from National crossing Block, 26 advanced progenies from MAS programme, 24 shoot fly nursery lines and 21 germplasm accessions were evaluated for shoot fly resistance. About 21 lines selected which showed on a par performance with resistant check IS 2312 (DSR, 2010).

Recently, in one of the novel approaches, field observations have suggested that shoot fly susceptible sorghum varieties emit attractive volatiles. The efforts are under way to identify plant-derived attractants for shoot fly in relation to breeding sorghum varieties less attractive to this pest (Padmaja *et al.*, 2010).

Stem Borer

Sources for stem borer resistance were reported by Trehan and Butani (1949) and Pant *et al.*, 1961. About 3953 germplasm lines from world collection were systematically screened by Sing *et al.* (1968). Plant height, tassel percentage, stem thickness, number of leaves, leaf length, leaf width, leaf thickness, and leaf strength are negatively correlated with deadheart formation (Khurana and Verma 1985). Days to panicle initiation and shoot length are associated with resistance to stem borers. Genotypes with early panicle initiation (IS 12308 and IS 13100) escape deadheart formation due to inability of the larvae to reach the growing point. Faster internode elongation is also associated with borer resistance. Shoot length, moisture content, plant growth rate or seedling vigor, leaf glossiness, and ligular hairs are associated with resistance (Sharma and Nwanze 1997). Antibiosis is the major factor associated with stem borer resistance. A number of biochemical factors such as amino acids, sugars, tannins, phenols, neutral detergent fiber, acid detergent fiber, lignins, and silica content also contribute for resistance mechanism. Information on these factors is very important for any breeding programme. On the basis of general (GCA) and specific combining ability (SCA) estimates revealed the additive type of gene action in governing resistance to spotted stem borer (Sharma *et al.*, 2007).

Sorghum Midge

The sorghum midge, *Contarinia sorghicola* (Coquillett), is probably the most common insect pest of sorghum in most of the sorghum growing countries. Resistance sources have been reported from several countries (Jotwani *et al.*, 1971; Parodi *et al.*, 1974). Midge resistance is polygenically inherited trait and governed by both additive and non-additive gene action. Knowledge about the traits attributing resistance is important for developing screening techniques. Most of these traits are not present in single genotypes so there is need to club them in a common background to increase the resistance levels. Use of random-mating populations appears to be ideal for accomplishing this task. Patil and Thombre (1983) reported that both gca and sca effects were important for midge resistance. They found additive genetic variance greater than non-additive genetic variance. Also the resistance has to be present in both parents of a potential hybrid.

Breeding for Disease Resistance

Grain Mold

Grain mold is one of the most important biotic challenges to address in *Kharif* (rainy season) sorghum growing areas. Overall the sorghum grain per hectare has increased but the total area under cultivation has been declining. There are several reasons for this decrease such as low demand for sorghum, limited commercialization, limited yield gains, and grain mold susceptibility (Rana *et al.*, 1997). The grain molds are caused by complexes of fungi, predominantly by *Fusarium moniliforme, Curvularia lunata, F. semitectum*, and *Phoma sorghina* (Forbes *et al.*, 1992). Breeding grain mould resistant sorghum hybrids and varieties requires identification of useful resistance gene(s) in the germplasm or defining other sources of such genes (Sharma, 1994). Highly significant correlations between measures of grain mould and seed hardness, seed phenol content in acid methanol extract, and glume colour indicated that they strongly affected grain mould response. Harder grain, higher levels of seed phenols, and darker glumes contributed to grain mould resistance. Weaker and less consistent correlations between measures of grain mould and seed colour, seed flavan-4-ol content, glume phenol and flavan-4-ol contents, and glume cover indicated relatively less effect of these traits on grain mould response.

Success in developing grain mould resistance in high yielding genotypes was limited because of the fact that grain mould resistance is governed by multiple mechanisms of resistance (Audilakshmi *et al.*, 1999), and many agronomically undesirable traits are associated with the resistance (Audilakshmi *et al.*, 2000; Aruna and Audilakshmi, 2004). Moreover, resistance should be incorporated in the genotype with acceptable grain quality, which is governed by several major and minor genes showing significant genotype (G) x environment (E) interactions (Audilakshmi *et al.*, 2005b). Due to this breeding efforts in about three decades to develop grain mould resistance in high yielding genotypes have not paid many dividends.The complex genetics of mould resistance is due to the presence of different mechanisms of inheritance from various sources. Evaluation of segregating population for resistance and selection for stable derivatives in advanced generations in different environments was found effective. The grain mould occurring before physiological maturity (PM) is influenced by genetics and to some extent by environment. Efforts were made to assess the host plant resistance at pre-PM stage (Audilakshmi *et al.*, 2005a). The new strategies were developed to breed for grain mould resistance in high yielding backgrounds. Audilakshmi *et al.*, Ambekar *et al.*, 2011 developed superior lines by crossing elite lines and grain mould resistant genetic stocks. These lines were advanced and evaluated across nine locations in India. Improved lines GMN 21, 52, 59 and 63 performed on par with resistant check B 58586 for grain mould resistance. These were selected in early generation at PM in multi-locations to get superior lines. Most of these lines are high yielding and on par with elite check, C43 for grain yield.

Foliar Diseases

In sorghum apart from grain mould, downy mildew, anthracnose and ergot are major diseases during *Kharif* season. Sorghum crop is also infected by several foliar diseases like leaf spots, chlorotic stripe virus and rust (Rust) are prevalent under warm humid conditions and are highly destructive(Das, 2008). These diseases reduce the amount of green leaf area available for photosynthesis and affect the quality of fodder by reducing the protein, zinc and IVDMD (in vitro dry matter digestibility) (Rana *et al.*, 1999). The use of host plant resistance is considered to be more practical and reliable for managing foliar diseases. To develop resistant cultivars, plant breeders require a detailed knowledge of the inheritance of resistance. Resistance can be complex, controlled by a single or several genes depending on the source of resistance, plant development stage and the pathotype used. Recently, genes for anthracnose resistance in sorghumhave been mapped to chromosomes SBI-05 and SBI-08 (Singh *et al.*, 2006; Perumal *et al.*, 2008). Murali Mohan *et al.*, 2010 reported QTLs (quantitative trait loci) conferring resistance to foliar diseases. This information generated will be useful for transferring resistance in to elite susceptible cultivar through marker assisted selection (MAS).

Charcoal Rot

Charcoal rot caused by *Macrophomina phaseolina* Tassi (Goid) is a stress related disease. The disese is prevalent in entire *Rabi* (post rainy) sorghum growing tracts. Post flowering lodging and poor grain filling indicate that the crop is infected by this disease. Studies on inheritance of this disease by Shinde *et al.* (1981) noticed the dominance of susceptibility governed by three major, one basic complementary and two duplicate complementary genes. Reports of partial dominance (Indira and Rane, 1983), non-additive with a high degree of dominance (Indira *et al.*, 1984), polygenic threshold character governed by partial dominance, duplicate epistasis and low heritability among the crosses involving resistant and susceptible parents have been studied. Venkata rao and Shinde (1988) reported polygenic nature of susceptibility, with dominance in the F_1 generation. On further investigations it revealed non-allelic interactions, like additive × dominance and dominance × dominance with major role in an inheritance of this trait. Dominance gene action was observed for resistance with over-dominance and low heritability values for number of nodes crossed and length of stem infected by the pathogen (Garud and Borikar, 1985). Phenolic compounds produced by sorghum plants also attributes resistance against pathogens by inhibiting spread of fungus during the dry period thus making sorghum plant tolerant to pathogen (Anahosur and Naik, 1986).

Genetic Improvement of Sorghum Grain Quality

Sorghum is an important drought tolerant grain cereal majorly consumed as food in Maharashtra, Karnataka and Andhra Pradesh states of India. Among the *Kharif* and *Rabi* sorghum types cultivated in India, the *Rabi* sorghum grains are the ones predominantly utilized for food purpose. The *Rabi* sorghum possess excellent grain quality and are characterized by bold and round grains and have a characteristic lustre (Murty and Subramanian, 1981; Reddy *et al.*, 2012). Among various sorghum food preparations, the unleavened roti and bhakri are the prepared and consumed traditionally. Compared to hybrids, landraces/varieties possess superior quality traits and are more suited for food purposes Dough and *roti* properties of hybrids were much inferior to those of the local cultivars (Rao *et al.*, 1964; Madhava Rao, 1965; Anantharaman, 1968). Viraktamath *et al.*, 1972 reported differences among varieties for culinary properties in sorghum and demonstrated that the traditionally grown cultivars possessed relatively superior culinary properties over recently developed cultivars. During sixties in India, several hybrid combinations superior for yield were developed however they were they were reported to non-acceptable (Rao, *et al.*, 1964). Anantharaman (1968) observed considerable variation for *roti* quality among hybrids with pearly white grains. M 35-1 (*Maldandi*) a sorghum cultivar is known for its good quality of *roti* due to having pearly white grain color, its flour having higher water holding capacity, and good organoleptic taste. However, this cultivar is low yielder. To evolve sorghum high yielding genotype coupled with this good *roti* qualities, systematic breeding program was planned and executed at MPKV, Rahuri. The local land races RSLG 428-1, RSLG 1238, RSLG 1275 and the genotypes RSV 290, RSV 292, RSV 858, RSV 859, RSV 861, RSV 868, RSV 894, RSV 985, RSV 992, RSV 995, RSV 999 were found to be promising for protein, sugar, water absorption, and soluble protein content and were suggested for further improvement in nutritional quality through breeding program. Among the latest cultivars, Phule Vasudha (RSV 423), CSV 22, and Phule Chitra (SPV 1546) were found to be most promising for *roti* quality (Chavan *et al.*, 2009). The bhakri prepared from *Rabi* sorghum is white and more palatable than that of *Kharif* sorghum (Salunkhe, 1984). Apart from roti, several other food preparations like popping are prepared and consumed in traditional sorghum growing states. The popped sorghum grains were found to have much flavor and be as nutritious as popcorn (Subramanian 1956). Popped sorghum grains are used in the preparation of sweet snacks, which are commonly consumed in Maharashtra. It was reported that popping varieties of sorghum belong to the *Talavirchina* group (S. *roxburghii* var *hians*) characterized by small grain with a dense and corneous endosperm (Ayyangar and Ayyer 1936). *Pelalu jonna* belonging to the Snowden species S. *membranaceum* was considered to be good for popping (Reddy, 1957).

The quality of the sorghum grain is differentiated at physical and biochemical level. The physical quality which refers the colour, size, texture and lustre of the grain gives more appealing grain appearance and fetches more market prices to the produce as well as plays a major role in milling. From inheritance studies for grain quality traits Ghorade and Kadam (1985) reported three genes were involved in the inheritance of hilum spot giving a 54: 10 of spotted to spotless condition. Three genes, one basic gene for cream seed colour (GrC) and a pair of complementary genes for white seed colour (Wha Whb), were involved in the inheritance of seed colour. Size of seed was governed by three complementary genes. Inheritance of starchy endosperm was governed by basic pair (EnSta, EnStb) with duplicate complementary action of genes (EnSt2 EnSt2). Grain hardness is governed by four complementary genes. The inheritance of threshability, is governed by basic gene (Grtha and three complementary genes, Grthb Grthe Grtha). The duplicate dominant genes Fr1 and Fr2 govern the inheritance of full fertility. Altogether 23 genes were found for the eight characters studied. The genes for oblong seed shape was linked 42.33 per cent cross over units with one of the complementary gene for grain hardness. The same gene is linked with complementary gene for grain thereshability with 27.10 per cent cross over distance. Genetic analysis also revealed that grain size is governed by dominant genes that are polygenic in nature. Predominance of dominance and epistatic interactions indicates that selection for higher grain size would be more effective if *the dominance and epistatic effects are first reduced by a few generations of selfing* (Audilakshmi *et al.*, 2005). Round grain shape is governed by a single dominant gene and grain lustre by two complementary recessive genes. It suggests that developing a sorghum hybrid with bold, round grain is feasible provided either of the parents has bold and/or round grain. However, for the hybrid to be lustrous, both parents need to be lustrous and homozygous for the alleles conferring grain lustre at a common locus. Majority of the food items especially in India are prepared with white/yellow white pericarp grains and are referred to as food grade sorghums. Yellow-white colored pearly grains fetches high market price. Although white or yellow white grain colour is highly preferred for food purposes over colored sorghum. Yellow endosperm textured grains do exist and are consumed in some parts of the Andhra Pradesh, India. Yellow endosperm textured grains are believed to be nutritionally superior, possessing medicinal properties but reports claiming the medicinal properties are poorly documented. The endosperm texture is the major physical grain attribute that determines its end-use (Murty and Subramanian, 1981). Endosperm texture relates to the proportion of the floury to corneous endosperm in the kernel and are broadly classified as vitreous (hard), intermediate and floury (soft) types. Govil and Murty, 1980 studied the genetics of the endosperm characteristics and sorghum and revealed significant role played by both GCA and SCA in controlling expression of those traits. Hard endosperm textured grains are generally associated with smaller seed size, contain higher proportion of corneous endosperm (densely packed starch granules), more resistant to milling, contain more protein content but less protein digestibility. Floury endosperms have lower seed weight and thereby have low grain yield potential. Most of the cultivars grown for food purposes during the *Rabi* season are intermediate endosperm textured (Murty and Subramanian, 1981). The endosperm texture is also associated with the quality and recovery of semolina food product. Higher semolina recovery (semi-processed food) yield was observed in hard endosperm types as compared to floury types.

The biochemical property also assesses its quality of sorghum grains as well in determining its end use. With regard to biochemical characteristics of sorghum grain, the starch and protein quality are the two important characters that enhances the value of sorghum as food and/or industrial crop. Several researchers have investigated the digestibility of sorghum starch and sorghum protein for its use as food or feed and the same is discussed below. The biochemical property which refers to the starch, protein, fat, vitamins, minerals, fiber and other proximate parameters have major role on nutritional properties. Sorghum grains are important source of dietary proteins, carbohydrates, minerals, and B group vitamins particularly to the vegetarian diets in India (Salunkhe *et al.*, 1984). Considering nutritional point of view, protein quality is one of the major focuses among the researchers for its improvement. Protein digestibility is an important aspect determining nutritional quality of sorghum. Murty *et al.*, 1967 studied the protein quality in Indian types and compared with world collections. Indian types especially M 35-1 Nandyal, BS 81were reported to contain less protein and lysine (Virupaksha and Shastry, 1968). Subsequently many studies were initiated to improve the protein and lysine content in the crop. Govil and Murthy, 1973 analyzed the nature of gene action using diallel analysis for grain yield and quality. The study revealed both additive and non-additive gene action for protein content, lysine content. Further the study also revealed role of non-allelic interactions for the quality traits. Nanda and Rao, 1975 analyzed the variability and nature of gene action in the inheritance of nutritional attributes like protein, aminoacids, β-carotene and yield attributes, yield per plant, seed size and grain hardness and the association between nutritional and yield characters using diallel mating system involving exotic, Indian and derived lines of sorghum. Adequate variability has been observed by them for essential amino acids and β-carotene and also reported that the protein, lysine, threonine, tryptophan and β-carotenen, grain yield, seed size and grain hardness are predominently additive in inheritance. Again the lysine, methionine and β-carotene were associated with grain yield indicating opportunities for recombining high yields with high nutritional quality. Khanure *et al.*, 1997 studied the variability in grain protein content among *Rabi* sorghum genotypes and observed range between 11.41 to 16.01 per cent and identified genotypes AGP-289, AGP 457, E 36-1, Afzalpur local, SPV 488, bhogapur local, Phul mallige, kempgund, RJ-12, RJ-14, RJ 16 to contain higher crude protein content. Tripathi *et al.*, 1971 reported protein content to be negatively correlated with lysine and threonine per cent of protein and positively correlated with leucine per cent of grain and leucine per cent of protein. Grain hardness was positively correlated with protein percentage, lysine per cent of grain, threonine per cent of grain, leucine per cent of grain and leucine per cent of protein. Among the different groups separated on the basis of physical characters the non-floating grains had a higher mean for protein percentage than the floating grain group. For protein digestibility, sufficient variability among the breeding lines for protein digestibility have been observed. Two improved lines SPV 1775 and SPV 1758 were found to possess high protein digestibility of about 70 per cent (Ganapathy *et al.*, 2010). Up to now no commercial cultivar has been released combining quality protein as that of maize wherein many Quality protein maize cultivars were developed and released with superior protein quality. Combining high protein digestibility with superior yield has been always a challenge among the breeders. Recently owing to the malnutrition problems, much attention has been undertaken to improve the mineral content in cereal grains. In sorghum Hariprasanna *et al.*, 2012 assessed the genetic variation for grain iron (Fe) and zinc (Zn) content across growing regions. The Fe and Zn in grain ranged from 28.9–34.9 mg/kg and 20.4–25.7 mg/kg, respectively among 10 cultivars.

Apart from its major use as food, the crop is finding potential for alternative uses such as livestock and poultry feed, potable alcohol, and starch and ethanol production. In India, the rainy season produce is most commonly used for industrial purposes due to inferior quality grains infected by molds. Mold affected grains are generally not used for food purposes but recent efforts are being made to remove mold affected parts by decortication or pearling techniques and are being tested for its feasibility for food uses. The grain deterioration due to grain mould has been investigated by various earlier workers (Narasimhan and Rangaswamy, 1969; Koteswara Rao and Poornachandrudu, 1971; Ravindranath, 1976). Both saprophytic and pathogenic genera are known to infect the grains and cause deterioration, the more commonly reported fungi being, *Curvularia, Fusarium, Helminthosporium, Phoma and Alternaria*. Rana *et al.*, 1977 made crosses using exotic and Indian lines for studying the genetics of resistance to grain deterioration in sorghum. Tan plant type, grain hardness and water absorption rate were reported useful for breeding resistant types. Further the studies revealed that the hardness of seed and rate of water absorption are predominantly additive and selection for these characters could be effective. Hard × hard crosses within tan types are likely to yield quick results. The male steriles 296A, 2277A and 2947A, the males CSV 4 (CS 3541) and CSV 5 (168) and amongst germplasm materials ESGPG-35, an African *zerazera*, are identified as most promising materials resisting grain deterioration. Kulkarni *et al.*, 1981 reported 'H-142' high yielding and resistant to head mould, both under natural and artificial infestation. The tan plant type progenies were found to be mould resistant and can form criteria for selecting progenies resistant to moulds. Any improvement in the quality of rainy season sorghum grain would make it more competitive in the market and fetch a remunerative price to the farmer. Audilakshmi *et al.*, 1999 examined various grain characteristics and their association with grain mould resistance. Harder grains, higher levels of seed phenols and darker glumes were reported to be associated with grain mould resistance. Highly significant correlations between measures of grain mould and seed hardness, seed phenol content in acid methanol extract, and glume colour indicated that they strongly affected grain mould response. Harder grain, higher levels of seed phenols, and darker glumes contributed to grain mould resistance. Weaker and less consistent correlations between measures of grain mould and seed colour, seed flavan-4-ol content, glume phenol and flavan-4-ol contents, and glume cover indicated relatively less effect of these traits on grain mould response. Genotype means indicated that combinations of several traits are required to achieve resistance. Germplasm lines, including coloured-seeded lines IS 14375, IS 14387, IS 18144, and IS 18528, and white-seeded lines IS 21443, IS 24495 and IS 25017, showed greatest grain mould resistance. Improved lines generally had poorer grain mould resistance than these landraces. However, the best improved lines were comparable in resistance to white-seeded landraces. B58586, IS 14375 and IS 14387 are hard-seeded guinea sorghum lines that can be used as sources of grain mould resistance for West Africa. SP 33316, SP 33349 and GM 15018 are agronomically elite lines that can be used as sources of grain mould resistance for further improvement of white-seeded sorghum for South Asia and other regions. Audilakshmi *et al.*, 2007 explored various approaches to manage grain mould problem in Maharastra, Karnataka, Andhra Pradesh and Tamil Nadu. The study suggested cultivation of good quality high yielding varieties followed by harvesting grains at physiological maturity with artificial drying and storing the solarized produce in metal bins as best package of technology for improving the grain quality of the rainy season produce. The study also identified CSH 16 was identified for its good quality mould tolerant grain. Its bold, round and lustrous grain fetched up to 21 per cent increased market price than those of other cultivars. The major challenges facing future sorghum research and development is to deliver technologies that will enable the agricultural sector to affect transformation of "subsistence farming" to sustainable and market oriented. This can be achieved through reassessment of crop research in terms of current and future demand, resolving specific production constraints by improving productivity, development of post-harvest processing and value-addition technologies, creation of demand, marketing strategies and policies that would result in additional farm income and employment without sacrificing overall goal of attaining sustainable food and nutritional security.

References

1. Agarwal M, Singh R and Shrotria PK. 2005. combining ability analysis for forage yield and quality characters in sorghum [*Sorghum bicolor* (L.). Moench]. *Forage Research* 31 (1): 9-11.
2. Ahluwalia M and Rao V M J. 1980. Fodder quality in sorghum. *Indian Journal of Genetics and Plant Breeding* **40**(3): 554-557.
3. AICSIP (All India Coordinated Sorghum Improvement Project). (2004). Sorghum physiology. All India Coordinated Sorghum Improvement Project Annual Progress Report for 2003"2004. AICSIP Pub. No. 7/Phy/2004, pp. 50. National Research Centre for Sorghum (NRCS), Rajendranagar, Hyderabad, AP, India.
4. AICSIP (All India Coordinated Sorghum Improvement Project). (2005). Physiology and sweet sorghum. All India Coordinated Sorghum Improvement Project Annual Progress Report for 2004"2005. AICSIP Pub. No. 8/Physiology and Sweet Sorghum/2005, pp. 80. National Research Centre for Sorghum (NRCS), Rajendranagar, Hyderabad, AP, India.
5. AICSIP (All India Coordinated Sorghum Improvement Project). (2006). All India Coordinated Sorghum Improvement Project Annual Progress Report for 2005"2006. AICSIP Tech. Pub. No– 3/Sweet Sorghum and Physiology/2006 (Book 3 of 3-agm06), pp. 100. National Research Centre for Sorghum (NRCS), Rajendranagar, Hyderabad, AP, India.
6. AICSIP (All India Coordinated Sorghum Improvement Project). (2007). Sweet sorghum and physiology. All India Coordinated Sorghum Improvement Project Annual Progress Report for 2006-2007. AICSIP Tech. Pub. No. 3/Sweet Sorghum and Physiology/2007 (Book 3 of 3-agm07 pre-meet), pp. 102. National Research Centre for Sorghum, Rajendranagar, Hyderabad, A.P., India.
7. AICSIP (All India Coordinated Sorghum Improvement Project). (2008). Sweet sorghum physiology. All India Coordinated Sorghum Improvement Project Annual Progress Report for 2007-2008. AICSIP Tech. Pb. No. 3/Sweet Sorghum and Physiology/2008 (Book 3 of 3-agm07 pre-meet), pp. 100. National Research Centre for Sorghum, Rajendranagar, Hyderabad, AP, India.
8. AICSIP (All India Coordinated Sorghum Improvement Project). (2010-11). Sweet sorghum report. DSR publication no. 1/2011 pp. 6-8 Directorate of Sorghum Research, Rajendranagar, Hyderabad, AP, India.

9. Ali MA, Jabran K, Awan SI, Abbas A, Ehsanullah, Zulfikar M. Tuba Acet, Farooq J and Rahman A. 2011. Morphophysiological diversity and its implications for improving drought tolerance in grain sorghum at different growth stages. *Australian J Crop Sci* 5(3): 308-317.
10. Ali MA, Niaz S, Abbas A, Sabir W, Jabran K. 2009. Genetic diversity and assessment of drought tolerant sorghum landraces based on morph-physiological traits at different growth stages. *Plant Omics Journal* 2: 214-227.
11. Almodares A, Sepahi A and Shirvani M. 1997. Sweet sorghum cultural practices in Iran. Pages 175"183 in Proceedings of the First International Sweet Sorghum Conference, 14"19 September 1997, Beijing, China (Li Dajue, ed.). Beijing, China: Institute of Botany, Chinese Academy of Sciences.
12. Anahosur, KH, Patil SH 1981. Date of planting and incidence of charcoal rot in sorghum. *Sorghum Newslet.*, 24: 118-119.
13. Anantharaman PV. 1968. Grain quality studies in sorghum. Sorghum Newsletter 11: 49-53.
14. Annual report. 2006-07. National Research Centre for Sorghum, Hyderabad, India. pp. 34-35.
15. Anonymous 2007. NRCS-Perspective Plan-Vision 2025, National Research Centre for Sorghum, Rajendranagar, Hyderabad 500 030, A.P. India. pp. 48.
16. armar H P and Tikka S B S. 2005. Heterosis and combining ability analysis in forage sorghum [*Sorghum bicolor* (L.). Moench]. *Forage Research* 30(4): 209-212.
17. Aruna C, Bhagwat VR, Sharma Vittal, Hussain T, Ghorade RB, Khandalkar HG, Audilakshmi S, Seetharama N (2011). Genotype x environment interactions for shoot fly resistance in sorghum (*Sorghum bicolor* (L.). Moench): Response of recombinant inbred lines. *Crop Protection* 30 : 623-630
18. Aruna C, Bhat B V, Umakanth A V, Shyamprasad G, Das I K and Patil J V. 2011. Forage Sorghum. Directorate of Sorghum Research, Rajendranagar, Hyderabad, India. pp. 52.
19. Aruna C, Shrotria P, Pahuja S, Umakanth A V, Bhat B V, Devender A V and Patil J V. 2012. Fodder yield and quality in forage sorghum- Scope for improvement through diverse male sterile cytoplasms. *Crop and Pasture Science* 63: 1114–1123.
20. Aruna C. and S. Audilakshmi. 2008. A strategy to identify potential germplasm for improving yield attributes using diversity analysis in sorghum. *Plant genetic resources: Characterisation and utilization*. 6: 187-194.
21. Aruna, C, Audilakshmi, S (2004). Genetic architecture of grain hardness-a durable resistance mechanism for grain moulds in sorghum [*Sorghum bicolor* (L.). Moench]. *Indian J. Genet. Plant Breed*. 64, 35-38.
22. Ashok Kumar. A. Reddy B. V. S. and Kaul, S. L. 2008. Alternate cytoplasmic male sterility system in sorghum and their utilization in '*Sorghum Improvement in the New Millenium* edited by Reddy Belum VS, Ramesh S, Ashok Kumar A and Gowda CLL, pp. 132-144.

23. Audilakshmi S and Aruna C (2005). Genetic analysis of physical grain quality characters in sorghum. The J Agric Sci 143: 267-273 Audilakshmi S, Aruna C, Solunke RB, Kamatar MY, Kandalkar HG, Gaikwade P, Ganesh Murthy K, Jayaraj K, Ratnavathi CV, Kannababu N, Indira S, Seetharama N (2007). Approaches to grain quality improvement in rainy season sorghum in India. *Crop Prot* 26: 630–641.

24. Audilakshmi S, Aruna C and Kiran V V S. 2003. Utilization of germplasm for improvement of varieties in India. In: Report of the AICSIP Coordinating Team, 2002–03. AICSIP Annual Meeting of the Sorghum Group, 1–3 April, held at Sorghum Research Station (GAU), Athwa Farm, Surat, Gujarat, pp. 47–53.

25. Audilakshmi S, Mall AK, Swarnalatha M and Seetharama N. 2010. Inheritance of sugar concentration in stalk (brix), sucrose content, stalk and juice yield in sorghum. *Biomass Bioenergy* 34: 813-820.

26. Audilakshmi S, Stenhouse JW, Reddy TP and Prasad MVR. 1999. Grain mold resistance and associated characters of sorghum genotypes. *Euphytica* 107: 91–103.

27. Audilakshmi, S, Aruna, C, Garud, TB, Nayakar, NY, Atale, SB, Veerabadhiran, P, Dayakar Rao, B, Ratnavathi, CV, Indira, S (2005b). A technique to enhance the quality and market value of rainy season sorghum grain. *Crop Prot*. 24, 251-258.

28. Audilakshmi, S, Stenhouse, JW, Reddy, TP (2000). Genetic analysis of grain mould resistance in coloured sorghum genotypes. *Euphytica* 16, 95-103.

29. Audilakshmi, S, Stenhouse, JW, Reddy, TP (2005a). Genetic analysis of grain mold resistance in white seed sorghum genotypes. *Euphytica* 145, 95-101.

30. Audilakshmi, S, Stenhouse, JW, Reddy, TP, Prasad, MVR (1999). Grain mold resistance and associated characters of sorghum genotypes. *Euphytica* 107: 91–103.

31. Audilakshmi, S. and Aruna, C., 2005. Genetic analysis of physical grain quality characters in sorghum. The J. Agric. Sci., 143: 267-273 Ayyangar GNR and Ayyer MAS (1936). Sorghum for popping. *Madras Agric J* 25: 323- 328

32. Bala Ravi S, Biswas PK and Ratanavathi CV. 1997. Genetic variability and stability for major traits of sweet sorghum. In: Li Dajue, editor. Proceedings of First International Sweet Sorghum Conference, Beijing, China; September 14–19, 1997. pp. 298–304.

33. Bangarwa K S, Grewal R P S, Lodhi G P and Thakral N K. 1989. Association analysis for some quantitative traits in forage sorghum. *Indian Journal of Heredity* **21**(1-2): 25-28.

34. Basnayake J, Cooper M, Ludlow MM, Henzell RG and Snell PI. 1995. Inheritance of osmotic adjustment to water stress in three-grain sorghum crosses. *Theoretical and Applied Genetics* 90(5): 675-682.

35. Belum VS Reddy, P Sanjana Reddy, AR Sadananda, Elango Dinakaran, A Ashok Kumar, SP Deshpande, P Srinivasa Rao, HC Sharma, Rajan Sharma, L

Krishnamurthy and JV Patil. 2012. Postrainy season sorghum: Constraints and breeding approaches, *SAT eJournal*.10: 1-12.

36. Bhale NL, Khidse SR and Borikar ST. 1982. Heterosis for proline accumulation in sorghum. *Indian Journal of Genetics and Plant Breeding* 42: 101.
37. Bhatt A and Baskheti D C. 2011. Combining ability studies for green forage yield and its component characters in sorghum (*Sorghum bicolor* L. Moench). through line × tester analysis. G.B. Pant University of Agriculture and Technology, Pantnagar, India. *Pantnagar Journal of Research* 9(1): 57-60.
38. Bhavsar VV and Borikar ST. 2002. Combining ability studies in sorghum involving diverse CMS steriles. *Journal of Maharashtra Agricultural Universities* 27: 35-38.
39. Bhavsar VV and Borikar ST. 2002. Combining ability studies in sorghum involving diverse cytosteriles. *J. Maharashtra Agric. Univ*. 27: 35.
40. Bichkar RP. 2005. Study of combining ability and inheritance of physiological traits and grain yield in *Rabi* sorghum. Thesis abst. *M.Sc.(Agri), thesis* submitted to M.P.K.V, Rahuri,(Maharashtra).
41. Bidinger FR, Mahalakshmi V, Rao GDP 1987. Assessment of drought resistance in pearl millet [*Pennisetum americanum* (L.). Leeke]. I. Factors affecting yield under stress. *Aust J Agric Res* 38: 37-48
42. Blum, A. and Sullivan, C. Y., The comparative drought resistance of land races of sorghum and millet from dry and humid regions. *Ann. Bot.*, 1986, 57, 835–846.
43. Boora K S. 2003. Molecular markers of three foliar diseases of sorghum for marker-assisted breeding. *Forage Research* 29 (1): 44-48.
44. Borikar, ST, Chopde, PR. 1980. Inheritance of shoot fly resistance under three levels of infestation in sorghum. *Maydica* 25: 175-183.
45. Borrell AK, Bidinger FR and Sunitha K. 1999. Staygreen trait associated with yield in recombinant inbred sorghum lines varying in rate of leaf senescence. *International Sorghum Millet News Letter* 40: 31-34.
46. BS Rana, Jaya Mohan Rao V, Tripathi DP and Rao NGP 1977. Genetic analysis of exotic × Indian crosses in sorghum xvii : resistance to grain deterioration. *Ind J Genet* 37 (3): 480 - 487.
47. Chakraborty S, Thakare I, Ravikiran R, Nikam V, Trivedi R, Sasidharan N and Jadeja GC. 2011. Assessment of diversity using RAPD and ISSR markers in Sorghum varieties across Gujarat, India. *Electronic Journal of Plant Breeding* 2(4): 488-493.
48. Chand K, Grewal R P S, Dutt Y and Singh P. 2005. Combining ability analysis for fodder traits in sorghum [*Sorghum bicolor* (L.) Moench]. *Forage Research* 31 (2): 134-135.
49. Chavana UD, Patil JV and Shinde MS 2009. Nutritional and roti quality of sorghum genotypes Indonesian *J Agric Sci* 10(2): 80-87.

50. Chinna BS and Paul PS. 1988. Heterosis and combining ability studies in grain sorghum under irrigated and moisture stress environment. *Crop Improvement* 15: 151-155.
51. Das, IK (2008). Lecture delivered at Model Training Course on "Value-addition and commercialization through alternate uses of sorghum and sweet sorghum for bio-ethanol production" 23-30 September. Directorate of Sorghum Research (DSR), Hyderabad, India.
52. Dayakar Rao B, Ratnavathi CV, Karthikeyan K, Biswas PK, Rao SS, Vijaya Kumar and Seetharama N. 2004. Sweet sorghum cane for Biofuel production: A SWOT analysis in Indian context. NRCS Technical Report no.21/2004. National Research Centre for Sorghum, Rajendranagar, Hyderabad 500 030, (AP), India.20 pp.
53. Deshpande, VP, Kamatar, MY, Kathnalli, DS, Malleshappa, SM, Nayakar, NY (2003). Screening of sorghum genotypes against shootfly, *Atherigona soccata* (Rondani). *Indian J. Plant Protect*. 31, 90-93.
54. Dhillon, MK, Sharma, HC, Folkertsma, RT, Chandra, S (2006). Genetic divergence and molecular characterization of sorghum hybrids and their parents for reaction to *Atherigona soccata* (Rondani). *Euphytica* 149, 199-210.
55. Dhillon, MK, Sharma, HC, Ram Singh, Naresh, JS. 2005. Mechanisms of resistance to shoot fly, Atherigona soccata in sorghum. *Euphytica* 144, 301-312.
56. Dhoble MV and Kale UV. 1988. Recovery of drought resistance in sorghum genotypes. *Journal of Maharashtra Agricultural University* 13(1): 118.
57. Dhole VJ. 2004. Genetic analysis and multiple trait selection indices based on selection methods in *Rabi* sorghum (*Sorghum bicolor* (L.). Moench). Thesis abst. Thesis submitted to M.P.K.V. Rahuri, (Maharashtra).
58. Donald, C. M., The breeding of crop ideotypes. *Euphytica*, 1968, 17, 385–403.
59. DSR (2010). Directorate of Sorghum Research Annual Report- 2009-10. Rajendranagar, Hyderabad 500030, Andhra Pradesh, India pp.93.
60. Forbes, GA, Frederiksen, RA, Seitz, LM (1989). Assessment of sorghum grain mould: disease severity and crop loss. *Seed Sci and Technol* 17: 297–307.
61. Gadakh SR, Shinde MS, Gaikwad AR and Patil VR (2012). Phule Revathi : a new *Rabi* sorghum variety for irrigated condition. *Crop Res* 43 (1, 2 and 3): 24-27.
62. Ganapathy KN, Audilakshmi S and Patil JV (2010). Sorghum as feed for poultry and animals: Performance of elite sorghum genotypes for protein digestibility. In: Research and Development in Millets: Present status and future strategies in "National Seminar on Millets" organized by Directorate of Sorghum Research at Hyderabad during 12th November 2010. III-20, pp: 123.
63. Ganga Kishan, A and Borikar, S. T. 1989. Genetic relationship between some cytoplasmic male sterility systems in sorghum, Euphytica. 42: 259-269.
64. Garud TB, Seetharama N, Deshpande SP, Syed Ismail and Dadke MS. 2002. Usefulness of non-senescent parents for charcoal rot resistance breeding in sorghum. *International Sorghum and Millets Newsletter* 43: 63-65.

65. Garud, TB, Borikar, ST. 1985. Genetics of charcoal rot resistance in sorghum. *Sorghum Newslet.*, 13 : 269–276.

66. Ghorade DS and Kadam BS. 1985. Inheritance of grain characters in Sorghum. *Ind J Genet* 45 (1): 38-44.

67. Gite, B.D.; Kahate, P.A.; Ratnaparkhi, R.D.; Ghodpage, R.M.; Anokar, D.N. 2006. Identification of some new *Rabi* sorghum hybrids for shoot fly resistance and grain yield. *Annals of Plant Physiology* 20(1): 154-155.

68. Govil GN, Murty BR (1980). Genetic variance for endosperm characteristics in sorghum. *Ind J Genet* 40 (2): 366-374.

69. Grewal R P S, Lodhi G P and Paroda R S. 1987. Inheritance of field resistance to oval leaf spot. *Indian Journal of Genetics* 47(1): 41-45.

70. Grewal R P S, Lodhi G P and Sabharwal P S. 1996. Forage sorghum germplasm: Evaluation in the past two decades. *Indian Journal of Plant Genetic Resources* 9: 287-293.

71. Grewal R P S, Pahuja S K, Yadav R, Jindal Y, Yadav K K, Rana D S and Pundir S R. 2005. Multi-cut forage sorghums–where do we stand? *Forage Research* 31(3): 170-184.

72. Grewal R P S.1988. Genetic basis of resistance to zonate leaf spot disease in forage sorghum. *Theoretical and Applied Genetics* 76(4): 550-554.

73. Hariprasanna K, Agte V, Prabhakar and Patil JV (2012). Genotype × Environment Interactions for Grain Micronutrient Contents in Sorghum [*Sorghum bicolor* (L.). Moench]. *Ind J Genet* 72(4): 429-434.

74. Het Ram and Lodhi G P. 1992. Stem borer resistance in sorghum. *Forage Research* 18: 6-8.

75. Indira, S, Rana, BS (1983). Host plant resistance to Macrophomina phaseolina in F1 varietal crosses of sorghum. *Tropl. Pl. Diseases*, 1 : 187-192.

76. Indira, S, Rana, BS, Rao, NGP (1984). Studies on host-plant resistance to charcoal rot rust and head moulds in sorghum. Paper Presented at the Annual Workshop of AICSIP, held at Dharwad from 3-6, May 1984.

77. Iyanar K and Khan A K F. 2005. Combining ability analysis in forage sorghum for multi-cut habit. *Crop Research* **29** (1): 129-133.

78. Jain S K and Patel P R. 2012. Genetic variability in land races of forage sorghum [*Sorghum bicolor* (L.). Moench] collected from different geographical origin of India. *International Journal of Agriculture Sciences*, 4(2): 182-185.

79. Jain S K, Elangovan M and Patel N V. 2009. Genetic and environmental variability in dual sorghum (*Sorghum bicolor* (L.). Moench). for yield and related traits. *Forage Research* 34(4): pp. 201-204.

80. Jayanthi, P.D.K. 1997. Genetics of shoot fly resistance in sorghum hybrids of cytoplasmic male sterile lines. Thesis submitted to the Acharya N.G. Ranga Agricultural University, Rajendranagar, and Hyderabad.

81. Jhansi Rani K, Rana BS, Swarnalatha Kaul, Rao SS and Ganesh M. 2007. Genetic analysis of certain morpho-physiological characters in *Rabi* sorghum. *Indian Journal of Genetics* 67(3): 281-283.

82. Jotwani, MG, Singh, SP, Chaudhari, S (1971). Relative susceptibility of some sorghum lines to midge damage. Pages 123-130 in Investigations on insect pests of sorghum and millets (1965-1970): final technical report (Pradhan, S. *et al.*, Investigators). New Delhi, India: Indian Agricultural Research Institute.

83. Kadam DE, Patil FB, Bhor TJ and Harer PN 2001. Line x tester analysis in sweet sorghum hybrids. *J Maharashtra Agri Univ*.25: 318–9.

84. Kamatar, MY, Salimath, PM (2003). Morphological traits of sorghum associated with resistance to shootfly, *Atherigona soccata* Rondani. *Indian J. Plant Protect*. 31, 73-77.

85. Kandekar. 2008. Genetics of some physiological parameters related to drought tolerant in *Rabi* sorghum (*Sorghum bicolor* (L.). Moench). *M.Sc. (Agri.). Thesis* submitted to M.P.K.V, Rahuri, (Maharashtra).

86. Kannababu,N., Prabhakar, MY Samdur and AV Gadewar. 2013. Variability for seed setting potential in *Rabi* sorghum parental lines, presented at Global Consultation on Millets Promotion for Health and Nutritional Security, held at Directorate of Sorghum Research, Hyderabad from 18 20, December, 2013, India.

87. Khanure SK, Barigidad H, Kuruvinashetti MS (1997). variability in grain protein content of sorghum genotypes across *Rabi* environments. *Karnataka J Agric Sci* 10(4): 1217-1219.

88. Khatri R S, Lodhi G P, Grewal R P S and Pahuja S K. 2001a. Inheritence of regrowth in forage sorghum [*Sorghum bicolor* (L.). Moench.]. *Forage Research* 27(2): 147-148.

89. Khatri R S, Pahuja S K, Avtar R and Phogat D S. 2001b. Inheritance of leafiness in forage sorghum [*Sorghum bicolor* (l.). Moench]. *Forage Research* 27(1): 35-38.

90. Kholova J, McLean G, Vadez V, Craufurd P and Hammer GL. 2013. Drought stress characterization of post-rainy season (*Rabi*). sorghum in India. *Field Crops Research* 141: 38-46.

91. Khot KB. 2008. Generation mean analysis and molecular assay of yield and yield components for drought tolerance in *Rabi* sorghum. *Ph.D. Thesis* submitted to M.P.K.V, Rahuri, (Maharashtra).

92. Khurana AD, Verma AN (1985). Some physical plant characters in relation to stemborer and shoot fly resistance in sorghum. *Indian J Entom* 47: 14–19.

93. Koteswara Rao G and Poornachandrudu D (1971). Isolation of head moulds and assessment of mould grains in certain sorghum varieties. *Andhra Agric J* 18: 153-156.

94. Kulkarni N, Sahib K. Hussain, Murty KN (1981). Genetics of mould tolerance in sorghum. *Ind J Genet* 41 (1): 84-87.

95. Kumar, C.V.S., Shreelakshmi, Ch. and Shivani, D. 2010. Genetic diversity analysis in *Rabi* sorghum (*Sorghum bicolor* (L.). Moench). local genotypes. *Electronic Journal of Plant Breeding* 1(4): 527-529.

96. Kumara Charyulu D, Bantilan MCS, Nedumaran S and Deb UK. 2011. Development and diffusion of improved sorghum cultivars in India: Impact on growth and variability in yield. Tracking of Improved Cultivars adoption in India, paper presented in the 7th ASAE Conference held at Vietnam (http://oar.icrisat.org/5452/7/Sorghum-paper-final-ASAE.pdf).

97. Lodhi G P and Dangi O P. 1981. Genetics of yield and quality characters in forage sorghum. Forage Research 7A: 57-71.

98. Lodhi G P and Grewal R P S. 1988. Forage Sorghum Production. Technical Bulletin of the Plant Breeding Department, Haryana Agricultural University, Hisar 125004, Haryana, India.

99. Ma ZH, Li D and Ning XB 1992. Study on brix degree, total sugar content and their relationship in the juice of sweet sorghum stem. *J Shenyang Agric Univ*. 23: 87–91.

100. Madhava Rao. 1965. Bread or roti making quality in sorghum hybrids. *Sorghum Newsletter* 8: 27-30.

101. Maiti, RK, Gibson, PT (1983). Trichomes in segregating generations of sorghum matings. II. Association with shoot fly resistance. *Crop Sci.* 23, 76-79.

102. Manickam S and Das L D V. 1995. Combining ability analysis for forage characters in sorghum (*Sorghum bicolor* (L.). Moench). *Annals of Agricultural Research* 16(1): 49-52.

103. Mathur P N, Rao K E P, Thomas T A, Mengesha M H, Sapra R L and Rana R S. 1991. Evaluation of Forage Sorghum Germplasm, Part-1: NBPGR-ICRISAT Collaborative Programme. NBPGR, New Delhi, India. pp. 269.

104. Mathur PN, KE Prasada Rao, IP Singh, RCAgrawal, MHMangesha and RS Rana (1992). Evaluation of Forage Sorghum Germplasm. NBPGR-ICRISAT Collaborative Programme, NBPGR, New Delhi, India. pp. 296.

105. Meshram MP, Atale SB, Murumkar RD and Raut PB. 2005. Heterosis and heterobeltiosis studies in sweet sorghum. *Ann Plant Physiol*. 19: 96–8.

106. Mittal M and Boora K S. 2005. Molecular tagging of gene conferring leaf blight resistanceusing microsatellites in sorghum [*Sorghum bicolor* (L.). Moench]. *Indian Journal of Experimental Biology* 43: 462-466.

107. Mohan M, Pahuja S K and Yadav R. 2007a. Combining ability studies for multi-cut traits in forage sorghum involving male sterile lines and testers. *Forage Research* 32(4): 249-253.

108. Mohan M, Pahuja S K, Yadav R and Avtar R. 2007b. Combining ability studies for fodder yield and components in forage sorghum involving male sterile lines and testers. *Forage Research* 33(1): 17-21.

109. Murali Mohan, S, Madhusudhana, R, Mathur, K, Chakravarthi, DVN, Sanjay Rathore, Nagaraja Reddy, R, Satish, K, Srinivas, G, Sarada Mani, N, Seetharama, N (2010). Identification of quantitative trait loci associated with resistance to foliar diseases in sorghum [*Sorghum bicolor* (L.). Moench]. *Euphytica*, 176: 199–211.

110. Murty BR, Arunachalam V and Saxena MBL (1967). Catalogue of World Collection of Sorghum. *Indian J. Genet* 27: Spl. Number: 1-312.

111. Murty DS and V Subramanian (1981). Sorghum Roti: I. Traditional Methods of Consumption and Standard Procedures for Evaluation In: Proceedings of the International Symposium on Sorghum Grain Quality held during 28-31 October 1981, ICRISAT Center, Patancheru, India.

112. Murty DS, Patil HD, and House LR (1979). Breeding for sorghum food quality at ICRISAT. Pages 38-39 in Proceedings of the eleventh biennial grain sorghum research and utilization conference, 28 Feb-2 Mar 1979, Wichita, Kansas, USA.

113. Murty, M.V.R., Piara Singh, Wani, S.P., Khairwal, I.S., Srinivas, K., 2007. Yield gap analysis of sorghum and pearl millet in India using simulation modeling. Global Theme on Agroecosystems Report no. 37. International Crops Research Institute for the Semi-Arid Tropics, Patancheru 502324, Andhra Pradesh, India, 82 pp.

114. Murumkar PN, Atale SB, Shivankar RS, Meshram MP and Bhandarwar AD. 2005. Combining ability studies in newly established 'MS' and 'R' lines in *Rabi* sorghum. *Indian Journal of Agricultural Research* 39(4): 263-268.

115. Nanda GS, Rao NGP (1975). Genetic analysis of some exotic × Indian crosses in sorghum. IX. nutritional quality and its association with yield. *Ind J Genet* 35 (1): 131-135.

116. Nandawankar KG. 1990. Heterosis studies for grain yield characters in *Rabi* sorghum. *Indian Journal of Genetics and Plant Breeding* 50: 83-85.

117. Narasimhan KS and Rangaswamy G (1969). Microflora of sorghum grain. *Mysore J Agric Sci* 3: 215-226.

118. Nerkar, Y.S. 1998. *Rabi* sorghum improvement for food and fodder security: Opportunities and strategies, Special lecture delivered at 29th Workshop of AICSIP, April 23-25, 1998 at University of Agricultural Sciences, Dharwad.

119. Nimbalkar VS and Bapat DR. 1991. Heterosis for shoot fly resistance in grain sorghum. *J. Maharashtra Agric. Univ*. 16(1): 7-9.

120. Nirmal S. V. and Patil J. V. 2008. A new drought tolerant genotype of *Rabi* sorghum - SPV 1546 (Phule. Chitra). *Annals of Plant Physiology*, 22: 165-168..

121. Padmaja PG, Woodcock CM, Bruce TJ. 2010. Electrophysiological and behavioral responses of sorghum shoot fly, *Atherigona soccata*, to sorghum volatiles. *J Chem Ecol*. (12): 1346-53.

122. Pahuja S K and Yadav R. 2008. Forage sorghum hybrids : prospectives and constraints. *Forage Research* 34(1): 9-13.

123. Pahuja S K, Aruna C, Shrotria P K, Kaur S, Ranwah B S and Patil J V (2013). Inducing variability in multi-cut forage sorghum through mutagenesis. *Plant Genetic Resources* 11: 114-120.

124. Pahuja S K, Yadav R and Grewal R P S. 2003. Genetics of fodder yield and its components in multi-cut x single cut forage sorghum crosses. *Forage Research* 29(3): 139-141.

125. Palanisamy S and Subramanian A. 1984. Genetic analysis of harvest index in sorghum. *Sorghum newsletter* 28: 39-40.

126. Pandey A K. 2010. Developing low-HCN producing forage sorghum (*Sorghum bicolor* [L.]). using anti-sense approach. *Ph.D. Thesis*. Jawaharlal Nehru Technological University, Hyderabad.

127. Pandey, S and Shrotria P K. 2009. Genetic parameters for hydrocyanic Acid content in forage sorghum [*Sorghum bicolor* (l.). Moench]. *Forage Research* 35(1): 17-19.

128. Parmar H P, Patel J R and Patel P C. 2004. Combining ability for quantitative traits in forage sorghum [*Sorghum bicolor* (L.). Moench]. *Forage Research* 29(4): 170-172.

129. Paroda R S and Lodhi G P. 1981. Genetic improvement in forage sorghum. *Forage Research* 7A: 17-56.

130. Parodi, RA, Gamba, RD, Scantamburlo, JL (1974). "Huerin INTA" grain sorghum variety tolerant to the sorghum midge (*Contarinia sorghicola* Coq.). *Sorghum Newsletter* 17: 1.

131. Patel DV, Makne VG, Mehta HD and Shete DM. 1993. Genetic architecture of grain yield and related in high energy sorghum. *Journal of Maharashtra Agricultural University* 18(2): 261-263.

132. Pathak H C and Sanghi A K. 1992. Genetic analysis of leaf and physiological traits in forage sorghum across environments. *Gujarat Agricultural University Research Journal* 17(2): 49-55.

133. Patil BS and RaviKumar RL. 2011. Osmotic adjustment in pollen grains: a measure of drought adaptation in sorghum. *Current Science* 100(3): 377-382.

134. Patil JV, Rakshit S and Khot KB. 2013. Genetics of post-flowering drought tolerance traits in post-rainy sorghum (*Sorghum bicolor* (L.). Moench). *Indian Journal of Genetics and Plant Breeding* 73(1): 44-50.

135. Patil, RC, Thombre, MV (1983). Inheritance of earhead midge incidence in sorghum. *Sorghum Newsletter* 27: 90.

136. Perumal R, Menz MA, Mehta PJ, Katile´ S, Gutierrez-Rojas LA, Klein RR, Klein PE, Prom LK, Schlueter JA, Rooney WL, Magill CW (2008). Molecular mapping of Cg1, a gene for resistance to anthracnose (*Colletotrichum sublineolum*). in sorghum. *Euphytica* 165: 597–606.

137. Ponnaiya, BWX (1951). Studies in the genus Sorghum. I. Field observations on sorghum resistance to the insect pest, *Atherigona indica* M. *Madras Agricultural Journal* 21: 96-117.

138. Poor SN and Rezai A. 1996. Estimates of genetic parameters for grain yield and related characters in sorghum. *Iranian Journal of Agricultural Science* 27(2): 77-87.

139. Prabhakar and M. S. Raut. 2010. Exploitation of heterosis using diverse parental lines in *Rabi* Sorghum. *Electronic Journal of Plant Breeding*, 1(4): 680-684.

140. Prabhakar, Kannababu N., Samdur M. Y., and Bahadure D. M., 2013. Exploitation and assessment of heterosis in *Rabi* sorghum hybrids across environments presented at Global Consultation on Millets Promotion for Health and Nutritional Security, held at Directorate of Sorghum Research, Hyderabad from 18 20, December, 2013, India.

141. Prabhakar, M Elangovan and MS Raut. 2010. Stability of new varieties for yield components in *Rabi* sorghum. *Indian J. Agric, Sci*. 80 (2): 106-109.

142. Prabhakar. 2002. Combining ability studies in *Rabi* Sorghum. *Ann. Agric. Res.* New series: -23 (3): 500-502.

143. Rajarajan K and Ganesamurthy K. 2011. Genetic diversity analysis of sorghum (*Sorghum bicolor* L. Moench). genotypes for drought tolerance using SSR markers *Indian J. Genet*. 71(1): 17-24.

144. Rajguru AB, Kashid NV, Kamble MS, Rasal PN and Gosavi AB. 2004. Combining ability analysis for yield and its components in *Rabi* sorghum. *Crop improvement* 31(2): 195-200.

145. Rajvanshi AK and Nimbkar N. 1996. Sweet sorghum R&D at the NARI. In Proceedings of the First European Seminar on Sorghum for Energy and Industry, Toulouse, France, 1"3, April 1996.

146. Rana BS, Rao MH, Indira S, Singh BU, Chari Appaji, Tonapi VA (1999). Technology for increasing sorghum production and value addition. National Research Center for Sorghum, Rajendranagar, Hyderabad.

147. Rana BS, Singh BU, Rao NGP (1984). Breeding for Shoot Fly and Stem Borer Resistance in Sorghum. In proceeding of international sorghum entomology workshop 15-21 July 1984. Texas A and M University, College Station, USA. pp. 347-360.

148. Rana BS, Swarnlata Kaul and Rao MH. 1997. Impact of genetic improvement on sorghum productivity in India. Pages 142-165 in the Proceedings of the International Conference on Genetic Improvement of Sorghum and Pearl millet, 22-27 September 1996, Holiday Inn Plaza, Lubbock, Texas.

149. Rana, B. S. and Kaul, S. L. 2005. Final Report on NATP project on Development of *Rabi* sorghum Hybrids,National Research Centre for Sorghum, Hyderabad, pp. 1-44.

150. Rana, B.S., Jaya Mohan Rao, V., Reddy, B.B., and Rao, N.G.P. 1985. Overcoming present hybrid yield in sorghum. Pages 48-59 In Genetics of heterotic system in crop plants: Proceedings of Precongress Symposium XV International Genetics Congress, 7-9 Dec 1983, Tamil Nadu Agricultural University, Coimbatore, India.

151. Rana, BS, Jotwani, MG, Rao, NGP (1981). Inheritance of host plant resistance to shoot fly in sorghum. *Insect Science and its Application* 2: 105-109.

152. Rana, BS, Rao, MH, Indira, S, Singh, BU, Rao, SS, Lodhi GP (1997). Introduction. In: XXVII Annual Group Meeting held at Tamil Nadu Agricultural University, Coimbatore, April 28– 30, 1997. All India Co-ordinated Sorghum Project, Hyderabad, India.

153. Rana, BS, Tripathi, DP, Balakotaiah, K, Damodar, R, Rao, NGP (1975). Genetic analysis of some exotic x Indian crosses in sorghum. XI. Selection for shoot fly resistance. *Indian Journal of Genetics and Plant Breeding* 35: 350-355.

154. Rani Ch and Umakanth AV.2012. Genetic variation and trait inter-relationship in F1 hybrids of sweet sorghum (*Sorghum bicolor* (L.). Moench). *Journal of Tropical Agriculture* 50 (1-2): 80-83, 2012.

155. Rao DG and Khanna SKCSR. 1999. Comparative performance of sorghum hybrids and their parents under extreme water stress. *Journal of Agricultural Science* 133(1): 53-59.

156. Rao N G P and Rana B S. 1982. Selection in temperate-tropical crosses of sorghum. Pages 403-419. In: Sorghum in the eighties vol. 1: Proceedings of the International symposium on sorghum, 2-7 Nov 81, ICRISAT, Patancheru, AP, India.

157. Rao N G P. 1972. Sorghum Breeding in India: Recent Developments. Pages 101-158. In: Sorghum in seventies. Rao N G P and House L R (Eds).

158. Rao SBP, Bhatia HP, and Dabir VN (1964). Quality tests on grain sorghum hybrids. *Sorghum Newsletter* 7: 34-45.

159. Rao SS. 2005. Collaboration on sweet sorghum big mill test –NRCS and Sagar Sugars. *Jowar Samachar*. 1 (2): 2.

160. Rao, N.G.P. 1970. Genetic analysis of some exotic × Indian crosses in sorghum. 1. Heterosis and its interaction with seasons. *Indian J. Genet*. 30: 347-361.

161. Rao, N.G.P. 1982. Transforming traditional sorghum in India. Pages 39-59 In Sorghum in the Eighties: Proceedings of the International Symposium on Sorghum, 2-7 Nov 1981, ICRISAT, India. Vol. 1. Patancheru, Andhra Pradesh 502 324, India: International Crops Research Institute for the Semi-Arid Tropics.

162. Rao, N.G.P., Jaya Mohan Rao, V., and Reddy, B.B. 1986. Progress in genetic improvement of *Rabi* sorghums in India. *Indian J. Genet*. 46(2): 348-354.

163. Ratanavathi CV, Dayakar Rao B and Seetharama N. 2003. Sweet sorghum stalk: A suitable raw material for fuel alcohol production. DSR/NRCS Report Number-12/2003. NATP (DSR). Series No. 1. Rajendranagar, Hyderabad 500 030, Andhra Pradesh, India: National Research Center for Sorghum (NRCS). pp. 8.

164. Ratnavathi CV, Dayakar Rao BD, Padmaja PG, Ravi Kumar S, Shashidhar Reddy Ch, Vijay Kumar BS, Pallavi M, Komala VV, Gopala Krishna D and Seetharama N. 2005. Sweet sorghum-the wonder crop for biofuel production. NRCS Technical Report no.27/2005. National Research Centre for Sorghum, Rajendranagar, Hyderabad, AP 500 030.

165. Ravikumar, R. L., Patil, B. S. and Salimath, P. M., Drought tolerance in sorghum by pollen selection using osmotic stress. *Euphytica*, 2003, 133, 371–376.

166. Ravindranath, V. (1976). Head mould on grain sorghum. Paper presented at Annual Sorghum Workshop, M. K. V., Parbhani.

167. Reddy B V S, Ahok Kumar A, Sanjana Reddy P and Elangovan M. 2008. Sorghum germplasm: diversity and utilization in 'Sorghum Improvement in the New Millenium edited by Reddy Belum VS, Ramesh S, Ashok Kumar A and Gowda CLL, pp. 153-169.

168. Reddy BVS and Prasada Rao KE. 1993. Genetic Diversification. ICRISAT Annual Report 1992. International Crops Research Institute for the Semi-Arid Tropics, Patancheru, Andhra Pradesh 502 324, India.

169. Reddy BVS, Ramesh S, Ashok Kumar A., Wani SP and Ortiz R. 2008. Biofuel crops research for energy security and rural development in developing countries. Bioenergy Research 1, 248–258. pp. 24.

170. Reddy BVS, Ramesh S, Sanjana Reddy P, Ramaiah B, Salimath PM and Rajashekar Kachapur. 2005. Sweet sorghum–A potential alternative raw material for bioethanol and bio-energy. *Intl. Sorghum Millets Newslett*. 46: 79–86.

171. Reddy BVS, Seetharama N, Maiti RK, Bidinger FR, Peacock JM and House LR. 1980. Breeding for drought resistance in sorghum at ICRISAT. Paper presented at the AICSIP workshop. 12-14 May 1980. Coimbatore, India.

172. Reddy BVS. 1997. Development, production, and maintenance of male-sterile lines in sorghum. Pages 12-21 in Training Manual.(Singh F, Rai KN, Reddy BVS and Diwakar B, eds.). ICRISAT, Patancheru 502 324, Andhra Pradesh, India.

173. Reddy PS, Patil JV, Nirmal SV and Gadakh SR (2012). Improving post-rainy season sorghum productivity in medium soils: does ideotype breeding hold a clue?. *Current Science* 102 (6): 1-5.

174. Reddy TV (1957). Pelalu jonna—a popping variety of sorghum of Andhra. *The Andhra Agric J*. 4: 90.

175. Reddy. B.V.S., Ahok Kumar, A., Sanjana Reddy, P. and Elangovan, M. 2008. Sorghum germplasm: diversity and utilization in 'Sorghum Improvement in the New Millenium edited by Reddy Belum VS, Ramesh S, Ashok Kumar A and Gowda CLL, pp. 153-169.

176. Rooney WL. and Smith CW. 2000. Techniques in developing new cultivars. In Sorghum: Evolution, History, Production and Technology. C.W. Smith and R.A.Frederiksen eds., John Wiley Press pp. 329-347.

177. Saballos A, Vermerris W, Rivera L and Ejeta G. 2008. Allelic association, chemical characterization and saccharification properties of brown midrib mutants of sorghum [*Sorghum bicolor* (L.). Moench]. *Bioenerg. Res*. 1: 193–204.

178. Sajjanar, G.M., Biradar, B.D., Biradar, S.S., 2011. Evaluation of crosses involving *Rabi* landraces of sorghum for productivity traits Karnataka. *J. Agric. Sci.* 24 (2), 227–229.

179. Salih AA, Ali IA, Lux A, Luxona M, Cohen Y, Sugimoto Y, Inanaga S. 1999. Response of sorghum to water stress under greenhouse conditions. Phenology, water consumption through evapotranspiration and transpiration and yield. *Crop Science* 39: 168-173.

180. Salunkhe DK, JK Chavan and SJ Jadhav (1984). *Nutritional and processing quality of sorghum*. Oxford and IBH Publishing Co. New Delhi. p. 275.

181. Salunkhe K (1984). Sorghum grain quality: problems and prospects. In: *Nutritional and Processing Quality of Sorghum* (Ed). D. K. Salunkhe, J.K. chavan and S. J. Jadha, Oxford and IBH Publishing Co.

182. Sanghi A K. 1982. Combining ability for fodder yield and other quantitative characters in forage sorghum. *Indian Journal of Genetics and Plant Breeding* **42**: 311-313.

183. Sanjana Reddy. P, J. V. Patil1, S. V. Nirmaland S. R. Gadakh.2013. Improving post-rainy season sorghumproductivity in medium soils: does ideotype breeding hold a clue?, *Current Science*: 102 (6): 904-908.

184. Sankarapandian R, Ramalingam J, Pillai MA and Vanniarajan C. 1994. Heterosis and combining ability studies for juice yield related characteristics in sweet sorghum. *Ann. Agric. Res.* 15: 199"204.

185. Sankarpandian R and Bangarusamy U. 1996. Stability of sorghum genotypes for certain physiological characters and yield under water stress conditions. *Crop Improvetment* 23(1): 61-65.

186. Sankarpandian R, Krishnadass D, Muppidathi N and Chidambaram S. 1993. Variability studies in grain sorghum for certain physiological characters under water stress conditions. *Crop Improvement* 20(1): 45-50.

187. Seetharama N, Bidinger FR, Rao KN, Gill KS and Madhuri M. 1987. Effect of pattern and severity of moisture deficit stress on stalk rot incidence in sorghum. I. Use of line source irrigation technique and the effect of time of inoculation. *Field Crops Research* 15(34): 289-308.

188. Seetharama N, Reddy BVS, Peacock JK and Bidinger FR. 1982. Sorghum improvement for drought resistance, In: Drought resistance in crop plants with emphasis on rice, IRRI, Philippines, 317-356.

189. Seetharama, N., Singh, Sardar and Reddy, B. V. S., Strategies for improving postrainy sorghum productivity. *Proc. Indian Natl. Sci. Acad.*, 1990, 56, 455–467.

190. Selvi B and Palanisamy S. 1990. Heterosis and combining ability for grain yield in sweet sorghum. *Madras Agric J.* 77: 493-496.

191. Sharma HC, Dhillon MK, Pampapathy G, Reddy BVS (2007). Inheritance of resistance to spotted stem borer, Chilo partellus, in sorghum, *Sorghum bicolor*. *Euphytica* 156 (1-2): 117-128.

192. Sharma, HC, Nwanze, KF, Subramanian, V (1997). Mechanisms of resistance to insects and their usefulness in sorghum improvement. In: Sharma, HC, Singh, F,

Nwanze, KF (Eds.), *Plant Resistance to Insects in Sorghum*. ICRISAT, Patancheru, India, pp. 81-100.

193. Sharma, HC, Nwanze, KF, Subramanian, V (1997). Mechanisms of resistance to insects and their usefulness in sorghum improvement. In: Sharma, H.C., Singh, F., Nwanze, K.F. (Eds.), *Plant Resistance to Insects in Sorghum*. ICRISAT, Patancheru, India, pp. 81e100.

194. Sharma, JR (1994). *Principles and Practice of Plant Breeding*. Tata McGraw Hill Publishing Company Limited, New Delhi, India.

195. Shinde V K, Sudewad S M, Ambekar S S, Rodge B M and Shete D M. 1994. A new forage sorghum variety - MKV Chari-1. *Journal of Maharashtra Agricultural Universities* 19(1): 73-74.

196. Shinde, VK, Nandanwankar, KG, Ambekar, SS (1981). Heterosis and combining ability for grain yield in *Rabi* sorghum. *Sorghum Newslet.*, 26 : 19.

197. Shukla GK, Gupta SK, Lakhendra Singh, Rao SS, Ratnavathi CV and Dayakar Rao B 2006. Successful pilot production of bio-ethanol from sweet sorghum in sub-tropical north India. *Jowar Samachar*. 2 (1): 1.

198. Singh M, Chaudhary K, Boora KS (2006). RAPD-based SCAR marker SCA 12 linked to recessive gene conferring resistance to anthracnose in sorghum [*Sorghum bicolor* (L.). Moench]. *Theor Appl Genet* 114: 187–192.

199. Singh R K and Shrotria P K. 2008. Combining ability analysis for forage yield and its components in forage sorghum [*Sorghum bicolor* (L.). Moench] *Forage Research* 34(2): 79-82.

200. Singh S, Dwivedi VK, Shrotria PK and Pandey S. 2010. Combining ability analysis for fodder yield and its components in forage sorghum. *Forage Research* 36(1): 4-8.

201. Singh, BU, Padmaja, PG, Seetharama, NS (2004). Stability of biochemical constituents and their relationship with resistance to shoot fly, Atherigona soccata (Rondani). in seedling sorghum. *Euphytica* 136, 279-289.

202. Singh, SR, Vedamoorty, G, Thobbi, VV, Jotwani, MG, Young, WR, Balan, JS, Srivastava, KP, Sandhu, GS, Krishnananda, N (1968). Resistance to stem borer, Chilo zonellus (Swinhoe). and stem fly, *Atherigona soccata* Rond. in the World Sorghum Collection in India. *Memoirs of the Entomological Society of India* 7: 79 pp.

203. Sinha SK and Khanna R. 1975. Physiological, biochemical and genetic basis of heterosis. *Adv. Agron*. 27: 123.

204. Srinivasa Rao P, Rao S S, Seetharama N, Umakanth A V, Sanjana Reddy P, Reddy BVS and Gowda CLL. 2009. Sweet sorghum for biofuel and strategies for its improvement. Information Bulletin No. 77. Patancheru 502 324, Andhra Pradesh, India: International Crops research Institute for the Semi-Arid Tropics. 80 pp. ISBN: 978-92-9066-518-2 Order Code: IBE 077.

205. Srivastava, A., Naresh Kumar, S., Aggarwal, P.K., 2010. Assessment on vulnerability of sorghum to climate change in India. *Agric. Ecosyst. Environ*. 138, 160–169.

206. Subramanian A (1956). Jowar pops are a real delicacy. *Indian Farming* 6: 25-26.

207. Sumalini K, Desale J S and Gulia S K. 2005. Genetics of grain and forage traits in *Sorghum bicolor* (L.). Moench. *Forage Research* 31(1). : 30-32.

208. Sunil S Gomashe, Misal, MB, Mehtre, SP, Sujay Rakshit, Ganapathy, KN (2012). Assessment of parental lines and crosses for shoot fly resistance mechanism in sorghum (*Sorghum bicolor* (L.). Moench). *Indian Journal of Genetics and Plant Breeding*. 72 (1): 31-37.

209. Swarnalatha Kaul and Rana BS. 1988. Combining ability for biological yield and harvest index in sorghum. *Indian Journal of Genetics* 48(2): 149- 153.

210. Talwar HS and Vincent Vadez. 2011. Improving sorghum varieties to meet growing grain and fodder demand in India. South Asia Newsletter, ACIAR Australia, pp. 22-24.

211. Talwar HS, Prabhakar, Elangovan M, Aruna Kumari, Rao SS, Mishra JS and Patil JV. 2010. Strategies to improve post-flowering drought tolerance in *Rabi* sorghum for predicted climate change scenario. *Crop Improvement* 37(2): 93-98.

212. Talwar, H. S, Prabhakar, Elangovan, M, Aruna Kumari, Rao, SS, Mishra and J.V., Patil JV. 2011. Strategies to improve post-flowering drought tolerance in *Rabi* sorghum for predicted climate change scenario, *Crop Improvement* 37 (2): 93-99.

213. Talwar, H. S., Surwenshi, A. and Seetharama, N. (2009). Use of SPAD chlorophyll meter to screen sorghum (*Sorghum bicolor*). lines for postflowering drought tolerance. *Indian Journal of Agricultural Sciences* 79 (1): 432-37.

214. Thul AV. 2007. Genetics of traits associated with shootfly and drought tolerance in *Rabi* sorghum. (*Sorghum bicolor* (L.). Moench). Ph.D.(Agri.). Thesis submitted to M.P.K.V, Rahuri (Maharashtra).

215. Tonapi V A, Elangovan M, Kannababu N, Rao K V R, Kulkarni R, Reddy D C S, Meena R S, Gawali H S and Seetharama N. 2009. Sorghum cultivars of India: Hybrids, Varieties and Parental lines. National Research Centre for Sorghum, Rajendranagar, Hyderabad 500 030, Andhra Pradesh, India. ISBN 81-89335-23-5. 144 pp.

216. Trehan, KN, Butani, DK (1949). Notes on the life history, bionomics and control of Chilo zonellus in Bombay Province. *Indian Journal of Entomology* 11: 47-59.

217. Tripathi BK and Gupta YP and House LR (1971). Selection for high protein and amino acids in grain sorghum. *Ind J Genet* 31 (2). : 275-282.

218. Umakanth AV, Madhusudhana R, Latha KV, Kumar PH and Swarnlata Kaul. 2002. Genetic architecture of yield and its contributing characters in post rainy season sorghum. *International Sorghum and Millets Newsletter* 43: 37-40.

219. Umakanth AV, Patil JV, Rani Ch, Gadakh SR, Siva Kumar S, Rao SS and Tanmay Vilol Kotasthane. 2012. Combining Ability and Heterosis over Environments for Stalk and Sugar Related Traits in Sweet Sorghum (*Sorghum bicolor* (L.). Moench.). Sugar Tech (July-September 2012). 14(3): 237–246.

220. Vedansh, Singh S K, Kerkhi S A, Singh A, Kumar M and Kumar V. 2010. Variability, heritability and genetic advance for forage yield and quality traits in forage sorghum (*Sorghum bicolor* L. Moench). *Progressive Agriculture* 10(2): 400-401.

221. Venkata Rao, DN, Shinde, VK (1988). Inheritance of charcoal rot and other qualitative characters in sorghum, *Sorghum bicolor* (L.). Moench. Sorghum Newslet., 24 : 98.

222. Vermerris W, Saballos A, Ejeta G, Mosier NS, Ladisch MR and Carpita NC. 2007. Molecular breeding to enhance ethanol production from corn and sorghum stover. *Crop Sci.* 47(S3): S142–S153.

223. Vietor D. and Miller F. 1990. Assimilation, partitioning, and nonstructural carbohydrates in sweet compared with grain sorghum. *Crop Science*. 30: 1109–1115.

224. Viraktamath CS, Raghavendra G and Desikachar HSR (1972). Varietal differences in chemical composition, physical properties and culinary qualities of some recently developed sorghum. *J Food Sci Tech* (Mysore). 9(2): 73-76.

225. Virupaksha TK and Shastry LVS (1968). The protein content and amino acid composition of some varieties of grain sorghum. *J. Agric Fd Chew* 16: 199-203.

226. Warkad Y N, Potdukhe N R, Dethe A M and Kahate P A. 2008. Analysis of genetic variation and trait inter-relationship in sorghum [*Sorghum bicolor* (L.). Moench] *Forage Research* 33(4): 199-202.

227. Yadav R and Pahuja S K. 2007. Combining ability for fodder yield and its components in forage sorghum. *Forage Research* 32(4): 220-223.

228. Yadav R, Grewal R P S and Pahuja S K. 2002. Assessment of variability for fodder yield and its component traits in forage sorghum. *Indian Journal of Agricultural Sciences*. 72: 428-430.

229. Zhou H, Zuhu D, Jun L and Zhang M. 2005. Analysis on combining ability for the energy sugarstalk breeding. *Acta Agriculturae* Universitatis Jiangxiensis. 27: 14–20.

Millets: Ensuring Climate Resilience and Nutritional Security (2015) *Pages* **197–231**
Editors: **Vilas A. Tonapi and J.V. Patil**
Published by: **DAYA PUBLISHING HOUSE, NEW DELHI**

Chapter 4

Genetic Improvement in Pearl Millet

P. Sanjana
ICAR-Indian Institute of Millets Research (IIMR)
Rajendranagar, Hyderabad – 500 030

Pearl millet (*Pennisetum glaucum* (L.) R. Br.) is the sixth important cereal crop under cultivation in the world after rice, wheat, maize, barley and sorghum. About 3,000 years ago, pearl millet crossed the Indian Ocean and got domesticated in India. It is India's fourth most important cereal after rice, wheat, and sorghum. India accounts for >80 per cent of the 12 million ha in Asia under pearl millet cultivation (Rai *et al.*, 1999). *Bajra,* as it is called, is currently grown on almost 10 per cent of India's food-grain area, and it yields about 5 percent of the country's cereal food. Rajasthan, Maharashtra, Gujarat, and Uttar Pradesh account for nearly 80 percent of the 14 million hectares planted and 70 percent of the 5 million tons of pearl millet grain produced each year.

It is a predominantly cross-pollinated crop with 75–80 per cent outcrossing. As it is C_4 species with high photosynthetic efficiency and dry matter production ability, it performs well under marginal conditions. However, the grain yields largely under subsistence farming are low (500–700 kg ha^{-1}) for pearl millet. Low soil fertility, rain-fed farming characterized by erratic and inadequate rainfall (annual precipitation as low as 300 mm), negligible external inputs, continued use of mostly unimproved cultivars with a low harvest index (<20 per cent), and the prevalence of diseases and insect pests are the major causes of low productivity. Under these conditions, yields of maize become too uncertain, and pearl millet stands out as the most reliable food cereal. Pearl millet, in general, occupies drier parts of the semi-arid tropics and extends into the arid zone, where soils tend to be sandier, the growing season shorter, and drought stress more frequent. With regard to nutritional quality, it is equivalent to maize and superior to sorghum in protein content and quality, protein efficiency ratio, metabolizable energy levels and does not contain tannins. Though it is deficient in essential amino acids, it contains 35 per cent more lysine than sorghum (Rooney and McDonough 1987). It contains 5 to 6 per cent oil (Jambunathan and Subramanian 1988) and is a rich source of iron and zinc. Apart from being a main staple in Western Rajasthan and Gujarat, its use for poultry feed is on the increase. It is also an important source of fodder and only source of feed in dry months (Hash *et al.*, 2003).

Based on seed shape, cultivated pearl millet has been classified into four basic races (*typhoides*, *nigritarum*, *globosum*, and *leonis*) (Brunken *et al.*, 1977), although intermediates between any two of these basic types also exist. The Indian subcontinent is the secondary center of diversity. About 93 per cent of the total pearl millet germplasm accessions deposited in ICRISAT gene bank from Asia belongs to India. Owing to allogamous breeding behavior, landraces of pearl millet display much larger within population genetic variation.

Pearl millet breeding involves recurrent selection (within populations formed with incorporation of various desirable characters) and pedigree selection aimed at developing fixed lines. Combinations of fixed lines can be used to form synthetic varieties and composite varieties can be developed by maintaining mixture of genotypes in bulk. This has influenced the extent of germplasm utilization, the breeding methods used, prioritization of target traits, and a consideration of alternative cultivar options.

Germplasm Utilization

Selection by plant breeders within pearl millet landraces, which have far greater intrapopulation variability has produced improved OPVs, but these have had limited gains in productivity. These improved landraces have been used by more traditional farmers.

Pearl millet germplasm has been widely used in developing composites, which include a wide range of germplasm and improved breeding lines, mostly from western and southern Africa (Rai and Anand Kumar 1994). Pearl millet composites were primarily intended for breeding OPVs. Most programs dedicated to breeding hybrid parents have used pedigree breeding. This approach continues to be successful, although the genetic base of hybrids and hybrid parents has narrowed considerably, and broadening the genetic base of elite hybrid parents by expanded use of new germplasm is essential for long-term progress in hybrid development.

Iniadi germplasm from the Togo – Ghana –Burkina Faso – Benin region of western Africa is being most extensively used in pearl millet breeding programs worldwide (Andrews and Anand Kumar, 1996). Typical characteristics of this germplasm are large-sized grain (15–18 g 1000^{-1}) of globular shape and dark grey color with bright outer exposed surface, starchy endosperm, compact and conical panicles, relatively less photoperiod-sensitive, earliness (70– 85 days to mature), rapid grain-filling (23–25 days relative to 25–32 days for others), good downy mildew resistance, and good combining ability for grain yield. A majority of hybrids in India include some degree of

Iniadi germplasm in their parental lines, and almost all the hybrids under advanced testing in the U.S.A. include a great deal of it. This germplasm is being increasingly introgressed runs the risk of disrupting heterotic patterns, if any, in improved breeding materials and, hence, reduces the probability of producing high-yielding hybrids in the future.

Photoperiod sensitive germplasm accessions have been little used in breeding programs outside their area of adaptation. This is because (*i*) the trait needs have been largely met from within the vast less-sensitive group, of which only a small fraction has been used so far, and (*ii*) the high attrition rate in segregating generations derived from sensitive × insensitive crosses has led to few successful outcomes.

The utilization of wild germplasm has received the least attention. However, a modest attempt at Tifton, Georgia to use wild pearl millet germplasm has produced interesting results, as revealed in the utilization of rust and leaf spot resistance and cytoplasmic male sterility from *Pennisetum glaucum* ssp. *monodii* accessions, greater stalk strength and male fertility restorer genes from *P. purpureum*, and apomictic genes from *Pennisetum squamulatum* (Hanna 1992). It has been shown that wild sorghum germplasm can contribute useful variability to highly adapted grain hybrid parents and that introgression through populations is a more effective way of moving this variability than through elite inbred lines (Bramel-Cox and Cox 1989; Menkir *et al.*, 1994).

Breeding Methods

Recurrent selection and pedigree breeding are the two most common methods used in the genetic improvement of pearl millet. The basic requirement of random mating for the application of recurrent selection in pearl millet is met through its natural outbreeding behavior, which is caused by protogynous flowering. Pearl millet is amenable to all forms of recurrent-selection schemes. However, a comparative study in pearl millet found no significant difference among various recurrent selection procedures and concluded that the choice of a recurrent- selection scheme should be determined largely by the objective of the program and the availability of resources (Singh *et al.*, 1988). Pedigree breeding from populations that have mostly been derived from single crosses is the most common method in breeding hybrid parents (Andrews *et al.*, 1997). This approach continues to be quite successful, although it has obvious effects in that it narrows the genetic base of hybrid cultivars. Pedigree breeding using composites as base populations has the potential to (*i*) accelerate the genetic gains for yield in hybrids and (*ii*) broaden the genetic base of hybrid cultivars that would contribute to yield stability. However, composites will be used for pedigree breeding of hybrid parents only when they fulfill the basic traits required of hybrid parents. The development of composites specifically targeted for breeding hybrid parents is a relatively recent development in pearl millet. However, a diverse range of existing composites, developed primarily for breeding OPVs, can serve as base material for deriving partial inbreds, some of which can be used directly as hybrid parents, but most of which will find use in crosses to further diversify hybrid-parent breeding programmes.

Cultivar Development

Both OPVs and hybrids are commercially grown in pearl millet. The highest-yielding hybrids have about 15–30 per cent grain yield advantage over the best varieties of comparable maturity.

Open Pollinated Varieties

An OPV in pearl millet is a highly heterozygous and heterogeneous population, open to significant genetic changes. Pearl millet OPVs, bred from single progenies (full-sib or S1–S2), do not perform well and, hence, are not under cultivation. Wider genetic variability is needed to maintain vigor over generations. Thus, pearl millet varieties are often developed by balanced random mating of 5–10 full-sib or S1–S2 progenies. Several OPVs developed on the basis of this procedure have been released (*e.g.*, WC-C75, ICTP 8203, and Raj 171) and are widely grown. Provided that some characters of greatest visual appeal, such as flowering, plant height, panicle size and shape, and grain size, shape, and color, can be kept sufficiently uniform for varietal identification purposes, even larger numbers of progenies (>50) can be used in breeding productive OPVs of pearl millet (*e.g.*, ICMV 155, ICMV 221, and ICMV 88908 (= Okashana 1)). Within a pearl millet variety there is continual reorganization of variability generating a narrow range of phenotypes within a recognizable norm. Mild mass selection during the seed increase of a variety is a useful exercise to maintain this norm. Genetic shift is likely to occur when a pearl millet variety is grown in a different environment. A corollary of this is that a pearl millet variety can be fine-tuned for its adaptation and productivity to a range of related environments, but seed stocks for each of these environments would then have to be maintained separately. Pearl millet varieties have seed yields 2–3 times higher than the inbred seed parents. Intrapopulation variability in pearl millet helps in the genetic management of several diseases. For instance, genetic heterogeneity confers durable resistance to downy mildew, and a variation in flowering confers pollen based escape from ergot and smut infection; this contrasts well with frequent downy mildew epidemics on single-cross hybrids of pearl millet and their greater vulnerability to ergot and smut. In consequence, improved pearl millet OPVs are readily acceptable to farmers and are easier to multiply and, hence, have carved a niche for themselves even in India, where hybrids are the preferred cultivars. In part, this may have to do with the memory of downy mildew epidemics on hybrids in India.

Hybrids

All the commercial grain hybrids in pearl millet are single cross hybrids based on A_1 CMS system. Two other types of hybrids can also be produced. The first one is a top cross hybrid, which is produced by crossing an OPV with an inbred-seed parent. This type of hybrid achieves a compromise between the positive attributes of both OPVs (yield stability and durable disease resistance) and single-cross hybrids (phenotypic uniformity and higher grain yield). The second type is an inter population hybrid, also called an inter varietal hybrid. This type of hybrid can be produced commercially by crossing an OPV with a male-sterile population. Inter population hybrids would be phenotypically as variable as OPVs but could produce 32–45 per cent greater yields (Ouendeba *et al.*, 1993). The higher seed yields and greater durability in downy mildew resistance of population seed parents would make them economically more attractive proposition for commercial utilization of inter population hybrids than inbred seed parents. One top cross hybrid, ICMH 88088, based on a CMS inbred line is under commercial production in Maharashtra State in India (Talukdar *et al.*, 1996) and another one, JBH 1 (= GICH 501), has been released by the Madhya Pradesh Variety Release Committee (Hash *et al.*, 1997). The protogynous flowering of pearl millet essentially makes its panicle equivalent to a male-sterile until it sheds its own pollen, by which time it would be pollinated by pollen from other plants. This trait can be effectively used to breed top cross hybrids that might be more desirable in ergot- and smut prone areas than top cross hybrids developed on CMS-based inbred seed parents (Andrews *et al.*, 1993). Inter population hybrids have not been commercialized so far, but this now appears possible, as the feasibility of breeding male-sterile populations using the A_4 CMS system has been demonstrated (Rai and Rao, 1995).

The first commercial hybrid HB-I (Tift 23A × BIL-3B) by Athwal (1965), developed at Punjab Agricultural University (PAU), Ludhiana and released in India in 1965, had twice the yield of OPVs (Athwal 1965). Since then private seed companies in India emerged as the most dominant and reliable force in the commercialization of hybrid breeding products. Hybrid programs in India accorded highest prority to grain yield and downy mildew resistance combined with maturity duration mostly in the range of 75-85 days, as per agro-ecological requirements. Apart from evident quality traits both in the grain (size, shape, color) and stover (less rust, thinner stem, lodging resistance) that receive some attention, consideration of any other quality traits has been negligible.

CMS Systems in Pearl Millet

Burton (1958) first reported the CMS system in pearl millet and designated it as A_1. Different sources of MS cytoplasm such as A_2, A_3 from genetic stocks and their derivatives (Athwal 1961, 1966), Av and A_4 CMS sources from *Pennisetum glaucum* ssp *monodii* (=*violaceum*) accessions (Marchais and Pernes 1985, Hanna 1989), Ghana and Botswana sources (Appa Rao *et al.*, 1989), A_{egp} and A_5 CMS sources from gene pools (Sujata *et al.*, 1994, Rai 1995), Gero, Maiwa (Aken'ova, 1982) have been discovered. Based on differential male fertility restoration patterns, it has been established that the A_1, A_2, A_3, A_v, A_4 and A_5 were distinctly different CMS systems.

Although all the commercial hybrids are based on A_1 CMS system, it has been observed that all the A-lines involved in these hybrids produce low frequency of pollen shedders. About 40 years of research with the A_2 and A_3 CMS systems, mostly at PAU, Ludhiana, failed to produce an A-line with stable male-sterility comparable to A_1-system A-lines. In the iso-nuclear background of 81, male sterility of A1 and Av had similar low frequency of pollen shedders, while A4 and A5 had no pollen shedders. These two new CMS systems also appear to have more stable male sterility than the A_1 CMS system. Also for these two CMS systems, much more of the germplasm is in the maintainer class than in the A_1 system, offers the opportunity to accelerate the diversification of the genetic base of pearl millet seed parents (Hanna 1989; Rai and Rao 1996; Rai *et al.*, 1996). Recently Kumar and Sagar (2009) compared five systems of male sterility in pearl millet and did not find the cytoplasm to be associated with downy mildew susceptibility.

Breeding for seed yield in pearl millet parental lines, especially seed parents, is a higher priority and a greater challenge due to existence of inbreeding depression. The hybrid parents should have good fertility restoration ability of R lines, complete male sterility maintenance ability in B lines, and acceptable levels of resistance to downy mildew. On top of the above features, there should also be adequate diversity for morpho-physiological and grain traits among the lines to satisfy the diverse needs of farmers. It is only after this has been achieved that evaluation for general combining ability starts, mostly at the F5–F6 or S4–S5 stage. Performance of lines per se and their general combining ability for seed yield in pearl millet, have been found to be either uncorrelated or positively correlated.

In Maharashtra State, with a large number of private seed companies and an aggressive state seed corporation, had about 18 improved pearl millet cultivars under various scales of cultivation during the mid-1990s compared with no more than three during the mid-1980s. Similar changes in pearl millet cultivar diversity have occurred in Gujarat State. These two leading states now have 85–90 per cent of the pearl millet hectarage under HYV cultivation; in Gujarat State this consists mostly of hybrids but, in Maharashtra State, a substantial proportion is still an OPV (ICTP 8203). In contrast, Rajasthan State, which has about 40 per cent of the total pearl millet area in India, much of it in unfavorable (dry and low soil fertility) environments, grows HYVs on <25 per cent of the area, because there is limited cultivar development and seed production effort directed toward this area.

Development of hybrids of pearl millet in the last forty years has led to its increased productivity and stability, largely in the regions with relatively better environments, while regions like western Rajasthan with poor environments still suffer from low productivity of about 470 kg ha^{-1}. This is because most of the hybrids recommended for this region were developed elsewhere and lacked the desired adaptability and characteristics required for this region (Kelley *et al.*, 1996).

Production Constraints

Poor productivity of unimproved cultivars is related to their low harvest index, numerous biotic and abiotic stress factors.

Downy Mildew

Pearl millet downy mildew, caused by *Sclerospora graminicola* (Sacc.) J. Schröt, presents a clear contrast to drought. Its impact on pearl millet production, in a global context, is secondary to that of drought, even though downy mildew epidemics on single-cross hybrids have occurred repeatedly in India and can be expected to occur on such cultivars in Africa as well. Breeding for resistance to downy mildew, however, has been accorded the highest priority among the biotic stress factors (Hash *et al.*, 1997), and it is an integral part of a majority of the breeding programs in India (especially at ICRISAT—Patancheru) and Africa, because (*i*) a large number of resistance sources (including elite breeding lines) have been identified in diverse genetic backgrounds and effective screening techniques have been developed (Singh *et al.*, 1997) and (*ii*) inheritance of resistance is better understood, with most of the evidence in favor of resistance being dominant and displaying continuous variation in segregating populations (Singh, 1995). Over the years, many centers in the National Agricultural Research Systems have initiated or stepped up their downy mildew resistance breeding activities.

Other biotic constraints have received relatively low priority in breeding programs targeted to the semi-arid tropical regions of Asia and Africa. These can be grouped into two categories. The first category includes ergot (*Claviceps fusiformis* Lovelesse) and smut (*Moesziomyces penicillariae* (Bref.) Vánky), diseases of pearl millet; for these, good resistance sources and effective screening techniques have been developed (Sharma *et al.*, 1991, Sharma *et al.*, 1993; Thakur *et al.*, 1992, Thakur *et al.*, 1993; Pande *et al.*, 1994; Ejeta *et al.*, 1997). However, these have been shown not to have as large an impact on yield, on as large a geographical scale. Also, some of these can, to some extent, be managed by other means, for instance, by breeding for improved male fertility restoration in hybrids and by resorting to alternative cultivar options (*e.g.*, open-pollinated varieties (OPVs), topcross hybrids, and protogyny-based single-cross hybrids (seed produced using protogynous flowering rather than by male sterile seed parents) can be effective in reducing ergot and smut severity in pearl millet). The second category includes stem borers of pearl millet (*Coniesta ignefusalis* Hampson); and head miner (*Heliocheilus albipunctella* de Joannis) and seedling thermo tolerance of pearl millet, for which confirmed sources of good resistance are not available and (or) the trait inheritance is too complex to permit its effective utilization in breeding. This category has received little attention in breeding programs in Asia and Africa.

Progenies are periodically evaluated for downy mildew resistance during the course of inbreeding and selection for DM resistance runs concurrent to agronomic evaluation finally producing B-lines and R-lines resistant to this disease. It has been found that selection for resistance to one pathotype is effective, to some extent, in selecting for resistance to another diverse pathotype. Integrated conventional and marker-assisted backcrossing to introgress DM resistance QTLs from sources IP 18293 and P 1449-2 into the genetic backgrounds of maintainer lines 81B, 843B, and PT 732B advanced by a generation following screens against highly virulent pathogen isolates from Patancheru and New Delhi.

Selection Criteria in B-lines

At maturity plants with desirable traits like DM resistance, good tillering, dwarf to medium dwarf plant height, lodging resistance, medium to long panicles, compact panicles, having complete panicle exertion, acceptable seed size, shape and color were selected for further pedigree breeding of B lines. The complete and stable male-sterility of a-lines, high levels of male fertility restoration of R-lines, good pollen shedding in B-lines and r-lines, flowering synchrony between A-lines and R-lines are also considered. The d_2 dwarf plant height has emerged as the most dominant plant type in seed parents breeding as it reduces danger of lodging, allows for rapid detection and rouging of pollen shedders and off types in A-lines, provides option of breeding hybrids with varying heights.

Selection Criteria in R-lines

Restorers must produce profuse pollen that should remain viable at temperatures as high as 42-44°C. Also they should produce highly fertile hybrids that also reduces losses due to ergot and smut. They should be productive, 150-180 cm height but no taller than the A-line, resistant to lodging and downy mildew.

Drought

Pearl millet is usually grown in areas receiving less than 500mm of rainfall annually. Pearl millet is one of those cereal grasses which has strong development of underground organs and tends to have efficient adaptive mechanisms to cope with drought (Winkel and Do, 1992; Winkel *et al.*, 1997). Because of its tolerance to high temperature and better ability to withstand drought and to grow even in low soil fertility conditions, pearl millet is best suited for arid and semi-arid regions of the country (Khairwal 2007).

It is usually planted at the start of the rains in the arid semi-desert areas of north-east India (Rajasthan) (Bidinger *et al.*, 1987a,b). Because the duration of the rains is normally shorter than the duration of the crop, the stress that millet commonly experience is a terminal drought, where by seed filling occurs with plants depending on the moisture available in the soil. Drought is the most devastating production constraint that can occur at any crop growth stage. The phenotyping of terminal drought tolerance uses three water regimes: full irrigation; early stress imposed by stopping irrigation at booting; and late stress imposed by stopping irrigation at flowering (Serraj *et al.*, 2005). Postflowering (terminal or end of season) drought stress is of greater significance causing a reduction in grain yield by about 45 – 49 per cent with large variation among the genotypes (Fussell *et al.*, 1991). Even under favorable conditions, pearl millet tends to have a shorter crop cycle than other cereals because it has a "built in" drought escape mechanism (early flowering) inherited from its wild progenitors, having evolved in semi-desert environments with adapted short life cycles. Therefore, pearl millet is short cycled, has a short grain-filling period and has small seed sizes. Its growth potential is no match for other longer-duration cereals growing in favorable environments. Yet, it enjoys a high crop growth rate that confers a fairly high growth potential under optimal conditions (Begg, 1965), relating in particular to its being a C4 plant, with a large leaf area index(LAI) due to its erect type (Craufurd and Bidinger, 1989), and high radiation-use efficiency (RUE; Squire *et al.*, 1986).

The study of the influence of the drought using osmotic solutions is one of the methods in the study of resistance during the germinal phase. Several literatures have indicated the superiority of the germination capacity of pearl millet compared to that of sorghum as well as its remarkable resistance in the drought (Van Den Abeele and Vandenput, 1956; Gaudy, 1957; White and Cooper., 1959; Martin and Leonard, 1967). Radhouane (2007) studied seeds of pearl millet from six provenances of Tunisia were subjected to germination and shoot and root length traits on filter paper treated with polyethylene glycol 6000 (PEG 6000) solutions and found PEG useful for early screening. Early and rapid elongation of roots is important indication of drought resistance. Ability of continued elongation of root under situation of water stress was remarkable character of most crop species and root length is more affected to drought condition than shoot length (Kulkarni and Deshpande, 2007).

Panicle threshing percentage under stress conditions is a good indicator of terminal drought tolerance in pearl millet. In conjunction with grain yield potential and earliness, it accounts for >80 per cent of the genotypic variation in grain yield under terminal drought stress (Bidinger and Mahalakshmi, 1993). However, published information confirming the existence of adequate variability for drought-tolerance parameters and adequate repeatable heritability of the trait(s) associated with drought resistance is still lacking. Also, the effectiveness of the screening techniques available for selecting for drought tolerance has not been demonstrated convincingly in terms of how material selected for superior performance in a drought nursery at one location will perform under drought stress at another location.

Breeding Strategies for Drought Tolerance

The overall goal of a breeding programme for drought stress is, ultimately, an improved genetic yield, or a more stable yield, under drought conditions. These two objectives are not necessarily related. The latter, the stabilization of yield across environments in drought-prone areas, is very important because of the large differences in the coefficient of variation of pearl millet production at the all-India level(26 per cent) compare to that in Rajasthan state (53 per cent), which is characterized by very low and erratic rainfall. There are different ways to assess what is commonly called "drought tolerance," and this depends mostly on how close the assessed trait/parameter is from the final target–an increased genetic yield. Therefore, the approaches to drought tolerance vary.

Three categories can broadly be defined, with advantages and drawbacks as highlighted below:

- ☆ Drought tolerance is seen purely as a higher and more stable yield under drought conditions, which is fully in line with the ultimate goal. However, in almost all cases, this is related to a large GEI because yield is the integration of many different processes, each of them having a close interaction with the environment.
- ☆ Drought tolerance is considered as the maintenance of different development and growth processes, such as leaf expansion, at levels that are close to control well-watered plants. Here, we assume that these would remain well-linked to yield performance. This approach is straight forward and may be easier to capture than yield itself. However, some of these traits can be cumbersome to measure, which may not allow time to assess large numbers of accessions and progenies.
- ☆ Drought tolerance can be seen as more upstream, at the organ or cell level, and can be seen as the capacity to sustain certain biological mechanisms, such as maintaining leaf turgor, close to the level of well-watered plants. Measuring such traits requires screening under controlled environment conditions where better management and reproducibility of environmental variation can help reach low levels of GEI. However, the main drawback of

this approach is that the traits may be loosely related to the final yield under stress.

- ABA levels under well-watered conditions were higher in tolerant genotypes, including NIL-QTLs, than in sensitive genotypes, and ABA did not increase under water stress. Well-watered Tr was lower in tolerant than in sensitive genotypes at all VPD levels. Except for one line, Tr slowed down in tolerant lines above a breakpoint at 1.40-1.90 kPa, with the slope decreasing >50 per cent, whereas sensitive lines showed no change in that Tr response across the whole VPD range. It is concluded that two water-saving (avoidance) mechanisms may operate under well-watered conditions in tolerant pearl millet: (i) a low Tr even at low VPD conditions, which may relate to leaf ABA; and (ii) a sensitivity to higher VPD that further restricts Tr, which suggests the involvement of hydraulic signals. Both traits, which did not lead to TE differences, could contribute to absolute water saving seen in part due to dry weight increase differences. This water saved would become critical for grain filling and deserves consideration in the breeding of terminal drought-tolerant lines (Kholova *et al.*, 2010).

Over the past decade, ICRISAT and its partners have made substantial investments in developing mapping populations (Hash and Witcombe, 1994) and in DNA-based molecular marker systems including restriction fragment length polymorphism (RFLP; Liu etal.,1994), sequence-tagged sites (STS; Devos *et al.*, 1995), amplified fragment length polymorphism(AFLP), and simple sequence repeat (SSR) markers (Qi *et al.*, 2000; Allouis *et al.*, 2001), and a bacterial artificial chromosome (BAC) library (Allouis etal.,2001) for pearlmillet. These genetic tools have been used to develop a DNA marker-based linkage map for pearlmillet (Liu etal.,1994), and to map quantitative trait loci (QTLs) conferring resistance to biotic stresses (Jonesetal.,1995,2002;Morgan *et al.*, 1998) and tolerance to terminal drought stress (Yadav *et al.*, 2002b). Levels of DNA marker polymorphism in pearl millet are very high, even between elite in bred parental lines of hybrids adapted to growth in India. The current pearl millet DNA marker based genetic linkage map covers about 700cM(Haldanefunction) distributed across the expected seven linkage groups for this diploid ($2n = 2x = 14$) species, and at least one free-floating pair of linked RFLP markers. These DNA marker-based linkage groups have not been definitively linked with the chromosome map of this species (Minocha and Sidhu,1981;Kaul and Sidhu,1997), which has been developed over the past 35years using morphological markers(Anand Kumar and Andrews,1993) and conventional cytogenetic methods(Jauhar and Hanna,1998). Compared to most other grasses, the pearl millet genome appears to have undergone a large number of structural rearrangements (Devos and Gale, 2000).

Tillering and developmental plasticity: This trait is derived from the wild parents. Pearl millet develops primary tillers and then the secondary tillers from the primary ones. The secondary tillers will compensate yield in case of mid season drought (Mahalakshmi and Bidinger, 1986). As the period between floral initiation and flowering is similar, the plant has tillers in all stages of apical development at all times (Craufurd and Bidinger, 1988a,b). It is often considered that pearl millet is not affected very much by mid-season drought, provided that moisture is available for the end of the season (Mahalakshmi and Bidinger, 1985a, b).

Early flowering is important as a drought escape mechanism. However, it appears that the prospect of breeding for earliness is limited because of the often poor predictability of rainfall events in the semi-arid tropics. Therefore, there seems to be an optimal time for flowering, suited to the average season length. It is within that particular range of flowering times for any particular environment that other traits likely to improve performance under water-limited conditions must be found.

Rooting

Pearl millet is known to be deep and profusely rooted, with the ability to match its rooting to water availability in a very plastic manner, leading to a highly varying root growth to shoot growth ratio, depending on the intensity of water limitation (Squire *et al.*, 1987). Roots appear to play an important role in pearl millet genotypes that differ in the presence or absence of a major terminal drought tolerance QTL (Vadez *et al.*, 2005). Further efforts are needed to clarify the extent of the role of the root in the drought tolerance of pearl millet.

Drought Resistance Index

It has been found that about 50 per cent of yield variation under drought stress conditions could be explained by differences in the yield potential of genotypes and their flowering time (Bidingeretal., 1982, 1987a). This led to develop an index, the"drought resistance index"(DRI), in which the effect of yield potential and drought escape (flowering time) are removed by assuming that yield under stress is a function of yield potential (control yield in the test environment), drought escape (proxied by time to flowering), and a residual that accounts for drought tolerance/susceptibility. So that:

$$\hat{}\,Ys = aYc + bFl + Residual$$

where,

^ Ys is the predicted yield under stress based on the yield under control conditions, respectively, the flowering time (Fl) and a residual. This residual variation in grain yield under stress that is not explained by either the potential yield (Yc) or by the flowering time (Fl) represents the DRI. The value of the residual (= DRI) is obtained as follows: DRI = Ys – ^ Ys, where Ys is the actual grain yield under stress conditions. Therefore, the DRI represent the deviation in grain yield under stress from a baseline yield that depends on the yield potential and flowering time and it there- fore allows comparing genotype's performance regardless of their yield potential and flowering time. The DRI approach has been used in a selection programme for improved drought tolerance (see below), using the panicle harvest index (PNHI), *i.e.*, the ratio of grain yield to panicle yield on a plot basis, as a proxy to assess the DRI.

DRI in both midseason and terminal stress treatments was unrelated to yield component structure in the irrigated control treatment, indicating that selection for plant type under non-stressed conditions will not influence drought response. The correlation of DRI and individual panicle yield was of sufficient magnitude for the latter to serve as a selection index in terminal stress. (Bidinger *et al.*, 1987).

Water use efficiency: Being a C4 plant, pearl millet already has high transpiration efficiency (TE). However, it seems that the major strategy of pearl millet is to maximize carbon fixation as long as water is available. Therefore, stomatal movements adapt in such a way that the transpiration rate is kept as high as possible (Squire, 1979; Henson and Mahalakshmi,1985). It also appears that stomata are sensitive to the vapor pressure deficit(VPD), particularly during the pre-flowering stage. It also appears that stomata are sensitive to the vapor pressure deficit (VPD), particularly during the pre-flowering stage, this being related to differences in the abscicic acid (ABA) content of the leaves (Henson and Mahaklakshmi, 1985). In any case, there have been no studies to assess the range of variation in TE across a diverse range of pearl millet cultivars and lines, nor on the sensitivity of stomata to VPD.

QTL for Terminal Drought Tolerance

In most of the environments where pearl millet is grown, the crop is facing stress during the grain-filling period, in particular in Northern India (van Oosterom *et al.*, 1996 a,b). Therefore, work has focused on identifying QTLs for terminal drought tolerance using the PNHI as a selection criterion.

Number of grains per panicle and the 100-grain weight were the yield components most affected under terminal drought conditions, leading to a decrease in the PNHI. DRI represents the share of the variation in yield across a set of genotypes that cannot be explained either by differences in yield potential or time to flowering, and is closely related to yield under stress conditions. Bidinger *et al.* (1987a,b) have also shown DRI was closely related to the PNHI and therefore, a high DRI was closely related to a higher percentage grain set and better grain filling, which are the major components of the PNHI. In subsequent works, the PNHI has then been used as an indirect proxy for DRI, and is readily and cost-effectively measured.

Trait-Based Conventional Approach

This approach was initially used to select genotypes achieving a high PNHI under terminal drought conditions. PNHI was initially tested in hybrid parent breeding, where it was used as a selection criterion by the following procedure:

1. Conduct bidirectional selection for combining ability for high and low PNHI in replicated potential maintainer(B) and restorer(R) line test cross nurseries (three testers each) grown in managed terminal drought stress environments.
2. Cross parents selected for high and low PNHI under stress conditions on three different A or R line testers from those used in the original test cross nurseries in which selection was carried out.
3. Evaluate these test crosses for general combining ability (GCA) for PNHI, grain yield and yield components, in both fully irrigated control environments and in managed stress environments.

Trait-Based Molecular Breeding Approach in Current Use

This is the current approach to pearl millet breeding for drought tolerance. It is based on the fact that PNHI remains a highly complex trait for which a molecular approach can increase precision during the selection process. We have shown earlier that the yield under terminal drought was a function of the yield potential under no stress, a drought escape mechanism, and DRI per se. Therefore, it is not advisable to cross parents with large variations in yield potential or flowering time if the intention is to develop a RIL population to map terminal drought tolerance. This approach has been used successfully for the identification of terminal drought tolerance QTLs (Yadav *et al.*, 2000, 2002b, 2003), and the introgression of a terminal drought tolerance QTL into the background of the popular pearl millet hybrid HHB67to create the new hybrid HHB67-improved.This terminal drought tolerance QTL has a major effect, explaining over 30 per cent of the yield variation under terminal drought. It is located on linkage group 2 (LG2). Further efforts are still needed to reduce the size of that QTL to improve the precision of its introgression.

Phenotyping Traits

Traits should be relevant to the target area

- ☆ The rate of water loss per unit of leaf area and time was lower in PRLT/89-33, a terminal drought tolerant parent compared to H77/833-2, a terminal drought sensitive genotype (Vadez *et al.*, 2007). These differences were found under well-watered conditions and were consistently found across experimental seasons, at both the pre-flowering and post-flowering stages.
- ☆ All terminal drought tolerant materials had profuse rooting in the deep soil layers than did sensitive materials under water stress (Vadez *et al.*, 2007). In contrast, there seemed to be little difference under well-watered conditions. Therefore, rooting appears to be an adaptive trait that tolerant pearl millet genotypes "develop" under stress conditions. However, the phenotyping of root trait differences would better be based on the volume of water uptake during the grain-filling period.
- ☆ QTL on LG 2 remains large in size and, therefore, relatively difficult to introgress. Hence secondary traits for the high PNHI of the terminal drought tolerant lines would be needed to refine the QTL interval and facilitate its use in modern breeding
- ☆ Identification of traits in different drought environments

Seedling Heat Tolerance

Pearl millet [*Pennisetum glaucum* (L.) R. Br.] is the staple cereal grain and fodder crop grown by subsistence farmers in the hottest, driest regions of the Sahelian zone of sub-Saharan Africa and the Indian subcontinent where rainfed crop production is possible. Early cessation of rains is the single most important factor contributing to low mean grain and straw yields and poor yield stability of this crop. Failure of seedling establishment due to extreme high soil temperatures at the start of the growing season result in the need to re-sow and an increased likelihood that the crop will be exposed to the damaging effects of post-flowering drought. Some of the pearl millet populations were shown to have good seedling emergence and survival at soil surface temperatures exceeding 62°C (Peacock *et al.*, 1993). Large pearl millet area is under cultivation in Gujarat which is the best testing site for screening for seedling heat tolerance.

Biotech Integration

Molecular MAS has been applied as a tool to enhance the efficiency of breeding for DM resistance, drought tolerance and stover quality in pearl millet. The MAS for drought tolerance and stover quality is in the strategic research phase. The MAS for DM resistance has been integrated with breeding. More than 56 quantitative loci (QTLs) for DM resistance against 14 diverse pathotypes (7 Indian and 7 African) have been identified.

Grain Quality Studies

Golden millet with about 1.37 ppm of beta-carotene, which is comparable to beta-carotene in golden rice and 3-5 times more than traditional pearl millet identified. This is in photosensitive background and yellow endosperm acts as a proxy to beta-carotene. Parental lines and populations with up to 120 ppm Fe and up to 68 ppm Zn were identified and high positive correlation (r= 0.84**) exists between Fe and Zn density.

Alternate Uses of Pearl Millet

There is a huge demand for pearl millet towards extraction of alcohol, provided the starch content in pearl millet is increased from the present level of 55 per cent to 65 per cent. Presently, most of the distilleries are using broken rice in the distilleries, as rice has high starch content and the unit cost of starch from rice is cheaper than starch from the pearl millet. Hence, the demand for pearl millet from the distilleries depends on the relative prices of broken rice and pearl millet (Nagaraj *et al.*, 2011).

Researchable Areas

1. **Utilization of alternate CMS systems:** The A_1, A_4 and A_5 CMS resources are well characterized for their commercial potential. The utilization of A_4 and A_5 as the more stable and useful CMS systems compared to the widely used A_1 CMS system should be enhanced. The pace of shift to these two new CMS sources depend on the pace of their restorer development, for which various approaches (pedigree breeding, backcross breeding and population approaches) have been shown to be effective.
2. **Hybrid options:** Considering the increasingly dominant role of the private sector in hybrid development and seed production, single cross hybrids will remain the only commercial option. Hence, inbred seed parent development with high yield potential, DM resistance and adaptation to the regions concerned should be focused. Hybrid development should focus on better-endowed environments and strategic research should be taken up for marginal environments.

Table 1: Hybrids and Varieties Suitable for India

Hybrid	*Parentage*	*Release Year*	*Notification Number and Date*	*Bred at*	*Area of Adaptation*	*Maturity Duration and Salient Features*
A. Hybrids						
KBH 108 (MH 1737)	KMPS-74A x KPR-9288R	2013	S. O. 244 (E) 24.01.2014	Krishna Seed (P) Ltd., Agra	Rajasthan, Gujarat, Haryana, Punjab, Delhi, UP and MP	Late maturing, tall plant height, purple anther colour, cylindrical very compact ear heads, obovate grey seed, resistant to downy mildew, blast and smut
GHB-905 (MH-1655)	ICMA 04999 x J- 2454	2013	S. O. 2815 (E) 19.09.2013	AICPMIP, MRS, Jamnagar	Rajasthan, Gujarat, Haryana, Punjab, Delhi, UP and MP	Medium maturing, medium height, yellow anther colour, compact cylindrical earheads with bristles, resistant to downy mildew, globular grey brown seeds
Nandi-72 (MSH-238) (NMH 75)	NMS 31A x NMP 75	2013	S. O. 2815 (E) 19.09.2013	New Nandi Seeds Corporation, Ahmedabad	Summer growing areas of Gujarat, Maharashtra, Rajasthan and Tamil Nadu	Medium tall plant height, purple coloured anthers lanceolate thick compact earheads, resistant to downy mildew, globular deep grey seed.
86M89 (MH 1747)	M163F x M164R	2013	S. O. 2815 (E) 19.09.2013	Pioneer Overseas Corp., Hyderabad	Rajasthan, Gujarat, Haryana, Punjab, Delhi, UP and MP	Late maturing, tall height, conical very compact earheads, yellow anther colour, resistant to downy mildew, obovate deep grey coloured seed
MPMH 17 (MH 1663)	ICMA 04999 x MIR 525-2	2013	S. O. 952 (E) 10.04.2013	AICPMIP, Jodhpur	Rajasthan, Gujarat, Haryana, Punjab, Delhi, UP and MP	Medium maturing, medium height, yellow anther colour, compact lanceolate earheads with bristles, resistant to downy mildew, grey brown seeds
HHB 234 (MH 1561)	HMS 7A x H77/833-2- 202	2013	S. O. 952 (E) 10.04.2013	AICPMIP CCS HAU Hisar	Western Rajasthan and drier part of Gujarat and Haryana	Early maturing, candle shaped earheads with small bristles, medium seed size and tolerant to downy mildew

Contd...

Table 1–*Contd...*

Hybrid	*Parentage*	*Release Year*	*Notification Number and Date*	*Bred at*	*Area of Adaptation*	*Maturity Duration and Salient Features*
Kaveri Super Boss (MH 1553)	KBMS 329 x KBR 621	2012	S. O. 2125 (E) 10.09.2012	Kaveri Seed Co. Ltd., Secundrabad	Rajasthan, Gujarat, Haryana, Punjab, Delhi, UP and MP Maharashtra, Karnataka, Andhra Pradesh, Tamil Nadu	Late maturing, tall height, long compact cylindrical earheads, purple anther colour, globular grey coloured seed
Bio 70 (MH 1632)	11A x R 207	2012	S. O. 2125 (E) 10.09.2012	Bioseed Research India Pvt. Ltd., Hyderabad	Western Rajasthan and drier part of Gujarat and Haryana	Early maturing, conical compact earheads, yellow anther colour, globular grey seed and tolerant to downy mildew
Bio 448 (MH 1671)	13A x R 210	2012	S. O. 2125 (E) 10.09.2012	Bioseed Research India Pvt. Ltd., Hyderabad	Rajasthan, Gujarat, Haryana, Punjab, Delhi, UP and MP	Late maturing, cylindrical compact earheads, yellow anther colour, globular grey seed and tolerant to downy mildew
Nandi-70 (MSH 224)	NMS 31A x NMP 55	2012	S. O. 2125 (E) 10.09.2012	New Nandi Seeds Corporation, Ahmedabad	Summer growing areas of Gujarat, Maharashtra, Rajasthan and Tamil Nadu	Medium plant height, cylindrical thick compact earheads, globular deep grey seed.
Pratap (MH 1642)	NB 101A x NB 152R	2012	S. O. 2125 (E) 10.09.2012	Nuziveedu Seeds Pvt. Ltd., Hyderabad	Maharashtra, Karnataka, Andhra Pradesh, Tamil Nadu	Medium maturity, medium plant height, cylindrical semi compact earheads, globular grey seed.
PKV-Raj (BBH 3)	BMS-5-23A x BR-333	2012	S. O. 2125 (E) 10.09.2012	Dr. PDKV, ARS, Buldana	Maharashtra	Medium maturing, Medium height, conical compact bristled earheads, grey coloured seed
MP-7872 (MH-1610)	M002A x M004R	2012	S. O. 1708 (E) 26.07.2012	Metahelix Life Science Ltd., Bangalore	Rajasthan, Gujarat, Haryana, Punjab, Delhi, UP and MP	Late maturing, medium height, yellow anther colour, spindle earheads, grey coloured globular seed
MP-7792 (MH-1609)	M001A x M004R	2012	S. O. 1708 (E) 26.07.2012	Metahelix Life Science Ltd., Bangalore	Rajasthan, Gujarat, Haryana, Punjab, Delhi, UP and MP	Late maturing, medium height, yellow anther colour, cylindrical earheads, grey coloured globular seed

Contd...

Table 1–*Contd...*

Hybrid	*Parentage*	*Release Year*	*Notification Number and Date*	*Bred at*	*Area of Adaptation*	*Maturity Duration and Salient Features*
86M86 (MH 1684)	M128F x M138R	2012	S. O. 1708 (E) 26.07.2012	Pioneer Overseas Corp., Hyderabad	Rajasthan, Gujarat, Haryana, Punjab, Delhi, UP and MP Maharashtra, Karnataka, Andhra Pradesh, Tamil Nadu	Late maturing, Medium to tall plant height, conical very compact earheads, purple anther colour, grey hexagonal seeds
CO 9	ICMA 93111 x PT 6029- 30	2012	S. O. 1708 (E) 26.07.2012	AICPMIP, TNAU, Coimbatore	Tamil Nadu	Medium maturing, medium height, Candle compact earheads, greyish yellow seed colour
Shine (VBBH 3040) (MH 1578)	VBBA 310089 x VBBR 330585	2012	S. O. 456 (E) 16.03.2012	Vibha Agrotech Ltd., Hyderabad	Maharashtra, Karnataka, Andhra Pradesh, Tamil Nadu	Medium maturing, medium height, red node pigmentation, yellow anther colour, compact spindle shaped earheads resistant to downy mildew, grey bold seeds
86M66 (MH 1617)	M124F x M118R	2011	S.O. 2326 (E) 10.10.2011	Pioneer Overseas Corp., Hyderabad	Rajasthan, Gujarat, Haryana, Punjab, Delhi, UP and MP	Late maturing, medium height, conical compact earheads, brownish yellow anthers, broad leaves, purple node colour, grey seed colour, resistant to downy mildew
PAC 909 (MH 1435)	110057 x 130453	2011	S.O. 2326 (E) 10.10.2011	Advanta India Ltd., Secundrabad	Maharashtra, Karnataka, Andhra Pradesh, Tamil Nadu	Medium maturing, medium height, medium thick compact cylindrical earheads, light yellow coloured anthers, grey seed colour, resistant to downy mildew
HHB-226 (MH 1479)	ICMA 843- 22 x HBL-11	2011	S.O. 632 (E) 25.3.2011	AICPMIP CCS HAU Hisar	Western Rajasthan and drier part of Gujarat and Haryana	Medium maturing, medium height, dark green leaves, candle shaped bristled earheads, resistant to downy mildew

Contd...

Table 1–*Contd...*

Hybrid	*Parentage*	*Release Year*	*Notification Number and Date*	*Bred at*	*Area of Adaptation*	*Maturity Duration and Salient Features*
RHB-177 (MH-1486)	ICMA 843- 22 x RIB 494	2011	S.O. 632 (E) 25.3.2011	AICPMIP, SKRAU, Jaipur	Western Rajasthan and drier part of Gujarat and Haryana	Early maturing, medium tall, cylindrical bristled earheads, resistant to downy mildew, light yellow anthers
RHB-173 (MH-1446)	ICMA 93333 x RIB 192 S/99	2011	S.O. 632 (E) 25.3.2011	AICPMIP, SKRAU, Jaipur	Rajasthan, Gujarat, Haryana, Punjab, Delhi, UP and MP	Medium maturity, medium to tall plant height, compact cylindrical ear heads, resistant to downy mildew
86M64 (MH 1540)	M096F x M 117R	2011	S.O. 283 (E) 7.2.2011	Pioneer Overseas Corp., Hyderabad	Maharashtra, Karnataka, Andhra Pradesh, Tamil Nadu	Late maturing, conical compact earheads, obovate light grey seeds, resistant to downy mildew
86M53 (MH 1541)	M096F x M 119R	2010	S.O. 283 (E) 7.2.2011	Pioneer Overseas Corp., Hyderabad	Maharashtra, Karnataka, Andhra Pradesh, Tamil Nadu	Late maturing, conical compact earheads, obovate grey bold seeds, resistant to downy mildew
86M64 (MSH 203)	M096F x M 117R	2011	S.O. 283 (E) 7.2.2011	Pioneer Overseas Corp., Hyderabad	Summer growing areas of Gujarat Maharashtra, Rajasthan and Tamil Nadu	Late maturing, conical compact earheads, obovate light grey bold seeds, resistant to downy mildew
RHRBH 9808	RHRB 13A x RHRBI 1314	2010	S.O.2137 (E) 31.8.2010	AICPMIP MPKV, Dhule	Maharashtra	Medium maturing, medium tall plant height, dark green leaves, cylindrical earheads, yellow anthers
Nandi 65 (MH 1549)	NMS 24A x NMP 75	2010	S.O.2137 (E) 31.8.2010	New Nandi Seeds Corp., Ahmedabad	Rajasthan, Gujarat, Haryana, Punjab, Delhi, UP and MP	Late maturing, cylindrical shaped non bristled compact ear heads, yellow anther colour, bold grains resistant to downy mildew
Nandi 61 (MH 1548)	NMS 24A x NMP 64	2010	S.O.2137 (E) 31.8.2010	New Nandi Seeds Corp., Ahmedabad	Rajasthan, Gujarat, Haryana, Punjab, Delhi, UP and MP	Late maturing, conical shaped non bristled compact earheads, bright yellow anther colour, bold grains resistant to downy mildew

Contd...

Table 1–*Contd...*

Hybrid	*Parentage*	*Release Year*	*Notification Number and Date*	*Bred at*	*Area of Adaptation*	*Maturity Duration and Salient Features*
HHB 223 (MH 1468)	ICMA 94555 x HBL 11	2010	S.O.211 (E) 29.1.2010	AICPMIP CCS HAU Hisar	Rajasthan, Gujarat, Haryana, Punjab, Delhi, UP and MP	Medium maturing, conical earheads with long purple bristled, resistant to downy mildew, tolerant to drought
HHB 216 (MH 1421)	HMS 37A x HTP 3/13	2010	S.O.211 (E) 29.1.2010	AICPMIP CCS HAU Hisar	Western Rajasthan and drier part of Gujarat and Haryana	Medium maturing, resistant to downy mildew, candle shaped medium long to togo earheads with brownish long bristled
Nandi 64 (MSH 199)	NMS 2-11A x NMP 4-1	2010	S.O.211 (E) 29.1.2010	New Nandi Seeds Corp., Ahmedabad	Summer growing areas of Gujarat Maharashtra, Rajasthan and Tamil Nadu	Late maturity, conical shaped earheads, light purple anther colour, light reddish plant base, resistant to downy mildew
RHB 154 Western	ICMA 95444 x RIB 57 S/05	2009	S.O.2187 (E) 27.8.2009	AICPMIP ARS Durgapura, Jaipur	Rajasthan and drier part of Gujarat and Haryana	Early maturing, medium tall, resistant to downy mildew, tolerant to lodging, light yellow anthers, cylindrical earhead
JKBH 676	JKMS 20A x JKR 6136	2009	S.O.2187 (E) 27.8.2009	JK Agri genetics Hyderabad	Rajasthan, Gujarat, Haryana, Punjab, Delhi, UP and MP	Late maturity, Medium tall plant height, highly compact earheads, stay green fodder
GK 1051	PM 678A-II x PM 1081 R-I	2009	S.O. 454 (E) 11.02.09	Ganga Kaveri Hyderabad	Maharashtra, Karnataka, Andhra Pradesh, Tamil Nadu	Medium maturity, medium tall, cylindrical shaped ear heads with pearly grey grain
B 2095 (MH 1257)	B 0009A x B 5220R	2009	S.O. 449 (E) 11.02.09	Zuari Seeds Ltd Bangalore	Rajasthan, Gujarat, Delhi, UP and MP	Medium maturity, medium tall, compact candle earheads, globular grey grains
PB 727 (Proagro 9555)	PSP 51 x PP 38		S.O. 1108 (E) 8.05.08	Bayer Bio Science Hyderabad	Summer growing areas of Gujarat Maharashtra, Rajasthan and Tamil Nadu	Late maturity, medium tall, compact cylindrical earheads with yellow anthers, obovate grey grains
PHB 2168	ICMA 92333 x PIB 686	2008	S.O. 72 (E) 10.01.08	AICPMIP PAU Ludhiana	Rajasthan, Gujarat, Haryana, Punjab, Delhi, UP and MP	Late maturity, medium tall, compact cylindrical shaped earheads with yellow anthers, obovate grey grains

Contd...

Table 1–*Contd...*

Hybrid	*Parentage*	*Release Year*	*Notification Number and Date*	*Bred at*	*Area of Adaptation*	*Maturity Duration and Salient Features*
HHB 197	ICMA 97111 x HBL 11	2008	S.O. 72 (E) 10.01.08	AICPMIP CCS HAU Hisar	Rajasthan, Gujarat, Haryana, Punjab, Delhi, UP and MP	Early maturity, medium tall, dark green leaves, cylindrical medium Togo earheads with long bristles, highly resistant to down mildew disease
GHB 732	ICMA 96222 x J 2340	2008	S.O. 72 (E) 10.01.08	AICPMIP MRS Jamnagar	Rajasthan, Gujarat, Haryana, Punjab,Delhi, UP and MP	Late maturity, medium tall, compact lanceolate earheads, purple anthers, globular grey brown bold grain
GHB 744	ICMA 98444 x J 2340	2008	S.O. 72 (E) 10.01.08	AICPMIP MRS Jamnagar	Rajasthan, Gujarat, Haryana, Punjab, Delhi, UP and MP	Medium maturity, medium tall, medium thick stem with basal pigmentation, compact cylindrical shaped earheads with yellow anthers, globular grey brown grains
GHB 757	ICMA 92777 x J 2467	2008	S.O. 72 (E) 10.01.08	AICPMIP MRS Jamnagar	Western Rajasthan and drier part of Gujarat and Haryana	Early maturity, medium tall, compact cylindrical shaped earheads with purple anthers, globular grey brown grains
GHB 719	ICMA 95222 x J 2454	2007	S.O. 122 (E) 06.02.07	AICPMIP MRS Jamnagar	Western Rajasthan and drier part of Gujarat and Haryana	70-75 days maturity, fully exerted conical shaped, compact and bristled earheads, globular, medium in size, grey coloured grains, tolerant to drought.
NMH 68 (Nandi 62)	ICMA 97444 x NMP 48	2007	S.O. 1703 (E) 5.10.07	New Nandi Seeds Corp., Ahmedabad	Rajasthan, Gujarat, Punjab, Delhi, UP and MP	Late maturity, tall, pubescent nodes, compact cylindrical earheads that are slightly tapering towards apex, yellow anthers, globular grey grains
B 2301	B 0009A x B 5103R	2007	S.O. 1703 (E) 5.10.07	Zuari Seeds Ltd. Bangalore	Maharashtra, Karnataka, Andhra Pradesh, Tamil Nadu	Late maturity, medium tall, compact spindle shaped earheads, globular grey bold grains

Contd...

Table 1–*Contd...*

Hybrid	*Parentage*	*Release Year*	*Notification Number and Date*	*Bred at*	*Area of Adaptation*	*Maturity Duration and Salient Features*
HHB 67 Improved	843-22 x H77/ 833-2-202	2005	S.O. 1566 (E) 05.11.05	Hisar	Western Rajasthan and drier part of Gujarat and Haryana	Extra early maturity, highly resistant to moisture stresses, resistance to downey mildew The first commercial cultivars developed using marker-assisted selection in India.
GHB 538	94555A x J2340	2004	S.O. 1177 (E) 25.08.05	Jamnagar	Western Rajasthan and drier part of Gujarat and Haryana	Early maturity, highly resistant to moisture stresses, resistance to downey mildew and lodging
PB 180	PSP 41 xPP 29	2004	S.O. 161 (E) 4.02.04	Proagro	Summer cultivation areas of Rajasthan Haryana, UP, MP, Punjab and Delhi	85 days, tall, high tillering, pubescent green nodes, compact candle shaped earheads, yellow anthers, obovate dark grey grains Medium maturity, resistant ot downey mildew, smut, ergot.stay green fodder till harvest
GHB 577	JMS A 101 x J 2405	2003	S.O. 161 (E) 4.02.04	AICPMIP MRS Jamnagar	Rajasthan, Haryana, Gujarat, MP, UP	80-85 days, medium thick stem with basal pigmentation, semi-compact cylindrical earheads with slightly incomplete exertion, globular grains
HHB 117	HMS 7 A x H77/29-2		S.O. 161 (E) 4.02.04	AICPMIP CCS HAU Hisar	Haryana 74	78 days, medium tall, yellow anthers, compact candleshaped medium thick earheads, obovate grey grains, possesses stay
HHB 146	95222 A xHTP 94/54	2003	SO 283(E) 12.3.03	Hisar	Rajasthan, Haryana, Gujarat, MP, UP	75-80 days, tall, violet nodes, hairy leaf sheath, long well-filled compact earheads, slow senescence

Contd...

Table 1–*Contd...*

Hybrid	*Parentage*	*Release Year*	*Notification Number and Date*	*Bred at*	*Area of Adaptation*	*Maturity Duration and Salient Features*
GHB 558	94555A x J 2290	2002	SO 283(E) 12.3.03	Jamnagar	All India	75-80 days, broad leaves, long thick conical earheads, yellow anthers, obovate dark grey bold grains
7688	PH 03 xPH 05	2001	SO 92(E) 2.2.01	Pioneer	Rajasthan, Haryana, Gujarat, MP, UP	78-80 days, hairy green internode, light purple anthers
PB 106	PSP 41 xPP 6	2001	SO 92(E) 2.2.01	ProAgro	All India	76-80 days, tall, basal pigmentation purple, purple anthers, conical semi-compact earheads with tan bristles
PB 112	PSP 35 xPP 1	2001	SO 1134(E) 15.11.01	ProAgro	Rajasthan, Haryana, Gujarat, MP, UP, Delhi	85-88 days, tall, good tillering, glabrous green nodes, compact conical earheads, yellow anthers, bold and globular grey grains
RHB 121	89111A x RIB 3135-18	2001	SO 1134(E) 15.11.01	Jaipur	Rajasthan, Haryana, Gujarat, MP, UP, Delhi	85 days, medium tall, compact thick earheads, yellow anthers, long purple bristles, globular grey brown grains
Pusa 605	841A xPPMI 69	1999	SO 425(E) 8.6 99	Delhi	Rajasthan, Haryana, Gujarat, MP, UP	74-80 days, medium tall, yellow anthers, compact cylindrical earheads, obovate grey grains
Nandi 32	NMS 7A x NMP 24	1999	SO 1050(E) 26.10.99	New Nandi	Rajasthan, Haryana, Gujarat, MP, UP, Punjab, Delhi	78 days, pink base of stems, long cylindrical earheads, medium sized grey grains
RHB 90	81A x RIB 3135-18	1999	SO 821(E) 13.9.2000	Jaipur	Rajasthan	85 days, medium tall, yellow anthers, purple bristles, compact cylindrical earheads, obovate yellow brown grains
Nandi 8	NMS 5A x NMP 23	1998	SO 425(E) 8.6 99	New Nandi	Rajasthan, Haryana, Gujarat, UP	80 days, long cylindrical earheads, medium sized cream yellow grains

Contd...

Table 1–*Contd...*

Hybrid	*Parentage*	*Release Year*	*Notification Number and Date*	*Bred at*	*Area of Adaptation*	*Maturity Duration and Salient Features*
Pusa 322 (MH 322)	841A x PPMI 301	1993	SO 615(E) 17.8.93	Delhi	All India	80 days, medium tall, thick cylindrical semi-compact earheads with sterile tip, globular grey brown grains
ICMH 356	ICMA 88004x ICMR 87003	1993	SO 615(E) 17.8.93	ICRISAT	All India	75-80 days, medium tall, semi-compact thick conical earheads, yellow anthers, obovate dark grey grains
RHB 30	843A x RIB 335/74	1991	SO 360(E) 1.5.97	Jaipur	Rajasthan	75 days, medium tall, loose conical earheads, purple anthers, globular dark grey grains
RHB 58	81A xRIB 20K-86	1991	SO 408(E) 4.5.95	Jaipur	All India	85 days, medium tall, semi-conical compact earheads, purple anthers, obovate yellow brown grains
HHB 67	843A xH77/ 833-2	1990	SO 386(E) 15.5.90	Hisar	All India	60-62 days, medium height, thin stem, medium narrow leaves, semi-compact spindle shape earheads, yellow anthers, grey globular grains, escapes drought and tolerates salt stress, fits well in inter- and multiple cropping
Pusa 23 (MH 169)	841A xD 23	1987	SO 834(E) 18.9.87	Delhi	All India	82 days, medium tall, glabrous leaves, yellow anthers, compact cylindrical earheads, light grey grains

Contd...

Table 1–*Contd...*

Variety	*Release Year*	*Notification Number and Date*	*Bred at*	*Area of Adaptation*	*Maturity Duration and Salient Features*
B. Open pollinated varieties					
ABPC-4-3 (MP 484)	2012	S. O. 2125(E) 10.09.2012	AICPMIP, Aurangabad	Maharashtra	Late maturing, medium plant height, lanceolate earheads, globular greyseeds
Mandor Bajra Composite 2 (MBC 2) (MP 489)	2011	S.O. 2326(E) 10.10.2011	SKRAU, A.R.S. Mandor, Jodhpur	Rajasthan, Harayana, Gujarat	Early maturing, medium height, medium long semi compact cylindrical earheads, obvate grey coloured seed
Pusa Composite 612 (MP 480)	2011	S.O. 632 (E) 25.03.11	AICPMIP IARI, New Delhi	Maharashtra, Karnataka, Andhra Pradesh, Tamil Nadu	Medium maturity, medium to tall plant height, compact cylindrical earheads
Pusa Composite 443 (MP 443)	2009	S.O. 449 (E) 11.02.09	AICPMIP IARI, New Delhi	Rajasthan, Harayana, Gujarat	Early maturity, medium tall, rod shaped earheads with bold grain
PCB 164	2007	S.O. 1178(E) 20.07.07	AICPMIP PAU Ludhiana	Punjab	85-88 days, dual purpose variety with broad leaves, thick stalks, cylindrical earheads, bred from sevenelite population and 27 diverse inbred lines
JBV 4 (MP403)	2007	S.O. 1178(E) 20.07.07	AICPMIP COA Gwalior	MP	Mature in 75 days,medium tall, thick stemmed. Plants with many leaves, slate greyseeds.
PPC-6 (Parbhani Sampada)	2005	S.O. 122 (E) 02.02.05	AICPMIP RRS NARP Aurangabad	Maharashatra	75-80 days, medium tall, light pink nodes, bold grains, bred from 8 inbredlines
CoCu 9	2005	S.O. 1177(E) 28.08.05	AICPMIP TNAU Coimbatore	Rainfed/irrigated conditions in Tamil Nadu	80-85 days, tall, semicompact to compact, candle/cylindrical shaped earheads, grey grains with yellow base.

Contd...

Table 1–*Contd...*

Variety	*Release Year*	*Notification Number and Date*	*Bred at*	*Area of Adaptation*	*Maturity Duration and Salient Features*
CZP 9802	2002	SO 283(E) 12.03.03	CAZRI	Dry areas of Rajasthan, Gujarat and Haryana (zone A1)	70-75 days, good tillering, narrow leaves, thin candle shaped earheads, thin stem, yellowish grains of medium size, drought tolerant, very high stover of good quality
Pusa Composite 383	2001	SO 1134(E) 15.11.01	Delhi	Zone A	82 days, tall, thick stems and panicles, resistant to lodging and downy mildew
JBV 3 (GICKV 96752)	2000	SO 92(E) 2.2.01	Gwalior	Zone A	82 days, tall, yellow anthers, long compact cylindrical heads, obovate grey grains, developed from Cycle 3 of SRC II
Pusa Composite 334	1999	SO 1050(E) 26.10.99	Delhi	Zone A	75-80 days, tall, thick semi-compact cylindrical earheads, developed from selected lines and elite inbreds
JBV-2(GICKV 93191)	1999	SO 425(E) 8.6.99	Gwalior	Zone A	79 days, tall, compact-cylindrical earheads, yellow anthers, obovate brown grains, developed from Early Composite 91
Pusa Bajri 266	1996	SO 360(E) 1.5.97	Delhi	Zone A	80 days, tall, thick semi -compact cylindrical earheads, yellow anthers, obovate grey brown grains
CZP-1C 923	1996	SO 360(E) 1.5.97	Jodhpur	Zone A	72-80 days, thick stem, long oblanceolate thick panicles, light grey seed with yellow base, light yellow to brown anthers
ICMV 155	1991	SO 527(E) 16.8.91	ICRISAT	All India	85-89 days, thick semi compact cylindrical earheads, yellow anthers, obovate grey grains, dual-purpose variety bred from 59 plants of NELC C_4 selected at Patancheru

Contd...

Table 1–*Contd...*

Variety	*Release Year*	*Notification Number and Date*	*Bred at*	*Area of Adaptation*	*Maturity Duration and Salient Features*
Raj 171 (RCB-IC-9)	1990	SO 814(E) 4.11.92	Jaipur	All India	80-85 days, tall, medium thick stems, long thin semi compact earheads, obovate grey brown grains, resistant to downy mildew, bred from Intervarietal Composite
Raj Bajra Chari-2	1990	SO 386(E) 15.5.90	Jobner	All India	Suitable for green fodder, broad and succulent leaves, suitable for saline soils

3. **Trait focus:** Grain yield and downy mildew resistance should be given top priority. Emphasis should be given for dual purpose hybrids for northern India, and medium maturing hybrids for central and southern India. Breeding for earliness and forage hybrids should also be taken up. Farmer preferred traits such as panicle compactness, panicle size, lodging resistance, grain size and color should be considered while selection.
4. **Exploitation of germplasm and breeding lines:** Past breeding efforts have introgressed large seed size, panicle length and girth and tillering ability. Apart from them new germplasm sources for high biomass yield, panicle compactness combined with good panicle size should be utilized. Breeding lines with complementary traits for generating productive lines should be used.
5. **Marker-assisted breeding:** Marker assisted backcross breeding should be followed for DM resistance gene deployment. Based on funding, other traits such as drought tolerance, stover quality, drought tolerance, soil salinity tolerance and heat stress at flowering, heterotic grouping of elite breeding lines and germplasm should be taken up.

D. Biochemistry

- ✰ Grain quality analysis of advance coordinated trials for identification of promising genotypes in respect of protein and fat content
- ✰ Screening of germplasm for desired character (better micronutrients availability, Low phytate, high phytase, high protein and starch digestibility, high carotenoids
- ✰ Shelf life of flour (Purification and characterization of isoenzymes of peroxidase, lipase and Lipoxygenase. Screening of germplasm for rancidity determinants)
- ✰ Reduction or degrading of anti nutritional elements and increase of shelf life.
- ✰ Dehulling, Milling, Malting Blanching Acid treatment, Dry heat treatmen and Parboiling

References

1. Aken'ova M. E. 1982. Male-sterility in Nigerian bulrush millets (*Pennisetum americanum* (L.) K. Schum.]. *Euphytica*, 31: 161-165.
2. Allouis,A., Qi,X., Lineup,S., Gale, M. D., and Devos,K.M. (2001). Construction of a BAC library of pearlmillet, Pennisetum glaucum. *Theory. Appl. Genet.* 102, 1200–1205.
3. Anand Kumar, K., and Andrews, D. J. (1993). Genetics of qualitative traits in pearlmillet: a review. *Crop Sci.* 33, 1–20.
4. Andrews, D.J., and Anand Kumar, K. 1996. Use of the West African pearl millet landrace *Iniadi* in cultivar development. *In* Plant Genetics Resources Newsletter

No. 105. The Crop Ecology and Genetic Resources Unit, Plant Production and Protection Division of the Food and Agriculture Organization of the United Nations (Rome, Italy) and the International Board for Plant Genetic Resources. pp. 15–22.

5. Andrews, D.J., Ejeta, G., Gilbert, M., Goswami, P., Anand Kumar, K., Maunder, A.B., Porter, K., Rai, K.N., Rajewski, J.F., Reddy, B.V.S., Stegmeier, W., and Talukdar, B.S. 1997. Breeding hybrid parents. *In* Proceedings of an International Conference on the Genetic Improvement of Sorghum and Pearl Millet, held at Lubbock, Texas, 22–27 September 1996. International Sorghum and Millet Research (INTSORMIL) – International Crops Research Institute for the Semi-arid Tropics (ICRISAT). pp. 173–187.

6. Andrews, D.J., Kiula, B., and Rajewski, J.F. 1993. The use of protogyny to make hybrids in pearl millet. *In* New crops. *Edited by* J. Janick and J. Simpson. John Wiley and Sons, New York. pp. 208–210.

7. Appa Rao S, Mengesha, MH and Rajagopal Reddy C. 1989. Development of cytoplasmic male-sterile lines of pearl millet from Ghana and Botswana germplasm. Pages 817"823 *in* Perspectives in cytology and genetics (Manna GK and Sinha U, eds.). India: Kalyani.

8. Athwal D. S. 1965. Hybrid bajra-1 marks a new era. *Indian Farming*, 15: 6-7.

9. Athwal DS. 1961. Recent developments in the breeding and improvement of bajra (pearl millet) in the Punjab. *Madras Agricultural Journal* 48: 18"19. (Abstract.)

10. Athwal DS. 1966. Current plant breeding research with special reference to *Pennisetum*. *Indian Journal of Genetics and Plant Breeding* 26A: 73"85.

11. Begg, J.E. 1965. The growth and development of a crop of bulrush millet (*Pennisetum typhoid's* S andH). *J. Agric. Sci.* 65, 341–349.

12. Bidinger FR, Mahalakshmi V and Rao GDP. 1987. Assessment of drought resistance in pearl millet (*Pennisetum americanum* (L.) Leeke). II. Estimation of genotype response to stress. *Australian Journal of Agricultural Research* 38(1) 49 – 59.

13. Bidinger, F. R., Mahalakshmi, V., and Rao, G. D. P. (1987a). Assessment of drought resistance in pearl millet (*Pennisetum americanum* (L) Leeke).I Factors affecting yields under stress. *Aust. J. Agric. Res.* 38, 37–48.

14. Bidinger, F. R., Mahalakshmi, V., and Rao, G.D.P. (1987b). Assessment of drought resistance in pearl millet(*Pennisetum americanum* (L) Leeke). II Estimation of genotype response to stress. *Aust. J. Agric. Res.* 38, 49–59.

15. Bidinger, F. R., Mahalakshmi, V., Talukdar, B.S., and Alagarswamy, G. (1982)."Improvement in drought resistance in pearl millet," in *Drought Resistance in Crops with Emphasis on Rice*, (LosBaños, CA: IRRI), 357–376.

16. Bidinger, F.R., and Mahalakshmi, V. 1993. Selection for drought tolerance. *In* The 1992 annual report on the cereals program of the International Crops Research Institute for the Semi-arid Tropics (ICRISAT). ICRISAT, Patancheru, India. pp. 57–59.

17. Bidinger, F.R., Mahalakshmi, V. and Rao, G.D.P. 1987a. Assessment of drought resistance in pearl millet (*Pennisetum americanum* (L) Leeke). I. Factors affecting yields under stress. *Aust. J. Agric. Res.* 38, 37–48.

18. Bidinger, F.R., Mahalakshmi, V. and Rao, G.D.P. 1987b. Assessment of drought resistance in pearl millet (*Pennisetum americanum* (L) Leeke). II. Estimation of genotype response to stress. *Aust. J. Agric. Res.* 38, 49–59.

19. Bramel-Cox, P.J., and Cox, T.S. 1989. Use of wild germplasm in sorghum improvement. *In* Proceedings of the 43rd Annual Corn and Sorghum Industry Research Conference, held at Chicago, Ill. American Seed Trade Association, Washington, D.C. pp. 13– 26.

20. Brunken, J. N. 1977. A systematic study of *Pennistum* Sect. *Pennisetum* (Gramineae). Am. J. Bot. 64: 161-76.

21. Burton G. W. 1958. Cytoplasmic male-sterility in pearl millet [*Pennisetum glaucum* (L.) R. Br.]. *Agron. J.*, 50: 230-231.

22. Burton G. W. and Athwal D. S. 1967. Two additional sources of cytoplasmic male-sterility in pearl millet and their relationship to Tift 23A. *Crop Sci.*, 7: 209-211.

23. Burton, G.W. 1965. Pearl millet Tift 23A released. Crops and Soils 17: 19.

24. Craufurd, P.Q. and Bidinger, F.R. 1989. Potential and realized yield in pearl millet (*Pennisetum americanum*) as influenced by plant population density and life-cycle duration. *Field Crops Res.* 22, 211–225.

25. Craufurd, P.Q., and Bidinger, F.R. (1988a). Effect of the duration of the vegetative phase on crop growth, development and yield in two contrasting pearl millet hybrids. *J. Agric. Sci.* 110: 71–79.

26. Craufurd, P.Q., and Bidinger, F.R. (1988b). Effect of the duration of the vegetative phase on shoot growth, development and yield in pearl millet (*Pennisetum americanum*(L) Leeke). *J. Exp.Bot.* 39: 124–139.

27. Craufurd, P.Q., and Bidinger, F.R. (1989). Potential and realized yield in pearl millet (*Pennisetum americanum*) as influenced by plant population density and life-cycle duration. *FieldCrops Res.* 22, 211–225.

28. Devos, K.M., and Gale, M.D. 2000. Genomic relationships: the grass model in current research. *Plant Cell* 12, 637–646.

29. Devos, K.M., Pittaway, T. S., Busso, C. S., Gale,M. D.,Witcombe, J. R., and Hash, C. T. (1995). Molecular tools for the pearl millet nuclear genome. *Int.SorghumMilletsNewslett.* 36, 64–66.

30. Ejeta, G., Butler, L.G., Hess, D.E., Obilana, T. and Reddy, B.V. 1997. Breeding for *Striga* resistance in sorghum. *In* Proceedings of an International Conference on the Genetic Improvement of Sorghum and Pearl Millet, held at Lubbock, Texas, 22–27 September 1996. International Sorghum and Millet Research (INTSORMIL) – International Crops Research Institute for the Semi-arid Tropics (ICRISAT). pp. 504–516.

31. Fussell, L.K., Bidinger, F.R., and Bieler, P. 1991. Crop physiology and breeding for drought tolerance: research and development. Field *Crops Res.* 27: 183–199.

32. Gaudy M. 1957. Manuel d'Agriculture tropicale. La maison Rustique. Paris. 139-140.

33. Hanna W. W. 1989. Characteristics and stability of a new cytoplasmic-nuclear male-sterile source in pearl millet. *Crop Sci.,* 29: 1457-1459.

34. Hanna, W.W. 1992. Utilization of germplasm from wild species. *In Desertified grasslands: their biology and management*. *Edited by* G.P. Chapman. Academic Press, London. pp. 251–257.

35. Hash C. T., Bhasker Raj A. G., Lineup S., Sharma A., Beniwal C. R., Folkertsma R. T., *et al.* (2003). Opportunities for marker-assisted selection (MAS) to improve the feed quality of crop residues in pearlmillet and sorghum. *Field Crops Res.* 84, 79–88

36. Hash, C.T. and Witcombe, J.R. 1994. "Pearl millet mapping populations at ICRISAT, "in *Use of Molecular Markers in Sorghum and Pearl Millet Breeding for Developing Countries*, eds J.R. Witcombe and R.R. Duncan (London: Overseas Development Administration), 69–75.

37. Hash, C.T., Witcombe, J.R., Thakur, R.P., Bhatnagar, S.K., Singh, S.D., and Wilson, J.P. 1997. Breeding for pearl millet disease resistance. *In* Proceedings of an International Conference on the Genetic Improvement of Sorghum and Pearl Millet, held at Lubbock, Texas, 22–27 September 1996. International Sorghum and Millet Research (INTSORMIL) – International Crops Research Institute for the Semi-arid Tropics (ICRISAT). pp. 337–372.

38. Hash, C.T., Witcombe, J.R., Thakur, R.P., Bhatnagar, S.K., Singh, S.D. and Wilson, J.P. 1997. Breeding for pearl millet disease resistance. *In* Proceedings of an International Conference on the Genetic Improvement of Sorghum and Pearl Millet, held at Lubbock, Texas, 22–27 September 1996. International Sorghum and Millet Research (INTSORMIL) – International Crops Research Institute for the Semi-arid Tropics (ICRISAT). pp. 337–372.

39. Henson, *I.E.*, and Mahaklakshmi, V. 1985. Evidence for panicle control of stomatal behavior in water-stressed plants of pearl millet. *Field Crops Res.* 11, 281–290.

40. Jambunathan R., Subramanian V. (1988). Grain quality and utilization in sorghum and pearl millet, in Workshop on Biotechnology for Tropical Crop Improvement, (Patancheru: ICRISAT;), 1330–1339

41. Jauhar, P.P., and Hanna, W.W. 1998. Cytogenetics and genetics of pearl millet. *Adv. Agron.* 64, 1–26.

42. Jones, E.S., Breese, W.A., Liu, C. J., Singh, S.D., Shaw, D.S., and Witcombe, J.R. 2002. Mapping quantitative trait loci for downy mildew resistance in pearl millet: field and glass house screens detect the same QTL. *Crop Sci.* 42, 1316–1323.

43. Jones, E.S., Liu, C.J., Gale, M.D., Hash, C.T., and Witcombe, J.R. 1995. Mapping quantitative trait loci for downy mildew resistance in pearl millet. *Theor. Appl. Genet.* 91, 448–456.

44. Kaul, J., and Sidhu, J.S. 1997. Establishment of chromosome map of pearl millet through the use of interchanges. *J. Cytol. Genet.* 32, 113–123.

45. Kelley TG, Rao PP, Weltzien RE and Purohit ML. 1996. Adoption of improved cultivars of pearl millet in an arid environment: Straw yield and quality considerations in western Rajasthan. *Experimental Agriculture* 32: 161-171.

46. Khairwal IS. 2007. Coordinator's review. Pages 1–11 in Annual report 2006–2007. All India Coordinated Pearl Millet Improvement Project (Indian Council of Agricultural Research), Mandor, Jodhpur 342 304, Rajasthan, India.

47. Kholova J, Hash CT, Kumar PL, Yadav RS, Kocova M and Vadez V. 2010. Terminal drought-tolerant pearl millet [*Pennisetum glaucum* (L.) R. Br.] have high leaf ABA and limit transpiration at high vapour pressure deficit. *J. Exp. Bot.* 61(5): 1431-40.

48. Kulkarni, M and U. Deshpande. 2007. *In Vitro* screening of tomato genotypes for drought resistance using polyethylene glycol. *African Journal of Biotechnology*, 6: 691-696

49. Kumar R. and Sagar P. 2009. Effect of cytoplasm on downy mildew vulnerability in pearl millet (*Pennisetum glaucum* (L.) R. Br.). *Indian J. Genet.*, 69: 115-121.

50. Liu, C.J., Witcombe, J.R., Pittaway, T.S., Nash, M., Hash, C.T., Busso, C. S., *et al.*, 1994. An RFLP- based genetic map of pearl millet (*Pennisetum glaucum*). *Theoy. Appl. Genet.* 89, 481–487.

51. Mahalakshmi, V. and Bidinger, F. R. 1985a. Flowering response of pearl millet to water stress during panicle development. *Ann. Appl. Biol.* 106, 571–578.

52. Mahalakshmi, V. and Bidinger, F.R. 1986. Water deficit during pan- icle development in pearl millet (*Pennisetum americanum*): yield compensation by tillers. *J. Agric. Sci.* 106, 113–120.

53. Mahalakshmi,V., and Bidinger, F.R. 1985b. Water deficit during panicle development in pearl millet: yield compensation by tillers. *J. Agric. Sci.* 106, 113–119.

54. Marchais L and Pernes J. 1985. Genetic divergence between wild and cultivated pearl millets (*Pennisetum typhoides*).I. Male sterility. Zeitschrift für Pflanzenzüchtung 95: 103"112.

55. Martin J.H. and Leonard W.H. 1967. Principles of field crop production, 2nded. The Mac Millan Co. New York. pp. 526-528.

56. Menkir, A., Bramel-Cox, P.J., and Witt, M.D. 1994. Comparisons of methods for introgressing exotic germplasm into adapted sorghum. Theor. Appl. Genet. 89: 233–239.

57. Minocha, J.L., and Sidhu, J.S. 1981. Establishment of Trisomic Stocks and Linkage Groups in Pennisetum Typhoides. Final research report of PL480 project, Punjab Agricultural University, Ludhiana, India.

58. Morgan, R.N., Wilson, J.P., Hanna, W.W., and Ozias-Akins, P. 1998. Molecular markers for rust and pyricularia leaf spot disease resistance in pearl millet. *Theor. Appl. Genet.* 96, 413–420.

59. Nagaraj N., Basavaraj G. and Rao PP. 2011. Policy Brief on Future Outlook and Options for Target Crops: The Sorghum and Pearl millet economy of India, ICRISAT.

60. Ouendeba, B., Ejeta, G., Nyquist, W.E., Hanna, W.W., and Kumar, K. 1993. Heterosis and combining ability among African pearl millet landraces. *Crop. Sci.* 33: 735–739.

61. Pande, S., Thakur, R.P., Karunakar, R.I., Bandyopadhyay, R. and Reddy, B.V.S. 1994. Development of screening methods and identification of stable resistance to anthracnose in sorghum. *Field Crops Res.* 38: 157–166.

62. Qi, X., Pittaway, T., Allouis, S., Lineup, S., Liu, H., Gale, M.D., *et al.*, 2000."Development of simple sequence repeats (SSRs) markers from small and large insert libraries in pearl millet," in *Abstracts, Plant and Animal Genome VIII Conference*, (San Diego,CA). Available online at: http://www.intl-pag.org/8/abstracts/pag8875.html

63. Radhouane L. 2007. Response of *Tunisian autochthonous* pearl millet (*Pennisetum glaucum* (L.) R. Br.) to drought stress induced by polyethylene glycol (PEG) 6000. *African Journal of Biotechnology*, 6: 1102-1105

64. Rai K. N. 1995. A new cytoplasmic-nuclear male sterility system in pearl millet. *Plant Breed.*, 114: 445- 447.

65. Rai KN, Murty DS, Andrews DJ and Bramel-Cox PJ. 1999. *Genetic enhancement of pearl millet and sorghum.*

66. Rai, K.N. and Rao, A.S. 1996. A new CMS system and its restorers in pearl millet. *In* Proceedings of an International Crop Science Congress, held at New Delhi, India, 17–24 November 1996. Abstract P5-0018.

67. Rai, K.N., and Anand Kumar, K. 1994. Pearl millet improvement at ICRISAT—an update. *International Sorghum and Millets Newsletter* No. 35. pp. 1–29.

68. Rai, K.N., and Rao, A.S. 1995. Commercial feasibility of growing interpopulation hybrids in pearl millet. *International Sorghum and Millets Newsletter* No. 36. pp. 49–50.

69. Rai, K.N., Virk, D.S., Harinarayana, G. and Rao, A.S. 1996. Stability of male-sterile sources and fertility restoration of their hybrids in pearl millet. *Plant Breed.* 115: 494–500.

70. Rooney L. W., McDonough C. M. (1987). Food quality and consumer acceptance in pearl millet, in *Proceedings of the International Pearl Millet Workshop*, 7–11, eds Witcombe J. R., Beckerman S. R., editors. (Patancheru: ICRISAT;), 43–61

71. Serraj, R., Hash, C.T., Rivzi, S.M.H. Sharma, A., Yaday, R.S., and Bidinger, F.R. 2005. Recent advances in marker-assisted selection for drought tolerance in pearl millet. *Plant Prod. Sci.* 8, 334–337.

72. Sharma, H.C., Agrawal, B.L., Abraham, C.V., Vidyasagar, P. and Nwanze, K.F. 1993. Identification and utilization of resistance to sorghum midge, *Contarinia sorghicola*. Coq. *Crop Prot*. 12: 343–350.

73. Sharma, H.C., Taneja, S.L., Leuschner, K. and Nwanze, K.F. 1991. Techniques to screen sorghum for resistance to insects. International Crops Research Institute for the Semi-arid Tropics (ICRISAT) Information Bulletin No. 32. ICRISAT, Patancheru, India.

74. Singh SD. 1995. Downy mildew of pearl millet. *Plant Disease* 79: 545"550.

75. Singh, P., Rai, K.N., Witcombe, J.R., and Andrews, D.J. 1988. Population breeding methods in pearl millet improvement (*Pennisetum americanum*). *Agron. Trop*. 43: 185–193.

76. Singh, S.D., Wilson, J.P., Navi, S.S., Talukdar, B.S., Hess, D.E. and Reddy, K.N. 1997. Screening techniques and sources of resistance to downy mildew and rust in pearl millet. *International Crops Research Institute for the Semi-arid Tropics (ICRISAT)* Information Bulletin No. 48. ICRISAT, Patancheru, India.

77. Squire, G. R., Marshall, B. and Ong, C. K. 1986. Development and growth of pearl millet (*Pennisetum typhoid's*) in response to water supply and demand. *Exp. Agric.* 22, 289–299.

78. Squire, G.R. 1979. The response of the stomata of pearl millet (*Pennisetum typhoid's* S and H) to atmospheric humidity. *J. Exp. Bot.* 30, 925–933.

79. Squire, G.R., Ong, C.K., and Montecito, J.L. 1987."Crop growth in semi-arid environments," in *Proceedings of the International Pearl Millet Workshop*, 7–11, eds J. R. Witcombe and S.R. Beckerman (Patancheru: ICRISAT), 219–231.

80. Sujata V, Sivaramakrishnan S, Rai KN and Seetha K. 1994. A new source of cytoplasmic male sterility in pearl millet: RFLP analysis of mitochondrial DNA. *Genome* 37: 482-486.

81. Talukdar, B.S., Singh, S.D., and Prakash Babu, P.P. 1996. Prospects of topcross hybrids in increasing and stabilizing grain yield in pearl millet. *Crop Improv*. 23: 147–150.

82. Thakur, R.P., King, S.B., Rai, K.N. and Rao, V.P. 1992. Identification and utilization of smut resistance in pearl millet. International Crops Research Institute for the Semi-arid Tropics (ICRISAT) Research Bulletin No. 16. ICRISAT, Patancheru, India.

83. Thakur, R.P., Rai, K.N., King, S.B. and Rao, V.P. 1993. Identification and utilization of ergot resistance in pearl millet. *International Crops Research Institute for the Semi-arid Tropics (ICRISAT) Research Bulletin* No. 17. ICRISAT, Patancheru, India.

84. Vadez, V., Kashiwagi, J., Krishnamurthy, L., Serraj, R., Sharma, K.K., Devi, J., *et al.*, 2005. "Recent advances in drought research at ICRISAT: using root trait sandrd29a: DREB1Ato increase water use and water use efficiency in drought-

prone areas," in *Abstract P5.88 in: Abstracts of Lectures and Posters, Inter drought II: The 2nd International Conference on Integrated Approaches to Sustain and Improve Plant Production Under Drought Stress,* (Rome, Italy), 24–28. Available online at: http://www.plantstress.com/ID2/Abstract per cent 20book per cent 20updated.pdf

85. Vadez,V., Krishnamurthy, L., Kashiwagi, J.W., Kholova, J., Devi, J.M., Sharma,K.K., etal. (2007). Exploiting the functionality of root systems for dry, saline, and nutrient deficient environments in a changing climate. *J. Semi-Arid Tropic Agric. Res.* 4. Available online at: http://ejournal.icrisat.org/special project/sp10.pdf

86. Van den Abeele M. and Vandenput R. 1956. Les principales cultures du Congo Belge. Ministère des colonies, Bruxelles. pp. 179-183.

87. VanOosterom, E.J., Mahalakshmi, V., Bidinger, F.R, and Rao, K.P. 1996b. Effect of water availability and temperature on the genotype by environment interaction of pearl millet in semi arid tropical environments. *Euphytica* 89, 175–183.

88. VanOosterom, E.J., Whitaker, M. L., and Weltzien, R.E. 1996a. "Integrating genotype by environment interaction analysis, characterization of drought patterns, and farmer preferences to identify adaptive plant traits in pearl millet,"in: *Plant Adaptation and Crop Improvement* eds M. Cooper and G.L. Hammer (Wallingford: CAB International), 383–402.

89. White R.O. and Cooper J.P. 1959. Les graminées en agriculture. FAO, Rome. pp. 418-419.

90. Winkel, T. and F. Do.1992. Caracteres morphologiques et physiologiques de resistance du mil. [*Pennisetum glaucum* (L.) R. Br.] a la secheresse. Agro. Trop. 46: 330-351.

91. Winkel, T., Renno, J. N. and W. Payne.1997. Effect of timing of water deficit on growth, phenology and yield of pearll millet. (*Pennisetum glaucum* (L.) R. Br.] growth in Sahelian conditions. *J. Exp. Bot.*, 48: 1001-1.

92. Yadav, O.P., Weltzien-Rattunde, E., Bidinger, F.R., and Mahalakshmi, V. 2000. Heterosis in landrace-based topcross hybrids of pearl millet across arid environments. *Euphytica* 112, 285–295.

93. Yadav, R.S., Bidinger, F.R., Hash, C.T., Yadav, Y.P., Yadav, O.P., Bhatnagar, S. K., etal. 2003. Mapping and characterization of QTL × E interactions for traits determining grain and stover yield in pearl millet. *Theory. Appl. Genet.* 106, 512–520.

94. Yadav, R.S., Hash, C.T., Cavan, G. P., Bidinger, F.R., and Howarth, C. J. 2002b. Quantitative trait loci associated with traits determining grain and stover yield in pearl millet under terminal drought stress conditions. *Theory. Appl. Genet.* 104, 67–83.

95. Burton, G.W. 1974. Factors affecting pollen movement and natural crossing in pearl millet. *Crop Sci.* 14: 802-805.

96. Burton, G.W. and Athwal, D.S. 1967. Two additional sources of cytoplasmic male sterility in pearl millet and their relationship to Tift 23A. *Crop Sci.* 7: 209-211.

97. Hanna, W.W. 1989. Characteristics and stability of a new cytoplasmic nuclear male sterile source in pearl millet. *Crop Sci.* 29: 1457-1459.

98. Rai, K.N. 1995. A new cytoplasmic-nuclear male sterility system in pearl millet. *Plant Breed.* 114: 445-447.

99. Scheifele, G.L., Whitehead, W. and Rowe, C. 1970. Increased susceptibility to Southern leaf spot (Helminthosporium maydis) in inbred lines and hybrids of maize with Texas male sterile cytoplasm. *Plant Dis. Rep.* 54: 501-503.

100. State of Indian Agriculture. 2012-13. Department of Agriculture and Cooperation, Ministry of Agriculture, Government of India.

101. Virk, D.S., 1988. Biometrical analysis in pearl millet – a review. *Crop Improv.* 15: 1-29.

102. Yadav O.P., Rai, K.N., Rajpurohit, B.S., Hash, C.T., Mahala, R.S., Gupta, S.K., Shetty, H.S., Bishnoi, H.R., Rathore, M.S., Kumar, A., Sehgal, S. and Raghvani, K.L. 2012. Twenty-five Years of Pearl Millet Improvement in India. All India Coordinated Pearl Millet Improvement Project, Jodhpur, India. 122 pp.

103. www.agricoop.nic.in. Department of Agriculture and Cooperation, Ministry of Agriculture, Government of India.

Millets: Ensuring Climate Resilience and Nutritional Security (2015) *Pages* **233–259**
Editors: **Vilas A. Tonapi and J.V. Patil**
Published by: **DAYA PUBLISHING HOUSE, NEW DELHI**

Chapter 5

Genetic Improvement in Small Millets

A. Seetharam

Former Project Coordinator (Millets) and Emeritus Professor, ICAR, Bangalore

Small millets have their origin chiefly in Asia and Africa. The most important domestication areas are East Asia, Indian sub-continent and the regions from Southern margin of Sahara to the Ethiopian high lands of Africa (Table 1). Different kinds of millets were domesticated in these two continents. Millets such as, Sorghum, pearl millet and finger millet represent African millets; while foxtail millet, common millet, Japanese barnyard millet and Indian barnyard Millet originated in Eurasia. Some small millets are grown only in specific regions even though they have a long history of cultivation. For *e.g.*, kodo millet and little millet are domesticated and predominantly cultivated only in India. On the other hand finger millet, foxtail millet, common millet and barnyard millet have much wider distribution in Asia, Africa and Europe. It is evident from the above that small millets as a group are quite important in areas of their areas of their production as dry land crops and for mountain or hill agriculture providing staple food for the people of the region. Many small millets having much longer history of cultivation than major food grains were rated highly in the past, playing an important role in our traditional food culture and farming systems (Seetharam, 1998).

Characteristics of Small Millets

They being component of traditional agricultural systems are grown with low inputs and accompanied less productivity. There is a trend to replace them with commercial crops such as maize or rice or oilseeds in more productive systems. This is true in many countries of South Asia resulting in decline of acreage. It must be

Table 1: Small Millets and their Place of Domestication

Crop	*Scientific Name*	*Chromosome Number*	*Place of Domestication*
Finger millet	*Eleusine coracana* (L.) Gaertn.	2n =36 (4x)	East Africa
Foxtail millet	*Setaria italica* (L.) Beauv.	2n =18 (2x)	Central Asia-India
Common millet	*Panicum miliaceum* L.	2n =36 (4x)	Central Asia-India
Indian barnyard millet	*Echinochloa frumentacea* (Roxb.) Link.	2n =54 (6x)	India
Japanese barnyard millet	*Echinochloa utilis* Ohwi *et* Yabuno.	2n =54 (6x)	East Asia
Little millet	*Panicum sumentranse* Roth.ex Roem. and Schultz.	2n =36 (4x)	India
Kodo millet	*Paspalum scrobiculatum* L.	2n =40 (4x)	India

recognized that there are situations in which major cereals do have advantage and difficult to arrest the area decline under millets. Nevertheless, small millets have a number of merits as a consequence of unique traits possessed by them and this should make their cultivation attractive even to the improved agricultural systems. Some of the unique features possessed by small millets are:

1. They can grow well even in areas where soil, climate and other conditions are less favorable for crop growth. They are fast maturing which is so important to fit them into more intensive cropping systems as a catch or relay crop.
2. Their grains are small and yields are stable. The grains store well for long periods ensuring food supply in years of crop failure and in lean seasons.
3. In view of small grain size they often require less cooking time which could be an important factor for women who is required to look after many farm related responsibilities.
4. Many method of using millets have been established as materials of traditional staple foods. There are also ways of processing millet into novel preparations as well. This can be a factor in increasing market demand for them. Millets are used as materials for making local drinks too. This application is closely related to the farm practices, dietary culture and agricultural rituals of rural communities.
5. Small millet grains in general have excellent nutritious properties with high levels of essential minerals such as iron, calcium, zinc, dietary fiber, quality protein, nutraceuticals etc. Finger millet is especially known for its excellent malting qualities which should make it an important ingredient for the preparation of several naval, high value foods.
6. These crops often have dual use and grown for grain and quality fodder/ straw and both are equally important in mixed farming (Subba Rao *et al*,1995; Seetharam *et al.*, 1994).

Small Millets in Indian Agriculture

There are six different Small millets grown in India which include finger millet (ragi), foxtail millet (kangni), kodo millet (kodo), proso millet (cheena), barnyard millet (sawan) and little millet (kutki). The area under these crops during the last 6 decades has significantly shrunk from 8 million in 1949-50 to around 2.3 million in 2012-13 This is also reflected in diminishing production, from around 4 million tones produced in late forties to around 2.5 million tones in 2011-12 (Tables 2–4).

Table 2: Area (m.ha) of Small Millets *vis-à-vis* other Millet Crops in India*

Crop/Year	*1955-56*	*1965-66*	*1975-76*	*1985-86*	*1995-96*	*2005-06*	*2012-13*
Jowar	17.36	17.68	16.09	16.10	11.33	8.68	6.18
Bajra	11.34	11.97	11.57	10.65	9.32	9.58	7.20
Ragi	2.30	2.70	2.63	2.41	1.77	1.53	1.11
Small millets	5.34	4.56	4.67	3.16	1.66	1.06	0.75

**Source*: Agricutural Cences, Directorate of Economics and Statistics, Department of Agriculture and Cooperation, Government of India.

Table 3: Production of small Millets(million tonnes) *vis a vis* other Millet Crops in India *

Crop/Year	*1955-56*	*1965-66*	*1975-76*	*1985-86*	*1995-96*	*2005-06*	*2012-13*
Jowar	6.73	7.58	9.50	10.20	9.33	7.63	5.33
Bajra	3.43	3.75	5.74	3.66	5.38	7.68	8.74
Ragi	1.85	1.33	2.80	2.52	2.50	2.35	1.59
Small millets	2.07	1.56	1.92	1.22	0.78	0.47	0.43

**Source*: Agricutural Cences, Directorate of Economics and Statistics, Department of Agriculture and Cooperation, Government of India.

Table 4: Productivity of Small Millets (kg/ha) *vis a vis* other Millet Crops in India *

Crop/Year	*1955-56*	*1965-66*	*1975-76*	*1985-86*	*1995-96*	*2005-06*	*2012-13*
Jowar	387	429	591	633	823	880	863
Bajra	302	314	496	344	577	802	1214
Ragi	800	492	1064	1049	1410	1534	1428
Small millets	388	341	412	386	469	443	571

**Source*: Agricutural Cences, Directorate of Economics and Statistics, Department of Agriculture and Cooperation, Government of India.

The loss of area is very severe in all small millets other than finger millet. However, in the last 15 years, the finger millet also has lost ground and area has come down from 2.4 million to 1.2 million ha and likely to loose further in the coming years. Despite reduction in area total production has not been much affected (Table 3). As regards productivity is conserved, finger millet has kept pace with most other major dry land crops in compound growth rates(CGR) for yield while the other small millets have shown little progress (Table 4). By and large, the low productivity of these crops is largely due to meager attention received in terms of inputs; water and technology back up which is further compounded by low value status of grains.

Small Millets and Food Security

The bulk of the millets produced in the country are consumed at the farm level and very little finds its way to organized market. So, it is quite obvious that the real value of these crops and the important role they play inn filling the food basket at home/farm level has not been appreciated. Small millets provide the much needed food and feed security especially to the vulnerable groups. However, it is pertinent to note that food security at National level will only be effective when regionally important crops such as small millets are allowed to play their due role in meeting food and fodder needs of the region and thus avoiding the undesirable dependence on other regions as well as a fewer number of food crops for meeting food needs. This calls for correction and more ideal situation would be producing the required food within the state/region itself by promoting and developing locally/regionally important rain fed crops. Small millets being eco-friendly crops are suitable for fragile and vulnerable ago-ecosystems and should be the preferred crops for sustainable and green agriculture. The promotion of these crops can lead to much efficient natural resource management and ultimately to a more holistic approach in sustaining precious agro-biodiversity.

Small Millet Production Systems in India

In most parts of India, small millet is grown with limited attention and inputs for local consumption. The marketing avenues are extremely limited in areas where millet is the main food crop. Live stock is an important component of most millet production systems and millet crop residues contribute significantly to fodder supplies. In Indian sub continent, millet cropping systems tend to be extensive with limited application of improved technologies except in parts of South India where intensive cultivation practices are followed in finger millet production. But for these exceptions, these crops are grown without irrigation or manures on light, marginal and even gravelly soils on hill slopes that are poor in native soil fertility. Because of these reasons the millet yield tends to be lower with significant year to year and location to location variation.

The bulk of small millet production in India is of finger millet (80 per cent) and the remaining from kodo millet, little millet, foxtail millet, barnyard millet and proso millet in that order. Small millets in India are grown in Karnataka, Andhra Pradesh, Maharashtra, Tamil Nadu, Odisha, Jharkhand, Chhattisgarh, Madhya Pradesh and Uttarkhand. They being the components of traditional farming system one could see well evolved cropping systems. Mixed cropping of small millets with as many as nine crops (Navadhanya) is not uncommon even today in several parts of India. The intercrops grown especially with finger millet are - sorghum (fodder), pearl millet, cowpea, field bean, niger, pigeonpea, mustard, soybean and horsegram. Rotation with several grain legumes like greengram/blackgram/rice bean/soybean/

groundnut/horse gram is practiced in different regions (Seetharam, 1998). Small millet production suffers from a number of constraints. Poor soil fertility, low and erratic rainfall, diseases like blast, *Helminthosporium* blight and *Cercospora* leaf spot (on finger millet); grain and head smuts (in small millets) and shoot fly and stem borer among pests are the major causes. The population pressures along with replacement of millet by other cash crops have pushed small millets into more marginal lands resulting in stagnation of yield levels and lowering of production. Lack of marketing avenues and lower prices has tended to keep the use of cash inputs especially fertilizers and pesticides as well as level of crop management at a minimum.

Utilization of Grain

Grain Morphology

Small millet grains show considerable diversity in physical and morphological features. They differ in colour, shape, size as well as in certain nutritional features. The basic grain structure is more or less similar in all millet grains - with pericarp, germ and endosperm as principal anatomical components (Table 5). Finger millet possesses a very thin layer of pericarp and fairly thick/thin bran as seed coat similar to wheat. Foxtail millet, proso millet, kodo millet, barnyard millet and little millet possess husk and bran similar to rice. So, they require dehusking and debraning prior to utilization.

Table 5: Structural Features of Kernels of Small Millets

Grain	*Type*	*Shape*	*Nature of Seed Coat*	*Colour*
Finger millet	Utricle	Globose	Thin pericarp and bran	White, red, copper brown, violet
Proso millet	Utricle	Oval	Husk and bran	Grey, brown, golden yellow, straw white
Foxtail millet	Utricle	Oval	Husk and bran	Red, black, white, yellow
Kodo millet	Utricle	Spherical	Husk and bran	Grey brown, brown,dark brown
Barnyard millet	Utricle	Oval	Husk and bran	Straw white, dull white
Little millet	Utricle	Oval	Husk and bran	Grey, straw white

Endosperm forms the largest component of millet grain which is the major storage tissue. It is composed of an aleurone layer and peripheral corneous and floury zones. The aleurone layer consisting of single layer of cells found just below the seed coat and testa is rich in minerals, B-complex, vitamins and oil. The peripheral endosperm contains starch granules and protein. The proportion of corneous and floury endosperm determines the texture of millet and millets vary widely for the kernel texture from floury to corneous types. Finger millet grains are soft textured and were floury in nature. On the other hand, other small millet grains contain less floury endosperm and are of a hard corneous texture (Seetharam and Krishne Gowda, 2008).

Nutritional Composition of Grains

Like other cereal grains, millets are predominantly starchy. The protein content is more or less equal and comparable to that of wheat, rice and maize (Table 6). Finger millet has slightly lower protein content but is in fact nutritionally superior because the protein quality is generally as good as or better than other cereals. Finger millet contains the lowest fat. Barnyard millet has the lowest carbohydrate content and energy value.

One of the characteristic features of the grain composition of millet is their high ash content (mineral composition). They are relatively rich in iron and phosphorus. Finger millet has the highest calcium content among all the food grains. High fibre content and lower digestibility of nutrients is the other characteristic feature of millet grains which influence their consumer acceptability. The whole grains are important sources of B-complex vitamins which is mainly present in the outer bran layer of the grain. Millets do not contain vitamin 'A'. However certain yellow endosperm varieties contain some amount of 'β' carotene, a precursor of vitamin 'A'. Considerable variation in grain composition of these millets has been reported.

Genetic factors play a major part in determining grain composition and thus screening of germplasm will be useful in identifying nutritionally superior varieties. Environmental factors also have a role in deciding nutrient make up of the grain. Often an inverse relationship has been observed between grain yield and protein content. Grain colour also has same relationship with nutrient composition especially protein content. The millet protein has well balanced amino acid profile and good source of methionine, cysteine and lysine. These essential amino acids are of special benefits to those who depend on plant food for their protein nourishment. The millet grain contains about 65-70 per cent carbohydrate, a high proportion of which is in the form of non starchy polysaccharides and dietary fibre. Millet grains are also rich in important vitamins *viz.,* thiamine, riboflavin, folin and niacin. Millet grains are known for good shelf life. The grains dried to 10 -12 per cent moisture stored in farm houses show less store pest damage. There are reports of finger millet kept for more than two decades without spoilage. The other small millets also store well when whole grains are kept without dehusking. The millet flour and their products also show good shelf life.

Traditional Foods from Millets

Millet is a high energy nutritious whole some food especially recommended for children, convalescents and the elderly. Several traditional food preparations are made from millet which differs between different regions of a country. Milled small millets (other than ragi) is similar to rice in several cooking properties and used in making several kinds of food items. The cooking style of millets in India varies from state to state and many different kinds of traditional foods are made and they form the staple diet for many rural and urban households. Ragi is eaten in the form of mudde (dumpling or stiff porridge) and/or roti. Many other traditional foods are made from popped ragi flour mixed with sugar/jaggery/ghee/milk/butter milk and salt. In several rural households, a vast variety of traditional snacks are made from

Table 6: Nutrient Composition of Small Millets and other Cereals (Per 100 g edible portion @ 12 per cent moisture)

Food	*Protein (g)*	*Fat (g)*	*Ash (g)*	*Crude Fibre (g)*	*Carbo-hydrate (g)*	*Energy (kcal)*	*Ca (mg)*	*Fe(mg) (mg)*	*Thiamin (mg)*	*Riboflavin (mg)*	*Niacin (mg)*
Rice (brown)	7.9	2.7	1.3	1.0	76.0	362	33	1.8	0.41	0.04	4.3
Wheat	11.6	2.0	1.6	2.0	71.0	348	30	3.5	0.41	0.1	5.1
Maize	9.2	4.6	1.2	2.8	73.0	358	26	2.7	0.38	0.2	3.6
Sorghum	10.4	3.1	1.6	2.0	70.7	329	25	5.4	0.38	0.15	4.3
Pearl millet	11.8	4.8	2.2	2.3	67.0	363	42	11.0	0.38	0.21	2.8
Finger millet	7.7	1.5	2.6	3.6	72.6	336	350	3.9	0.42	0.19	1.1
Foxtail millet	11.2	4.0	3.3	6.7	63.2	351	31	2.8	0.59	0.11	32
Common millet	12.5	3.5	3.1	5.2	63.8	354	8	2.9	0.41	0.28	4.5
Little millet	9.7	5.2	5.4	7.6	60.9	329	17	9.3	0.3	0.09	32
Barnyard millet	11.0	3.9	4.5	13.6	55.0	300	22	18.6	0.33	0.10	4.2
Kodomillet	9.8	3.6	3.3	5.2	66.6	353	35	1.7	0.15	0.09	2.0

Source: Hulse, Laing and Pearson, 19.

ragi and other small millets. However, very little information about the food uses of small millets has been documented. There is a need for collection and dissemination of detailed information about the traditional uses and processing methods of small millets.

Millet Processing for Value Addition

The millet grains offer many opportunities for diversified utilization and in adding value. With proper processing, it is possible to make many different kinds of food products by adopting appropriate milling, popping and other technologies. Milled millet can be processed today at cottage level as well as in large scale processing. Milled millet can be further processed towards various food uses such as flakes, quick food cereals, ready to eat snacks, supplementary foods, extrusion cooking, malt based products, weaning foods and more importantly health foods. Finger millet flour is easy to make since the endosperm and bran are pulverized freely and in such flour, fibre content is normally higher. However, it is possible to reduce fibre content by adopting simple sieving methods. Millets are well adaptable to a wide range of ecological and unfavourable agro-climatic conditions. There is a great potential for their enhanced production, especially in the arid regions. Development of several technologies for their diversified uses and preparation of value-added foods from small millets have opened up new opportunities for their wide spread utilization and thereby improving demand for these grains.

Research Progress

Research efforts in India have been in progress since the beginning of the 20 the century (Seetharam, 1998). But, the launching of coordinated crop improvement programs during late 1950s and 60s has contributed significantly by way of developing new superior varieties and concomitant production and protection technologies in all small millets. The release of these varieties and production packages for general cultivation has helped in 5 fold increase in grain production from 50 million to 250 million tones in the country. It is generally seen that this increase has largely come from two major crops –rice and wheat- and less from dry land crops such as millets and more so from small millets. The small millets have been the last priority crops in the agriculture developmental agenda in the country. Finger millet among small millets has received a little more attention than the rest especially in the southern states. An attempt has been made here to trace the progress especially in the field of crop improvement during the last 100 years.

Crop Improvement Efforts in Pre-coordinated Project Era

Before the launching coordinated project in 1950s and 60s; the crop improvement in small millets was confined to a fewer states such as Tamil Nadu, Andhra Pradesh, Karnataka and Uttar Pradesh. The emphasis was on varietal improvement through selection of better types from local cultivars. In Tamil Nadu, Millet Research Station was established in 1923 at Coimbatore under the erstwhile Madras Presidency and the work later extended to Anakapalle (A.P.) and Hagari (Karnataka). Finger millet work in Karnataka dates back to 1900, initiated at Bangalore especially on finger millet and in Uttar Pradesh at Kanpur and Gorakhpur in 1944. As a pre requisite to millet improvement, the inheritance of qualitative and quantitative characters, period of anthesis, pollination and crossing methods were studied at Coimbatore.

The first finger millet variety released in the country was H 22 as early as 1918 in Karnataka. The other finger millet varieties released were Co 6 (1935); R 0870, ES13, K1, ES11 (1939); Hagari1 (1941), Co1, Co2, Co3, Co4 (1942), VZM 1, VZM 2 (1958) and T36 B (1949). Interest in finger millet improvement got a fillip in Karnataka during 1950-60 and several new varieties such as Aruna, Udaya, K1, Purna, ROH 2 and Cauvery were released. Similarly, many varieties were released in other small millets also in many states (Seetharam, 1998). This included little millet variety Co 1 (1954); foxtail millet varieties Co1, Co2, Co 3 (1943), H1, H2 (1948), T 4 (1949); kodo millet varieties PLR 1(1942, T 2 (1949), Co 1 (1953), proso millet variety Co 1 (1954) and barnyard millet varieties T 46, T25 (1949).

During fifties, with food production remaining stagnant and with raising population the importance of millet crops to Indian agriculture started gaining recognition, as they formed the important constituents of dry land agriculture. Project for intensification of research on cotton, oil seeds and millets was launched during this period with several centres working on millets, The importance of genetic resources as primary raw material for crop improvement was recognized prior to initiation of coordinated project and the first attempt to collect the germplasm of millets in the country was made in 1961 under the PL 480 project on "storage, Maintenance and distribution of millets germplasm " resulted in the collection of nearly 3000 genetic stocks of various small millets _718, finger millet,584 in kodo millet,431 9 in little millet,615 in foxtail millet 250 in proso millet and 399 in barnyard millet.

Crop Improvement Efforts during Coordinated Project Era

Small millets have always been of local and regional importance and as a result have attracted little attention both at national and International level. Millets in general started receiving with attention with the launching of All India Coordinated Millets Improvement Project (AICMIP) in 1969. In this project small millets also started receiving some attention at a selected few centres. Small millets improvement received the major boost during 1978-79 with the establishment of five crop specific lead research centres in the country under IDRC assistance. They were Almora in Uttarakhand (barn yard millet), Dholi in Bihar (proso millet), Dindori in Madhya Pradesh (Kodo millet), Semiliguda in Odisha (Little millet) and Nandyal in Andhra Pradesh (foxtail millet.) in this project finger millet was not included as it was already receiving some attention in the Coordinated Millets Improvement Project. The IDRC project continued till 1985 and the "All India Coordinated Small Millets Improvement Project" (AICSMIP) was established in the year 1986 with head quarters at The University of Agricultural Sciences, Bangalore. The centres that were functioning under IDRC project became part of AICSMIP. With the inception of separate AICRP, research on small millets has been getting focused attention for developing varieties and other agro production and protection technologies suitable to different regions. There are 14 centres functioning under AICSMIP spread all over the country to address to the research needs of small millets. Small millets are known for their suitability to dry land areas, hill and tribal agriculture and contribute to food and nutritional security of the disadvantaged regions. The research in the project is focused to state/ regional needs from the point of developing appropriate varietal and agro production technology for maximizing production/productivity. The work is multi-disciplinary and applied in nature.

Germplasm Availability

In the past small millets scientists hardly had access to germplasm and worked with a handful of local collections which lacked diversity. This blunted the opportunities for of yield improvement through breeding. This situation was to some extent rectified in the 1960 when first attempts were made by ICAR to pool the collections under PL 480 project as mentioned earlier. The conservation activities further gained momentum with NBPGR, New Delhi, playing a key role in augmenting the small millets collection. Recognizing the importance and conservation and easy access to germplasm, AICMIP established a separate germplasm unit at Bangalore in 1979. This unit since then has been making efforts to collect as well as pool the available germplasm from various sources and make it available to breeders in the country. This unit is also recognized as National Active Germplasm Site (NAGS) by ICAR/NBPGR and has the mandate to assist in collection, conservation, evaluation and documentation of small millets germplasm in the country (Seetharam, 2006). Presently the unit at Bangalore is maintaining one of the largest collections of more than 15000 accessions of 6 small millets (7122 in finger millet,2821 foxtail millet,1537 kodo millet,939 proso millet,1657 little millet and 988 barnyard millet) in the country (AICSMIP,2012).

Utilization of Germplasm

The full utilization of germplasm depends on two factors: (1) evaluation and characterization and (2) identification of useful gene sources. These two areas have received considerable attention during the last 25 years and majority of the accessions have been screened for agronomic, physiological, pathological and even important grain quality parameters. The breeding value of many accessions has been judged by growing in field trials more than once. There is good data base available for most accessions and germplasm catalogues have been brought out (Seetharam *et al.*, 2006). In order to improve the efficiency for utilization of germplasm, core subsets have been formed and made available to breeders working at different centres. Selected germplasm have also been evaluated in the all India testing network and a number of superior accessions were identified and a couple of them have been released for general cultivation in different parts of the country. The exotic collections especially from Africa in finger millet have been largely used in recombination breeding resulting in release of many superior high yielding varieties in many states. The African germplasm have thick stem, dark leaves, robust growth, large ears, and high grain density and source of resistance to blast disease (Naik *et al.,* 1993). Hybridization between African and Indian elite varieties has been highly rewarding and has resulted in the release of many high yielding varieties in the country (Seetharam, 1998).

Several useful genetic stocks have been identified in all small millets accessions possessing higher protein, desirable agronomic attributes with high carbon dioxide fixation and low leaf area suitable for rain fed situations, and genotypes which can germinate under limited moisture under hard soil crust have been identified (Sashidar *et al*,1983,1986; Seetharam *et al*,1984; B.T.S.Gowda *et al*.1986). Long glume types with higher ear photosynthesis reflecting in higher seed size and weight will be of interest in improving yield of finger millet in the coming years (Sashidar *et al.*, 1983). Accessions capable of producing higher biomass, dual purpose types with superior stover quality are available for improving grain and Stover yield of cultivars. Identification several sources of stable resistance to blast disease of finger millet and their deployment in breeding research has been highly rewarding in evolution of high yielding blast resistant cultivars in finger millet in the country (B.T.S.Gowda *et al*,1986; Ravikumar *et al*,1990,1991; Seetharam and Ravikumar,1993; Byre gowda *et al*,1998,1999). There are finger millet germplasm with significantly higher grain calcium and protein and useful in breeding for improving quality parameters. In the secondary gene pool, *Eleusine coracana* sub sps *africana* will be of interest from the breeding point of view. This could be a useful gene source for improving tillering ability, fodder yield and quality, drought tolerance and even finger number and length. A care fully planned pre breeding is required for introgression of characters from *E.africana* to *E. coracana* for deriving lines useful in regular breeding programmes.

In foxtail millet, new sources of dwarfing controlled by oligo genes have e been identified (Dinesh Kumar *et al.*, 1992). These accessions are useful in breeding dwarf foxtail millet similar to the ones available in wheat. The variability available for protein content (7.176-15.73) and seed fat content (4.0-7.1) in foxtail millet is enormous and can be exploited directly and use in breeding (Laxminarayana *et al.*, 1983). The optimum use of utilization of germplasm in small millets is the most important in future breeding activities for making notable advances in the productivity of these crops and also to make small millets competitive *vis- a- vis* other crop options. Based on the evaluation and geographical origin, core set of germplasm have been formulated in finger, foxtail and proso millets for improving their utilization in the crop improvement. There is variability available for Stover quality parameters as well as quantity in the germ plasm of various small millets offering scope for the improvement of feed value of crop residues (Schiere *et al.*, 2004; Subba Rao *et al.*, 1995; Seetharam *et al.*, 1993).

Outcome of Crop Improvement

The All India Coordinated Small Millets Improvement Project was established in the year 1986 with head quarters at University of Agricultural Sciences, Bangalore and with 14 centres spread over the country to address research needs of small millets. Small millets are known for their suitability to dry land areas, hill and tribal agriculture and contribute to food and nutritional security of the disadvantaged regions. The research in the project is focused to state/regional needs from the point of developing appropriate agro production technology for maximizing production/ productivity. The work is multi-disciplinary and applied in nature. The crop improvement is mainly aimed at developing high yielding varieties with resistance to blast disease quality fodder, early and medium maturity and white seed in finger millet, resistance to head smut in kodo millet and resistance to shoot fly in both proso and little millets. So far, a total of 236 varieties in 6 small millets have been released in the country (Table 7).

Table 7: No. of Varieties Released in Various Small Millets in Pre and Post-coordinated Project Era

State	*Finger Millet*	*Foxtail Millet*	*Kodo Millet*	*Little Millet*	*Proso Millet*	*Barnyard Millet*	*Total*
Andhra Pradesh	18	9	–	–	4	–	31
Karnataka	37	5	1	–	2	–	45
Tamil Nadu	20	9	7	8	11	4	59
Odisha	6	–	–	5	–	–	11
Madhya Pradesh	–	–	17	4	–	–	21
Chattissgarh	1	–	1	–	–	–	2
Uttar Pradesh	4	1	2	–	1	3	11
Bihar	1	–	–	–	3	2	6
Jharkand	4	–	–	1	–	–	5
Rajastan	–	5	–	–	1	1	7
Gujarat	5	–	2	1	–	1	9
Maharashtra	2	1	1	1	2	–	7
Uttarkhand	14	2	–	–	–	6	22
Total	**112**	**32**	**31**	**20**	**24**	**17**	**236**

Out of this 86 varieties were released before 1986 (pre coordinated era and 151 during 1986-2012 (post coordinated project era (Table 8).

Table 8: Break Up of No. of Varieties Released during Pre (before 1986) and Post-coordinated Project Era (after 1986)

Crop	*No of Varieties Released*		*Total*
	Before 1986	*After 1986*	
Finger millet (1918-2012)	45	67	112
Foxtail millet (1942-2012)	12	20	32
Little millet (1954-2012)	6	14	20
Proso millet (1954-2012)	8	16	24
Barnyard millet (1949-2012)	4	17	21
Kodo millet (1942-2012)	11	20	31
Total	**86**	**154**	**240**

The break up for various small millets are 112 in finger millet (45 in pre and 67 in post coordinated project era respectively); 32 in foxtail millet (12 in pre and 20 in post coordinated project era respectively); 20 in little millet (6 in pre and 14 in post coordinated project era respectively); 31 in kodo millet (11 in pre and 20 in post coordinated project era respectively) and 21 in barnyard millet (4 in pre and 17 in post coordinated project era respectively). Pure line selection has been the approach so far in little and kodo millet and as a result the genetic gain made has been very limited and the varieties are less diverse too. On the contrary recombination breeding has been the approach especially in finger millet resulting in creation highly diverse and productive varieties (Tables 9 and 10).

Table 9: No. of Varieties Released from Various Selection Methods in different Small Millets

Selection Method	*Foxtail Millet*	*Little Millet*	*Kodo Millet*	*Proso Millet*	*Barnyard Millet*	*Total*
Pure line selection	24	18	30	17	14	103
Bulk/Pedigree method	7	1	–	7	3	18
Mutation Breeding	1	1	1	–	–	3
Total	**32**	**20**	**31**	**24**	**17**	**124**

Table 10: No. of Varieties Released from Various Selection Methods in Finger Millet

Selection Method	*<1930*	*1931-50*	*1951-70*	*1971-80*	*1981-2000*	*2001-12*	*Total*
Pure line selection	1	8	8	7	20	6	50
Bulk/Pedigree method	–	–	5	6	19	20	50
Mutation Breeding	–	2	–	1	6	–	9
MAS	–	–	–	–	–	1	1
Total	**1**	**10**	**13**	**14**	**45**	**27**	**110**

The search for genetic male sterility in finger millet has been successful and now available for use. The modern finger millet varieties has the genetic potential if 5- 6 tones/ha under optimum management. On the same lines recombination breeding has resulted in release of many superior varieties in foxtail, proso and barnyard millets also. With the launching of AICSMIP, the variety development and release has gained immense momentum in all major small millet growing states resulting in identifying varieties to different ecologies in the country. The large scale verification variety tasting on farmers fields in participatory mode (Gowda *et al.*, 2000) and through large number of Front Line Demonstrations has helped in fine tuning recommendation domains and in popularization of high yielding varieties among farmers especially in finger millet. The varieties of released during the last 15 for cultivation in different states are given in Table 11.

The research centres have undertaken the production and supply of breeder seed indented by the Department of Agriculture and Co-operation, GOI. However, large scale seed production and distribution which is the key to success in spread of HYV's is very week in many states especially in crops other than finger millet in the entire country. This has led to opportunity deprivation of benefits of improved varieties to farmers in most parts of the country. The harnessing of yield advantages from these improved varieties is the need of the hour in order to make small millets

Table 11: Varieties of Small Millets Released for Cultivation during the Last 15 Years

Sl.No.	*Name of the Variety*	*Pedigree*	*Year of Release*	*Maturity*	*Average Yield*	*Recommended Domain*
			FINGER MILLET			
1.	Champavathi (VR 708)	Pure line selection	1998	90-95	2000-2500	AP, UP, TN, Karnataka, Odisha
2.	MR 1	Hamsa x IE 927	1998	125-130	3500-4000	Karnataka
3.	PR 230 (Maruthi)	Pure line selection	1998	90-100	2500-3000	Andhra Pradesh
4.	GPU 28	Indaf 5 x IE 1012	1998	110-115	3500-4000	Karnataka, TN, Chattisgarh, Odisha, Jharkhand
5.	BM 9-1	Mutation breeding	1999	105-110	2500-3000	Karnataka, Andhra Pradesh, Odisha, Madhya Pradesh, Maharashtra
6.	OUAT 2	Mutant of Co 9	1999	120-125	2800-3000	All ragi growing areas of the country
7.	L 5	Malawi x Indaf 5	1999	120-125	3500-4000	Karnataka
8.	GPU 26	(Indaf 5x Indaf 9) x IE 1012	2000	100-105	3000-3500	Karnataka
9.	GPU 45	GPU26 x L5	2001	95-100	2700-2900	Madhya Pradesh, Gujarat, Jharkhand, Karnataka, Maharashtra
10.	Chilika (OEB 10)	GE 68 x GE 156	2001	120-125	2600-2700	Odisha, Madhya Pradesh, Gujarat, AP, TN
11.	VL 315	SDFM 69 x VL 231	2004	105-110	2600-2700	Uttarakhand
12.	TNAU 946 (CO 14)	Malavi 1305 x Co 13	2004	105-110	2500-2600	Tamil Nadu
13.	GPU 48	GPU 26 x L5	2005	100-105	3000-3500	Karnataka
14.	GN 4	Pure line selection	2006	105-115	2800-3000	Gujarat
15.	VL 332	VL 127 x IE 1213824	2006	100-105	2300-2400	All ragi growing areas of the country
16.	PRM 1	Pure line selection	2006	110-115	2000-2500	Hills of Uttarakhand
17.	VL 324	VL 162 x IE 3808	2006	110-120	2000-2200	Lower and mid hills of Uttarakhand
18.	VR 762 (Bharathi)	Pure line from VMEC 134	2006	110-115	2600-3000	Andhra Pradesh

Contd...

Table 11–*Contd...*

Sl.No.	Name of the Variety	Pedigree	Year of Release	Maturity	Average Yield	Recommended Domain
19.	Paiyur 2	VL 145 x Selection 10	2008	110-115	3000-3500	Tamil Nadu
20.	ML 365	IE 1012 x Indaf 5	2008	110-115	5000-5500	Zone 5 of Karnataka
21.	GN 5	Pure line selection	2009	120-125	3000-3200	Gujarat
22.	Srichaithanya (VR 847)	GPU 26 x L 5	2009	110-115	2600-2800	Andhra Pradesh
23.	GPU 66	PR 202 x GPU 28	2009	110-115	3500-4000	Karnataka
24.	KMR 301 (Gowri)	MR 1 x GE 1409	2009	120-125	4000-4500	Southern dry zone of Karnataka
25.	GPU 67	Selection from GE 5331	2009	115-120	3000-3200	Chhattisgarh, Jharkhand, Karnataka, Maharashtra, Tamil Nadu and Uttarakhand
26.	PRM 2	Pure line selection	2010	105-110	2500-2800	Hills of Uttarakhand
27.	VL 347	VR 708 x VL 149	2010	95-100	2000-2200	All ragi growing areas of the country
28.	KOPN 235 (Phule Nachni)	Pure line selection	2011	115-120	2500-2600	Sub mountain and Western Ghat zones of Maharashtra
29.	OEB 526	SDFM 30 x PE 244	2011	110-115	2400-2600	Odisha, Bihar, Chattisgarh, Karnataka, TN
30.	OEB 532	GPU 26 x L 5	2012	110-115	2200-2500	Odisha, Chattisgarh, Karnataka, Maharashtra, TN
31.	KMR 204	GPU 26 x GE 1409	2012	100-105	3000-3500	Karnataka
32.	VR 936 (Hima)	IE 2695 x PR 202	2012	115-120	2800-3000	Andhra Pradesh
33.	Vakula (PPR 2700)	KM 55 x U 22/B	2012	105-110	2500-3000	Andhra Pradesh
34.	Indira Ragi 1	HR 911 x GE 669	2012	120-125	2500-2600	Chattisgarh
35.	VL 352	VR 708 x VL 149	2012	95-100	3300-3500	All ragi growing areas of the country

Contd...

Table 11–*Contd...*

Sl.No.	Name of the Variety	Pedigree	Year of Release	Maturity	Average Yield	Recommended Domain
			LITTLE MILLET			
1.	Paiyur 2	Pure line selection	2000	95-100	800-1000	Tamil Nadu
2.	Kolab (OLM 36)	Mutant of SS 81-1	2001	95-100	2800-3000	MP, Odisha, Chattisgarh, Bihar, Karnataka, Gujarat
3.	Tarini (OLM 203)	Selection from KL 2	2001	105-110	2500-3000	Odisha, AP, Bihar, TN
4.	Sabara (OLM 20)	Mutant of SS 81-1	2003	75-80	1100-2000	All little millet growing areas of the country
5.	Co 4	Co 2 x MS 1684	2005	75-80	1600-2000	Tamil Nadu
6.	JK 36	Pure line selection	2009	75-80	1000-1200	Madhya Pradesh
7.	Soura (OLM 208)	Pure line selection	2009	100-105	1400-1500	Odisha, Gujarat, Chattisgarh
8.	OLM 217	Pure line selection	2009	105-110	1500-1600	Odisha, Gujarat, Chattisgarh
			PROSO MILLET			
1.	GPUP 8	S 7 x L111	2001	85-90	2500-2600	Karnataka and Tamil Nadu
2.	GPUP 21	GPUP 14 x K 1	2003	65-75	1500-1600	Karnataka and Tamil Nadu
3.	Pratap Cheena	Pure line selection	2006	65-70	1500-1700	All proso millet growing areas of the country
4.	Co 5	PV 1403 x GPUP 21	2007	70-75	1800-2300	Tamil Nadu
5.	TNAU 145	PV 1454 x TNAU 96	2007	70-75	1800-2000	All proso millet growing areas of the country
6.	PRC 1	Selection from GPMS 519	2008	70-75	1000-12000	Uttarakhand hills
7.	TNAU 151	TNAU 96 x PV 1673	2008	70-75	1800-2000	Tamil Nadu
8.	TNAU 164	TNAU 137 x Co 4	2009	70-75	1800-2000	Andhra Pradesh, Bihar, Karnataka, Maharashtra, Tamil Nadu and Uttarakhand
9.	TNAU 202	PV 1453 x GPUP 16	2011	70-75	1800-2000	Dry lands of India

Contd...

Table 11–*Contd...*

Sl.No.	Name of the Variety	Pedigree	Year of Release	Maturity	Average Yield	Recommended Domain
			FOXTAIL MILLET			
1.	TNAU 186	Co 5 x SiA 326	1998	80-85	1800-2000	Andhra Pradesh, Tamil Nadu, Karnataka
2.	PS 4	Mutation of SiA 2616	1998	80-85	1800-2000	All foxtail millet growing areas of the country
3.	Meera (SR 16)	Pure line selection	2000	75-80	1500-1700	Rajasthan
4.	Srilakshmi	Pure line selection	2002	80-85	2300-2500	All foxtail millet growing areas of the country
5.	Prathap Kangani (SR 1)	Pure line selection	2003	65-70	1600-1800	Rajasthan
6.	Co 7 (TNAU 196)	Co 6 x Ise 247	2005	85-90	1800-1900	Tamil Nadu
7.	HMT 100-1	RS 118 x PS 3	2008	90-95	2000-2500	Karnataka
8.	SiA 3085	Pure line from 2644	2011	75-80	2000-3000	All foxtail millet growing areas of the country
9.	Suryanandi	Pure line from 1244	2012	70-75	2000-2500	All foxtail millet growing areas of the country
10.	SiA 3156	Pure line selection from 2871	2012	85-90	2000-2500	Andhra Pradesh, Bihar, Gujarat, Karnataka, Madhya Pradesh Tamil Nadu and Uttarakhand
			KODO MILLET			
1.	KK 1	Pure line selection	1999	80-85	1800-2000	Uttar Pradesh
2.	JK 155 (RBK 155)	Selection from GPLMP 251	2000	80-100	2000-2200	Karnataka and Madhya Pradesh
3.	JK 48 (DPS 48)	Pure line selection	2001	95-100	2600-2700	Madhya Pradesh
4.	KK 2	Pure line selection	2002	80-85	2000-2300	Uttar Pradesh
5.	JK 439	Pure line selection	2004	100-110	2200-2300	Chattisgarh and Madhya Pradesh
6.	JK 13	Mutant of JK 76	2007	95-100	2200-3000	Madhya Pradesh, Chattisgarh, Uttar Pradesh, Andhra Pradesh, Karnataka and Tamil Nadu

Contd...

Table 11–*Contd...*

Sl.No.	*Name of the Variety*	*Pedigree*	*Year of Release*	*Maturity*	*Average Yield*	*Recommended Domain*
7.	JK 106	Pure line selection	2009	100-105	1900-2000	Madhya Pradesh
8.	JK 65	Pure line selection	2009	105-110	2500-3000	Madhya Pradesh, Gujarat, Uttar Pradesh and Tamil Nadu
9.	JK 36	Pure line selection	2009	75-80	1000-1200	Madhya Pradesh
10.	JK 98	Pure line selection from 317	2010	100-105	2500-3000	Madhya Pradesh
11.	DPS 9-1	Pure line selection	2011	105-110	2600-2800	Madhya Pradesh
12.	Indira Kodo 1	Pure line selection	2012	105-110	2200-2500	Chattisgarh
13.	TNAU 86	Pure line selection	2012	95-110	2700-2300	Andhra Pradesh, Chattisgarh, Gujarat, Karnataka, Madhya Pradesh, Tamil Nadu and Uttar Pradesh
14.	RK 390-25	Mutant of RK 390	2012	100-105	2500-2800	Andhra Pradesh, Chattisgarh, Gujarat, Karnataka and Madhya Pradesh
			BARNYARD MILLET			
1.	VL Madira 172	EF 2 x VHC 5205	2000	75-80	2200-2300	Uttar Pradesh, Gujarat, Karnataka
2.	Sushrutha (RAU 11)	Pure line selection	2000	75-80	2000-2200	Zone 8 of Karnataka
3.	VL Madira 181	ECC 27 x VL 60	2001	70-80	1600-1800	Bihar, Karnataka, Madhya Pradesh and Tamil Nadu
4.	PRJ 1	Pure line selection from IEC 542	2003	115-120	2300-2500	Uttarakhand
5.	Pratap Sawan 1/ER 64	Pure line selection	2008	85-90	1500-1700	Rajasthan
6.	VL Madira 207	VL 172 x GECH 506	2008	80-90	1600-1800 and Gujarat	All barnyard millet growing states except Tamil Nadu
7.	CO 2	Pure line selection from EF 79	2009	95-100	2100-2200	Tamil Nadu

cultivation competitive and economically viable. The success of GPU-28 variety of ragi in Karnataka is a best example in harnessing the benefits of a good seed production and distribution program involving line departments like Karnataka State Seed Corporation Limited and Karnataka State Department of Agriculture. The package of practices for cultivation different small millets such as time of sowing/planting, choice of varieties, time and method of application of fertilizers have been developed for different regions, of the country. Management practices for aberrant weather conditions for mitigating early, mid and late season drought have been worked out. Remunerative cropping systems involving different pulse crops in millet for different regions have been evolved. Plant protection measures to control economically important diseases and pests have been evolved.

Crop Production

Small millets are mainly grown under rain fed conditions during *Kharif*. Therefore conservation of rain water is very essential. Early tillage, deep ploughing at least once, land smoothening, frequent inter-cultivation upto 40-45 days from sowing are essential. Line sowing is beneficial in inter-culturing and control of weeds effectively. Maintenance of optimum plant population is essential for realizing higher yields. Two to three inter -cultivations followed by one hand weeding minimizes weed problem. These crops respond well to added fertilizers. Judicious blending of organic manure with 20-40 kg N, 20-40 kg P_2O_5 and 20-25 kg K_2O kg ha^{-1} brings higher yield. Nitrogen in two splits helps in efficient use of applied fertilizers. Profitable small millets based intercropping systems have been identified for different regions.

Crop Protection

Research activities in plant protection discipline are concerned with identifying resistant sources, forecasting, loss estimation, epidemiological studies and chemical control measures for major diseases and pests. Some of salient findings are:

- Finger millet genotypes highly resistant to both neck and finger blast were identified.
- New resistant sources for all major pests and diseases have been located for further use in crop improvement research.
- Finger millet Genotypes highly resistant to both neck and finger blast were identified.
- Spraying of SAAF @ 0.2 per cent at 50 per cent flowering and one more need based spray after 10 days are effective in controlling neck and finger blast.
- Choosing resistant variety and treating seeds with carbendazim @ 2 g/kg seed effectively controlled the blast incidence and resulted in higher yield of finger millet;
- In foxtail millet sowing early in July minimized the incidence of blast and rust.
- Isolation, purification and characterization of phytotoxin and phytoalexin have helped in rapid diagnostic methods for resistance evaluation.

- ☆ The role of phenols, PAL, PPO, Chitinase, β-13-gluconase activities has been established in disease resistance.
- ☆ Varieties resistant to insect pests were identified in all small millets.
- ☆ Intercropping/sprinkling of niger/mustard help in enhancing the population of parasites and predators in various small millets and thus help in effective bio control of major pests.
- ☆ Early sowing with the onset of monsoon helps in escaping the spray schedules for the control of shoot fly in little, kodo and proso millets.
- ☆ Seed treatment with chlorpyriphos @ 12 ml/kg seed was effective in warding of shoot fly damage up to 25-30 days
- ☆ The host plant resistance studies on shoot fly in little millet have revealed that plant characters like Trichome density on leaves and other plant parts and silica content in plants are responsible in conferring resistance to shoot fly.

Crop Physiology

- ☆ A quick screening technique for assessing genotypes for drought resistance based on determination of crop canopy air temperature differences (CCATD) has been standardized. Using these techniques two types of genotypes namely negative CCATD (cooler canopy) and positive CCATD (warmer canopy) have been identified. Negative CCATD types are superior in water extraction and they possess higher root volume and root length.
- ☆ Studies on harvest index in finger millet indicated that the variability for this character was large from 0.23 to 0.35. In different varieties grain yield showed significant positive relationship with total biomass and mean ear weight but non-significant relationship with productive tillers. In hills, the biomass production significantly increased in the crop sown in May. This was also reflected in higher yield. Tissue Culture-Protocols for plantlet regeneration from various explants have been standardized. Transformation protocols using agro-bacteria mediated transfer has been developed. Screening of finger millet germplasm for higher water use efficiency and root traits has helped in identifying are superior genotypes. Similarly saline tolerant varieties have also been identified.
- ☆ Technology transfer attempted through frontline demonstrations on the farmers' field and on large scale station demonstrations have helped in narrowing down the yield gap that existed between farmers field, demonstration plot and research station trials. Participatory selection of varieties has also been helpful in fine tuning variety recommendation domain by narrowing down to identifying varieties with high degree of location specificity.

Future Prospects

- ☆ After years of neglect, small millets are receiving some attention in agricultural R&D agenda in the country. They are viewed as important for health and well ness of people and can help in preventing many kinds of diseases related to modern life style including obesity, diabetes. A couple of elite food chains have begun selling millets and millets based products on their shelves as health food. The productivity could be increased by more than 50 per cent by adopting improved production practices as demonstrated by the results of frontline demonstrations. Since the seed rate is low for these crops (10 to 12 kg/ha) the extra expenditure incurred on seeds of high yielding varieties is not high and could be considered as low cost technology. In lieu of climate change contingency planning is required in case of excessive, scanty or delayed rains, wherein the versatile small millets would fit in more eminently.

- In recent years milling technology has been improved to enhance the grain quality. Millet mills are available for cottage level and large scale processing. Small millets can be further processed towards various foods such as flakes, quick food cereals, ready to eat snacks, supplementary foods, extrusion cooking, malt based products, weaning foods and more importantly health foods. Most of the small millet traded internationally is imported by the pet food industry in the developed countries for use as bird feed, which needs to be explored. There is a greater demand for organically produced small millets which can be further stepped up so that the demand for these crops is increased for the benefit of millet farmers.
- The importance of regular food use of nutrient dense millet for achieving a holistic food and nutritional security is widely recognized. Intensification of efforts to secure the millet growing areas by offering improved germplasm, more assured market through social, industrial and marketing back up is very much needed now. The farmers who had shifted from millets to other crops are keen to go back to millets in view of the stable harvests ensured, easy crop production, drought resistance, and eco friendly production. The higher prices being offered in some parts of the country to small millets duly recognizing their unique nutritional features is making small millets a remunerative proposition vis a vis with other crop options in the region.
- The R&D efforts made in the area of grain processing and value addition through the development of novel diversified foods especially in finger millet and other small millets are opening new avenues for expanding consumer base, enhanced absorption for food use. However, several challenges specific to region are being faced in many key areas of production to consumption system which need close attention. Intensification of research for genetic improvements and improved farming practices are important to increase and stabilize farm yields and make these crops remunerative.
- Securing the existing millets cultivation area is a matter of concern. In the past 5 decades the area under small millets has gone down by more than 75 percent. Area loss has been very steep in the last 20 years and this trend need to be arrested and efforts are necessary to secure the existing area. The support essential for protecting small millets is to be carefully assessed

and organized with out further delay. The tendency of discouraging small millets for making way to other crops such as mono cropping of oils seeds and pulses should be done away with.

- ☆ The rich diversity of small millets crops has made these crops well suited for contingency crop planning and also to address the issues of climate change. The plasticity exhibited has made them flexible for - apparent early as well as delayed planting, very low and high rainfall areas, various elevations and different soil regimes. These positive features have not been duly recognized and exploited in the country.
- ☆ The Government of India is implementing national food security to all people in the country. This will be feasible only by protecting and promoting regionally important food grains especially small millets and providing necessary technological and marketing back up as is being presently done for rice and wheat.
- ☆ There is need for the R&D to address to whole range of activities from conservation to final consumption. In this process, the existing market potential and likely channels for enhanced absorption need to be effectively met. Rising awareness on the nutrient richness of small millet grains and identifying technology and policy options should receive attention. Some of these millets, being the first crops domesticated and with cultivation history of more than 5000 years have many reasons from Social, Cultural and Health considerations necessitating their continued production and consumption especially in the present juncture of providing effective food and nutritional cover to all in the country. Besides, India is the primary/ secondary centre of diversity for many small millets such as finger millet, kodo millet, little millet and Indian barnyard millet and even sorghum and pearl millet. Three of the small millets-kodo, little millet and Indian barnyard millet - have originated and domesticated in India. Hence, small millets deserve to be declared as "Heritage Crops" of India.

The immediate needs of small millet promotion are:

- ☆ Sustaining and increasing production of small millets by using nutritional superiority of grains and expanded opportunities for value addition as incentives.
- ☆ Conserving genetic resource base and providing superior germplasm possessing the required attributes keeping in view the food, feed and diversified utilization needs.
- ☆ Assisting to overcome the constraints in production by providing technological inputs appropriate to the region.
- ☆ Creating awareness among general public, policy makers, traders, farming communities, consumers, entrepreneurs and donors of the role of nutrient dense small millets in ensuring better health (nutritional security) and there by increasing demand and use.

Keeping the above in view the research activities are to be restructured and should have very high degree of location specificity. The R&D should not be viewed from crop and productivity angles only; but holistically on a wider perspective of other advantages accruing through small millet based cropping systems such as ensuring greater ecological balance with environment, efficient soil and moisture conservation and utilization, adjustments for climate change, minimum pest/disease load, assured harvests and more so the food, nutrition and health security to people at large.

Promotion of Small Millets and Enhanced Utilization: Areas for Action

1. Optimum management of the crop is essential for higher productivity. Awareness creation on improved production practices among farmers is necessary. Small millets are grown in remote areas where Govt. machinery is not in extension programmes. This has to be overcome by involving Civil Society Organizations to demonstrate the potentials available for enhancing the productivity. Crop Insurance has to be extended to dry land crops especially to millets.
2. In small millets organized seed supply systems are absent barring a few states where limited seed production of ragi and its supply is undertaken. This being a critical issue needs holistic action plan. State Dept of Agriculture should compulsorily take up seed production and distribution
3. Extending liberal seed subsidy for millets needs to be considered. In small millets 75 per cent of cost of seed should be subsidized. Since many of the public/organised private seed producing agencies are not be present in most places, genuine and capable NGO's be engaged in seed production and supply to govt.,
4. Encouraging Institutions for forming Seed Cooperatives be addressed by providing support for infrastructure and for production of seeds. Local seed supply systems on the lines of seed banks successfully functioning in some areas need to be replicated.
5. There are critical agricultural implements available especially for sowing operations for line sowing suitable for small millets. These implements should be distributed at subsidy to millets farmers for supporting proper cultivation practices.
6. Special schemes for improving productivity of small millets are to be launched. The scheme should focus on seed and input supply, implements, credit support, insurance, price support and procurement etc.
7. Even in crops like small millets, scarcity of farm labour is increasingly felt. This has a direct effect on increasing cost of production and in turn profitability to the farmers. Selective Mechanization of key operations such as planting, inter cultivation and harvesting is the need of the hour. The research on these be initiated in AICRPs

8. Considerable research efforts have gone in the area of value addition and grain processing in the last one decade. Many novel food products have been developed with millets grains especially in ragi and other small millets. This is offering great opportunities for commercialization of value added products. As a promotional incentive special schemes are to be launched for encouraging entrepreneurs to start small millets based food processing industries.
9. Ragi and other small millets are known to have several neutraceutical properties. Owing to this they are ideally suited for developing functional foods. This calls for close attention of food technologists. Greater understanding of bio availability of nutrients and minerals of millet grains will further help in enlarging food value of millets.
10. Market channels are totally absent for small millets. It is necessary to provide market support for sustaining their production and consumption. Declaring MSP does not help these crops unless procurement facilities as is done in rice and wheat are established.
11. Sto*Rabi*lity of millets grains is one of the often asked questions. This needs systematic analysis and study in order to develop proper storage systems.
12. In spite of fascinating qualities that small millets have, they are condemned to be called as *coarse* grains, poor men's food etc. The attitude towards grains has to change especially in the urban population in order main stream small millets as superior food grains. There re is need for initiating a focused and intensive publicity/awareness campaigns at the national level for popularizing these grains.
13. Policies relating to incentives on bonus for climate change, biodiversity and ecological balance may have to be formulated as small millet farming leads to saving on the current subsidies on chemical fertilizers and power. These policies can rescue farmers from their current reliance on high external input agriculture and move them to low input organic and bio-diverse farming thereby finding a lasting solution for the agrarian crisis.
14. There are many innovations in the processing of small millet grains. Ready to eat products have been developed at many institutions. Institutional linkages must be established to harness their potentials. SHG's can be supported with the machinery required and can be engaged in supplying the processed millet foods for ICDS programme and mid day meal programme. This not only creates demand for the products, but also helps in reducing the malnourishment among the children and pregnant women.
15. Strategies that promote small millet farming not only encourage rain fed farming, but also improve the livelihoods of section of agricultural work force living in these areas. So, ways and means are to be found for synergizing with the several rural development programs especially under MGNREGS.
16. Major fodder needs in the country especially in rain fed areas is met from crop residues and all millets being nutritious are important in this context.

Mixed forming is an important source of livelihood for small farmers. Millet based mixed farming system has good potential to address the food and feed security as well as poverty alleviation in the country.

Conclusions

Small millets are the oldest foods known to humans and possibly the first cereal grain to be used for domestic use. They have been in food use in India since 2500 BC. India has been the native home for several small millets such as little millet, Indian barnyard millet and kodo millet and these were first domesticated in Indian sub continent. The antiquity of cultivation of millets has led evolution of time tested millet based cropping systems and ethnic food culture thus ensuring stable well orchestrated food, feed and nutritional security. These crops despite no support what so ever from the developmental agencies have survived and have withstood repeated onslaughts from so many competing crops. This is testimony of their value to harsh farming situations and their resilience ability to adjust to changing climates and yet producing reliable and dependable harvests. It is time that these crops are brought back to main stream and allowed to play their due role in contributing to national food basket.

References

1. AICSMIP, Bangalore. 2012. *Annual Progress Report 2012-13.*Project Coordination Cell, All India Coordinated Small Millets Improvement Project, University of Agricultural Sciences, Bangalore.
2. Byre Gowda M, Shankare Gowda B T and Seetharam A. 1998. Relationship of Proteins, Phenols and Tannins with blast disease in finger millet (*Eleusine coracana* G.) *Crop Improvement*. 25(1): 43-47.
3. Byre Gowda M, Shankare Gowda, B.T. and Seetharam A. 1999. Selection for combining grain yield with high protein and blast resistance in finger millet (*Eleusine coracana.* G.), *Indian. J. Genet*. 59(3): 345- 349.
4. Byre Gowda M, Seetharam A and Shankare Gowda BT. 2000. Possibilities of combining high protein content with grain yield in finger millet, *Ind. J. Gent*. 60: 387-390.
5. Dinesh Kumar S P, Sashidhar V R, Ravikumar R L, Seetharam A and Gowda B T S. 1992. Identification of true dwarfing genes in foxtail millet (*Setaria italica* Beauv.) *Euphytica* 60: 207-212.
6. Gowda B T S, Seetharam A, Viswanath S and Sannegowda S.1986. Incorporation of blast resistance to Indian elite finger millet cultivars from African cv. lE 1012, *SABRAO J*. 18: 119-120.
7. Gowda B T S, Halaswamy B H, Seetharam A, Virk D S and Witcombe J R. 2000. Participatory approach in varietal improvement: a case study in finger millet in India, *Cur. Sci.,* 79 (3): 366-368.
8. Naik, B J, B.T.S. Gowda B T S, and Seetharam A. 1993. Pattern of variability in relation to domestication of finger millet in Africa and India, (*In*) *Recent Advances*

in Small Millets. (Eds. K. W. Riley, S.C. Gupta, A. Seetharam and J. Moshanga). Proc. Second Intl. Small Millets Workshop. Oxford-IBH Publishing Company. pp. 347- 364.

9. Paroda R S.1989. Research advances in crop sciences. (In) 40 years of Agricultural Research and Extension in India, pp. 35-81.Indian council of Agricultural Research, New Delhi.

10. Ravikumar R L, Seetharam A and Gowda B T S.1990. Identification of sources of stable resistance to finger millet (*Eleusine coracana* Gaertn.) blast, SABRAO J. 22: 117-121.

11. Ravikumar R L, Dinesh Kumar S P and Seetharam A.1991. Association of seed colour with seed protein, phenol, tannin and blast susceptibility in finger millet (*Eleusine coracana* Gaertn.) SABRAO J. 23: 91-94

12. Sashidhar V R. Prasad T G, Seetharam A and Udayakumar M. 1983. Ear photosynthesis in long and normal glumed genotypes of finger millet (*Eleusine coracana* Gaertn.), *Indian J. Plant Physiol*. 4: 359-362.

13. Sashidhar V R, Gurumurthy B R, Prasad T G, Udayakumar M, Seetharam A and Sastry K S K.1986. Genotypic variation in carbon exchange rate, functional leaf area and productivity in finger millet (*Eleusine coracana*): An approach to identify desirable plant types for higher water use efficiency under rainfed conditions, *Field Crops Res*. 13: 133-146

14. Schiere J B, Joshi A L, Seetharam A, Costing S J, Goodchild A V, Deinum Band Van Keulen H. 2004. Grain and straw for whole plant value: Implications for crop management and genetic improvement strategies. *Expl Agric.*, Vol. 40, pp 277-294

15. Seetharam A, Aradhya K M, and Laxrninarayana.M R. 1983.Variation for seed oil content in a world collection of foxtail millet germplasm, *SABRAO J. 15*: 99-102.

16. Seetharam A, Aradhya K M, Sashidhar V R, Mahishi D M and Gowda.B T S. 1984. Protein content in white and brown seeded finger millet genotypes, *SABRAO I.* 16: 65-67.

17. Seetharam A, Gowda B T S, Subba Rao A, and Prabhu U H. 1993. Genetic parameters and relation of morphological characteristics associated with fodder yield and quality in finger millet (*Eleusine coracana*), (*In*) *Feeding of ruminants on fibrous crop residues*. Proc. Int. Workshop, (Eds. Kiran Singh, J.B. Schiere). ICAR, New Delhi and Agriculture University, Wageningen. The Netherlands. 359-364-370.

18. Seetharam A. 1998. Small millets -Achievements during 1947-97, *Indian J Agric.* Sci. 68 {Suppl): 47-54.

19. Seetharam A. 2006. Millets, In: *Handbook of Agriculture*, ICAR, New Delhi, pp. 892-912.

20. Seetharam A, Patel D P, and Halaswamy B H. 2006. Small millets (*In*) *Plant Genetic Resources-Food Grain Crops*. (Eds: B.S. Dhillon, S. Saxena, A. Agrawal and R.K.Tyagi) Narosa publishing house, Delhi, pp. 204-223.

21. Seetharam A,, and. Ravikumar R L. 1993. Blast resistance in finger millet -Its inheritance and biochemical nature, (*In*) *Recent Advances in Small Millets* (Eds. K. W. Riley, S.C. Gupta, A. Seetharam and J. Moshanga). Proc. of Second IntI. Small Millets Workshop. Bulawayo (Zimbabwe). Oxford-IBH Publishing Company, New Delhi. pp 449-466.

22. Seetharam A, Gowda B T S, Prabhu U H and Subba Rao A. 1994. Genetic and managemental influence on stover quality and quantity of finger millet (*Eleusine coracana*), (*In*) *Variation in the quantity and quality of fibrous crop residues*. Proc. Nat. Seminar, BAIF Development Research Foundation, Pun6;.ICAR, New Delhi and Agriculture University, Wageningen. The Netherlands (Eds. A. L. Joshi, P.T Doyle and S.J. Oosting), pp. 1-8.

23. Seethatram A, and Krishne Gowda K T. (2008).Production and utilization of small millets in India, (*In*) *Food uses of small millets and avenues for further processing and value addition*. Project coordination cell, All India Coordinated Small Millets Improvement Project, ICAR, G.K.V.K. Campus, Bangalore.

24. Subba Rao A, Prabhu U H, Seetharam A, Gowda B T S,. Oostings S T, Sam path K R and Avinash Kumar. 1995. Finger millet straw, (*In*): *Hand Book for Straw Feeding Systems*. ICAR, New Delhi and Agriculture University, Wageningen. The Netherlands (Eds. Kiran Singh and J. B. Schiere), pp. 355-366, ICAR, New Delhi.

Millets: Ensuring Climate Resilience and Nutritional Security (2015) *Pages* **261–275**
Editors: **Vilas A. Tonapi and J.V. Patil**
Published by: **DAYA PUBLISHING HOUSE, NEW DELHI**

Chapter 6

Genetic Improvement in Finger Millet

K.N. Ganapathy

ICAR-Indian Institute of Millets Research (IIMR) Rajendranagar, Hyderabad – 500 030

Finger millet (*Eleusine coracana* (L.) Gaertn.) is an important and staple food crop of Africa and Southern Asia. Among the finger millet growing countries, India is the major producer of finger millet contributing about 60 per cent of the global production (Haradari *et al.*, 2012). The crop ranks third next to sorghum and pearl millet both in area and production in India. Cultivation of this crop extends from sea level (southern India) to hilly regions of Himalayas and is adapted to wide range of soil and environmental conditions but thrives best under good arable, well drained, loamy or clay loamy soils. The crop can also tolerate certain degree of salinity. Among the finger millets growing states, Karnataka contributes major share of about 60 per cent of the area and production. Other major growing states are Maharashtra, Uttarakhand, Tamil Nadu, Andhra Pradesh and Uttarakhand, Odisha and Andhra Pradesh. Grains of finger millet are highly nutritive and is known for its highest amount of calcium iron, zinc, dietary fiber and essential amino acids (Singh *et al.*, 2012; Shobana *et al.*, 2013). The grains are highly resistant to storage insect pests and under minimum attention the crop can stay in good condition upto 50 years (Iyengar *et al.*, 1945). The stover after harvest of grains is source of nutritive fodder to animals and is highly preferred due to sweet smelled stalks. Comparing different small millets, genetic improvement is much pronounced in finger millet but not exploited to its fullest extent like in sorghum and pearl millet mainly due to smaller sized florets owing to difficulties in hybridization. However the yield potential, nutritional superiority and its ability to tolerate drought make it a promising crop for the future. Research efforts should aim at using a combination of conventional and molecular approaches for genetic improvement of grain and forage yield, nutritional parameters, biotic and abiotic stresses, identification of end use specific genotypes and possibilities of exploitation of heterosis in finger millet crop.

Area, Production and Productivity

Globally finger millet planting area is estimated to be around 4-4.5 million ha with a production of about 5 million tons of which 2-2.2 million tones is contributed by India, about 2 million tons by Africa (Prajapati *et al.*, 2014) and rest by countries of south Asia. The important finger millet growing countries in eastern and southern Africa are the sub-humid regions of Ethiopia, Kenya, Malawi, Tanzania, Uganda, Zaire, Zambia and Zimbabwe. Similarly in South Asia, the crop is largely cultivated in India, Nepal and to certain extent in Bhutan and Sri Lanka. It is also reported to be grown in certain areas of China and Japan. In India, the crop is majorly grown in Karnataka, Uttarkhand, Odisha, Maharastra, Andhra Pradesh and Tamil Nadu.

There is a growing concern about the decline in area of finger millet. The area under finger millet during 1951-55 were around 2.2 million ha. Area gradually increased and reached to about 2.5 million ha during 1961-66. Thereafter the area started declining and reached to 2.3 million ha during 1986-90. From 1990 on-wards there was a steep decline in area and reached to 1.3 million ha during 2010. The area during 2011-12 was 1.17 million ha. Out of the total area 6 lakh ha area is occupied by Karnataka state, followed by Maharashtra (1.3 lakh ha), Uttarkhand (1.25 lakh ha), Tamil Nadu (0.8 lakh ha) and Andhra Pradesh (0.42 lakh ha). In Karnataka the crop is grown during *Kharif*, *Rabi* and summer season with 85-90 per cent during *Kharif* and rest 10-15 per cent during *Rabi* and summer season. Similarly in Odisha state the crop is grown during both *Kharif* and *Rabi* season with more than 90 per cent area during *Kharif*. List of potential finger millet growing districts is given in Table 1.

In India the production was about 1.6 million tons during 1951-55. The production gradually increased and reached to about 2.2 million tones during 1971-75. During 1976-80 highest production of 2.6 million tones was recorded and this was due to increase in area during the period. The production gradually declined after 1980 and reached to around 2.0 million tones. During 2011-12 the production was estimated to be around 2.0 million tones. During 2011-12 the highest production is obtained from Karnataka (1.2 million tones), followed by Tamil Nadu (0.224 million tones), Uttarkhand (0.174 million tonnes), Maharastra (0.138 tones) and Andhra Pradesh (0.040 million tones). State wise area and production and given in the Figure 2.

Productivity of finger millet in India showed continuous improvement. The productivity in India was 704 kg/ha during 1951-55. The productivity improved gradually and reached to 1021 kg/ha during 1976-80. The productivity further increased and reached to 1641 during 2011-12. Over the years, the yield has increased more than two folds from 704 kg/ha (1951-55) to 1641kg/ha (2011-12) at national level. The highest productivity increase is contributed by the improved varieties by crossing the Indian types with the exotic types especially the African gene pool which were otherwise isolated in Indian breeding programme. Highest yield gains

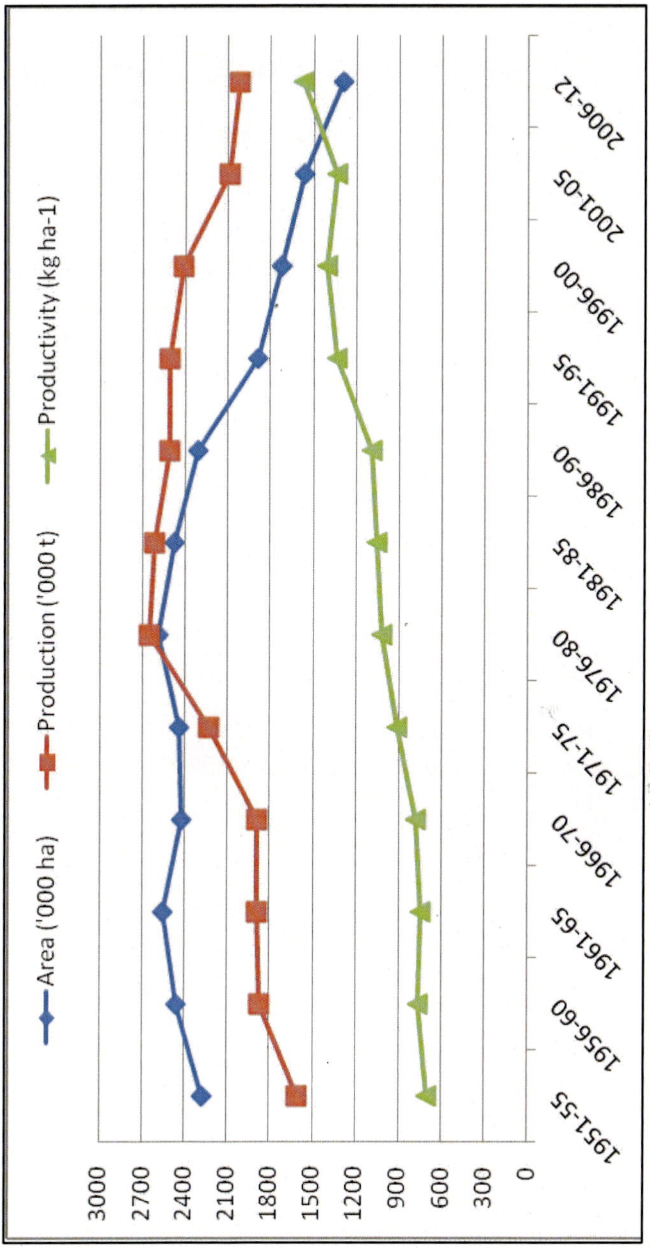

Figure 1: Quinquinnal Area, Production and Productivity of Finger Millet in India.

Source: AICSMIP, UAS, Bengaluru.

Table 1: Potential Finger Millet Growing Districts of India

Sl.No.	*State*	*District with an Average 10,000 ha Area (upto 2010)*
1.	Karnataka	***Kharif***: Bengaluru Rural, Bengaluru Urban, Chamrajnagar, Chitradurga, Davangere, Hassan, Kolar, Mandya, Mysore Tumkur, Chickmaglur and Shimoga
		Rabi/summer: Chamrajnagar, Chitradurga, Mandya and Tumur (with an average of 1000 ha or above)
2.	Tamil Nadu	Dharmapuri, Erode, Salem, Thiruvannammalai, Vellore and Villupuram
3.	Maharastra	Thane, Raigarh, Ratnagari, Nasik, Dhule, Ahmednagar, Pune, Satara and Kolhapur
4.	Andhra Pradesh	Vizianagaram, Vishakhapatnam, Chittoor, Mahaboobnagar and Rangareddy
5.	Uttarkhand	Chamoli, Pauri Garhwal, Rudra Prayag, Tehri Garhwal, Uttar Kashi, Almora, Bageshwar, Champawat and Pithoragarh
6.	Odisha	Kalahandi, Koraput, Malkangiri, Nowrangpur, Raigada, Ganjam and Gajpati
7.	Chattisgarh	Bastar
8.	Jharkhand	Gumla
9.	Gujarat	Dang and Valsad
10.	West Bengal	Darjeeling

Source: Status paper on millets 2010, Directorate of millets development, India

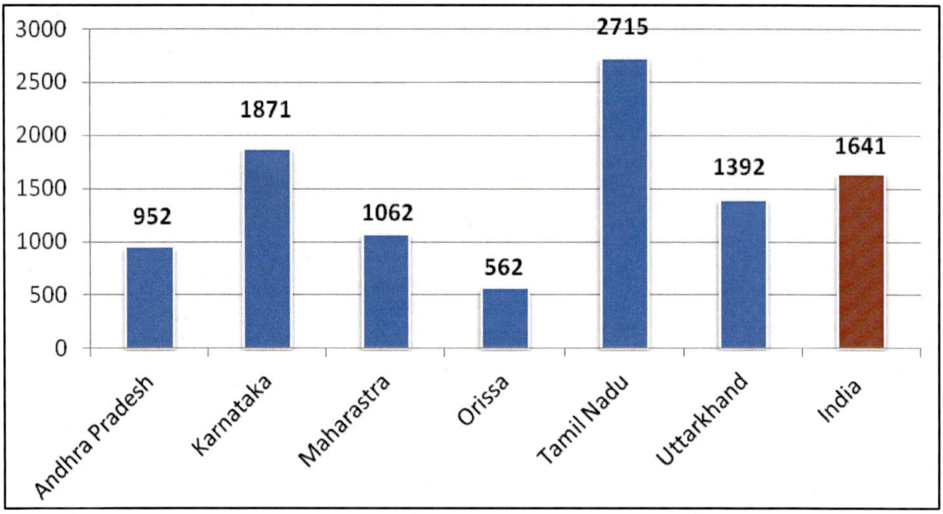

Figure 2: Productivity of Major Finger Millet Growing States during 2011-12.
Source: AICSMIP, UAS, Bengaluru.

was recorded by Tamil Nadu (2715 kg/ha) followed by Karnataka (1871 kg/ha). One of the reason for highest productivity in Tamil Nadu is due to its cultivation in *Kharif* season while in Karnataka the crop is cultivated in *Kharif, Rabi* and also in summer. GPU 28 a blast resistant high yielding variety has made a major contribution to yield improvement in Karnataka.

Status of Genetic Improvement

(a) Genetic Variability for Economically Important Traits

For genetic improvement of economically important traits it is imperative that we know the economically important traits and the variability existing for these traits. Early maturing varieties combining high grain yield, quality and Stover yield are ideally preferred by farmers. National Active Germplasm Collection Site (NAGS), Project Co-ordination cell, All India Coordinated Millets Improvement Project (AICSMIP) Bangalore made efforts to assemble large collections of germplasm at global level. Presently, the centre has total collections of 7070 germplasm. Both the wild (*Eleucine coracana* subsp. *africana*) and cultivated (*Eleucine coracana* subsp *coracana*) are considered important for germplasm collection and conservation forming primary gene pool. The cultivated finger millet has been divided into races and sub-races which are mainly based on inflorescence morphology.

Table 2: Cultivated Races and Sub-races

Sl.No.	*Races/Sub-races*	*Characteristics*
1.	Race: Elongata Sub-races: *laxa*, *reclusa* and *sparsa*	Long and slender inflorescence, 10-24 cm long branches, arranged digitately, spikes spreading and curved outward during maturity
2.	Race: *Plana* Sub-races: *seriata*, *confundere* and *grandigluma*	8-15 cm large spikelets arranged in 2 or more rows along with rachis giving inflorescence branch a flat ribbon like appearance
3.	Race: *compacta*	Spikelets composed of 9 or more florets, curved at top to give large fist like appearance
4.	Race: *vulgaris* Sub-races: liliaceae, stellata, incurvata and digitata	Four to eight florets in a spikelet giving semi-compactness appearance with tip incurved at maturity

The Indian and African germplasm are highly diverse and a very few systematic studies have been conducted to compare the diversity. The Indian germplasm are reported to be diverse for grain and stover yield. The African germplasm exhibited larger variation for tallness, stout plant stature, long narrow finger, higher number of spikelets, poor threshability, late maturing and poor harvest index. The African germplasm are said to possess higher level of resistance to blast- the most devastating disease in finger millet.

For effective utilization of the germplasm for genetic improvement program, core and mini core were developed both by the AICSMIP and ICRISAT (Gowda *et al.*, 2007; Upadhyaya *et al.*, 2010; Haradari *et al.*, 2012). Haradari et and his co-workers reported a core collections of 77 germplasm made from set of 1000 accessions of finger millet belonging to both India and Africa. ICRISAT developed a core collection of 622 genotypes (Upadhyaya *et al.*, 2010). The 622 core collections were evaluated for 15 quantitative and 5 qualitative traits for constituting mini core collection of 80 accessions. Wide variability were reported for economically important traits like days to flowering (51.24-93.73), plant height (72.66-113.31 cm), length of longest finger (49.79 to 139.73 mm), finger per ear (6.13-9.41) and grain yield (691-2430 kg/ha) in mini-core collections. Subsequently efforts were made by different centers to study the variability and identify location specific genotypes for different economic traits such as days to flowering, plant height, number of ears, ear length, productive tillers, days to maturity, grain yield, grain quality, 1000 grain weight for utilization in their breeding programmes.

(a) Varietal Development

Various breeding methods such as mass selection, pure line selection, recombination breeding and mutation breeding are the widely used approached for genetic improvement in finger millet. Mass selection has been extensively used for purification of the cultivars (landraces) and multiplication of varieties bred by pure line or pedigree systems. Mass selection resulted in marginal increase in the performance of the farmer's finger millet varieties like *gidda ragi*, *hullubele* of Karnataka, *saluchodi* of Andhra Pradesh, *udumalpet* ragi and *guddapah* rajampet ragi of Tamil nadu and *murky* and *nangkatna* of Sikkim. Pure line selection has also been widely practiced in finger millet improvement. Single plants were selected from landraces (germplasm) and improved farmer's varieties and the superior progenies mostly for earliness, pest and disease resistance and grain yield were tested under multilocational trials and released as varieties. Pure line resulted in development and release of 41 varieties in finger millet (Harinarayana *et al.*, 1086).

Progress through hybridization has received greater attention in finger millet. Until 1950's the finger millet genetic improvement revolved around the locally adapted lines by centres located in the states. Subsequent millet improvement was carried forward after establishment of The Project for Implementation of Regional Research on Cotton, Oilseeds and Millets (PIRRCOM) during late 1950's and later formulation of the All India Coordinated Millets Improvement Project during 1965. Germplasm of finger millet and other millets were assembled during 1960's. The exotic germplasm mainly from Africa was introduced to India during 1970's. The African germplasm had high vigour, large ears, more grain density and thick robust stems with broad dark green leaves (Seetharam, 1982). Hybridization was performed among indigenous, exotic and indigenous with exotic. Pedigree selections were followed and selections were made from Indian x Indian, Indian x exotic or exotic x Indian and exotic x exotic cross for early (less than 100 days), mid-late (100-110 days) and late (more than 110 days). Among the exotic x exotic crosses, varieties HR 374 (early) and Co-6 (late) were developed and had productivity of less than 2000 kg/ha. Large number of varieties were developed from Indian x exotic crosses. Among them, Indaf 9 for early, Co-9 and Indaf 5 for mid-late and Indaf 1, Indaf 3, K5, Indaf 7, Indaf 8, Indaf 6, HR 911 and Indaf 11 as late maturing varieties were released. The Indian x exotic crossed has productivity of more than 2500 kg/ha. Varieties from Indian x Indian crosses such as udaya, K 7 (early), Purna and Annapurna (mid-late) and Cauvery, Shaktii and HPB 7-6 (late) were developed and released having intermediate productivity. Hybridization among the Indian x exotic cross provided real breakthrough in breaking yield barriers and improving productivity of finger millet. This aided in achieving by more than 50 per cent increased productivity in Karnataka (Seetharam, 1982) and by

60 per cent in Tamil Nadu (Nagarajan and Raveendran, 1983). Subsequent large scale efforts were made by major finger millet growing states to develop location specific varieties combining high productivity, blast resistance and drought tolerance. GPU 28 is a medium duration variety maturing in 110-115 days released during 1996. The variety is suited for delayed sowing under terminal drought conditions. It is also resistant to neck and finger blast which is a major pest to finger millet production. The potential yield of the variety is 35 -40 qtl/ha. However under farmers field condition the productivity increased from 10 qtl/ha to 20 qtl/ha by introduction of this variety and presently the variety is grown on a larger area (about 70 per cent area) in Karnataka. Joint efforts were made by the Department of agriculture in each of the finger millet growing states, state seed corporation, state agricultural universities and private seed sectors for augmenting the availability of quality seed of GPU 28. GPU 28 was released during 1998. During the initial stages of the about 1100 kgs certified seed was produced during 1998-99 but during 2008-09 about 15548 of the total certified seeds were produced out of which 93 per cent is of GPU 28 and has become the most popular variety of finger millets in Karnataka state. Other varieties released under the national programme, varietal characteristics and their area of adaptation is given in Table 3. Farmer's participatory varietal selection was carried out at UAS, Bangalore under the MC Knight Foundation funded project Development of high yielding disease resistant and drought tolerant finger millet genotypes suited for southern part of Karnataka. Twenty eight superior recombinant inbred lines derived from IE 6012 x Indaf 5 were evaluated through farmer's participatory varietal selection. The study identified ML 31, ML 322 and ML 365 as more preferred genotypes based on grain yield, fodder yield and quality, cooking quality and nutritional parameters.

(c) Blast Disease Resistance Improvement

Blast disease caused by *Magnaporthe grisea* is a major problem and destructive disease in India as well as In Africa causing substantial yield losses in the crop. Average loss has been reported to be around 28-36 per cent (Vishwanath *et al.*, 1986, Nagaraja *et al.*, 2007) and yield losses in endemic areas is as high as upto 80-90 per cent (Rao *et al.*, 1990). Although number of cultural and chemical methods are available for management of blast disease, it is mostly tacked by host plant resistance as numbers of resistance sources are available in germplasm. Sources of blast resistance are reported in finger millet and efforts were made to incorporate resistance into improved cultivars/breeding lines. A good number of varieties/genotypes GPU 28, GPU 45, VL 315, GPU 48, OEB 259, VL 340, PRM 9809, HR 374, VL 149 are reported to be blast resistant. Efforts to screen the core collection to identify sources for blast resistance. Blast screening was carried out in minicore collections and reported 58 accessions resistant to blast and represented one accession from wild (Africana) and four cultivated races (*plana*, *vulgaris*, *elongata* and *compacta*) originating from 13 countries of Africa and Asia and exhibiting considerable diversity for agronomic traits such as maturity, plant height and panicle type (Kiran Babu *et al.*, 2013). Bijendra Kumar and Kumar, 2011 reported varietal mixture as an option for controlling blast in finger millet. For understanding of the genetics, identification of genomic loci associated for blast disease resistance RIL populations (PR 202 x GPU 48, VR 708 x GPU 48, GPU 28

Table 3: List of Varieties Recently Developed along with its Characteristics

Sl.No	Crop/Variety	Pedigree	Institute where Developed	Year of Release	Maturity (Days)	Yield Level Kg/ha	Area of Adaptation
1.	GPU 66	PR 202× GPU 28	PC Unit, UAS, Bangalore	2013	115-120	4000-4500	Karnataka, Jharkhand, Maharastra, Tamil Nadu, Uttrakhand Chattisgarh
2.	KMR 301	MR-1× GE-1409	Mandya	2009	120-125	3000-3500	Karnataka
3.	VR 847	GPU 26 × L5	Vizayanagaram	2009	110-115	3000-3500	Andhra Pradesh
4.	GPU 67	PLS from GE 5331	PC Unit, UAS, Bangalore	2009	115-120	4000	Karnataka, Chattisgarh, Jharkhand, Maharastra, Tamil Nadu and Uttarakhand.
5.	GPU 48	GPU 26 x L 5	PC Unit, UAS, Bangalore	2005	100-105	3000-3500	Karnataka
6.	TNAU 946	Malavi 1305 x Co 13	TNAU, Coimbatore	2004	105-115	2000-2500	Tamil nadu
7.	VL 315		UPKAS, Almora	2004	105-115	2600-2800	Uttaranchal
8.	GPU 45	GPU 26 x L 5	PC Unit, UAS, Bangalore	2001	104-109	2700-2900	Gujarat, Jharkhand, Karnataka, Madhya Pradesh, Maharashtra
9.	Chilika (OEB 10)	GE 68 x GE 156	OUAT, Bhubaneshwar, Odisha	2001	120-125	2600-2700	Odisha, Madhya Pradesh, Gujarat, Andhra Pradesh and Tamil Nadu
10.	GPU 26	(I-5 x I-9) IE 1012	PC Unit, UAS, Bangalore	2000	100-105	3000-3500	Karnataka
11.	BM 9-1	Mutant from Budha Mandia	OUAT, Berhampur	1999	103-105	2500-3000	Karnataka, Andhra Pradesh, Odisha, Madhya Pradesh and Maharashtra
12.	L-5	Malawi x Indaf 9	UAS, Bangalore	1999	120-125	3500-4000	Karnataka
13.	Champavathi (VR 708)	Puerline selection	ANGRAU, ARS, Vizianagaram	1998	90-95	2000-2500	Andhra Pradesh, Uttar Pradesh, Tamil Nadu, Karnataka, Odisha
14.	MR 1	Hamsa x IE 927	UAS, Bangalore	1998	125-130	3500-4000	Karnataka
15.	PR 230 (Maruthi)	Pure line selection	ANGRAU, Paleru	1998	90-100	2500-3000	Andhra Pradesh (Telangana region)
16.	GPU 28	Indaf 5 x IE 1012	PC Unit, Bangalore	1996	110-115	3500-4000	Karnataka

Contd...

Table 3–*Contd...*

Sl.No	Crop/Variety	Pedigree	Institute where Developed	Year of Release	Maturity (Days)	Yield Level Kg/ha	Area of Adaptation
17.	Birsa marua 2	Pureline selection	BAU, Ranchi	1996	110-115	2000-2500	Chotanagapur regions of Jharkand
18.	VL-146	Selection from germplasm	APAU, Perumallapalli	1995	100-105	2500-3000	Andhra Pradesh
19.	BM 2	Selection	BAU, Ranchi	1995	105-110	2400-2600	Bihar
20.	Dapoli 1	Selection from mutant No. 50-1	KVK, Dapoli	1994	100-110	1500-2000	Konkan regions of Maharashtra
21.	Suraj (VR 520)	Pureline selection	ANGRAU, Vizianagaram (AP)	1994	90-95	2200-2800	All over India
22.	KM 65	Selection from exotic germplasm	CSAUA and T, Kanpur	1994	98-102	1800-2100	Uttar Pradesh
23.	A 404	Selection from germplasm	BAU, Ranchi	1993	110-115	2200-2500	Bihar
24.	Gautami (PR 1158-9)	PR 202 x U22	MRS, Vizianagaram	1993	115-120	2800-3000	Andhra Pradesh
25.	GN 3	KM 13 x GN 2	GAU, Gujarat	1993	130-136	2200-2500	Gujarat
26.	Padmavathi (PPR 2350)	Pureline selection	ANGRAU, Perumalapalli (AP)	1993	110-115	2500-3000	Coastal Andhra Pradesh
27.	Indaf 15	IE 927 x IE 67	UAS, Bangalore	1991	110-115	3500-4000	Karnataka
28.	VL 149	VL 204 x IE 882	VPKAS, Almora	1991	98-102	2000-2500	All states
29.	Saptagiri (PR 2614)	MR 1 x Kalyani	APAU, Perumalapalli	1990	105-110	2500-3000	Tamil Nadu, Maharashtra, Odisha, Andhra Pradesh
30.	KM 13	Pureline selection	CSAUA and T, Kanpur	1989	95-110	2500-3000	Uttar Pradesh, Madhya Pradesh, Odisha
31.	PES 400	Pureline selection	GBPUAT, Pantnagar	1989	98-102	1800-2000	Hills of U.P
32.	Co 13	Co 7 x TAH 107	TNAU, Coimbatore	1989	110-120	2500-3000	Tamil Nadu

Contd...

Table 3–*Contd...*

Sl.No	*Crop/Variety*	*Pedigree*	*Institute where Developed*	*Year of Release*	*Maturity (Days)*	*Yield Level Kg/ha*	*Area of Adaptation*
33.	TRY 1	Selection from HR 374	TNAU, Coimbatore	1989	100-105	2000-2500	Tamil Nadu
34.	VL 124	Selection from local germplasm	VPKAS, Almora	1989	95-100	2000-2500	Hills of U.P
35.	RAU 8	BR 407 x Ranchi Local	RAU, Dholi	1989	105-110	2200-2500	Bihar and other states
36.	Gujarat nagli 2 (NS 109)	Pureline selection	GAU Sardar Krishinagar, Gujarat	1988	110-115	2500-3000	Gujarat
37.	Indaf 9	KI x IE 98 R	UAS, Bangalore	1988	100-105	3000-3500	Karnataka
38.	HR 911	UAS 1 x IE 927	PC Unit, Bangalore	1986	110-115	4000-5000	Karnataka
39.	Indaf 8	Hullu bele x IE 929	UAS, Bangalore	1986	115-120	3500-4000	Karnataka

x KM 252, PR 202 x GPU 28) of about 500 progenies in F3/F4 stage are developed (AICSMIP Annual Report).

(d) Progress in Genetic Male Sterility Development

Gupta (1999) reported genetic male sterile line in finger millet, INFM 95001 using EMS mutagenesis from the finger millet germplasm line IE 3318. The source of male sterility is a single recessive gene and reported to produce fully fertile F1 progeny when pollinated by any parent homozygous for male sterility. The GMS line is easily distinguishable from male sterility at anthesis. The INFM 95001 genetic male sterile lines was increasingly used in the finger millet improvement programs and due the recessive allele governing male sterility maintenance of lines requires perpetuation through heterozygotes which is tedious and requires skilled manpower. To circumvent the difficulties associated with this, a new partial sterile lines was isolated from GPU 28 through EMS mutagenesis at AICSMIP Bangalore (DARE/ICAR Annual Report 2013-14). The identified male sterile source would facilitate crossing for development of finger millet hybrid progenies to generate new recombinants in finger millet breeding programmes. However, for effective utilization of male sterility and exploitation of heterosis in crop breeding programs there is need to search for cytoplasmic genetic male sterility system.

(e) Mutation Breeding Achievements

Mutation breeding was effectively used in finger millet for development of early maturing types, for generating polygenic variability and for development of complete/partial male sterile lines. Physical, chemical and combinations of mutagens were used for this purpose. Using gamma irradiation, Hamsa an early mutant of finger millet with increased finger number and grain bearing area was obtained (Nayar *et al.*, 1979). Mutation breeding led to release of early maturing Dibya singh (90 days) originated from variety Sarada (110 days) from Odisha, and Co-3 (110 days) originating from CO-1 (120 days) in Tamil Nadu. Eswari *et al.*, 2014 investigated the mutagenic efficiency in TRY 1 variety and identified desirable mutants for plant height, tillers/plant, productive tillers, finger number, finger length and 1000 grain weight and suggested 0.15 per cent concentration to be ideal for increasing the frequency of obtaining desirable segregants. The genetic male sterile line in finger millet, INFM 95001 was done using EMS mutagenesis from the germplasm line IE 3318. The partial sterile line (PS 1) was isolated through EMS mutagenesis of GPU 28.

Breeding Strategies for Enhancing Productivity and Utilization of Finger Millet

1. *Germplasm Evaluation and Utilization*: Valuation of genetic diversity collected and conserved in gene banks needs urgent attention in order to identify accessions possessing specific traits. In addition to earliness and yield attributing traits, there is increasing need to identify germplasm and develop cultivars with disease and insect pest resistance, cultivars adaptive to climate change and nutritional superior cultivars.

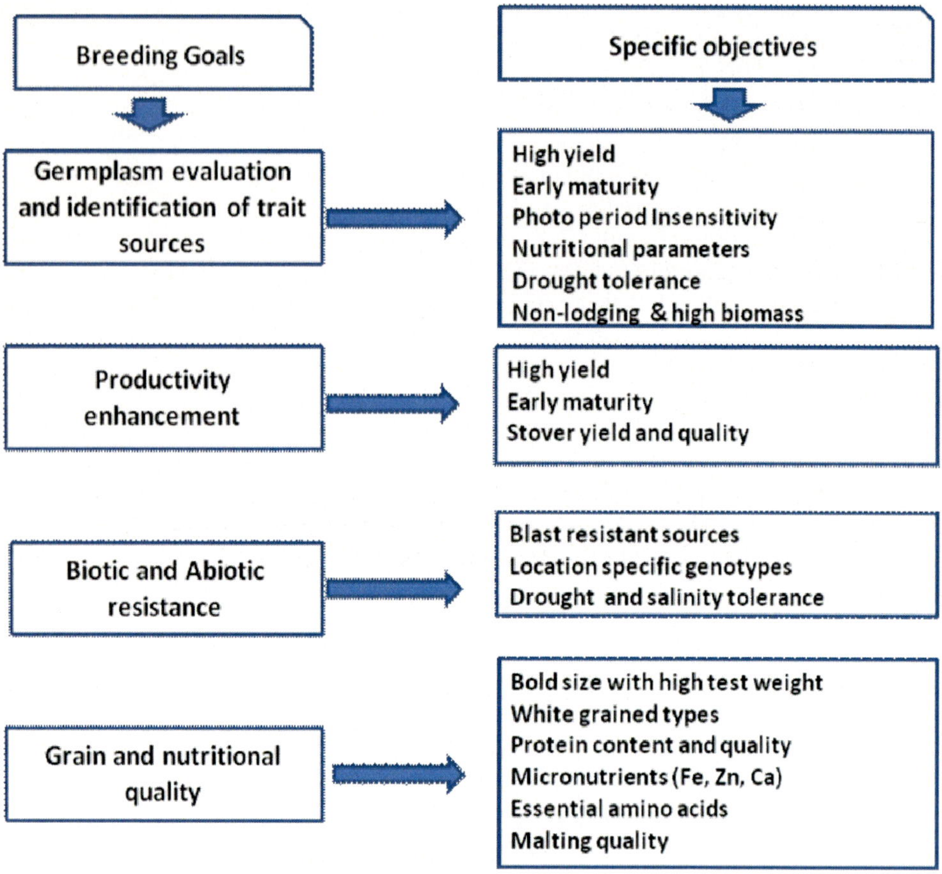

Figure 3: Breeding Goals and Objectives in Finger Millet.

2. *High yielding white finger millet varieties:* Only a fraction of available germplasm is being used in crop improvement programmes. The economic value and usefulness of large amounts of conserved germplasm available in the gene banks is still to be assessed and hybridization not fully exploited due to problems associated with floral biology and breeding behavior in finger millet. This activity needs to be strengthened and should receive priority.
3. *Development of early maturing, photo-insensitive varieties for different cropping systems:* From germplasm screening germplasm especially in core and minicore collections, early germplasm with 50-52 days flowering are available and there is need to utilize these germplasm for development of early maturing superior genotypes which can fit in different cropping systems.
4. *Genetic improvement for drought tolerance:* Systematic studies to identify diverse drought tolerant genotypes, understand mechanisms of drought

tolerance from multilocational screening as well as to identify key surrogate/adaptive traits requires major attention.

5. *Stover yield and quality improvement:* Although finger millet crop is predominantly grown for grain/food purpose, the stover after harvest of crop is excellent source of nutritive fodder to cattle. Varieties with good grain and fodder yield and blast resistance are highly desired by farmers. The tall types are known to be susceptible to lodging and therefore the recent released high yielding cultivars are dwarf types and medium maturing. There is need to identify non-lodging tall types with superior grain and fodder yield. Also very few isolated reports are available on the nutritive value and feeding trials of fodder and therefore there is need to strengthen these research gaps.
6. *Nutritional improvement:* Finger millet is highly nutritious and thus calls for intensive evaluation of germplasm to assess its nutritional qualities. Attempts to breed finger millet for enhanced grain nutrients are still in its infancy. Evaluation of finger millet core germplasm for grain nutrients and agronomic traits revealed a substantial genetic variability for grain Fe, Zn, Ca and protein contents. Therefore there is need to use the identified accessions for its use in strategic research for development of nutritionally rich cultivar development in finger millet.
7. *Breeding for blast resistance and other pests:*Substantial progress has been made to introgress blast resistance in the elite genotypes/varieties such as GPU 28, GPU 45 and GPU 48. Among various pests, blast disease is major problem in finger millet affecting the crop at all stages of the crop growth and number of varieties are susceptible. Understanding the pathogen diversity in the geographical area and development of suitable blast resistance and other disease and insect resistant genotypes based on knowledge of the strains is desirable.
8. *Interspecific hybridization and search for novel traits:* Three primary wild relatives (*E. Africana*, *E. indica* and *E. Kigeziensis*), secondary wild relatives (*E. tristachya*, *E. Floccifolia* and *E. intermedia*) and tertiary wild relatives (*E. Jaegeri*, *E. multiflora* and *E. compressa*) are recognized in finger millet. There is need for search of novel traits in wild species especially disease resistance, nutritional parameters, male sterility for introgression into cultivated types.

Conclusion

Among various millets, finger millet is one of the most nutritious crops for quality protein, minerals (calcium and iron) and amino acids (methionine). The crop has unique property of slower digestibility for long sustenance and also has an excellent malting property which shows scope for wide utility in food processing and value addition. The crop is productive in wide range of environments especially in water limited conditions where other cereals cannot be grown profitably. In spite of all these advantages, the crop is neglected both at national and global levels. Although productivity is high in finger millet, there is scope for further genetic improvement of this crop through combination of conventional and molecular approaches for genetic improvement of grain and forage yield, nutritional parameters, biotic and abiotic stresses and identification of cultivars for specific end uses.

References

1. Chandrashekhar H, Jayarame G and Jayashree U (2012). Formation of core set in Indian and African finger millet (*Eleusine coracana* (L.) Gaertn) germplasm accessions. *Indian J of Genet* 72 (3) 358-363.
2. Shobana S, Krishnaswamy K, Sudha V, Malleshi NG, Anjana RM, Palaniappan L, Mohan V (2013). Finger millet (Ragi, *Eleusine coracana* L.): a review of its nutritional properties, processing, and plausible health benefits. *Adv Food Nutr Res* 69: 1-39.
3. Gowda J, Suvarna, Somu G, Bharathi S and Mathur PN (2007). Formation of core set in finger millet (*Eleusine coracana* (L.) Gaertn) germplasm using geographical origin and morpho-agronomic characters. *J Plant Genet Res* 20 (1): 38-42.
4. Singh P and Raghuvanshi RS (2012). Finger millet for food and nutritional security. *African J Food Sci* 6(4) : 77 - 84.
5. Iyengar KG, Doraisami LS, Iyengar RS (1945). Ragi (*Eleusine coracana*). *Mysore J Agricultural.*
6. Prajapati VP, Sabalpara and Pawar DM (2014). Disease on finger millet crop phenology. *Trends in Biosciences* 7 (2): 140-142.
7. Seetharam A (1982). Finger millet improvement. *Indian Farming* 32(3): 3-6.
8. Harinarayana G (1986). Breeding and varietal improvement of small millets in India. In: (Eds. Seetharam A, Riley KW and Harinarayana G) *Small millets in global agriculture*: proceedings of the First International small millets workshop, Bangalore, India during October 29 - November 2, 1986, Bangalore, India.
9. Nagarajan C and T.S. Raveendran (1985). Germplasm mobilization and utilization in finger millet in Tamil Nadu. *National Seminar on Finger Millet. Genetics and Breeding*, UAS, Bangalore.
10. Nagaraja A, Jagadish PS, Ashok EG and Krishne Gowda KT (2007). Avoidance of finger millet blast by ideal sowing time and assessment of varietal performance under rain fed production situations in Karnataka. *J Mycopathological Res* 45(2): 237-240.
11. Vishwanath S, Sanne Gowda S, Seetharam A and Shankare Gowda BT (1986). Reaction to blast disease of released and pre-released varieties of finger millet from different states. *Millet Newsletter* 5: 31.
12. Rao ANS (1990). Estimates of losses in finger millet (*Eleusine coracana*) due to blast disease (*Pyricularia grisea*). *J Agric Sci* 24: 57-60.
13. Kiran Babu, T and Thakur, R P and Upadhyaya, H D and Reddy, P N and Sharma, R and Girish, A G and Sarma, N D R K (2013). Resistance to blast (Magnaporthe grisea) in a mini-core collection of finger millet germplasm. *European Journal of Plant Pathology*, 135 (2). pp. 299-311. ISSN 1573-8469.

14. Bijendra Kumar, Kumar J and Krishne Gowda KT (2010). Reaction of finger millet recombinant inbred lines (RILs) to blast]. *Crop Res.* 39 (1, 2 and 3) : 120-122.
15. Gupta SC (1999) Inheritance of genetic male sterility in finger millet. *African Crop Sci J.* 7(2): 125-128.
16. Nayar KMD, G. Pushpa and BGS Reddy (1979). Open earhead Hamsa true breeding early mutants of finger millet. *Curr Res.* 8: 212-213.
17. Eswari K, Gogulan G, Hari Prathab KA and Sakila M (2014). Development of early maturing mutants in finger Millet. *Res J Agric Forestry Sci* 2(6): 1-9.

Millets: Ensuring Climate Resilience and Nutritional Security (2015) Pages 277–304
Editors: Vilas A. Tonapi and J.V. Patil
Published by: DAYA PUBLISHING HOUSE, NEW DELHI

Chapter 7
Genetic Improvement in Foxtail Millet

K. Hariprasanna
ICAR-Indian Institute of Millets Research (IIMR)
Rajendranagar, Hyderabad – 500 030

Foxtail millet [*Setaria italica* (L.) Beauv.] is one of the oldest of the cultivated millets. It is cultivated in about 23 countries in Asia, Africa and America, especially where the climatic hazards do not permit the cultivation of other cereals. It is a self pollinating cereal (2n = 2x = 18), belonging to Family *Poeceae* and Subfamily *Panicoideae*. It is known by different names in different parts of India (Table 1). The foxtail millet is valued as a crop of short duration, which is good as food, feed and fodder. Foxtail millet ranks second in the total world production of millets and continues to have an important place in the world agriculture providing approximately six million tonnes of food to millions of people, mainly on poor or marginal soils in southern Europe and in temperate, subtropical and tropical Asia (Marathee, 1993). Its grain is used for human consumption and as feed for poultry and cage birds. In China, foxtail millet is next to rice and wheat in importance. It provides approximately 15-17 per cent of the

Table 1: Common Names of Foxtail Millet

English	Foxtail millet, Italian millet	Malayalam	Thina
Hindi	Kangni, Kakum	Marathi	Kang, Rala
Sanskrit	Kanguni	Gujarati	Kang
Kannada	Navane	Bengali	Kaon
Tamil	Tenai	Oriya	Kanghu, Kangam, Kora
Telugu	Korra, Korralu	Punjabi	Kangni
		Kashmiri	Shol

Table 2: Nutritive Value of Foxtail Millet vs. Rice and Wheat (per 100 g)

Item	*Foxtail Millet*	*Wheat (Whole)*	*Rice (Raw, milled)*
Carbohydrates (g)	60.9	71.2	78.2
Protein (g)	12.3	11.8	6.8
Fat (g)	4.3	1.5	0.5
Energy (KCal)	331	346	345
Crude fibre (g)	8	1.2	0.2
Mineral matter (g)	3.3	1.5	0.6
Amylose (per cent)	17.5	25	12-19
Amylopectin (per cent)	82.5	75	88-81
Ca (mg)	31	41	10
P (mg)	290	306	160
Essential Amino acids (mg/g of N)			
Arginine	220	290	480
Histidine	130	130	130
Lysine	140	170	230
Tryptophan	60	70	80
Phenyl Alanine	420	280	280
Tyrosine	–	180	290
Methionine	180	90	150
Cystine	100	140	90
Threonine	190	180	230
Leucine	1040	410	500
Isoleucine	480	220	300
Valine	430	280	380
Vitamins (mg/100 g)			
Thiamin	0.59	0.41	0.41
Niacin	3.2	5.1	4.3
Riboflavin	0.11	0.1	0.04
Vit A (carotene) (μg/100g)	32	64	0
Vit B6	–	0.57	–
Folic Acid	15	36.6	8
Vit B5	0.82	–	–
Vit E	31	–	–
Micronutrients (mg/100 g)			
Fe	2.8	5.3	0.7
Zn	2.4	2.7	1.4
Mg	81	138	90
Na	4.6	17.1	–

Contd...

Table 2–*Contd...*

Item	*Foxtail Millet*	*Wheat (Whole)*	*Rice (Raw, milled)*
K	250	284	–
Cu	1.4	0.68	0.14
Mn	0.6	2.29	0.59
Mb	0.07	0.051	0.058
Cr	0.03	0.012	0.004
Su	171	128	–
Cl	37	47	–
Fatty acids			
Palmitic	6.4	24.5	15
Palmoleic	–	0.8	–
Stearic	6.3	1	1.9
Oleic	13	11.5	42.5
Linoleic	66.5	56.3	39.1
Linolenic	–	3.7	1.1

Source: Nutritive value of Indian foods, NIN, 2007.

total food consumed in china. China still ranks first in foxtail millet production in the world (4.6 m t). In the US, foxtail millet was introduced from the old world about the middle of the nineteenth century as a forage crop (Baltensperger, 2002).

Because of the drought tolerance, it was once an indispensable crop of vast rainfed areas in semi-arid regions in India. At present, foxtail millet is cultivated on a limited area in Andhra Pradesh, Karnataka, Maharashtra, Tamil Nadu, Rajasthan, Madhya Pradesh, Uttar Pradesh and north eastern states. The area under foxtail millet in India has come down by more than half during 1990's mainly due to introduction of more remunerative crops like sunflower and soybean in black soils. At present foxtail millet is grown in an area of 0.98 lakh ha (2005-06) producing about 0.56 lakh t of grains with an average productivity of 565 kg/ha, and the important growing states are Andhra Pradesh (0.17 lakh t), Karnataka (0.14 lakh t), Rajasthan (0.05 lakh t), Maharashtra (0.05 lakh t), Arunachal Pradesh (0.05 lakh t), Tamil Nadu (0.01 lakh t) and Madhya Pradesh.

Foxtail millet is mostly grown to meet the domestic needs of the rural people. It is widely used as an energy source for pregnant and lactating women, and also for sick people and children, and especially for diabetics (Sema and Sarita, 2002). It has been suggested that foxtail millet protein be used as a food component to fight type 2 diabetes and cardiovascular diseases (Choi *et al.*, 2005). Like other small millets, foxtail millet is highly nutritious and even superior to rice and wheat in certain constituents (Table 2). Foxtail millet is highly drought tolerant crop and its water use efficiency (WUE) has been found to be higher than maize, wheat and sorghum (Zhang *et al.*, 2007). Its drought tolerance has been accredited to the association between increased WUE and its several morphological characteristics such as dense root system, thick cell walls, epidermal cell arrangements and minuscule leaf areas (Li, 1997). For producing 1 g of dry biomass, foxtail millet requires only 257 g of water which is much lower than maize (470 g) and wheat (510 g) (Diao, 2007; Li and Brutnell, 2011). Foxtail millet because of its small diploid genome along with the release of its genome sequence by Beijing Genomics Institute, China and US Department of Energy-Joint Genome Initiative, is on a fast-track to transforming into a model for studying functional genomics and to probe plant architecture, genome evolution, drought tolerance, and physiology in the bioenergy grasses (Dekker, 2003; Diao, 2011; Lata *et al.*, 2013).

Origin and Distribution

Foxtail millet has the longest history of cultivation among the millets, having been grown in China since sometime in the sixth millennium BC. Some records from China indicate that foxtail millet originated there, while other varieties arrived in China from Africa. Carbonized foxtail millet has been identified in archaeological sites in China, with various reports of origination date. Its planting has been mentioned in Chinese records as early as 2700 BC (Vinall, 1924). Foxtail millet was the most important plant food in the neolithic culture in China, and its domestication and cultivation was the earliest identifiable manifestation of this culture, the beginning of which has been estimated at over 4,000 years ago (Chang, 1968). In Russia, foxtail millet has been cultivated since ancient times and there is evidence that it was grown as long as 1,500 years ago (Movsisyants, 1950). According to Vavilov (1926), the principal center of diversity for foxtail millet is East Asia, including China and Japan. Though several hypotheses concerning the origin and domestication of foxtail millet have been proposed, a multiple domestication hypothesis (de Wet *et al.*, 1979) is widely accepted. From China or Central Asia it spread to India and European countries (Oelke *et al.*, 1990).

Foxtail millet can grow in altitudes from sea level to 2000 m. It is fairly tolerant of drought, but it cannot tolerate water logging. It has a low water requirement, though it does not recover well from drought conditions because it has a shallow root system. It is adapted to a wide range of elevations, soils and temperatures.

Taxonomy

The genus *Setaria* consists of approximately 125 species extensively dispersed in warm and temperate regions worldwide (Dwivedi *et al.*, 2012). The genus *Setaria* is widely distributed in warm and temperate areas. Foxtail millet is the most economically valuable of the genus (Baltensperger, 1996). Malm and Rachie (1971) thoroughly reviewed the domestication of foxtail millets and the taxonomy. They report the use of subspecies by Scholz (1960) based on papillae on the epidermal cells of the glumes (*moharia*) and no papillae (*maximas*).

The geographical origin of foxtail millet based on cytological studies indicated that wild ancestor of foxtail millet is *S. viridis* (Li *et al.*, 1945). Phylogenetic analyses using both chloroplast and nuclear genes show foxtail and green millet (*S. viridis*) as close relatives (Giussani *et al.*, 2001; Doust *et al.*, 2007), consistent with the hypothesis that foxtail millet is a domesticated version of green millet (Li *et al.*, 1945; deWet and Harlan, 1975; Wang *et al.*, 1995; Le Thierry d'Ennequin *et al.*, 1999). Three main groups of cultivated foxtail millet gene pool were suggested namely, Chinese (from China, Japan and Korea), tropical (from Taiwan, India and Kenya), and European group, on the basis of isozyme studies in accessions of *S. italica* and *S. viridis*, respectively (Jusuf and Pernes 1985; Panaud, 2006).

The accessions were classified into races and subraces on the basis of inflorescence morphology (Prasada Rao *et al.*, 1987). On the basis of inflorescence morphology foxtail millet is classified into two species *S. pumila* and *S. italica*. The species *S. italica* is classified into two subspecies *viridis* and *italica*. The subspecies *italica* is classified into three races and ten sub-races. The race *Moharia* (common in Europe, southeast Russia, Afghanistan and Pakistan) is divided into sub-races *Aristata, Fusiformis, Glabra;* race *Maxima* [common in eastern China, Georgia (Eurasia), Japan, Korea, Nepal and northern India (it has also been introduced in the USA)] is divided into sub-races *Compacta, Spongiosa, Assamense;* and race *Indica* (remaining parts of India and Sri Lanka) is divided into sub-races *Erecta, Glabra, Nana* and *Profusa*.

Morphology and Floral Biology

Foxtail millet is an annual grass. Successful production is due almost entirely to its short growing season. It matures in 65-70 days. Foxtail millet can be planted when it is too late to plant most other crops. It forms a slender, erect, leafy stem varying in height from 1-5 ft. The inflorescence is a spike with short side branches bearing spikelets and bristles. The compressed hairy panicle 5-30 cm long resembles yellow foxtail, green foxtail, or giant foxtail. Each spikelet consists of a pair of glumes that embrace two minute flowers (about 1 mm in length); the lower one sterile and the upper one bisexual, with three stamens and a long oval smooth ovary with two long styles, that terminate in a brush like stigma (Hector, 1936). From one to three bristles develop at the base of each spikelet (Vinall, 1924). Anthesis in foxtail millet generally takes place near midnight and in the morning, but varies significantly with the environment (Malm and Rachie, 1971). The rate of anthesis is generally favoured by low temperature and high humidity.

The small seeds, around 2 mm in diameter, are encased in a thin, papery hull which is easily removed after threshing. The seed coat and husk of foxtail are generally of single entity with glossy appearance. The grains are very similar to paddy rice in grain structure. They contain an outer husk, which needs to be removed in order to be used. Seed colour varies greatly between genotypes, and usually yellowish in colour.

Morphology and anthesis behaviour make foxtail millet one of the most difficult species to cross pollinate. Foxtail millet is largely a self-pollinating species, with cross pollination averaging about 4 per cent (Li *et al.*, 1935). Out-crossing rates have been estimated from 0.0 to only 1.4 per cent for plants separated by 0.30 m (Till-Bottraud *et al.*, 1992), although Li *et al.* (1935) reported rates as high as 5.6 per cent for some varieties under certain conditions. Siles *et al.* (2001) described an artificial technique of hybridization that resulted in 67.5 per cent hybrid seed set per flower crossed.

Genetic Studies

Information on the inheritance of important agronomic traits of foxtail millet, including susceptibility to key diseases, is limited. Most of the previous work has focused on estimating broad-sense heritabilities and realized genetic gains, with little attention directed to measuring levels of heterosis or to assessing the relative importance of different types of gene action (Athwal and Singh, 1966; Singh and Athwal, 1966; Gill and Randhawa, 1975; Vishwanatha *et al.*, 1981; Gurunadha Rao *et al.*, 1984; Prasada Rao *et al.*, 1985). Darmency *et al.* (1987) reported that most of 19 morphological and reproductive traits were probably under the control of nonadditive genetic components, but this research was conducted on an interspecific cross between foxtail millet and its wild relative *S. viridis*. Also, information on the importance of genotype x environment interaction for this species is lacking.

Important agronomic traits in foxtail millet include not only grain yield, days to heading, days to maturity, and plant height, but also number of tillers and spike length. When foxtail millet is used as a forage, tillering is a desirable trait. However, nontillering cultivars are preferred for use in producing birdseed because the seeds typically are larger. Spike length also is correlated with seed size. Correlation and path coefficient analysis revealed that straw yield per plant and harvest index were the major determining characters for grain yield among foxtail genotypes (Murugan and Nirmalakumari, 2006). The total number of tillers, number of productive tillers, panicle length, plant height, days to 50 per cent flowering and days to maturity exhibited highly positive correlations with grain yield per plant. Correlation and path analysis in exotic genotypes revealed positive significant correlation of days to 50 per cent flowering, plant height, days to maturity, number of productive tillers per plant, ear length, ear weight and straw weight with yield per plant. Characters such as flag leaf area and 1000-grain weight were also observed to influence yield. The association of protein content and calcium content with grain yield was in negative direction. The association of carotene with grain yield was negative. So simultaneous improvement of these traits along with grain yield may not be possible (Lakshmi Prasanna *et al.*, 2013).

Varietal Development

Small millets as a whole received less priority in the agricultural development in the past both at National and International level. The ultimate goal of breeding small millets remains improvement of grain yield including maximization of biomass and the harvest index. Most of the small millets particularly foxtail millet mature early, and therefore, provide one first harvest for human consumption. Genotypes need to be tailored for maturity-early, mid-late and late, depending on the location specific requirements of soil, rainfall, temperature, humidity, day length and cropping patterns (Harinarayana, 1989). Being low or no-input crops, selection needs to be effected for fertilizer use efficiency, particularly nitrogen, both native and applied. Breeding of dwarf varieties is an objective of intensive cultivation. Water-use efficiency of these crops deserves breeding efforts. Post-harvest technology demands free threshing grain for minimizing human drudgery (Harinarayana, 1989).

The research needs of small millets to a limited extent were looked after under the erstwhile PIRCOM of Govt. of India. Though the crop improvement work on small millets in India under the coordinated programme started in 1964, the launching of a separate coordinated millet improvement project in 1969 helped in giving greater attention to these crops with a few centres established in different states for specific millets. Five crop specific lead centres were established in 1978-79 with IDRC assistance; at Dholi in Bihar for proso millet, Dindori in Madhya Pradesh for kodo millet, Nandyal in Andhra Pradesh for foxtail millet, Semiliguda in Odisha for little millet, and at Almora in Uttar Pradesh for barnyard millet. The establishment of these centres has greatly assisted in building the necessary infrastructure and manpower for these crops. These efforts have started bearing fruit and many new technologies and varieties are now undergoing critical evaluation in different parts of the country (Harinarayana, 1989). The major goal during the VII Five-Year Plan was to further consolidate the outcome of these research efforts and stabilize the productivity of small millets at a higher level. As an important step in this direction a separate All India Coordinated Small Millets Improvement Project (AICSMIP) was launched in 1986 with its headquarters at UAS, Bangalore, and the IDRC centres became part of AICSMIP. The AICSMIP centre at Regional Agricultural Research Station ANGRAU, Nandyal, is exclusively working on foxtail millet. Other cooperating centres are Palem (ANGRAU, Hyderabad), Dahod (AAU, Anand), Hanumanamatti (UAS, Dharwad), Hagari and Gulbarga (UAS, Raichur).

With the establishment of AICSMIP the varietal development in small millets gained momentum. Pure line selection remained the prime breeding method for improving performance, particularly grain yield of small millets. As foxtail millet is largely self-pollinated and out-crossing is very much limited, and because of the anthesis behaviour recombination breeding could not gather much research efforts compared to other crops. Progress through hybridization has been extremely limited, in India. This has been chiefly due to difficulties in emasculation and pollination, in identification of true hybrids, limited heterosis in intervarietal crosses (Srivastava and Yadav, 1977), and the availability of unexploited genetic resources. Identification of natural hybrids by contact method [the panicles of selected plants are enclosed in a parchment bag before flowering to enhance the chances of natural cross pollination (Ayyangar, 1934)] has resulted in a low frequency (1.5 per cent) of true hybrids in foxtail millet (Mahishi *et al.*, 1982). Use of marker plant types like foxtail millet dwarfs or other distinctly distinguishable lines in germplasm and breeding nurseries for identification of natural hybrids (Harinarayana *et al.*, 1984) through grow out tests met with limited success.

Some of the early efforts in crop improvement of foxtail millet have resulted in the release of improved varieties as early as 1940s. For example Co 1, Co 2 and Co 3 were released in 1943; H1 and H2 were released in 1948; and T4 was released in 1949. Since 1978, the centre for the improvement of small millets established at Nandyal (ANGRAU) by ICAR with the assistance of International Development Research Centre, Canada is devotedly working towards the improvement of foxtail millet. Till date (1942-2012) about 32 varieties have been released, of which 12 were released before establishment of AICSMIP (1986) and 20 after 1986. Pure line selection has resulted in the development and release of maximum number of foxtail millet varieties - till 1985, 13 varieties were developed (Harinarayana, 1989) and between 1986-2011 eight varieties were developed. One hybrid derivative of foxtail millet, CO 5 (Co 1 x A 113/2) was released in Tamil Nadu (Raveendran *et al.*, 1979) and notified at the central level in 1982. Between 1986-2011, five varieties were released following recombination breeding and one through mutation breeding.

A number of varieties with high yield potential have been released for different states (Table 3). Some of the recent varieties with yield potential, area of adaptation and special features are listed in Table 4. One of the very early variety released in India is Birsa Gundli (Haider and Birsa, 1991), but the present status of cultivation of this variety is not available. Super early maturing varieties (50-60 days) have been developed at Japan and China (Liu *et al.*, 2006). Till date 27 foxtail millet varieties have been notified by Govt. of India, of which 13 are central releases and rest are state releases. The list of notified varieties is given in Table 5. New short duration varieties SiA 3085 and SiA 3088 are becoming popular in the major cultivated areas. Because of the short duration nature, instead of fallow followed by chickpea cropping system, foxtail-chickpea has increased income significantly in Kurnool. In paddy fallows also, short duration foxtail is becoming popular in Kurnool in view of limited water supply. New varieties have a yield potential of 20-25 q/ha.

Table 3: State-wise High Yielding Foxtail Millet Varieties

State	*Foxtail Millet*
Andhra Pradesh	SiA 326, Chitra, PS 4, TNAU 186, Narasimharaya, Krishnadevaraya
Bihar	SiA 326, PS 4
Jharkhand	SiA 326, PS 4
Karnataka	SiA 326, PS 4, TNAU 186
Maharashtra	SiA 326, PS 4
Madhya Pradesh	SiA 326, PS 4
Chattisgarh	SiA 326, PS 4
Tamil Nadu	Co 5, K 3, Co 4, PS 4, K 2, TNAU 186, TNAU 43
Uttar Pradesh	Nischal, PS 4
Uttarakhand	PRK 1
Rajastan	SR 51, SR 11, SR 16

Source: http://smallmillets.res.in/

Under AICSMIP, the foxtail varietal evaluation trials are taken up at about a dozen locations across the major growing states of Andhra Pradesh (Nandyal, Vizianagaram, Palem and Chintapalle), Bihar (Dholi), Karnataka (Bangalore, Mandya, Hiriyur, Dharwad, Hanumanamatti, Gulbarga and Hagari), Madhya Pradesh (Rewa), Jharkhand (Ranchi), Tamil Nadu (Coimbatore) and Uttarakhand (Pantnagar). The latest national checks are SiA 326 (Pure line selection, Nandyal) and SiA 3156 (Selection from SiA 2871, Nandyal).

Table 4: Improved Foxtail Millet Varieties with Characteristics

Sl.No.	Name of Variety	Pedigree	Institute where Developed	Year of Release	Maturity (Days)	Av. Yield (kg/ha)	Area of Adaptation	Special Features
1.	Nischal	Selection from accessions of Basti district	CSAUA&T, Kanpur	1988	–	1100-1500	Uttar Pradesh	High seed yield
2.	K 3	Selection from SiA 2567	TNAU, Coimbatore	1989	85-90	1500-1800	Tamil Nadu	High seed yield
3.	AK 132-1	Pureline selection	APAU, ARS Ananthpur	1990	82-85	1800-2000	Andhra Pradesh	Drought tolerance, high seed yield, suitable for Rayal Seema
4.	Krishna-devaraya	Selection from SiA 2593	APAU, ARS Nandyal	1993	85-96	2000-2200	Andhra Pradesh	High seed yield
5.	TNAU 43	Co 5 x ISe 30	TNAU, Coimbatore	1994	80-85	1500-1600	Tamil Nadu	–
6.	Gavari (SR 11)	Selection from local germplasm	RAU, Rajasthan	1995	75-80	1500-1600	Rajasthan	High seed yield
7.	PS 4	SiA 2616 (0.2 per cent EMS induced mutant)	GBPUA&T, Pantnagar, U.P	1998	80-85	1800-2000	All States	Wider adaptability and high yield
8.	TNAU 43	Co 5 x ISe 30	TNAU, Coimbatore	1998	80-85	1500-1600	Andhra Pradesh, Tamil Nadu and Karnataka	
9.	TNAU 186	Co-5 x SiA 326	TNAU, Coimbatore	1998	85-90	1800-2000	Tamil Nadu, Andhra Pradesh, Karnataka	
10.	Meera (SR 16)	Pure line selection	MPUA&T, Udaipur	2001	75-80	1500-1700	Rajasthan	Stay green character, tolerant to downey mildew
11.	Sri Lakshmi	Pure line selection	ANGRAU, Hyderabad	2002	75-80	2000-2300	Andhra Pradesh	High seed yield
12.	Prathap kangni 1 (SR-1)	Selection from local germplasm	MPUA&T, Udaipur	2003	65-70	1800-2000	Rajasthan	Extra early maturity

Contd...

Table 4–*Contd...*

Sl.No.	*Name of Variety*	*Pedigree*	*Institute where Developed*	*Year of Release*	*Maturity (Days)*	*Av. Yield (kg/ha)*	*Area of Adaptation*	*Special Features*
13.	SR 51	Pure line selection	MPUA&T, Udaipur	2003	62-68	1800-2000	Rajasthan	Bold seeds
14.	PRK 1	Selection from Tehri collection	GBPU&T, Ranihauri	2000	75-80	1900-2000	Uttar Pradesh	–
15.	TNAU 196	Co 5 x ISE 248	TNAU, Coimbatore	2005	85-90	1800-2000	Tamil Nadu	–
16.	HMT-100-1	PS 3 X RS 118	ARS, Hanumanamatti	2007	90	2000-2500	Karnataka	Resistant to lodging
17.	SiA 3085	Selection from SiA 2644	RARS, ANGRAU, Nandyal	2012	75-82	3000-3200	Andhra Pradesh, Karnataka, Tamil Nadu, Bihar, Madhya Pradesh and Uttarakhand	Resistant to blast and downy mildew

Source: Seed Division, DAC; http://smallmillets.res.in/

Table 5: List of Notified Foxtail Millet Varieties in India

Sl.No.	Variety Name	Year of Release	Notification Date	Notification No.	State of Release
1.	Co.3	1942	01/01/1944	–	Tamil Nadu
2.	Co.2	1942	01/01/1944	–	Tamil Nadu
3.	Co.1	1942	01/01/1944	–	Tamil Nadu
4.	H.2	1948	01/01/1950	–	Karnataka
5.	H.1	1948	01/01/1950	–	Karnataka
6.	Arjun	1969	01/01/1971	–	Andhra Pradesh
7.	RS-118	1976	02/02/1976	S.O.786	Karnataka
8.	RC-128	–	02/02/1976	786	Central
9.	K-221-1	1970	02/02/1976	786	Karnataka
10.	Co-4	1973	19/12/1978	19	Central
11.	K.2	1978	14/01/1982	19(E)	Central
12.	Co-5	1979	14/01/1982	19(E)	Central
13.	Chitra	1982	01/01/1984	–	Andhra Pradesh
14.	Nellamale Korra	1985	09/04/1985	295(E)	Central
15.	SIC-3(548-79K)	1985	01/01/1987	–	Central
16.	Nischal	–	13/04/1989	280(E)	Central
17.	K-3	–	31/07/1989	599(E)	Central
18.	Lepakshi (AK-132-1)	–	17/08/1990	639(E)	Central
19.	Krishna Devaraya	–	04/05/1995	408(E)	Central
20.	Gavari (SR-11)	1995	01/01/1996	1(E)	Central
21.	TNAU-43	1994	01/05/1997	360(E)	Tamil Nadu
22.	Pant Setaria-4 (PS 4)	1998	08/06/1999	S.O. 425(E)	Central
23.	Meera (SR-16)	1999	02/02/2001	92(E)	Rajasthan
24.	Pratap Kangni-1 (SR-51)	2003	04/02/2004	161(E)	Rajasthan
25.	Co(Te) 7	2005	25/04/2006	599(E)	Tamil Nadu
26.	HMT-100-1	2007	11/02/2009	S.O. 449(E)	Karnataka
27.	SiA 3085	2011	10/04/2013	S.O. 952(E)	Central

Source: Seed Variety Management System, http://seedvariety.dacnet.nic.in/

Heterosis Breeding and Male Sterility

For exploiting heterosis in foxtail millet, artificial hybridization is very difficult because of the small flower, and chemical emasculation technology is still in exploratory stage. Although foxtail millet is mainly autogamous (less than 2 per cent outcrossing), pollen from a 5 m diameter plot could fertilize a fertile variety up to 24 m away, and a male-sterile variety up to 40 m (Wang *et al.*, 1997). Wind velocity and direction resulted in variations of gene flow intensity. Pollen competition for available target ovules was observed up to 2 m away from a 1 m diameter pollen source. The percentage of pollen grains that fertilized ovules outside the pollen donor source was 1.4 per cent of the total pollen grain production.

The ability to hybridize foxtail millet opens other options to breeders besides selecting from landraces. If non-additive gene action is important, then either mid- and/or high-parent heterosis may be sufficient to justify the production and use of varietal crosses (Siles *et al.*, 2004). In a study to determine whether sufficient high-parent heterosis is expressed in foxtail millet for grain yield and other key traits to justify the development and use of varietal crosses, on the basis of the estimate of average heterosis the expected yield of the F1 generation was 68 per cent greater than the average yield of the parental cultivars. This high level of heterosis for grain yield suggested that varietal crosses or other types of cultivars in which there exists relatively high amount of heterozygosity would provide a significant yield benefit over nonhybrid cultivars. Although significant heterotic effects were observed for each of the other traits, additive effects were more important. Significant correlations between traits of estimates of additive and/or variety heterosis effects suggested that at least some of the genes controlling grain yield, plant height, and spike length were either the same or in coupling phase linkage (Siles *et al.*, 2004).

Foxtail millet heterosis utilization research has made a number of breakthroughs in China since 1967 (Tian *et al.*, 2009). Various male-sterility lines have been identified such as dominant and recessive genic highly male-sterility lines and photo-(thermo-) sensitive nuclear lines (GMS) (Cui *et al.*, 1979; Wang, 1991; Wang *et al.*, 1993, 2002; Zhao *et al.*, 1996; Hao *et al.*, 2009), gene interaction male-sterility lines (GIMS) (Hu *et al.*, 1986), cytoplasmic male sterility (CMS) (Zhu *et al.*, 1991), and cytoplasmic-nuclear male sterile type (CGMS) (Zhi *et al.*, 2007). Recently, Wang *et al.* (2010) identified a highly genic sterile line Gao146A adapting to spring-sowing late maturing ecological zones of China, showing 95 per cent sterile rate and not sensitive to light and temperature, and the F_1 generation showed fertile reaction and the segregation ratio of fertility to sterility was 3:1 in F2 generation, which showed that the sterile trait of Gao146A was mainly controlled by a single recessive gene. Wang *et al.* (2002) considered the male-sterile line 1066A as a genic highly male-sterile line, and its inbred seeding rate is about 5 per cent, and the sterility trait was controlled by a recessive gene located on the chromosome VI using primary trisomics positioning method. On the chromosome VI, a genic highly male-sterile and a photo-(thermo-)sensitive male-sterile genes have been reported by Wang *et al.* (2002) and Hao *et al.* (2011), respectively. Breeding of male-sterile lines has become the mainstream for the heterosis utilization in foxtail millet, but the genetic basis of most male-sterile lines used for the hybrid production is still an area to be elucidated. In highly male-sterile line Gao146A, using F2 population derived from cross Gao146A/K103, one recessive gene controlling the highly male sterility, tentatively named as *ms1*, which linked to SSR marker *b234* with genetic distance of 16.7 cM, was mapped on the chromosome VI. These results not only laid the foundation for fine mapping of this highly male-sterile gene, but also helped to accelerate the improvement of highly male-sterile lines by using molecular marker assisted breeding method (Wang *et al.*, 2013).

Exploitation of heterosis has been proven to be the most important genetic factor in increased foxtail millet yields in China like rice and many other crops. A hybrid cultivar, Zhangzagu5, released from Zhangjiakou Academy of Agricultural Sciences, Hebei Province, China, yielded up to 12,159 kg/ha compared to conventional cultivars ranging from 4500 to 6000 kg/ha in 2007. However, most of the currently released Chinese spring foxtail millet male sterile lines were derived from Chang10A, whose cytoplasm was contributed by Qinyuanmujizui (Liu *et al.*, 2011; Wang *et al.*, 1998), while most summer foxtail millet male sterile lines were derived from Huangmi1A with the cytoplasm from Dahuanggu (Liu *et al.*, 1996; Liu *et al.*, 2006). Thus, lack of genetic diversity in cytoplasmic sources is a serious problem for effective utilization of heterosis in foxtail millet.

Because of maternal inheritance of cytoplasm, using a single source of cytoplasm in male sterile lines will inevitably lead to all hybrid cultivars carrying a single source cytoplasm. A specific cytoplasm-related pathogen race may quickly develop, like that observed in case of maize T-type sterile cytoplasm in hybrid maize production, and become the predominant race under high selection pressure, and epidemics of the disease can cause devastating losses in hybrid crop production. Thus, utilization of novel cytoplasmic types in male sterile lines is urgently needed, and determination of cytoplasm genetic diversity and classification of foxtail millet germplasm is the foundation for discovery and deployment of novel types of cytoplasm in breeding programmes. Several attempts have been made on construction of evolutionary trees to investigate the evolution and genetics of foxtail millet germplasm (Jia *et al.*, 2009a, 2009b; Wang *et al.*, 2006), and on assembling heterosis groups through population structure analysis (Liu, 2010; Liu *et al.*, 2011). These reports were mainly aimed at the nuclear genes, and analysis on cytoplasmic genetic diversity is limited, especially through comparative analysis of mtDNA evolutionary trees and female-parent derivative pedigree trees to classify the germplasm.

Fukunaga and Kato (2003) used mtDNA to identify the geographical origination and the center of diversity of foxtail millet cytoplasm types. To explore genetic diversity of Chinese foxtail millet cytoplasm, 23 pairs of mtDNA primers that showed a high level of polymorphism among selected 14 genetically diverse foxtail millet accessions were identified after screening 34 pairs of consensus mtDNA primers that were derived from various plant species. The 23 pairs of primers were used to analyze genetic diversity of cytoplasm in 111 Chinese elite foxtail millet accessions. The results showed that the genetic diversity of foxtail millet cytoplasm was low, which makes Chinese foxtail millet germplasm vulnerable to the infection by cytoplasm-related diseases. The 111 tested accessions were classified into eight cytoplasmic types: Qinyuanmujizui, Huangruangu, Riben60ri, Dahuanggu, Yingsuigu, Heizhigu, MissingI, and MissingII. The eight cytoplasmic types matched with the geographic and ecological distribution pattern of the most tested accessions. The cytoplasmic types of foxtail millet germplasm from China follow certain geographic and ecological distribution patterns, which is different from the result derived from a study on the nucleus genome (Liu *et al.*, 2011). These results can help in rationally utilizing the germplasm lines, diversifying cytoplasmic types in male sterile lines, and further improving hybrid foxtail millet cultivars (Liu *et al.*, 2014). As in breeding programmes, locally adapted cultivars are usually used as the female parent in crosses, and introduced cultivars from other regions are usually used as male parents to contribute to the nucleus genome, to diversify cytoplasmic types in breeding programmes, introduced cultivars should also be used as female parents in breeding crosses (Liu *et al.*, 2014).

Cytological Studies

A complete set of nine primary trisomics (2n+1) for cv.'Yugu No. 1'of foxtail millet was identified cytologically from progenies derived from crosses between autotriploids (2n= 2x=27) and their diploid counterparts. Five autotriploid plants were identified from 2100 seedlings derived from 4x-2x crosses; the reciprocal crosses (2x-4x) failed to produce autotriploids. Autotriploids grew vigorously and were morphologically very similar to diploids. Theprimary trisomics (2n=2x=19) constituted @ 32.5 per cent of the total progeny from the 3x-2x crosses, whereas 59.8 per cent of the descendants were aneuploids with chromosome numbers ranging from 20 (double trisomics and tetrasomics) to 37 (2n=4x+1; or autotetraploid with one additional chromosome). The nine primary trisomics identified were self-fertile; seven had characteristic morphology, whereas trisomics VIII and IX resembled the disomics. The seed set for trisomic V was the lowest (20 per cent), and trisomic VIII the highest (74 per cent). Other aneuploids with 20 or more somatic chromosomes were either self-sterile or partially fertile with various, but low, levels of seed set. Each of the primary trisomics showed its unique transmission rate when self-pollinated; trisomic IX had the highest (45.8 per cent), whereas trisomic V had the lowest (19.6 per cent) transmission rate (Wang *et al.*, 1999).

In another study, autotetraploid material was induced by colchicine treatment of a pure line of foxtail millet and an interspecific hybrid between foxtail millet and giant green foxtail. Compared with diploid material, tetraploids were smaller, flowered later, and had a two-fold reduced fertility (number of grain per cm of spike). Grain weight increased by 20 per cent with polyploidy, but total grain yield decreased by 46 per cent. Hence, scope of breeding polyploidy lines of foxtail millet that are of agronomic use is limited (Ahanchede and PoirierHamon, 2004).

Germplasm Status

There is wide genetic diversity available in foxtail millet, and characterizing these resources is a prerequisite for the genetic improvement of its cultivars. More than 30,000 germplasm of foxtail millet have been conserved in different germplasm consortium all over the world *viz.*, Chinese National Gene Bank (CNGB), (26,670accessions), International Crops Research Institute for the Semi-Arid Tropics (ICRISAT), Patancheru (1,535 accessions), National Institute of Agro-biological Sciences (NIAS), Japan (1,279 accessions), and the Plant Genetic Resources Conservation Unit (PGRCU), USDA (766 accessions) (Doust *et al.*, 2009). At the CGIAR genebanks 9494 accessions of foxtail millet have been conserved, while at Svalbard genebank, Norway 1442 accessions have been conserved (Padulosi *et al.*, 2009). At National Genebank at NBPGR, New Delhi at total of 4473 accessions including 94 exotic accessions have been conserved, and catalogue has been published (Umesh Chandra *et al.*, 1987).

At UAS, Bangalore National Active Germplasm Site (NAGS) was established for small millets in 1992. A total of 2821 accessions of foxtail millet, mainly landraces and cultivars from 13 states in India, Afghanistan, USA, USSR, Turkey, Taiwan and African countries are being conserved at UAS, Bangalore. Out of these, catalogue on evaluation of 1482 accessions for 14 and 11 qualitative and quantitative descriptors, respectively, has been published.

At TNAU, Coimbatore considerable diversity was observed among 741 accessions maintained for all the agronomic characters. Highest heritability, genotypic coefficient of variation and genetic advance as per cent of mean was recorded in grain yield per plant and the lowest was recorded in days to 50 per cent flowering. Days to 50 per cent flowering, plant height (cm), total number of tillers, number of productive tillers, panicle length (cm) and days to maturity exhibited highly significant positive correlations with grain yield. The path analysis revealed that direct effect of days to 50 per cent flowering on grain yield was positive and negligible. Direct effect of number of productive tillers on grain yield was positive and high. Panicle length showed moderate positive direct effect on grain yield and negligible indirect effects through other traits. Hence, yield advantage could be achieved by using germplasm with more productive tillers, medium panicle length and medium duration. Clustering pattern shows wide range of variability in the population (Nirmalakumari and Vetriventhan, 2010).

Core Collection

As the grain yield levels of foxtail millet are low in comparison with other staple cereals, greater use of diverse germplasm in breeding is suggested as a means to improve the productivity of the crop. To utilize the diversity concealed in the ICRISAT foxtail millet collection from 26 countries in research, a core collection (10 per cent of the entire collection) was established using the taxonomic and qualitative traits data (Upadhyaya *et al.*, 2008). The germplasm accessions were stratified into three taxonomic races (Indica, Maxima and Moharia). Principal coordinate analysis was performed on 12 qualitative traits for each of the biological races, separately that resulted in the formation of 29 clusters. From each cluster, 10 per cent of the accessions were selected to constitute a core collection of 155 accessions. Different statistical analyses indicated that the diversity from the entire collection was optimally represented in the core collection. From the core collection 25 foxtail millet accessions with better grain yield and early maturity have been identified. Twenty-three accessions flowered significantly early (25-40 days) and 19 (990-1760 kg/ha) produced greater grain yields than control SiA 326 (48 days; 980 kg/ha). Cluster analysis based on first five principal components indicated that the selected accessions were diverse than the control cultivars.

Multi-location evaluation of core collection has resulted in identification of a number of diverse germplasm accessions with agronomically (earliness and high grain yield) and nutritionally (high seed protein, calcium, iron and zinc) superior traits. The hierarchical cluster analysis of data using phenotypic distances resulted in 25 clusters, from each cluster, ~10 per cent or a minimum of one accession was selected to form a mini core, which comprised of 35 accessions. Statistical tools revealed that the mini core indeed captured adequate variability from the core collection. This mini core collection is an ideal pool of diverse germplasm for studying population structure and diversity, and identifying new sources of variation for use in breeding and genomics studies in foxtail millet (Upadhyaya *et al.*, 2011). Some of the promising accessions of foxtail millet are given in Table 6.

Table 6: Promising Germplasm Accessions and other Entries of Foxtail Millet

Trait	*Promising Accessions*
High grain yield	ISe 710, ISe 969, ISe 1820, ISe 388, ISe 842, ISe 49, ISe 1888, ISe 90, ISe 364, ISe 1767, ISe 362, ISe 1808
Early flowering	ISe 1151, ISe 1161, ISe 1161, ISe 1201, ISe 1227, ISe 1227, ISe 1234, ISe 1254, ISe 1258, ISe 1286, ISe 1286, ISe 1312, ISe 1320, ISe 1335, ISe 1647, ISe 1658
Tall	ISe 769 and ISe 1434
High tillering (More basal tillers)	ISe 1433 and ISe 1434, ISe 796, ISe 1009, ISe 1026, ISe 1134, ISe 1408, and ISe 1892
High Panicle exertion	ISe 719, 827, 1151, 1161, 1163, 1227, 1286, 1312, 1320, 1547, 1563, 1593, and 1655
Longer inflorescence length and grain yield	ISe 785, ISe 1780, ISe 1433 and ISe 1434, ISe 1593, ISe 701
High protein (≅17 per cent)	ISe 1312, ISe 1227, ISe 1789, ISe 1254, ISe 1541, ISe 827, ISe 748, ISe 1305, ISe 1647, ISe 1335, ISe 751, ISe 1118, ISe 1134
High Ca (≥200 mg/kg Ca)	ISe 1227, ISe 1181, ISe 1059, ISe 1419, ISe 827, ISe 751, ISe 1474, ISe 1685, ISe 1277, ISe 1418, ISe 1161
High Fe (≥62 mg/kg Fe)	ISe 1151, ISe 1286, ISe 1400, ISe 1305, ISe 1332, ISe 1059, ISe 1581, ISe 1320, ISe 748, ISe 1387, ISe 1511
Zn (≥55 mg/kg Zn)	ISe 1286, ISe 748, ISe 1387, ISe 195, ISe 1134, ISe 1408, ISe 1419, ISe 1161, ISe 900, ISe 1820, ISe 1511
Blast (leaf spot) resistance	ISe 1181, ISe 1547, ISe 1067, ISe 1575
Sheath blight resistance	RFM 29, RFM 84, GS 162, GS 1204, SiA 1827, SiA 2566, SiA 2667, SiA 2863, ISe 1488
Shoot fly tolerant	SIA 1538, SIA 1533, SIA 1507, SIA 1581, SIA1566, SIA 1549
Other Promising accn.	ISe 941, ISe 905, ISe 781, ISe 1363, ISe 426, ISe 1135, ISe 989, ISe 1013, ISe 1042, ISe 1077, ISe 184, ISe 275, ISe 941, ISe 905, ISe 781, ISe 1363, ISe 426, ISe 1135, ISe 989, ISe 1013, ISe 1042, ISe 1077, ISe 148, ISe 275

Fingerprinting of 155 core collection accessions using 84 markers (81 SSRs and 3 EST-SSRs) showed the presence of greater molecular diversity in the foxtail millet core collection. The 84 markers detected a total of 1356 alleles with an average of 16.14 alleles (4–35) per locus. Of these, 368 were rare alleles, 906 common alleles and 82 the most frequent alleles. Sixty-one unique alleles that were specific to a particular accession and useful for germplasm identification were also detected. The genetic diversity of foxtail millet was fairly correlated well with racial classification, and the race Indica showed a greater genetic distance from the races Maxima and Moharia. Two hundred and fifty pairs of genetically most diverse accessions were identified.

This large molecular variation observed in the core collection could be utilized effectively by breeders or researchers for the selection of diverse parents for breeding cultivars and the development of mapping populations (Vetriventhan *et al.*, 2012). In another study, high degree of molecular diversity among the 155 core collection accessions was found using 72 SSRs, with an average of 16.69 alleles per locus. STRUCTURE analyses classified the accessions into four subpopulations (SP) based on SSR allelic diversity. The Neighbour joining clustering and the principal coordinate analysis were in accordance with the racial classification. The distribution of molecular genetic variation among and within the four SP and three races showed high degree of variability within each group, and low level of genetic distance (GD) among the groups (Vetriventhan *et al.*, 2014).

Linkage Mapping and QTLs

For foxtail millet, the only available genetic linkage map was constructed by Wang *et al.* (1998) using RFLP markers, which provided foundation for genetic analysis of some traits. Doust *et al.* (2004, 2005) used the map to locate QTL for vegetative branching and inflorescence architecture. Furthermore, Devos *et al.* (1998, 2000) did comparative genomic research among foxtail millet, rice and pearl millet. They found that some homologous chromosomal fragments existing among these genomes, indicating close relationships of the gramineous crops. Development of EST-SSR in foxtail millet for polymorphism and transfe*Rabi*lity study was reported by (Jia *et al.*, 2007). From 1213 EST sequences, 30 SSRs were obtained and primers were designed for 26 SSRs. Among them, four pairs of SSR primers amplified polymorphic products in 12 foxtail millet cultivars and one accession of *S. viridis*, with 10 alleles detected for the four loci and 2.5 alleles per locus.

Microsatellite markers or SSRs, which are distributed widely and throughout plant genomes, are valuable genetic markers because of their high polymorphism and simple operation procedures (Tautz and Renz, 1984; Matsuoka *et al.*, 2002). SSR markers were developed from two genomic libraries enriched for (GA)n and (CA)n of foxtail millet and a foxtail millet SSR linkage map was constructed by integrating 81 new developed SSR markers with 20 RFLP anchored markers. The 81 SSRs covered nine chromosomes of foxtail millet. The length of the map was 1,654 cM, with an average interval distance between markers of 16.4 cM. The 81 SSR markers were not evenly distributed throughout the nine chromosomes, with Ch.8 harbouring the least (3 markers) and Ch.9 harbouring the most (18 markers) (Jia *et al.*, 2009).

A number of QTLs for agronomic traits have been identified in foxtail millet by linkage mapping approach, including basal branching (tillering) and axillary branching (Doust *et al.*, 2004), inflorescence branching (Doust *et al.*, 2005), plant height (PH), panicle neck length, stem node number, panicle length (PcL) and seed shattering (Mauro-Herrera *et al.*, 2013). Following association mapping approach, 184 foxtail millet accessions from diverse geographical locations were genotyped using 50 SSR markers representing the nine chromosomes of foxtail millet. The genetic diversity within these accessions was examined using a genetic distance-based and a general model-based clustering method. The model-based analysis using 50 SSR markers identified an underlying population structure comprising five subpopulations which corresponded well with distance-based groupings. The linkage disequilibrium analysis considering population structure and relative kinship identified eight SSR markers ($p<0.01$) on different chromosomes showing significant association ($R^2 = 18$ per cent) with nine agronomic traits. Four of these markers were associated with multiple traits (Gupta *et al.*, 2014).

Genomics and Genome Sequencing

Foxtail millet, because of its small diploid ($2n = 2x = 18$) genome size (1C genome size = 515 Mb), comprising of relatively low quantity of repetitive DNA and conserved genome structure, inbreeding nature, abiotic stress-tolerance, is considered to be a model species (Devos *et al.*, 1998; Doust *et al.*, 2009; Lata *et al.*, 2013; Lata and Prasad, 2013) with its genome sequenced by the US Department of Energy - Joint Genomic Institute (JGI) (Bennetzen *et al.*, 2012) and BGI (formerly the Beijing Genomics Institute), China (Zhang *et al.*, 2012) for studying functional genomics and to probe plant architecture, genome evolution, drought tolerance, and physiology in the bioenergy grasses (Dekker, 2003; Diao, 2011; Lata *et al.*, 2013).

Though the foxtail millet is regarded as a model crop for studying functional genomics of biofuel grasses, its genetic improvement to some extent was limited due to the non-availability of molecular markers, particularly the SSRs and the saturated genetic linkage map. In an attempt to generate a significant number of microsatellite markers in cultivar 'Prasad', 256 clones were sequenced to generate 41.82-kb high-quality sequences retrieved from genomic library enriched with dinucleotide repeat motifs. Microsatellites were identified in 194 (76 per cent) of the 256 positive clones, and 64 primer pairs (pp) were successfully designed from 95 (49 per cent) unique SSR-containing clones. All the 64 microsatellite markers displayed high level of cross-species amplification (~67 per cent) in 10 millets and non-millets species (Gupta *et al.*, 2013).

Development of genomic resources in foxtail millet and its breeding had made great progress, resulting in significant improvement of yield potential in recent years (Diao, 2011; Pandey *et al.*, 2013; Kumari *et al.*, 2013; Muthamilarasan *et al.*, 2013). Setaria genome sequencing efforts uncovered several distinctive attributes of Setaria genome that may help in understanding its physiology, evolution and adaptation. This will not only aid in comparative genomics studies of Setaria and related crops including bioenergy grasses but also help in rapid advancements of genomics information for developing varieties with superior traits either through marker-assisted selection (MAS) or using transgenic approaches in these crops (Lata and Prasad, 2013).

Major Yield Limiting Factors

Blast disease caused by pathogenic fungus, *Magnaporthe grisea*, is one of the major constraints for the production of foxtail millet, especially in northern China and India (Nakayama *et al.*, 2005). Blast affects both forage and grain production of foxtail millet. Symptoms of the disease appear as circular spots with straw-coloured centers on leaf blades. The spots are small and scattered, and are 2–5 mm in diameter surrounded by a dark brown margin. When the disease appears in severe form during humid weather conditions, especially with a dense plant stand, the leaves wither and dry. Plants are infected at all growth stages (Gaikwad and D'Souza, 1986); lower leaves are the most severely affected. Node, neck and stem tissues are also infected in addition to leaf sheaths. The pathogen is variable, but highly specialized in its host range. Thus, *M. grisea* strains from rice or any other hosts do not infect foxtail millet and *vice versa*. The evaluation of core collection has resulted in identification of two accessions, ISe 1547 and ISe 1181, of race *maxima* resistant to all the four isolates of *M. grisea* indicating a high level of resistance in race *maxima* to different populations of *M. grisea* adapted to foxtail millet (Sharma *et al.*, 2014). ISe 1419, which is resistant to sheath, neck and head blast against all the four isolates and has leaf blast resistance to two isolates. Blast resistant accessions ISe 1059, ISe 1181, and ISe 1685, and ISe 1059 and ISe 1187 were among the trait specific accessions identified for Ca and Fe content, respectively (Upadhyaya *et al.*, 2011).

Other frequently occuring diseases in foxtail millet are sheath blight, downey mildew and rust. Some of the promising accessions (<10 per cent) for sheath blight are RFM 29, RFM 84, GS 162, GS 1204, SiA 1827, SiA 2566, SiA 2667, SiA 2863 and ISe 1488. Another important disease affecting foxtail millet is leaf spot, which is caused by at least three species of *Helminthosporium*, ie., *H. setariae* Sawada, *H. turcicum* Pass., and *H. carbonum* Ullstrup (Haenseler, 1941; Robert, 1962).

Among the insect pests, shoot fly (*Atherigona atripalpis* wiede) is one of the important pests in foxtail millet. shoot fly do not prefer the foxtail millet genotypes possessing lower moisture, crude protein and total sugar content for its oviposition and dead heart infestation (Sanjeev *et al.*, 2011).

Lodging is another important yield and quality reducing factor depending on the environmental conditions, and weak stems and poorly-developed anchored roots always result in crop lodging. Studies have demonstrated that lodging is associated with stem quality rather than plant height and internode length at the stem base. The most important trait that directly impacted on lodging coefficient was mechanical strength of the stem and plant height in the landraces, while the weights of the above- and underground tissues, in combination with mechanical strength of the stem, were the most important contributors to lodging coefficient in the improved cultivars (Tian *et al.*, 2010).

Research Needs, Critical Gaps, Emerging Research Areas, Strategies and Action Plan

Though many high yielding varieties have been released with satisfactory genetic enhancement the average yield remains one-fourth to one-third of the yield levels observed in the advanced varietal trials of AICSMIP. The spread of high yielding varieties in different parts of the country is very much limited as organized seed production and supply is absent in most states. Spread of improved varieties is evident only in southern states. Varieties with high degree of specific adaptation are needed in remote areas. FLDs have helped in spread of HYVs and demonstrating their superiority under all situations as well as fine tuning recommendation domains. The yield levels achieved under FLDs are 52 per cent more than the farmers' practice (1882 kg/ha vs.1239 kg/ha, averaged over the period of 2000-13).

The yield gaps resulting due to crop management constrains that mask the available genetic yield potential under on-farm conditions in different states need to be addressed on priority. Use of exotic germplasm in recombination breeding needs to be strengthened for creating more variability and identifying superior segregants adapted to low inputs, and moisture and other biotic stresses. Introduction and study of behavior of Chinese male sterile lines, and utilization in hybrid development or transfer of male sterile trait to locally well adapted high yielding cultivars for new parental line development should gain momentum.

Development of core collection and identification of trait specific donors, association mapping and identification of markers linked to agronomic traits, genomic resources, QTL mapping, etc. are some of the recent developments in foxtail millet. With the availability of complete genome sequence study of functional genomics and probing plant architecture, genome evolution, drought tolerance, and other basic physiological mechanisms shall be the focus areas in the future. Some of the strategies and action plan for future are:

- ✰ Study of genetics of yield and yield components
- ✰ Use of exotic germplasm in recombination breeding
- ✰ Development of blast and sheath blight tolerant varieties
- ✰ Heterosis breeding – utilization of Chinese CMS lines and crossing with local adapted cultivars, identification of maintainers and restorers for hybrid development
- ✰ Mapping population development and QTL mapping for important agronomic traits, especially seed size

References

1. Ahanchede A and PoirierHamon S. 2004. Why no tetraploid cultivar of foxtail millet?. *Genetic Resources and Crop Evolution* 51(3): 227-230.
2. Athwal DS and Singh G. 1966. Variability in Kangni-1. Adaptation and genotypic and phenotypic variability in four environments. *Indian J. Genet. Plant Breed.* 26(2): 142–152.
3. Ayyangar GNR. 1934. Recent work in the genetics of millets in India. *Madras Agric. J.* 22: 16-26.
4. Baltensperger DD. 1996. Foxtail and proso millet. In: *Progress in new crops*. Janick J. (Ed.). ASHS Press, Alexandria, VA. pp. 182-190.
5. Baltensperger DD. 2002. Progress with proso, pearl and other millets. In: *Trends in new crops and new uses*. Janick J. and Whipkey A. (Eds.). ASHS Press, Alexandria, VA. pp. 100-103.
6. Bennetzen JL, Schmutz J, Wang H, Percifield R, Hawkins J, Pontaroli AC, Estep M, Feng L, Vaughn JN, Grimwood J *et al.*, 2012. Reference genome sequence of the model plant Setaria. *Nat Biotech*. 30: 555–561.
7. Chang K. 1968. Archeology of ancient China. *Science* 162: 519-526.

8. Choi Y, Osada K, Ito Y, Nagasawa T, Choi M and Nishizawa N. 2005. Effects of dietary protein of Korean foxtail millet on plasma adinopectin, HDL-cholesterol and insulin levels in genetically type 2 diabetic mice. *Biosci. Biotechnol. Biochem*. 69: 31-37.

9. Cui WS, Du G, Zhao ZH. 1979. The selection and utilization of male sterile line Shanxi 280 of foxtail millet. *Scientia Agricultura Sinica* 12: 43-46. (in Chinese).

10. Darmency H, Quin, and Pernes J. 1987. Breeding foxtail millet (*Setaria italica*) for quantitative traits after interspecific hybridization and polyploidization. *Genome* 29: 453–456.

11. de Wet JMJ and Harlan JR. 1975. Weeds and domesticates: evolution in the man-made habitat. *Econ. Bot*. 29: 99–107.

12. de Wet JMJ, Oestry-Stidd LL and Cubero JI. 1979. Origins and evolution of foxtail millet (*Setaria italica*). J. d'Agric. *Trad. et de Bota*. 26: 53-64.

13. Dekker J. 2003. Evolutionary biology of the foxtail (Setaria) speciesgroup. In: Inderjit K (Ed) *Principles and practices in weed management: weed biology and management*. Kluwer Academic Publishers, The Netherlands, pp. 65–114.

14. Devos KM, Pittaway TS, Reynolds A and Gale MD. 2000. Comparative mapping reveals a complex relationship between the pearl millet genome and those of foxtail millet and rice. *Theor. Appl. Genet*. 100: 190–198.

15. Devos KM, Wang ZM, Beales J, Sasaki T and Gale MD. 1998. Comparative genetic maps of foxtail millet (*Setaria italica*) and rice (*Oryza sativa*). *Theor. Appl. Genet*. 96: 63–68.

16. Diao XM. 2011. Current status of foxtail millet production in China and future development directions. In: *The industrial production and development system of foxtail millet in China*. Chinese Agricultural Science and Technology Press, Beijing. pp. 20–30.

17. Diao X. 2007. Foxtail millet production and future development direction in China. In: *Reports on Minor Grain Development in China*. Chai Y and Wan SH. (Eds.). Beijing: Chinese Agricul Sci and Tech Press, 32–43.

18. Doust AN, Devos KM, Gadberry MD, Gale MD and Kellogg EA. 2004. Genetic control of branching in foxtail millet. *Proc. Natl. Acad. Sci. USA*. 101: 9045–9050.

19. Doust AN, Devos KM, Gadberry MD, Gale MD and Kellogg EA. 2005. The genetic basis for inflorescence variation between foxtail and green millet (Poaceae). *Genetics* 169: 1659–1672.

20. Doust AN, Kellogg EA, Devos KM and Bennetzen JL. 2009. Foxtail millet: a sequence-driven grass model system. *Plant Physiol*. 149: 137–141.

21. Doust AN, Penly AM, Jacobs SW and Kellogg EA. 2007. Congruence, conflict, and polyploidization shown by nuclear and chloroplast markers in the monophyletic "Bristle clade" (Paniceae, Panicoideae, Poaceae). *Syst. Bot*. 32: 531–544.

22. Dwivedi S, Upadhyaya H, Senthilvel S, Hash C, Fukunaga K, Diao X, Santra D, Baltensperger D and Prasad M. 2012. Millets: Genetic and Genomic Resources. In: *Plant Breed Rev.* Janick J (Ed.), Vol. 35. USA: John Wiley and Sons, Inc., pp. 247–375.

23. Fukunaga K and Kato K. 2003. Mitochondrial DNA variation in foxtail millet, *Setaria italica* (L.) P. Beauv. *Euphytica* 129: 7–13.

24. Gaikwad AP and D'Souza TF. 1986. Studies on blast disease of Rala Setaria italica. *Journal of Maharashtra Agricultural Universities* 11: 108-109.

25. Gill AS and Randhawa AS. 1975. [*Setaria italica* (L.) P. Beauv.] Heritable variation and interrelationships in foxtail millet. *Madras Agric. J.* 62(5): 253–258.

26. Giussani LM, Cota-Sanchez JH, Zuloaga FO and Kellogg EA. 2001. A molecular phylogeny of the grass subfamily Panicoideae (Poaceae) shows multiple origins of C4 photosynthesis. *Am. J. Bot.* 88: 1993–2012.

27. Gupta S, Kumari K, Muthamilarasan M, Parida SK, Manoj Prasad. 2014. Population structure and association mapping of yield contributing agronomic traits in foxtail millet. *Plant Cell Rep.* 33: 881–893.

28. Gupta S, Kumari K, Muthamilarasan M, Subramanian A and Manoj Prasad. 2013. Development and Utilization of Novel SSRs in Foxtail Millet [*Setaria italica* (L.) P. Beauv.]. *Plant Breeding* 132(4): 367-374.

29. Gurunadha Rao Y, Anjanappa M and Appa Rao P. 1984. Genetic variability in yield and certain yield components of foxtail millet (*Setaria italica* Beauv.). *Madras Agric. J.* 71(5): 332–333.

30. Haenseler CH. 1941. Helminthosporium leaf spot on millet in New Jersey. *Plant Dis. Rep.* 25: 486.

31. Haider ZA and Birsa AS. 1991. Gundil 1 little-millet for Bihar's tribals. *Indian Farming*, 41: 28.

32. Hao XF, Wang JZ, Wang GQ, Wang LY and Wang XY. 2009. AFLP analysis of photo sensitive male sterile gene in millet. *Journal of Shanxi Agricultural Sciences* 37: 3-5, 10. (in Chinese).

33. Hao XF, Wang ZM, Wang GQ, Wang GR, Wang XY, Wang LY and Wang JZ. 2011. SSR analysis of photosensitive male sterile gene of millet. *Acta Agronomica Boreali-Sinica* 26: 112-116. (in Chinese).

34. Harinarayana G, Navale PA and Mungse HB. 1984. Dwarf genotypes of foxtail millet. *J. Maharashtra Agric. Univ.* 9(3): 343-344.

35. Harinarayana G. 1989. Breeding and varietal improvement of Small Millets in India. In: *Small millets in global agriculture*. Proc. First Int. Small millets workshop. Bangalore. Seetharama A, Riley KW and Harinarayana G. (Eds.) Oxford and IBH publishing Co. Pvt. Ltd. New Delhi. pp. 59-70.

36. Hector JH. 1936. Introduction to the botany of field crops. Millets. In: *Cereals.* Vol. I. Central New Agency Johannesburg, South Africa. p. 307–319.

37. Hu HK, Ma SY and Shi YH. 1986. The discovery of a dominant male-sterile gene in millet (Setaria italica). *Acta Agronomica Sinica* 12: 73-78. (in Chinese).

38. Jia X, Shi Y, Song Y, Wang G, Wang T and Li Y. 2007. Development of EST-SSR in Foxtail Millet (*Setaria italica*). *Genetic Resources and Crop Evolution* 54(2): 233-236.

39. Jia X, Zhang Z, Liu Y, Zhang C, Shi Y, Song Y, Wang T and Li Y. 2009. Development and Genetic Mapping of SSR Markers in Foxtail Millet [*Setaria italica* (L.) P. Beauv.]. *Theoretical and Applied Genetics* 118: 821-829.

40. Jia XP, Tan XJ, Li YX, Wang TY and Li Y. 2009a. A study on the genetic diversity of foxtail millet cultivars by SSR markers. (In Chinese.) *Acta Agric. Uni. Jiangxi*. 31: 633–638.

41. Jia XP, Zhang ZB, Liu YH, Zhang CW, Shi YS, Song YC, *et al.*, 2009b. Development and genetic mapping of SSR markers in foxtail millet [*Setaria italica* (L.) P. Beauv.]. *Theor. Appl. Genet*. 118: 821–829.

42. Jusuf M and Pernes J. 1985. Genetic variability of foxtail millet (*Setaria italica* (L.) P. Beauv): Electrophoretic study of five isoenzyme systems. *Theor. Appl. Genet*. 71: 385–391.

43. Krishne Gowda KT, 2006. Status of research and development of small millets in India. In: *Strategies for Millets Development and Utilization: Inaugural souvenir of Society of Millets Research* (Eds. N. Seetharama and Vilas A Tonapi), Society for Millets Research, Directorate of Sorghum Research, Rajendranagar, Hyderabad. pp. 81-92.

44. Kumari K, Muthamilarasan M, Misra G, Gupta S, Subramanian A, Parida SK, Chattopadhyay D and Prasad M. 2013. Development of eSSR-markers in *Setaria italica* and their applicability in studying genetic diversity, cross-transfe*Rabi*lity and comparative mapping in millet and non-millet species. PLoS ONE 8: e67742.

45. Lakshmi Prasanna P, Samba Murthy JSV, Ramakumar PV and Srinivasa Rao V. 2013. Studies on correlation and path analysis in exotic Genotypes of Italian millet [*Setaria italica* (L.) Beauv]. *Electronic Journal of Plant Breeding*, 4(1): 1080-1085.

46. Lata C and Prasad M. 2013. Setaria genome sequencing: an overview. *J. Plant Biochem. Biotech*. 22: 257–260.

47. Lata C, Gupta S and Prasad M. 2013. Foxtail millet: a model crop for genetic and genomic studies in bioenergy grasses. *Crit. Rev. Biotech*. 33: 328–343.

48. Le Thierry d'Ennequin M, Toupance B, Robert T, Godelle B and Gouyon PH. 1999. Plant domestication: a model for studying the selection of linkage. *J. Evol. Biol*. 12: 1138–1147.

49. Li YM. 1997. Breeding for foxtail millet drought tolerant cultivars (in Chinese). In: *Foxtail millet breeding*. Li Y, (Ed.). Beijing: Chinese Agr Press, 421–446.

50. Li HW, Li CH and Pao WK. 1945.Cytological and genetical studies of the interspeciÞc cross of the cultivated foxtail millet, *Setaria italica* (L.) Beauv. and the green foxtail millet, S. viridis L. *J. Am. Soc. Agron*. 37: 32-54.

51. Li HW, Meng CJ and Liu TN. 1935. Problems in the breeding of millet (*Setaria italica* (L.) Beauv.). *J. Am. Soc. Agron*. 27: 963–970.

52. Li P and Brutnell TP. 2011. Setaria viridis and Setaria italica, model genetic systems for the Panicoid grasses. *J. Exp. Bot*. 62: 3031–3037.

53. Liu Z, Sun S, Cheng R, Huang W, Liu J, Qu Z, Xia X and Shi Z. 2006. Development of a New Foxtail Millet Germplasm with Super Early Maturity and High Iron Content. *Agricultural Sciences in China* 5(7): 558-562.

54. Liu Z, Zhang T, Li C and Bai G. 2014. Genetic diversity and classification of cytoplasm of Chinese elite foxtail millet [*Setaria italica* (L.) P. Beauv.] Germplasm. *Crop Sci*. 54: 659–666.

55. Liu ZL, Bai GH, Zhang DD, Zhu CS, Xia XY, Cheng RH and Shi ZG. 2011. Genetic diversity and population structure of elite foxtail millet [*Setaria italica* (L.) P. Beauv.] germplasm in China. *Crop Sci*. 51: 1655–1663.

56. Liu ZL, Cheng RH and Li XY. 1996. The pedigree analysis and evaluation of north China summer millets. (In Chinese.) *Crops*. 5: 24.

57. Liu ZL. 2010. Establish method of heterotic group in foxtail millet and its research progress. (In Chinese.) *J. Hebei Agric. Sci*. 14: 102–104.

58. Mahishi DM, Aradhya KM, Seetharam A and Gowda BTS. 1982. Efficacy of contact method for hybridization in foxtail millet. *MILWAI Newsletter* 1: 16.

59. Malm NR and Rachie KO. 1971. Setaria millets: A review of the world literature. S.B. 513. University of Nebraska, Lincoln. pp. 19–29.

60. Marathee JP. 1993. Structure and characteristics of the world millet economy. In: *Advances in small millets*. Riley KW, Gupta SC, Seetharam A and Mushonga JN. (Eds.). New Delhi, India: Oxford and IBH. pp. 159–178.

61. Matsuoka YS, Mitchell E, Kresovich S, Goodman M and Doebley J. 2002. Microsatellites in Zea: Variability, patterns of mutations, and use for evolutionary studies. *Theor. Appl. Genet*. 104: 436–450.

62. Mauro-Herrera M, Wang X, Barbier H, Brutnell TP, Devos KM and Doust AN. 2013. Genetic control and comparative genomic analysis of flowering time in Setaria (Poaceae). G3 3: 283–295.

63. Movsisyants AP. 1950. Foxtail millet-A valuable forage crop. Sotsialist. *Zhivotn*. 12: 60-63.

64. Murugan R and Nirmalakumari A. 2006. Genetic divergence in foxtail millet [*Setaria italica* (L.) Beauv.]. *Indian J. Genet*., 66(4): 339-340.

65. Muthamilarasan M, Suresh BV, Pandey G, Kumari K, Parida SK and Prasad M. 2013. Development of 5123 Intron length polymorphic (ILP) markers for large-scale genotyping applications in foxtail millet [Setaria italica (L.)]. *DNA Res*. doi: 10.1093/dnares/dst039.

66. Nakayama H, Nagamine T and Hayashi N. 2005. Genetic variation of blast resistance in foxtail millet (*Setaria italica* (L.) P. Beauv.) and its geographic distribution. *Genet. Resour. Crop Ev*. 52: 863-868.

67. Nirmalakumari A and Vetriventhan M. 2010. Characterization of foxtail millet germplasm collections for yield contributing traits. *Electronic Journal of Plant Breeding*, 1(2): 140-147.

68. Oelke EA, Oplinger ES, Putnam DH, Durgan BR, Doll JD and Undersander DJ. 1990. Millets. In: Alternative Field Crops Manual, University of Wisconsin, Cooperative Extension and University of Minnesota, Center for Alternative Plants and Animal Products and Minnesota Extension Service. (https://www.hort.purdue.edu/newcrop/afcm/millet.html).

69. Padulosi S, Bhag Mal, Bala Ravi S, Gowda J, Gowda KTK, Shanthakumar G, Yenagi N and M Dutta. 2009. Food Security and Climate Change: Role of Plant Genetic Resources of Minor Millets. *Indian J. Plant Genet. Resour*. 22(1): 1-16.

70. Panaud O. 2006. Foxtail millet. In: *Genome mapping and molecular breeding in plants*. Kole C. (Ed.). 1: 325–332.

71. Pandey G, Misra G, Kumari K, Gupta S, Parida SK, Chattopadhyay D and Prasad M. 2013. Genome-wide development and use of microsatellite markers for large-scale genotyping applications in foxtail millet [*Setaria italica* (L.)]. *DNA Res*. 20: 197–207.

72. Prasada Rao GP, Nagaraja Rao M and Anjanappa M. 1985. Genetic variability in setaria. *Andhra Agric. J*. 32(1): 34–56.

73. Prasada Rao KE, de Wet JMJ, Brink DE and Mengesha MH. 1987. Intraspecific variation and systematics of cultivated Setaria italica (Gramineae). *Economic Botany* 41: 108–116.

74. Raveendran TS, Nagarajan C, Natarajan US and Appadurai R. 1979. Tenai CO 5-a high yielding rust resistant variety suitable for Tamil Nadu. *Madras Agric. J*. 66: 771-773.

75. Robert AL. 1962. New hosts for three Helminthosporium species from corn. *Plant Dis. Rep*. 46: 321–324.

76. Sanjeev U, Jagadish PS, Mahendra R, Santosh DT and Shashidhara N. 2011. Biochemical basis of resistance to shoot fly (*Atherigona atripalpis* wiede) in foxtail millet. *Int. Jr. of Agril., Env. and Biotech*. 4: 221-222.

77. Scholz H. 1960. A contribution to knowledge of cultivated plants. *Ber. Dtsch. Bot. Ges*. 73: 471-476.

78. Sema A and Sarita S. 2002. Suitability of millet-based food products for diabetics. *J. Food Sci. Technol*. 39: 423-426.

79. Sharma R, Girish AG, Upadhyaya HD, Humayun P, Kiran Babu Talluri, Veeranki Panduranga Rao and Thakur RP. 2014. Identification of blast resistance in a core collection of foxtail millet germplasm. Plant Disease (DOI: http://dx.doi.org/10.1094/PDIS-06-13-0593-RE).

80. Siles MM, Baltensperger DD and Nelson LA. 2001. Technique for artificial hybridizatioof foxtail millet [*Setaria italica* (L.) Beauv.]. *Crop Sci*. 41: 1408–1412.

81. Siles MM, Ken Russell W, Baltensperger DD, Nelson LA, Johnson B, Dale Van Vleck L, Jensen SG and Hein G. 2004. Heterosis for grain yield and other agronomic traits in foxtail millet. *Crop Sci.* 44: 1960–1965.

82. Singh G and Athwal DS. 1966. Variability in Kangni-2 genotype x environment interaction, heritability and genetic advance. *Indian J. Genet. Plant Breed.* 26(2): 153–161.

83. Srivastava DP and Yadav A. 1977. Heterosis in *Panicum miliare* Lam. *Curr. Res.* 6(4): 66-67.

84. Tautz D and Renz M. 1984. Simple sequences are ubiquitous components of eukaryotic genomes. *Nucleic Acids Res.* 12: 4127–4138.

85. Tian B, Wang J, Zhang L, Li Y, Wang S and Li H. 2010. Assessment of resistance to lodging of landrace and improved cultivars in foxtail millet. *Euphytica* 172: 295–302.

86. Tian G, Wang YW, Li HX, Wang GH and Shi QX. 2009. Breeding and studies of foxtail millet hybrid Changzagu 2 with herbicide-resistance. *Journal of Agricultural Science and Technology*, 11(S2), 138-141. (in Chinese).

87. Till-Bottraud I, Reboud X, Brabant P, Lefranc M, Rherissi B, Vedel F and Darmency H. 1992. Outcrossing and hybridization in wild and cultivated foxtail millets: Consequences for the release of transgenic crops. *Theor. Appl. Genet.* 83: 940–946.

88. Umesh Chandra, Thomas TA, Sapra RL and Paroda RS. 1987. Catalogue on Foxtail Millet (*Setaria italica* L.) Germplasm. NBPGR Publication, New Delhi: 80 p.

89. Upadhyaya HD, Pundir RPS, Gowda CLL, Reddy VG and Singh S. 2008. Establishing a core collection of foxtail millet to enhance the utilization of germplasm of an underutilized crop. *Plant Genet. Resour.* 7: 177–184.

90. Upadhyaya HD, Ravishankar CR, Narasimhudu Y, Sarma NDRK, Singh SK, Varshney SK, Reddy VG, Singh S, Parzies HK, Dwivedi SL, Nadaf HL, Sahrawat KL and Gowda CLL. 2011. Identification of trait-specific germplasm and developing a mini core collection for efficient use of foxtail millet genetic resources in crop improvement. *Field Crops Research* 124: 459–467.

91. Vavilov NI. 1926. Studies on the origin of cultivated plants. *Inst. Appl. Bot. Plant Breed., Leningrad.* p. 248.

92. Vetriventhan M, Upadhyaya HD, Anandakumar CR, Senthilvel S, Parzies HK, Bharathi A, Varshney RK and Gowda CLL. 2012. Assessing genetic diversity, allelic richness and genetic relationship among races in ICRISAT foxtail millet core collection. *Plant Genetic Resources* 10 (3): 214-223.

93. Vetriventhan M, Upadhyaya HD, Anandakumar CR, Senthilvel S, Varshney RK and Heiko K Parzies. 2014. Population structure and linkage disequilibrium of ICRISAT foxtail millet (*Setaria italica* (L.) p. Beauv.) core collection. Euphytica 196(3): 423-435.

94. Vinall HN 1924. Foxtail millet: its culture and utilization in the United States. USDA. Farmers Bull. 793.

95. Vishwanatha JK, Mallanna KN, Aradhya KM, Gowda MVC and Shanker RU. 1981. Genetic variability in a world collection of germplasm of foxtail millet, Setaria italica Beauv. *Mysore J. Agric. Sci.* 15: 234–238.

96. Wang J, Wang Z, Yang H, Yuan F, Guo E, Tian G, An Y, Li H, Wang Y, Diao X and Guo P. 2013. Genetic analysis and preliminary mapping of a highly male-sterile gene in foxtail millet (*Setaria italica* L. Beauv.) using SSR markers. *Journal of Integrative Agriculture* 12(12): 2143-2148.

97. Wang JZ, Hao XF, Wang GQ, Wang LY and Sun MR. 2006. Study of genetic diversity of millet germplasm resources by molecular markers. (In Chinese.) *Biotechnology* 16: 12–16.

98. Wang R, Gao J and Liang GH. 1999. Identification of primary trisomics and other aneuploids in foxtail millet. *Plant Breeding* 118(1): 59-62.

99. Wang RL, Wendel JF and Dekker JH. 1995.Weedy adaptation in Setaria spp. I. Isozyme analysis of genetic diversity and population genetic structure in *Setaria viridis. Am. J. Bot.* 82: 308–317.

100. Wang RQ, Gao JH, Mao LP, Du RH, Diao XM and Sun JS. 2002. Chromosome location of the male-sterility and yellow seedling gene in line 1066A of foxtail millet. *Acta Botanica Sinica* 44: 1209-1212. (in Chinese).

101. Wang T, Du RH and Hao F. 1993. Studies and utilization of highly male sterility of summer millet (*Setaria italica*). *Scientia Agricultura Sinica* 26: 88. (in Chinese).

102. Wang T. 1991. Studies and utilization of a high nuclear male sterile line in foxtail millet. J. *Hebei Agric. Univ*. 14: 60–66.

103. Wang TY, Chen HB and Darmency H. 1997. Pollen-mediated Gene Flow in an Autogamous Crop: Foxtail Millet (*Setavia italica*). *Plant Breeding* 116(6): 579-583.

104. Wang YW, Li HX, Tian G and Shi QX. 2010. Study on innovation and application of highly-male-sterile line with high outcrossing rate in millet. *Scientia Agricultura Sinica* 43: 680-689 (in Chinese).

105. Wang YW, Li HX, Wang GH and Tian G. 1998. Breeding of foxtail millet highly sterile line "Chang10A". (In Chinese.) *Gansu Agr. Sci. and Techn*. 12: 12–13.

106. Zhang G, Liu X, Quan Z, Cheng S, Xu X, Pan S, Xie M, Zeng P, Yue Z and Wang W. 2012. Genome sequence of foxtail millet (*Setaria italica*) provides insights into grass evolution and biofuel potential. *Nat. Biotech*. 30: 549–554.

107. Zhang JP, Liu TS, Zheng J, Jin Z, Zhu Y, Guo JF and Wang GY. 2007. Cloning and characterization of a putative 12-oxophytodienoic acid reductase cDNA induced by osmotic stress in roots of foxtail millet. *DNA Seq* 18: 138–144.

108. Zhao ZH, Cui WS, Du G and Yang SQ. 1996. The selection of millet photo-(thermo-) sensitive line 821 and study on the relation of sterility to illumination and temperature. *Scientia Agricultura Sinica* 29: 23-31. (in Chinese).

109. Zhi H, Wang YQ, Li W, Wang YF, Li HQ, Lu P and Diao XM. 2007. Development of CMS material from intra-species hybridization between green foxtail and foxtail millet. *Journal of Plant Genetic Resources* 8: 261-264.

110. Zhu GQ, Wu QM and Ma YT. 1991. Breeding of Ve type CMS in foxtail millet. *Shaanxi Agricultural Sciences* 1: 7. (in Chinese).

Millets: Ensuring Climate Resilience and Nutritional Security (2015) *Pages* **305–317**
Editors: **Vilas A. Tonapi and J.V. Patil**
Published by: **DAYA PUBLISHING HOUSE, NEW DELHI**

Chapter 8

Genetic Improvement in Kodo Millet

K. Hariprasanna

ICAR-Indian Institute of Millets Research (IIMR) Rajendranagar, Hyderabad – 500 030

Kodo millet (*Paspalum scrobiculatum* L.) is an annual grain that is grown primarily in India, but also in the Philippines, Indonesia, Vietnam, Thailand and in West Africa. It is harvested as a wild cereal in West Africa and India. It is grown as a minor food crop in most of these areas, with the exception of the Deccan plateau in India where it is grown as a major food source. It is a self pollinating grass (2n = 4x = 40) belonging to Family *Poaceae* and Subfamily *Panicoideae*. It grows wild as a perennial in the west of Africa, where it is eaten as a famine food. Often it grows as a weed in rice fields, and farmers do not weed it out as it can be harvested as an alternative crop if their primary crop fails. In the Southern United States and Hawaii, it is considered to be a noxious weed. It is a very hardy crop, drought tolerant and can survive on marginal soils where other crops may not survive, and can supply 450–900 kg of grain per hectare (Heuzé *et al.*, 2012).

In the Indian subcontinent and on Socotra (Miller and Morris, 2004) it is used as secondary cereal. In India, kodo millet, known by different names (Table 1), is wide spread and grown in about 2.44 lakh ha (2005-06) in many states in south, west, central and north producing about 0.73 lakh t of grains with a productivity of about 312 kg/ha. Madhya Pradesh (0.50 lakh t) and Chattisgarh (0.17 lakh t) accounts for nearly 80 per cent area under this crop, and other important states are Tamil Nadu (0.12 lakh t), Maharashtra (0.08 lakh t), Uttar Pradesh (0.07 lakh t) and Gujarat (based on mean production during 2001-06). Its cultivation is mostly confined to tribal areas and it is grown under poor environment. It is also used as fodder and forage for

Table 1: Common Names of Kodo Millet

English	Kodo millet	Malayalam	Varagu
Hindi	Kodon	Marathi	Kodra
Sanskrit	Kodara	Gujarati	Kodra
Kannada	Harka	Bengali	Kodo
Tamil	Varagu	Oriya	Kodua
Telugu	Arikelu, Arika	Punjabi	Kodra

Table 2: Nutritive Value of Kodo Millet vs. Rice and Wheat (per 100 g)

Item	*Kodo Millet*	*Wheat (Whole)*	*Rice (Raw, milled)*
Carbohydrates (g)	65.9	71.2	78.2
Protein (g)	8.3	11.8	6.8
Fat (g)	1.4	1.5	0.5
Energy (KCal)	309	346	345
Crude fibre (g)	9	1.2	0.2
Mineral matter (g)	2.6	1.5	0.6
Amylose (per cent)	24	25	12-19
Amylopectin (per cent)	76	75	88-81
Ca (mg)	27	41	10
P (mg)	188	306	160
Vitamins (mg/100 g)			
Thiamin	0.15	0.41	0.41
Niacin	2.0	5.1	4.3
Riboflavin	0.09	0.1	0.04
Folic Acid	23.1	36.6	8.0
Micronutrients (mg/100 g)			
Fe	0.5	5.3	0.7
Zn	0.7	2.7	1.4
Mg	147	138	90
Na	4.6	17.1	–
K	144	284	–
Cu	1.6	0.68	0.14
Mn	1.1	2.29	0.59
Cr	0.02	0.012	0.004
Su	136	128	–
Cl	11	47	–

Source: Nutritive value of Indian foods, NIN, 2007.

livestock. The leaves are antiseptic in action and their paste is applied externally in cutaneous afflictions. On the basis of ethnobotanical records this species is also used for carbuncle, diabetes, intoxication, narcotics, opthalmia, parturition and sores (Knees and Gupta, 2013).

The fiber content of the whole grain is very high. The grain varies in colour from light red to dark grey and is enclosed in a tough husk that is difficult to remove. The seed coat is multiple layered. Kodo millet has around 11 per cent protein, and the nutritional value of the protein has been found to be slightly better than that of foxtail millet but comparable to that of other small millets. It is however deficient in the amino acid tryptophan. It is also reasonably low in fat with high fiber content. Due to high antioxidant content, it is beneficial in protecting against oxidative stress and maintaining glucose levels in type-2 diabetes (Taylor, 2004). As with other food grains, the nutritive value of kodo millet protein could be improved by supplementation with legume protein.

In India, kodo millet is ground into flour and used to make pudding (Heuzé *et al.*, 2012). In Africa it is cooked like rice. It is also a good choice for animal fodder for cattle, goats, pigs, sheep, and poultry. The grains are larger than foxtail and little millet grains. When cooked as rice, the grains keep their shape and can be used to make lime rice, coconut rice and any other such dish where one wants grain integrity to be maintained. Kodo millet is very easy to digest, it contains a high amount of lecithin and is excellent for strengthening the nervous system. Kodo millets are rich in B vitamins, especially niacin, B6 and folic acid, as well as the minerals such as calcium, iron, potassium, magnesium and zinc (Table 2). Kodo millet contain no gluten and is good for people who are gluten intolerant. Regular consumption of kodo millet is very beneficial for postmenopausal women suffering from signs of cardiovascular disease, like high blood pressure and high cholesterol levels (Malathi *et al.*, 2012). Kodo millet has the highest total phenolic content for whole grain and hulls among the different types of millets (Chandrasekara *et al.*, 2012). It has been reported that the DPPH radical scavenging activity was also high in kodo millet hulls compared to whole or dehulled or cooked grains vs. other millets. The roasting or boiling of kodo millet reduces the free radical quenching activity (Hegde and Chandra, 2005) there by resulting in a reduction in antioxidant activity. Insoluble bound fraction phenolic compounds of kodo millet showed the highest phenolic content as well as antioxidant activity in the *in vitro* test systems (Chandrasekara and Shahidi, 2011a).

Origin and Distribution

Kodo millet is indigenous to Indian subcontinent and known to be in cultivation for more than 3000 years. The kodo millet, also known as cow grass, rice grass, ditch millet, Native Paspalum, or Indian Crown Grass originated in tropical Africa, and it is estimated to have been domesticated in India 3000 years ago (Malleshi and Hadimani, 1994). Kodo millet is widely distributed in damp or humid habitats of tropics and subtropics across the Old World. Kodo millet is variable but lacks racial differentiation. Hybridization with wild *P. scrobiculatum* obscures racial differentiation. Weedy kodo is harvested with the crop, making it difficult to distinguish wild and cultivated complexes (de Wet *et al.*, 1983). The wild form is common across tropical Africa (as well as across wetter parts of the Asian tropics from Indonesia to Japan). It is often abundant along paths, ditches, and low spots, especially where the ground is disturbed (which accounts for the reason it is sometimes called ditch millet).

Taxonomy

Paspalum scrobiculatum var. *scrobiculatum* is grown in India as an important crop, while *P. scrobiculatum* var. *commersonii* is the wild variety indigenous to Africa (Heuzé *et al.*, 2012). The species is recognised, as demonstrated by Clayton (1975), to include the African *P. orbiculare*, *P. polystachyum*, and *P. cartilagineum* with usually smaller spikelets (1.8-2.5 mm long) than those of Indian wild specimens (2.5-3.5 mm). These taxa merge completely not only spikelet morphology, but also in overall inflorescence morphology and vegetative traits (de Wet *et al.*, 1983). On the basis of inflorescence morphology the species *P. scrobiculatum* is divided into three races *Regularis*, *Irregularis* and *Variabilis*.

Morphology and Floral Biology

Kodo millet is a vigorous, annual tufted (up to 60 cm diameter) and slender perennial grass. Plants are slender to stout, up to 90 cm tall, and often root from the lower nodes. The roots are rather shallow and the stems are ascending, branched and somewhat succulent. Leaf blades are linear, glabrous or pubescent, up to 40 cm long, with the basal leaf sheaths glabrous or pilose. Leaf blades are 15-40 cm long, 5-12 mm wide, pale green. Leave-sheath and leaves are glabrous (Galinato *et al.*, 1999). Inflorescences are composed of rarely more than 5 racemes (3-7) that are alternately arranged on a short to elongated primary axis. Racemes are up to 13 cm long, with the subsessile spikelets arranged in 2-3 rows along one side of a flattened rachis. Spikelets are glabrous, orbicular or broadly elliptic in outline, conspicuously plano-convex, 1.8-3.5 mm long, and never awned. The lower glume is absent, and the upper glume is as long as the spikelet. The lower lemma is almost flat, more or less membranaceous and without a palea or floret. The upper lemma is crustaceous, often brown and shiny when grains are mature, and embraces the crustaceous palea. The grain is elliptic-orbicular in outline and 1.5-2.5 mm long (de Wet *et al.*, 1983).

The grain is enclosed in hard, corneous, persistent husks that are difficult to remove. The grain may vary in colour from light red to dark grey. The crop matures in 3-4 months with average yield varying from 250 to 1000 kg/ha (Hulse *et al.*, 1980) and a potential yield of 2000 kg/ha (Harinarayana, 1989). It has a 1000 grain weight of 6.7 g. Kodo millet prefers full light for optimal growth, but can tolerate some partial shading. Its ideal temperature for growth is 25-27°C (Heuzé *et al.*, 2012).

Kodo millet is cleistogamous, but protogyny has been observed in some cultures like IPs 147, IPs 197, IPs 427, etc. Protogynous types have been selected, and crosses have been made. Inter-planting of protogynous kodo millets in germplasm or in breeding nurseries or in association with chosen lines enhances the prospects of natural hybrids. Controlled pollination could also be effected in protogynous types. In the Indian wild types the stigmas protrude from the spikelets. The observation by de Wet *et al.* (1983), on the lack of racial differentiation after 3,000 years of domestication, suggests that it could be a very interesting crop with which to work.

Varietal Development

Like other small millets kodo millet also received less priority in the agricultural research. The pure line selection remained the prime breeding method for improving performance, particularly grain yield. Pure line selection has resulted in the development of many varieties in kodo millet (Harinarayana, 1989). Since 1978, the centre of excellence for the improvement of small millets established at Dindori (JNKVV) by ICAR with the assistance of International Development Research Centre (IDRC), Canada is devotedly working towards the improvement of kodo millet. With the establishment of All India Coordinated Small Millets Improvement Project (AICSMIP) in 1986 at UAS, Bangalore the IDRC centre became part of AICSMIP and varietal development has gained momentum. The AICSMIP centres located at Tribal Agricultural Research Station, JNKVV, Dindori, and Agricultural College, JNKVV, Rewa are exclusively working on kodo and little millets.

Some of the early efforts in crop improvement of kodo millet have resulted in the release of improved varieties as early as 1940s. For example PLR1 was released in 1942 and T2 in 1949. Improved variety Co 1 was released from TNAU during 1953. Till date (1942-2012) about 31 varieties have been released, of which 11 were released before establishment of AICSMIP (1986) and 20 after 1986. Pure line selection has resulted in the development and release of maximum number of kodo millet varieties - till 1985, 13 varieties were developed (Harinarayana, 1989) and between 1986-2011, 17 varieties were developed. Mutation breeding has resulted in the development of one variety.

A number of varieties with high yield potential have been released for different states (Table 3). Some of the recent varieties with yield potential, area of adaptation and special features are listed in Table 4. Some of the important varieties include RBK 155, DPS 48, GPUK 3, APK 1, JK 41, JK 48, JK 76, JK 106, JK 65, JK 98, GK 1, GK 2, DPS 9-1, JK155, K1, Co 2, Co 3, APK 1 and Vamban. Till date 27 kodo millet varieties have been notified by Govt. of India, of which 13 are central releases and rest are state releases. The list of notified varieties is given in Table 5.

Table 3L State-wise High Yielding Kodo Millet Varieties

State	*Kodo Millet*
Andhra Pradesh	PSc 1, GPUK 3, DPS 48
Gujarat	GK 2, DPS 48
Karnataka	GPUK 3, RBK 155, DPS 48
Maharashtra	GPUK 3
Madhya Pradesh	JK 62, JK 76, JK 41, GPUK 3, RBK 155, DPS 48
Chattisgarh	JK 62, JK 76, JK 41, GPUK 3, RBK 155, DPS 48
Tamil Nadu	GPUK 3, KMV 20, APK 1
Uttar Pradesh	GPUK 3, KK1, KK2

Source: http://smallmillets.res.in/

Table 4: Improved Kodo Millet Varieties with Characteristics

Sl.No.	*Name of Variety*	*Pedigree*	*Institute where Developed*	*Year of Release*	*Maturity (Days)*	*Av. Yield (kg/ha)*	*Area of Adaptation*	*Special Features*
1.	JK 76	Selection from local germplasm	JNKVV, Rewa	1989	85-90	1600-1800	Madhya Pradesh and Chattisgarh	Earliness
2.	JK 62	Selection from local germplasm	JNKVV, Rewa	1989	100-105	1800-2000	Madhya Pradesh and Chattisgarh	Earliness and high yield
3.	GPUK 3	Selection from germplasm GPLM 826	P.C. Cell, Bangalore	1991	100-105	1800-2000	All states	Yield, earliness and resistance to head smut
4.	APK 1	Selection from PSC 5	ARS, Aruppukotai	1993	100-102	1800-2000	Tamil Nadu	High seed yield
5.	GK 2	Selection from GPUK 5	GAU, Gujarat	1993	110-112	1500-2000	Gujarat	
6.	KMV 20 (Vamban 1) (Varugu 1)	Pureline selection from entry Pali	TNAU, Coimbatore	1996	95-100	1000-1750	Tamil Nadu	Tolerant to sheath blight
7.	Jawahar kodo 155 (RBK 155)	Selection from germplasm no. GPLM 251	JNKVV, Jabalpur	2000	105-110	2000-2200	Madhya Pradesh, Karnataka	Resistant to head smut and shoot fly
8.	Jawahar kodo 48 (JK 48)	Pure line selection from local germplasm 48	JNKVV, Jabalpur	2001	110-115	2600-2700	Andhra Pradesh, Madhya Pradesh, Chattisgarh, Karnataka, Gujarat	Tolerance to head smut and high grain yield
9.	KK 2	Selection from genetic collections of Ballia Dist.	CSAUA&T, Kanpur	2002	110-115	2000-2300	Uttar Pradesh	Resistant to drought and lodging, suitable for saline condition
10.	Jawarhar Kodo-439	Selection from germplasm RPS-123	JNKVV, Jabalpur	2002	89-110	2000-2200	Madhya Pradesh	Moderately resistant to smut and shoot fly
11.	Jawahar Kodo 13 (JK-13)	Mutant plant No. 13 of cv JK 76	JNKVV, Rewa	2006	79-129	2200-2300	All states	Moderately resistant to head smut, Resistant to shoot fly

Contd...

Table 4–*Contd...*

Sl.No.	*Name of Variety*	*Pedigree*	*Institute where Developed*	*Year of Release*	*Maturity (Days)*	*Av. Yield (kg/ha)*	*Area of Adaptation*	*Special Features*
12.	Jawahar Kodo-65 (RK 65-18)	Selection from local germplasm of Rewa	JNKVV, Rewa	2008	104	2300	Madhya Pradesh, Gujarat and Uttar Pradesh	–
13.	Jawahar Kodo-106	Selection from local germplasm no. 106	JNKVV, Rewa	2008	100	1947	Madhya Pradesh	Resistant to head smut and shoot fly
14.	Indira Kodo-1 (BK-1)	Selection from kodo genetic stock	IGAU, SGCA, Jagdalpur	2012	103-105	2437-2440	Chhattisgarh	Moderately susceptible to head smut

Source: Seed Division, DAC; http://smallmillets.res.in/

Table 5: List of Notified Kodo Millet Varieties in India

Sl.No.	*Variety Name*	*Year of Release*	*Notification Date*	*Notification No.*	*State of Release*
1.	PLR-1	1942	01/01/1944		Tamil Nadu
2.	CO.1	1953	01/01/1955		Tamil Nadu
3.	CO.2	1970	01/01/1972		Tamil Nadu
4.	Niwas-1	1971	01/01/1973		Madhya Pradesh
5.	JNK-101	1976	01/01/1978		Madhya Pradesh
6.	JNK-364	1977	01/01/1979		Madhya Pradesh
7.	K-1	1979	14/01/1982	19(E)	Tamil Nadu
8.	Jawahar Kodo-2	1976	14/01/1982	19(E)	Madhya Pradesh
9.	Jawahar Kodo-1	1976	14/01/1982	19(E)	Madhya Pradesh
10.	CO-3	1980	14/01/1982	19(E)	Tamil Nadu
11.	Gujarat Kodra-1	1977	09/04/1985	295(E)	Central
12.	PSC-1		14/05/1986	258(E)	Central
13.	Jawahar Kodo-62		14/05/1986	258(E)	Central
14.	Jawahar Kodo-41		14/05/1986	258(E)	Central
15.	Jawahar Kodo-76		16/08/1991	527(E)	Central
16.	GPUK-3		16/08/1991	527(E)	Central
17.	APK-1		17/08/1993	615(E)	Central
18.	GK-2		04/05/1995	408(E)	Central
19.	Vamban-1 (KMV-20)	1995	01/05/1997	360(E)	Tamil Nadu
20.	KK-1	1998	08/06/1999	425(E)	Uttar Pradesh
21.	Jawahar Kodo-155 (RBK-155)		13/09/2000	821(E)	Central
22.	Jawahar Kodo-48 (DPS-48)	2001	15/11/2001	1134(E)	Central
23.	Jawarhar Kodo-439	2002	05/11/2005	1566(E)	Central
24.	Jawahar Kodo 13 (JK-13)		06/02/2007	122(E)	Central
25.	Jawahar Kodo-65 (RK 65-18)		11/02/2009	S.O. 449(E)	Central
26.	Jawahar Kodo-106	2008	11/02/2009	S.O. 449(E)	Madhya Pradesh
27.	Indira Kodo-1 (BK-1)	2012	10/09/2012	S.O. 2125(E)	Chhattisgarh

Source: Seed Variety Management System, http://seedvariety.dacnet.nic.in/

Under AICSMIP, the kodo millet varietal evaluation trials are taken up at different testing centres, identified across the country based on kodo millet growing regions. Centres are located in Andhra Pradesh (Nandyal), Chattisgarh (Jagdalpur and Ambikapur), Gujarat (Dahod), Jharkhand (Ranchi), Karnataka (Bangalore), Madhya Pradesh (Dindori and Rewa) and Tamil Nadu (Coimbatore, Pudukottai and Aruppukottai). The latest national checks are GPUK 3 (selection from germplasm

GPLM 826, Bangalore), RK 390-25 (Mutant of RK 390, Rewa) and TNAU 86 (selection from IPs 85, Coimbatore).

Genetic Studies

Information on the inheritance of important agronomic traits of kodo millet, including susceptibility to major diseases and insect pests, is limited. Some of the limited previous work has focused on estimating broad-sense heritabilities and genetic gains, with little attention directed to measuring levels of heterosis or to assessing the relative importance of different types of gene action because of difficulty in crossing. Estimation of genetic parameters in the context of trait characterization is an essential component for targeted trait based crop improvement. High genotypic variance, phenotypic variance, GCV and PCV were observed for dry fodder yield, plant height and grain yield per plant in a study of 188 accessions of kodo millet at TNAU, Coimbatore. Broad sense heritability ranged from 97 per cent for dry fodder yield to 57 per cent for number of tillers. The highest genetic advance as percentage of mean was recorded by dry fodder yield followed by grain yield per plant and plant height. High heritability coupled with high genetic advance observed for grain and dry fodder yield indicates the influence of additive gene action (Subramanian *et al.*, 2010).

Germplasm Status

About 5457 accessions of kodo millet have been conserved in main *ex situ* genebanks around the world (Padulosi *et al.*, 2009). At the CGIAR genebanks around 2418 accesions have been conserved while at ICSRISAT 658 accessions of kodo millet have been conserved. A core collection of the available germplasm has been developed at ICRISAT recently. At the Svalbard genebank at Norway 108 accessions have been conserved. At National Genebank at NBPGR, New Delhi at total of 2273 accessions including 3 exotic accessions have been conserved and catalogue has been published (Umesh Chandra *et al.*, 1987). At UAS, Bangalore National Active Germplasm Site (NAGS) was established for small millets in 1992. A total of 1537 accessions of kodo millet, comprising landraces from 8 Indian states and USA are being conserved at UAS, Bangalore. Of these, catalogue on evaluation of 1038 accessions for 16 and 11 qualitative and quantitative descriptors, respectively, has been published. At TNAU, Coimbatore 427 accessions have been maintained. Some of the promising accessions of kodo millet are given in Table 6.

Table 6: Promising Germplasm Accessions and Other Lines of Kodo Millet

Trait	*Promising Accessions*
High grain yield	IPs 149, IPs 147-1
Drought tolerance	IPs 583
Protein stability	IPs 147, IPs 200, IPs 198, IPs 112, IPs 66, IPs 138, IPs 115
Head smut	KOPKM 1, KOPKM 10, RPS 973
Other promising accessions	IPs 199, IPs 270, IPs 681, IPs 3417, IPs 669, IPs 585, IPs 79, IPs 346, IPs 570, IPs 325, IPs 135, IPs 663, IPs 679

Major Yield Limiting Factors and Constraints

Head smut incited by *Sorosporium paspalithunbergione* is an important disease of kodo millet, which causes economic loss in grain yield (Viswanath and Seetharam, 1989). In Madhya Pradesh, up to 32.9 per cent loss in grain yield was reported (Jain and Yadava, 1997). Resistant sources for head smut have been identified (Jain, 2005). In a study at ZARS, Mandya, 12 varieties namely RK31, RK65-18, RK87-9, RK162, RK106, DPS486, DPS516, DPS542, DPS672, DPS700, DPS727 and ICK769 were observed free from smut incidence and showed resistance reaction to smut (Ramappa *et al.*, 2004). Some of the promising accessions against head smut are KOPKM 1, KOPKM 10 and RPS 973.

Among the other diseases, kodo millet is susceptible to ergot. Hardened masses of this fungus will grow in place of the millet grain, called sclerotia. These compact fungi growths contain a chemical compound that is poisonous to humans and livestock if consumed, and potentially fatal. It causes damage to the central nervous system, causing excitability in animals and eventually loss of muscle control. If the symptoms are caught early and the animals are removed from the infected food, they have a good chance of recovery. Cleaning the seeds by winnowing them before storage may remove the fungal spores. Farmers believe that kodo millet is poisonous after a rain. It is known to produce unconsciousness or delirium with violent tremors of the voluntary muscles. There are reports that elephants have died from eating kodo millet. To prevent poisoning, the grains are carefully removed from the glume, lemma and palea before cooking. It is possible that poisoning results from a smut that invades and eventually replaces the developing grain (de Wet *et al.*, 1983).

Kodo millet grains are also found infected with fungal pathogens in humid environment or if crop moistened before harvesting and threshing. Fungal growth in grains and fodder causes a decrease in nutritional value and also results in health hazards locally known as *Malona* in Madhya Pradesh and Chattisgarh (Pall *et al.*, 1980; Ansari and Shrivastava, 1991). It produces unconsciousness, vomiting, giddiness, difficulty in swallowing and sometimes death of human being and cattle after their use as food and feed (Jain and Singh, 2004). Kodo millet grains are frequently infested by *Aspergillus tamarii* Kita which produces cyclopiazonic acid (CPA), a mycotoxin producing acute hepatotoxicty in humans and animals (Antony *et al.*, 2003). Cyclopiazonic acid is a toxic indoletetramic acid that has been isolated from numerous species of *Aspergillus* and *Penicillium*.

Among the insect pests, like in many other small millets, shoot fly (*Atherigona simplex*) is a serious pest in kodo millet and can cause significant yield reduction in late sown crops. Early sowing with the onset of monsoon helps in escaping the shoot fly damage. At some isolated pockets *Striga* spp., a phanerogamic semi-root parasite are predominant in kodo millet (Jain and Tripathi, 2002). The kodo millet is prone to lodging at maturity, causing loss of grain (de Wet *et al.*, 1983). To prevent this, limited fertilization is recommended. While plenty of fertilizer dramatically improves yields, there is the risk of lodging accompanying vigorous growth. The basic problem of kodo millet is milling for the grains, milling drudgery associated with upper husk sticking endosperm tightly that reduces the efficiency of grain recovery from each spikelet.

Research Needs, Critical Gaps, Emerging Research Areas, Strategies and Action Plan

The crop improvement efforts have achieved only average genetic enhancement in kodo millet so far though many improved varieties have been released for cultivation. The average yield remains one-fifth to one-sixth of the yield levels observed in the advanced varietal trials of AICSMIP. The spread of high yielding varieties in different parts of the country is very much limited as organized seed production and supply is absent in most states. Varieties with specific adaptation are needed in remote tribal areas. FLDs have helped in spread of improved varieties to some extent and demonstrating their superiority. The yield levels achieved under FLDs are 65 per cent more than the farmers' practice (1628 kg/ha vs. 988 kg/ha, averaged over the period of 2000-13). The yield gaps resulting due to crop management constrains under farmers' fields in different states need to be addressed on priority.

As genetic variability is essential for any crop improvement programme, the creation and management of genetic variability becomes a focal point, more so in crops like kodo millet, in which the available genetic variability is very limited owing to complete self pollination due to its cleistogamous nature (Harinarayana, 1989). Among the approaches to create genetic variability, induced mutation is an important approach (Subramanian *et al.*, 2011). Many of the entries tested under advanced yield trials of AICSMIP have resulted from mutagenesis. Recombination breeding efforts are still in infancy because of difficulty in crossing due to flowering habit. Emasculation and crossing techniques have to be standardized for effective deployment of trait-specific germplasm available in gene banks for genetic enhancement following recombination breeding. Development of core collection and identification of trait specific donors, development of molecular markers and association mapping, identification of markers linked to agronomic traits, etc. are some of the areas that need urgent efforts. Some of the strategies and action plan for future are:

- ☆ Study of genetics of yield and yield components
- ☆ Standardization and fine tuning of emasculation and crossing techniques and recombination breeding
- ☆ Development of head smut and shoot fly tolerant varieties
- ☆ Identification of nutrient-responsive germplasm and tolerant to lodging
- ☆ Development of appropriate package of practices to realise the yield potential of improved varieties

References

1. Ansari AA and Shrivastava AK. 1991. Susceptibility of minor millets to *Aspergillus flavus* for aflatoxin production. *Indian Phytopath*. 44: 533-534

2. Antony M, Shukla Y and Janardhanan KK. 2003. Potential risk of acute hepatotoxicity of kodo poisoning due to exposure to cyclopiazonic acid. *Journal of Ethnopharmacology* 87: 211–214

3. Chandrasekara A, Naczk M and Shahidi F. 2012. Effect of processing on the antioxidant activity of millet grains. *Food Chemistry* 133: 1–9.
4. Chandrasekara A and Shahidi F. 2011a. Inhibitory activities of soluble and bound millet seed phenolics on free radicals and reactive oxygen species. *Journal of Agricultural and Food Chemistry*, 59: 428–436.
5. Clayton WD. 1975. The *Paspalum scrobiculatum* complex in Tropical Africa. *Kew Bull.* 30: 101-105.
6. de Wet JMJ, Prasada Rao KE, Mengesha MH and Brink DE. 1983. Diversity in Kodo Millet, *Paspalum scrobiculatum*. *Economic Botany* 37(2): 159-163.
7. Galinato MI, Moody K and Piggin CM. 1999. Upland rice weeds of south and southeast Asia. International Rice Research Institute.
8. Harinarayana G. 1989. Breeding and varietal improvement of Small Millets in India. In: Small millets in global agriculture. *Proc. First Int. Small millets workshop*. Bangalore. Seetharam A, Riley KW and Harinarayana G (Eds.). Oxford and IBH Publishing Co. Pvt. Ltd. New Delhi. pp. 59-70.
9. Hegde PS and Chandra TS. 2005. ESR spectroscopic study reveals higher free radical quenching potential in kodo millet (*Paspalum scrobiculatum*) compared to other millets. *Food Chemistry*, 92, 177–182.
10. Heuzé V, Tran G and Giger-Reverdin S. 2012. Scrobic (*Paspalum scrobiculatum*) forage and grain. Feedipedia.org. A programme by INRA, CIRAD, AFZ and FAO. http://www.feedipedia.org/node/401
11. Hulse JM, Laing EM and Pearson OE. 1980. *Sorghum and the Millets: their composition and nutritive value*. Academic Press, New York.
12. Jain AK and Singh RP. 2004. Current status of biotic stresses of kodo millet production in Madhya Pradesh. Proceedings of Third National Seminar on Millets research and development- Future policy options in India (Volume III: Small Millets), 2006 (Editors: KT Krishnegowda, N Seetharama, IS Khairwal, Vilas A. Tonapi, S Ravikumar, B. Jayaramegowda, KV Raghavendra Rao), March 11-12,2004. Organized by All India Coordinated Pearl Millet Improvement Project, Agricultural Research station, Mandor, Jodhpur 342304, Rajasthan, India. p.61-64.
13. Jain AK. 2005. Stable sources of resistance for head smut in kodo millet. *Indian Phytopath.* 58 (1): 117
14. Jain AK and Tripathi SK. 2002. Occurrence of *Striga* and its influence on yield of kodo millet. *Crop Res.* 23(3): 532-535.
15. Jain AK and Yadava HS. 1997. Recent approaches in disease management of Small millets. In National Seminar on *Small Millets: Current Research Trends and Future Priorities* as *Food, Feed and in Processing for Value Addition,* TNAU, Coimbatore. April 23-24, 1997.
16. Knees SG and Gupta AK. 2013. *Paspalum scrobiculatum*. The IUCN Red List of Threatened Species. Version 2014.2. <www.iucnredlist.org>. Downloaded on 30 July 2014.

17. Malathi D, Thilagavathi T and Sindhumathi G. 2012. Traditional Recipes from Kodo millet. Tamil Nadu Agricultural University, Coimbatore - 641 003. p. 14.

18. Malleshi NG and Hadimani NA. 1994. Nutritional and technological characteristics of small millets and preparation of value-added products from them. In: Riley KW, Gupta SC, Seetharam A and Mushonga JN (Eds). *Advances in Small Millets*. New York: International Science Publisher. pp. 271-287.

19. Miller A and Morris M. 2004. *Ethnoflora of Soqotra Archipelago*. Royal Botanic Garden Edinburgh, Huddersfield, Scotland. 759 pp.

20. Padulosi S, Bhag Mal, S Bala Ravi, J Gowda, KTK Gowda, G Shanthakumar, N Yenagi and M Dutta. 2009. Food Security and Climate Change: Role of Plant Genetic Resources of Minor Millets. *Indian J. Plant Genet. Resour*. 22(1): 1-16.

21. Pall BS, Jain AC and Singh SP. 1980. *DIseases of lesser millets*. JNKVV, Jabalpur, M.P., India. p. 66.

22. Ramappa HK, Ravishankar CR and Prakash P. 2004. Reaction of kodo millet and barn yard millet entries to smut disease. Proceedings of Third National Seminar on Millets research and development- Future policy options in India (Volume III: Small Millets), 2006 (Editors: KT Krishnegowda, N Seetharama, IS Khairwal, Vilas A. Tonapi, S Ravikumar, B. Jayaramegowda, KV Raghavendra Rao), March 11-12,2004. Organized by All India Coordinated Pearl Millet Improvement Project, Agricultural Research station, Mandor, Jodhpur 342304, Rajasthan, India. p. 73-74.

23. Subramanian A, Nirmalakumari A. and Veerabadhiran P. 2010. Trait based selection of superior Kodo millet (*Paspalum scrobiculatum* L.) genotypes. *Electronic Journal of Plant Breeding*, 1(4): 852-855.

24. Subramanian A, Nirmalakumari A. and Veerabadhiran P. 2011. Mutagenic Efficiency and Effectiveness in Kodomillet (*Paspalum scrobiculatum* L.). *Madras Agric. J.*, 98 (1-3): 22-25.

25. Taylor R. 2004. Causation of type 2 diabetes - the Gordian knot unravels. *N Engl J Med*. 350(7): 639-641.

26. Umesh Chandra, Thomas TA, Ranjana Nagpal and Paroda RS. 1987. Catalogue on Kodo Millet (*Paspahum scrobiculatum* L.) Germplasm. NBPGR Publication, New Delhi: 66p.

27. Viswanath S and Seetharam A. 1989. Diseases of small millets and their management in India. In: *Small millets in Global Agriculture*. Seetharam A, Riley KW and Harinarayana G (Eds). Oxford and IBH Publishing Co. New Delhi. pp. 237-253.

Millets: Ensuring Climate Resilience and Nutritional Security (2015) *Pages* **319–331**
Editors: **Vilas A. Tonapi and J.V. Patil**
Published by: **DAYA PUBLISHING HOUSE, NEW DELHI**

Chapter 9

Genetic Improvement in Proso Millet

Sunil S. Gomashe

ICAR-Indian Institute of Millets Research (IIMR)
Rajendranagar, Hyderabad – 500 030

Proso millet (*Panicum miliaceum* L.) also called as common millet, broomcorn millet, French millet, Hog millet, white millet, hersey millet in different regions of the world. It is one of the most important dry land crops. It was domesticated more than 10000 years ago in Central and Eastern Asia. By 5000 BC its cultivation extended from China to the Black Sea region of Europe (Lu *et al.*, 2009; Lawler, 2009; Hunt *et al.*, 2014). There have been various opinions on the possible origin of proso millet, like central Asia or China (Zohary and Hopf, 2000; Zhao 2006). Recently the archaeological evidences collected from Cishan Site located on the north terrace of of the Ming River in China have revealed and confirmed that the loess-soil belt of northern China is likely the location where proso millet was domesticated (Lu *et al.*, 2009).

This crop is under cultivation in China, India, former USSR, Iran, Iraq, Syria, Turkey, Afghanistan and Romania. It is grown both in the tropics and sub-tropical climates at altitudes as high as 2700 feet as a minor crop by the resource poor marginal farmers. Agronomically it is a short duration crop with very low water requirement and highly tolerant to heat and drought. This makes it highly preferable crop for extreme soil and climatic conditions where it can give reasonable yield when other crops succumb. Nutritionally it is very rich in protein, vitamins, minerals, and micro nutrients like iron, zinc, copper and manganese when compared to staple cereals. In India it called by various names in different states and languages.

Table 1

Bengali	*Gujarati*	*Hindi*	*Kannada*	*Marathi*	*Oriya*	*Punjabi*	*Tamil*	*Telugu*
Cheena	Cheno	Chena, Barri	Baragu	Vari	China Bachari bagmu	Cheena	Pani - varagu	Variga

Botany and Taxonomy

Proso millet is considered as a short-day plant and usually an erect annual, 30 to 100 cm tall with few tillers and an adventitious root system. Proso millet stems and leaves are covered with slight hairs. The leaves have short ligule with out auricles and may grow up to 30 cm long (Cobley, 1976). The plant produces a 10 to 45 cm long drooping panicle that may be open or compact (Hulse *et al.*, 1980). Proso millet ($2n$ = 36) is predominantly a self-pollinated crop, but natural cross-pollination may exceed 10 per cent (Popov, 1946). Proso millet seeds are smaller than major coarse grain cereals (Sorghum and pearl millet), generally oval in shape and about 3 mm long and 2 mm wide (Cardenas *et al.*, 1983). The seeds vary in color as white, cream, yellow, orange, red, black to brown. Cultivated types of *P. miliaceum* are commonly subdivided into five subspecies (Lyssov, 1975). These are mainly recognized as races without any taxonomic validity. Races have no ecogeographic unity.

Table 2

Race	*Inflorescence Morphology*	*Characteristics*
Miliaceum	Large, open inflorescences with sub-erect branches that are sparingly subdivided Resembles wild *P. miliaceum* in inflorescence morphology	These two races occur across the range of broomcorn millet cultivation from Eastern Europe to Japan
Patentissimum	its slender and diffused panicle branches is often difficult to distinguish from race *miliaceum*	
Contractum	Compact, drooping inflorescences	Highly evolved cultivars of broomcorn millet have more or less compact inflorescences
Compactum	Cylindrical shaped inflorescences that are essentially erect	
Ovatum	Cultivars with compact and slightly curved inflorescences that are ovate in shape	

Growth and Development

Growth and development in proso millet can be divided mainly in to three distinct phases (Cardenas *et al.*, 1983) *viz.*, vegetative, reproductive, and maturity/ripening, which may be further sub-divided into physiologically distinct stages. The period from germination to panicle initiation is the vegetative phase. Depending on the cultivar used and climate in the area may be completed 16 to 20 days after sowing. This phase is specifically characterized by an increase in number of leaves, tiller buds, and plant height. The period from panicle differentiation to flowering of the main culm (20 to 25 days) is the reproduction phase. This phase initiates when the panicle primordium is greater than 0.5 mm. Rapid elongation of stem internodes and an increase in leaf area accompanied by more tillers are recorded in this phase. The maturity/ripening phase starts at flowering and continues to the end of physiological maturity, which covers a period of 20 to 30 days. Throughout this period, the plant actively accumulates dry matter, particularly in the grains.

Nutritional Composition and Food Value

Proso millet is high in protein content (>14 per cent), rich in minerals, vitamins and other nutritional attributes which are better than commonly eaten cereals (Table 3). The protein quality is also very superior when compared with other cereals (Dikshit and Sivaraj, 2013).

Table 3: Nutrient Composition of Proso Millet and Cereals (per 100 g)

Food Gain	*Protein (g)*	*Carbohydrates (g)*	*Fat (g)*	*Crude fibre (g)*	*Mineral Matter (g)*	*Calcium (mg)*	*Phosphorous (mg)*	*Fe (mg)*
Proso millet	12.5	70.4	3.1	7.2	1.9	14	206	10.0
Wheat	11.8	71.2	1.5	1.2	1.5	41	306	5.3
Rice	6.8	78.2	0.5	0.2	0.6	45	160	0.7

Area, Production and Productivity in India

Proso millet occupies substantial area in Tamil Nadu, Karnataka, Bihar, Andhra Pradesh, Maharashtra, Madhya Pradesh and Odisha as an important constituent of dry land agriculture. In Bihar, it is cultivated throughout the year, as a catch crop before the main *Kharif*, or after the *Rabi* crop harvest. Two quick crops may be taken during the summer in March-June. In India, it is grown on 0.49 lakh ha area with production of about 0.24 lakh tones. Overall productivity is about 490 kg/ha (Anonymous, 2010). Maharashtra (63 per cent) has maximum area under proso millet followed by Bihar and Rajasthan. Tamil Nadu recorded highest productivity followed by Bihar and Odisha (Figures 1a and b).

Potential Yield Gaps

The limited number of varieties developed. These varieties have remarkably shown higher yield than the state average productivity. Yield potential of recently released proso millet variety (GPUP-8) is more than 2500 kg per ha but the present maximum state average yields are very low around 387 kg per ha (Maximum State average yield of UP). The crop and variety wise yield potential and state average yield of reveals a larger yield gap which could be exploited for making these nutritious crops more profitable. Front line demonstrations (FLDs) conducted over five years (2007-08 to 2012-13) in major growing states revealed yield gap of 60 per cent and 27 per cent in Karnataka and Tamil Nadu over control plot. Thus, suggesting wide scope for genetic improvement of this crop in India.

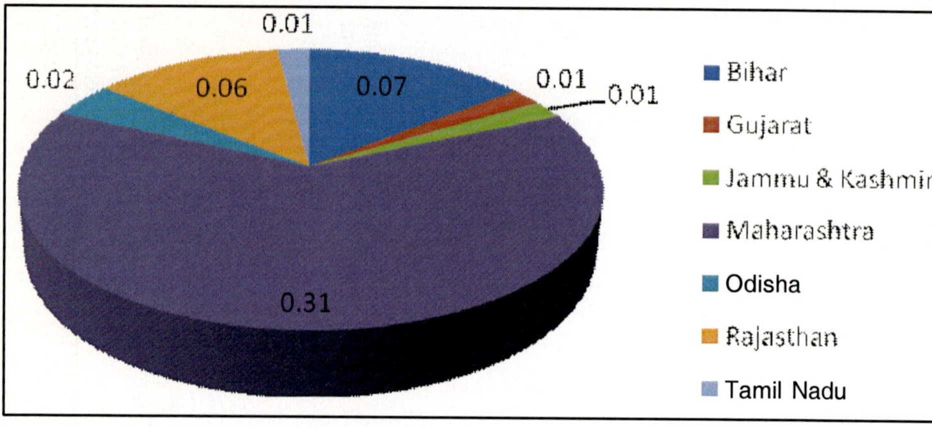

Figure 1a: Proso Millet Area (Lakh ha).

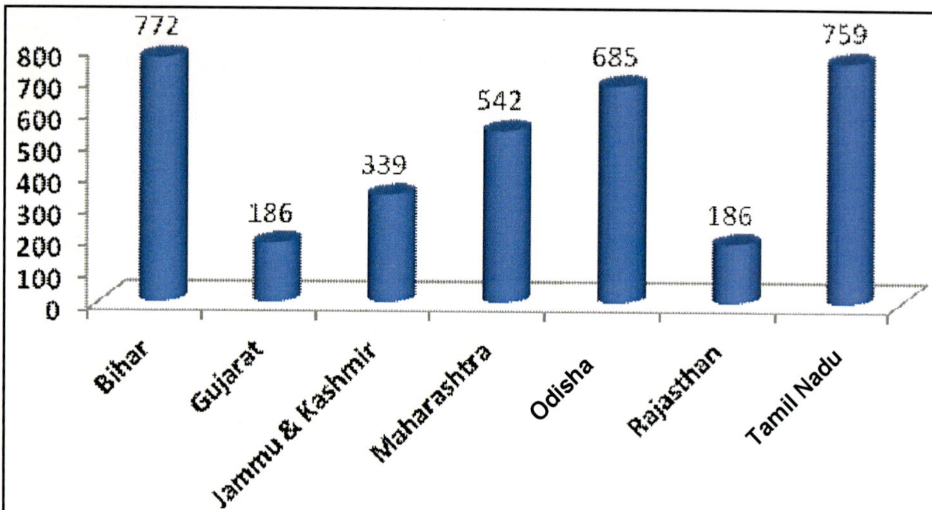

Figure 1b: Proso Millet Productivity (Kg/ha).

Genetic Improvement: Present Status

All India Coordinated Small Millets Improvement Project (AICSMIP) and SAUs have been screening the accessions/germplasm lines since inception of research for yield and other traits improvement. Few improved varieties (Table 4) of proso millet have been developed using conventional plant breeding techniques like pure line selection, pedigree breeding and back crossing.

Genetic Resources and Utilization

The utilization of germplasm in small millets as in most crops is most important. In the case of small millets, the utilization of germplasm is poor and has been drastically restricted by difficulties in artificial hybridization. Systematic use of these

Table 4: Improved Varieties Released in Proso Millet

Sl.No.	Variety	Pedigree	Developing Centre	Year	Duration	Grain Yield (kg/ha)	State	Salient Features
1.	Ram Cheena	A pure line selection	–	1960	65-95	2000-2500	Bihar	–
2.	Shyam Cheena	A pure line selection	–	1960	80	1200-1300	Bihar	–
3.	V-27	–	Andhra Pradesh	1965	–	–	Andhra Pradesh	–
4.	V-306	–	Andhra Pradesh	1965	–	–	Andhra Pradesh	–
5.	K2	A selection from PV 1685	TNAU, Coimbatore	1980	68-70	1334	Tamil Nadu	Tolerant to major pests and diseases
6.	K 1	Selection from Local	TNAU, Coimbatore	1982	70-75	900-1000	Tamil Nadu	–
7.	BR-7	Pure line selection	–	1982	80-95	2200-2500	Bihar	Moderately Resistant to *Helminthosporium* and Shoot fly
8.	CO-2	Pure line selection from a local collection from *Manjanaikanur* in *Annamalai* block Coimbatore	TNAU, Coimbatore	1985	70-75	1600	Central	Shining yellow bold grain; densely pubescent
9.	Co-3	Pure line selection from the local collection maintained at TNAU	TNAU, Coimbatore	1985	–	1280	Central	Shining golden yellow grain drought tolerant, leaf pubescent
10.	Co 4	Pure line selection	TNAU, Coimbatore	1989	70-75	1200-1500	Tamil Nadu	–
11.	Nagarjuna	Pure line selection	ANGRAU, Hyderabad	1989	60-65	1500-1800	Andhra Pradesh	Early Maturity
12.	Sagar	Selection from Local	ANGRAU, Hyderabad	1989	80-87	1800-2000	Andhra Pradesh	High seed yield
13.	Bhawna	Pure line selection	CSAUAT, Kanpur	1992	70-75	1200-1500	Uttar Pradesh	Early Maturity
14.	Bhawna	Selection from genetic collections of *Deoria* district	CSAUAT, Kanpur	1992	70-75	1200-1500	Central	Plants remain stout, erect and semi dwarf. All productive tiller come out

Contd...

Table 4–*Contd...*

Sl.No.	Variety	Pedigree	Developing Centre	Year	Duration	Grain Yield (kg/ha)	State	Salient Features
15.	GPUP 8	S 7 x L 111	UAS, Dharwad	2001	70-75	1500-1600	Karnataka	Resistant to brown spot
16.	GPUP 21	GPUP 14 x K 1	PC Unit, UAS, Bangalore	2003	67-75	1500-1800	Karnataka, Tamil Nadu	High yield, moderately tolerant to shoot fly
17.	Pratap Chenna-1(PR-18)	Pure line selection	MPUA and T, Udaipur	2006	75-85	1500-1700	Central	Recommended for cultivation in sub-marginal/shallow, undulating and hilly agro-ecological situations in Rajasthan State
18.	CO (PV) 5	Cross derivative of PV 1403 x GPUP 21	TNAU, Coimbatore	2007		2300-2400	Tamil Nadu	Resistant to brown spot and tolerant to rust and grain smut
19.	PRC-1	Selection from GPMS 519	GBPUA&T, Pantnagar	2008	90	1125	Uttarakhand	Resistant to *helminthosporium*
20.	TNAU 145	cross derivate of PV 1454 x TNAU 96	TNAU, Coimbatore	2009	88	1800-1900	Central	Tolerant to rust and shoot fly
21.	TNAU 151	TNAU 96 x PV 1673	TNAU, Coimbatore	2009	72	1800	Central	Tolerant to rust and shoot fly
22.	TNAU 164	Derivate of TNAU 137 x CO4	TNAU, Coimbatore	2010	72	1890	Central	Resistant to rust and grain smut disease incidence
23.	TNAU 202	PV 1453 x GPUP 16	TNAU, Coimbatore	2011	75	1800-2000	Central	Recommended for cultivation in dry lands of India

Table 5: Number of Varieties Developed with different Breeding Methods

Breeding Method	*Pure Line Selection*	*Bulk/Pedigree Method*	*Mutation Nreeding*
Varieties developed	17	7	–

Table 6: Proso Millet Varieties Recommended for Cultivation in India

State	*Varieties Recommended (Last 15 years)*	*Most Popular Varieties in State*
Andhra Pradesh	TNAU 151 AND TNAU 164	
Bihar	TNAU 151 AND TNAU 164	
Karnataka	GPUP 8, GPUP 21, TNAU 145, TNAU 151, TNAU 164	GPUP 8 and GPUP 21
Maharashtra	TNAU 164	
Rajasthan	Pratap chena 1	
Tamil Nadu	GPUP 21, CO (Pv) 5, TNAU 145, TNAU 151 and TNAU 164	K1, K2, CO2, CO3, CO4 and CO(Pv)5
Uttarakhand	PRC 1 AND TNAU 164	

resources for breeding can help in improving grain and fodder yield, grain quality with better composition, biotic and abiotic stress tolerance.

Table 7: Genetic Resources Identified for Various Traits

Trait	*Promising Genotypes*	*Reference*
Days to flower	IPM 40, IPM 2125, IPM 2589, IPM 84, IPM 273, IPM-982, JPM 44, IC345024	NBPGR Annual Report, 1986, 1987
Grain yield (g)	IC 41847, IPM 36, IPM 33, IC 52792, IPM 364, IC 28829, EC 24114, EC 24113	NBPGR Annual Report, 2012
Panicle length	Shai Local, Rampur Local	NBPGR Annual Report, 2011
Blast, Sheath blight, Leaf blight, Grain smut	TNAU 218, TNAU 164, DHPRMV 2769, TNAU 148	Anonymous, 2012

Grain Quality Improvement

Worldwide proso millet is gaining more attention as a health food. To develop various products using proso millet grains some novel end-use characteristics such as waxy starch, seed size, grain lustre, appearance etc. can bring better market attention. Heyduck *et al.*, 2008, evaluated 18 experimental F_5 waxy lines derived from a cross of 'Huntsman' and PI436626 across seven locations to improve the existing elite lines with introgression of waxy endosperm trait. *Chao et al.*, 2014 also examined the starch composition four waxy proso millet varieties and found that they are rich in nutritional composition when compared to non- waxy genotypes.

Protein Content and Quality

Among the small millets proso contains highest protein content. The highest identified content of protein in proso millet grains is about 17 per cent. This mainly depends on variety and growing conditions. However usual range is from 11.3 to 12.7 per cent. The dehulled grains contain 16 per cent of protein in dry matter. In an experiment involving eight varieties of proso millet for assessing protein content and quality Kalinova and Moudry (2006) highlighted the richness proso protein in essential amino acids (leucine, isoleucine, methionine) which was better than wheat. The proso grain contained about 3.3 g kg"1 of the limiting amino acid-lysine. Significant differences in protein and its quality were found among the evaluated proso varieties. Dikshit and Sivaraj (2013) evaluated eighteen diverse accessions of proso millet germplasm varying in morpho-agronomic traits. Found high variability (Coefficient of variation) for protein content, inflorescence length and days to maturity in proso millet germplasm collected from the Southern region of Ratnagiri district using DIVA-GIS grid maps.

Variability Studies and Character Associations

Variability studies give a broad picture about the differences among the available germplasm which is primary requirement for any breeding programme. The different studies have revealed high CV with regards to morphological and yield attributing traits in proso millet (Manoharan and Sivasubramanian, 1982; Natrajan *et al.*, 1978; Panwar and Kapila, 1992; Reddy *et al.*, 2007).

Salini *et al.*, 2010 evaluated 364 proso millet germplasm accessions to study genetic variability, diversity, correlation and path analysis of yield and its components. High variability observed for most of the characters. Highest variability was observed for grain yield per plant and total number of productive tillers and with high heritability. High genetic advance as per cent of mean was recorded for most of the characters. High heritability estimates in addition to high genetic advance was observed for grain yield, number of productive tillers and total number of tillers. All characters exhibited highly significant positive correlations with grain yield. Plant height, number of productive tillers and hundred grain weight exhibited positive direct effect indicating suitability of direct selection for these characters to improve the grain yield in proso millet.

Hawlader (1991) reported high genotypic and phenotypic coefficients for grain yield and total number of tillers, Chidambaram and Palanisamy (1995) for grain yield and Prasad *et al.* (1995) for total number of tillers. Medium genotypic and phenotypic coefficients of variation were observed for plant height; hundred grain weight and panicle length which indicated variation for these characters were medium among the genotypes studied. Heritability was more than 80 per cent for all the characters studied which indicated that these characters were less influenced by environmental conditions and selection would be effective on the basis of phenotype alone with equal probability of success. Sasamala *et al.*, 2011 recorded moderate to high estimates of heritability along with moderate to high genetic advance (GA) and genotypic coefficient of variation (GCV) for grain yield and all the component traits except plant height and panicle number. Significant positive association of grain yield with plant height, flag leaf length, flag leaf area, panicle length, panicle weight, panicle yield, biological yield, harvest index and 1000-grain weight was observed. Path analysis pooled over environments revealed that flag leaf length, panicle weight, 1000-grain weight and plant height had maximum contribution towards grain yield.

Research Needs and Strategies

Development and Evaluation of Core Sets for Biotic and Abiotic Stresses as well as Quality Traits

Presently the AICSMIP is holding about 939 accessions of proso millet (AICSMIP, 2012). Out of these many germplasm accessions there is need to identify the set of genotypes representing the overall diversity. The development of core collections and mini-cores will certainly boost the future breeding efforts for making notable advances in productivity. In the recent developments in this area have come up with development of core collection using geographic and morphological data (Upadhyaya *et al.*, 2011) which would provide a new avenue to access the germplasm for important agronomic traits and sources for various abiotic and biotic stresses. There is variability available for stover quality and quantity in the germplasm offering scope for the improvement of feed value of this crop.

DUS Characterization of Varieties

There are about 24 varieties released in the proso millet through national and state programmes. As this crop is gaining importance as health food the private sector will definitely come up for commercialization aspect through development of value added products thereby increasing the seed demand for product specific varieties. To protect the existing varieties these needs to be registered with the PPVFRA and for that purpose DUS testing is mandatory. DUS testing also useful for identifying novel traits in the existing cultivars and it will restrict the development of similar varieties and will give boost for identifying new varieties with distinct traits. Development of detailed descriptors for this crop is necessary for DUS testing. IBPGR *Panicum* descriptors can be of great use as base for further refining and developing new DUS testing guidelines in this crop.

Identification of Location and Trait Specific Germplasm for Utilization in Crop Improvement

Genotypes requires to be designed and developed for different maturity duration groups depending upon the location specific climatic parameters such as soil, rainfall, temperature, humidity, day length and cropping patterns. Few localities more prefer a variety with a particular trait such grain colour and lustre, increased plant height, stover quality. In such conditions preferred varieties can be developed and supplied to the niche areas. Some proso millet varieties (Table 6) perform better under specific location making them popular amongst the farmers. All varieties may not produce significantly higher yield across locations. This is expected since variety development mostly targeted for dryland production.

Development of Varieties with Abiotic and Biotic Stress Tolerance/ Resistance

Varieties with high level of tolerance to the abiotic stresses *viz.*, drought, heat, salinity, cold etc. need to be evolved to sustaining the crop yields. The varieties with such resistance level s would find the place where the present cash crop fails to sustain the profitability. Among the insects shoot fly is the only major pest reported in proso millet. Germplasm and the newly developed varieties in AICSMIP are evaluated for the shoot fly resistance. In the recent evaluation studies under AICSMIP entries DHPVMV 21-64 and DHPVMV 27-21 exhibited high resistance shoot fly over five locations in the year 2013-14 (AICSMIP, 2014). Blast, Sheath blight, leaf blight and grain smut are the important diseases recorded so far on this crop. Use of resistant germplasm to improve host plant resistance in the high yielding varieties is the key strategy to minimize losses.

Developing Lodging and Shattering Resistant Varieties

After attaining maturity ripening in proso millet starts from upper part of the panicle and it usually not uniform most of the time seeds in the upper portion of the panicle ripe while lower seeds remain green (Matz 1986). If we delay harvesting keeping in mind to dry grains sufficiently then there is possibility of considerable losses on account of shattering. In Indian conditions this is not of much concern during the rainy season crop but for post rainy and summer crop this may affect the overall yields drastically. Threshing can be delayed until the grain is below 13 per cent moisture (Baltensperger *et al.*, 1995).

Development of Varieties with Early Vigour and Short Duration

One of the most critical periods in growing small millet is the two weeks after planting. Generally small millets grow very slowly during the first few weeks and they are relatively poor weed competitors resulting in unsatisfactory yield on weedy land. There is need to identify germplasm with early vigour and highest per day productivity. Short duration genotypes will generally have high early vigour.

Identification of Elite Germplasm and Varieties with Superior Nutritional Traits and Bio-fortification of Existing Elite Lines

Human consumption of proso millet is high in Asia and it is also increasing in developed countries. The waxy proso millet genotypes are preferred in such markets because of their glutinous nature and greater properties of adhesion after cooking. Waxy forms of proso millet exist in the germplasm and these can be used to develop the waxy proso millet cultivars. In addition to different grain quality traits the increase in seed size is of great value to diversify the utilization. In addition to this trait the brightness of the endosperm which contains carotenoids and resistance to *Melanos* disease (Blackening of endosperm due to bacterial infection) are most important. The yellowish, bright and lustrous traits are directly related to the content of the carotenoids. Red coated varieties are preferred mainly because they offer protection from light degradation of carotenoids under field conditions.

Inducing Cytoplasmic Genetic Male Sterility

Availability of male sterility will offer a new gateway to exploit heterosis and thereby increasing the yield potential of the crop. As of now there are no reports of stable male sterility in proso millet. Some efforts have been made to use the chemical mutagens such as N-ethyl-*N*-nitrosourea and diethyl sulfate for producing mutagenic populations (Konstantinov *et al.*, 1991). They isolated mutants resistant to smut disease in elite Ukrainian varieties.

Wide Hybridization

The cultivated proso millet has very narrow genetic base. To increase the yield potential and resistance levels for various biotic and abiotic stresses there is necessity to introgress new genes in to the improved varieties. The wild species/wild relatives of this species can play an important role for exploiting the novel genes. The proso millet probable progenitors *Panicum capillare or Panicum repens* can be effectively used to broaden the genetic basis and identify novel adaptive genotype combinations (Hunt *et al.*, 2014).

Improved Plant Type with High Harvest Index (HI)

Plant harvest index, the ratio of grain weight to total plant weight, is an important trait associated with the dramatic increases in crop yields that have occurred in the 20th century specifically in major cereal crops wheat, rice, maize, Sorghum etc. Harvest index reflects the partitioning of photosynthates between the grains and the vegetative plant. Harvest index of modern cultivars of these intensively cultivated grain crops fall within the range of 0.40 to 0.60. The scenario in small millets is quite different and it is around 0.25 to 0.35. Thus, provides a large scope for improving the varieties of small millets to achieve HI of about 0.40.

Breeding Varieties for Intensive Cultivation and Suitable for Mechanical Harvesting

The availability of agriculture labour has decreased substantially and it is continuously declining making it very difficult to sustain farm production/ profitability with increased cultivation cost. One of the solutions is mechanization of the labour intensive farm operations. Therefore there is great necessity to develop varieties amenable to mechanical harvesting. Developing plant ideotype suitable for mechanical harvesting has many fold advantages. It would increase yield as new plant architecture will assimilate more food as grains with increased photosynthetic activity. Erect genotypes with optimum height and well arranged branching and panicle morphology is very useful for mechanical harvesting.

References

1. AICSMIP (2014). Annual report 2013-14- All India Coordinated Small Millets Improvement Project (AICSMIP), Bangaluru.
2. Anonymous (2010). Status paper on millets. Directorate of Millets Development, Department of Agriculture and Cooperation, Ministry of Agriculture, Government of India.

3. Anonymous (2014). Status paper on coarse cereals. Directorate of Millets Development, Department of Agriculture and Cooperation, Ministry of Agriculture, Government of India.

4. Baltensperger D, Lyon D, Anderson R, Holman T, Stymiest C, Shanahan J, Nelson L, DeBoer K, Hein G and Krall J (1995). Producing and marketing Proso millet in the High Plains. University of Nebraska Cooperative Extension EC 95-137-C.

5. Cardenas A, Nelson L and Neild R (1983). *Phenological stages of proso millet*. MP45. Univ. Nebraska, Lincoln.

6. Chao G, Gao J, Liu R, Wang L, Li C, Wang Y, Qu Y and Feng B (2014). Starch physico-chemical properties of waxy proso millet (*Panicum miliaceum* L.). *Starch* 66: 1005-1012.

7. Chidambaram S, Palanisamy S (1995). Dry matter production and harvest index in relation to grain yield in Panivaragu - proso millet (*Panicum miliaceum* L.). *Madras Agric J*. 82: 13-15.

8. Cobley LS (1976). *An introduction to the botany of tropical crops*. 2nd ed. Longman, London.

9. Dikshit N and Sivaraj N (2013). Diversity for Protein and Morpho-Agronomical Characteristics in Proso Millet Germplasm Collections of Ratnagiri District, Maharashtra, India. *Vegetos*. 26 (2): 164-170.

10. Hawalder MSH (1991). Genetic variability and correlation study in proso millet (*Panicum miliaceum* L.). *Ann. Bangadesh Agri.* 1(2): 61–64.

11. Heyduck RF, Baltensperger DD, Nelson LA, Graybosch RA (2008). Yield and Agronomic Traits of Waxy Proso in the Central Great Plain. *Crop Science* 48(2): 741-748.

12. Hulse JH, Laing EM and Pearson OE (1980). *Sorghum and the millets. Their composition and nutritional value*. Academic Press. New York.

13. Hunt HV, Badakshi F, Romanova O, Howe CJ, Jones MK and Pat Heslop-Harrison JS (2014). Reticulate evolution in *Panicum* (Poaceae): the origin of tetraploid broomcorn millet, *P. miliaceum*. *Journal of Experimental Botany*, Vol. 65, No. 12, pp. 3165–3175.

14. Kalinova J and Moudry J (2006). Content and Quality of Protein in Proso Millet (*Panicum miliaceum* L.). Varieties. *Plant Foods for Human Nutrition* 61: 45–49.

15. Konstantinov SI, Linnik VM, Shapina LYa and Grigorashchenko LV (1991). Breeding proso millet for resistance to diseases. In Urozhai i adaptivnyi potentsial ekologicheskoi sistemy polya. pp. 112-117 (ISBN 5-7778-0012-2).

16. Lawler A.2009. Bridging East and West: Millet on the Move. Science, 325(5943): 942-943.

17. Lu H, Zhang J, Liu K, *et al.* (2009). Earliest domestication of common millet (*Panicum miliaceum*) in East Asia extended to 10,000 years ago. *Proceedings of the National Academy of Sciences, USA,* 106: 7367–7372.

18. Manoharan V and Sivasubramanian S (1982). Variability studies in Proso millet (*Panicum miliaceum* L.). *Madras Agric J.* 69(5): 343–345.

19. Matz SA (1986). Millet, wild rice, adlay, and rice grass. pp. 225-229. In *Cereal science.* AVI Press, Westport, CT.

20. Natarajan US, Raveendran TS and Appadurai R (1978). A path coefficient analysis of yield and yield components in proso millet (*Panicum miliaceam* L.). *Madras Agric J* 65(7): 430 – 434.

21. NBPGR (1986). National Bureau of Plant Genetic Resources, India. *Annual Report.* 1986.

22. NBPGR (1987). National Bureau of Plant Genetic Resources, India. *Annual Report.* 1987.

23. NBPGR (2012). National Bureau of Plant Genetic Resources, India. *Annual Report.* 2012-13.

24. Panwar KS and Kapila RK (1992). Variation and character association in proso millet. *Crop Improv* 19: 130-133.

25. Popov GI (1946). The Importance of diversity in millet. *Agrobiologiya* 2: 28-43.

26. Prasad SG, Nagaraja TE, Seetharam A and Gowda BTS (1995). Genetic variability and character association studies in proso millet. *Crop Improv.* 22: 225–227.

27. Reddy VG, Upadhyaya HD and Gowda CLL (2007). Morphological characterization of World's proso millet germplasm collection. *SAT e-Journal.icrisat.org.* 3: 1-4.

28. Salini K, Nirmalakumari A, Muthiah AR and Senthil N (2010). Evaluation of proso millet (*Panicum miliaceum* L.). germplasm collections. *Electronic J Plant Breed* 1(4): 489-499.

29. Sasamala AC, Sahoo LP, Mahapatra KC and Pandey S(2011). Genetic variability and character association in common millet germplasm of Odisha. *Indian Journal of Plant Genetic Resources.* 24(1). : 82-86.

30. Upadhyaya HD, Sharma S, Gowda CLL, Reddy VG and Singh S (2011). Developing proso millet (*Panicum maliaceum* L.). core collection using geographic and morpho-agronomic data. *Crop and Pasture Science* 62: 383-389.

31. Zhao ZJ (2006). Domestication of millet-paleoethanobotanic data and ecological perspective. Archaeology in China and Sweden, eds Institute of Archaeology Chinese Academy of Social Sciences and the Institute of Archeology Swedish National Heritage Board (Science Press, Beijing), pp. 97-104.

32. Zohary D and Hopf M.2000. *Domestication of Plants in the Old World*, 3rd edn. Oxford University Press, Oxford.

Millets: Ensuring Climate Resilience and Nutritional Security (2015)
Editors: Vilas A. Tonapi and J.V. Patil
Published by: DAYA PUBLISHING HOUSE, NEW DELHI *Pages* 333–344

Chapter 10

Genetic Improvement in Barnyard Millet

Sunil S. Gomashe

ICAR-Indian Institute of Millets Research (IIMR)
Rajendranagar, Hyderabad – 500 030

Barnyard millet has two major cultivated species *Echinochloa frumentacea* (Indian barnyard millet, sawa millet, or billion dollar grass) and *E. utilis* (Japanese barnyard millet). This millet is widely grown as a minor cereal in India, China, Japan, Pakistan, Africa and Nepal. It is cultivated on marginal lands where rice and other crops will not grow well. In India, it is largely grown in harsh and fragile environments, with minimal use of agricultural inputs due to its remarkable ability to withstand erratic rainfall and varying weather conditions it is a regular crop up to 2,300 m msl during *Kharif* season in Uttarakhand, Tamil Nadu and form main stay of agricultural diet and cultural system of hill people (Kumar *et al.*, 2007). Like other small millets barnyard millet is also gaining attention as health food. Traditionally it was used as a substitute for rice in northern India. Traditional foods prepared from barnyard millet such as idli, dosa and muruku are very popular in southern India. Recently it is being used for developing various value added products like biscuits, sweets, vermicili, ready mixes, multi-grain atta (Sood *et al.*, 2015). As a health food it is highly suitable and recommended for diabetics, infants and pregnant women. In addition to human food it has an important place as cattle feed due to high palatability of its fodder which can be used for making hay or silage.

Table 1

Bengali	*Gujarati*	*Hindi*	*Kannada*	*Marathi*	*Oriya*	*Punjabi*	*Tamil*	*Telugu*
Shyama	Samo, Moriyo	Sanwa	Oodalu	Bhagar, Varai	Khira	Swank	Kuthiraivali	Udalu, Kodisama

Taxonomy

The majority of the species (total 35) in the genus *Echinochloa* are spread over the warmer regions of the world. Most of them are aggressive weeds of cultivated farms. Out of 35 *Echinochloa* species *E. frumentacea* Link and *E. Utilis* Ohwi and Yabuno are the cultivated species. Indian barnyard millet (*E. frumentacea* Link) revealed a simultaneous evolutionary path both in India and Africa. Its wild progenitor is the *E. colona* (L.) Link is the tropical grass popularly named Jungle rice. The exact information about the domestication is not available (Sood *et al.*, 2015). Japanese barnyard millet [*E. esculenta* (A. Braun) H. Scholz; syn *E. utilis* Ohwi et Yabuno] is the habitant of temperate region (De Wet *et al.*, 1983) in East Asian countries. Its wild ancestor is barnyard grass (*E. crus-galli* (L.) Beauv.) from which was domesticated ca. 4000 years ago in Japan (Doggett, 1989).

It is grown in India from Himalayan region in the north to Deccan plateau in the south. It is generally cultivated in hill slopes and undulating fields of hilly, tribal or backward areas, where few options exist for crop diversification. Easy cultivation, voluminous fodder and assured crop harvest have made it an indispensable crop of N. W. Himalayan region.

Table 2: Wild Species of Barnyard Millet

Wild Species	*Chromosome No.*	*Characteristics*
E. crus-galli	2n= 48	Annual tufted grass grown as fodder
E. stagnina	2n = 54	Perennial weed, grown as fodder, made into hay
E. pyramidalis	2n = 54, 72	A weed like perennial, excellent fodder and pasture
E. colona	2n = 36, 48, 54, 72	Jungle rice: a good fodder at all stages.

Source: Joel *et al.*, 2006.

Table 3: Barnyard Millet Species, Subspecies and Races

Cultivated Species	*Subspecies*	*Races*
Indian barnyard millet	*Colona*	–
[*E. colona* (L.) Link]	*Frumentacea*	*Stolonifera, intermedia, robusta and laxa*
Japanese barnyard millet	*Crus-galli*	*Crus-galli and macrocarpa*
[*E. crus-galli* (L.) Beauv]	*Utilis*	*Utilis and intermedia*

Botany

Barnyard millet is freely tillering, herbaceous annual. The culm is smooth and slender. Internodes hollow. The first formed roots are seminal roots and subsequently fibrous root system develops. Leaves are flat, glabrous or slightly hairy with ligule. Inflorescence identified as a digitate panicle, contracted or pyramidal with many spikes. Thickened with densely packed in 3-5 rows, turgid, glumes two, lower floret sterile, upper floret fertile, hermaphorodite, stamens three, ovary superior, styles two with plumose stigma. Fruit a caryopsis enclosed with in lemma and palea. Seed small, smooth, shinning and rounded at base and pointed at apex.

Nutritional Composition and Food Value

Barnyard millet is also an appropriate food for patients intolerant to gluten causing celiac disease or other forms of allergies/intolerance of wheat, as none of the millets are closely related to wheat, so they are appropriate foods for such patients. Grains of the crop are very nutritious. Its grain contains 11.2g/100g Proteins, 10.1g/ 100g crude fibre, 4.4g/100g minerals and 15.2mg/100g iron. (Anonymous, 2010)

Table 4: Nutrient Composition of Barnyard Millet and Cereals (per 100 g)

Food Gain	*Protein (g)*	*Carbohy drates (g)*	*Fat (g)*	*Crude fibre (g)*	*Mineral Matter (g)*	*Calcium (mg)*	*Phospho-rous (mg)*	*Fe (mg)*
Barnyard millet	6.2	65.5	2.2	9.8	4.4	11	280	15.0
Wheat	11.8	71.2	1.5	1.2	1.5	41	306	5.3
Rice	6.8	78.2	0.5	0.2	0.6	45	160	

Area, Production and Productivity in India

The area under small millets has been steadily sliding down during last 3 decades and in recent years the pace of decline has been much faster. These crops in recent years have been pushed further to marginal and sub marginal areas where no other crop can be grown. Barnyard millet is an important minor millet crop well adapted to low and moderate rainfall areas (500-700 mm) due to its early maturity character. It is the second most important small crop after finger millet among *Kharif* crops both in terms of acreage and production in the hill region of the Uttarakhand. Uttarakhand has maximum acreage under barnyard millet followed by Madhya Pradesh, Karnataka, Arunachal Pradesh and Nagaland (Figures 1 a and b). Uttarakhand also ranks first in productivity followed by Nagaland, Meghalaya and Arunachal Pradesh (Anonymous, 2010, 2014).

Table 6: Important Barnyard Millet Growing Districts in different States of India

State	*Important Districts*
Chhattisgarh	Dantiwada, Sarguja and Korea
Madhya Pradesh	Chhindwara, Seoni, Sidhi, Jhabua and Betul
Maharashtra	Nasik
Tamil Nadu	Madurai
Uttar Pradesh	Sonbhadra
Uttrakhand	Chamoli, Pauri Garhwal, Tehri Garhwal, Uttar Kashi and Almora

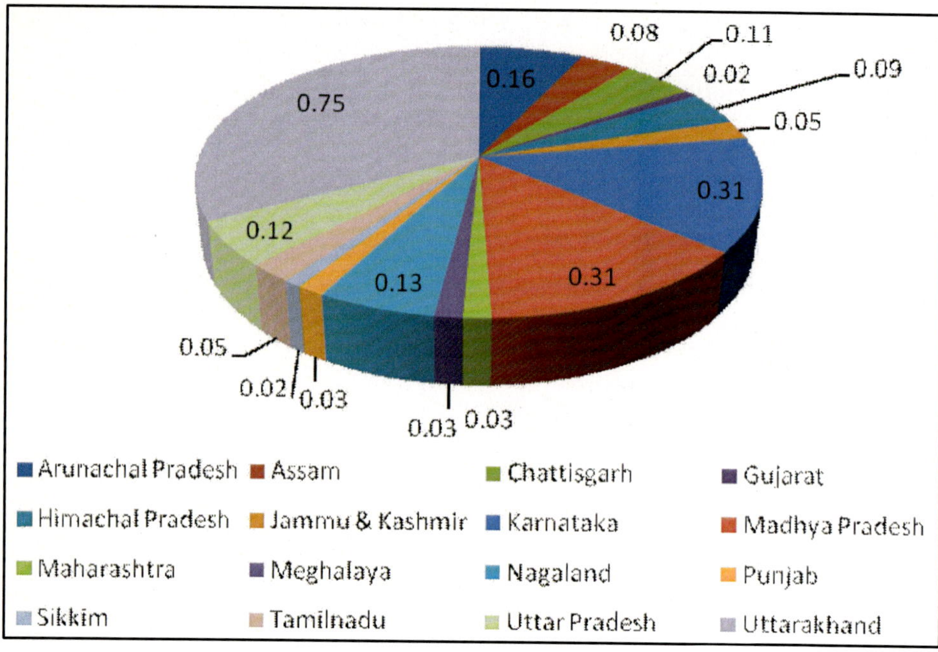

Figure 1a: Barnyard Millet Area (Lakh ha).

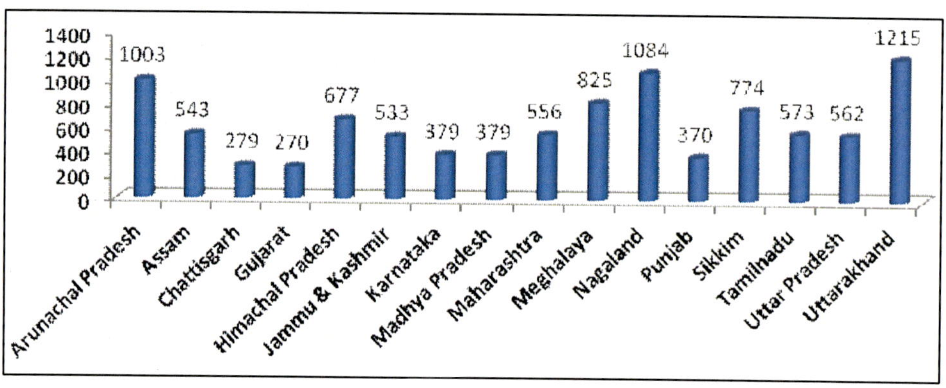

Figure 1b: Barnyard Millet Productivity (kg/ha).

Potential Yield Gaps

In India, the cultivation of barnyard millet is mainly confined to Tamil Nadu, Andhra Pradesh, Karnataka and Uttarakhand. The yield level of barnyard millet is as high as 10 t/ha in Japan, where as in India it is 1.5 to 2 t/ha. So there is a greater scope for exploiting its potential in Indian condition (Channappagoudar *et al.*, 2008). Yield potential of popular released barnyard millet variety (PRJ-1) is more than 2500 kg per ha but the present maximum state average yields are very low around 1138 kg per ha (Maximum State average yield of Uttarakhand). The crop and variety wise yield potential and state average yield of reveals a larger yield gap which could be

Table 7: Barnyard Millet Released Varieties in India

Sl.No.	*Variety*	*Pedigree*	*Institute where Developed*	*Year*	*Maturity (days)*	*Av. Yield (kg/ha)*	*Area of Adaptation*	*Salient Features*
1.	K-1	A pureline selection from Tenkasi local	TNAU	1970	90-110	1000	Tamil Nadu	Panicles loose and spear shaped. Grains dull. White in colour
2.	Kudiraivali K2 (Kovilpatty-2)	A pure line selection from a local cultivated type	TNAU	1978	90	2300-3000	Tamil Nadu	–
3.	VL Madira-8	Isolated from local collection	VPKAS, Almora	1978	90-120	1800	Uttar Pradesh	Grains light grey, smooth with white endosperm
4.	Anurag	Selection from genetic collection of Gorakhpur	ARS	1983	75	1700-2000	Uttar Pradesh	Resistant to pink borer and shoot fly
5.	Gujarat banti-1	F-2 is a selection from the segregating material of Banti	AICSIP, GAU, Surat	1984	85-90	2000	Gujarat	The grain colour of this variety is yellowish white and preferred by
6.	Rau-3	Pure line selection	IDRO-minor millets improvement project, RAU, Dholi, Muzaffarpur (Bihar)	1985	80-50	1800	Bihar	Moderately resistant to *helminthosporium frumentace*
7.	VL Madira-21	Local collection from Garhwal region of UP	VPKAS, Almora	1988	85-100	2500-2600	Central	Very low incidence of smut and leaf spot diseases
8.	VL Madira-29	Location collection (vhc 5360-1) from Kumaon region of UP	VPKAS, Almora	1988	88	2500	Central except Tamil Nadu	Very low incidence of smut disease
9.	Chandan	Selection from genetic collections of Ballia district	–	1989	75-85	1200-1600	Central	Plants remain erect with thick and strong stem synchrony in tillering spike
10.	Kanchan	Selection from germplasm	CSAUAT, Kanpur	1993	85-90	1500-1800	Uttar Pradesh	

Contd...

Table 7–*Contd...*

Sl.No.	*Variety*	*Pedigree*	*Institute where Developed*	*Year*	*Maturity (days)*	*Av. Yield (kg/ha)*	*Area of Adaptation*	*Salient Features*
11.	VL Madira 172	EF 2 x VHC 5205	VPKAS, Almora	2000	90-95	2100-2300	Uttar Pradesh, Gujarat and Karnataka	Tolerant to grain smut
12.	Sushrutha (RAU 11)	Introduction	RAU, Pusa and UAS, Dharwad	2000	90-95	2000-2200	Karnataka	
13.	VL Madira 181	ECC 27 x VL 60	VPKAS, Almora	2001	90-95	1600-1700	Bihar, Karnataka, Madhya Pradesh, Tamil Nadu	
14.	Prj-1	Selection from germplasm accession IEc 542 obtained from ICRISAT, India	GBPU&T, Uttaranchal	2003	118	2500	Uttarakhand	Resistant to grain smut disease
15.	Praptap sanwa-1 (ER-64)	Selection from local germplasm of Agro-climatic zone IVa (Udaipur)	MPUA&T, Udaipur, Rajasthan		2004	85-91	1500-1700	Rajasthan Moderately resistant to the diseases like grain smut and brown Spot.
16.	VL Madira 207	Vl Madira 172 x GECH 506	VPKAS, Almora	2008	85-100	1800-1900	All the states except Gujarat and Tamil Nadu	Tolerant to grain smut
17.	Co(KV)2	Pure line selection from a genetic accession ef 79	TNAU, Coimbatore	2009		2100		
18.	Co-1	A selection from the germplasm collection from *palladam* area in Coimbatore district	TNAU, Coimbatore	2009	105-110	1600	Tamil Nadu	

exploited for making these nutritious crops more profitable. Front line demonstrations (FLDs) conducted over five years (2007-08 to 2012-13) in major growing states revealed yield gap of 52 per cent and 27 per cent in Uttarakhand and Tamil Nadu over control plot. Thus, suggesting wide scope for genetic improvement of this crop in India.

Genetic Improvement

VPKAS, Almora in the hills of Uttarakhand is the main centre working on this crop. This centre has developed many improved cultivars resistant various biotic stresses. Barnyard millet is quite important in Uttarakhand, where it occupies some 75000 ha, almost equally divided between hills and plains. Within India, it is also cultivated in Madhya Pradesh, Maharashtra, and Tamil Nadu. Besides being an important crop very little improvement work has been carried out in barnyard millet. Most of the varieties (Tables 7 and 8) developed and released are selections from local germplasm of *E.frumentacea* and are susceptible to various diseases and are low yielding. Only in recent years some exotic introduction of barnyard millet has been evaluated and one of the successful variety PRJ-1 was released for the cultivation in Uttarakhand (Yadav *et al.*, 2010).

Table 8: Number of Varieties Developed with different Breeding Methods

Breeding Method	*Pure Line Selection*	*Bulk/Pedigree Method*	*Mutation Nreeding*
Varieties developed	14	4	–

Table 9: Varieties Recommended for Cultivation in different States

State	*Varieties Recommended (Last 15 years)*
Bihar	VL Madhira 181, VL Madhira 2007
Gujarat	VL Madhira 172
Jharkhand	VL Madhira 181, VL Madhira 2007
Karnataka	VL Madhira 172, VL Madhira 181, VL Madhira 2007
Madhya Pradesh	VL Madhira 181
Rajasthan	Pratap sanwa 1, VL Madhira 2007
Tamil Nadu	VL Madhira 181, CO1, K1, K2
Uttar Pradesh	Kanchan, VL Madhira 172
Uttarakhand	Kanchan, VL Madhira 172, VL Madhira 2007, PRJ1

Variability, Heritability and Character Association Studies

Joshi *et al.*, 2013, observed considerable variability among the evaluated germplasms *viz.*, days to maturity (73-90 days), number of productive tillers (2-15), number of leaves on the main tiller (2-18), ear length (9-21 cm), finger length (21-55 mm), number of grains/finger (26-117), grain yield/ear (0.13-18 g) and grain yield/plant (0.35- 26g). Prakash and Vanniarajan (2013) in an experiment comprising 65 barnyard germplasms of indigenous and exotic origin observed wide range of variability both at genotypic and phenotypic level with little influence of environment on the expression of the most of the agronomic traits. High heritability coupled with high genetic adavance recorded for ear head width, grain weith per ear head, single ear head weight, ear head length and grain yield per plant over the four locations. Lal and Maloo (2006) recorded that main panicle weight had maximum direct effect on seed yield followed by plant height and primaries per panicle.

Correlation and path coefficient analysis studies by Mehta *et al.*, 2007 indicated that heat units at maturity (HUM), photothermal units at 50 per cent flowering (PTUF), plant height, days to 50 per cent flowering and weight of panicle had high direct effect on grain yield and could serve as useful selection criterion during the segregating generations for developing short, high yielding genotypes. However as per Gupta *et al.*, 2009, correlation studies revealed that flag leaf width, number of racemes along with internode length should be considered while performing selection in segregating generations.

Genetic Resources and Utilization

Table 10: Trait Specific Barnyard Millet Germplasm

Trait	*Promising Genotypes*	*Reference*
Grain smut, head smut, sheath blight	HCBM 1016, HCBM 1018, HCBM 1019, HCBM 1020	Anonymous, 2012
Grain smut	TNAU 25, TNAU 63, VL 10, PRJ 1, PRB 903, DHBMV 56-6, DHBMV93-3, RBM 82, RBM 45, RBM 78, RBM 85	Anonymous, 2012 and 2013
Shoot fly	VL 207, VL 172, TNAU 151, TNAU 155, KOPBM 46	Anonymous, 2012
Days to 50 per cent flowering	IEC 336, IEC 343, IEC 344, IEC 330, IEC 217, JEC 340 and, JEC 338, IC257799	NBPGR Annual Report 1986, 87
Days to maturity	IEC I57, IEC 159, IEC 71, IEC 78, IEC 257, IEC 271, IEC 158, IEC 156, IEC 282, IC426595, IC436926, IC426594, IC257799	NBPGR Annual Report 1986, 87
Basal tillers	IC257799, IC24848, VL 29, IC52701, IC338896	NBPGR Annual Report 2011-12
Length of inflorescence	IC426595, IC97034, VL 29, IC97031	NBPGR Annual Report 2011
Width of inflorescence	IC426594	NBPGR Annual Report 2011
1000 seed weight	IC326725, KL I	NBPGR Annual Report
Yield per plant	IC97034, VL 29 IC338960, IC97031	NBPGR Annual Report
Ear head length	IC548696, IC548635, IC548697	NBPGR Annual Report
Plant height	IC548681, IC548696, IC548658	NBPGR Annual Report
100 seed weight	IC548641, IC538089, IC548697	NBPGR Annual Report

Research Needs and Strategies

Development and Evaluation of Core Collections

Core collections comprising of about 10 per cent of the available accessions represents nearly all the available diversity within the germplasm. Core collections comprising 50 (Gowda *et al.*, 2009) and 89 (Upadhyaya *et al.*, 2014) genotypes have been developed in barnyard millet. These core collections can be evaluated in different locations to identify location specific desirable genotypes and to isolate other trait specific stable genotypes which can be great vale for over all genetic improvement.

Exploitation of the Wild Relative Gene Pool

Introgression of desirable traits from wild relatives and progenitors can open new arena for genetic improvement. Inter-specific hybridization between Indian and Japanese barnyard millets can bring mutual genetic improvement of the two species. There is also need to evaluate existing variability identify genetic resources for better agronomic performance, disease resistance, nutritional qualities.

Breeding for Waxy Endosperm Genotypes

Recently demand for barnyard millet has increased due to its nutritious grains with high levels antioxidative compounds. Most of the released cultivars are non-waxy (non-glutinous) which give a poor texture to the food products. In Asian countries including India and Japan a glutinous texture as like rice is preferred so incorporating waxy trait in to the elite cultivars thereby imparting glutinous texture to the food preparations would be highly beneficial to make the barnyard popular. Fully waxy stable mutant lines have been developed using Gamma irradiation of low amylase land race 'Nogehie' (Hoshino *et al.*, 2010).

Breeding for Pest and Diseases

Grain smut is most important disease in barnyard millet. It is caused by *Ustilago panici-frumentacei* Brefeld fungus. While head smut is caused by *Ustilago crus-galli* Tracy and Earle (Nagaraja *et al.*, 2007). Phenotypic screening is the most important to identify the sources for resistance. The accessions belonging to *E. esculenta* mostly show immune reaction to this disease. The two accessions of this species PRB 9402 and PRB 9602 were observed to be immune for the grain smut (AICSMIP 2001). Among the insect pest pink stem borer (*Sesamia inferens*) and shoot fly (*Atherigona falcate*) are the major pests damaging this crop. The accessions identified can be used to develop cultivars with increased resistance levels.

Breeding for Dual Purpose Genotypes (Grain and Stover)

Barnyard millet stover is as important as grains. It is considered superior to rice and oat straw because of high protein and calcium content (Yabuno 1987). While breeding for increasing grain yield there is need to look for the high straw yield as well. Thus the genotypic selection of barnyard millet could be practiced on the basis of its performance on both grain and fodder yield while, selection of genotypes on the basis of grain yield alone might not be suitable for the promotion of dual purpose cultivars at higher elevation of hills as in this region fodder has great value. In

Uttarakhand, barnyard millet contributes 11.5 per cent of the total fodder consumption of the state (Singh and Singh 2005)

Breeding for Genotypes Suitable for Mechanical Harvesting and Post Harvest Processing

Breeding for optimum plant is required to develop genotypes for mechanical harvesting. Genotypes with plant height of about 120-130 cm could also minimize the problem of lodging will give required strength to the culm (Sood *et al.*, 2015). The reduction in the fodder yield due to reduced height can be compensated by increasing number of basal tillers. Breeding genotypes for easy dehulling is much needed to reduce the drudgery involved in the post harvest processing. Recently Gupta *et al.*, 2010 identified easy dehulling barnyard millet genotype B 29 from the germplasm maintained at Vivekananda Parvatiya Krishi Anusandhan Sansthan, Almora. Dehulling capacity of B 29 along with two other released varieties was evaluated using Vivek Thresher 1. The genotype B 29 dehulled 42–146.4 per cent faster over varieties VL *Madira* 172 and PRJ 1, respectively. It was observed that B 29 has more extended lemma after kernel embedded in it, which resulted in splitting of lemma more easily by impact force sheared by the machine.

References

1. AICSMIP (2014). Annual report 2013-14- All India Coordinated Small Millets Improvement Project (AICSMIP), Bangaluru.
2. AICSMIP (2001). Annual Report 2000–2001. All India Coordinated Small Millet Improvement Project (ICAR), Bangalore, p. 6.
3. Anonymous (2010). Status paper on millets. Directorate of Millets Development, Department of Agriculture and Coöperation, Ministry of Agriculture, Government of India.
4. Anonymous (2014). Status paper on coarse cereals. Directorate of Millets Development, Department of Agriculture and Cooperation, Ministry of Agriculture, Government of India.
5. Anonymous (2010). Annual Report 2009–2010. Vivekananda Parvatiya Krishi Anusandhan Sansthan (ICAR) Almora-263601, Uttarakhand, 23.
6. Channappagoudar BB, Hiremath SM, Biradar NR, Koti RV and Bharamagoudar TD (2008). Influence of morpho-physiological and biochemical traits on the productivity of barnyard millet. *Karnataka J. Agric. Sci.*, 20(3) : 477-480.
7. De Wet JMJ, Rao KEP, Mangesha MH, Brink DE (1983). Domestication of sawa millet (*Echinochloa colona*). *EcoBot* 37: 283–291.
8. Doggett H (1989). Small millets-a selective overview. In: Seetharam A, Riley KW and Harinarayana G (eds), *Small Millets in Global Agriculture*, pp. 3-18. Oxford and IBH, New Delhi.
9. Gupta A, Mahajan V and Gupta HS (2010). Genetic resources and varietal improvement of small millets for Indian Himalaya. In: Tewari LM, Pangtey YPS and Tewari G (eds), *Biodiversity Potentials of the Himalaya*, pp. 305-316. Gyanodaya Prakashan, Nainital, India.

10. Gupta A, Mahajan V, Gupta HS, Singh P and Bisht GS (2009). Kagazi madira (B 29) (IC568707; INGR09023), a Barnyard millet (*Echinochloa frumentacea*) germplasm, easy de-hulling type. *Indian J. Plant Genet. Resour*. 22: 281-317.

11. Gupta A, Sood S, Agrawal PK, Bhatt JC (2015). B 29: An easy dehulling Barnyard Millet (*Echinochloa frumentacea* Link) genotype. *National Academy Science Letters*. 38(1): 21-24.

12. Gupta AMahajan V, Kumar M and Gupta HS (2009). Biodiversity in the barnyard millet (*Echinochloa frumentacea* Link, Poaceae) germplasm in India. *Genet. Resour. Crop Evol*. 56, 883-889.

13. Hoshino T, Nakamura T, Seimiya Y, Kamada T, Ishikawa G, Ogasawara A, Sagawa S, Satto M, Shimizu H, Nishi M, Watanabe M, Takeda J and Takahata Y (2010). Production of a full waxy line and analysis of waxy genes in the allohexaploid crop, Japanese Barnyard millet. *Plant Breed*. 129, 349-355.

14. Joel JA, Raveedran TS, Vanniarajan C, Gopalan A (2006). *Fundamentals of crop botany*. Directorate of open and distance learning, Tamil Nadu Agricultural University (TNAU), Coimbatore. ISBN: 81-902081-9-5.

15. Joshi V (2013). Assessment of Genetic Variability and identification of genotypes fordifferent traits in Barnyard millet (*Echinochola* spp.) *International Journal of Agriculture and Food Science Technology (IJAFST)*, 4(2): 65-67.

16. Kumar J, Kumar B and Yadav VK (2007). *Small millets research* at G.B. Pant University, pp. 3-17.

17. Mehta H, Tyagi PC, and Mohapatra KP (2007). Genetic analysis of morpho-physiological and yield component traits in barnyard millet (I). *Crop Improvement* 34 (1): 82-85.

18. Nagaraja A, Kumar J, Jain AK, Narasimhudu Y, Raghuchander T, Kumar B and Hanumanthe Gowda B. (2007). Compendium of Small Millets Diseases, 80pp. All India Coordinated Small Millets Improvement Project, GKVK, Bangalore.

19. NBPGR (1986). National Bureau of Plant Genetic Resources, India. *Annual Report*. 1986.

20. NBPGR (1987). National Bureau of Plant Genetic Resources, India. *Annual Report*. 1987.

21. NBPGR (2012). National Bureau of Plant Genetic Resources, India. *Annual Report*. 2012-13.

22. Prakash R and Vanniarajan C (2013). Genetic variability for panicle characters inindigenous and exotic barnyard millet (*Echinochloa frumentacea* (Roxb.) Link) germplasm over environment. *Vegetos* 26(2): 297-306.

23. Singh H S and Singh K (2005). Status and needs of pasture and fodder management in Uttaranchal. In: J. Bisht K and Srivastava A K (eds), *Road Map for Pasture and Fodder Development in NWHR for Livestock Sustenance*, pp. 39-64. Vivekananda Parvatiya Krishi Anusandhan Sansthan (Indian Council of Agriculture Research), Almora, Uttarakhand, India.

24. Sood S, Khulbe RK, Gupta AK, Agrawal PK, Upadhyaya HD and Bhatt JC (2015). Barnyard millet- a potential food and feed crop of future. *Plant Breeding*. DOI: 10.1111/pbr.12243.

25. Upadhyaya H D, Dwivedi SL, Singh SK, Singh S, Vetriventhan M and Sharma S 2014. Forming core collections in Barnyard, Kodo, and Little Millets using morphoagronomic descriptors. *Crop Sci.* 54, 2673—2682.

26. Yabuno T 1987. Japanese barnyard millet (*Echinochloa utilis,* Poaceae) in Japan. *Econ. Bot*. 41,: 484-493.

Millets: Ensuring Climate Resilience and Nutritional Security (2015) *Pages* **345–357**
Editors: **Vilas A. Tonapi and J.V. Patil**
Published by: **DAYA PUBLISHING HOUSE, NEW DELHI**

Chapter 11

Genetic Improvement in Little Millet

K.N. Ganapathy

ICAR-Indian Institute of Millets Research (IIMR)
Rajendranagar, Hyderabad – 500 030

Little millet (*Panicum sumatrense* Roth. Ex. Roem and Schultz) is one of the important small millets indigenous to Indian subcontinent. In India, the crop is grown predominantly in Madhya Pradesh, Chhattisgarh, Tamil Nadu, Karnataka, Odisha, Andhra Pradesh, Jharkhand and Bihar. The crop is hardy crop and provides reasonable harvest even in degraded soils and unfavorable weather conditions. It is cultivated both in the tropics and sub tropics and even at an altitude of 7000 ft above mean sea level. Little millet is well known for its drought tolerance and is considered as one of the least water demanding crops. Little millet is known for highest content of crude fibre in its grains. It is also rich source of protein and fat, minerals and vitamins and requires consideration as essential food for nutritional security. The crop is reported to be grown in the eastern parts of India where it formed part of tribal agriculture. Little millet locally known by different vernacular names.

Table 1: Vernacular Names of Little Millet

English	Little millet	Marathi	Sava
Scientific name	*Panicum sumatrense*	Oriya	Suan
Gujarati	Gajrao/Kuri	Punjabi	Swank
Hindi	Kutki/Savan	Telugu	Samalu
Kannad	Same/Save	Tamil	Samai
Malayalam	Sama		

Area, Production and Productivity

Area under little millet has drastically declined from about 5 lakh hectares during 2001-02 to presently about 3- 4 lakh hectares and production of about 1 lakh tonnes. However exact figures on current area, production and productivity is not available. Madhya Pradesh occupies about 50 per cent of the area under little millet followed by other states Chhattisgarh, Tamil Nadu, Karnataka, Maharashtra, Odisha, Andhra Pradesh and Jharkhand (Table 2). Concerning productivity, it is highest in Tamil Nadu (769 kg/ha) and Karnataka (664 kg/ha) while in other states it is less than 450 kg/ha. Chhattisgarh and Gujarat recorded least productivity in range of 165 and 186 kg/ha respectively. The list of potential districts for little millet cultivation is given in Table 3.

Table 2: State-wise Average (2001-06) Area, Production and Productivity of Little Millet

Sl.No.	*State*	*Area (Lakh ha)*	*Production (Lakh tones)*	*Productivity (Kg/ha)*
1.	Andhra Pradesh	0.23	0.10	425
2.	Chhattisgarh	0.67	0.12	186
3.	Gujarat	0.02	0.00	165
4.	Jharkhand	0.24	0.11	445
5.	Karnataka	0.31	0.20	664
6.	Madhya Pradesh	1.63	0.37	229
7.	Maharashtra	0.25	0.09	371
8.	Odisha	0.25	0.10	422
9.	Tamil Nadu	0.41	0.32	769
	All India	4.01	1.41	352

Source: Status Paper on Millets, 2010.

Table 3: List of Potential Districts of Little Millet in India

Sl.No.	*State*	*Districts*
1.	Andhra Pradesh	Vishakhapatanam and Chitoor
2.	Chhattisgarh	Raipur, Rajnanadgaon, Durg, Bilaspur, Raigarh
3.	Jharkhand	Ranchi, Palamu, Garhwa and Gumla
4.	Karnataka	Belgaum, Chitradurga, Dharwad, Haveri and Tumkur
5.	Madhya Pradesh	Jabalpur, Balaghat, Chhindwara, Seoni, Mandla, Sidhi, Khargone and Betul
6.	Maharasthra	Nasik
7.	Odisha	Bolangir, Koraput, Rayagada, Mayurbhanj and Sundergarh
8.	Tamil Nadu	Vellore, Thiruvannamalai, Dharmapuri and Krishnagiri

Source: Status Paper on Millets, 2010.

Based on the yield performance data of front line demonstration (FLDs) in little millet published in Status paper on millets, 2010 there exists large yield gap over the control plot/farmer practice. The largest yield gap has been recorded in Madhya Pradesh (116 per cent as intercrop and 104 per cent as sole crop) and Odisha (110 per cent) with least gap in Karnataka (22 per cent) and Tamil Nadu (31 per cent).

Genetic Improvement in Little Millet

Little millet is a reliable fast-growing crop that is early maturing and resistant to pests and diseases and tolerant to adverse agro-climatic conditions. Grains are nutritionally rich and used for food purpose while the stover is suitable for cattle. ICRISAT's gene bank maintains 473 little millet accessions from 5 countries for use in research and development. Compared to other cereals, reports on genetics improvement in little millet is scanty. Research on genetic improvement is carried out mainly under the All India Coordinated Small millets improvement program at different SAU centres/stations such as Berhampur (Odisha), Coimbatore (TN), Dindori (MP), Jagdalpur (Chhattisgarh), Kolhapur (MH), Ranchi (Jharkhand) and Rewa (MP). AICSMIP, Bangalore is the lead center working on basic and applied aspects of little millet and other small millets. The Lead centre coordinates and monitors research work in different centres and organizes research programmes relevant to regional needs. Other activities of the centre include development of improved varieties and refine production and protection technologies, conserve, evaluate and document germplasm resources and to promote their utilization in the research programmes in different states. AICSMIP is also aimed at strengthening research on value addition and grain processing for expanding utility range by finding alterative uses jointly with Central Food Technological Research Institute, located Mysore. During the last decade, the MS Swaminathan Research Foundation (MSSRF) in partnership with several Agriculture Research Institutes and universities, NGOs, GOs and International Agencies have been conducting a participatory research work on millets, aimed at reviving, conserving and enhancing their sustainable use. The activities included establishing community Institutions for promotion of millets, strengthening local millet seed supply system, increasing yield through improving millet cultivation practices and participatory selection of better varieties

Status of Genetic Improvement

Germplasm Screening and Identification Traits/Genotypes of Economic Importance

Channappagoudar *et al.*, 2007 attempted to identify important traits influencing grain yield potential in little millet. Taller genotypes were found to be high yielding capacity and shorter genotypes are medium to low yielding. Genotypes DPI 1865 (152cm) and TNAU 63 (121 cm) was recorded as taller genotypes. Other parameters are number of leaves and tillers per plant positively contributing towards higher grain yield. Leaf area Index and leaf area duration were also important growth parameters for improving the yield and total dry matter as it indicates the efficiency of the photosynthetic system. Overall, the study reported TNAU 63 (20.2 q/ha) and DPI 1869 (18.3 q/ha) and TNAU 18 (16.9 q/ha) were high yielding with high values for most morpho-physiological parameters. Golagi *et al.*, 2013 analyzed the biophysical basis of productivity in little millet genotypes and observerd higher transpiration rate in low yielding genotypes and low in high yielding genotypes. They also observed that high yielding genotypes had higher stomatal conductance at 60 DAS which could be because of higher stomatal frequency on abaxial surface leading to enhanced canopy photosynthesis. There was a minimum interveinal distance and high vein load frequency in high yielding genotypes.

Salini *et al.*, 2014 evaluated 105 germplasm of little millets for 17 quantitative and 12 qualitative traits and selected 12 promising lines as parents and generated a total of 132 cross combinations. Gene action was studied for 11 characters and they reported non-additive gene action for all the characters except plant height where additive and non-additive played equal role. High CV was observed for grain yield per plant and number of basal tillers per plant. Heritability estimates were high for all the characters except flag leaf sheath length. Inheritance of qualitative characters indicated monogenic simple dominance inheritance for most traits except grain colour. Stability analysis was carried out for 12 parents in four environments for eight characters and identified IPmr 1046 and IPmr 889 as stable genotypes for grain yield.

M.S. Swaminathan Research Foundation (MSSRF), Chennai have taken major initiatives to collect, conserve and evaluate little millet in Tamil Nadu for improving utilization of little millet. In Kohli regions of Tamil Nadu, little millet were preferred by tribal farmers as they provided sustainable benefits. But during the last two decades active introduction of high yielding varieties of other food and commercial crops like cassava changed the focus to commercial production relegating cultivation of millets to an obscure background leading to genetic erosion of millet diversity. MSSRF attempted to bring back cultivation of the millets and also to revitalize their conservation of local landraces and cultivars. The participatory rural appraisal conducted with region farmers revealed abiding interest of the farmers in millets, but are getting eroded due to the poor productivity of landraces that are under traditional cultivation. Ideal approaches are to introduce scientific steps to optimize their cultivation practices under site specific constraints. Under a program of Swiss Development Corporation, optimization of cultivation practices was taken up under a participatory plant breeding activity. A two season operation of these initiatives has shown the following. (a) Long discontinuation of small millet cultivation has not yet destroyed the inherent genetic diversity, (b) The genetic integrity of short and long duration landraces is still maintained, (c) Even out of the seven prevalent landraces sufficient genetic diversity existed giving four distinctly different genetic groups and (d) Introduction of improved practices into farmer method of cultivation has improved the productivity up to 200 per cent at no extra cost. Arunachalam *et al.*, 2005 from MSSRF studied the ecological stability of genetic diversity among landraces of little millet in south India. Genetic divergence studies using Tocher's and Principle component analysis revealed high diversity among set of landraces studies analyzed for two sites for two seasons. Days to maturity and flowering time contributed most to the genetic differentiation. The study confirmed the sustained availability of the distinct divergence in little millet in Kolli hills of Tamil Nadu. They observed site specific expression of traits among the landraces and further suggested that breeding and selection to be executed at the site. It was expected that the high level of diversity would be preserved at kolli hills as long as the micro (including traditional cultivation practices) and macro (like edaphic and climatic factors, soil status and fertility) environments are left largely undisturbed.

Chaturvedi *et al.*, 1990 analyzed the genetic variability among 83 genotypes for days to flowering and maturity, number of basal and effective tillers, plant height, ear length, number of ears per plant, 100 grain weight, grain yield, grains per plant and per ear, biological yield and harvest index. Study revealed high GCV for days to 50 per cent flowering, grain yield, number of grains per plant and ear. High heritability with high genetic advance as percentage of mean were recorded for plant height, grain yield, number of grain per plant and ear. Positive association of grain yield with number of grains per plant and ear, ear length, plant height, number of basal and effective tillers, and days to 50 per cent flowering was observed. Grain yield exhibited high degree of co-heritability with plant height, ear length, number of grains per plant and per ear and days to 50 per cent flowering. Selection for grain yield had marked increase in plant height, ear length, 100-seed weight, number of grains per plant and per ear and days to 50 per cent flowering. In another similar study, Yadav Abinash and Srivastava, 1976 reported positive correlation of grain yield with days to maturity, number of tillers, number of panicles, length of panicle and yield of straw. Path-coefficient analysis showed that number of panicles per plant had the highest positive direct effect on grain yield, followed successively by 100-grain weight, length of panicle and days to maturity. The regression of grain yield on number of panicles per plant and 100-grain weight was significant indicating these traits may be given due improvement of improvement of grain yield in little millet. Patel *et al.*, 1990 analysed eight genotypes for grain stability during 1981-86 at Wagjao, Gujarat State and observed genotype V15-5 had above average stability in unfavorable environments. GV1, V2-2 and V2-5 were better adapted to favourable environments (Patel *et al.*, 1990)

Varietal improvement: Varietal development has received less attention as the case in other small millets. Most of the varieties released were developed through mass selection or pureline selection methods. Breeding for new varieties using hybridization techniques are limited owing to difficulties encountered in crossing due to tiny spikelets on brittle pedicels (Nirmalakumari *et al.*, 2007). Mutation breeding was used as complement to conventional breeding methods for genetic improvement of little millet. A number of high yielding varieties have been developed and released for cultivation in different little millet growing states. The improved varieties are able to meet the specific requirements of different regions. Although yield has been main criterion for development of new varieties, the varieties OLM 20 possesses drought tolerance, OLM 36 for brown spot and sheath blight, OLM 203 possesses blast and grain smut. Birsa Gundli 1, selection from local developed from BAU, Ranchi is very early and matures in 55 – 60 days with reasonable grain yield. The varieties released in different states and popular in different little millet growing states are given in Tables 4 and 5.

Table 4: List of Varieties Released and Popular in Little Millet different States

State	*Varieties Recommended (Last 15 years)*	*Most Popular Varieties in State*
Andhra Pradesh	Tarini	–
Bihar	Tarini and Kolab	–
Chhattisgarh	Kolab and Sabara	JK 1, VG 1 and TNAU 63
Gujarat	Kolab and TNAU 63	–
Jharkhand	Tarini	–
Karnataka	Tarini, Kolab and TNAU 63	TNAU 63, OLM 20, OLM 3 and OLM 203
Madhya Pradesh	Kolab, sabara and Jawahar Kutki 36	JK 8, JK 36, Co 2 and PRC 3
Odisha	Tarini, Kolab and sabara	–
Tamil Nadu	TNAU 63 and Co3	K1, Co 2, Co3, Co 4, Paiyur 1 and Paiyur 2

Source: Status Paper on Millets, 2010.

Improvement in Grain Smut Resistance

For management of grain smut in little millet, Jain and Tripathi, 2007 analyzed different approaches such as varietal differing for maturity, and chemical methods for management of grain smut (*Macalpinomyces sharmae*) in little millet. Average grain smut incidence ranged between 5.9 per cent to 31.0 per cent and reduction in grain yield ranging from 9.8 per cent to 53.5 per cent in different cultivars. Early maturing cultivars (upto 70 days) namely OLM 20, OLM 23, OLM 45, DLM 80, DLM 93, DLM 314, DLM 369, RLM 36 and K 8 were observed susceptible to grain smut and substantiated that the weather conditions prevailing during flowering and floral characteristics as reasons for high incidence. Cultivars with more than 100 days duration (Varisukdhara, RLM 13, RLM 14, OLM 203, VMLC 281, VLMC 296, OLM 40) were reported to be free from grain smut. Seed treatment with carboxin @2g/kg of seed + one foliar spray of carbendazim (0.05 per cent) as most effective for control of grain smut.

Nutritional Improvement

Nambi *et al.*, 2012 from MSSRF studied the proximate and mineral composition among 18 little millet genotypes collected from millet growing regions of Tamil Nadu (Pachamalai hills, Javadu hills, Yelagiri hills, Megamalai hills and Dharmapuri dist). The accessions varied for seed colour. Analysis was carried out with advanced quantification techniques such as Inductively Coupled Plasma (ICP) spectroscopic method. Among proximate compositions, protein content ranged between 6.36 to 9.15 g/100g, carbohydrate (61.48 to 68.57 g/100g), ash (3.15 to 9.07 g/100g), fibre (5.54 to 7.84g/100g), fat (3.61 to 5.56 g/100g). Among mineral contents potassium ranged between 199.2 to 3.3.4 mg/100g, phosphorous 184.3 to 3.8.9 mg/100g, magnesium 131.3 to 186.6 mg/100g, iron content 7.84 to 90.8 mg/100g, calcium 22.6 to 56.4mg/100g and zinc 1.52 to 2.14 mg/100g. Overall analysis of the mineral composition showed the minerals contained in the following order

Table 5: List of Improved Varieties and their Characteristics

Sl.No.	Variety Name	Central/ State	Date of Notification	Parentage	Adaptation	Maturity Duration (Days)	Pests and Disease Reaction	Grain Yield (Kg/ha)
1.	Saura (OLM-208)	Central	2010	Selection from Lajigada local	Recommended for release and cultivation in the states of Odisha, Chhattisgarh and Gujarat under rainfed upland conditions	71-137	Resistant to Blast and moderately resistant to rust, grain smut, sheath blight and shoot fly	1413-3407
2.	Co 4	Tamil Nadu	2007	Co2 x MS 1684	Low rainfall and low fertile soils in the marginal and sub-marginal, Rainfed dry lands, hill and tribal areas of Tamil Nadu	75-80	–	1500
3.	Sabara (OLM-20)	Central	2004	Mutant of SS 81-1	Recommended for release and cultivation for *Kharif* season in Uplands of Odisha, Madhya Pradesh and Chattishgarh	46-77	Moderately resistant to shoot fly. Moderately susceptible to grain smut; moderately resistant to Sheath blight; brown spot.	1196-2100
4.	Kolab (olm-36)	Central	2001	Mutant of SS81-1	Recommended for cultivation in MP, Odisha, Chattishgarh, Bihar, Karnataka and Gujarat under rainfed upland condition	77	Moderate resistant to Grain smut and Shootfly, Resistant to brown spot and shoot fly	1091
5.	Tarini (OLM-203)	Central	2001	Selection from a local cultivar (kl-2) of Koraput district	Recommended for release and cultivation under rainfed upload conditions in the states of Karnataka, Andhra Pradesh, Odisha, Bihar and Tamil Nadu	105-115	Resistant to blast and grain smut and moderately resistant to brown spot and shoofly	–

Contd...

Table 5–*Contd...*

Sl.No.	Variety Name	Central/ State	Date of Notification	Parentage	Adaptation	Maturity Duration (Days)	Pests and Disease Reaction	Grain Yield (Kg/ha)
6.	Paiyur-2	Tamil Nadu	2000	Pure line selection from the germplasm accession pm 295	Recommended for cultivation in marginal and sub marginal dry land of Tamil Nadu	80-85	Lesser incidence of Grain smut as compared to Co2, Co3 and Paiyur-1	744-800
7.	Co-3	Tamil nadu	1997	Selection from germplasm	Recommended for cultivation in rainfed conditions in Tamil Nadu	80-85	Tolerant to major diseases and pests.	1066
8.	TNAU-63	Central	1997	Single plant selection from germplasm MS 2369	Low rain fall and marginal soil areas of Karnataka, Gujarat and Tamil Nadu	69-104	Shoot fly incidence (8 per cent), aphids (14 per cent), brown spot (0.5 per cent grade), Smut Incidence (3.4 per cent)	1152
9.	Birsa gundli-1	Central	1993	Selection from local germplasm	Upland, rainfed situation of plateau region of Bihar	55-60	–	700-800
10.	Paiyur-1	Central	1989	Pureline selection from Sathur local	Tropics of South India		–	873
11.	Jawahar Kutki-2	Central	1987	Selection from local material named Gariya-band	–	75-80	–	700-1000
12.	Jawahar Kutki-8	Central	1987	Pedigree - single plant selection from germplasm no.8.	–	80	Moderately resistant to shootfly	

Contd...

Table 5–*Contd...*

Sl.No.	Variety Name	Central/ State	Date of Notification	Parentage	Adaptation	Maturity Duration (Days)	Pests and Disease Reaction	Grain Yield (Kg/ha)
13.	PRC-3	Central	1986	Single plant selection from GPM 97	Recommended for release for cultivation in MP, Odisha, Bihar, Maharashtra and Karnataka	61-93	Moderately resistant to shootfly	–
14.	Gujarat Vari-1	Gujarat	1985	A selection from local strain vari 8-3	–		–	1500
15.	Gariya Band (Dindori-2)	Madhya Pradesh	1984	–	–	75-80	–	700-1000
16.	K-1	Tamil Nadu	1982	Developed from culture no. Pm-368	–	90 days	–	1285
17.	Co-2	Central	1978	A pure line selection from a local (pm-296) collection of AP	–	80-85 days	Fairly tolerant to pests under field conditions	
18.	Dindori-1	Madhya Pradesh	1973	Mutant of no. 50-1	Heavy rainfall tract of Maharashtra state especially Konkan Region	125-130	No serious pest or disease observed	1500-2000
19.	Co1	Tamil Nadu	1956	Pureline selection from Anantpur selection	–	90-100	The incidence of shoot fly is similar to that of other varieties	400-500

Contd...

Table 5–*Contd...*

Sl.No.	*Variety Name*	*Central/ State*	*Date of Notification*	*Parentage*	*Adaptation*	*Maturity Duration (Days)*	*Pests and Disease Reaction*	*Grain Yield (Kg/ha)*
20.	Jawahar Kutki-36 (2009)	Madhya Pradesh		Selection from local germplasm no. 36	Suitable for sole and inter-cropping, lodging medium, resistant to drought, responsive to NPK application, seed rate of 10 kg/ha	76	Tolerant to grain smut and shoofly	4600
21.	V-15	Central		Pure line selection	Popular in south Bihar and parts of Odisha	78-80	Moderately resistant to diseases	1400-1600
22.	V-17	Central		Pureline selection	Popular in south Bihar and parts of Odisha	78-80	Moderately resistant to disease	1400-1600

Source: AICSMIP, small millets, Bengaluru and http://smis.dacnet.nic.in/

Potassium>Phosphorous> Magnesium>Iron>Calcium>Zinc. Overall high variability was recorded for all the proximate and mineral contents analyzed indicated its utilization in breeding programs.

Mutational Approaches for Genetic Improvement

Mutation breeding has been widely used for creation of new variability in millets including little millet. Most of the varieties released were developed through mass selection or pureline selection methods. Breeding for new varieties using hybridization techniques are limited owing to difficulties encountered in crossing due to tiny spikelets on brittle pedicels (Nirmalakumari *et al.*, 2007). Therefore mutation breeding was used as complement to conventional breeding methods for genetic improvement of little millet. Nirmalakumari *et al.*, 2007 carried out induced mutagenesis in CO3 and CO4 cultivars using gamma rays. They recorded maximum variability for plant height and total number of tillers in both the cultivars. High heritability with high genetic advance was recorded for total number of tillers, lodging and grain yield in the M2 generation. Mutants in 500 and 600 Gray exhibited maximum variability, heritability and genetic advance for plant height, total number of tillers, number of nodes, inter nodal length, culm thickness and grain yield indicating efficiency of selection for these characters for development of improved genotypes. Nirmalakumari *et al.*, 2010 evaluated 109 germplasm of little millet and observed wide variability, heritability and genetic advance for grain yield, productive tillers per plant, total number of tillers per plant, flag leaf length and flag leaf sheath length indicating additive gene effects for these characters. The study also observed strong correlation of most of the component characters with grain yield indicating possibility of simultaneous improvement of these characters by selection. Ganapathy *et al.*, 2008 studied the mutagenic effectiveness and efficiency in two little millet varieties Co-3 and Co 4 using Gamma rays. The mutagen dose of 500 Gray registered highest mutation rate both on M_1 panicle family basis and on M_2 seedlings basis in both varieties. They also observed that the effectiveness decreased with the increase in dose of the mutagen in both genotypes.

Strategies for Genetic Improvement

1. Systematic evaluation and unlocking the genetic potential in the germplasm should be the primary objective for genetic improvement.
2. Grain yield has always been an important trait for genetic improvement. Efforts towards improvement for yield enhancement have led to marginal improvement. In addition to yield per se, selection for component traits such as compact panicle types and large seed size should be done.
3. Although millets are grown on poor soils under dry conditions, genotypes/ cultivars responsive to high input conditions should also be identified.
4. Varieties with drought tolerance better regenerative capacity on reversal of dry spell for harsh conditions drought prone areas should be of focus.
5. Improved cultivars possessing high nutritional value would make a great way in increasing the demand for the crop.

6. Pests and diseases infecting little millet are meager. However earlier studies have indicated shoot-fly and grain smut causing economic losses. Host plant resistance should be the tried for development of the disease/insect resistance.
7. Stem lodging and grain shattering are the other important agronomic traits which should receive due attention while improving for grain yield.
8. Breeding for specific end-uses should receive priority as this attracts the private sector industries to invest in millets.
9. Little millet is known to mature in 60- 80 days. However early maturing genotypes with substantial high yield and photo-insensitive genotypes suiting different cropping systems should be developed.

References

1. Abinash Y and Srivastava, DP (1976). Path analysis in *Panicum miliare. Indian J Genet* 36 (1) 64-68.
2. Arunachalam, V., Rengalakshmi, R. and Kubera Raj, M. (2005). Ecological stability of genetic diversity among landraces of little millet (*Panicum sumatrense*) in south India. *Genet Res Crop Evol* 52: 15- 19.
3. Channappagoudar BB, Hiremath N SM, Biradar R, Koti, RV and D Bharamagoudar T (2007. Morpho-physiological traits influencing the grain yield potential in little millet. *Karnataka J Agric Sci* 20(3): 473-476.
4. Chaturvedi VK, Mishra Y, Singh CB and Kharche SK (1990). Genetic variability, correlated response and association analysis in little millet (*Panicum miliare* L.). *Mysore J Agric Sci* 24: 163-167.
5. Ganapathy S, Nirmalakumari A, Senthil N, Souframanien J and Raveendran TS (2008). Isolation of macromutations and mutagenic effectiveness and efficiency in Little Millet varieties. *World J Agric Sci* 4: 483-486.
6. Gollagi SG, Chetti MB and Ashvathama VH (2013). Biophysical basis of productivity in little millet (*Panicum miliare* L). *Int J Agric Sci* 9(1) 117-120.
7. Jain AK and Tripathi SK (2007). Management of grain smut (*Macarpinomyces sharmae*) in little millet. *Indian Phytopath* 60(4): 467-471.
8. Nambi VA, Eganathan P, Philip PI Maria (2012). Proximate and Mineral Composition Analysis of Little Millet Collected from Some Millet Growing Areas in Tamil Nadu. *Indian J Plant Genet Res* 25 (2): 189-191.
9. Nirmalakumari A, Arulselvi S, Ganapathy S, Souframanian J, Senthil N and Devan P (2007). Gamma ray induced variation for lodging resistance and its associated characters in littlemillet (*Panicum sumatrense* Roth Ex-roem and schult). *Madras Agric J* 94 (7-12): 151-155.
10. Nirmalakumari A, Salini K and Veerabadhiran P (2010). Morphological Characterization and Evaluation of Little millet (*Panicum sumatrense* Roth. ex. Roem. and Schultz.) Germplasm. *Electronic Journal of Plant Breeding* 1(2): 148-155.

11. Patel RH, Desai KB, Desai MS and Modi ND (1990). Study on stability analysis in little millet (*Panicum miliare* L.). *Madras Agric J* 77 (5-6) 252-254.

12. Salini K, Nirmala Kumari A and Senthil N (2014). Genetic analysis of yield and its components in Little Millet (*Panicum sumatrense* Roth ex Roem. and Schultz). In : *2nd International Conference on Agricultural and Horticultural Sciences* Hyderabad, India during February 03-05, 2014 pp. 177.

Part II
Physiology and Agronomy

Millets: Ensuring Climate Resilience and Nutritional Security (2015)
Editors: Vilas A. Tonapi and J.V. Patil
Published by: DAYA PUBLISHING HOUSE, NEW DELHI

Pages 361–376

Chapter 12
Physiology of Pearl Millet

S.S. Rao and J.V. Patil
ICAR-Indian Institute of Millets Research (IIMR)
Rajendranagar, Hyderabad – 500 030

Pearl millet (*Pennisetum gluacum* (L.) R. Br.) is a C_4 tropical grass adapted to arid and semi-arid tropical regions of the world under harsh climatic and edaphic conditions, where other crops such as sorghum and maize usually fails to produce economic yields (Yadav and Rai 2013). In India, pearl millet is entirely grown as rainfed crop for food, feed and fodder on sandy often shallow soils in the states of Rajasthan, Gujarat, Haryana, Maharashtra, Uttar Pradesh, Tamil Nadu, and Andhra Pradesh over 9.0 million ha with a production of 9.3 million tonnes with an average grain yield of about 1106 kg ha^{-1}, while, 60 per cent of its area is confined to Rajasthan alone (Yadav and Rai, 2013). Regions growing pearl millet have little option to choose high value crops on account of marginal lands and short growing period.

Yield growth in the last decades was relatively more in pearl millet as compared to other crops. Globally Pearl millet is grown on 26.0 million ha and ranks sixth in global cereal hectare (Rai *et al.*, 1999). The pearl millet area did not decline in the most arid regions of northwestern India, which receives less than 400 mm rainfall (accounts for 30 per cent of the pearl millet area in India (A1 zone) as its grain serves as a major staple and provide food and fodder security to the smallholder farmers in rural households. This could be due to its relatively more tolerance to abiotic stresses particularly drought and high temperatures compared to other cereal grains

Soils and Agro-climatic Environment

Pearl millet growing regions of northern and peninsular India receiving annual rainfall from <400 to 600 mm year $^{-1}$ (AICPMIP, 2014). These zones where pearl millet is grown are generally characterized by short rainy seasons (2-4 mo), high mean temperatures(>35°), high potential evapotranspiration rates, and shallow, sandy soils (Kowal and Kassam 1978). Inter- and intra seasonal variability in available soil moisture is the major hazard to pearl millet production. Rainfall distribution is highly erratic as well as low with soils having poor water-holding capacity (Bidinger *et al.*, 1982). The pearl millet growing areas (eg. Jodhpur) receives lower mean weekly rainfall than mean potential evapotranspiration and the probability of receiving at least 20 mm of rainfall/week during the rainy season does not exceed 50 per cent (Bidinger *et al.*, 1982). Pearl millet has been found to be far more tolerant to acidic soils (pH < 5.0) than sorghum (Clark *et al.*, 1990; Flores *et al.*, 1991).

Comparison of Pearl Millet in Relation to other Crops

The area and production of major cereals and millets for the world and India are presented in Table 1 (Prasad and Venkateswarlu, 2013).

Table 1: Area and Production Statistics of major Cereals and Millets

Crop	*World (million ha/tonnes)*		*India (million ha/tonnes)*	
	Area	*Production*	*Area*	*Production*
Rice	163	718	42	152
Maize	176	875	8.4	21
Wheat	216	674	29	94
Sorghum	37	58	6.3	6.0
All millets	33	25	11.5	13.2
Pearl millet	–	–	9.0	9.3
Finger millet	–	–	1.8	2.0

However, the productivity of most pearl millet is very low (0.6 to 1.4 t ha^{-1}). The main reason for low productivity is due to absence of improved inbred/varieties/composites/hybrids, stress tolerant genotypes and improved crop technologies. In addition, pearl millet is often grown on marginal and unfertile soils with minimum inputs. Murty *et al.* (2007) quantified that total yield gap (simulated rainfed potential yield - farmers' yield) in production zones ranged from 680 to 1040 kg ha^{-1} for pearl millet suggesting that yield can be increased by 1.8 to 2.3 times from their current levels of productivity.

The crop in this region is continuously subjected to recurrent pre-and postflowering drought, and heat stresses besides downy mildew (DM) disease threats, and there is as need to utilize both conventional and molecular tools to overcome these production problems. Efforts of AICPMIP resulted in enhancing the pearl millet productivity in India from a 530 kg/ha in 1988 to 1106 kg/ha in 2013 (Yadav and Rai, 2013).

Rationale

The pearl millet crop is primarily grown during rainy season. The crop is currently extending to summer season in northwestern India under intensive cropping cultivation and grain yield of 4 to 5 t ha^{-1} and stover yield of 10 to 15 t ha^{-1} can be realized with available hybrids in 80 to 85 days. High temperature (often reaches > 42°C) coinciding with flowering, and hence the crop suffers reproductive sterility leading to drastic reductions in grain yield leading to poor seed set. Thus, there is urgent need to identify heat tolerant hybrid parents to increase hybrid options for farmers. For pearl millet to be competitive, further research should be focused on yield improvement, tolerance to multiple stresses, value added products and nutritional quality of grain and forage.

In the advent of climate change, it is predicted that several regions in the semi-arid tropics will experience longer dry-spells and warmer temperatures. These changes will provide niche areas where pearl millet will be more productive than other cereals. Besides, where limited water is available, with one or two supplemental irrigations, pearl millet can perform better than sorghum and maize in regions with high temperatures, sandy soils with low soil fertility.

Production Constraints and Research Gaps

- Low, and unpredictable rainfall and growing crop on sandy shallow soils.
- Low native soil fertility with limited use of applied nutrients
- Erratic rainfall in the production environment resulting in early, mid-season and terminal drought stress.
- Lack of high temperature (>42°C) or heat stress tolerance at seedling and flowering stages.
- Lack of pearl millet genotypes that tolerate drought and high temperature stresses occurring during seedling establishment and early vegetative phase
- Lack of effective phenotying or screening techniques for selecting for drought and heat tolerance in the different soil types and across locations.

Review of Status of Research on the Pearl Millet Crop Physiology Research

Growth, Botany and Biology

Pearl millet is a robust annual grass usually 1.5-3.0 m tall, but it can grow up to 5.0 m. The plant produce 2-3 tillers depending on the environmental conditions include primary, basal, secondary, or nodal. Leaves are usually 90-100 cm long and about 5-8 cm wide. The growth and development of pear millet was described by Prasad Maiti and Bidinger (1981).

Bhatnagar *et al.* (1998) reviewed the Indian research progress on pearl millet which includes cultivar development, cytoplasmic diversification, crop management and identification of sources for both biotic and abiotic stresses. Yadav (2011) showed that in peal millet genotypes, both early flowering and high tillering had a significant advantage over late flowering and low tillering material and selecting materials that produce smaller but many panicles is much more important than producing a few larger panicles under drought conditions. The selection criteria should aimed at developing both high biomass and harvest index for improving both grain and stover yields in pearl millet targeted in stressful conditions. Bidinger and Rao (1990) reviewed the progress of genetic and cultural improvement in the production of pearl millet and proposed new research avenues for increased yield and sustainability under stress conditions in India and sub-Saharan Africa.

Effect of Drought Stress on Phenology, Growth, Biomass and Grain Productivity and Relative Tolerance

Phenology, Biomass Accumulation, Components of Grain and Stover Yields

Though pearl millet is a drought tolerant crop, the yield is decreased if the drought stress occurs during reproductive stages (panicle development, and flowering). Genetic improvement for drought stress tolerance in pearl millet is focused during flowering and grain-filling stages because stress during these stages decreases number of grains and individual grain weight leading to lower grain yield. However, drought stress early in the season (seedling stage) and during vegetative growth can decreases leaf area, reduce tillering, plant height and biomass biomass production leading to decreased yields (Prasad and Venkateswarlu, 2013).

Blum and Sullivan (1986) indicated that land-races of sorghum pearl millet which evolved along geographical gradients of rainfall in Africa and India, differ in their drought resistance. Drought resistance, in terms of relatively less growth inhibition under stress, was higher in races from dry regions than in races from humid regions. The results showed that the general trend for greater drought resistance in sorghum than in pearl millet, indicating that the commonly observed adaptation of the pearl millet to dry environments may be due to other factors, such as drought escape or heat tolerance. Mahalakshmi *et al.* (1987) reported that in pearl millet when stress was imposed late in crop growth and not relieved, the time of initiation of stress was directly related to the extent of loss in grain yield. Water stress relieved at anthesis or early grain filling had little adverse effect on the grain yield because additional tillers produced panicles.

Bidinger *et al.* (1987b) reported that pearl millet yield potential was generally positively related to yield under stress, but accounted for a much smaller fraction of yield variation than time to flowering, while these two factors together accounted for more than 50 per cent of the total variation in grain yield under stress, suggesting that genotype drought response (drought resistance/susceptibility) was a secondary rather than a primary factor in differences in yield under stress among the pearl millet genotypes field tested. Sivakumar *et al.* (1987) and Bidinger *et al.* (1987b) have shown that 50 per cent of the variation in grain yield of pearl millet lines under drought stress treatments could be attributed to variation in yield potential and time of flowering that was used to develop a drought-response index (DRI) based on the residual variation in grain yield, adjusted for experimental error. DRI was positively correlated to measured yield in the drought treatments, and independent of both yield potential and time to flowering. They suggested the use of DRI as a component of breeding for better adaptation to stress conditions in pearl millet.

Matching the phenology of the crop to the expected water supply is the most important drought adaptation trait under drought situation. Traits that enhance leaf survival are more important in intermittent than in terminal stress, and more important in subsistence than in modern agriculture (Ludlow and Muchow, 1990). Postflowering (terminal or end of the season) drought stress is of greater significance and is much better understood than preflowering drought stress in pearl millet. Crop growth and grain production that occurs during the postflowering period in pearl millet was shown to be the most sensitive to water deficits (Mahalakshmi and Bidinger 1985), while terminal drought had reduced grain yield by 45–49 per cent in pearl millet genotypes (Fussell *et al.*, 1991).

Panicle threshing percentage or panicle harvest index (PNHI) under stress conditions was proposed as a good indicator for terminal drought tolerance in pearl millet (Bidinger *et al.*, 2000). In conjunction with grain yield potential and earliness, PNHI accounts for >80 per cent of the genotypic variation in grain yield under terminal drought stress (Bidinger *et al.*, 2000). van Oosterom *et al.* (1995) reported that pearl millet grain yield potential was related to grain yield under stress when pre-flowering drought was combined with low temperatures. They showed that escape was the predominant factor if temperature was high, except if combined with pre-flowering drought, in which case tolerance became more important. van Oosterom *et al.* (1996) compared cumulative rainfall with three different possible outputs derived from a simple soil water budget analysis (Frere and Popov, 1979) as descriptors of the water resource availability for a pearl millet multi-environment (MET) trials covering 24 different environments, many of which were affected by drought stress. Evaluation of pearl millet landrace germplasm for drought tolerance traits revealed that drought response index (-8.5 to 13.9) was accounted for 73 per cent of variation in grain yield obtained under stress conditions, while higher grain number per panicle and harvest index were associated with drought tolerance (Yadav *et al.*, 2003).

Molecular Markers and QTLs Linked to Drought Tolerance in Pearl Millet

Yadav *et al.* (1999) identified flanking molecular markers for QTLs associated with increased harvest index, improved grain-fill rate, and maintenance of grain yield under terminal drought stress in pearl millet from an RIL population (H77/833-2 and PRLT 2/89-33) differing in their drought response. Yadav *et al.* (2001) too evaluated various selection indices for drought stress and non-stress environment and found that earliness was useful under stress conditions. Previous work showed that drought indices such as stress susceptibility and drought response were highly useful in identifying the pearl millet cultivars performing well under drought conditions especially when days to flowering was differing considerably (Yadav *et al.*, 2001).

Investigation on pearl millet RILs population for grain and stover yield associated with terminal drought tolerance revealed that a QTL associated with grain yield *per se* and for the drought tolerance of grain yield mapped on linkage group 2 and explained up to 23 per cent of the phenotypic variation, while, some QTLs were common across stress environments and others were specific to only a particular stress environment (Yadav *et al.*, 2002). Yadav *et al.* (2004) reported that genomic regions associated with grain yield and post-flowering drought tolerance in pearl millet were harvest index and panicle harvest index that co-mapped with grain yield suggesting that increased drought tolerance and yield of pearl millet that mapped to these regions was realized due to increased partitioning of stem reserves to the grains. Serraj *et al.* (2005) reported the use QTLs as possible selection criteria for improving stress tolerance of an elite hybrid cultivar and found that several QTL introgression lines had a significant positive general combining ability for grain yield under terminal stress and significantly out-yielded testcross hybrids made with the original recurrent parent both in unrelieved terminal drought and in gradient stress conditions.

Bidinger *et al.* (2005) reported that field evaluation of drought tolerance QTL effects on phenotype and adaptation in pearl millet, the QTL-based hybrids were significantly, but modestly, higher yielding in a series of both absolute and partial terminal stress environments. They concluded the effectiveness of the putative drought tolerance QTL on LG 2, but suggested that it may enhance drought tolerance by favoring a particular phenotype with adaptation to terminal stress. Kholová *et al.* (2012) reported that in pearl millet, two H77/833-2 and one PRLT2/89-33 alleles co-mapped to a previously identified major terminal drought tolerance quantitative trait locus (QTL), although in a much smaller genetic interval. They suggest that different alleles influencing plant water use and have close interactions with one another and with the environment, so that different ideotypes for plant water use exist or could be designed from specific allele combinations conferring particular physiological characteristics for specific adaptation to a range of terminal drought conditions.

Pearl Millet Ideal Plant Type for Increased Drought and Heat Stress Tolerance

Gregory (1982) suggested that pearl millet grain yield could be increased i) by increasing the size of the root system to exploit deeper water reserves, ii) by managing the existing water reserves more efficiently, and iii) by increasing the quantity of assimilates moved to fill grain. van Oosterom (2003) proposed yield architecture model for pearl millet landraces include small size of the main shoot panicle, as it minimized (1) the loss of productive tillers during stem elongation; (2) the delay in anthesis if mid-season drought occurs; and (3) the reduction in panicle productivity of the basal tillers under stress. On the other hand, low-tillering varieties are suited to end-of-season drought stress, provided anthesis is early. In a path analysis study, Maman *et al.* (2004) showed that number of kernels per panicle and kernel weight were significantly associated with grain yield under dryland temperate conditions.

Rai *et al.* (2006) reviewed the progress of pearl millet research on yield improvement, biotic and abiotic stress tolerance at ICRISAT and summarized the strategies and approaches that need to be followed to enhance the productivity. They emphasized the importance of developing pearl millet cultivars that set seeds under air temperatures exceeding 46°C during flowering and need to identify heat-tolerant hybrid parents especially to promote summer cultivation. Haussmann *et al.* (2012) indicated that breeding strategy should aimed at traits that potentially enhance adaptive phenotypic plasticity or yield stability in variable climates which include photoperiod-sensitive flowering, plastic tillering, flooding tolerance, seedling heat tolerance and phosphorus efficiency.

Phenotyping Plant Traits Associated with Mechanism(s) of Drought Tolerance, Water Use Efficiency, TE, Biomass and Components of Grain Yields in Pearl Millet

Austin *et al.* (1982) reported that in pearl millet, yields from drought plots of five genotypes, expressed as a percentage of control yields, were positively correlated (r=0.95) with accumulation of ABA. The water-use efficiency of pearl millet was lower than that sorghum for both grain and total dry matter due to shorter growth duration of pearl millet than sorghum besides higher crop water use and grain yield of sorghum than pearl millet (Kanemasu *et al.*, 1985; Seetharama *et al.*, 1984). In contrast, Singh and Singh (1995) showed less reduction of in dry matter yield in pearl millet than sorghum under drought conditions and also recorded greater WUE than sorghum.

Drought stress effects during vegetative period reduce yields less than drought in later stages in pearl millet (Lahiri and Kumar 1966). Craufurd and Bidinger (1988) reported that both African and Indian pearl millet phenotypes are equally productive under good agronomic conditions, while, increased GS 1 duration effected the increase in drymatter production and decreased in HI. Traore *et al.* (1989) compared the glossiness, trichomes and cuticular transpiration among sorghum, maize and pearl millet and showed that pearl millet exhibited leaf glossiness later than sorghum, while, cuticular transpiration of glossy leaves was often more than double that of non-glossy leaves.

Alagarswamy *et al.* (1988) reported that nitrogen use efficient pearl millet genotypes maintained thicker leaves and were generally less responsive in terms of leaf N concentrations and leaf CO_2 exchange rate to the N available in solution as compared to the N-inefficient ones, and suggested that nitrogen use efficiency was more related to the partitioning of N resources available into additional leaf area. Muchow (1989) in a comparative study on productivity of maize, sorghum and pearl millet demonstrated that productivity varied significantly with sowing date in both sorghum and pearl millet but stable in maize. Average grain-yield realized over the three sowing dates was 2.9 t ha^{-1} for millet. High yield was associated with high biomass production both at maturity and during grain-filling, high harvest index, and an increase in stover weight during grain-filling, while, high biomass-accumulation was too associated with long growth-duration, especially the duration of grain-filling and thus high cumulative radiation interception, and with high radiation-use efficiency (Muchow, 1989).

Subramanian and Maheswari (1989) showed that leaf water potential, stomatal conductance, photosynthesis, transpiration and transpiration efficiency of flag leaves decreased more in sorghum than in pearl millet under drought, while, pearl millet recovered quickly than sorghum after rewatering. They concluded that based on physiological responses, pearl millet is adapted to drier regions compared to sorghum. Zegada-Lizarazu, and Iijima (2005) reported that water uptake from deep soil layers by pearl millet and barnyard millet was highest for drought and water logging tolerance, respectively. Based on the WUE the pearl millet had the highest and lowest resistances to drought and water logging conditions, respectively, while barnyard millet was the most stable under both conditions. The drought resistance of pearl millet can be explained by higher WUE but not by increased water uptake efficiency compared to barnyard millet, another drought-resistant millet species.

Kholova (2010) characterized the mechanism of terminal drought tolerance in pearl millet and concluded that terminal drought tolerance of tolerant pearl millet genotypes was linked to their lower transpiration (Tr) rate. It appears that low Tr of these genotypes might be contributed to saving the water in the soil profile enabling more water available for grain filling stage (drought avoidance mechanism).They also reported that It is further discussed that Tr could be influenced by the level of leaf ABA and the hydraulic properties of plant tissues. However, these "water saving" drought tolerance mechanisms seems to be specific to the environmental conditions in which plants` development took place. Kholova *et al.* (2010) too demonstrated that constitutive traits in pearl millet controlling leaf water loss under well-watered conditions correlated with the terminal drought tolerance and selection of such traits may lead to more water being available for grain filling under terminal drought.

In lysimetric experiment on pattern of water uptake in pearl millet, Vadez *et al.* (2013) showed that tolerant genotypes extracted less water prior to anthesis, and more water after anthesis and yield was positively related to the amount of water extracted during week three after panicle emergence. They concluded that increased water uptake after anthesis, which results from earlier water conservation during pre-anthesis had increased grain yield under terminal drought. Burman *et al.* (2011) quantified the effect of terminal drought on pearl millet and found that drought stress decreased plant water potential and relative water content that led to a significant decline in the rate of net photosynthesis (PN), and its components, nitrate reductase activity and grain yield. Cv. Pusa-266 and CZP-9802 showed lower reduction in plant water status and were superior for drought tolerance.

Effect of High Temperature Stress on Phenology, Growth, Biomass and Grain Productivity and Relative Tolerance

In India, pearl millet is generally grown as rainfed crop and often experiences both drought and high temperature stress or in combination. Uneven or delay in rainfall accompanied by high temperature during seedling stage effects seedling establishment resulting in decreased crop stand and grain yield. In addition, pearl millet is often exposed to high temperature stress during reproductive stages of crop development, particularly when grown in off-season. Research on pearl millet has shown that reproductive stages of crop development were relatively more sensitive to high temperatures compared to vegetative stages (Prasad, 2013). High temperature stress during male or female gametogenesis results in pollen or ovule sterility, which causes decreased percentage seed-set and leads to decreased number of grain number per panicle. Besides, high temperature during flowering decreases pollen shed, pollen germination, inhibits pollen tube growth and decreases fertilization and seed-set. Studies on genetic mechanism for high temperature tolerance or susceptibility and its association with grain yield and other related traits would help in the development of tolerant genotypes.

Investigation on pearl millet germplasm for high temperature stress tolerance during reproductive stages revealed that certain genotypes of pearl millet had higher reproductive success (seed-set percent) at daytime temperatures about 40°C. Compared to other cereal crops, pearl millet had higher threshold temperature (maximum temperature, T_{max}, temperature beyond which physiological or growth processes does not occur) for yield. Pollen viability and germination is generally related to seed-set percentages. In several crops, such as rice, peanut and sorghum, pollen viability and pollen production begin to reach zero at a T_{max} of 39°C. However, pearl millet pollen showed T_{max} value of >42°C (Prasad, 2013).

Soman and Peacock (1985) developed a laboratory technique for screening seedling emergence of sorghum and pearl millet under high soil temperature and observed significant differences for emergence in both crops. Mortlock and Vanderlip (1989) reported that medium-sized pearl millet seed showed higher germination over a wider temperature range than small or large seed. Pearl millet germination rate and plumule emergence had optima at 37–38°C, while germination at 48 h was lower for large seed than small seeded. High-density pearl millet seeds had higher emergence than low-density ones at a constant temperature of 43°C.

Rachie and Majmudar (1980) indicated that pearl millet adaptation to harsh stress conditions might be due to a combination of short duration and heat tolerance than to specific (physiological) drought resistance mechanisms. Peacock *et al.* (1993) developed a field technique to screen pearl millet genotypes for their emergence and survival at high soil surface temperatures. They showed that genetic variation in seedling emergence and survival largely due to tolerance of high temperatures rather than tolerance of soil moisture deficit. The pearl millet cultivars tested had a relatively low degree of leaf thermal injury (< 20 per cent after treatment at 48° C for 1 hour) compared to sorghum (30-38 per cent injury under the same conditions) as reported by Sullivan *et al.* (1972). Sivaramakrishna *et al.* (1990) reported that pearl millet produced more heat shock proteins with greater apparent molecular weights than sorghum under seedling stage heat stress (35-45°C). Yadav *et al.* (2011) and Yadav *et al.* (2013) had compared different screening procedures for supra-optimal temperature tolerance in pearl millet and found that methods such as seedling thermo-tolerance index (STI) and seed to seedling thermo-tolerance index (SSTI) may be employed for identifying best genotypes, while, membrane thermostability (MTS) may be used for screening large number of genotypes or germplasm accessions for heat tolerance in pearl millet.

Strategies/Research Needs for Improving Pearl Millet Drought and Heat Stress Tolerance

- ☆ Exploit differences in developmental plasticity and recovering ability after a stress period, which include i) delay in development during severe stress, ii) degree of asynchrony in tiller development which allows some tillers to escape a stress period that damages others, and iii) rapid release and growth of secondary tillers after stress damage to the main tillers.

- ☆ Develop pearl millet genotypes tolerant to both drought and high temperature stresses that occur during field germination, seedling establishment and early vegetative phase.
- ☆ Identify and develop germplasm and hybrid parents possessing midseason (panicle initiation to flowering) and terminal (flowering to physiological maturity) drought and heat tolerance combining both high grain and stover yields and quality.
- ☆ Phenotype pearl millet for key traits (eg. phenology, canopy architecture, biomass accumulation, TE, WUE, root traits, etc.,) conferring adaptation to future climate change and developing high throughput phenotyping capabilities for selecting these key traits.
- ☆ Integrate phenotyping of key traits, and modeling of climate change/adaptation.
- ☆ Evaluate the relative sensitivities of male and female gamete to high temperature using male sterile lines.
- ☆ Phenotype and genotype pearl millet germplasm for putative traits, QTLs and genes associated with drought, heat and combination of both these stresses.

Action Plan

Theme

Phenotyping Pearl millet resilience for climate change adaptation traits (pre- and post-flowering drought and heat stress tolerance); Climate change impacts on vulnerability of pearl millet to drought and heat stress will be done through agro-climatic analyses and using crop growth modeling framework; Key plant traits contributing adaptation to elevated CO_2, heat and drought stresses and their combinations will be identified and characterize in field and controlled growth chambers.

Facilities Needed

To achieve the above objectives, there is a need for availability of modern facilities such as automatic rainout shelters, movable heat tents, controlled environment green houses, plant growth chamber etc at DSR.

Outcome

Phenotyping pearl millet germplasm for drought and heat tolerance/susceptibly will help understanding on mechanism of tolerance and identification of suitable donors, besides developing RILs for identifying QTLs governing drought tolerance.

References

1. AICPMIP, 2014. Progress in Pearl millet research. In: Proc. 49th Annual Pearl Millet Workshop, All India Coordinated Research Project on Pearl Millet, March 13-15, 2014, RARI (SKNAU), Durgapura, Jaipur, 47pp. (www.aicpmip.res.in).

2. Alagarswamy,G., Gardner, J.C., Maranville J. W., and Clark, R.B. 1988. Measurement of instantaneous nitrogen use efficiency among pearl millet genotypes. *Crop Science*. 28: 681-685.

3. Austin, R. B., I. E. Henson, and S. A. Quarrie. 1982. Abscisic acid and drought resistance in wheat, millet, and rice. In International Rice Research Institute. *Drought resistance in crops, with emphasis on rice*. Los Baños. Philippines.

4. Bhatnagar, S. K., Yadav, O. P., and Gautam, R. C. 1998. Research achievements in pearl millet (*Pennisetum glaucum*). *Indian Journal of Agricultural Sciences* 68 (8, Special issue): 423-30.

5. Bidinger FR, Mahalakshmi V, and Rao GDP. 1987a. Assessment of drought resistance in pearl millet [*Pennisetum americanum* (L.) Leeke]. I. Factors affecting yield under stress. *Aust.J. Agric. Res*. 38: 37-48.

6. Bidinger FR, Mahalakshmi V, and Rao GDP (1987b). Assessment of drought resistance in pearl millet (*Pennisetum americanum* (L.) Leeke). II. Estimation of genotype response to stress. *Australian Journal of Agricultural Research* 38, 49–59.

7. Bidinger, F R and Mahalakshmi, V and Talukdar, B S and Alagarswamy, G 1982. Improvement Of Drought Resistance In Pearl Millet. Pages 357-375. In: *Drought resistance in crops with emphasis on rice* IRRI Symposium, 4-8 May, 1981, Los Banos, Philippines.

8. Bidinger, F. R., Rao, P. P., Sinha, S. K., Sane, P. V., Bhargava, S. C., and Agrawal, P. K. (1990). Genetic and cultural improvements in the production of pearl millet. Pages 194-206. In : *Proceedings of the international congress of plant physiology*, New Delhi, India, 15-20 February 1988. Volume 1. Society for Plant Physiology and Biochemistry.

9. Bidinger, F. R., Chandra, S., and Mahalakshmi, V. (2000). Genetic improvement of tolerance to terminal drought stress in pearl millet (*Pennisetum glaucum* (L) R Br). In *Molecular approaches for the genetic improvement of cereals for stable production in water-limited environments. A Strategic Planning Workshop held at CIMMYT, El Batan, Mexico* (pp. 59-63). Mexico: CIMMYT.

10. Bidinger, F. R., Serraj, R., Rizvi, S. M. H., Howarth, C., Yadav, R. S. and Hash, C. T. 2005. Field evaluation of drought tolerance QTL effects on phenotype and adaptation in pearl millet [*Pennisetum glaucum* (L.) R. Br.] top cross hybrids. *Field Crop. Res*. 94, 14–32. (doi: 10.1016/j.fcr.2004.11.006).

11. Blum, A., and Sullivan, C. Y. (1986). The comparative drought resistance of landraces of sorghum and millet from dry and humid regions. *Annals of Botany*, 57(6), 835-846.

12. Burman, U., Garg, B. K., Yadav, O. P., and Kathju, S. (2011). Effect of terminal water stress on growth, plant water status and yield of pearl millet genotypes. *Indian Journal of Plant Physiology*, 16(3-4), 276-284.

13. Chaudhuri, U.N.and Kanemasu ET. 1985. Growth and water use of sorghum (*Sorghum bicolor* (L.) moench) and pearl millet (*Pennisetum americanum* (L.) leeke). *Field Crop Research:* 10: 113-124.

14. Chaudhuri, U.N.and Kanemasu ET. 1985. Growth and water use of sorghum (*Sorghum bicolor* (L.) moench) and pearl millet (*Pennisetum americanum* (L.) leeke). *Field Crop Research*: 10: 113-124.

15. Clark, R.B., Flores, C.I., Gourley, L.M., and Duncan, R.R. 1990. Mineral element concentrations and grain yield of sorghum (*Sorghum bicolor*) and pearl millet (*Pennisetum glaucum*) grown on acid soil. *J. Plant. Nutr*. **13**: 391–396.

16. Flores, C.I., Clark, R.B., and Gourley, L.M. 1991. Genotypic variation of pearl millet for growth and yield on acid soil. *Field Crops Res*. **26**: 347–354.

17. Craufurd, P. Q., and Bidinger, F. R. (1988). Effect of the duration of the vegetative phase on crop growth, development and yield in two contrasting pearl millet hybrids. *The Journal of Agricultural Science*, 110(01), 71-79. ICRISAT, Hyderabad.

18. Frere, M. and Popov, G.F. (1979) *Agrometeorological Crop Monitoring and Forecasting*.FAO Plant Production and Protection Paper No. 17. FAO, Rome, Italy.

19. Fussell, L. K., Bidinger, F. R., and Bieler, P. (1991). Crop physiology and breeding for drought tolerance: research and development. *Field Crops Research*, 27(3), 183-199.

20. Gregory. P. J.1982. Interaction of shoot and root characteristics in the response of millet to drought. Pages 135-144. In: *Drought resistance in crops with emphasis on rice* IRRI Symposium, 4-8 May, 1981, Los Banos, Philippines.

21. Haussmann, B. I. G., Fred Rattunde, H., Weltzien-Rattunde, E., Traoré, P. S. C., vom Brocke, K., Parzies, H. K. (2012), Breeding Strategies for Adaptation of Pearl Millet and Sorghum to Climate Variability and Change in West Africa. *Journal of Agronomy and Crop Science*, 198: 327–339. doi: 10.1111/j.1439-037X.2012.00526.x, University of Hohenheim, Institute of Plant Breeding, Seed Science and Population Genetics, Stuttgart, Germany; Email: bettina.haussmann@uni-hohenheim.de.

22. Kanemasu *et al*., 1984. International Crops Research Institute for the Semi-Arid Tropics. 1984. *Agrometeorology of Sorghum and Millet in the Semi-Arid Tropics: Proceedings of the International Symposium*, 15-20 Nov 1982, ICRISAT Center, India. Patancheru, A.P. 502324, India: ICRISAT.

23. Kholova J. 2010 b. Constitutive water-conserving mechanisms are correlated with the terminal drought tolerance of pearl millet [*Pennisetum glaucum* (L.) *R*. Br.] Journal of Experimental Botany, Vol. 61, No. 2, pp. 369–377.

24. Kholova J. (2010a) Understanding of terminal drought tolerance mechanisms in pearl millet (*Pennisetum glaucum* (L.) R. Br.). Charles University in Prague, Faculty of Science, *PhD Thesis*, pp. 107.

25. Kholová, J., Nepolean, T., Hash, C. T., Supriya, A., Rajaram, V., Senthilvel, S.,. and Vadez, V. (2012). Water saving traits co-map with a major terminal drought tolerance quantitative trait locus in pearl millet [*Pennisetum glaucum* (L.) R. Br.]. *Molecular Breeding*, *30*(3), 1337-1353.

26. Kowal JM and Kassam AH.1978 *Agricultural ecology of savanna. A study of West Africa*. Clarendon Press, Oxford, 403 pp.

27. Lahiri, A. N., and V. Kumar. 1966. Studies on plant water relationships. 111. Further studies on the drought mediated alterations in the performance of bulrush millet. *Proc. Nat'l Inst. Sci. India* B. 32: 116-129.

28. Ludlow MM, Muchow RC. 1990. Critical evaluation of the traits for improving crop yields.

29. Mahalakshmi V, Bidinger FR, Raju DS. 1987. Effect of timing of water deficit on pearl millet (*Pennisetum americanum*). *Field Crops Res.* 15: 327-339.

30. Mahalakshmi, V., and Bidinger, F. R. (1985). Water stress and time of floral initiation in pearl millet. *The Journal of Agricultural Science*, *105*(02), 437-445.

31. Maiti, R K and Bidinger, F R (1981) *Growth and Development of the Pearl Millet Plant*. Research Bulletin no. 6. Research Report. International Crops Research Institute for the Semi-Arid Tropics, Patancheru, Andhra Pradesh, India. Oxford and IBH, New Delhi.

32. Maman, N; Mason, S C,Lyon,D.J.and Dhungana,P. 2004. Yield Components of Pearl Millet and Grain Sorghum across Environments in the Central Great Plains. Crop Sci.44: 2138-2145.

33. Mortlock M.Y., and Vanderlip R.L. 1989. Germination and establishment of pearl millet and sorghum of different seed qualities under controlled high-temperature environments. *Field Crops Research*.22: 195-209.

34. Muchow R.C. 1989. Comparative productivity of maize, sorghum and pearl millet in a semi-arid tropical environment I. Yield potential. *Field Crops Research*. 20: 191-205.

35. Murty, M.V.R., Piara Singh, Wani, S.P, Khairwal, I.S. and Srinivas, K. 2007. Yield gap analysis of sorghum and pearl millet in india using simulation modeling. Global Theme on Agro-ecosystems Report no. 37. Patancheru 502 324, Andhra Pradesh, India: International Crops Research Institute for the Semi-Arid Tropics. 82 pp.

36. Peacock J. M., Soman P., Jayachandran R., Rani A. U., Howarth C. J. and Thomas A. (1993). Effects of High Soil Surface Temperature on Seedling Survival in Pearl Millet. *Experimental Agriculture*. 29: 215-225.

37. Prasad PVV and Staggenborg, SA 2009. Growth and production of sorghum and millets, in Soils, Plant Growth and Crop Production, [Ed. Willy H. Verheye], in Encyclopedia of Life Support Systems (EOLSS), Developed under the Auspices of the UNESCO, Eolss Publishers, Oxford,UK, [http: / /www.eolss.net] [Retrieved February 19, 2011] ©UNESCO-EOLSS.

38. Prasad and Venkateswarlu, 2013. Importance of millets in global food and nutritional security in changing climates. Proc. Global Consultation on Millets Promotion for Health and Nutritional Security, December 18-20, 2013. Compendium of Papers and Abstracts, Directorate of Sorghum Research, Hyderabad, 6 pp.

39. Rachie, K. O., and J. V. Majmudar. 1980. *Pearl millet*. Pennsylvania State University Press, University Park, Pa. 307 pp.

40. Rai, K. N., Kulkarni, V. N., Thakur, R. P., Haussmann, B. I. G., and Mgonja, M. A. (2006). Pearl millet hybrid parent s research: approaches and achievements. Pages 11-74. In: *Hybrid parents research at ICRISAT*. Gowda CLL, Rai KN, Reddy Belum VS and Saxena KB. (eds.). Patancheru 502 324, Andhra Pradesh, India: International Crops Research Institute for the Semi-Arid Tropics. 212 pp. ISBN 92-9066-489-4.

41. Rai, K.N. and Murty, D.S. and Andrews, D.J. and Bramel-Cox, P.J. (1999) Genetic enhancement of pearl millet and sorghum for the semi-arid tropics of Asia and Africa. *Genome* 42: 617–628 (1999), 42. pp. 617-628.

42. Seetharama N *et al.*, 1984. International Crops Research Institute for the Semi-Arid Tropics. 1984. Agrometeorology of Sorghum and Millet in the Semi-Arid Tropics: *Proceedings of the International Symposium*, 15-20 Nov 1982, ICRISAT Center, India. Patancheru, A.P. 502324, India: ICRISAT.

43. Seetharama, N., Mahalakshmi, V., Bidinger, F.R. and Singh S. 1984. Response of Sorghum and Pearl Millet to Drought Stress in Semi-Arid India. Pages159-173. *In* Agrometeorology of Sorghum and Millet in the Semi-Arid Tropics: Proceedings of the International Symposium, 15-20 Nov. 1982, ICRISAT Center, India. Patancheru, A.P. 502324, India: ICRISAT.

44. Serraj, R., Tom Hash, C., Rizvi, S. M. H., Sharma, A., Yadav, R. S., and Bidinger, F. R. (2005). Recent advances in marker-assisted selection for drought tolerance in pearl millet. *Plant Production Science*, *8*(3), 334-337.

45. Singh, B.R. and Singh D.P. 1995. Agronomic and physiological responses of sorghum, maize and pearl millet to irrigation. *Field Crop Research*. 42: 57-67.

46. Sivakumar, MVK., Singh P, and Williams JH. 1987. Agroclimatic Aspects in Planning for Improved Productivity of Alfisols. Pages15-30. In : Alfisols in the semi-arid tropics. Proceedings of the Consultants' Workshop on the State of the Art and Management Alternatives for Opt imizing the Productivity of SAT Alfisols and Related Soils. 1-3 December 1983, ICRISAT Center, India. Patancheru, A.P. 502324. India: ICRISAT.

47. Sivaramakrishna, S.,Patell VZ and Soman, P.1990. Heat Shock Proteins of Sorghum [*Sorghum bicolor* (L.) Moench] and Pearl Millet [Pennisetum glaucum (L.) R.Br.] Cultivars with Differing Heat Tolerance at Seedling Establishment Stage. *Journal of Experimental Botany* 41: 249-254.

48. Soman P. and. Peacock J. M (1985). A Laboratory Technique to Screen Seedling Emergence of Sorghum and Pearl Millet at High Soil Temperature.

49. Subramanian VB; and Maheswari M. 1989. Comparison of physiological responses of pearl millet and sorghum to water stress. Proceedings of the Indian Academy of Sciences, *Plant-Sciences*. 99: 517-522; 13 ref.

50. Sullivan, C. Y. 1972. Mechanisms of heat and drought tolerance in grain sorghum and methods of measurement. Pages 247-269 in N. G. P. Rao and L. R. House, eds. Sorghum in the seventies.

51. Traore, M., Sullivan, C. Y., Rosowski, J. R., and Lee, K. W. (1989). Comparative leaf surface morphology and the glossy characteristic of sorghum, maize, and pearl millet. *Annals of Botany*, 64: 447-453.

52. Vadez, V., Kholova, J., Yadav, R.S., and Hash CT (2013). Small temporal differences in water uptake among varieties of pearl millet (*Pennisetum glaucum* (L.) R.Br.) are critical for grain yield under terminal drought. *Plant and Soil 371: 447-462.* doi: 10.1007/s11104-013-1706-0.

53. van Oosterom E J, Bidinger F R and Weltzien R E. 2003. A yield architecture framework to explain adaptation of pearl millet to environmental stress. *Field Crops Research* 80: 33–56.

54. van Oosterom E J, Mahalakshmi V, Arya G K, Dave H R, Gothwal B D, Joshi A K, Joshi P, Saxena M B L, Singhania D L and Vyas K L. 1995. Effect of yield potential drought escape and drought tolerance on yield of pearl millet (Pennisetum glaucum) in different stress environments. *Indian Journal of Agricultural Sciences* 65: 629–35.

55. van Oosterom EJ, Bidinger FR, Mahalakshmi V, Rao KP. 1996. Effect of water availability pattern on yield of pearl millet in semi-arid tropical environments. *Euphytica* 89: 165-173.

56. Yadav O P, Weltzien R E and Bidinger F R. 2003. Genetic variation in drought response of landraces of pearl millet (*Pennisetum glaucum*). *Indian Journal of Genetics and Plant Breeding* 63: 37–40.

57. Yadav R.S. and Howarth C.J., Hash CT and Bidinger FR.1999. QTL analysis and marker-assisted breeding of traits associated with drought tolerance in pearl millet. Pages 211-223. In : Ito O, O'Toole J, Hardy B(eds). *Genetic improvement of rice for water-limited environments*. Proc. of the Workshop on Genetic Improvement of Rice for Water-Limited Environments, 1-3 December 1998, Los Baños, Philippines. Los Baños (Philippines): International Rice Research Institute. 353.

58. Yadav, A. K., Narwal, M. S., and Arya, R. K. 2011. Genetic dissection of temperature tolerance in pearl millet (*Pennisetum glaucum*). *Indian Journal of Agricultural Sciences* 81 (3): 203–13,.

59. Yadav, A. K., Narwal, M., and Arya, R. K. 2013b. Evaluation of pearl millet (*Pennisetum glaucum*) genotypes and validation of screening methods for supra-optimal temperature tolerance at seedling stage. *Indian Journal of Agricultural Sciences* 83 (3): 260–271.

60. Yadav, O P. 2011. Breeding dual-purpose pearl millet (*Pennisetum glaucum*) for north-western India: understanding association of biomass and phenotypic traits. *Indian Journal of Agricultural Sciences* 81 (9): 816–20.

61. Yadav, O. P., and Bhatnagar, S. K. (2001). Evaluation of indices for identification of pearl millet cultivars adapted to stress and non-stress conditions. *Field Crops Research*, 70(3), 201-208.

62. Yadav, O.P., and Rai, K.N. 2013a. Genetic Improvement of Pearl Millet in India. *Agricultural Research* 2: 275-292.

63. Yadav, R. S., Hash, C. T., Bidinger, F. R., Cavan, G. P., and Howarth, C. J. (2002). Quantitative trait loci associated with traits determining grain and stover yield in pearl millet under terminal drought-stress conditions. *Theoretical and Applied Genetics*, 104(1), 67-83.

64. Yadav, R. S., Hash, C. T., Bidinger, F. R., Devos, K. M., and Howarth, C. J. (2004). Genomic regions associated with grain yield and aspects of post-flowering drought tolerance in pearl millet across stress environments and tester background. *Euphytica*, 136(3), 265-277.

65. Zegada-Lizarazu, W., and Iijima, M. (2005). Deep root water uptake ability and water use efficiency of pearl millet in comparison to other millet species. *Plant production science*, 8(4), 454-460.

Millets: Ensuring Climate Resilience and Nutritional Security (2015) *Pages* **377–404**
Editors: **Vilas A. Tonapi and J.V. Patil**
Published by: **DAYA PUBLISHING HOUSE, NEW DELHI**

Chapter 13

Physiology of Small Millets

H.S. Talwar

ICAR-Indian Institute of Millets Research (IIMR)
Rajendranagar, Hyderabad – 500 030

Small millets consist of a about a dozen distinct species of small-seeded grasses that are grown for grain, each with their own unique traits and value. The most economically significant of these at present is Finger millet, but the other small millets (including five more, Foxtail millet (*Setaria italica*), kodo millet (*Paspalum scrobiculatum*), Little millet [*Panicum sumatrense* Roth. ex. Roem. and Schult.], Proso millet (*Panicum miliaceum*), and Barnyard millet (*Echinocloa* sp., *Echinochloa colona* (L.) Link] and [*Echinochloa crus-galli* (L.) P. Beauv.), are each in their own way important to the farmers who grow them. About 90 per cent of global millet production is utilized in the developing countries (FAO, 1990) and 38.6 per cent of total world millet is produced by India alone (FAO, 1995).

Small millets provide staple food grain for millions of poor smallholders and households in the developing world's harshest, most food insecure regions, such as the Sahel in Africa and in South Asia's semi-arid zone. They also provide feed grain and fodder for livestock. Millets in general provide many essential vitamins and micronutrients that can bolster nutrition for those living in dryland areas, particularly women and children. Small millets are especially rich in iron, zinc and calcium, and have other dietary qualities that can help stave off anemia, celiac disease, and diabetes.

Small millets are extraordinarily tolerant of drought and other abiotic stresses, such as high temperatures and poor soils. This makes them "climate smart", and a good source for genetic traits that can strengthen the resilience of other crops in the face of climate change. As climates get hotter and drier, small millets and the more dominant dryland cereals will become increasingly well suited for production in areas where other crops are now grown. The high nutritional value – coupled with the impressive hardiness of small millets – makes them desirable food security crops, as well as a good sources of fodder and feed in mixed crop/livestock systems.

Finger Millet

Finger millet (*Eleusine coracana* (L.) Gaertn) is the world's fourth most important food crop belonging to Poaceae family, and is widely grown in Africa and Asia (Upadhyaya *et al.*, 2007; www.fao.org/ag/AGP/AGPC/doc/Gbase/safricadata/eleucor. htm). Finger millet constitute approx. 12 per cent of th global millet area and is cultivated in more than 25 countries in Asia and Africa (Hema *et al.*, 2014). In the Indian sub-continent, Finger millet is known locally as Ragi in the, is mostly cultivated as a base crop in a mixed cropping. Seeds of local landraces may be consciously or unconsciously mixed along with the seeds of high yielding varieties of other species and these combinations ensure a high degree of stability of the whole cropping system in variable environments. It is rich in calcium, dietary fibre and known for its health benefits. It is known for its high degree of tolerance against drought, salinity and blast disease (Shailaja and Thirumeni 2007; Agarwal *et al.*, 2011). In recent times, in view of higher incidence of diabetes and obesity, it is considered a wholesome food to diabetics (Kumari and Sumathi, 2002). Finger millet is highly nutritious as its grain contain 65-75 per cent carbohydrates, 5-8 per cent protein, 15-20 per cent dietary fibre and 2.5-3.5 per cent minerals (Chetan and Malleshi, 2007). Considering its importance in food and fodder security, adequate information on genetic variability between yield and its attributes is meagre in finger millet. In India, it is the third important crop among millets in area and production after sorghum and pearl millet. It is cultivated mostly as a rainfed crop for its valued food grains and its adaptability to wide range of geographical areas and agro-ecological diversity in India. It is cultivated over nearly 1.6 million ha, with an annual production of 2.4 million tonnes and a productivity of approximately 1,534 kg/ha (Government of India, 2009). Important finger millet growing states are Karnataka, Tamil Nadu, Andhra Pradesh, Maharashtra, Odisha, Jharkhand, Chhattisgarh and Uttarakhand.

Finger millet is reported to be resistant to flood, drought, pest and diseases. The crop can be cultivated across a range of soil moisture availabilities. Lack of high yielding varieties adapted to diverse agro-ecological conditions is the major reason of low productivity. Though more than 200 finger millet varieties have been released in India, only very few varieties show adaptability to wide range of environments, while most others show better adaptability to specific agro-ecological or agro-management condition. Evaluation of interaction of genotypes with locations and other agro-management conditions would help in getting information on adaptability and stability of performance of genotypes.

Genetic Variability in Morphological, Physiological and Yield Traits

Previous and limited studies have demonstrated considerable variability for different morphological, physiological and yield traits in Finger millet. Reddy *et al.* (2013) evaluated eighteen elite entries of finger millet in a field study to assess the magnitude of genetic variability, heritability and genetic advance for yield and yield contributing traits. A wide range of variation was recorded for days to 50 per cent flowering, plant height, ear head length, chlorophyll content, grain yield and straw yield. The phenotypic coefficient of variation was greater than genotypic coefficient of variation for all the characters studied which shows the influence of the environmental effect on the characters. High values for phenotypic coefficient and genotypic coefficient was recorded for number of fingers per ear head. High heritability and high genetic advance was recorded for ear head yield, ear head length and number of fingers per ear head indicating that these characters were controlled by additive gene effects. Selection based on these characters would be effective for future finger millet crop improvement program. Moderate heritability coupled with moderate genetic advance was observed for grain yield and straw yield. Threshing percentage showed low heritability as well as low genetic advance.

Das *et al.* (2013) observed that grain yield per plant has exhibited significant positive association with total no. of basal tillers per plant, productive tillers per plant, total fingers on the main ear, finger length and finger width. Path analysis revealed that maximum positive direct effect on grain yield per plant was exhibited productive tillers per plant followed by Finger length, Finger width. Therefore, it is emphasized to lay attention on traits like productive tillers per plant followed by Finger length, Finger width while selecting for improvement in grain yield of finger millet.

Sashidhar (1986) investigate the genotypic differences in stomatal frequency (F), leaf area (LA), leaf area duration (LAD), total stomatal number per plant (LA× F), and the relationship of these parameters to biomass production, grain yield and photosynthetic rate in 100 medium-duration genotypes of Finger millets. The differences in stomatal number per plant amongst genotypes was, as expected, influenced more by the leaf area than by the stomatal frequency of leaves at any canopy position. There were several genotypes with a consistently low leaf area but having a productivity under rainfed conditions that was equivalent to that of the high leaf area types. Mean biological yield and grain productivity were not significantly different between the two groups, although the differences in both LAD and stomatal number per plant were significant. Genotypes with a consistently low leaf area per plant associated with high biomass and grain yield were selected on the broad assumption that these genotypes may also show smaller canopy water loss under rainfed conditions, due to a low transpirational area. Photosynthetic $^{14}CO_2$ fixation and gas exchange studies showed that the rate of leaf carbon fixation was two-fold higher in genotypes that had a consistently low leaf area but were able to produce dry matter and grain yield in the same amount as the high leaf area types. Such genotypes are desirable for rainfed or drought situations

Shankar *et al.* (1990). Germplasm lines of finger millet (400 lines) with diverse genetic background were screened to identify high carbon exchange rate (CER) and crop growth rate (CGR) genotypes. Considerable variation was noticed in LA, biomass and all the growth and yield attributes but range in stomatal frequency was relatively less. The difference in plant stomatal number was more due to variation in LA but not the frequency. Since high TDM is a primary criteria for achieving higher grain yield, from amongst the high biomass types, genotypes differing in LA/DM ratios but with similar DM and HI were selected. In six genotypes each from low and high LA/DM group which showed stability across the environment, gas exchange characteristics were studied. The CER in low LA/DM types was 45 per cent higher. A significant negative relationship (r = -0.87) exists between CER and LA/DM ratios. Higher net assimilation rate in these genotypes therefore could be attributed mainly due to CER rather than the dark respiration. An approach to identify high CER types with high CGR and low plant conductances from the germplasm lines has been proposed. These conceptual types will have high CGR but with low functional LA, thus low transpiration leaf surface. These conceptual types will have specific advantage under rainfed water limited conditions.

Abiotc Stress Tolerance

Finger millet is a crop which is widely grown in semi-arid and Arid tropics of Africa and Asia. It has a high production potential reaching up to 4-t tonnes per hectare under optimum conditions. The yield levels achieved are far below its actual potential because is is predominantly grown under rainfed conditions. Among the abiotic stresses drought stress severely limit the yield of finger millet although it is reputed one of the most resistant crops to drought. Even short period of drought during some critical growth stages markedly reduce the yield (Uday Kumar, 1986). Various agronomic practices during sowing and plant establishment in the field (Kalarani *et al.,* 2001a, Kalarani 2001b and Yadav *et al.,* 1999), have not improved the productivity of finger millet (Hebbar *et al.,* 1992). Hence, developing plants having higher intrinsic stress tolerance is necessary for yield improvement.

Drought Tolerance

Finger millet (*Eleusine coracana* L.) is considered a drought-tolerant crop. Because of high tolerance to drought, finger millet is preferred in drought prone areas with less rainfall (Goud and Laxmi, 1972). Since the major area of cultivation of finger millet is rain-fed, the potential yields are affected by severe drought stress (Hebbar *et al.,* 1992, Sashidhar *et al.,* 1986, Reddy and Patil, 1983, Shivkumar *et al.,* 1996) and drought-induced oxidative stress (Bhatt *et al.,* 2011). In addition, salinity stress also reduces potential yields (Shailaza and Thirumeni, 2007). More specifically, seed germination and seedling establishment stages of finger millet are highly susceptible to drought and salinity stress (Hebbar *et al.,* 1992, Shailaja and Tirumeni, 2007). However, within this species, there is considerable genotypic variation in tolerance to this environmental stress. Assefa, A and Fetene, M. (2013) screened 96 finger millet accessions collected from different agro-ecological zones of the Ethiopia for drought tolerance. Data were collected for plant height (PH), green leaf number (GLN), green leaf area (GLA), ear number (EN), ear length (EL) and relative water content (RWC). Significant difference ($P<0.05$) were obtained between accessions for most selected physiological, morphological and yield related traits such as Chlorophyll Content Index (CCI), RWC, and yield related parameters, such as tiller number (TN), productive tillers (PT), seed weight per head and per plant. Based on high RWC, CCI, GLN, PT, EL, and EN, Grain Yield per head and per plant, a total of 23 accessions were categorized as drought stress tolerance and promoted to the next intensive physiological and yield related evaluation. Moreover, AAUFM-7, AAUFM-2, AAUFM-

44 were the first top three accessions with the highest grain yield of 77.5, 72g/plant and 65.4g/plant, respectively. Overall, the higher genetic variability observed among accession in this study should be further utilized for finger millet improvement.

In India several studies have demonstrated considerable variability for drought and tolerance to water stress in a multiple of traits in finger millet. Krishnasastry *et al.* (1982) evaluated varieties of finger millet were in two seasons under rainfed conditions and for one season (Summer) under imposed intermittent stress. Growth and yield analysis revealed that in all the seasons and when grouped according to duration, a significant positive correlation was obtained between shoot dry weight and yield. Even under intermittent stress a similar relationship was observed. Av rage ear weight per unit area was significantly correlated with yield but not e ar number. Under stress the number of late formed tillers enhanced but their contribut on to yield was less. It is suggested that high dry matter production, lesser l af area coupled with higher photosynthetic efficiency and larger average ear weight are desirable features in finger millet genotypes suited to rainfed conditio s.Various physiological approaches are available to improve the stress olerance. One of the approaches that can be applied in Finger Millet by developi g finger millet plants that can accumulate novel osmolytes is useful for improving stress tolerance. From this perspective, utilizing the existing genetic variability among finger millet germplasm for increased osmolyte production and genetic engineering for a particular osmolyte are approaches that have been attempted in finger millet (Bhat *et al.,* 2011, Rai *et al.,* 2009, Uma *et al.,* 1995, Dinesh Kumar *et al.,* 1987 and Sashidhar *et al.,* 1986). However, incorporating traits from a stress tolerant germplasm line into cultivated variety by conventional breeding is time consuming (Jaiprakash *et al.,* 1992). Finger millet is known to accumulate proline under stress (Kandpal and Rao, 1985; Prasad *et al.,* 1980) but, mannitol production has not been reported in this plant.

Kandpal and Rao (1981) showed that in polyethyleneglycol (PEG)-treated ragi (*Eleusine coracana*) leaves, the level of free proline increased by 85-fold and this increase was probably due to an enhanced activity of the enzymes involved in the synthesis of proline and marked inhibition of the activity of enzymes degrading proline (Kandpal *et al.,* 1981). The activity of ornithine aminotransferase (EC 2.6.1.13), an enzyme of proline biosynthesis, increased by about 3 fold in water-stressed ragi leaves. In their subsequent study, they showed that the increase in the activity of ornithine aminotransferase was not due to the presence of an activator but due to alteration in the properties of the enzyme isolated from water stressed leaves (Kandpal and Appaji Rao, 1982).

Kandpal and Rao (1984) Water stress resulted in a specific response leading to a large and significant increase (80-fold) in free proline content of ragi (*Eleusine coracana*) leaves and seedlings. L-Proline protected ornithine aminotransferase, an enzyme in the pathway for proline biosynthesis, isolated from normal and stressed ragi leaves against heat inactivation and denaturation by urea and guanidinium chloride. The protection of the stressed enzyme by L-proline was much more complete than that of the enzyme isolated from normal leaves. While L-ornithine, one of the substrates, protected the stressed enzyme against inactivation, it enhanced the rate of inactivation of the normal enzyme. α-Ketoglutarate protected both the normal and stressed enzyme against inactivation and denaturation. These results support the suggestion that ornithine aminotransferase has undergone a structural alteration during water stress. In view of the causal relationship between elevated temperature and water stress of plants under natural conditions, the protection afforded by proline against inactivation and denaturation of the enzyme from stressed leaves assumes significance. These results provide an explanation for a possible functional importance of proline accumulation during water stress.

Genetic differences in water deficit stress tolerance offer the unique opportunity to compare the changes in metabolic processes in plants under water stress that might be involved in water deficit stress tolerance. Genotypic diversity existing in finger millet is well known and hence is the primary reason that holds promising for further improvement for water deficit tolerance. Rai *et al.* (2009) studied the effect of mannitol -induced water deficit stress on seed germination, early seedling growth, proline, MDA and hydrogen peroxide content in ten finger millet genotypes. The seedlings of PRM-6107, VL-283 and VL-328 recorded relatively better drought tolerance in terms of growth and oxidative damage. While, PES-400, VR-708, VL-149 and VL-146 showed susceptibility as per cent reduction in germination, shoot growth, root growth, MDA content and hydrogen peroxide level was relatively higher in comparison to tolerant group. Rest of the three genotypes *viz.* VL-315, PR-202 and PRM-9802 showed intermediary character in between these two groups.

Maqsood and Azam Ali (2007) conducted studies on two finger millet landraces in glasshouses under two moisture regimes (fully irrigated and subjected to drought) to investigate the effects of environmental stress on the growth, development, radiation use efficiency and yield of finger millet (*Eleucine coracana*), in two landraces *viz.*, TZA-01 and TZM-01. The drought treatment was imposed at 28 DAS beyond which no irrigation was applied to the droughted treatment. Growth and development were monitored between 21 DAS and 105 DAS. Soil moisture had an effect on the growth of both the landraces. Drought reduced leaf area, dry matter accumulation, seed weight, radiation use efficiency and yield of finger millet. Drought significantly reduced the number of leaves per plant, leaf area per plant from 42 DAS to 105 DAS. LAI increased significantly in all the treatment between 28 DAS and 84 DAS, thereafter there was a sharp decline in LAI for TZM-01 under both irrigation and drought. Maximum LAI values (5) in irrigated TZA-01 at 105 DAS and minimum LAI value (2) was recorded in droughted TZA-01 and TZM-01 at 105 DAS. Drought had significant ($p<0.009$) effect on grain yield of two finger millet landraces. The maximum grain yield (4.88 t ha-1) was recorded under irrigated TZA-01 followed by irrigated TZM-01, where (3.22 t ha^{-1}) grain yield was recorded. The minimum grain yield (1.92 t ha-1) was recorded in droughted TZM-01. Biomass was affected significantly ($p<0.028$) by drought. Maximum radiation use efficiency was recorded as 3.11g MJ-1 of the accumulated intercepted radiation

Neshamba (2011) reported variation in 12 putative morphological and agronomic traits of drought tolerance in 203 accessions in 2 environments, one with and the other without stress, on an Acrisol in northern Zambia. One sample T-test of differences in attributes of traits showed no significant difference between test environments in grain weight (GW, p = 0.744). The test, however, showed significant differences in spike length (SL); highly significant differences in biomass (BW) and chaff weight (CW), days to 50 per cent flowering (DTP), pest and disease susceptibility (PDS), plant height (PH), number of productive tillers (NPT), spike weight per plot (SY) and stay-green characteristic (SGC). Withholding water for 5 days during flowering reduced SL by 3.4 per cent, BW by 3.1 per cent, SGC by, CW by, PDS by 10.5 per cent, NPT by 40.0 per cent), and SY by 48.5 per cent and increased PH and DTP by 3.1 and 4.3 per cent, respectively. One-way analysis of variance detected significant differences among accessions in SY under optimal conditions. The same analysis detected significant differences in number of spikes per panicle and SGC; and highly significant differences in PDS under stress. Eighty-three accessions outweighed the best check in SY under optimal conditions. Accession ZM 3813 had the highest SY and ZM 203 the lowest in the environment. Under stress 92 accessions exceeded the best check in SN. ZM 3825 had the largest attribute and ZM 193 the smallest. In SGC 77 accessions were scored better than the best check, FMM 165. ZM 225, ZM 245, and ZM 112 were scored the best and 37 others the poorest in the trait. And in PDS 31 accessions had better scores than the best check. The least susceptible accession was ZM 3860 and the most susceptible was ZM 3652. Overall ranking of accessions according to tolerance to stress in 9 traits showed 16 accessions namely, ZM 3825, ZM 3920 ZM 1459, ZM 3906, ZM 3819, ZM 3834, ZM 3834, ZM 203, ZM 40122, ZM 3816, ZM 229, ZM 3685, ZM 154, ZM 153, and ZM 3824, to be more tolerant to stress than the best check, Nyika. These accessions occurred across a wide range of clusters. Six of them were collected from the high, and 5 from the low rainfall region. Five were collected from Kawambwa showing the district to be an important source of tolerance. Nearest neighbour analysis using the dissimilarity coefficient, Euclidean Distance, generated 3 dendrograms depicting relationships among the accessions based on the traits evaluated 1 dendrogram for each of the test environments and 1 for the most tolerant accessions. Accessions could be consigned to several clusters. All clusters could be read between Euclidean Distance 0.84 and 1.00. The pattern of clustering in iv the 2 environments was similar. Under optimal conditions ZM 153 and ZM 3813 were the most divergent. Under stress maximum cluster distance was between the most tolerant accession (ZM 3825) and the recently released mutant variety, FMM 175. Among the most tolerant accessions ZM 3834 and ZM 40 122 were the most closely related, and ZM 154 and ZM 203 the most divergent. There was obvious differentiation at both inter-cluster and intra-cluster levels. In the dendrogram of the most tolerant accessions ZM 3825, ZM 3824, ZM 3819, ZM 3834, ZM 3783 and ZM 3816 collected from Luapula did not occur in the same cluster at short distances indicating wide divergence among the most tolerant accessions. This also showed extensive divergence between the most and less tolerant accessions. The accession may be used to introgress drought tolerance in the mutant variety

Hema *et al.* (2014) studied performance of transgenic finger millet with mtnl gene under drought, salinity and oxidative stress was studied at plant level in T_1 generation and in T_1 and T_2 generation seedlings. Results from these experiments showed that transgenic finger millet had better growth under drought and salinity stress compared to wild-type. At plant level, transgenic plants showed better osmotic adjustment and chlorophyll retention under drought stress compared to the wild-type.

Salinity Tolerance

Finger millet known for its resilience and ability to provide assured harvest even under environmentally fragile habitats, viz saline (Kaliappan *et al.*, 1967) and alkaline (Rachie and Peters, 1982) could be considered as a potential crop to alleviate the problem of salinity. Enormous work has been done on various crops with reference to salinity, however, little information is known regarding the effect of salinity on finger millet. Progressive salinization of soil is a major limiting factor to crop growth and productivity across the globe. Agarwal *et al.* (2011) reported a gradual and significant reduction in per cent germination at all salinity levels except in cv. Local Hills at 3 EC. Length, fresh and dry weights of root and shoot were also decreased with increase in concentration in saline solutions. Cv. VL-315 registered least reduction (4.7-56.2 per cent) in seedling dry matter whereas cv. Local Hills recorded highest reduction (12.2-71.8 per cent) at all salinity levels. We concluded that cv. VL-315 performed better and proved to be the most promising variety as it had SSI value > 0.8 whereas cv. Local Hills recorded as most sensitive (SSI value < 1.0). Shailaja and Thirumeni (2007) evaluated the finger millet genotypes at seedling stage and reported significant difference was observed among the genotypes at different stress levels. All the 5 traits, viz germination, root length, shoot length, seedling length and vigour index declined with increasing salt stress in all the genotypes. However, 'CO 12', 'GPU 48', 'Indaf 5' were highly tolerant, while 'CO 11 ' was highly susceptible. Similar results were obtained by Kaliappan *et al.* (1967), Panigarh *et al.* (1978), Onkware (1993), Anantharaju (2001) and Mahadeviah (2004) while screening finger millet genotypes for salt-tolerance at seedling stage. Rahman *et al.* (2014). attempted to unravel the physiological and molecular basis of salinity tolerance in two contrasting finger millet genotypes *viz.*, CO 12 and Trichy 1. Physiological studies revealed that the tolerant genotype Trichy1 had lower Na+ to K+ ratio in leaves and shoots, higher growth rate (osmotic tolerance) and ability to accumulate higher amount of total soluble sugar in leaves under salinity stress. These workers sequenced the salinity responsive leaf transcriptome of contrasting finger millet genotypes and identified salinity responsive genes and genotype specific responses.

Uma *et al.* (1995) reported marked differences among 28 finger millet genotypes (*Eleusine coracana* Gaertn.) in acquired tolerance to osmotic stress as assessed by the recovery of root growth from severe stress [-1·2 MPa polyethylene glycol, (PEG) or 400 mM NaCl]. However, these differences in tolerance were observed only when the seedlings were subjected to a preceding mild induction stress (-0·6 MPa PEG or 200 mM NaCl). In two contrasting genotypes, synthesis of stress-induced proteins was studied. Proteins with apparent molecular weight of 70-72, 52, 37, 34 and 23 kDa were synthesized in the highly responsive genotype (GE 415) and poorly responsive (VL 481) genotype following a mild induction stress (200 mM NaCl). However, GE-415 synthesized a 54 kDa protein that was not observed in VL-481. Addition of abscisic acid (ABA) to the induction medium containing 200 mM NaCl enhanced the acquired tolerance of finger millet seedlings over those without ABA in association with the appearance of several ABA-responsive proteins. GE-415 required much less

ABA than VL-481 to obtain the same response. With 10 μM ABA + 200 mM, A NaCl induction stress, GE-415 had significantly higher endogenous ABA. In association with higher levels of ABA, GE-415 had greater recovery root growth following severe stress from 600 mM NaCl. Pretreatment with 10 μM ABA + 200 mM NaCl induced several proteins with apparent molecular weights of 70-72, 54, 45, 36, 29 and 21 kDa in both genotypes. Qualitatively, GE-415 synthesized a unique 23-24 kDa protein and quantitatively there was significantly more of the 21 kDa protein in GE-415 compared to VL-481. The results indicate that the synthesis of stress proteins is correlated with the observed variation in acquired tolerance of the two genotypes.

Aarati (2003) developed an optimum induction protocol for salinity stress for finger millet seedlings. Seedlings induced with 200 mM NaCl showed substantially high recovery growth compared to non-induced seedlings. The addition of ABA during induction markedly increased recovery growth when challenged with a very high concentration of NaCl (up to 600 mM). A number of stress responsive proteins were expressed upon induction. Among them, 21 kDa protein was the most prominent. ^{35}S methionine incorporation studies showed that >60 per cent of the total label in the heat stable protein fraction was in the 21 kDa protein. The expression of this protein was seen within 2 h of induction stress. Further, the 21 kDa protein is induced under desiccation, low temperature, ABA and ABA together with NaCl, indicating that it is ABA responsive. This protein showed considerable accumulation in the mature seed and exhibited a strong cross-reaction with the antibodies raised against the conserved sequence of LEA 1 group of proteins, suggesting that the 21 kDa protein belongs to LEA 1. The relevance of this protein was examined in species varying in stress tolerance (finger millet, setaria, proso millet and cucumber). Cucumber, a salinity sensitive species, did not show any detectable level of expression of this protein. The relatively tolerant species, such as finger millet and proso millet, showed higher levels of expression compared to setaria, which is a sensitive species, signifying the relevance of the 21 kDa protein in stress tolerance.

Anjaneyulua (2014) subjected both, controls and transgenic plants to 100 and 200 mM NaCl and certain biochemical and physiological parameters were studied. Relative water content (RWC), plant height, leaf expansion, finger length and width and grain weight were severely reduced (50–70 per cent), and the flowering period was delayed by 20 per cent in control plants compared to transgenic plants under salinity stress. With increasing salt stress, the proline and chlorophyll contents as well as the enzyme activities of superoxide dismutase (SOD), catalase (CAT), ascorbate peroxidase (APX), guaiacol peroxidase (GPX) and glutathione reductase (GR) increased by 25–100 per cent in transgenics, while malondialdehyde (MDA) showed a 2–4-fold decrease. The increased activities of antioxidant enzymes and the reduction in the MDA content suggest efficient scavenging of reactive oxygen species (ROS) in transgenics and, as a consequence, probably alleviation of salt stress. Also, the leaf tissues of the transgenics accumulated 1.5–2.5-fold higher Na^+ and 0.4–0.8-fold higher K^+ levels.

Foxtail Millet

Foxtail millet (*Setaria italica* (L.) P. Beauv) has been a very important cereal since ancient times in Eurasia and has contributed greatly to human civilization both in

Asia and Europe (Li and Wu, 1996). Foxtail millet is a minor millet belonging to the family poaceae and is one of the oldest domesticated diploid C4 Panicoid crops having a comparatively small genome size of approximately 515 Mb, short life cycle, and inbreeding nature (Doust *et al.*, 2009). The most recent archaeological evidence demonstrates that foxtail millet is one of the most ancient crops as its domestication in China dates back to 8,700 years ago (Lu *et al.*, 2009). According to Bennetzen *et al.* (2012), foxtail millet was domesticated from Setaria viridis in Northern China around 6000 BC. Green foxtail (*Setaria viridis* (L.) Beauv.), a weed distributed worldwide, is the presumed wild progenitor of domesticated foxtail millet, based on cytological evidence and RAPD, AFLP, and other markers (Fukunaga and Kato 2003, Fukunaga *et al.*, 1997). But the number of domestication centers and the age of domestication of foxtail millet remains controversial (Fukunaga *et al.*, 2006).

In its long history of cultivation, foxtail millet has been adapted to the conditions of low and moderate rainfall area due to early maturity period. Because of its outstanding tolerance to drought stress foxtail millet is well suited to arid and semi-arid areas of Asia, North Africa, South and North America and has the potential to become a new model organism. Foxtail millet is closely related to the major crops of maize and sorghum as well as to the biofuel crop switchgrass (Li and Brutnell 2011; Bennetzen *et al.*, 2012; Zhang *et al.*, 2012). It requires warm weather and minimal water for rapid maturation during the hot and dry months of the year. It comes second after pearl millet in the global millets production (FAOSTAT 2005; http://faostat.fao/org/). It is commonly known as Italian millet or German millet or Russian millet or Chinese millet or Hungarian millet. With the rapid development of maize and other modern crops, foxtail millet has gradually become a minor crop in the last 80 years, but is nonetheless still widely cultivated in Asia, Europe, North America, Australia and North Africa as grain food or forage (Austin, 2006). Foxtail millet is a staple crop used extensively for food and fodder in parts of Asia and Africa. In India, the cultivation of fox millet is confined to Andhra Pradesh, Karnataka, and Tamil Nadu. The potentiality of foxtail millet is not yet exploited properly in India. The yield levels in China is 11 t/ha, whereas in India it is just ranging between 0.4-0.8t/ha suggesting a greater scope for exploitation of this millet under Indian conditions (Jiaju, 1986).

Genetic Diversity

As revealed by earlier studies, genetic diversity is low in foxtail millet (Li *et al.*, 1998), yet these analyses were based on genetic markers that only represent a subset of the possible information that can be gained regarding genomic polymorphism (Wright and Gaut, 2005). However, research at the sequence level is very limited in foxtail millet and its close relatives. although sequencing of the foxtail millet genome is now nearly completed (Doust *et al.*, 2009). Its two species, *Setaria italica* (domesticated) and *Setaria viridis* (wild progenitor), have characteristics that classify them as excellent model systems to examine several aspects of architectural, evolutionary, and physiological importance in Panicoid grasses especially the biofuel crops such as switchgrass (*Panicum virgatum*), and napiergrass (*Pennisetum purpureum*).

Physiological Studies

The low yield in foxtail millet is generally attributed to genetic, physiological and agronomic factors. All the factors influencing growth and development of crop plants are to be integrated at an optimum level for maximum production potential. The total dry matter production by crop depends on size of photosynthetic system and activity, length of growth period during which photosynthesis continues. The present investigation was carried out to find out the variation in morpho-physiological, biophysical and biochemical characters and their relationship with the productivity in foxtail millet genotypes.

The number of tillers per plant is considered as one of the important character related to yield in foxtail millet (Prasad Rao *et al.*, 1985) and little millet (Chidambaram and Palanisamy, 1996). Hanumanthrao *et al.* (1987) also reported the correlation between total dry meter accumulation and grain yield in foxtail millet. Assimilatory surface area over unit ground area gives a fairly good idea of photosynthetic capacity of the plant. These observations are in conformity with the findings of Udapudi *et al.* (1998) and Bhoite (2000) in foxtail millet

Channappagoudar *et al.* (2008): In an attempt to study the physiological basis of yield variation in 20 foxtail millet genotypes and reported that Plant height, number of tillers per plant, LAI, LAD at post an thesis period and TDM at harvest varied significantly among the genotypes. Photosynthetic rate, total chlorophyll content and NRase at flowering also varied significantly. Plant height, photosynthetic rate and number of tillers were found to have positive correlation with TDM and grain yield. Hence, these parameters can be considered as major determinants of grain yield in foxtail millet

Development pattern in little millet is an important aspect, as the phenological characters *viz.* rate of flowering, length of flowering vary with the genotypes. Therefore, the study of these aspects and the relationship with yield would help in understanding the basis of genotypic differences in productivity. The study of growth analysis would help in understanding contributions of various growth processes in accumulation of dry matter and yield.

Abiotic Stress Tolerance

Foxtail millet accessions are mostly tolerant to abiotic stresses, and further exploiting these traits can enhance the utility of foxtail millet for dry and neglected lands and would also be further useful in development of stress tolerant cultivars (Li and Wu, 1996; Dekker, 2003). Foxtail millet has one of the largest collections of cultivated as well as wild-type germplasm rich with phenotypic variations and hence provides prospects for association mapping and allele-mining of elite and novel variants to be incorporated in crop improvement programs. Most of the foxtail millet accessions can be primarily abiotic stress tolerant particularly to drought and salinity, and therefore exploiting these agronomic traits can enhance its efficacy in marker-aided breeding as well as in genetic engineering for abiotic stress tolerance. In addition, the release of draft genome sequence of foxtail millet would be useful to the researchers worldwide in not only discerning the molecular basis of biomass production in biofuel crops and the methods to improve it, but also for the introgression of beneficial agronomically important characteristics in foxtail millet as well as in related Panicoid bioenergy grasses

Drought Tolerance

During the long process of its domestication and improvement, foxtail millet has gradually adapted to arid and semiarid climates. Its morphological and anatomical features, such as thick cell walls, densely distributed roots and small leaf areas, give foxtail millet the strong drought tolerance. Strikingly, foxtail millet has been reported to have gained higher water use efficiency than maize, sorghum and wheat (Diao 2005, 2007). Further for 1 g of dry biomass, foxtail millet requires only 257 g of water which is much lower than maize and wheat which require 470 and 510 g water, respectively (Diao, 2007; Li and Brutnell, 2011). The prominent drought resistance features of foxtail millet, as well as its small sequenced diploid genome, inbreeding nature, short growing duration and extensive germplasm collection, make foxtail millet an excellent candidate model system for investigating abiotic challenges. With the severe environmental challenges facing agriculture today, it is of considerable significance to understand how transcriptome-level changes mediate plant drought response and tolerance. Suppression subtractive hybridization (SSH) and cDNA-amplified fragment length polymorphism (cDNA-AFLP) methods have been used to study the stress-induced transcriptome in foxtail millet (Jayaraman *et al.*, 2008; Lata *et al.*, 2010). However, these studies suffered from a lack of comprehensive and extensive datasets. Here, we used deep sequencing technology (Nobuta *et al.*, 2007; Wang *et al.*, 2009; Li *et al.*, 2010; Lu *et al.*, 2010; Metzker 2010; Trapnell *et al.*, 2010; Wang *et al.*, 2010a; Kakumanu *et al.*, 2012) to investigate the genome wide transcriptome reconfiguration of foxtail millet challenged by polyethylene glycol-simulated drought stress in a high-throughput manner. We found that numerous protein-coding genes as well as partial ncRNAs exhibited drought-regulated expression patterns. To better understand the biological functions of drought-responsive genes, the enriched Gene Ontology (GO) categories, drought responsive metabolic pathways and transcriptional regulation were examined. Furthermore, subsets of genes that were conserved in foxtail millet, maize, sorghum and A*Rabi*dopsis were characterized. The impact of drought responsive siRNAs and lncRNAs on gene expression under stimulated water deficit conditions was investigated extensively. In addition, to further explore the conservation of lncRNAs, syntenic blocks in foxtail millet and sorghum were identified and analyzed.

Dai *et al.* (2012) studied the changes of net photosynthetic rates (P_N), the contents of chlorophyll, soluble protein, malondialdehyde (MDA), superoxide radical (O_2^{-*}) and hydrogen peroxide (H_2O_2), and the activities of superoxide dismutase (SOD; EC1.15.1.1) and catalase (CAT; EC 1.11.1.6) in the leaves of two foxtail millet varieties (05-61 and Jingu3) under drought stress from 14 days after anthesis to maturity. P_N, the contents of chlorophyll and soluble protein and the activities of SOD and CAT increased and then declined as the plants aged, whereas the accumulation of MDA, O_2^{-*} gradually increased with senescence. Although the leaves of two varieties generally shared a similar pattern of senescence, their rates of aging differed. The activities of SOD and CAT in leaves of 05-61 declined more quickly than Jingu3, particularly in the late stages of grain filling. Taken together, the results implicated that the alleviation of leaf senescence played an important role in promoting grain filling and enhancing the yield and quality of 05-61 in the rain-fed agriculture area

Foxtail millet is as an elite drought tolerant crop due to its high WUE and short life cycle (Zhang *et al.*, 2007c; Lata *et al.*, 2011b). A total of 17,799 foxtail millet accessions including 17,313 landraces and 486 elite cultivars, were categorized into five grades of drought tolerance using seedling survival following repeated drought stress, where grade 1 accessions were the most drought tolerant and comprised more elite cultivars (Li, 1991, 1997). Similarly, a rapid and easy screen for drought tolerance using polyethylene glycol (PEG-6,000) and mannitol was developed in China and relative water content and germination rates were identified as markers of drought tolerance during seedling stage (Zhang *et al.*, 2005; Zhu *et al.*, 2008). Zhang *et al.* (2005) reported that foxtail millet seedlings can be screened for drought tolerance using 20 per cent PEG-6000. More recently, we have also screened a core set of 107 Indian foxtail millet cultivars at seedling stage for their drought tolerance using 20 per cent PEG-6000 (Lata *et al.*, 2011a). On the basis of primary and secondary screening, two highly tolerant (cv. IC-403579 and cv. Prasad) and two highly sensitive (cv. IC-480117 and cv. Lepakshi) foxtail millet cultivars were identified. The identified group of dehydration-tolerant foxtail millet cultivars may be used for plant breeding purposes. Even though the drought-tolerance capacity of foxtail millet was ascribed to its cellular and morphological characteristics (Li, 1997), there have been very less study on differentially expressed genes that might impart drought tolerance to foxtail millet. Zhang *et al.* (2007c) examined differential gene expression in the drought tolerant cultivar 'Mar51', following stress treatment (PEG-6000) and reported several notable changes in gene expression particularly those involved in metabolism, proteolysis, and stress signaling. In our laboratory, we have examined transcriptome changes of tolerant cv. 'Prasad' at 0.5 and 6 h of dehydration stress and established that two very distinct waves of transcriptional changes take place at early and late stages of stress treatment in foxtail millet (Lata *et al.*, 2010). The expression analysis of nine up-regulated transcripts (above 2.5-fold variations) between drought tolerant and susceptible cultivars upon dehydration stress suggests their function in dehydration adaptation and tolerance in foxtail millet.

Xin *et al.* (2013) used a deep sequencing approach to generate a genome-wide transcriptome of foxtail millet after exposure to simulated drought stress to understand the processes that contribute to drought resistance. A large number of differentially expressed genes were characterized; in particular and examined the roles of small interfering RNAs (siRNAs) and long non-coding RNAs (lncRNAs) in response to a water-deficit condition. These RNAs have remained largely unexplored in previous studies of stress-induced transcriptomes. These authors found that the reduced levels of 24-nt siRNA flanking genes were associated, for the most part, with proximal up-regulated genes, indicating a potential effect of 24-nt siRNAs on drought-regulated gene expression. Several lncRNAs that responded to the simulated drought stress were also identified and found that one of them shared sequence conservation and co-linearity with its counterpart in sorghum (*Sorghum bicolor*). These findings provide new insights into drought-induced changes in the foxtail millet transcriptome.

Salinity Tolerance

In order to study salinity tolerance, Zhi *et al.* (2004) screened 260 foxtail millet accessions using relative germination rate at 1.0 and 1.5 per cent NaCl concentration respectively, and detected a wide array of variations ranging from 0 to ~100 per cent. Sreenivasulu *et al.* (1999) established differences in salt tolerance among two foxtail millet cultivars, categorizing cv. Prasad as salt tolerant and cv. Lepakshi as salt sensitive. To understand the genetics of salt tolerance, Sreenivasulu *et al.* (2004) isolated differentially expressed ESTs in the above mentioned salt tolerant and sensitive cultivars using a 711 cDNA inserts macro array filter representing 620 unigenes of a barley EST collection, and found that at high salinity, the transcripts of H2O2 scavenging enzymes along with those of some cellular metabolism genes were particularly induced in the tolerant cultivar. Veeranagamallaiah *et al.* (2007) correlated proline accumulation (increased GS and P5C reductase activities) with salinity tolerance in foxtail millet. In order to get a better perspective of the salinity stress responses at molecular level, the temporal changes in total protein profile of cv. 'Prasad' seedlings were examined under different salt stress conditions (Veeranagamallaiah *et al.*, 2008) which led to the identification of 29 differentially expressed salt responsive proteins (both up- and down-regulated). Further, in our own laboratory, the cDNA-AFLP technique was used to evaluate gene expression profiles between the above mentioned two cultivars (Jayaraman *et al.*, 2008). In this study, 27 non-redundant differentially expressed transcripts were identified which were involved in various cellular responses in the tolerant cultivar cv. 'Prasad'. More recently, macroarray analysis of the transcripts identified in cv. 'Prasad' and cv. 'Lepakshi' revealed 159 (63.9 per cent) transcripts to be differentially expressed by e"two-fold in response to 250 mM salt stress (Puranik *et al.*, 2011a). A comparison of drought (20 per cent PEG 6,000) and salinity (250 mM NaCl) induced transcripts in the tolerant cultivar revealed that only 18 genes (10 per cent) were common between the two stresses, suggesting that the drought and salinity stress perception and/or responses are mediated by distinct gene sets in foxtail millet (Puranik *et al.*, 2011a).

Karyudi and R.J. Fletcher (2003) studies the association of osmoregulation of F3 families of accession 108046 (demonstrated low osmoregulation) and the accession 108541 (demonstrated high osmoregulation) with yield components and morphological characters. F3 families of foxtail millet demonstrated variability in osmoregulation. Osmoregulation was not associated with any morphological characters but was highly associated with grain yield and harvest index. F3 families within the high osmoregulation class produced an average of 1.58 times more yield than the F3 families in the low osmoregulation class.

Tolerance to other Abiotic Stresses

Foxtail millet can be tolerant to abiotic stresses other than drought and salinity also such as low temperature, lodging and water logging. The researchers in China have developed an extremely cold tolerant foxtail millet cultivar "Liggu No. 26", that can be cultivated 385 km further north to 54°, thus lengthening the northern limit of its cultivation in China (Chen and Qi, 1993; Dwivedi *et al.*, 2012). Foxtail millet has also been reported to bear lodging constraint and thus incurring substantial losses in crop yield as well as grain quality. Tian *et al.* (2010) reported that a lodging coefficient based on stem and root traits of foxtail millet could be an apt pointer for lodging resistance in field conditions. It was also suggested that mechanical strength of the stem, heights and weights of the above and under-ground plant tissues were the most significant donors to lodging coefficient in the wild-type as well as elite cultivars (Tian *et al.*, 2010; Dwivedi *et al.*, 2012). Several lodging resistant landraces and cultivars such as 'Longgu 28' and 'Nenxian 13' have been developed in China (Dwivedi *et al.*, 2012). Water logging is another important abiotic constraint in foxtail millet. A foxtail millet cv. 'Lugu No.7' has been reported to be tolerant to water logging (Chen and Qi, 1993). Thus, detection and exploitation of unidentified variations for abiotic stress tolerance could add to the adaptation of foxtail millet to adverse environmental conditions. Accurate phenotyping is the only solution to discover, exploit and introduce new genes for abiotic stress tolerance in any crop. Further, improved understanding of the physiological and molecular basis of abiotic stress tolerance mechanisms in foxtail millet will contribute towards developing more stress-tolerant cultivars.

Proso Millet [*Panicum miliaceum* L.]

Proso millet was domesticated in Manchuria and Transcaucasia about 7,000 years ago, and introduced in Europe about 3,000 years ago, followed by its introduction to the Near East and India. Proso millet is well adapted to many soil and climatic conditions. Being a short season crop with a low water requirement, it grows further north than the other millets and also adapts well to plateau conditions and high elevations. Proso millet is found high in the mountains of the former USSR (up to 1,200 masl) and in India (up to 3,500 masl). Proso millet generally matures between 60 and 90 days after sowing and can be grown successfully in poor soils and hot dry weather. It is an easy crop to grow and it seems to be better adapted than most to primitive agricultural practices. It requires very little water, possibly the least of any cereal, and converts water very efficiently to dry matter and grain. This is not because of its drought tolerance but because of its short growing season. Proso millet is a type of short-day C4 plant with thermophilic and photophilic responses, and it can be adapted to various types of soils, particularly sandy soils with low fertility. As the resistance of broomcorn millet to salt, alkali, and drought stress, as well as its shorter growth cycle, it is usually grown for exploration of new wastelands and deserts, or as a remediation crop in the event of natural disasters.

Physiological Studies

Channappagoudar (2007) studied the physiological basis of yield variation in nine proso millet genotypes, field studies were carried out for two years. The productivity of proso millet ranged from 1.5 to 2.75 q/ha. Significant differences in plant height, number of green leaves and tillers per plant, LAI, LAD and TDM were noticed. The TDM was further found to be dependent on photosynthetic rate, total chlorophyll content and NR'ase activity. Test weight and partitioning efficiency (HI) formed major determinants of grain yield. The genotype GPUP-8 had higher plant height (91.0 cm), number of tillers/plant (5.3), LAI (6.14), TDM (1200.5 g/m2) along with higher photosynthetic rate (16.7 μ mol/m2/s), NR'ase (505.5 μ mol/g fresh wt/hr), total chlorophyll content (3.48 mg/g fresh wt).

This crop can survive with sub-optimal rate of fertilizer and without organic manures (Dai *et al.*, 2008; Dai *et al.*, 2009). Water and nutrients are the limiting factors for getting good harvest. Little information is known about efficient water management in combination with nutrient management (Dai *et al.*, 2008). Leaf senescence and ROS metabolism is very important for adaptation to drought. But very little is known about this in proso millet. Dai *et al.* (2011) studied the physiological and biochemical mechanisms underlying the synergetic senescence between root and leaves with Yumi 3 as the experiment material. The results indicated that the vigour of root and the leaves activity of SOD, CAT and POD fluctuated from the jointing stage to maturing stage, namely, rose up to the peak and then declined. The soluble protein content of root and leaves decreased gradually whilst their membrane lipid peroxidation product MDA increased substantially. Senescence of root and leaves was closely related. Compared with leaves, the antioxidative enzyme (SOD, CAT and POD) activity of root was a little weak and the level of its soluble protein was relatively low. Based on the above comprehensive analysis, enhancing rhizosphere nutrition and root vigour may be conducive to delay leaves senescence after the jointing stage of broomcorn millet. These authors further studied the changes in the contents of chlorophyll and soluble protein, malondialdehyde (MDA) and the activities of anti-oxidative enzymes: superoxide dismutase (SOD, EC1.15.1.1), catalase (CAT, EC 1.11.1.6) and peroxidase (POD,EC 1.11.1.7) in the leaves of three broomcorn millet varieties (Ning-Mi 13,Nei-Mi 5 and Yu-Mi 3) from anthesis to mature stage (Dai *et al.*, 2011). The results showed that the contents of chlorophyll and soluble protein and the activities of SOD, CAT and POD in leaves of three broomcorn millet varieties rose and then declined with the progression of leaf senescence, the content of activities of antioxidative enzymes of flag leaf was high and declined slowly, and the content of MDA was low and rose gradually with senescence. Although the leaves of the three varieties generally presented a similar tendency in their senescence, their senescence progresses significantly differed. Among the varieties, the activities of SOD, CAT and POD reduced slowly in Ning-Mi 13, followed by Nei-Mi 5 and Yu-Mi 3. These changes were observed significantly in the middle and late stage of grain filling. Taken together, the results implicated a vitally important role played by leaf senescence alleviation in promoting grain filling and enhancing the yield and quality of Ning-Mi 13 in the rain-fed agriculture area.

Karyudi and Fletcher (2002) conducted experiments involving 14 accessions of *Panicum miliaceum* L. (Proso millet) and 11 accessions of *Setaria italica* L. (Foxtail millet) have demonstrated variability in the degree of osmoregulative capacity among these accessions. Birdseed millet is generally claimed to be sensitive to drought stress, apparently because of a shallow root system. Accessions with high osmoregulative capacity demonstrate at least some drought tolerance. Osmoregulative capacity was measured on ag leaves of headed millet plants in pots undergoing water stress in a controlled environment chamber. Osmoregulative capacity was determined from the relationship between osmotic potential and leaf water potential; and the logarithmic relationship between osmotic potential and relative water content. The group of accessions of *S. italica* showed an overall level of osmoregulative capacity which was greater than that observed for the group of *P. miliaceum* accessions. Four accessions of

S. italica (108042, 108463, 108541 and 108564) and one accession of *P. miliaceum* (108104) demonstrated high osmoregulative capacity. Differences of 1.05 MPa or more between observed and estimated osmotic potential were found at relative water contents of 80 per cent among these accessions. The extent of osmoregulative capacity was associated with osmotic potential at full turgor and the rate of decline in osmotic potential as leaf water potential declined.

Barnyard Millet [*Echinochloa colona* (L.)]

Barnyard millet is an important minor millet crop well adapted to low and moderate rainfall areas (500-700 mm) due to its early maturity character. In India, the cultivation of Barnyard millet is mainly confined to Tamil Nadu, Andhra Pradesh, Karnataka and Uttar Pradesh. The grains of barnyard millet are low in phytic acid and rich in iron and calcium (Sampath *et al.*, 1986). The yield levels in Japan is more than 10 t/ha whereas, in India it is 15-20 q/ha thereby indicating a greater scope for exploitation of this millet under Indian condition. The low yield in barnyard millet is generally attributed to genetic, physiological and agronomic factors. The factors influencing growth and development of crop plants are to be integrated at an optimum level for maximum production potential. Growth and yield analysis in crop plants help in understanding the contribution of various growth processes in dry mater accumulation and yield. The present investigation was carried out to study the genotypic variation in morop-physiological traits and biochemical parameters and their relationship with productivity and to identify the suitable ideotype for higher productivity.

Channappagoudar *et al.* (2008) studied the influence of morpho-physiological and biochemical traits on the productivity of barnyard millet genotypes. Positive association between plant height, number of green leaves and tiller number per plant with TDM was noticed. The grain yield was found to be dependent on total dry matter production at harvest and its partitioning efficiency. The genotypes RAU-11, VL-29 and K-1 recorded higher LAI at flowering (2.69, 2.64 and 2.64, respectively), LAD during post anthesis period (53.8, 52.7 and 50.7 days, respectively), TDM at maturity (7265.0, 7147.4 and 7119.5 g/m2, respectively), number of tillers per plant (9.5, 9.4 and 9.3, respectively), photosynthetic rate (34.1, 31.0 and 31.6μmol/m2/s, respectively), chlorophyll content (4.30, 4.40 and 4.27, mg/g fresh wt., respectively), NR'ase (600.0, 590.0 and 578.5, μ mol/g fresh wt/hr, respectively) and grain yield (14.30, 14.15 and 14.00 q/ha, respectively). These genotypes also pocessed higher assimilates translocation efficiency (19.6-19.8 per cent).

Little Millet (*Panicum miliare* L.)

Little millet (*Panicum miliare* L.) is an important minor millet of low and moderate rainfall areas. Besides India, it is cultivated in Nepal and Western Burma. The potentiality of little millet has not been exploited in India and the yield levels are very low there by indicating a greater scope for exploitation of this millet under Indian condition. Nutritionally the grains are rich in proteins, minerals and vitamins.

In little millet very little work has been done with respect to physiological indices associated with productivity. Thus, it becomes imperative to find out variations in morpho-physiological, biophysical and biochemical traits among the genotypes and relationship of these parameters with productivity. Extensive studies have been made in other cereals which emphasized the importance of such traits in understanding the process of yield formation based on which plant ideotypes have been defined. However, such studies have been meagre in millets in general and little millet in particular.

Channappagoudar *et al.* (2007) conducted studies to find out the relation of morpho-physiological traits with yield potential of little millet in 14 genotypes. Genotypes differed significantly for plant height, number of green leaves and number of tillers per plant, LAI and LAD. The photosynthetic rate also differed significantly among the genotypes leading to significant variation in total dry mater accumulation. However, the productivity was not only dependent on higher values of physiological traits but also on partitioning efficiency and test weight. Chlorophyll content and Nitrate reductase activity was higher in TNAU-63 (2.78 mg/g fresh weight and 528 µ mol/g fresh weight/hr) followed by TNAU-18 (2.71 mg/g fresh weight and 510 µ mol/g fresh weight/hr). The genotypes TNAU-63, DPI-1864 and TNAU-18 had higher values for most of the physiological traits and yield components leading to higher grain yield in these genotypes as compared to others.

Kodo Millet [*Paspalum scrobiculatum* L.]

Kodo millet was domesticated in India almost 3,000 years ago. It is found across the old world in humid habitats of the tropics and subtropics. It is a minor grain crop in India and most important in the Deccan plateau. The fibre content of the whole grain is very high. Kodo millet has around 11 per cent protein, and the nutritional value of the protein has been found to be slightly better than that of foxtail millet, but comparable to that of other small millets. As with other food grains, the nutritive value of kodo millet protein can be improved by supplementation with legume protein. Madhya Pradesh (MP) is the top producing state in India with total production of 41200 tonnes (Padulosi *et al.*, 2009).

Kodo millet is a C4 plant with virtual absence of photorespiration and having certain specialised biochemical feature resulted in higher optimum temperature for photosynthesis makes it more adaptive to tropics and subtropics. Kodo millet with high LAI, CGR and NAR results in high grain yield (Sharma, 1988). Nirmala devi *et al.* (1988) suggested that LAI at all the growth stages, dry matter and CGR at flowering and grain filling, and NAR at flowering contributed are likely to be effective as direct selection for grain yield in Kodo millet. Rao (1989) suggested an ideotype of Kodo millet should be taller with long internodes, long panicle and high dry matter accumulation in all plant parts. The dough stage is the most critical stage of crop growth with maximum influence on the yield potential of kodo millet genotypes.

Kodo millet is cultivated in extreme conditions of soil, climate, and rainfall where moisture deficiency is the recurrent feature. The genotypes possessing tolerance to drought thus produces higher grain yield. So far not much studies were conducted and nothing much is known about the mechanisms of drought/abiotic stresses tolerance in Kodo Millert. There is a need of identifying morpho-physiological characters related to drought tolerance and better seedling establishment under early season moisture stress. A through investigations are needed to understand the drought resistance mechanism in Kodo Millet which will certainly help the breeders the high yielding drought tolerant lines of this crop.

Research Gaps and Physiological Research Agendas in Small Millets

- Small millet serves as an excellent grass genetic model system for studying abiotic stress tolerance mechanisms, physiology of C4 photosynthesis and biomass production, domestication process, population structure, comparative mapping and evolution.
- More efforts are needed to collect and characterize landraces in small millet to identify potential germplasm resources for physiological traits to use in crop improvement programs.
- Comprehensive investigations on physiological, biophysical and biochemical traits and their association with productivity of small millets particularly of Foxtail, Proso, Kodo, Little and Barnyard *are* needed
- Through investigations are needed to understand the drought resistance mechanism in Small Millet which will certainly help the breeders the high yielding drought tolerant lines.
- Cloning and characterization of some of the known and unknown up-regulated genes in foxtail millet to understand the molecular basis of drought tolerance.
- More importantly with the completion of the draft genome sequence of foxtail millet (*S. italica*) and as well as the availability of efficient transformation systems, the genetic and molecular studies will progress swiftly.
- By comparing the genome sequences of important cereal and bioenergy crops with that of foxtail millet, it would be easy to find sequence variation across species and relate these differences to useful traits.
- Moreover, the foxtail millet DNA sequence information available to researchers would enable identification of genes and functional molecular markers linked with various traits of agronomic importance to be utilized in crop improvement programmes.
- Development of DNA-based functional markers, genetic linkage maps of varying density and association of QTLs with various physiological/agronomic traits would accelerate marker-aided breeding.
- Breeding strategies could be practiced to incorporate abiotic stress tolerance traits into the improved genetic background of small millets.
- Explore the nutritional and medicinal properties of foxtail and millets as well as enhanced research and development to increase its production and productivity globally.

References

1. Aarati P, Krishnaprasad B.T., Ganesh kumar, Savitha M, Gopalakrishna R, Ramamohan G, Udayakumar M. (2003). Expression of an ABA responsive 21 kDa protein in finger millet (*Eleusine coracana* Gaertn.) under stress and its relevance in stress tolerance. *Plant Science* 164 (1): 25-34.

2. Agarwal S, Kumar A, Singh PK, Singh A (2011). Responses of some genotypes of finger millet (*Eleusine coracana* Gaertn.) for their salt tolerance. *Int J Curr Res* 3: 045–050.

3. Anantharaju P. (2001). 'Studies on genetic variability, association and diversity in finger millet (*Eleusine coracana* (L.) Gaertn.) for grain yield, yield attributes and salinity tolerance'. *M Sc (Ag) thesis*, CPBG, TNAU, Coimbatore.

4. Anjaneyulua, E., Palle Surender Reddyb, Merla Srilakshmi Sunitab, Polavarapu B. Kavi Kishorb, Balaji Merigaa (2014). Salt tolerance and activity of antioxidative enzymes of transgenic finger millet overexpressing a vacuolar H+-pyrophosphatase gene (SbVPPase) from *Sorghum bicolor*," *Journal of Plant Physiology* 171: 789–798.

5. Assefa, A and Fetene, M. (2013). Agro-morphological, Physiological and Yield related Performances of Finger Millet [*Eleusine coracana* (l.) Gaertn.] Accessions Evaluated for Drought Resistance under Field Condition. *Asian Journal of Agriculture and Rural Development*, 3(10) 2013: 709-720.

6. Austin DF (2006). Fox-tail Millets (Setaria: Poaceae)–Abandoned Food in Two Hemispheres. *Economic Botany* 60(2): 143-158.

7. Bennetzen JL, Schmutz J, Wang H, Percifield R, Hawkins J, Pontaroli AC, Estep M, Feng L, Vaughn JN, Grimwood J, Jenkins J, Barry K, Lindquist E, Hellsten U, Deshpande S, Wang X, Wu X, Mitros T, Triplett J, Yang X, Ye CY, Mauro-Herrera M, Wang L, Li P, Sharma M, Sharma R, Ronald PC, Panaud O, Kellogg EA, Brutnell TP, Doust AN, Tuskan GA, Rokhsar D, Devos KM (2012). Reference genome sequence of the model plant Setaria. *Nat Biotechnol* 30: 555–561.

8. Bhatt D, Negi M, Sharma P, Saxena SC, Dobriyal AK, *et al.* (2011) Responses to drought induced oxidative stress in five finger millet varieties differing in their geographical distribution. *Physiology and Molecular Biology of Plants* 17: 347–353.

9. Bhoite, A.G., 2000, Physiological indices for higher productivity in foxtail millet *Ph.D thesis*, University of Agricultural sciences, Dharwad, India.

10. Channappagoudar BB, S. M. Hiremath, N. R. Biradar, R. V. Koti And T. D. Bharamagoudar (2008). Physiological Basis Of Yield Variation In Foxtail Millet. *Karnataka J. Agric. Sci.*, 20(3): 481-486.

11. Channappagoudar, B. B., Hiremath S. M., Biradar, N. R., Koti, R. V., and Bharamagoudar, T. D. (2007). Morpho-physiological Traits Influencing the Grain Yield Potential in Little Millet. *Karnataka J. Agric. Sci.*, 20(3): 473-476.

12. Channappagoudar, B. B., S. M. Hiremath, N. R. Biradar, R. V. Koti and T. D. Bharamagoudar (2008). Influence of Morpho-physiological and Biochemical Traits on the Productivity of Barnyard Millet. *Karnataka J. Agric. Sci.*, 20(3): 477-480.

13. Channappagoudar; B B., S M Hiremath;,N R Biradar; R V Koti; T D Bharamagoudar (2007). Variation in Morpho-physiological Traits and Dry Matter Accumulation that Determine the Yield of Proso millet. *Karnataka Journal of Agricultural Sciences*; 20 (3): 469-472.

14. Chen J, Qi Y. 1993. Recent developments in foxtail millet cultivation and research in China. In: Riley KW, Gupta SC, Seetharam A, Mushonga JN, eds. *Advances in Small Millets*. New Delhi: Oxford.

15. Chetan S and Malleshi N G. (2007). Finger mille polyphenols: Characterization and their nutraceuticalpotential. *American J. Food Technol.* 2: 582–592.

16. Chidambaram, S. and Palanisamy, S., 1996, Dry matter production and harvest index in little millet (*Panicum sumatrense* Roth Ex. Roem and Schult). *Madras Agricultural Journal*, **83**: 15-17

17. Dai HP, Feng BL, Gao JF, Gao XL,Wang PK, ChaiY (2008) Senescence and activate oxygen metabolism of leaf in *Panicum miliaceum* L. *Agr Res Arid Areas*. 26(1): 217-220.

18. Dai HP, Jia GL, Feng BL, Qu JQ, Sun SM, Qin XW, Ren XM (2009) Accumulation of sucrose and starch contents during the grain filling stage of broomcorn millet in two cultural practices. *J China Agr Univ*. (Nat. Sci. Ed.) 6: 37–40.

19. Dai HP, Jia GL, Lu, Chao., Wei, A. Z., Feng B.L.and Zhang S. Q. (2011) Studies of synergism between root system and leaves senescence in broomcorn millet (*Panicum miliaceum* L.). *Food, Agriculture and Environment* 9(2) : 177-180.

20. Dai, Hui-Ping., Shan, Chang-juan.,Wei, An-Zhi., Yang, Tuxi., Sa, Wen-Qing., and Feng Bai-Li (2012). Leaf senescence and photosynthesis in foxtail millet [*Setaria italica* (L.) P. Beauv] varieties exposed to drought conditions. *Australian Journal of Crop Science* 6(2): 232-237.

21. Dai, Hui-Ping., Zhang, Pan-Pan., Lu, Chao., Jia, Gen-Liang., Song, Hui., Ren, Xue-Min., Chen, Jia., Wei, An-Zhi., Feng Bai-Li., and Zhang She-Qi. (2011). Leaf senescence and reactive oxygen species metabolism of broomcorn millet (*Panicum miliaceum* L.) under drought condition. *Australian Journal of Crop Science* 5(12): 1655-1660.

22. Das, R., Sujatha M., Pandravada, S.R., and Sivasankar, A. (2013). Trait relationship and path coefficient analysis in finger millet (*Eleusine coracana* L.). *Journal of Progressive Agriculture* 4(1): 81-84.

23. Diao Xianmin (2005) Advances in foxtail millet biotechnology and its future directions. *J Hebei Agric Sci* 9: 61–68.

24. Diao Xianmin (2007) Foxtail millet production in China and its future development tendency. In: Chai Y, Wan FS (eds) *The industrial development of*

China special crops. Chinese Agricultural Science and Technology Press, Beijing, pp. 32–43.

25. Dinesh-Kumar SP, Sashidhar VR, Prasad TG, Udayakumar M, Seetharam A. (1987) Solute accumulation, solute potential, germinability and seedling vigour of seeds of finger millet (*Eleusine coracana* Gaertn.) raised under rain-fed conditionsand under irrigation. *Plant, Cell and Environment* 10: 661–665.

26. Doust AN, Devos KM, Gadberry MD, Gale MD, Kellogg EA (2005). The genetic basis for inflorescence variation between foxtail and green millet (poaceae). *Genetics* 169(3): 1659-1672.

27. Doust AN, Devos KM, Gadberry MD, Gale MD, Kellogg EA(2004). Genetic control of branching in foxtail millet. *Proc Natl Acad Sci USA* 101(24): 9045-9050.

28. Doust AN, Kellogg EA (2006). Effect of genotype and environment on branching in weedy green millet (*Setaria viridis*) and domesticated foxtail millet (*Setaria italica*) (Poaceae). *Mol Ecol* 15(5): 1335-1349.

29. Doust AN, Kellogg EA, Devos KM, Bennetzen JL(2009). Foxtail millet: a sequence driven grass model system. *Plant Physiol* 149(1): 137-141.

30. Dwivedi S, Upadhyaya H, Senthilvel S, Hash C, Fukunaga K, Diao X, Santra D, Baltensperger D, Prasad M. 2012. Millets: Genetic and Genomic Resources. In. Janick J (ed) *Plant Breed Rev*, Vol. 35. USA: John Wiley and Sons, Inc., pp. 247–375.

31. Fukunaga K, Domon E, Kawase M (1997). Ribosomal DNA variation in foxtail millet, *Setaria italica* (L.) P. Beauv., and a survey of variation from Europe and Asia. *Theoretical and Applied Genetics* 95(5-6): 751-756.

32. Fukunaga K, Ichitani K, Kawase M (2006). Phylogenetic analysis of the rDNA intergenic spacer subrepeats and its implication for the domestication history of foxtail millet, *Setaria italica*. *Theor Appl Genet* 113(2): 261-269.

33. Fukunaga K, Kato K (2003). Mitochondrial DNA variation in foxtail millet, *Setaria italica* (L.) P. Beauv. *Euphytica* 129(1): 7-13.

34. Goud, J.V. and Laxmi, P.V. (1972). Ragi varieties for drylands. *Current Res.*, 1(1): 11-12.

35. Government of India (2009). Area, Production and Productivity of Millets in India. Directorate of Millets, Ministry of Agriculture, Jaipur, India.

36. Hanumantharao, G.V., Satyamarayana, C.L. (1987). Chorecter association and path analysis of physiological determinants of grain yield in certain millet genotypes. *Andhra Agricultural Journal*, 34: 300 -303.

37. Hebbar SS, Nanjappa HV, Ramachandrappa BK (1992). Performance of finger millet after harvest of early *Kharif* crops under rainfed conditions. *Crop Research* 5: 92–97.

38. Hema, R., Vemanna, R.S., Sreeramulu, S., Reddy, C. P., Senthil-Kumar, M. and Udayakumar, M. (2014). Stable Expression of *mtlD* Gene Imparts Multiple Stress

Tolerance in Finger Millet. *PLoS ONE* 9(6): e99110. doi: 10.1371/journal.pone.0099110.

39. Jayaprakash TL, Sashidhar VR, Prasad TG, Udaykumar M (1992). Photoperiodic response of advanced breeding lines and cultivars of finger millet under irrigated and rainfed conditions. *Plant Physiology and Biochemistry* 19: 49–54.
40. Jayaraman A, Puranik S, Rai NK, Vidapu S, Sahu PP, Lata C, Prasad M. 2008. cDNA-AFLP analysis reveals differential gene expression in response to salt stress in foxtail millet (*Setaria italica* L.). *Mol. Biotechnol* 40: 241–251.
41. Jiaju C (1986) Importance and genetic resources of small millets with emphasis on foxtail millet (*Setaria italica*) in China. In '*Small millets in global agriculture*'. (Eds A Seetharam, KW Riley, G Harinarayana) pp. 93–100. (Oxford and IBH Publishing Co Pty Ltd: New Delhi).
42. Kakumanu A, Ambavaram MM, Klumas CM, Krishnan A, Batlang U, Myers E, Grene R, Pereira A (2012) Effects of drought on gene expression in maize reproductive and leaf meristem tissue revealed by RNA-Seq. *Plant Physiol* 160: 846–867.
43. Kalarani MK, Sivakumar R, Mallika V, Sujatha KB (2001a) Effect of seed soaking treatments in late sown finger millet under rainfed condition. *Madras Agricultural Journal* 88: 509–511.
44. Kalarani MK, Thangaraj M, Sivakumar R, Vanangamudi M, Srinivasan PS (2001b) Ameliorants for late sown finger millet under rainfed condition. *Indian Journal of Plant Physiology* 6: 435–437.
45. Kaliappan R, Ramachandran M and Rajagopal B. 1967. Effect of salinity on the South Indian field crops: Germination and early vigour of ragi (*Eluesirze coracana* (L.) Gaertn.). *Madras Agricultural Journal 54: 619-23.*
46. Kandpal R. P. and Appaji Rao N (1985) Changes in the levels of polyamines in ragi (*Eleusine coracana*) seedlings during water stress. *Biochemistry International* 11: 365–370.
47. Kandpal R.P. and Appaji Rao N. (1984). Water stress induced alterations in ornithine aminotransferase of ragi (*Eleusine coracana*): Protection by proline against heat inactivation anddenaturation by urea and guanidinium chloride. *J. Biosci.,* 6(1): 61-67.
48. Kandpal, R. P. and Appaji Rao, N. (1982). *Biochemistery International,* 5, 297.
49. Kandpal, R. P., Vaidyanathan, C. S., Udaya Kumar, M., Krishnasastry, K. S. and Appaji Rao, N. (1981). *J. Biosci.,* **3,** 361.
50. Karyudi and R. J. Fletcher (2002). Osmoregulative capacity in birdseed millet under conditions of water stress. I. Variation in *Setaria italica* and *Panicum miliaceum*). *Euphytica* 125: 337–348.
51. Karyudi and R. J. Fletcher. (2003). Osmoregulation in birdseed millet under conditions of water stress II. Variation in F3 lines of *Setaria italica* and its relationship to plant morphology and yield. *Euphytica* 132: 191–197.

52. Krishnasastry, K.S., Udayakumar M., and Viswanath H.R. (1982). Desirable Plant Characteristics in Genotypes of Finger Millet (*Eleusine coracana* Gaertn.) for Rainfed Conditions. *Proc. Indian natn. Sci. Acad.* B48 (2) : 264-270.

53. Kumari LP and Sumathi S (2002). Effect of consumption of finger millet on hyperglycemia in non-insulin dependent diabetes mellitus (NIDDM) subjects. *Plant Foods Hum Nutr*. 57: 205–13.

54. Lata C, Bhutty S, Bahadur RP, Majee M, Prasad M. 2011a. Association of an SNP in a novel DREB2-like gene SiDREB2 with stress tolerance in foxtail millet [*Setaria italica* (L.)]. *J Exp Bot* 62: 3387–3401.

55. Lata C, Gupta S, Prasad M. (2013). Foxtail millet: a model crop for genetic and genomic studies in bioenergy grasses. *Critical Reviews in Biotechnology* 33(3): 328-43.

56. Lata C, Jha S, Dixit V, Sreenivasulu N, Prasad M. 2011b. Differential antioxidative responses to dehydration-induced oxidative stress in core set of foxtail millet cultivars [*Setaria italica* (L.)]. *Protoplasma* 248: 817–828.

57. Lata C, Sahu PP, Prasad M (2010) Comparative transcriptome analysis of differentially expressed genes in foxtail millet (*Setaria italica* L.) during dehydration stress. *Biochem Biophys Res Commun* 393(4): 720–727.

58. Li P, Brutnell TP. 2011. *Setaria viridis* and *Setaria italica*, model genetic systems for the Panicoid grasses. *J Exp Bot* 62: 3031–3037.

59. Li P, Ponnala L, Gandotra N, Wang L, Si Y, Tausta SL, Kebrom TH, Provart N, Patel R, Myers CR, Reidel EJ, Turgeon R, Liu P, Sun Q, Nelson T, Brutnell TP (2010) The developmental dynamics of the maize leaf transcriptome. *Nat Genet* 42(12): 1060–1067.

60. Li Y and Wu S (2006)Traditional maintenance and multiplication of foxtail millet (*Setaria italica* (L.) P. Beauv.) landraces in China. *Euphytica* 87(1): 33-38.

61. Li Y-M. 1991. A study on the identification of drought-resistance on millet germplasm (in Chinese. English abstract). *Acta Agril Boreali-Sinica* 6: 20–25.

62. Li Y-M. 1997. Breeding for foxtail millet drought tolerant cultivars (in Chinese). In: Li Y, ed., *Foxtail millet breeding*. Beijing: Chinese Agr Press, p. 421–446.

63. Lu H, Zhang J, Liu KB, Wu N, Li Y, Zhou K, Ye M, Zhang T, Zhang H, Yang X, *et al.* (2009) Earliest domestication of common millet (*Panicum miliaceum*) in East Asia extended to 10,000 years ago. *Proc Natl Acad Sci USA* 106(18): 7367-7372.

64. Lu T, Lu G, Fan D, Zhu C, Li W, Zhao Q, Feng Q, Zhao Y, Guo Y, Huang X, Han B (2010) Function annotation of the rice transcriptome at single-nucleotide resolution by RNA-seq. *Genome Res* 20(9): 1238–1249.

65. Mahadevaiah C. 2002. 'Heterosis, combining, ability and salt tolerance studies in finger millet (*Eluesine coracatla* (L.) Gaertn.). *M.Sc . (Ag.) Thesis*, TNAU, Coimbatore.

66. Maqsood, M. and S.N. Azam Ali, S.N. (2007). Effects of environmental stress on growth, radiation use efficiency and yield of finger millet (*Eleucine coracana*). *Pak. J. Bot.*, 39(2): 463-474.

67. Metzker ML (2010) Sequencing technologies—the next generation. *Nat Rev Genet* 11(1): 31–46.

68. Neshamba, Mubaiwa Saul (2010). Variability for Drought Tolerance in finger millet [*Eleusine coracana* (L.)] accessions from Zambia. *M.Sc. dissertation* submitted to the School of Agricultural Science, The University of Zambia, Lusaka.

69. Nobuta K, Venu RC, Lu C, Belo A, Vemaraju K, Kulkarni K, Wang W, Pillay M, Green PJ, Wang GL, Meyers BC (2007) An expression atlas of rice mRNAs and small RNAs. *Nat Biotechnol* 25(4): 473–477.

70. Onkware A O. 1993. The germination, growth and grain production of salt stressed finger millet (*Eleusine coracana* (L.) Gaertn.). *Indian Journal* of *Plant Physiology* 3 6: 77-8 1.

71. Padulosi S, Mal B, Bala Ravi S, Gowda J, Gowda KTK, Shanthakumar G, Yenagi N and Dutta M (2009). Food Security and Climate Change: Role of Plant Genetic Resources of Minor Millets. *Indian J. Plant Genet. Resour. 22(1):* 1-16

72. Panigarh A K, Misra M K and Misra B N. 1978. Effect of sodium chloride stress on seed germination and seedling growth of local varieties of *ragi* (*Eluesirie coracatla* (L.) Gaertn.). *lndian Journal of Ecology 5:* 98-1 01.

73. Patil, M.S.and Mohankumar, H.D. (1988). Studies on genetic variability for yield and its components in foxtail millet (*Setaria italica* L). *Karnataka Journal of Agricultural Sciences,* **2** : 165- 169.

74. Prasad SV, Rao GG, Rao GR (1980). Studies on salt tolerance of ragi (*Eleusine coracana* Gaertn): Germination and free proline accumulation. *Proceedings of the Indian Academy of Sciences* 89: 481–484.

75. Prasadarao, G. P. Nagorajara, M. And Anjannappa, M., 1985. Genetic variability in seteria. *Andra Agricultural Journal,* 32: 34-36.

76. Puranik S, Jha S, Srivastava PS, Sreenivasulu N, Prasad M. 2011a. Comparative transcriptome analysis of contrasting foxtail millet cultivars in response to short-term salinity stress. *J Plant Physiol*. 168: 280–287.

77. Rachie K V and Peters L V. 1982. *Tlie Eleusines,* pp 2-3. Land O. Lakes Ltd, Dodge Iwawe, USA.

78. Rahman, H., N. Jagadeeshselvam, R. Valarmathi, B. Sachin, R. Sasikala, N. Senthil, D. Sudhakar, S. Robin, and Raveendran Muthurajan (2014). Transcriptome analysis of salinity responsiveness in contrasting genotypes of finger millet (*Eleusine coracana* L.) through RNA sequencing. *Plant Mol Biol* (2014) 85: 485–503.

79. Rai S, Gaur A.K., Guru S.K., Agrawal S., and Arora S. (2009). *In vitro* evaluation of relative tolerance of *eleucine coracana* genotypes against water deficit stress. *Indian J. Plant Physiol.*, 14 (1) : 82-87.

80. Reddy VC, Patil VS (1983) Influence of soil moisture stress on yield and its components of finger millet. *Mysore Journal of Agricultural Sciences* 17: 325– 328.

81. Reddy, CVCM., Reddy, PVRM., Munirathnam, P. and Gowda, J. (2013). Studies of genetic variability in yield and yield attributing traits of Finger millet [*Eleusine coracana* (l.) Gaertn]. *Indian J. Agric. Res.*, 47 (6) : 549 - 552.

82. Sampath, T.V., Razvi, S.M., Singh, D.N. and Bandale, K.V.(1986). Small millets in Indian Agriculture. In: *Small Millets in Global Agriculture*. Ed. Seetaram, A.Riley, KU. And Hariyana, G.Oxford and IBH Publishing Co. Pvt. New Delhi, p.34.

83. Sashidhar VR, Gurumurthy BR, Prasad TG, Udayakumar M, Seetharam A, *et al.* (1986) Genotypic variation in carbon exchange rate functional leaf area and productivity in finger millet (*Eleusine coracana*) an approach to identify desirable plant types for higher water use efficiency under rainfed conditions. *Field Crops Research* 13: 133–146.

84. Shailaja H, Thirumeni S (2007). Evaluation of salt-tolerance in finger millet (Eleusine coracana) genotypes at seedling stage. *Indian J Agric Sci* 77: 672–674.

85. Shankar, A. G., Udaya Kumar M and T. G. Prasad (1990). Genetic variability for net photosynthesis in Finger Millet (*Eleusine coracana* Gaertn) genotypes an approach to identify high CER types. *Journal of Agronomy and Crop Science* 165 (4): 240–252.

86. Shivkumar BG, Yadahalli YH (1996) Performance of finger millet genotypes under late sown dryland conditions as influenced by intercrops and intercropping systems. *Indian Journal of Agricultural Research* 30: 173–178.

87. Sreenivasulu N, Miranda M, Prakash HS, Wobus U, Weschke W. 2004. Transcriptome changes in foxtail millet genotypes at high salinity: identification and characterization of a PHGPX gene specifically upregulated by NaCl in a salt-tolerant line. *J Plant Physiol* 161: 467–477.

88. Sreenivasulu N, Ramanjulu S, Ramachandra-Kini K, Prakash HS, Shekar-Shetty H, Savithri HS, Sudhakar C. 1999. Total peroxidise activity and peroxidase isoforms as modified by salt stress in two cultivars of fox-tail millet with differential salt tolerance. *Plant Sci141*: 1–9.

89. Tian B-H, Wang J, Zhang L, Li Y, Wang S, Li H. 2010. Assessment of resistance to lodging of landrace and improved cultivars in foxtail millet. *Euphytica* 172: 295–302.

90. Trapnell C, Williams BA, Pertea G, Mortazavi A, Kwan G, van Baren MJ, Salzberg SL, Wold BJ, Pachter L (2010). Transcript assembly and quantification by RNA-Seq reveals unannotated transcripts and isoform switching during cell differentiation. *Nat Biotechnol* 28(5): 511–515.

91. Udapudi, S.S., Hiremath, S.M. Chetti, M.B. And Awari, S.B., 1998, Influence of leaf area index and total dry matter on grain yield in fox tail millet genotypes. Paper presented at *National Seminar on role of plant physiology and bio-technology in Agriculture and Industry* Hyderabad, February, pp. 14-16.

92. Udaya Kumar, M., Sashadhar, V.R. and Prasad, T.G. (1986). Physiological approaches for improving productivity of Finger millet under rainfed conditions.

In: *Small millets in Global agriculture, proceedings of 1st international small millets workshop*, Bangalore 29th Oct. to 2nd Nov. 1986, Seetharam, A., Riley, K.B., Harinarayana, G. (eds.). pp 179-207. Oxford and IBH Publishing Co. Pvt. Ltd.

93. Uma S., Prasad T. G., and Udaya Kumar M. (1995). Genetic Variability in Recovery Growth and Synthesis of Stress Proteins in Response to Polyethylene Glycol and Salt Stress in Finger Millet. *Ann Bot*. 76 (1): 43-49.

94. Upadhyaya HD, Gowda CLL, Gopal Reddy V. (2007). Morphological diversity in finger millet germplasm introduced from southern and eastern Africa. *SAT e Journal ICRISAT*, p. 3.

95. Veeranagamallaiah G, Chandraobulreddy P, Jyothsnakumari G, Sudhakar C. 2007. Glutamine synthetase expression and pyrroline- 5-carboxylate reductase activity influence proline accumulation in two cultivars of foxtail millet (*Setaria italica* L.) with differential salt sensitivity. *Environ Expt Bot* 60: 239–244.

96. Veeranagamallaiah G, Jyothsnakumari G, Thippeswamy M, Reddy PCO, Surabhi G-K, Sriranganayakulu G, Mahesh Y, Rajasekhar B, Madhurarekha C, Sudhakar C. (2008). Proteomic analyses of salt responses in foxtail millet (*Setaria italica* L. cv. Prasad) seedlings. *Plant Sci* 175: 631–641.

97. Wang B, Guo G, Wang C, Lin Y, Wang X, Zhao M, Guo Y, He M, Zhang Y, Pan L (2010a). Survey of the transcriptome of *Aspergillus oryzae* via massively parallel mRNA sequencing. *Nucleic Acids Res* 38(15): 5075–5087.

98. Wang Z, Gerstein M, Snyder M (2009). RNA-Seq: a revolutionary tool for transcriptomics. *Nat Rev Genet* 10(1): 57–63.

99. Wright SI, Gaut BS(2005). Molecular population genetics and the search for adaptive evolution in plants. *Mol Biol Evol* 22(3): 506-519.

100. Xin Qi, Shaojun Xie, Yuwei Liu, Fei Yi, Jingjuan Yu (2013). Genome-wide annotation of genes and noncoding RNAs of foxtail millet in response to simulated drought stress by deep sequencing. *Plant Mol Biol* (2013) 83: 459–473.

101. Yadav DS, Goyal AK, Vats BK (1999) Effect of potassium in *Eleusine coracana* (L.) Gaertn. under moisture stress conditions. *Journal of Potassium Research* 15: 131–134.

102. Zhang G, Liu X, Quan Z, Cheng S, Xu X, Pan S, Xie M, Zeng P, Yue Z, Wang W, Tao Y, Bian C, Han C, Xia Q, Peng X, Cao R, Yang X, Zhan D, Hu J, Zhang Y, Li H, Li N, Wang J, Wang C, Wang R, Guo T, Cai Y, Liu C, Xiang H, Shi Q, Huang P, Chen Q, Li Y, Zhao Z (2012) Genome sequence of foxtail millet (*Setaria italica*) provides insights into grass evolution and biofuel potential. *Nat Biotechnol* 30(6): 549–554.

103. Zhang JP, Liu TS, Zheng J, Jin Z, Zhu Y, Guo JF, Wang GY. 2007c. Cloning and characterization of a putative 12-oxophytodienoic acid reductase cDNA induced by osmotic stress in roots of foxtail millet. *DNA Seq* 18: 138–144.

104. Zhang J-P, Wang M-Y, Bai Y-F, Jia J-P, Wang G-Y. 2005. Rapid evaluation on drought tolerance of foxtail millet at seedling stage (in Chinese. English abstract). *J Plant Genet Resour* 6: 59–62.

105. Zhi H, Diao X, Lu P, Li W, Akolova Z. 2004. Methodology analysis on screening of salt tolerant genotypes from foxtail millet and other Setaria species. *J Hebei Agr Res* 8: 15–18.

106. Zhu X-H, Song Y-C, Zhao Z-H, Shi Y-S, Liu Y-H, Li Y, Wang T-Y. 2008. Methods for identification of drought tolerance at germination period of foxtail millet by osmotic stress (in Chinese, English abstract). *J Plant Genet Resour* 9: 62–67.

Millets: Ensuring Climate Resilience and Nutritional Security (2015)
Editors: **Vilas A. Tonapi and J.V. Patil**
Published by: **DAYA PUBLISHING HOUSE, NEW DELHI**

Pages **405–430**

Chapter 14

Sorghum Agronomy for Higher Productivity

S. Ravikumar

Centre on Rabi Sorghum-IIMR, Shelgi, Solapur, Maharashtra

Sorghum has a major regional presence across Asia, Africa and the Americas, while on lower scale across Europe and Oceania. When we look at the changing production trends across these regions (Figures 1a and b), there is significant change in these statistics between 1964-65 and 2011-12.

African region witnesses an increasing trend from 11 m tons to 24 m tons of sorghum production, while Asian region looses ground from 17 m tons to 10 m tons over five decades. These changes can be attributed to proportional changes in area and productivity across these regions (Figures 2a and b). Mean productivity which was less than 800 kg ha^{-1} in Asia and Africa increased close to 1200 kg ha^{-1} in both these regions over a period of nearly four decades (1972-2010). Definitely area changes can be attributed primarily to the decreasing and increasing trends in sorghum production across these two regions, while productivity changes could be brought about by funding of international research. Productivity in general across millet crops can be modeled through environment (E) by genotype (G) by management (M) interactions.

Grain yields of dryland (nonirrigated) grain sorghum [*Sorghum bicolor* (L.) Moench], a major crop in the southern Great Plains of USA, more than tripled in studies at the USDA-ARS Conservation and Production Research Lab., Bushland, TX, during the period from 1939 to 1997 (Unger and Baumhardt, 1999). The study documents the yield increases that occurred and as well determine factors primarily

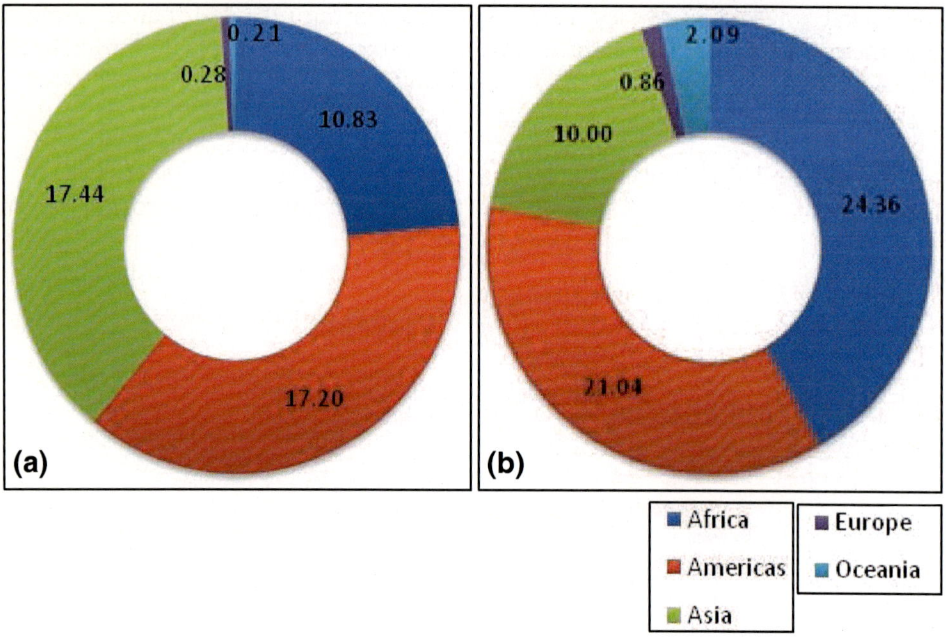

Figure 1: Sorghum Production (m. tons) Trends Across World Regions during (a) 1964-65 and (b) 2011-12.

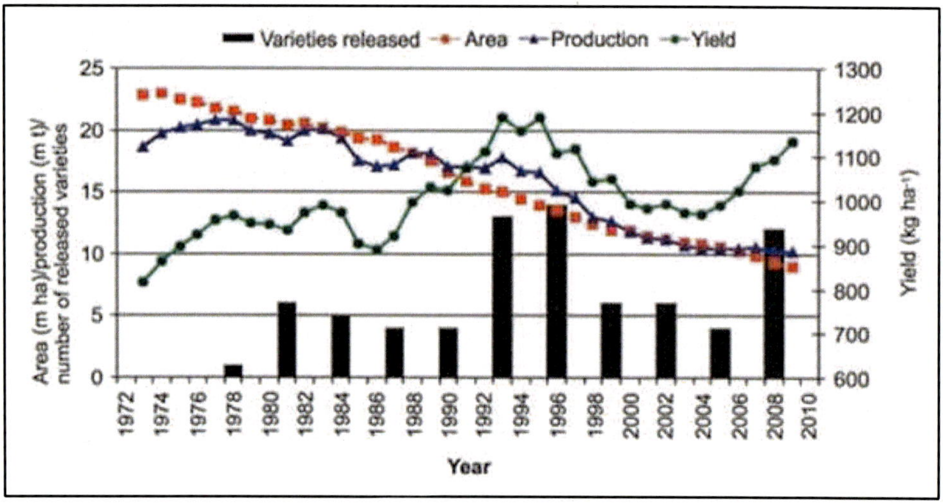

Figure 2: Sorghum Area, Production and Yield Trends in a. Asia and b. Western and Central Africa (1972-2010).

Source: ICRISAT (http://www.icrisat.org/crop-sorghum.htm).

responsible for the yield increases. Factors evaluated were annual precipitation, growing-season rainfall, soil water content at planting, soil water use, growing-season evapotranspiration, and year of record. For the report, 502 treatment-years of grain yield data from 37 studies were assembled. For the 1939–1997 period, grain yields increased about 50 kg ha^{-1} annually. Yields increased 139 per cent during the 1956–1997 period, with 46 of those percentage units resulting from use of improved hybrids, based on results of a uniformly managed 40-year study. The remaining 93 percentage units for that period were attributed to other factors, primarily to soil water at planting. Increases in soil water at planting resulted from changes in management practices with time, mainly the adoption of improved crop residue management practices after about 1970.

Sorghum in particular grown in rainfed production systems is totally dependent on environment, which influences yield both through above ground and below ground factors. Crop improvement brought about through changes in genotype influences yield both by genetic and physiological functional factors. Management the third component of the production system through experiential learning guides the decision process, so as to predict the consequences of both short and long term approaches to yield stability. This chapter uses this model of looking at the five decades of research progress across these three sub-system of E x G x M interaction.

Environmental Influences on Sorghum Yield

Early during 1960' and 1970's, environmental research was mostly dealt with above ground factors like carbon-dioxide and light intensity on yield components and yield. Eastin and Sullivan, 1969, monitored Carbon dioxide exchange rates in compact and semi-open heads of sorghum (*Sorghum bicolor* (L.) Moench). Highest CO_2 uptake in both head types occurred at the bloom stage when net uptake rates for semi-open and compact type heads were 3.9 and 1.2 mg CO_2 g dry wt^{-1} hr^{-1}, respectively. Beginning at the milk stage, a net CO_2 evolution of the order of 1 to 1.4 mg g dry wt^{-1} hr^{-1} occurred in light. Dark respiration decreased from about 4 to 1 mg CO_2 g dry wt^{-1} hr^{-1} from bloom to hard dough.

'Georgia 615' grain sorghum (*Sorghum bicolor* (L.) Moench) was shaded for weekly periods in three field experiments to determine if periods exist during the life cycle of the sorghum plant when low light intensities are deleterious to grain yield components (Pepper and Prine, 1972). Two stages of plant development most adversely affected by low light intensity were found. The first stage was approximately the two-week period just before 50 per cent anthesis. Shading during this period resulted in reduced number of seed per panicle. The second stage was approximately the three-week period just after 50 per cent anthesis. Shading during this period resulted in lighter seed with little or no change in number of seed per panicle.

In any dryland area, production and transpiration are directly related (Hanks *et al.*, 1969). This relation was established using, Lysimeters, which eliminated runoff and percolation below 90 cm and caused about 10 cm of additional water to be available for growth of grain sorghum in 1967. This additional water doubled yields, with an increase in evapotranspiration of only 50 per cent. Estimates of the amount of water evaporated from the soil within a crop allowed for estimation of transpiration from measurements of evapotranspiration. Similarly, another lysimeter study to determine the influence of contrasting soils (below ground factors) on grain sorghum (*Sorghum bicolor* Moench) growth, water use, and yield was conducted at Bushland,

TX containing monolithic soil cores of silty clay loam, silt loam, and fine sandy loam (Tolk *et al.*, 1997). The crop in the silt loam soil produced lower grain yield under high soil water conditions, but greater grain yield, total biomass, and seed number under reduced irrigation compared with the crop on the clay loam. The crop in the sandy loam consistently produced the lowest leaf areas and yield components in all irrigation treatments, possibly due to high soil bulk densities which may have restricted rooting. Crop in the silt loam had the highest water use in all treatments, and extracted water uniformly throughout the profile in both years. High strength silty clay and clay horizons and possibly a calcic horizon in the silty clay loam may have delayed or limited rooting, and affected crop growth and yield. The crop in the sandy loam consistently produced the lowest yield components in all irrigation treatments, possibly due to restricted rooting resulting from high bulk densities and also low water holding capacity.

During the 1980's Saeed and Francis, 1984, determined the relative contribution of several weather variables to variation in environment and genotype ' environment (GE) interactions of grain sorghum [*Sorghum bicolor* (L.) Moench]. Variation in temperature and rainfall accounted for more than half of the environment and genotype ' environment interaction sums of squares for yield, and seed number and seed weight components of yield. Temperature was the most important factor affecting environmental variability for yield and seed number. Generally, minimum temperature was more important than maximum temperature especially for the early and the late maturing genotypes. Preseason precipitation contributed more than seasonal rainfall to the GE interaction sums of square for yield and seed number, however, the opposite was observed for seed weight. Variation in weather factors contributed more to GE interaction for yield of the late maturing genotypes than the early and medium maturing genotypes.

The 1990's had greater emphasis of abiotic environmental influences like heat and drought stress on sorghum yield. The high temperature influences in field grown Sorghum with reference to linear rate of grain growth, the duration of the effective grain-filling period, the rate of increase in harvest index during grain-filling, the rate of development of black layer (as an indicator of maturity) and final grain yield were studied (Muchow, 1990). Contrary to previous findings from controlled environment experiments, the rate of grain growth increased linearly with temperature to 30°C. However, the duration of grain-filling and the rate of development of black layer were not closely related to mean temperature. Final grain yield was not necessarily adversely affected by high temperature during grain-filling. Also, high grain yield was not necessarily associated with long duration of grain-filling. Similar to the rate of grain growth, harvest index increased linearly during grain-filling. The rate of increase in harvest index was not significantly related to temperature, with a mean value across sowings of 0.0185 day^{-1}. An important implication of the linear increase in harvest index is that grain yield accumulation can be estimated from the crop biomass at any stage of grain growth without knowledge of grain number or the rate of grain growth.

Sorghum grain yield was affected by both the timing and the severity of the drought stress (Craufurd and Peacock, 1993). The largest reduction (87 per cent) in grain yield resulted from stress imposed during booting and flowering (late stress) in the early flowering lines; the same stress treatment on vegetative plants had no effect on grain yield. Increasing the duration of severe stress on vegetative plants (early plus late stress) reduced grain yield by 50–60 per cent. Grain yield was strongly and positively correlated with the number of grains m^{-2}. Variation in grain yield was associated with variation in total dry matter rather than with harvest index, which was only reduced when stress occurred at flowering. Treatment effects on thermal growth rates ($g\ m^{-2}\ °Cd^{-1}$) during the phase from booting to flowering confirmed that growth during this phase is the major determinant of yield potential (number of grains). The importance of sorghum phenology in research into drought resistance was emphasized by the authors.

Rainfed crop production in the subtropics is a risky enterprise due to high rainfall variability. When planting opportunities occur, farmers face risky choices because the consequences of decisions made at planting are uncertain. Simulation modeling offers an approach to generating the information required to assist in making planting decisions in climatically variable subtropical environments of Australia (Muchow, Hammer, and Vanderlip, 1994). The approach involved coupling a sorghum growth simulation model to long-term sequences of climatic data to provide probabilistic estimates of yield for the range of decision options, such as planting time and cultivar maturity, for a range of soil conditions. The likely change in the amount of stored soil water with delay in planting was also simulated to account for the decision option of waiting for a subsequent planting opportunity. Production risk varied with location, time of planting, soil water storage, and cultivar phenology. Yield responses to these factors were associated closely with differences in leaf area development and degree of depletion of the water resource. The probabilistic estimates of yield and change in stored soil water can be used in decision analysis or in computerized decision support, where decision-makers, and their risk preferences, can interact directly with the information.

Crop simulation models were also used to characterize the drought environment types (ET) that occur in the target population of environments (TPE) for dryland sorghum (Chapman, Hammer, Butler, and Cooper, 2000). The variable nature of rainfall confounds the process of selecting sorghum hybrids that are broadly adapted. Seventy seasons (1921–1990) of simulations of the yield of a sorghum genotype and the associated within season sequence of a stress index were run for a small TPE of 6 locations and also for a large TPE of 211 locations that attempted to represent the entire sorghum region in north-eastern Australia. The dataset was broken into a sequence of 16 multi-environment trials (METs), each of 3 years and 6 locations, the ratios of ET1 : ET2 : ET3 differed greatly among METs, *i.e.* any single MET was not randomly sampling the TPE. For any MET, the average yield (GVu) was not the same as the overall mean of the entire 70-year dataset.

If the trial yields were weighted according to the ratio of ET1 : ET2 : ET3 in the overall TPE, then GVw (s.d. = 0.13) for a single MET was much closer to the overall mean than was GVu (0.38). For different METs, the values of GVw were up to 30 per cent higher or 15 per cent lower than GVu. Across METs, the difference between GVu and GVw was positively correlated ($r = 0.88$, $n = 16$, $P < 0.05$) with the frequency of

ET1 ('low terminal stress') encountered within the MET and negatively correlated (r = –0.82) with the frequency of ET2.

The large TPE consisted of more than 15 000 simulations and was classified in 2 stages (within/among locations), repeated for each of 3 soil types. In years in which the simulation sowing criteria were met, the ratios of ET1 : ET2 : ET3 were about 4:2:4, 4:5:1, and 6:3:1 in the shallow, intermediate, and deep soils, respectively. Hence, over all soil types and locations, the sorghum TPE for northern Australia consists of at least 30 per cent each of low terminal stress (ET1) or severe terminal stress (ET2) and these environment types need to be sampled. The incidence and nature of the 'intermediate midseason/terminal stress' environment type (ET3) varies with soil type and location.

During the 21st century environmental issues were related to climate change, which still continues to be the current researcher's primary focus. Abiotic stress influences on sorghum studied during the earlier decades formed the basis of designing simulation models that could predict the outputs through input strategies that mimicked the anticipated climate changes. The vulnerability of the maize and sorghum crops to climate change were studied using crop simulation models while climate change scenarios were generated from Global Circulation Models (Chipanshi *et al.*, 2003). This study examines the sensitivity of maize and sorghum crops to global warming in Botswana, a country with arid climatic conditions and shortfalls in locally produced grain. Simulated yields indicated that rain-fed crop production under the observed climate was a small fraction of what could be produced under optimal conditions. The gap was attributed to both physical (especially lack of rain) and socio-economic constraints. Using the southern African core climate change scenario, simulated yields declined by 36 per cent in the case of maize and 31 per cent for sorghum in the sand veldt region. Yield reductions from the hard veldt region were in the order of 10 per cent for both maize and sorghum. The growing season became shorter, the average reduction in days in the sand veldt region being 5 and 8 days for maize and sorghum respectively, and correspondingly, 3 and 4 days over the hard veldt region. The food security option currently followed in Botswana was found to be a good adaptive strategy under a changed climate.

Apart from simulation models, correlation and regression studies used stress indices to predict the consequences of climatic factors on sorghum yield. Ajayi and Olufayo, 2004, examined the relationship between canopy temperature (T_c) based stress indices and grain yield and evapotranspiration (ET) of sorghum [*Sorghum bicolor* (L.) Moench]. Soil water content and T_c were measured, and relationships between T_c based stress indices [stress degree day (SDD), temperature stress day (TSD), and crop water stressed index (CWSI)] and yield as well as ET were examined. Although observations showed that T_c can be influenced by climatic condition, this study confirmed that it can serve as a useful indicator of water stress in the case of sorghum. High correlation found between T_c based stress indices TSD, SDD, and CWSI and ET as well as grain yield suggest the possibility of using these relationships for predictive purposes.

Modern tools like Remote Sensing (RS) and Geographical Information Systems (GIS) are being applied to relate designed indices with sorghum yield and monitor drought on a spatial mode. The distributed hydrologic model 'SWAT' was used to simulate soil moisture and evapotranspiration from daily weather data at a high spatial resolution (16 km^2) using GIS (Narasimhan and Srinivasan, 2005). Using this simulated data the drought indices Soil Moisture Deficit Index (SMDI) and Evapotranspiration Deficit Index (ETDI) were developed based on weekly soil moisture deficit and evapotranspiration deficit, respectively. SMDI was computed at four different levels, using soil water available in the entire soil profile, then soil water available at the top 2 ft. (SMDI-2), 4 ft. (SMDI-4), and 6 ft. (SMDI-6). ETDI and SMDI-2 had less auto-correlation lag, indicating that they could be used as good indicators of short-term drought. The developed drought indices showed high spatial variability (spatial standard deviation <"1.00) in the study watersheds, primarily due to high spatial variability of precipitation. The wheat and sorghum crop yields were highly correlated (r and gt; 0.75) with the ETDI and SMDI's during the weeks of critical crop growth stages, indicating that the developed drought indices can be used for monitoring agricultural drought.

Controlled environmental studies related to climate change still continue to be a source of knowledge that helps understand the physiological basis of sorghum yield. Plants of hybrid DK-28 E were grown in growth chambers at daytime maximum/ nighttime minimum optimum temperature (OT) of 32/22°C until 29 d after sowing (Prasad, Pisipati, Mutava, and Tuinstra, 2008). Thereafter, plants were exposed to OT or HT (40/30°C) or were reciprocally transferred at 10-d intervals (10 d before flowering, 0, 10, 20, and 30 d after flowering [DAF]) from OT to HT and vice versa. Continuous HT stress delayed panicle emergence and decreased plant height, seed set, seed numbers, seed yield, seed size, and harvest indices but did not influence leaf photosynthesis. Exposure to short (10-d) periods of HT stress at flowering and 10 d before flowering caused maximum decreases in seed set and seed yield, and HT stress during post flowering stages (10, 20, and 30 DAF) decreased seed yield, with a larger reduction at early stages of seed development.

Genotype Influences on Sorghum Yield

Grain yield of sorghum genotype was a function of rachis morphology and panicle weight components *i.e.*, number of branch whorls per panicle, number of branches per whorl, number of grains per branch and grain weight (Blum, 1967). High soil fertility and low plant competition increased the 1000-grain weight and number of grains per panicle. The latter component was affected by soil fertility through the number of branches per whorl and the number of grains per branch, and by plant competition through the number of whorls per panicle and the number of grains per branch. Higher soil fertility and lower plant competition promoted a longer panicle through the increase in rachis internode length and number of internodes, respectively. A negative correlation, which became stronger as the number of grains per panicle increased, was found between weight of grain and number of grains per branch.

The growth cycle of a sorghum genotype has been fully described in terms of a standard set of growth stages based on detailed studies of grain sorghum hybrids of different maturities (Vanderlip and Reeves, 1972). The ten stages of development have been defined and illustrated: emergence, three-leaf, five-leaf, growing-point, differentiation, final leaf visible in whorl, boot, half-bloom, soft dough, hard dough, and physiological maturity. These stages are suggested as standards to describe the timing of sampling or while managing a treatment in sorghum.

Source sink relation is an important determinant of grain yield in sorghum. Relative importance of head storage capacity and carbohydrate supply were examined in five sorghum varieties of different heights (Willey and Basiime, 1973). Varietal differences could be attributed to pre-anthesis competition between head and stem, but in the dwarfs a more important factor appeared to be pre-anthesis competition between head and leaf. It is suggested that for maximum efficiency of grain-yield production, maximum carbohydrate should be utilized for pre-anthesis head and leaf growth and that partitioning between these plant parts should be such that the supply of carbohydrate for grain filling is in balance with head storage capacity. Similarly under favourable growing conditions, the source-sink relationships in grain yield of sorghum were analysed in terms of capacities for net assimilation, of head storage, and of the transport system to move assimilates between source and sink (Muchow and Wilson, 1976). In four commercial hybrids (Dekalb E57, Pacific Goldfinger, Texas 610SR, and Texas 626) grown at three population densities (20.2, 40.4 and 80.8 plants m^{-1}), the assimilates supply was varied by increasing or decreasing the radiation available per plant (by thinning or shading), the potential grain storage capacity was decreased by spikelet removal, and the transport system was reduced by incision of the culm, all manipulations being performed at anthesis.

Decreasing the number of grains increased the size of those remaining in all cultivars at all population densities The degree of increase was greatest for T626 and T610 and least for E57 Thinning increased the grain size in all cultivars, but only sufficiently to cause a significant increase in grain yield in T610 and T626. Reduction in the assimilate supply by shading decreased the grain size and yield in all cultivars. Reduction in the transport system had no significant effect on grain yield. These results showed that there was surplus capacity for storage and transport in all cultivars. In T626, T610, and Goldfinger, all post-anthesis assimilate was stored as grain and grains could grow larger. The yield was therefore completely source-limited. In E57, however, not all post-anthesis assimilate was stored as grain, and these grains showed little capacity to grow larger, which suggests that the yield was partially limited by both source and sink.

Physiological responses to water stress and improving water use efficiency in sorghum genotypes were emphasized during the 1970's (Blum, 1974), (Blum and Naveh, 1976). The most drought susceptible genotype was 'Shallu' which under increasing soil moisture stress had relatively low leaf water potential, high leaf diffusion resistance, and the lowest total soil moisture extraction. 'Combine Kafir-60' and 'RS 610' hybrid had, under stress, comparatively high leaf water potential and low leaf diffusion resistance, and were therefore regarded as "drought avoidant." 'Feterita' had the greatest amount of soil moisture extraction, which was associated with very low leaf water potential, but relatively low leaf diffusion resistance. *'Durra'* also had low leaf diffusion resistance in spite of reduced leaf water potential. Appreciable inter-genotypic differences also were revealed in the amount of soil moisture extracted prior to heading as determined by percent of the total at maturity.

Plant competition was promoted by planting sorghum in double rows (DR), *i.e.*, 40 cm between rows within a pair, and 160 cm between row pairs, as compared with regular-spaced rows (RR) (100 cm between rows), with the same number of plants/ unit area. The èffect of competition in DR was observed beginning at 20 to 35 days after emergence, in reduced total stover dry matter/plant, and reduced tillering and LAI, as compared with the RR plots. In two out of four experiments, grain yield was significantly increased by DR configuration and was associated with an increased number of grains/panicle and 1,000-grain weight. During the boot to full bloom growth stages, less soil water was extracted in DR than in RR plots. Consequently, more soil water was available in DR plots during the grain-filling period.

Growth rate, leaf area index, net assimilation rate, and distribution of dry matter among aboveground plant parts were determined for sorghum [*Sorghum bicolor* (L.) Moench] hybrid 'RS 610' and its parents throughout the growing season in a replicated field experiment (Gibson and Schertz, 1977). The hybrid developed faster than its parents. RS 610 out yielded its parents in grain, and had more kernels per panicle than the female parent, but did not differ from the male parent in this characteristic. The yield superiority of RS 610 was attributed mainly to the combination of the high CGR of the female parent during grain filling with the effective conversion of dry matter to grain of the male parent.

A 2-year study was conducted to determine to what extent seed size and density influence germination and subsequent field performance (Maranville and Clegg, 1977). Large and small seed lots from the same genotype were compared to "more dense" and "less dense" seed lots which were separated using either urea-phosphate or sucrose solutions. The results indicated that seed lots with larger and denser seeds had a higher percent germination. The data also supported the conclusion that a higher percentage of viable seeds could be selected from seed lots with low average germinability by using specific gravity separations. However, the establishment of seedlings, final stands, and grain yields were not a function of size or density when the same number of viable seeds were planted in the field.

Drought, yield component analysis, resource use efficiency in terms of both water and nitrogen were the major focus during 1980' and 1990's. A field experiment was conducted using a modified line source sprinkler system to establish treatments that resulted in several different ET levels during each growth stage (Garrity, Watts, Sullivan, and Gilley, 1982). In two seasons the relationship between evapotranspiration (ET) and grain yield (Y) or dry matter was linear. Evapotranspiration reduction under all conditions tested incurred some yield reduction. For a given genotype the slope of the ET-Y relationship varied greatly depending on the growth period treatment. The lowest slopes were associated with a treatment in which stress built up steadily over the entire growing season. One hybrid had a lower slope of the ET-Y relationship than the other genotypes for the treatment that provided water stress of gradually increasing severity. Water use efficiency consistently decreased as ET declined below the maximum. It appeared that water stress, regardless of timing, tended to reduce water use efficiency. The reductions were smallest when stress increased gradually throughout the growing season. Similarly to identify the most sensitive phase, the effects of water stress occurring at various growth stages during the reproductive cycle of grain sorghum were evaluated in open field conditions under a Mediterranean climate (Mastrorilli *et al.*, 1995). During three soil drying cycles it was observed that: (1) the highest stomatal conductance occurred in correspondence with leaf water potential (Ψ values higher than "1.5 MPa; (2) stomata tended to close when Ψ becomes lower than "1.8 MPa. These values were retained in the second step for monitoring plant water status in four different plots: one never stressed and three stressed at different phenological stages (early flowering, seed-setting and ripening). An evaluation of the sensitivity of phenological stages submitted to identical water stress was obtained by comparing four yield components (above ground biomass, grain yield, 1000 grain weight and seed number) and WUE to those of the well-irrigated crop. The sensitivity was greatest at the early flowering stage.

A sorghum F1-hybrid, E57, with reputed drought resistance was compared with another F1-hybrid, TX- 671, under irrigation and dryland conditions of severe rainfall insufficiency but ample soil water at depth in a deep cracking clay soil (Wright, Smith, and McWilliam, 1983). Significant cultivar differences in grain yield occurred under rain-fed but not under irrigated conditions. Differences in yield were associated with different morphological responses to drought stress, with the variety E-57, in comparison with TX-671, having a lower level of floret abortion, the ability to maintain green leaf area longer during grain-filling and probably a higher root to shoot ratio. Similarly root development and plant water use was compared in early ('58M') and late ('90M') isogenic lines of sorghum (*Sorghum bicolor* L. Moench) grown in a root observation field installation at Temple, Texas (Blum and Arkin, 1984). The installation consisted of soil-filled steel chambers equipped with a glass panel and installed in retaining structures in the ground. Each chamber was lifted out of the ground for weighing and measuring roots visible on the glass panel. Transpiration was considerably less for the early than for the late genotype in relation to the respective leaf area of the two genotypes. Transpiration demand was met without apparent plant stress, as long as the soil moisture content of the total soil volume was greater than 20 per cent of available soil water. Below 20 per cent of available soil water, transpiration was controlled mainly by a reduction in leaf area through leaf senescence. Total cumulative root length per plant and root length density were larger and root length density per unit leaf area was smaller for the late than for the early genotype. Root length density per unit of green leaf area increased under conditions of soil moisture stress due mainly to the reduction in green leaf area. Root mortality of approximately two-thirds of the total cumulative root length produced by heading was recorded. Root mortality was not only associated with total plant senescence, but with plant age and rooting depth.

Relative importance of yield components and the effects of environmental factors in two growth periods, planting to bloom (Period 1) and bloom to physiological maturity (Period 2), on yield and its components in 46 sorghum hybrids was tested in contrasting field environments (Saeed, Francis, and Clegg, 1986). Number of seed was the major contributing component of yield, but the relative importance of seed weight to yield compared to number of seed increased from low night-temperature to high-night-temperature environments. Number of seeds head^{-1} was more important than number of heads in affecting variation in number of seed and yield among sorghum hybrids. Rate of grain fill day^{-1} accounted for more variation in seed weight than rate of fill per growing degree unit (GDU), but both were more important than the length of grain fill period. The relationship between the mean yield and the total GDU of environments was quadratic (R2=0.70), and GDU in the range of 1250 to 1350 favored development of high yields. Slow rate of GDU accumulation in Period 1 coupled with a relatively high rate in Period 2 was associated with high sorghum yields in high-night-temperature environments. In low-night-temperature environments, high GDU accumulation rate in both Period 1 and Period 2 favored high yields. Growing degree day unit (GDU) and rate of GDU have significant influence on accumulation to yield and yield component development across diverse environments, and it suggest the need to exploit genetic variation of metabolic efficiency in grain fill in breeding programs designed for stress environments.

Worldwide, nitrogen use efficiency (NUE) for cereal production (wheat, *Triticum aestivum* L.; corn, *Zea mays* L.; rice, *Oryza sativa* L. and *O. glaberrima* Steud.; barley, *Hordeum vulgare* L.; sorghum, *Sorghum bicolor* (L.) Moench; millet, *Pennisetum glaucum* (L.) R. Br.; oat, *Avena sativa* L.; and rye, *Secale cereale* L.) is approximately 33 per cent. The unaccounted 67 per cent represents a $15.9 billion annual loss of N fertilizer (assuming fertilizer-soil equilibrium). Loss of fertilizer N results from gaseous plant emission, soil de-nitrification, surface runoff, volatilization, and leaching. Grain sorghum, *Sorghum bicolor* (L.) Moench has been evaluated for its variability in nitrogen use efficiency by various researchers (Maranville, Clark, and Ross, 1980), (Muchow, 1998), (Raun and Johnson, 1999).

A 2-year study using 12 hybrids was conducted and nitrogen efficiency (NE) was calculated as: 1. total dry matter (DM) per unit N uptake (NE1), 2. total grain yield per unit N uptake (NE2), and 3. the product of NE2 and the grain N:stover N ratio (NE3). NE1 values increased with plant age and were significantly different among hybrids at all stages of growth. A 20 per cent difference in NE1 was found between the least and most efficient hybrids at maturity. A significant hybrid x year interaction showed that the environment strongly influenced hybrid ranking. Hybrids which ranked high in one year often did poorly the other year and visa versa for NE1, NE2, or NE3. Each N efficiency criterion ranked the hybrids differently, although NE1 and NE2 appeared to be the most closely related. Hybrids generally maintained satisfactory values for NE3 if one parent (either male or female) had a high NE3 value in hybrid combination. Differences were more dramatic for partitioning of N between grain and stover than for NE values among hybrids.

Comparative efficiency of utilization of N by maize and sorghum (defined as grain yield per unit N uptake) varied under different climatic, soil and management conditions. The maximum N-use efficiency (NUE) of sorghum was smaller than for maize (48 vs. 61 g grain$^{"1}$ g$^{"1}$ N absorbed), and was associated with a grain N concentration in sorghum. Cultivar differences in NUE in sorghum were related to grain N concentration. In both crops, NUE declined when N was supplied at high levels or when growth was limited by moisture supply.

Due to environmental influences, it appears that at least 3 or more years of data may be needed in order to establish consistent trends for N efficiency in grain sorghum. Increased NUE is unlikely, unless a systems approach is implemented that uses varieties with high harvest index, incorporated NH_4^-N fertilizer, application of prescribed rates consistent with in-field variability using sensor-based systems within production fields, low N rates applied at flowering, and forage production systems. Crop models can integrate the effects of abiotic factors on grain yield, and allow the development of strategies for improving the utilization of N to maximize grain yield and to minimise N losses from the production system.

During the 21st century major focus shifted to energy sorghums with increasing crude oil price and growing importance of renewable energy worldwide. Reducing the use of non-renewable fossil energy reserves together with improving the environment are two important reasons that drive interest in the use of bioethanol as an automotive fuel. Conversion of sugar and starch to ethanol has been proven at an industrial scale in Brazil and the United States, respectively, and this alcohol has been able to compete with conventional gasoline due to various incentives. First generation biofuel production concentrated on sweet sorghum as the feedstock, while the second generation biofuel emphasized the use of biomass as well as hydrogen production for future use.

Various studies examined making ethanol from the sugar extracted from the juice of sweet sorghum (Gnansounou *et al.*, 2005; Ratnavathi *et al.*, 2010; Zegada-Lizarazu and Monti, 2012). Sweet sorghum with its juicy stem containing sugars equivalent to that of sugarcane is a very good alternative for bio-ethanol production to meet the energy needs in India. It is important to identify superior genotypes for ethanol production in terms of percent juice brix, juice extractability, total fermentable sugars, ethanol yield and fermentation efficiency. Five Sweet Sorghum (*Sorghum bicolor* L. Moench) genotypes were evaluated for ethanol production from stalk juice (Keller, SSV 84, Wray, NSSH 104 and BJ 248). Maximum ethanol production of 9.0 per cent w/v ethanol was obtained with Keller variety (20 per cent sugar concentration was used), and decreased for other genotypes. A distiller's strain of Saccharomyces cerevisiae (gifted by Seagram Distilleries Ltd.) was employed for fermentation. The fermentation efficiency (FE) was 94.7 per cent for this strain. High biomass of yeast was obtained with BJ 248 variety. When the similar experiments were conducted with unsterile sweet sorghum juice (15 per cent sugar concentration) 6.47 per cent w/v ethanol was produced.

In general terms, ethanol production from sorghum bagasse appears very favorable, but other agricultural residues such as corn stover and rice hulls would likely provide a more attractive feedstock for making ethanol in the medium and long term due to their extensive availability in North China and their independence from other markets. Low input requirements, low production costs, drought resistance, versatility, and high yields gives to sweet sorghum a better energy balance compared to other competing energy crops, especially if bagasse is also processed to energy. Furthermore, the process for residue conversion was based on particular design assumptions, and other technologies could enhance competitiveness while considerations such as perceived risk could impede applications. In non-traditional potential growing areas (such as in temperate climates of Europe) where productivity/ adaptation improvements through genetically modified crops is not allowed, appropriate and sustainable agricultural practices constitute the most immediate option to improve yields. Based on scientific reports, research efforts seem particularly needed for harvesting techniques, and handling and storing.

Second generation biofuels focuses on the exploitation of sweet sorghum biomass as a source for hydrogen and methane (Antonopoulou *et al.*, 2008; Ntaikou *et al.*, 2008; Pedersen *et al.*, 2008; Ben *et al.*, 2010). Fermentative hydrogen production from the sugars of sweet sorghum extract was investigated at different hydraulic retention times (HRT). The subsequent methane production from the effluent of the hydrogenogenic process and the methane potential of the remaining solids after the extraction process were as well assessed. The highest hydrogen production rate (2550 ml H_2/d) was obtained at the HRT of 6 h while the highest yield of hydrogen produced per kg of sorghum biomass was achieved at the HRT of 12 h (10.4 l H_2/kg sweet sorghum). It has been proved that the effluent from the hydrogenogenic reactor is an ideal substrate for methane production with approximately 29 l CH_4/kg of sweet sorghum. Anaerobic digestion of the solid residues after the extraction process yielded 78 l CH_4/kg of sweet sorghum.

Four sorghum [*Sorghum bicolor* (L.) Moench] genetic stocks, BN611 (Reg. No. GS-140, PI 652943), AN612 (Reg. No. GS-141, PI 652944), BN612 (Reg. No. GS-142, PI 652945), and RN613 (Reg. No. GS-143, PI 652946), with stacked brown midrib genes bmr-6 and bmr-12 were developed jointly by the USDA-ARS and the Agricultural Research Division, Institute of Agriculture and Natural Resources, University of Nebraska, and were released in July 2006. The genetic stocks BN611, AN612 and BN612, and RN613 were tested as Atlas bmr-6 bmr-12, Wheatland bmr-6 bmr-12, and RTx430 bmr-6 bmr-12, respectively. The genetic stocks have immediate application for basic research involving lignin synthesis and in breeding strategies that are targeting second generation biofuels.

Double-cropping systems have been suggested as a way to increase annual dry matter production per hectare, while simultaneously delivering environmental benefits. Twelve sorghum varieties were grown as sole crops and within a double-cropping system with winter triticale [× Triticosecale Wittmack] in central and northern Iowa. When the chemical composition of the crops was considered, the theoretical ethanol yield of the single-cropping system was greater than the double-cropping system for all sorghum varieties. Although triticale was capable of offsetting any potential loss in dry matter, as an ethanol feedstock it was inferior to sorghum.

Most studies demonstrated that biohydrogen production can be very efficiently coupled with a subsequent step of methane production and that sweet sorghum could be an ideal substrate for a combined gaseous biofuels production.

Management Influences on Sorghum Yield

The 1960's and 1970's witnessed the date and rate type of experiments in Sorghum field management research. The sensitivity of genotype in terms of vegetative structures like plant height, tiller number, stalk size and reproductive features like head size, seed number, seed size as indicated in earlier sub-sections, led to the study of planting geometry, planting dates nutrient/manure application rates, that optimize the final yield.

Plant stand establishment studies in terms of planting density were conducted by various researchers (Stickler, 1964; Stickler and Wearden, 1965; Atkins *et al.*, 1968; Plaut *et al.*, 1969; Blum, 1970; Fischer and Wilson, 1975). Planting date of sorghum was another aspect that was initially standardized for a rainfed environment wherein rainfall was highly variable (Lingegowda *et al.*, 1971; Blum, 1972).

Highest grain yields were obtained with a 2.5-inch seed spacing. Where surface moisture availability was low, a small lister opener gave best stand establishment. Under favorable conditions, a tool-bar, unit type planter was best. In 1961 and 1962, press wheels increased stands by 14 per cent, and yield by 8 per cent. In 34 trials, grain yields in 20-inch rows exceeded those in 40-inch rows by 10 per cent in eastern Kansas and 7 per cent in central Kansas. Grain yields were remarkably constant across stand densities because of inter compensations among individual components of yield. Little tendency for row width and stand density interaction was noted. The performance of two short-stature grain sorghum (*Sorghum bicolor* (L) Moench) hybrids showed an 11 per cent advantage for 76-cm (30-inch) over 102-cm (40-inch) row spacings in Iowa. A within-row population of five plants/30 cm (1 foot) produced highest grain yields at each row spacing. With each row spacing, seeds/head and heads/plant decreased progressively as the within-row plant population increased from four through eight plants/30 cm. Days to mid-bloom, plant height, and 100-seed weight were not altered appreciably by row spacings and within-row plant populations. Interactions among the treatments did not identify differential performance patterns of marked practical consequence, and the interactions most often did not meet the tests for significance.

Effect of soil moisture regime and row spacing on grain yield of RS-610 sorghum (*Sorghum bicolor* L. Moench.) was investigated under semi-arid conditions in 1964 and 1965. Decreasing row spacing from 70 to 45 cm had hardly any positive effect on yield in irrigated sorghum. In one experiment, three experimental grain sorghum (*Sorghum bicolor* L. Moench.) hybrids (heading 55, 60, and 70 days from emergence) were planted at 12 plant densities (from 2.9 to 27.9 plants/m2) in a systematic design. Plants were grown under conditions of 302 mm available soil water, as determined at emergence, with no additional water supply. It was found that the grain yield of the late maturing hybrid was highest under the low plant density and that of an early maturing hybrid was highest under the high density. All grain yield components were operative in the determination of inter-hybrid differences under moderate competition. The expression of components was suppressed by increased competition, and at the highest plant density grain yield was determined only through the interaction between the number of grains per panicle and the weight per grain. Growth analysis was applied to grain sorghum (cv. RS610) grown at low, medium and high population densities, *i.e.* 14,352, 143,520 and 645,836 plants ha^{-1} respectively. The medium densities had two arrangements of plants, square (S) and rectangular (R). Differences in crop growth rate between populations in the early stages were attributed to leaf area development—specifically to the initial leaf area (dependent on seedling number) and not to differences in leaf growth rates. Peak crop growth rates were 15.0, 27.5, 26.0 and 45.8 g m^{-2} day^{-1} for the low, medium (S), medium (R) and high populations respectively. A maximum grain yield of 14,250 kg ha^{-1} was obtained at the high population density as a result of higher dry matter production, but a similar harvest index to that of the crops grown at the other densities. The maximum grain yield represents an efficiency of utilization of short-wave solar radiation during crop life of 2.5 x 10.6g cal^{-1}.

In trials in 1967-69 under rain-fed conditions with sorghum hybrid CSH-1 given 112 kg N + 67 kg P_2O_5 + 45 kg K_2O + 14 kg thimet [phorate] ha^{-1} and foliar applications of endrin at 15-day intervals to control pests, average grain yields decreased progressively from 3.71 to 0.38 t ha^{-1} with delay in sowing over the period from the 3rd week of June to the 3rd week of September. Average number of plants with dead growing points as a result of shoot fly [*Atherigona varia* soccata] incidence were 4.2, 104-192 and 39-77 with sowings in June, between July and early Aug. and between late Aug. and late Sept., respectively. Early sowing of sorghum with the onset of monsoon in India has been recommended to avoid shoot fly. It has been postulated that water regime was the major operative factor responsible for different yield response of dryland grain sorghum [*Sorghum bicolor* (L.) Moench] to different planting dates. For four consecutive years, four grain sorghum hybrids were planted on two planting dates (March 20–24 and April 12–16) in Israel under conditions of limited water supply as determined by the amount of stored soil moisture prior to planting. Early planting, as compared with late planting, increased grain yield through increased tillering and greater weight per grain. During the period from emergence to 51 days, early planted sorghum used about half as much water as late-planted sorghum. Early planted sorghum was less water stressed, as evidenced by lower leaf diffusive resistance prior to heading and smaller reduction in LAI (leaf desiccation) after heading, as compared with late-planted sorghum. Lower water use in early planted sorghum, during the period prior to heading, was ascribed to lower potential evapotranspiration, smaller LAI, and possibly slower root development, as compared with the respective period in late-planted sorghum. Total amount of water used at maturity did not differ between planting dates, apparently due to the longer period of growth and the larger amount of stover dry matter per plant in early planted sorghum, which served to offset the decreased water use during the earlier phase of growth. No appreciable differences were found between hybrids tested in terms of adaption to early planting or water use.

An understanding of the seasonal dry matter accumulation pattern of the sorghum [*Sorghum bicolor* (L.) Moench] plant under varying fertility, manure, compost levels is important for planning an efficient fertilization program, and while mulching to improve soil moisture retention. Field experiments were conducted to determine the effects of varying rates of nitrogen (N) and manures on the dry matter accumulation pattern, yield, grain N content and nutrient uptake of sorghum by various researchers (Campbell and Pickett, 1968; Onken and Sunderman, 1972' Singh and Choubey, 1972; Mays *et al.*, 1973; Roy and Wright, 1973; Singh and Bains, 1973; Roy and Wright, 1974; Jacques *et al.*, 1975; Day and Tucker, 1977; Mathers *et al.*, 1977).

The rate of dry matter accumulation in fertilized plants was rapid early in the season and continued almost linearly until maturity (91 days after sowing), but in unfertilized plants it began to drop after 70 days. In general, dry matter weight of leaves and stems reached a maximum by the 56th and 70th days, respectively, and then remained almost constant. The subsequent increase in total dry weight was due to increased head weight. Accumulation rate in the heads of fertilized plants at 70 and 77 days was very high in comparison to that of unfertilized plants. Dry matter yield of whole plant, grain yield, and grain N content were higher with 60 kg N/ha than with no nitrogen by 26, 47.6, and. 14.5 per cent respectively. They were also higher (9.2, 8.8, and 10.9 per cent respectively) with 120 N/ha. Eighteen inbred lines from the world sorghum collection and RS610 were tested with three levels of nitrogen fertilization. Nitrogen fertilization affected the protein production significantly, but variation among lines was much greater. Genotypic range for protein production was 11.7 to 16.3 per cent. Lysine as percent of protein was not affected significantly by nitrogen fertilization, but genotypes ranged from 1.64 to 2.52 per cent. There was a significant interaction between genotype and location for lysine, protein ' lysine, and grain weight per panicle, but not for protein production.

Multi-rate nitrogen studies were conducted for a 3-year period, on an irrigated clay loam soil at two locations, to determine the influence of applied and residual soil N on the yield of grain sorghum (*Sorghum bicolor* L. Moench). Applied and residual soil NO_3 N were found to influence grain yields. Regression analysis of the data showed highly significant relationships between (i) quantities of soil nitrates measured at upper depths to those measured at lower depths, and (ii) grain yield and applied N + residual NO_3 N. Best correlations were obtained when residual NO_3 N was included in the regression equation. Results indicated that soil samples taken to 15 or 30 cm would be sufficient for evaluation of residual nitrogen effects. In trials during the summer rainy seasons of 1968 and '69, sorghum cv. Vidisha 60-1, CSH 1, and Ujjain 6 were given 0, 40, 80 or 120 kg N ha^{-1}. Increase in rates of N increased plant height, length and width of ears, weight of ears/plant, 1000-grain weight and grain yields. Vidisha 60-1 gave the highest grain yields, 4.8 t in 1968 and 4.09 t ha^{-1} in 1969.

Field experiments were conducted to determine the effects of varying rates of N (0, 60, and 120 kg ha^{-1}) and P (0 and 26 kg ha^{-1}) on the N, P, and K accumulation pattern by various plant parts of a sorghum [*Sorghum bicolor* (L.) Moench] hybrid. Nitrogen, P, and K accumulation rates were higher during the 35th to 42nd day and 70th to 91st day which coincided with the peak vegetative growth period and the grain-filling stage, respectively. In unfertilized plants relatively higher translocation of N and P from the vegetative parts to the developing grain occurred. A much smaller percentage of total K was found in the head and more K accumulated in the stem than N and P. Application of 60 kg N ha^{-1} significantly increased N, P, and K yields of whole plant by 60.9, 49.4, and 19.1 per cent, respectively. Further respective increases were 19.0, 18.9, and 12.3 per cent, with 120 kg N ha^{-1}. Respective increases were 28.6, 49.0, and 17.1 per cent with 26 kg P ha^{-1} over no phosphorus.

Feedlot manure was incorporated into Pullman clay loam at rates of 0, 22, and 67 metric tons ha^{-1} annually (tons/ha/per year). Additional treatments were 67 tons ha^{-1} manure the first year only and 224-kg N/ha/per year as anhydrous ammonia. The time required for irrigation water to advance downgraded furrows was increased by the manure treatment. Also water intake and grain yields were higher on manure- than on ammonia-treated plots. A field study was conducted to determine the effects of straw mulch rates on soil water storage during fallow and on subsequent grain sorghum [*Sorghum bicolor* (L.) Moench] production. Wheat [*Triticum aestivum* (L.)] straw at rates of 0 (check), 1, 2, 4, 8, or 12 metric tons ha^{-1} was placed on field plots in July 1973, 1974, and 1975. A different area, previously cropped to dryland wheat, was used each year. Available soil water at planting averaged 12.3 and 21.4 cm for the 0- and 12-metric tons ha^{-1} mulch treatments, respectively. The additional water, along with greater infiltration and lower evaporation during the growing season, increased grain yields, which averaged 1,780 and 3,990 kg ha^{-1} for the respective treatments. Water-use efficiency increased from 55.6 kg/ha-cm for no mulch to 115.0 kg/ha-cm for 12 metric tons mulch ha^{-1}.

Experiments were conducted at Tucson, Arizona, to study effects of treated municipal waste water on growth, fiber, protein, and amino acid content of grain from sorghum (*Sorghum bicolor* (L.) Moench.). The plant growth response of sorghum to waste water, when grown in Grabe silt loam, was similar to that observed when the crop was produced in Comoro sandy loam. Municipal waste water may be an effective source of irrigation water and plant nutrients in the production of high yields of high quality sorghum grain for livestock feed in the irrigated areas of the southwestern U. S.

During the 1980's and 1990's too, many similar Agronomic studies related to planting density and nutrient (inorganic/organic) responses were repeated (Rees, 1986), (Wade and Douglas, 1990), (Jones and Johnson, 1991), (Schlegel, 1992), (Bellakki and Badanur, 1993), (Bellakki and Badanur, 1994), (Staggenborg *et al.*, 1999), but the focus moved on to long term implications and systems approach to sustainable sorghum yields in different cropping systems, tillage and fallow management practices.

Winter legume cover crops have been shown to provide significant amounts of N to subsequent non-leguminous crops. A 3 yr field study at two Georgia locations utilizing a randomized, complete-block, split-plot design with four replications was begun in 1985 to measure the equivalent fertilizer N supplied by winter annual legumes and to monitor changes in soil physical and chemical properties (McVay, Radcliffe, and Hargrove, 1989). Corn, *Zea mays* L. was grown on a Rome gravelly clay loam soil (fine-loamy, mixed, thermic Typic Hapludult) at the Limestone Valley location, and grain sorghum [*Sorghum bicolor* (L) Moench] on a Greenville sandy clay loam soil (clayey, kaolinitic, thermic Rhodic Paleudult) in the Coastal Plain. Main plots were cover crops of hairy vetch (*Vicia villosa* Roth.), crimson clover (*Trifolium incarnatum* L), berseem clover (*Trifolium alexandrinum* L), winter pea [*Pisum sativum* subsp. arvense (L) poir], wheat (*Triticum aestivum* L), and fallow. Subplots were broadcast NH_4NO_3 fertilizer of varying rates. Hairy vetch and crimson clover replaced the greatest amount of fertilizer N averaging 123 and 99 kg N ha^{-1}, respectively. More water-stable aggregates were found following cover crops than fallow in the 0- to 0.025-m depth at the Coastal Plain location. Higher infiltration rates were found following cover crops than fallow at both locations and infiltration rates were greater following hairy vetch than following wheat at the Coastal Plain site. An adapted winter legume cover crop can replace all of the fertilizer N necessary for optimum rain-fed grain sorghum and up to two-thirds of that required for corn production, and improve soil physical properties.

Particulate organic-matter (POM) and mineral-associated C was compared among three tillage treatments (20 yr under cultivation) and an undisturbed grassland at Sidney, NE (Cambardella and Elliott, 1992). The POM C in the native sod represented 39 per cent of the total soil organic C. Twenty years of bare-fallow, stubble-mulch, and no-till management reduced the C content in this fraction to 18, 19, and 25 per cent, respectively, of the total organic C. The mineral-associated organic-matter fraction showed no reduction in C content in the bare-fallow treatment compared with the grassland soil but increased in the no-till and stubble-mulch treatments. Nitrogen dynamics generally mirrored those observed for C. Analysis of the POM fraction for lignin and cellulose content indicated that this fraction was 47 per cent lignin and had a lignocellulose index of 0.7. The stable C-isotope composition of the POM fraction suggests that wheat-derived POM turns over faster than grass-derived POM. It was suggested that the POM fraction closely matches the characteristics of a SOM pool variously described as slow, decomposable, or stabilized organic matter.

The effects of tillage frequency (conventional, reduced and zero), primary tillage implement (disc, blade and chisel plough), stubble management (retention and removal), gypsum application, and paraplowing were examined with respect to soil water storage, soil nitrate accumulation, crop establishment, crop growth, grain yield and grain nitrogen content for 4 successive sorghum crops on a sodic, texture-contrast soil in south west Queensland (Gibson *et al.*, 1992). Retention of sorghum stubble (v. removal) produced an increase in mean yield of sorghum grain of 393 kg ha^{-1}, due to increased soil water extraction and increased water use efficiency by the following crop. The highest mean yield occurred after reduced blade tillage with stubble retained. Zero tillage with stubble removed gave the lowest mean grain yield. Zero tillage always had the lowest quantity of soil nitrate-nitrogen at sowing. In one fallow, increased aggressiveness of primary tillage (disc v. blade plough) increased the quantity of nitrate-nitrogen in the top 60 cm of soil at sowing. These effects on available soil nitrogen did not result in corresponding differences in grain nitrogen content. Results indicate that for optimum fallow management on this texture-contrast soil in south west Queensland, sorghum residues should be retained, tillage frequency should be reduced, but not to zero, blade ploughing should be preferred to disking, and gypsum application should not be practiced.

In Nebraska, six tillage and planting systems selected for evaluation were: no-till, no-till with row-crop cultivation, disk, double disk, chisel, and plow (Dickey, Jasa, and Grisso, 1994). After one crop rotation cycle, for the first 3 yr of yield measurements, no differences occurred in grain yield among the tillage and planting systems. After five additional years, differences in yield were measured, with no-till tending to have the greatest yield for both sorghum and soybean crops. Row-crop cultivation of no-till soybean did not result in any measurable yield differences, but for grain sorghum, row-crop cultivation resulted in an average yield decrease of 6 bu/acre. Soil organic matter tended to be greatest for the continuous no-till system and lowest for the plow system. The plow system had slightly less penetration resistance within the 4- to 8-in. depth than the other treatments, whereas, the double-disk system was slightly greater within the 2- to 6-in. depth. Draft and power requirements for planting in the selected tillage and planting systems were not different. The major difference among the tillage and planting systems was residue cover remaining after planting. No-till had the most residue cover, but there was no appreciable accumulation of residue over the 10 yr of continuous use of the tillage and planting systems. For the last 5 yr, no-till tended to have the greatest yield for both crops. Thus, for the soil and conditions evaluated, no-till yields were as good as the other systems during early years, and were better after 5 yr of continuous use.

Tillage and N management are important in dryland crop production of the west central Great Plains because of frequent periods of limited soil moisture (Thompson and Whitney, 1998). The study involved a wheat (*Triticum aestivum* L.)-sorghum [Sorghum biocolor (L.) Moench]- fallow (W-S-F) rotations under three tillage systems on a nearly level Harney silt loam soil (fine, montmorillonite, mesic Typic Argiustoll). The three tillage systems were clean-till (CT), reduced-till (RT), and no-till (NT). Nitrogen was broadcast preplant as ammonium nitrate on each crop at rates of 0, 20, 40, and 60 lb N/acre. As the level of soil moisture increased in each tillage system, there was a corresponding larger yield increase of wheat and sorghum to applied N. The correlation of grain yields of wheat and sorghum with soil profile N at all depths was highest for nitrate N and lowest for ammonium and total inorganic N. For all three tillage systems, sampling deeper than 6 in. resulted in little improvement in the coefficient of determination (R2) for grain yields regressed on soil nitrate N. Residual soil nitrate N was highest in the top 6 in., dropped significantly in the 6- to 12-in. depth, and remained relatively low thereafter throughout the 72-in. sampling depth. Data from this long-term study showed the optimum broadcast N rate was approximately 60 lb N/acre applied on each crop grown in a W-S-F rotation with the exact rate depending on soil moisture, fertilizer, and crop prices. Yields from CT were comparable with RT on this nearly level upland soil but failed to meet the residue requirements mandated in conservation compliance plans. Poorer stands, increased weed competition, and drier soils resulted in generally lower yields from NT plots. Considering all factors, RT systems for dryland wheat and sorghum production are recommended on upland fertile soils in the west central Great Plains.

A field experiment was conducted on deep vertisols of Bhopal, India to evaluate the manural potential of three organic manures: farmyard manure (FYM), poultry manure (PM), phosphocompost (PC) vis-a-vis 0 per cent, 75 per cent and 100 per cent recommended dose of fertilizer-NPK and to find out the most productive cropping system at various combinations of organic manures and chemical fertilizers (Ghosh *et al.*, 2004). Sorghum, both as sole and intercrop, responded more to PM while soybean to FYM. Application of 75 per cent NPK in combination with PM or FYM or PC to preceding rainy season crops (soybean and sorghum) and 75 per cent NPK to wheat produced significantly higher grain yield of wheat than those in inorganics and control indicating noticeable residual effect on the succeeding wheat crop and saving of 25 per cent fertilizer-NPK. The effect of PC on rainy season crops was not as prominent as those of FYM and PM, but its residual effect on grain yield of wheat was comparable to those two organic manures. Among the cropping systems, soybean as preceding crop recorded the highest seed yield of wheat and was on a par with that of soybean/sorghum intercropping system. The yield of wheat following sorghum was the lowest. The total system productivity (TSP) was the highest in sorghum + soybean–wheat system and the lowest in the soybean–wheat system.

During the 21st century sorghum yield prediction and forecasting through integration of modern tools like remote sensing (RS), geographic information systems (GIS) and trait specific breeding strategies using cropping system simulation modeling have been standardized.

A system analysis was undertaken to identify those plant traits that might be altered to improve crop yield in a water-limited environment (Sinclair and Muchow, 2001). A mechanistic crop model was used to simulate maize (*Zea mays* L.) yield over 20 yr at Columbia, MO, which had high inter annual variability in precipitation. Because sorghum [*Sorghum bicolor* (L.) Moench] is known to be better adapted to drier environments, a number of individual plant traits were adjusted in the maize model to represent a sorghum crop. Increasing the depth of water extraction resulted in increased yields and an increased ratio of grain yield to evapotranspiration. Combining sorghum-like traits in the maize model also increased yield and the ratio of grain yield to evapotranspiration when averaged for all years. For all seasons where simulated yield was less than approximately 550 g m^{-2}, grain yields were greater for a crop with sorghum-like traits than for a crop with maize traits.

Sorghum is the main dryland summer crop in NE Australia and a number of agricultural businesses would benefit from an ability to forecast production likelihood at regional scale (Potgieter, Hammer, Doherty, and de Voil, 2005). In this study we sought to develop a simple agro-climatic modelling approach for predicting shire (statistical local area) sorghum yield. Actual shire yield data, available for the period 1983–1997 from the Australian Bureau of Statistics, were used to train the model. Shire yield was related to a water stress index (SI) that was derived from the agro-climatic model. The model involved a simple fallow and crop water balance that was driven by climate data available at recording stations within each shire. Parameters defining the soil water holding capacity, maximum number of sowings (MXNS) in any year, planting rainfall requirement, and critical period for stress during the crop cycle were optimized as part of the model fitting procedure. Cross-validated correlations (CVR) ranged from 0.5 to 0.9 at shire scale. When aggregated to regional and national scales, 78–84 per cent of the annual variation in sorghum yield was explained.

High spatial resolution satellite imagery has the potential for mapping crop growth variability and identifying problem areas within fields (Yang, Everitt, and

Bradford, 2006). The objectives of this study were to use QuickBird satellite imagery for mapping plant growth and yield patterns within grain sorghum fields as compared with airborne multispectral image data. A QuickBird 2.8-m four-band image covering a cropping area in south Texas, USA was acquired in the 2003 growing season. Airborne three-band imagery with sub-meter resolution was also collected from two grain sorghum fields within the satellite scene. Yield monitor data collected from the two fields were resampled to match the resolutions of the airborne imagery and the satellite imagery. The airborne imagery was related to yield at original sub-meter, 2.8 and 8.4 m resolutions and the QuickBird imagery was related to yield at 2.8 and 8.4 m resolutions. The extracted QuickBird images for the two fields were then classified into multiple zones using unsupervised classification and mean yields among the zones were compared. Results showed that grain yield was significantly related to both types of image data and that the QuickBird imagery had similar correlations with grain yield as compared with the airborne imagery at the 2.8 and 8.4 m resolutions. Moreover, the unsupervised classification maps effectively differentiated grain production levels among the zones. These results indicate that high spatial resolution satellite imagery can be a useful data source for determining plant growth and yield patterns for within-field crop management.

Impacts of climate change on the sorghum production system in India are analysed using InfoCrop-SORGHUM simulation model (Srivastava, Naresh Kumar, and Aggarwal, 2010). In general, impact of climate change is projected to be more on winter crop in central (CZ) and south-central zones (SCZ), while in south-west zone (SWZ) the impacts are likely to be higher on monsoon crop. Climate change is projected to reduce monsoon sorghum grain yield to the tune of 14 per cent in CZ and SWZ and by 2 per cent in SCZ by 2020. Yields are likely to be affected even more in 2050 and 2080 scenarios.

Climate change impacts on winter sorghum are projected to reduce yields up to 7 per cent by 2020, up to 11 per cent by 2050 and up to 32 per cent by 2080. Impacts are projected to be more in SWZ region than in SCZ and CZ. But, the yield loss due to rise in temperature is likely to be offset by projected increase in rainfall. However, complete amelioration of yield loss beyond 2 °C rise may not be attained even after doubling of rainfall in south-central zone (SCZ) and in central zone (CZ). Results indicate that adaptation strategies like changing variety and sowing date can reduce the vulnerability of monsoon sorghum to about 10 per cent, 2 per cent and 3 per cent in CZ, SCZ and SWZ regions in 2020 scenario. Adaptation strategies reduced the climate change impacts and vulnerability of winter crop to 1–2 per cent in 2020, 3–8 per cent in 2050 and 4–9 per cent in 2080. This indicates that more low-cost adaptation strategies should be explored to further reduce the net vulnerability of sorghum production system in India.

References

1. Ajayi, A. E., and Olufayo, A. A. (2004). Evaluation of Two Temperature Stress Indices to Estimate Grain Sorghum Yield and Evapotranspiration. *Agron. J.*, *96*(5), 1282–1287.

2. Antonopoulou, G., Gavala, H. N., Skiadas, I. V., Angelopoulos, K., and Lyberatos, G. (2008). Biofuels generation from sweet sorghum: Fermentative hydrogen production and anaerobic digestion of the remaining biomass. *Bioresource Technology*, *99*(1), 110–119.
3. Atkins, R. E., Reich, V. H., and Kern, J. J. (1968). Performance of Short Stature Grain Sorghum Hybrids at Different Row and Plant Spacings. *Agron. J.*, *60*(5), 515–518.
4. Bellakki, M. A., and Badanur, V. P. (1993). Influence of Organic Residue Recycling on Crop Yield and Nutrient Uptake by *Rabi* Sorghum. *Karnataka Journal of Agricultural Sciences*, *6*(4), 339–344.
5. Bellakki, M. A., and Badanur, V. P. (1994). Effect of crop residue incorporation on physical and chemical properties of a vertisol and yield of sorghum. *Journal of the Indian Society of Soil Science*, *42*(4), 533–535.
6. Ben M Goff, Kenneth, J. M., Steven, L. F., and Emily, A. H. (2010). Double-Cropping Sorghum for Biomass. *Agron. J.*, *102*(6), 1586–1592.
7. Blum, A. (1967). Effect of Soil Fertility and Plant Competition on Grain Sorghum Panicle Morphology and Panicle Weight Components. *Agron. J.*, *59*(5), 400–403.
8. Blum, A. (1970). Effect of Plant Density and Growth Duration on Grain Sorghum Yield under Limited Water Supply. *Agron. J.*, *62*(3), 333–336.
9. Blum, A. (1972). Effect of Planting Date on Water-Use and Its Efficiency in Dryland Grain Sorghum. *Agron. J.*, *64*(6), 775–778.
10. Blum, A. (1974). Genotypic Responses in Sorghum to Drought Stress. I. Response to Soil Moisture Stress1. *Crop Sci.*, *14*(3), 361–364.
11. Blum, A., and Arkin, G. F. (1984). Sorghum root growth and water-use as affected by water supply and growth duration. *Field Crops Research*, *9*(0), 131–142.
12. Blum, A., and Naveh, M. (1976). Improved Water-use Efficiency in Dryland Grain Sorghum by Promoted Plant Competition1. *Agron. J.*, *68*(1), 111–116.
13. Cambardella, C. A., and Elliott, E. T. (1992). Particulate Soil Organic-Matter Changes across a Grassland Cultivation Sequence. *Soil Sci. Soc. Am. J.*, *56*(3), 777–783.
14. Campbell, A. R., and Pickett, R. C. (1968). Effect of Nitrogen Fertilization on Protein Quality and Quantity and Certain Other Characteristics of 19 Strains of *Sorghum bicolor* (L.) Moench. *Crop Sci.*, *8*(5), 545–547.
15. Chapman, S. C., Hammer, G. L., Butler, D. G., and Cooper, M. (2000). Genotype by environment interactions affecting grain sorghum. III. Temporal sequences and spatial patterns in the target population of environments. *Australian Journal of Agricultural Research*, *51*(2), 223–234.
16. Chipanshi, A. C., Chanda, R., and Totolo, O. (2003). Vulnerability Assessment of the Maize and Sorghum Crops to Climate Change in Botswana. *Climatic Change*, *61*(3), 339–360.

17. Craufurd, P. Q., and Peacock, J. M. (1993). Effect of Heat and Drought Stress on Sorghum (*Sorghum bicolor*). II. Grain Yield. *Experimental Agriculture*, *29*(1), 77–86.

18. Day, A. D., and Tucker, T. C. (1977). Effects of Treated Municipal Waste Water on Growth, Fiber, Protein, and Amino Acid Content of Sorghum Grain. *J. Environ. Qual.*, *6*(3), 325–327.

19. Dickey, E. C., Jasa, P. J., and Grisso, R. D. (1994). Long Term Tillage Effects on Grain Yield and Soil Properties in a Soybean/Grain Sorghum Rotation. *Journal of Production Agriculture*, *7*(4), 465–470.

20. Fischer, K. S., and Wilson, G. L. (1975). Studies of grain production in *Sorghum bicolor* (L. Moench). V.* Effect of planting density on growth and yield. *Australian Journal of Agricultural Research*, *26*(1), 31–41.

21. Garrity, D. P., Watts, D. G., Sullivan, C. Y., and Gilley, J. R. (1982). Moisture Deficits and Grain Sorghum Performance: Evapotranspiration-Yield Relationships. *Agron. J.*, *74*(5), 815–820.

22. Ghosh, P., Ramesh, P, Bandyopadhyay, K., Tripathi, A., Hati, K., Misra, A., and Acharya, C. (2004). Comparative effectiveness of cattle manure, poultry manure, phosphocompost and fertilizer-NPK on three cropping systems in vertisols of semi-arid tropics. I. Crop yields and system performance. *Bioresource Technology*, *95*(1), 77–83.

23. Gibson, G., Radford, B. J., and Nielsen, R. G. H. (1992). Fallow management, soil water, plant-available soil nitrogen and grain sorghum production in south west Queensland. *Australian Journal of Experimental Agriculture*, *32*(4), 473–482.

24. Gibson, P. T., and Schertz, K. F. (1977). Growth Analysis of a Sorghum Hybrid and its Parents. *Crop Sci.*, *17*(3), 387–401.

25. Gnansounou, E., Dauriat, A., and Wyman, C. E. (2005). Refining sweet sorghum to ethanol and sugar: economic trade-offs in the context of North China. *Bioresource Technology*, *96*(9), 985–1002.

26. Hanks, R. J., Gardner, H. R., and Florian, R. L. (1969). Plant Growth-Evapotranspiration Relations for Several Crops in the Central Great Plains. *Agron. J.*, *61*(1), 30–34.

27. Jacques, G. L., Vanderlip, R. L., and Ellis, R. (1975). Growth and Nutrient Accumulation and Distribution in Grain Sorghum. II. Zn, Cu, Fe, and Mn Uptake and Distribution. *Agron. J.*, *67*(5), 611–616.

28. Jones, O. R., and Johnson, G. L. (1991). Row Width and Plant Density Effects on Texas High Plains Sorghum. *Journal of Production Agriculture*, *4*(4), 613–621.

29. Lingegowda, B. K., Inamdar, S. S., and Kirshnamurthy, K. (1971). Effects of dates of planting on yield of sorghum CSH-1. *Indian Journal of Agronomy*, *16*(2), 155–156.

30. Maranville, J.W., Clark, R. B., and Ross, W. M. (1980). Nitrogen efficiency in grain sorghum. *Journal of Plant Nutrition*, *2*(5), 577–589.

31. Maranville, Jerry W., and Clegg, M. D. (1977). Influence of Seed Size and Density

on Germination, Seedling Emergence, and Yield of Grain Sorghum. *Agron. J.*, *69*(2), 329–330.

32. Mastrorilli, M., Katerji, N., and Rana, G. (1995). Water efficiency and stress on grain sorghum at different reproductive stages. *Agricultural Water Management*, *28*(1), 23–34.

33. Mathers, A. C., Stewart, B. A., and Thomas, J. D. (1977). Manure Effects on Water Intake and Runoff Quality from Irrigated Grain Sorghum Plots. *Soil Sci. Soc. Am. J.*, *41*(4), 782–785.

34. Mays, D. A., Terman, G. L., and Duggan, J. C. (1973). Municipal Compost: Effects on Crop Yields and Soil Properties. *J. Environ. Qual.*, *2*(1), 89–92.

35. McVay, K. A., Radcliffe, D. E., and Hargrove, W. L. (1989). Winter Legume Effects on Soil Properties and Nitrogen Fertilizer Requirements. *Soil Sci. Soc. Am. J.*, *53*(6), 1856–1862.

36. Muchow, R. C. (1990). Effect of high temperature on the rate and duration of grain growth in field-grown *Sorghum bicolor* (L.) Moench. *Australian Journal of Agricultural Research*, *41*(2), 329–337.

37. Muchow, R. C. (1998). 32 Nitrogen utilization efficiency in maize and grain sorghum. *Nutrient Use Efficiency in Rice Cropping Systems*, *56*(1–2), 209–216.

38. Muchow, R. C., Hammer, G. L., and Vanderlip, R. L. (1994). Assessing climatic risk to sorghum production in water-limited subtropical environments II. Effects of planting date, soil water at planting, and cultivar phenology. *Field Crops Research*, *36*(3), 235–246.

39. Muchow, R. C., and Wilson, G. L. (1976). Photosynthetic and storage limitations to yield in *Sorghum bicolor* (L. Moench). *Australian Journal of Agricultural Research*, *27*(4), 489–500.

40. Narasimhan, B., and Srinivasan, R. (2005). Development and evaluation of Soil Moisture Deficit Index (SMDI) and Evapotranspiration Deficit Index (ETDI) for agricultural drought monitoring. *Agricultural and Forest Meteorology*, *133*(1–4), 69–88.

41. Ntaikou, I., Gavala, H. N., Kornaros, M., and Lyberatos, G. (2008). Hydrogen production from sugars and sweet sorghum biomass using Ruminococcus albus. *International Journal of Hydrogen Energy*, *33*(4), 1153–1163.

42. Onken, A. B., and Sunderman, H. D. (1972). Applied and Residual Nitrate-Nitrogen Effects on Irrigated Grain Sorghum Yield. *Soil Sci. Soc. Am. J.*, *36*(1), 94–97.

43. Pedersen, J. F., Toy, J. J., Funnell, D. L., Sattler, S. E., Oliver, A. L., and Grant, R. A. (2008). Registration of BN611, AN612, BN612, and RN613 Sorghum Genetic Stocks with Stacked bmr-6 and bmr-12 Genes. *J. Plant Reg.*, *2*(3), 258–262.

44. Pepper, G. E., and Prine, G. M. (1972). Low Light Intensity Effects on Grain Sorghum at Different Stages of Growth. *Crop Sci.*, *12*(5), 590–593.

45. Plaut, Z., Blum, A., and Arnon, I. (1969). Effect of Soil Moisture Regime and Row Spacing on Grain Sorghum Production. *Agron. J.*, *61*(3), 344–347.

46. Potgieter, A. B., Hammer, G. L., Doherty, A., and de Voil, P. (2005). A simple regional-scale model for forecasting sorghum yield across North-Eastern Australia. *Agricultural and Forest Meteorology*, *132*(1–2), 143–153.

47. Prasad, P. V. V., Pisipati, S. R., Mutava, R. N., and Tuinstra, M. R. (2008). Sensitivity of Grain Sorghum to High Temperature Stress during Reproductive Development. *Crop Sci.*, *48*(5), 1911–1917.

48. Ratnavathi, C. V., Suresh, K., Kumar, B. S. V., Pallavi, M., Komala, V. V., and Seetharama, N. (2010). Study on genotypic variation for ethanol production from sweet sorghum juice. *Biomass and Bioenergy*, *34*(7), 947–952.

49. Raun, W. R., and Johnson, G. V. (1999). Improving Nitrogen Use Efficiency for Cereal Production. *Agron. J.*, *91*(3), 357–363.

50. Rees, D. J. (1986). Crop Growth, Development and Yield in Semi-arid Conditions in Botswana. I. The Effects of Population Density and Row Spacing on *Sorghum bicolor*. *Experimental Agriculture*, 22(02), 153–167.

51. Roy, R. N., and Wright, B. C. (1973). Sorghum Growth and Nutrient Uptake in Relation to Soil Fertility: I. Dry Matter Accumulation Patterns, Yield, and N Content of Grain. *Agron. J.*, *65*(5), 709–711.

52. Roy, R. N., and Wright, B. C. (1974). Sorghum Growth and Nutrient Uptake in Relation to Soil Fertility, II. N, P, and K Uptake Pattern by Various Plant Parts. *Agron. J.*, *66*(1), 5–10.

53. Saeed, M., Francis, C. A., and Clegg, M. D. (1986). Yield Component Analysis in Grain Sorghum. *Crop Sci.*, *26*(2), 346–351.

54. Schlegel, A. J. (1992). Effect of Composted Manure on Soil Chemical Properties and Nitrogen Use by Grain Sorghum. *Journal of Production Agriculture*, *5*(1), 153–157.

55. Sinclair, T. R., and Muchow, R. C. (2001). System Analysis of Plant Traits to Increase Grain Yield on Limited Water Supplies. *Agron. J.*, *93*(2), 263–270.

56. Singh, A., and Bains, S. S. (1973). Yield, grain quality and nutrient uptake of "CSH 1" and "Swarna" sorghum at different levels of N and plant population. *Indian Journal of Agricultural Sciences*, *43*(4), 408–413.

57. Singh, P., and Choubey, S. D. (1972). Effect of varying levels of nitrogen on the yield and yield attributes of some sorghum varieties. *Indian Journal of Agricultural Sciences*, *42*(4), 337–341.

58. Srivastava, A., Naresh Kumar, S., and Aggarwal, P. K. (2010). Assessment on vulnerability of sorghum to climate change in India. *Agriculture, Ecosystems and Environment*, *138*(3–4), 160–169.

59. Staggenborg, S. A., Fjell, D. L., Devlin, D. L., Gordon, W. B., and Marsh, B. H. (1999). Grain Sorghum Response to Row Spacings and Seeding Rates in Kansas. *Journal of Production Agriculture*, *12*(3), 390–395.

60. Stickler, F. C. (1964). Stand Establishment and Yield of Grain Sorghum as Affected by Method of Planting and Use of Press Wheels. *Agron. J.*, *56*(1), 53–56.

61. Stickler, F. C., and Wearden, S. (1965). Yield and Yield Components of Grain Sorghum as Affected by Row Width and Stand Density. *Agron. J.*, *57*(6), 564–567.

62. Thompson, C. A., and Whitney, D. A. (1998). Long-Term Tillage and Nitrogen Fertilization in a West Central Great Plains Wheat-Sorghum-Fallow Rotation. *Journal of Production Agriculture*, *11*(3), 353–359.

63. Tolk, J. A., Howell, T. A., Steiner, J. L., and Evett, S. R. (1997). Grain sorghum growth, water use, and yield in contrasting soils. *Agricultural Water Management*, *35*(1–2), 29–42.

64. Unger, P. W., and Baumhardt, R. L. (1999). Factors Related to Dryland Grain Sorghum Yield Increases: 1939 through 1997. *Agron. J.*, *91*(5), 870–875.

65. Vanderlip, R. L., and Reeves, H. E. (1972). Growth Stages of Sorghum [*Sorghum bicolor*, (L.) Moench.]. *Agron. J.*, *64*(1), 13–16.

66. Wade, L. J., and Douglas, A. C. L. (1990). Effect of plant density on grain yield and yield stability of sorghum hybrids differing in maturity. *Australian Journal of Experimental Agriculture*, *30*(2), 257–264.

67. Willey, R. W., and Basiime, D. R. (1973). Studies on the physiological determinants of grain yield in five varieties of sorghum. *The Journal of Agricultural Science*, *81*(03), 537–548.

68. Wright, G. C., Smith, R. C., and McWilliam, J. R. (1983). Differences between two grain sorghum genotypes in adaptation to drought stress. I. Crop growth and yield responses. *Australian Journal of Agricultural Research*, *34*(6), 615–626.

69. Yang, C., Everitt, J., and Bradford, J. (2006). Comparison of QuickBird Satellite Imagery and Airborne Imagery for Mapping Grain Sorghum Yield Patterns. *Precision Agriculture*, *7*(1), 33–44.

70. Zegada-Lizarazu, W., and Monti, A. (2012). Are we ready to cultivate sweet sorghum as a bioenergy feedstock? A review on field management practices. *Biomass and Bioenergy*, *40*(0), 1–12.

Millets: Ensuring Climate Resilience and Nutritional Security (2015) *Pages* **431–441**
Editors: **Vilas A. Tonapi and J.V. Patil**
Published by: **DAYA PUBLISHING HOUSE, NEW DELHI**

Chapter 15

Pearl Millet Agronomy for Higher Productivity

J.S. Mishra
ICAR-Indian Institute of Millets Research (IIMR)
Rajendranagar, Hyderabad – 500 030

Pearl millet [*Pennisetum glaucum* (L.) R. Br.], the fifth most important cereal crop in the world after rice, wheat, maize, and sorghum is widely grown in the arid and semi-arid regions of Africa and Southern Asia. The crop can be grown in areas with annual rainfall of 200 to 600 mm. Pearl millet accounts for almost half of global millet production, with 60 per cent of the cultivation areas in Africa, followed by 35 per cent in Asian countries. Globally, the crop is grown on 27 million ha in some of the harsh, arid and dry-semi-arid tropical environments of South Asia and sub-Saharan Africa (Amarender Reddy *et al.*, 2013). India is the largest producer of pearl millet in Asia, both in terms of area and production. It is the most widely cultivated cereal after rice and wheat. All time high production (12.57 mt) and productivity (1431 kg/ha) of pearl millet was achieved during 2011-12 (Project Coordinator's Report, 2013). However, the average productivity declined to 1138 kg/ha in 2012-13 due to drought conditions in pearl millet growing regions (Pearl Millet News, 2014). The major pearl millet growing state is Rajasthan which account for more than 50 per cent (57.6 per cent in current year) of pearl millet acreage in country with Maharashtra, Gujarat, Uttar Pradesh and Haryana together sharing another 40 per cent. It is mainly cultivated during *Kharif* (rainy) season across the country. However, it is also grown to a lesser extent during *Rabi* (post-rainy) season in Andhra Pradesh, Karnataka and Tamil Nadu. Summer pearl millet cultivation is popular in Gujarat, producing high yields and excellent grain quality. The productivity imbalance is due to erratic rains and shifting of pearl millet cultivation to marginal soils due to diversification of traditional areas to higher value crops across the country.

The area under pearl millet is marginally declining or stagnant and there is a probability of further decline in area under *Kharif* pearl millet due to expansion of area under more remunerative and high value crops such as pulses and oilseeds (greengram, blackgram, sesamum, cluster bean, etc). However, there is the scope of expanding the area under *Kharif* pearl millet in low rainfall regions with little scope for expanding irrigation facilities. Still there is a large pearl millet area under *Kharif* season in western India; hence, it is important to increase pearl millet production through all the best possible interventions in this season. Pearl millet is having a drought-escaping mechanism but still responds well to irrigation and improved agronomic management conditions. A majority of the subsistence farmers who typically cultivate this crop are unable to take advantage of high yield potential of pearl millet because of the lack of application of improved management practices. Pearl millet productivity can be increased by growing varieties/hybrids with improved tolerance to drought, resistance to diseases, and with fertilizer applications and soil moisture management.

Review of the Research Work Done on Production Technologies

Agronomic research conducted under All-India Co-ordinated Pearl Millet Improvement Project and elsewhere led to the establishment of detailed recommendations for pearl millet-growing areas. A brief resume of derived recommendations on agronomic aspect is given here.

Time of Sowing and Seeding Depth

Optimum time of sowing of crop plays an important role in determining the performance of a crop. In rainfed areas the sowing should be commenced with the onset of monsoon showers. Chances of disease and insect-pest incidence are also reduced in timely sown crop. Under delayed sowing, crop becomes more susceptible to rust as rust inoculums become more prevalent later in the season resulting in lower yields (Wilson *et al.*, 1995). Delay in sowing reduces the grain yield consistently. The rate of decline varied from 4 to 80 kg/day/ha under rainfed and 4 to 150 kg/day/ha under irrigated conditions (Batnagar *et al.*1998). They reported that the crop should be sown by first fortnight of July. The results of the experiments conducted under AICPMIP revealed that the proper sowing time for zone A1 was up to mid July, for zone A- last week of June to first week of July and for zone B-last week of June to mid July (AICPMIP, 2012). First fortnight of October is appropriate time for *Rabi* season in Tamil Nadu.

Gap filling should be done by transplanting seedlings after 2-3 weeks of sowing if scanty population exists. In Marathwada area of Maharashtra, dry sowing prior to first monsoon rains is recommended. Summer pearl millet should be sown in the last week of January to 1st week of February to obtain higher production of summer pearl millet in zone B. The crop should be seeded at 2.5-3 cm soil depth.

Transplanting or Direct Seeding

For delayed sowing, transplanting is recommended in pearl millet. Delayed direct seeding results in poor plant stand and poor grain setting thereby causing low grain yield. Transplanting was found as a better substitute of direct seeding under such cases. It is because 3-week-old transplanted seedlings produce more tillers and show better ear development resulting in higher yields.

Seed Rate and Plant Populations and Planting Geometry

The recommended seed rate is 4-5 kg/ha. The optimum plant population should be 1,75,000-2,00,000 plants/ha. However, plant population of 1,60,000/ha should be maintained in erratic rainfall areas. This plant population is achieved by maintaining 50 cm distance between rows and 10-12 cm distance between plants within rows. Widening the row distances decreases the pre-flowering moisture use and the conserved water can be utilized during later growth stages. For arid-western plain of Rajasthan, Haryana and Kutch of Gujarat (A1 zone), pearl millet should be planted in rows 60 cm apart, maintaining low plant population of 1.00 to 1.25 lac/ha. For the area receiving rainfall more than 450 mm (zone A and B), the crop should be planted at the spacing of 45 x 10-15 cm keeping plant population of 1.75 to 2.0 lakhs/ha. Plant stand establishment is critical to obtain higher grain yields. Soil surface crust formation after sowing is a major problem resulting in poor plant stand. Application of FYM in seed furrows has helped to establish plant stands and produce higher grain yields (1834 kg/ha) than *pora* sowing (1083 kg/ha) or *kera* sowing (1094 kg/ha) (AICMIP, 1977-1982).

Weed Management

Because pearl millet is slow to grow initially, weeds are problematic (Asbil, 2001) and early control is needed to optimize yield. As crop is grown predominantly in the hot and humid rainy season, weeds deprive pearl millet of vital nutrients and moisture. Like other rainy-season crops, pearl millet also faces severe weed competition leading to heavy reduction in grain yield. Competition from weeds could reduce yield by 25-50 per cent. Because weeds compete with crops for water, nutrients, air, and space, it is imperative to keep fields weed-free. Hand weeding is superior to herbicidal application, but is more expensive and labor intensive. As manually uprooting them becomes difficult because of soggy nature of soil during the crop season, chemical weed control was recommended. Triazine compounds, atrazine and propazine when applied @ 0.5 kg/ha at the pre-emergence stage controlled 71-96 per cent of the weeds thus increasing the yield by 30-40 per cent over the unweeded control. Application of atrazine @ 0.5 kg/ha superimposed with 1 hand-weeding 4 weeks after sowing was found most effective and economic method of controlling weeds. Chemical weed control was more economical than manual weeding. There was no adverse effect of atrazine applied in recommended doses in pearl millet in rainy season on the succeeding winter crops. Crop management practices, such as reduced row spacing (Dowler and Wright 1995; Limon-Ortega *et al.*, 1998; Asbil 2001), increased seed density (Asbil, 2001), using a false seedbed technique (Dowler and Wright 1995), or burn down with a broad spectrum herbicide prior to seeding (Asbil 2001), have been shown to reduce weed pressure in pearl millet.

Water Management

Pearl millet is generally grown during rainy season under rainfed condition in areas of low rainfall. Water requirement of the crop is much lower (250 mm-350 mm). However, the crop is most sensitive to water stress during flowering and grain filling stages (Hattendorf *et al.*, 1988). Under prolonged dry spells, irrigation should be applied at critical stages of crop growth *i.e.* tillering, flowering and grain developmental stage, if water is available. In summer, pearl millet should be irrigated at regular intervals (0.75-1.0 IW/CPE with 40 mm) as per need of the crop.

Moisture Conservation

Evaporation from the soil surface constitutes a large proportion of evapotranspiration of pearl millet fields. Conservation of moisture either in soil or plant constitutes an important practice in pearl millet cultivation. A widely spaced pearl millet crop grown in the hotter parts of the year can result in loss of considerable soil moisture. To have a control over evaporational losses, dust mulching (breaking of upper crust) or organic mulching (wheat or rice straw) @ 5 tonnes/ha is recommended which could increase the pearl millet yield by 25 per cent. This also decreased the soil temperature by 4°C and increased the soil-moisture content by 3 per cent.

Removing the upper one-third of the plant; or foliar application of chemicals such as borax, kaoline, atrazine; or pre-sowing treatments with 0.2 per cent solutions of KNO_3 or NaCl was effective in controlling leaf transpirational losses and increasing grain yield considerably. These chemical treatments were as good as straw mulch.

Nutrient Management

Next to quality seed of high-yielding varieties, nitrogen becomes critical, particularly under drought conditions. Rajasthan, Maharashtra, Gujarat, and Uttar Pradesh together account for 76.7 per cent of the area and 68.8 per cent of pearl millet production in the country. However, the total N, P, and K consumption of Rajasthan does not exceed 10 kg/ha with a total food grain production of less than 500 kg/ha (AICMIP 1983). As a no- or low-energy input crop, pearl millet is fertilizer- starved and fertilizer-hungry, but its response to fertilizer is tremendous and the grain yield increases many-fold. Fertilizers should be allocated to rainfed pearl millet, subsidized to cover any risks, and preferably applied on a community basis. This is a necessary "minimum" of dry farming technology if pearl millet is expected to contribute significantly to the national food basket. At the present productivity level, pearl millet removes 72 kg of N + P + K/ha/annum, whereas only 10-11 kg of these nutrients are being actually applied (Bhatnagar *et al.*, 1998). It is estimated that for the production of 100 kg grain, pearl millet requires 3.6 kg N, 0.8 kg P and 3.4 kg K. A crop producing grain yield of 3.0 t/ha removed 79.8, 28.2 and 97.8 kg NPK, respectively from soil (Khairwal *et al.*, 2007). Fertilizers not only increase above-ground growth but also enhance below-ground root proliferation. In this way, replenishment of the soil nutrients encourages more efficient water use.

Nitrogen

Of the 3 macro-nutrients, response to N has.almost been universal in all regions of the country where pearl millet is grown. The optimum rate of N fertilization in experiments averaged over 5 years and 30 locations ranged from 92 to 137 kg/ha at a response level of 6.0-10.5 kg grain/kg N applied. The output: input ratio of nitrogenous fertilizer ranged between 0.8 and 2.7 in different zones (De and Gautam 1987). Pearl millet usually receives no or low amounts of chemical fertilizer, but its response to fertilizer is tremendous. Even without fertilization, a high-yielding variety or a hybrid out yields local cultivars. Nitrogen application, however, widens this difference, indicating that the high-yielding varieties and hybrids use nitrogen more efficiently. Based on the response to nitrogen application, it is estimated that for every 1 kg N applied, the hybrids or improved varieties return 10-15 kg of grain at 30-60 kg/ha N as compared to no application. However, 40 kg/ha N was found to be profitable for most of the pearl millet growing regions, but higher nitrogen levels (60-80 kg/ha) are recommended in areas with assured rainfall such as Gujarat and Uttar Pradesh. The response was also higher in hybrids than in synthetics and composites (AICMIP, 1977-79).

The studies aimed at determining response to N fertilization of newly developed hybrids and populations under varying time of sowing showed that most of the cultivars were found responsive to N fertilization. Application of 40 kg N/ha in zone A and 60-80 kg N/ha in zone B were found optimum doses for both hybrids and populations under rainfed situation. Full dose of N should be applied in 2 equal splits, *i.e.* half at planting and other half as top-dressed 3-4 weeks after sowing, subject to the avail-ability of soil moisture. The field experiments, initiated in 1992 at several co-ordinated centres on the permanent site with the objective to determine the long-term effect of organic manures like farmyard manure on nitrogen economy in pearl millet cultivation and also on soil physical conditions revealed that application of 2.5 tonnes of farmyard manure for AI zone and 5 tonnes of farmyard manure/ha every year for A and B zones in combination with 40kg N/ha was more economical and conducive for yield stability in pearl millet. Addition of organic materials with low C:N ratio, such as cluster bean residue in conjunction with 40 kg N/ha to increased the pearl millet yield (Aggarwal *et al.*, 1997). Application of 50 per cent RDF (40:20 kg NP/ha) along with compost 2.5 t/ha and castor cake 500 kg/ha provided maximum yield and profits in pearl millet (Kanzaria *et al.*, 2010). Under dryland conditions with low rainfall, 30 to 40 kg of nitrogen/ha may be adequate for varieties of low yield potential. High/assured rainfall regions or irrigated areas may require 80 to 100 kg nitrogen/ha to get higher yields. Under prolonged dry spell, skipping of top dressing of N and spraying of 2 per cent urea is advised. Under excessive rain situation during vegetative phase, additional dose of nitrogen @ 20 kg/ha should be applied.

Phosphorus

The recommended rate of phosphorus application is 25 kg P_2O_5/ha for rainfed pearl millet and 45 to 60 kg P_2O_5/ha for irrigated pearl millet. The entire quantity of phosphorus is applied at the time of sowing. Phosphorus has to be applied as side dressing or by placement below the seed for its efficient utilization by the plants.

Potassium

Soils with medium or high potassium levels will not show a yield response to added potassium fertilizer. For soils low in potassium (less than 150 kg/ha), the recommended rates are 25 kg/ha of K_2O for rainfed pearl millet and 50 kg/ha of K_2O for irrigated pearl millet. Like phosphorus, potassium is also applied in one dose at the time of sowing along with nitrogen fertilizer.

Micronutrients

Gupta and Jain (2008) concluded that application of sulphur at 45 kg S/ha through single super phosphate in pearl millet was effective in increasing the productivity of the pearl millet- Indian mustard cropping system in the semi-arid tract of Rajasthan. In zinc deficient soils of the pearl millet growing areas of the country, soil application of 10 kg $ZnSO_4$/ha is recommended. To correct the zinc deficiency in standing crop, spray of 0.2 per cent $ZnSO_4$ at tillering to pre-flowering stage is recommended.

Biofertilization

Use of biofertilizer (*Azospirillum* and PSB) can economize the N and P fertilizer application. *Azospirillum brasilence* is a free-living bacterium which shows associated symbiosis when present in the rhizosphere. These bacteria are capable of fixing 10-28 mg/g N of calcium malate in pure culture. In experiments conducted at 7 locations over 3 years, inoculation of pearl millet seed with the local strains of the bacterium increased the yield markedly. Field studies aimed at quantifying the response of pearl millet to inoculation with *Azotobacter* and phosphate for economizing nutrient-solubilizing bacteria revealed that part of inorganic nitrogen could be saved with Azotobacter. Seed inoculation with *Azospirillum brasilence* in conjunction with N application was found to save nitrogen to the extent of 10-20 kg/ha. Sole application of phosphobacteria or with P_2O_5 did not produce as much visible effect as sole azotobacter or azotobacter with nitrogen.

Intercropping

Being a rainfed crop, pearl millet is also grown in mixed, cropping or intercropping systems, as an insurance against bad weather. Paired-row system of planting of pearl millet along with a row of legume as an intercrop not only increases the total production but also reduces the fertilzer N requirement of the crop. Pearl millet + pigeonpea (2:1 paired row 30/60 cm) or 4:2 intercropping system was found most productive and remunerative (Patil and Patil, 1989; Patel *et al.*, 1985; Gadhia *et al.*, 1993). Pearl millet + groundnut (2:4) was also found superior (Patel *et al.*, 1985). The highest net returns were obtained when pearl millet was intercropped with greengram 2:1(Singh and Chauhan, 1973; Hooda and Khippal, 2000) obtained. Paired row planting of pearl millet with greengram or cowpea (30/70 cm) increased the yield of pearl millet with additional yield of pulses. The other intercrops suited with pearl millet are blackgram, cowpea, dew gram, castor and cluster bean. Details of intercropping systems practiced in the country are listed in Table 1.

Table 1: Intercropping Systems in Pearl Millet Growing Regions (Khaïrwal *et al.*, 2007)

States	*Intercropping Systems*
Eastern Rajasthan	Pearl millet + Cluster bean
	Pearl millet + Cowpea
	Pearl millet + Greengram
Western Rajasthan	Pearl millet + Mothbean
	Pearl millet + Clusterbean
Maharashtra	Pearl millet + Pigeonpea/soybean
	Pearl millet + Greengram
	Pearl millet + Blackgram
	Pearl millet + Cowpea
	Pearl millet + Mothbean
	Pearl millet + Sunflower
Gujarat	Pearl millet + Greengram
	Pearl millet + Cowpea
	Pearl millet + Sesame
Haryana	Pearl millet + Clusterbean
	Pearl millet + Greengram
	Pearl millet + Cowpea
Uttar Pradesh	Pearl millet + Greengram
	Pearl millet + Cowpea
	Pearl millet + Sesame
Madhya Pradesh	Pearl millet + Pigeonpea
	Pearl millet + Cowpea
	Pearl millet + Soybean
Karnataka	Pearl millet + Pigeonpea
	Pearl millet + Soybean
	Pearl millet + Greengram
	Pearl millet + Sunflower
Andhra Pradesh	Pearl millet + Pigeonpea
	Pearl millet + Soybean
	Pearl millet + Greengram
	Pearl millet + Sunflower
	Pearl millet + Groundnut
Tamil Nadu	Pearl millet + Pigeonpea
	Pearl millet + Soybean
	Pearl millet + Greengram
	Pearl millet + Cowpea
	Pearl millet + Sunflower

Production Constraints

- ☆ Cultivation of pearl millet on marginal lands.
- ☆ Unreliable rainfall and frequent drought.
- ☆ Minimum use of inputs such as fertilizers and organic manures.
- ☆ Low soil moisture resulting in poor plant stand
- ☆ Crust formation on soil surface
- ☆ Crop losses due to occurrence of downy mildew.
- ☆ Bird damage.

Research Strategies

1. *Nutrient management in pearl millet-based cropping systems*: Pearl millet is grown under various crop sequences involving cereals like pearl millet-wheat, oilseeds like pearl millet-mustard/linseed and pulses like pearl millet-chickpea/pea/lentil. Also it is intercropped with various crops including legumes. The nutrient requirement of pearl millet-based cropping system as a whole need to be worked out for different regions and soil types. A long-term study need to be initiated in major pearl millet based cropping sequences in different ecological regions to quantify the role of legume crops on nutrient balance and soil health.
2. *Nutrient x genotype interaction studies:* Pearl millet genotypes exhibit differential responses to fertilizer input, and different production potentials at different fertility levels indicate that the genotypes have different threshold responses. These studies should be initiated in different types of soils and agro-ecological regions.
3. *Biofertilization:* Bacterial fertilization has been found to affect consumption of recommended inorganic nitrogen. *Azospirillum* application alone increased grain yield from 1115 to 1434 kg/ha (Harinarayana 1980). The effect varies depending on location, pH, C:N ratio, and supplemental nitrogen application. *Azospirillum* with 10 kg/ha N has yielded as much grain as 20 kg/ha N, indicating that10 kg/ha N could be saved. *Azospirillum* and *Azotobacter* production should be commercialized on the lines of *Rhizobium* and applied extensively to increase pearl millet grain yield with or without the addition of inorganic nitrogen. More efficient strains better compatible with specific genotypes, and factors, both plant and bacterial, responsible for this specificity should be identified for large-scale screening. The rhizosphere bacteria are influenced by mycorrhizae which increase the uptake of phosphorous, even in soils with threshold levels (Hubbell, 1987). Studies to identify associative mycorrhizae beneficial to pearl millet in poor soils would be useful.
4. *Secondary and Micronutrients:* Pearl millet growing soils are deficient in secondary elements like sulphur, and most of the micro-nutrients like Fe, Zn, etc. Hence, there is a need to initiate experiments on micronutrient doses and method of application.

5. *Cropping systems approach:* The ever-increasing demand for food puts a premium on increasing food production per unit area. The traditional mixed farming systems have been scientifically transformed into intercropping systems which include pearl millet. The total system needs modifications based on selection of genotypes for plant habit, plant height, maturity, manipulation of ratio and proportion of component crops, and management of production components such as fertilizer, water, weeds, etc.
6. *Managing soil moisture:* Rainfed pearl millet faces a soil-moisture deficit during grain formation and seed development. Soil moisture conservation through grading, land shaping, drainage, erosion control, etc., is pivotal for the success of any technology. This often transcends individual boundaries and involves community and group action. Large scale adoption requires huge capital investments for equipment and technical expertise. The necessary infrastructure and institutional support is often missing in the rainfed areas.
7. *Weed management:* More research is needed to identify herbicides and other weed management strategies that can safely control annual grasses and broadleaved weeds in pearl millet crops. Herbicides for weed management in sole crop of pearl millets are available; however, there is a need to evaluate safe and effective herbicides for pearl millet–based intercropping systems and crop sequences.
8. *Intercropping:* There is a need to evaluate the performance of newly released hybrids and varieties of pearl millet vis-à-vis the component crops for their compatibility for intercropping systems in different zones. More research is needed on nutrient management in pearl millet-based intercropping systems. The role of pearl millet intercropping/cropping system on suppression of insect-pests and diseases and weed management need to be quantified.
9. *Evaluation of pearl millet cultivars to different sowing window*: Under changing climate scenario where monsoon is getting late with erratic rainfall, there is a need to evaluate the pearl millet cultivars for delayed sowing with acceptable grain yield.
10. *Seed priming:* It involves soaking of seed in water or other nutrient solutions before sowing of crop to enhance the germination. As the pearl millet crop is grown under rainfed/dryland conditions, where moisture at sowing is critical, the studies on seed priming to enhance the germination need to be initiated.
11. *Site-specific nutrient management:* There is a need to prepare an inventory of the soils in pearl millet growing regions and nutrient dose may be recommended based on the soil nutrient status.
12. *Graded technology:* The adoption levels of improved technology vary depending on farm size, and farmer financial and technical resources. Field experience shows that technology based on low monetary inputs such as the introduction of high-yielding varieties and improved cultural practices

has been widely adopted by all classes of farmers. However, graded technology is necessary to suit the varying levels of adoption. The relative contribution of each of the inputs to farmers of divergent resource bases operating at different levels of skills and management should be determined.

References

1. NRC. 1996. Pearl millet. In: *Lost Crops of Africa. Volume 1: Grains.* National Academy Press, Washington DC, USA. pp. 77-126.
2. Yadav, O.P. 2011. Project Coordinator's Review (2010-11) on Pearl Millet Research. (http://www.aicpmip.res.in/pcr2011.pdf).
3. Gupta, A.K. and Jain, N.K. 2008. Sulphur fertilization in a pearl millet(*Pennisetum glaucum*)-Indian mustard (*Brassica juncea*) cropping system, *Archives of Agronomy and Soil Science*, 54: 5, 533-539, DOI: 10.1080/03650340802280435
4. Harinarayana, G. 1980.Prospects of millets research in1980's. Presented at the All India Coordinated Millets Improvement Project Workshop, 17-19 A p r 1980, Hisar, Haryana, India.
5. De, R. and Gautam, R.C.1987. Management priorities to increase and stabilize pearl millet production in India. In: Proceedings of International pearl millet Work shop, held during 7-11April, 1986 at the International Crop Research Institute for Semi-Arid. Tropics, Patancheru, Hyderabad,, pp. 247-254.
6. Amarender Reddy A., Parthasarathy Rao P., Yadav O.P., Singh I.P., Ardeshna N.J., Kundu K.K., Gupta SK, Rajan Sharma, Sawargaonkar G, Dharm Pal Malik, Shyam Moses D and Sammi Reddy K.2013. Prospects for *Kharif* (rainy season) and summer pearl millet in western India. Working Paper Series no. 36. Patancheru 502 324, Andhra Pradesh, India: International Crops Research Institute for the Semi-Arid Tropics. 24 pp.
7. AICPMIP, 2012. Proceedings of the 47th Annual Pearl Millet Workshop of the All India Coordinated Pearl Millet Improvement Project, held at SK RAU, ARS, Jaipur, March, 17-19, 2012.
8. Bhatnagar, S.K., Yadav, O.P. and Gautam, R.C. 1998. Research Achievements in pearl millet (*Pennisetum glaucum*). *Indian Journal of Agricultural Sciences*, 68 (8, Special issue): 423-30.
9. Wilson, J.P., Hanna, W.W., Gascho G. and Wilson, D.M. 1995. Pearl millet grain yield loss from rust infection. Page 54–56. *In: Proceedings of the First National Grain Pearl Millet Symposium*, 17–18 Jan 1995 (Teare ID, ed.). Tifton, Georgia, USA: University of Georgia and USDA
10. Asbil, W. 2001. Optimizing pearl millet and sorghum for forage and grain production in eastern Ontario. Pages 153-185 in Sorghum, pearl millet and chickpea, annual report 2001. Agriculture Environmental Renewal Canada Inc., Nepean,ON
11. Dowler, C. C. and Wright, D. L. 1995. Weed management systems for pearl millet in the southeastern United States. Pages 64-71 in I. D. Teare, ed. Proc. of the first

national grain pearl millet symposium, 17-18 Jan. 1995, University of Georgia, Tifton, GA.

12. Limon-Ortega, A., Mason, S. C. and Martin, A. R. 1998. Production practices improve grain sorghum and pearl millet competitiveness with weeds. *Agronomy Journal*, 90: 227-232.
13. Hattendorf, M.J., M.S. Dedelfs, B. Amos, L.R. Stone, and R.E. Given Jr. 1988. Comparative water use characteristics of six row crops. *Agronomy Journal*, 80: 80–85.
14. Aggarwal, R.K., Praveen Kumar and Power, J.F. 1997. Use of crop residue and manure to conserve water and enhance nutrient availability and pearl millet yields in an arid tropical region. *Soil and Tillage Research*, 41: 43-51.
15. Kanzaria, K.K., Sutaria, G.S., Akbari, K.N., Vora, V.D. and Padmani, D.R. 2010.Effect of integrated nutrient management on productivity of pearl millet and soil fertility of sandy loam soils under rainfed conditions. *An Asian Journal of Soil Science*, 5 (1): 154-456.
16. Khairwal IS, Rai KN, Diwakar B, Sharma YK, Rajpurohit BS, Bindu Nirwan and Ranjana Bhattacharjee. 2007. Pearl millet: Crop management and seed production Manual. Patancheru 502 324, Andhra Pradesh, India: International Crops Research Institute for the Semi-Arid Tropics. 104pp.
17. Patil, C.B. and Patil, B.R. 1989. Studies on cereal legume association on intercrop in pearl millet with different legumes under dry land conditions. *Indian Journal of Dryland Agriculture Research and Development*, 4(1): 37-42.
18. Patel, J.S., Khisteria, M.K., Paida, V.J., and Patel, J.C. 1985. Intercropping of pearl millet and pigeonpea in rainfed conditions. *Indian Journal of Agronomy*, 30 (3): 283-286.
19. Gadhia, D.S., Khanpara, V.D. and Patel, T.C. 1993. Production potential and economic returns of pearl millet-based inter cropping system with different grain legumes and oilseed crops under rainfed conditions. *Indian Journal of Agronomy*, 38 (2): 282-283.
20. Singh, K.B and Chauhan, R.S. 1973. Studies on intercropping with pearl millet. Progress Report, *Kharif* (1973). AICRP for Dryland Agriculture, RBS College, Agra. pp 5-7.
21. Hooda, R.S and Khippal, A. 2000. Comparative performance of different intercropping systems in pearl millet in relation to nutritional requirements. *Haryana Journal of Agronomy*, 16 (1 and 2): 70-72.

Part III

Disease and Pest Management

Millets: Ensuring Climate Resilience and Nutritional Security (2015) *Pages* **445–471**
Editors: **Vilas A. Tonapi and J.V. Patil**
Published by: **DAYA PUBLISHING HOUSE, NEW DELHI**

Chapter 16

Millet Diseases: Current Status and Future Research Needs

I.K. Das

ICAR-Indian Institute of Millets Research (IIMR)
Rajendranagar, Hyderabad – 500 030

In global perspective rice, wheat and maize are major cereals in terms of food security followed by barley, sorghum, pearl millet, oat and rye. The remaining cereals are minor millets which account for about one per cent of the food grain production in the world. These minor cereals are not important in terms of world food production, but essential as food crops in their respective agro-ecosystems. They are mostly grown in marginal lands or under agricultural conditions where other cereals fail to consistently produce an acceptable harvest. The principal members of the small millets group are finger millet or ragi (*Eleusine coracana*), Italian or foxtail millet (*Setaria italica*), common or proso millet (*Panicum miliaceum*), little millet (*Panicum miliare*) (renamed *P. sumatrense*) kodo millet (*Paspalum scrobiculatum*) and barnyard millet (*Echinochloa frurnentacea*) renamed *E. colona*). All these are cultivated in one or more states in India and occupy about 4.5 per cent of the total cultivated area. Globally, India is the leading producer of small millets with about 20 per cent of the area under these crops. Annual planting area under small millets is around 2.5 m ha. Due to poor management practices, crops succumb to various biotic stresses *viz.*, diseases like blast, leaf spot, brown spot, downy mildew, smut, rust, ergot and viral diseases. Sometimes new diseases occur or old disease reappears and all these have become major hindrance for the production of millets. A list of fungal diseases reported on millets has been given in Table 1.

Pearl Millet Diseases

Pearl millet is a crop that is grown under dry climate, yet it sufferers from infection of many diseases. Compared to fungal diseases incidence of bacterial and viral diseases are negligible in pearl millet. A list of fungal diseases of pearl millet is given in Table 1. However, all of these diseases are not economically significant in India. Downy mildew, leaf blast, ergot, smut and rust occur almost regularly in low to severe form in pearl millet growing areas across India.

Table 1: Fungal Diseases of Millets

Crop	*Disease*	*Pathogen*
Pearl millet	Downy mildews	*Sclerospora graminicola* (Sacc.) Schroet.
	Blast	*Pyricularia grisea* (Cke.) Sacc
	Ergot	*Claviceps fusiformis* Loveless; *Claviceps africana* Frederickson
	Smut	*Moesziomyces penicillariae* (Bref.) Vanky
	Rust	*Puccinia substriata* Ell. and Barth. indica Ramachar and Cumm
	Top rot	*Fusarium moniliforme* Sheldon
	Head mold	Various fungi
	Bipolaris leaf spot	*Bipolaris setariae* (Saw.) Shoem
	Cercospora leaf spot	*Cercospora penniseti* (Chupp)
	Curvularia leaf spot	*Curvularia penniseti* (Mitra) Boedijn
	Dactuliophora leaf spot	*Dactuliophora elongata* Leakey
	Drechslera leaf spot	*Drechslera dematioidea* (Bubak and Wroblewski) Subram. and Jain
	Exserohilum leaf blight	*Exserohilum rostratum* (Drechs.) K.J. Leonard and *E.G.* Suggs
	Myrothecium leaf spot	*Myrothecium roridum* Tode ex. Fr
	Phyllachora leaf spot	*Phyllachora penniseti* Syd.
	Phyllosticta leaf blight	*Phyllosticta penicillariae* Speg.
	Rhizoctonia blight	*Rhizoctonia solani* Kuhn; *Rhizoctonia zeae* Voorhees
	Southern blight	*Sclerotium rolfsii* Sacc.
	Zonate leaf spot	*Gloeocercospora sorghi* Bain and Edgerton
Finger millet	Blast	*Pyricularia grisea*
	Brown spot	*Helminthosporium nodulosum*
	Banded blight	*Rhizoctonia solani*
	Smut	*Melanopsichium eleusinis*
	Downy mildew	*Sclerophthora macrospora*
	Wilt or Foot rot	*Sclerotium rolfsii*
	Rust	*Puccinia substriata*

Contd...

Table 1–*Contd...*

Crop	Disease	Pathogen
Foxtail millet	Blast	*Pyricularia setariae*
	Rust	*Uromyces setariae italic*
	Smut	*Ustilago crameri*
	Downy mildew	*Sclerospora graminicola*
	Udbatta	*Ephelis* sp.
Kodo millet	Head smut	*Sorosporium paspali*
	Rust	*Puccinia substriata*
	Udbatta	*Ephelis* sp.
Little millet	Rust	*Uromyces linearis*
Barnyard millet	Head smut	Ustilago crus-galli
	Kernel smut	Ustilago paradoxa
Proso millet	Head smut	*Phacelotheca destruens*

Downy Mildew

Of all the above diseases downy mildew is considered number one important disease of pearl millet due to its wide spread occurrence sometimes in epidemic form, and widespread damaging effect on vegetative as well as reproductive parts restricting yield. In farmers' field the range was 0 to 90 per cent depending on location and specific cultivar. Gujarat, Rajasthan, Haryana in the North-West and Tamil Nadu and Karnataka in the South are the major downy mildew prone areas for pearl millet. Maharashtra, Madhya Pradesh and Andhra Pradesh generally record less incidence of downy mildew.

Leaf Spots

Leaf spots due to several pathogens like; *Pyricularia grisea, Bipolaris setariae, Cercospora penniseti, Curvularia penniseti, Drechslera dematioidea* and *Exserohilum rostratum* etc. are also taking toll on crop destroying foliage and thus reducing yield and yield attributes. Among foliar diseases blast or Pyricularia leaf spot is becoming more severe in the states of Rajasthan, Gujarat, Maharashtra and Madhya Pradesh. Once considered a minor disease of pearl millet, incidence of blast disease has increased at an alarming rate in the recent past, particularly on commercial hybrids in several states of India. Mean disease severity ranges from 10 to 30 per cent and sometimes may go upto 50 per cent on hybrids and varieties depending on season.

Rust

Rust is predominant in almost all areas where pearl millet is grown. Tamil Nadu, Karnataka, Rajasthan, Gujarat and Maharashtra experience the disease regularly. Mean rust severity in pearl millet across India varies from 10 to 45 per cent depending on season. Highly susceptible cultivars develop large pustules on leaf blade and leaf sheath.

Smut

Smut is another disease of pearl millet that occurs regularly across India. The disease commonly occurs in Rajasthan, Madhya Pradesh and Maharashtra and incidence varies from 2 to 20 per cent depending on season and cultivars. It causes direct loss of grain by replacing them with smut sori, and yield losses of up to 30 per cent have been reported. In recent years the disease has become more important on commercial F1 hybrids in India and on exotic early-maturing cultivars in countries of Africa.

Ergot

Ergot is observed in Karnataka, Maharashtra and Rajasthan. Disease incidence is relatively less (2 to 10 per cent) in zone A (North-west India) than zone B (around 25 per cent) (South India).

Small Millet Diseases

Blast

Blast disease is conspicuous on two small millets namely finger millet and foxtail millet. Blast caused by heterothallic ascomycete *Magnaporthe grisea* (Anamorph: *Pyricularia grisea*) is the most important constraint to finger millet production in most finger millet growing environments. The disease causes considerable grain loss ranging from 30 to 80 per cent in many regions depending on weather. The disease occurs almost every year during rainy season but the extent of crop loss depends on the severity and the time of onset of disease. The is common in all finger millet growing areas in Tamil Nadu, Karnataka, Andhra Pradesh, Maharashtra, Odisha, Bihar, Uttaranchal, Madhya Pradesh, Gujarat and Himachal Pradesh. It infects all stages of crop growth from seedling to flowering. Pathogen Infection of leaf, neck (peduncle) and fingers (panicle branch) are common causing huge loss to the crop. Crop loss is greater when the disease appears on the neck and ears during flowering and grain development phase. Weeds or some cereal plants act as collateral hosts from where primary inoculums spread to the crop. Seeds also have been found to carry the pathogen in the pericarp and endosperm. Systematic breeding programme are in practice for development of blast resistant varieties of finger millet. Gautami, GPU 28, BM 9-1, GPU 26, GPU 45, Chilika, VL 315 and GPU 48 are some of the blast tolerant/resistant varieties which are cultivated in India.

Blast

Blast disease is conspicuous on two small millets namely finger millet and foxtail millet. Blast caused by heterothallic ascomycete *Magnaporthe grisea* (Anamorph: *Pyricularia grisea*) is the most important constraint to finger millet production in most finger millet growing environments. The disease causes considerable grain loss ranging from 30 to 80 per cent in many regions depending on weather. The disease occurs almost every year during rainy season but the extent of crop loss depends on the severity and the time of onset of disease. The is common in all finger millet growing areas in Tamil Nadu, Karnataka, Andhra Pradesh, Maharashtra, Odisha, Bihar, Uttarakhand, Madhya Pradesh, Gujarat and Himachal Pradesh. It infects all stages of crop growth from seedling to flowering. Pathogen Infection of leaf, neck (peduncle) and fingers (panicle branch) are common causing huge loss to the crop. Crop loss is greater when the disease appears on the neck and ears during flowering and grain development phase. Weeds or some cereal plants act as collateral hosts from where primary inoculums spread to the crop. Seeds also have been found to carry the pathogen in the pericarp and endosperm. Systematic breeding programme are in practice for development of blast resistant varieties of finger millet. Gautami, GPU 28, BM 9-1, GPU 26, GPU 45, Chilika, VL 315 and GPU 48 are some of the blast tolerant/resistant varieties which are cultivated in India.

Blast of foxtail millet is caused by *Pyricularia setariae.* On foxtail millet the disease is not as severe as in finger millet. Only leaf blast phase is seen on foxtail millet and most damaging neck blast is not observed on it. The isolate from foxtail millet is reported to infect finger millet, pearl millet, wheat and *Dactyloctaenium aegyptium*.

Rust

Two different pathogens cause rust on small millets. Finger millet and kodo millet are infected by *Puccinia substriata.* Small yellow or white raised spots are developed on upper and lower leaf surfaces. Spots tend to be more numerous on lower leaf surface. Red-brown pustules may be surrounded by a yellow halo. This pathogen also infects pearl millet and cause considerable damage. Foxtail millet and little millet, on the other, hand is attacked by *Uromyces* spp. The *Uredia* are brown and erumpent. Brown colors urediospores germinate readily to cause infection. The teliospores are thick walled, smooth, globose, brown with persistant, long thick pedicels. Rust on millet is not a regular disease and having little economic importance. But it is known to be prevalent in all the states in India, wherever this crop is grown. In recent years there are reports of the rust becoming serious and destroying the crop before the ears have formed. Little information is available on systemic research on this disease on small millets.

Smut

All the millets are affected by various types of smuts. Grain smut is common on finger millet, foxtail millet and barnyard millet. Finger millet smut (*Melanopsichium eleusinis*) is gaining importance after the introduction of high yielding varieties especially in Karnataka. Air borne spores cause infection of the flower. Smut sori are developed in the grains, main rachis or in peduncle. The pathogen may also convert floral organs into saclike shaped smut sori. Smut disease in foxtail millet is caused by *Ustilago crameri* mainly in southern states of India. The fungus is externally seed borne and affects most of the grains in an ear. Certain amount of soil borne infection has also been observed. Head smuts are common in kodo millet, barnyard millet and proso millet. In kodo millet head smut (*Sorosporium paspali*) assumes significance as an important disease with considerable yield loss in Madhya Pradesh, Andhra Pradesh, Bihar, Tamil Nadu and Karnataka. The entire panicle is transformed into a long sorus. In some cases, it is enclosed in the flag leaf and may not emerge fully. The membrane bursts and exposes the black mass of spores. The disease is mainly seed-borne. The spores adhere to the surface of the grains and infect the seedlings.

Systematic breeding for development of smut resistance variety has not been common. Cultural practices are mainly followed for management of the disease and fungicidal seed treatment also used occasionally. Two types of smuts are reported on barnyard millet, head smut (*Ustilago crus-galli*) and kernel smut (*Ustilago paradoxa*). Both the diseases are seed-borne and can be managed through seed treatments.

Leaf Spots

Among all the small millet leaf spots the brown spot of finger millet caused by *Helminthosporium nodulosum* is the most important one. The pathogen also causes seedling blight when infection occurs in the emerging seedling. The disease is prevalent in Madhya Pradesh, Andhra Pradesh, Maharashtra, Uttar Pradesh, Bihar, Tamil Nadu and Karnataka. The disease affects all the parts of the plants like base of the plant, culms, leaf sheath, leaf blade, neck and fingers and cause considerable yield loss. Preliminary data on epidemiology is available. The primary infection is caused by the pathogen on the seed and secondary spread is through air borne conidia. Several other cereals are infected by this organism including *Cynodon dactylon, Echinochloa colonum, Echinochloa frumentacea, Eleusine indica, Panicum miliaceum, P. milliare, P. ramosum, P. repens, Setaria glauca, S. italic, Oryza sativa, Sorghum vulgare and Zea mays*. Seed treatment with any organomercurial compounds controls the pre-emergence damping off seedling blight. The secondary infection in the field can be reduced by protective sprayings with suitable fungicides. Studies have been conducted to identify genetic resistance in the germplasm collection.

Downy Mildew

Apart from pearl millet, downy mildew also observed on small millets. Pathogen *Sclerospora graminicola* infects both pearl millet and foxtail millet whereas, *Sclerophthora nacrospora* infects finger millet. Downy mildew incidence on small mileets is not as severe as in pearl millet. Affected finger millet plants are generally stunted with shortened internodes and profuse tillering. The plant assumes a bunchy and bushy appearance. The green ear manifests itself at the time of grain formation and completely converts the ear heads into green narrow leafy structures causing complete sterility. The disease is mainly observed in peninsular India particularly Tamil Nadu and Karnataka. The fungus is internally as well as externally seed borne. It has a wide host-range including *Eleusine indica,* maize, wheat, oat, *Eragrostis pectinacea* and *Digitaria marginata*. Physiologic specialization and races of the pathogen are reported.

Udbatta

Udbatta disease is reported on kodo millet and foxtail millet. The disease is caused by *Ephelis* sp. Udbatta disease earlier noticed as pandemic and of minor importance in certain paddy growing areas. The disease is noticed on several host plants such as grasses *Cynodon dactylon, Pennisetum* sp. and *Ergostis tenufolia*. The disease was found to occur in sporadically on Kodo millet in the experimental plots of All India Coordinated Research Project on Small Millets in Karnataka. Symptom first appears during panicle initiation stage, then the grains are infected and the entire panicle converts into hard silver color cylindrical spike resembling that of udbatta disease in paddy. No grains are formed on the affected panicle.

Millet Diseases of High Economic Significance

Like other cereals production of millets also are affected by many diseases caused by fungi, bacteria and viruses. Fungal diseases are of major significance in millet whereas bacterial and viral diseases are highly sporadic in nature and negligible in terms of economic importance. Among fungal diseases though many are reported to occur on millets, all are not economically major constraints. A few diseases are there which occur regularly in wide areas in moderate to severe form, and are considered important constraints for production of millet. Millets, their major growing areas and their important diseases in India are given in Table 2. Downy mildew of pearl millet and blast of finger millet are considered two most important diseases of millets which causes considerable yield and quality loss every year in these crops. Blast is another disease of pearl millet that is gaining significance in recent years. Other important diseases are smut and rust in pearl millet, *Helminthosporium* leaf spot in finger millet, blast in foxtail millet and head smuts in kodo millet.

Details about the disease, status of research works carried out till now, research gap and future research priorities are given in the next section for two most important disease of millet *viz.*, downy mildew of pearl millet and blast of finger millet.

Current Status and Future Research Needs for Downy Mildew of Pearl Millet in India

Pearl millet is the most widely cultivated cereal in India after rice and wheat. India produces more than half the world's pearl millet. Presently it is cultivated in an area of 7.2 million ha. Rajasthan is the major pearl millet growing states covering 50 per cent area. Around 40 per cent areas are in the states of Maharashtra, Gujarat, Uttar Pradesh and Haryana. Remaining 10 per cent area is spread over Madhya Pradesh, Karnataka, Andhra Pradesh and Tamil Nadu (Roy, 2013). Pearl millet used to be the staple food in arid and semi-arid regions in India particularly Rajasthan, but has been replaced by wheat and rice in last three decades. The extension of irrigation facilities and increased production of wheat and rice has been major reason for this shift. Preference of farmers also shifted towards more economic pulses and oilseed crops. India witnessed a major breakthrough in total production and productivity of pearl millet after the release of high yielding hybrids. However, this increase has been short lived as majority of the hybrids succumbed to downy mildew disease during mid seventies. In the present day hybrids the downy mildew disease is under control as the hybrids are based on diverse genetic background. Pearl millet can provide economical grain yield (600–700 kg/ha) under marginal and low management conditions. Hybrids with a grain yield potential of 4-5 t/ha maturing in 80-85 days in summer under irrigated and high management conditions is possible. Yields as high as 5.5 t/ha have been realized in summer cultivation in Gujarat. Wide range of hybrids and varieties are available to fit into diverse cultivation and consumption requirements. Lately emphasis has been towards developing genotypes with high nutritional value especially Fe and Zn. Dual purpose varieties have been developed to meet the demand of fodder.

Table 2: Millets, Growing Areas and their Important Diseases in India

Crop	*Area (M ha)*	*Cropping Zone*	*Diseases*		
			Most Important	*Important*	*Less Important*
Pearl millet (*Pennisetum glaucum*)	7.20	Rajasthan, Maharashtra, Gujarat, UP, Haryana, MP, Karnataka, AP and TN	Downy mildew	Blast, Rust	Smut, Ergot
Finger millet (*Eleusine coracana*)	1.53	TN, Karnataka, AP, Maharashtra, Odisha, Bihar, Uttaranchal, MP, Gujarat, HP	Blast	*Helminthosporium* leaf spot	Smut, Downy mildew, Banded blight, Wilt
Foxtail millet (*Setaria italica*)	0.10	AP, Karnataka, Maharashtra, TN, Rajasthan, MP, UP, NE states	–	Smut	Blast, Rust, Downy mildew, Udbatta
Kodo millet (*Paspalum scrobiculatum*)	0.24	MP, Chhattisgarh, Maharashtra, UP, Gujarat and TN	–	Head smut,	Rust, Udbatta
Little millet (*Panicum sumatrense*)	0.29	TN, Karnataka, AP, Maharashtra, Jharkhand, MP, Odisha and Gujarat	–	–	Rust
Barnyard millet	0.20	Bihar, TN, Maharashtra, MP, Bihar	–	–	Head smut
Proso millet (*Panicum miliaceum*)	0.04	Grown in patches from the Himalayas to TN in south	–	–	Smut

Downy Mildew of Pearl Millet

Downy mildew caused by Sclerospora graminicola is considered number one important disease of pearl millet due to its wide spread occurrence sometimes in epidemic form, and extensive damaging effect on vegetative as well as reproductive parts restricting yield. Systemic symptoms generally appear on the second or subsequent leaves and panicles, except in cases of recovery resistance where plants outgrow the disease (Singh and King 1988). The disease is present in all pearl millet growing regions in India and Africa. However, it has not yet been reported on pearl millet in USA. Gujarat, Rajasthan, Haryana in the North-West and Tamil Nadu and Karnataka in the South are the major downy mildew prone areas for pearl millet (AICPMIP, 2013-14). Maharashtra, Madhya Pradesh and Andhra Pradesh generally record less incidence of downy mildew. In farmers' field the range was 0 to 90 per cent depending on location and specific cultivar. Systematic research works have been started on the subject during 60's or 70's. Major focuses of research have been on pathogen biology and variability, disease epidemiology, host plant resistance, screening for resistance, inheritance of resistance and management of the disease. Details of researches on various aspects of pearl millet downy mildew up to 1992 had been reviewed by Singh *et al.* (1993). Considerable progress has been made in understanding host-pathogen interaction, refining disease screening methods, identifying and utilizing resistant sources, and breeding downy mildew resistant parental lines and hybrids (Thakur *et al.*, 2008).

Economic Importance

First pearl millets hybrid HB 1 (developed by PAU, Ludhiana) was introduced in India in 1965 and average grain yields become nearly doubled. But this advantage was short-lived because of heavy outbreak of downy mildew. Newly developed hybrids during 70's and 80's were worst affected due to severe outbreak of downy mildew. Downy mildew epidemics caused substantial yield losses in F1 hybrids during 1970 to 1976, during 1983 and 1984, and again in 1987 and 1988 (Singh *et al.*, 1993). Many popular hybrids were then withdrawn from cultivation. The epidemic was mainly due to narrow genetic base of parental lines of the hybrids. Presently the downy mildew disease is under check as the hybrids are based on diverse genetic background. About yield loss due to this disease various reports are available. Some reports suggested 45 per cent loss in Allahabad (Mitter and Tandon, 1930), and up to 27 per cent loss from 1962 to 1964 in Rajasthan (Mathur and Dalela, 1971). During last three seasons mean downy mildew incidence in research farms across India was around 5 per cent, and range was 0 to 90 per cent (AICPMIM, 2011-2014). In farmers' field the range was 0 to 90 per cent depending on location and specific cultivar. Gujarat, Rajasthan, Haryana in the North-West and Tamil Nadu and Karnataka in the South are the major downy mildew prone areas for pearl millet. Maharashtra, Madhya Pradesh and Andhra Pradesh generally record less incidence of downy mildew. The disease affects grain yield and there was highly significant ($P = 0.05$) negative linear correlation between downy mildew incidence ($r = 0.6272 - 0.9626$) and severity ($r = 0.7154 - 0.9148$) on yield (Gwary *et al.*, 2009). Apart from yield the disease can adversely affect the fodder quality. There may be significant reduction in total soluble sugars in diseased plants (26.2 per cent) compared to healthy plants (31.8 per cent) of pearl millet hybrids (Upasana *et al.*, 2010).

The Pathogen

Lot of information is available on pathogen biology and variability of downy mildew of pearl millet. *Sclerospora graminicola* is a biotroph that produce determinate type dichotomously branched sporangiophores. Sporangiophores emerge from systemically infected leaves through stomata and produce abundant sporangia usually during night. Oospores are produced in large numbers inside the leaf. They can survive from 8 months to 10 years under laboratory conditions (Nene and Singh 1976). Treatment of oospores with sodium hypochlorite has been reported to give about 76 per cent germination of the spore (Panchbhai *et al.*, 1991). Two mating types have been identified in *S. graminicola* and they are considered to be necessary for the production of oospores, suggesting the heterothallic nature of the pathogen (Michelmore *et al.*, 1982). Pushpavathi *et al.* (2006) observed low frequency of homothallic isolates of *S. graminicola* in Indian. Of the 70 single-zoospore isolates tested for fertility in terms of oospore production potential, 62 were self-sterile and 8 were self-fertile. Pathogenic variability in *S. gramillicola* was first reported in Karnataka by Bhat (1973). Later it has been shown that some genotypes differ in their downy mildew reactions between locations and pathogen populations (Werder and Ball, 1992; King *et al.*, 1989). Singh and Singh (1987) provided evidence that *S. graminicola* is highly cultivar-specific. They concluded that a pathogen population specific to a particular cultivar can disappear within 3 years if the cultivar is withdrawn from cultivation. Thakur and Rao (1997) observed highly significant ($P<0.001$) pathotype x host genotype interaction effects for virulence, aggressiveness, and virulence index. This further indicated the existence of host-pathogen specificity in the *S. graminicola* system. Sivaramakrishnan *et al.* (2003) studied variability of 15 isolates of *S. graminicola* collected from various pearl millet cultivars in India and classified them into five major virulence groups. Molecular characterization of the pathogen has been carried out using SSR, ISSR, RFLP, RAPD for studying variability (Sastry *et al.*, 1995; Sudisha *et al.*, 2009; Sharma *et al.*, 2010). A SCAR primer pair (UOM3-Sg-Path1-F/R) has been developed for rapid identification and specific detection of *S. graminicola* pathotype-1 (Sudisha *et al.*, 2009).

Disease Development

Considerable amount of information is available on epidemiology of pearl millet downy mildew. *S. graminicola* survives as oospores in the soil. Oospores germinate and infect seedling at the underground part and colonizes the growing point. Seedlings may show yellowing symptom within 6 days after sowing. Under humid conditions, systemically infected leaves produce abundant sporangia which cause spread of the disease within and among fields. Optimum temperature and humidity for sporangia production is 20-25°C and RH 95-100 per cent. Sporangia liberate zoospores that remain infective for about 4 h at 30°C, and for a longer period at lower temperatures (Singh and Gopinath, 1990). Sporangia are short-lived, and do not survive in the field during daytime but can survive for about 24h at low temperature (5-15°C). Sporangia remain infective to seedling for short period of time that lies between germination of seed to the 1-2 leaf stage. Thereafter, the susceptibility of seedling decreases sharply (Singh and Gopinath 1985). Zoospore germ tubes penetrate the leaves directly through the epidermal cells, with or without appressoria and very rarely through stomata. The pathogen colonized the leaf tissue inter- and intra-cellularly (Celia Chalam, 1996). Both nitrogen and phosphorus are reported to significantly increase incidence and severity of the disease (Zarafi *et al.*, 2005). Oospores are produced in infected leaves when compatible mating types of *S. graminicola* are present in the same tissue, or when homothallism is operative. Available information is not clear about role of alternate/collateral hosts are survival of pathogen. Although 14 graminaceous species belonging to eight genera have been reported as hosts of *S. graminicola* (Bhat 1973) there role in disease epidemiology is doubtful. Singh and Williams (1979) tested 23 hosts belonging to 11 genera and found none of them, except *P. glaucum*, to develop downy mildew symptoms. Although it is believed that the disease is seed-borne in the form of seed-carried oospores, internal seed transmission of the disease has not been accepted by all. It has been a subject of controversy as there are reports for and against seed transmission (Sundaram *et al.*, 1973, Safeeulla 1976, Thakur and Kanwar 1978, Shetty *et al.*, 1980).

Screening for Resistance

Development of standard screening technique is required for the identification of resistance. A field-screening technique which mainly utilizes sporangia as infection propagules was developed (Williams *et al.*, 1981). This technique deploys infector rows, test rows and indicator rows for development of disease and identification of resistance. Singh and Gopinath (1985) developed a technique which involves inoculation of potted seedlings in the coleoptile stage using a micro syringe. Greenhouse and laboratory screening techniques have been developed, for mass screening of breeding materials (Thakur, 2003).

Sources of Resistance

A number of sources of downy mildew resistance have been identified in India and Africa. More than 50 sources of stable resistance, primarily originating from Nigeria, had been identified, and seven of these possessed an extremely high degree of stable field resistance (Singh *et al.*, 1990; Zarafi, 2007). Several other lines and germplasm accessions, including P 310-17, P1449-3, IP 18292, IP 18293, IP 18294, IP 18295 and IP 18298 with high levels of resistance have been identified (Singh *et al.*, 1997, Thakur *et al.*, 2004). Singh *et al.* (1997) reported that there is enough geographical, morphological and genetic diversity in germplasm and breeding lines for DM resistance. Yadav *et al.* (2005) evaluated 13 diverse, exotic populations of pearl millet for their phenotypic characteristics including downy mildew resistance and use them for diversification of local landraces. Molecular breeding tools have been used for identification of resistance against pearl millet downy mildew. A number of quantitative trait loci for downy mildew resistance have been identified on different linkage groups, and some of them are specific to different pathotypes (Hash *et al.*, 1999; Hash and Witcombe 2001; Jones *et al.*, 2002). Several identified QTLs have been transferred to backgrounds of elite inbred parental lines of a popular single-cross hybrid HHB 67.

Genetics of Resistance

The information on the mode of inheritance of downy mildew resistance is unclear. There are several reports that say that inheritance of resistance to downy mildew is controlled by one or two dominant genes with some modifiers (Singh 1974; Appadurai *et al.*, 1975; Gill *et al.*, 1978). There is one report indicating that resistance is a recessive trait (Singh *et al.*, 1978). Deswal *et al.* (1998) observed both additive and non-additive gene effects were important. Epistatic interactions were present in low magnitude. Avirulence was dominant over virulence and a single gene pair (AA/Aa) controlled avirulence in isolates Sg 110-9, to a corresponding resistance gene Rsg1 in IP 18292 (Pushpavathi *et al.*, 2005). Cytoplasm effects were reported to be not responsible for downy mildew susceptibility (Yadav, 1996; Kumar and Sagar, 2009). This may be taken as an opportunity for exploitation of different cytoplasmic bases of the male-sterile lines for development of hybrids.

Mechanism of Resistance and Induced Resistance

Relatively little information is available on mechanism of downy mildew resistance in pearl millet. Available reports are based on studies on histopathology (Celia Chalam, 1996) or cell wall structure (Shailasree *et al.*, 2004) of resistant and susceptible cultivars. A series of reports from a laboratory showed that HRGPs are accumulated in cell wall of resistant cultivar of pearl millet in response to attack by downy mildew pathogen *S. graminicola*. Hydroxyproline-rich glycoproteins (HRGPs) are important plant cell wall structural components that are known to be induced in response to pathogen attack. These glycoproteins are composed of subunits of 27, 17, and 14 kDa (Shailasree *et al.*, 2004; Sujeeth *et al.*, 2010). It was suggested that cell wall strengthening in *S. graminicola*-infected resistant pearl millet was brought about by a combination of polypeptide cross-linking of isodityrosine as well as by the high content of hydroxyproline in HRGPs, and H_2O_2 (Deepak *et al.*, 2007). There was higher silicon accumulation in the resistant cultivar compared with the susceptible cultivar (Deepak *et al.*, 2008) suggesting that silicon might play role in induction of resistance. Later on many living or inert substances have been used for induction of resistance in pearl millet against downy mildew. Recently there are many reports are coming that have experimented for induction of resistance agaisnt this disease. Among chemical substances L-methionine, vitamins, synthetic jasmonate analogon, nitric oxide donor, mannitol, thiamin etc have been reported to induce resistance against downy mildew in pearl millet(Sarosh *et al.*, 2005; Deepak *et al.*, 2007; Manjunath *et al.*, 2008; Pushpalatha *et al.*, 2011). Substances or elicitors derived from living cells like Cellulysin crude cellulose from *Trichoderma*, chitosan, Trichoshield, elicitors derived from yeast, *Datura* extract, have also been used for induction of resistance (Niranjan-Raj *et al.*, 2005; manjunath *et al.*, 2008; Devaiah *et al.*, 2009; Hindumathy, 2012; Pushpalatha *et al.*, 2013). Most of these studies were to prove the role of those materials in induction of resistance. Only a few of these materials have been tested under field conditions.

Management of Downy Mildew

Use of Resistant Cultivars

For management of pearl millet downy mildew use of host plant resistant has been thought to be the best and economical method. Resistant sources have been used in the hybrid program for development of seed-parents. Present day hybrids having wider genetic base and hold a good degree of tolerance to downy mildew. Varieties show better tolerance to downy mildew than hybrids. In major pearl millet area like Rajasthan, varieties are preferred over hybrid. There has been considerable success in breeding for downy mildew resistance using conventional pedigree breeding, and a large number of disease resistant hybrids have been developed and deployed (Khairwal *et al.*, 2004). Many varieties including four open-pollinated varieties developed at ICRISAT (WC-C75, ICMS 7703, ICTP 8203, and ICMV 155 have been released for cultivation in India (Thakur *et al.*, 2008). A high yielding downy mildew resistant pearl millet hybrid GHB-732 has been released recently for Gujarat (Dangaria *et al.*, 2013). Efforts have been made to utilize recovery resistance in which systemically infected plants outgrow the disease and produce healthy panicles. Efforts have been made to develop transgenic against downy mildew. Transgenic pearl millet lines expressing *pin* gene-exhibiting high resistance to downy mildew pathogen, *Sclerospora graminicola* (Latha *et al.*, 2006).

Use of Chemicals and other Practices

Considerable research has been done to control downy mildew by protective and systemic fungicides. However, protective fungicides (Zineb, Maneb) are not as effective as systemic one. The systemic fungicide metalaxyl has been used successfully to control downy mildew in pearl millet. Seed treatment with metalaxyl at 2 g a.i./kg seed controlled the disease for about the first 35 days after sowing (Venugopal and Safeeulla 1978, Dang *et al.*, 1983). Foliar application of the fungicide at 125 mg a.i./L arrests further development of the disease in systemically infected plants, and if applied before floral initiation, disease-free heads are produced (Singh *et al.*, 1984). Shankara Rao *et al.* (1987) reported that seed treatment with metalaxyl, coupled with a single foliar application of metalaxyl + mancozeb was much superior to the seed treatment alone. Mani and Hepziba (2009) observed that seed treatment with Apron 2 g/kg + Ridomil 0.6 per cent foliar spray 20 DAS was by far the most effective in reducing the incidence of disease to a minimum of 4.7 per cent from a maximum of 66.1 per cent in the control. Anaso and Anaso (2010) reported that Apron Star 42 WS, a new formulation of metalaxyl, for seed dressing is cost-effective and sustainable in the downy mildew endemic areas of northern Nigeria. Biological control has not been tried much for control of pearl millet downy mildew. Few reports of experimentation with plant extracts (Deepak *et al.*, 2005), boicontrol agents (Mani and Hepziba, 2009), endophytic bacteria (Chandrashekhar *et al.*, 2007) and PGGR (Sudisha *et al.*, 2010) are there but they have not reached field level applications.

Research Gaps and Future Priority

Many aspects of pearl millet downy mildew have been researched intensively by national and international research bodies. Considerable progress has been made in understanding host-pathogen interaction, refining disease screening methods, identifying and utilizing resistant sources, and breeding downy mildew resistant parental lines and hybrids. Presently the disease is somewhat under control but still a major challenge towards realizing the high yield potential of hybrids. Continuous evolution of science has provided a new set of tools that can be used more efficiently and precisely for improvement of understanding of various critical aspects of downy mildew.

1. *S. graminicola* evolve quickly, because of sexual reproduction and create great variability in pathogen population and their virulence on new set of cultivars. This calls for continuous monitoring of virulence, characterization of the pathogen populations, and continuous identification and utilization of new sources of resistance.
2. Virulence mapping for pathotypes in different zones/states and strategic deployment of cultivar will be a continuous process for efficient management of pearl millet downy mildew. Regular screening of breeding lines against representative pathotypes from each zone and designing of hybrid parental lines for resistance to specific pathotypes should be part of strategic plan for management of downy mildew.
3. Study on genetics of resistance and mechanism of resistance need further emphasis that will lead to development of more efficient control method.
4. Recent advances in molecular techniques can be used for analyzing genomic variation in pathogen population and host-pathogen interactions. A saturated genetic linkage map based upon molecular markers has been developed for pearl millet that might assist breeders to identify markers in selecting individual plants carrying specific resistance alleles of interest. Molecular tools can also be used to better understand genetics of host-pathogen interaction.
5. Detailed and convincing information on oospores survival, germination, infection, colonization and their role in disease epidemiology is still required.
6. Similarly genuineness of seed transmission needs to be established keeping in view its quarantine significance.

Current Status and Future Research Needs for Blast of Finger Millet in India

Minor millets accounts for about one per cent of the food grain production in the world. In global perspective rice, wheat and maize are major cereals in terms of food security followed by barley, sorghum, pearl millet, oat and rye. Though minor millets are not important in terms of world food production, they are essential as food crops in their respective agro-ecosystems. They are mostly grown in marginal lands or under agricultural conditions where other cereals fail to consistently produce an acceptable harvest. They occupy about 4.5 per cent of the total cultivated area spread over many states. Globally, India is the leading producer of small millets with about 20 per cent of the area under these crops. Among small millets, finger millet (*Eleusine coracana*), is widely cultivated and covers around 60 per cent of the areas under small millets in India (acreage of small millets 2.5 m ha and finger millet 1.53 m ha). Globally finger millet is cultivated in Asia (India, Srilanka, Malaysia, China, Myanmar, Nepal and Japan) and Africa (Kenya, Uganda, Tanzania, Ethiopia, Eritrea, Rwanda, Democratic Republic of Congo, Zaire, Eritrea, and Somalia) main as staple food of millions of poor people. In India, the important finger millet growing states are Karnataka, Odisha, Maharashtra, Tamil Nadu, Andhra Pradesh, Uttarakhand, Uttar Pradesh and Bihar. Major growing state is Karnataka which covers around 50 per cent of the total finger millet area (Nagaraja *et al.*, 2008). Finger millet grains are nutritionally rich in minerals like calcium, iron etc and a good source of dietary fiber. For its rich nutritive value such as presence of 8-10 times more calcium than that of rice and wheat, high quality protein, plenty of minerals, dietary fiber, phytochemicals and vitamins it has become popular in human diet. Finger millet flour is used in various food preparations such as cakes, bread and other bakery products. The sprouted seeds are also nutritious and easily digested. The grain may also be malted and a flour of the malted grain is used as a nourishing food for infants (Mgonja *et al.*, 2007). The crop is attacked by many diseases caused by fungal, bacterial and viral pathogens. They are blast, seedling blight/brown spot, banded blight, smut, downy mildew, wilt or foot rot, rust, bacterial leaf spot, bacterial wilt, mottle streak and streak virus (Viswanath and Seetharam, 1986).

Blast of Finger Millet

Blast of finger millet is caused by the fungi *Pyricularia grisea*. It is heterothallic, ascomycete fungi whose sexual stage (Anamorph) is called *Magnaporthe grisea*. Blast is the most important constraint to finger millet production in most crop growing environments. In India, the disease was first reported from Tanjore delta of Tamil Nadu by McRae (1920). It is now common in all finger millet growing areas in Tamil Nadu, Karnataka, Andhra Pradesh, Maharashtra, Odisha, Bihar, Uttarakhand, Madhya Pradesh, Gujarat and Himachal Pradesh. The disease occurs almost every year during rainy season but the extent of crop loss depends on the severity and the time of onset of disease. It infects all stages of crop growth from seedling to flowering. Infection of leaf, neck (peduncle) and fingers (panicle branch) are common that inflicts huge loss to the crop. Depending on weather the grain loss ranging from 30 to 80 per cent is reported in many regions. Crop loss is greater when the disease appears on the neck and ears during flowering and grain development phase. Weeds or some cereal plants act as collateral hosts from where primary inoculums spread to the crop. Seeds also have been found to carry the pathogen in the pericarp and endosperm. Seed treatment combined with varietal resistance is practiced for management of the disease.

Economic Importance

Blast of finger millet is a serious disease that occurs regularly in all finger millet growing areas in the country such as Tamil Nadu, Karnataka, Andhra Pradesh, Maharashtra, Odisha, Bihar, Uttarakhand, Madhya Pradesh, Gujarat and Himachal Pradesh. The pathogen can infect the crop at all stages, from seedling to post flowering phase. Neck infection is the most serious phase of the disease causes significant loss in grain number, grain weight and increase in spikelet sterility (Rath and Mishra, 1975). The average loss has been reported to be around 28-36 per cent and as high as 80-90 per cent in endemic areas (Bisht, 1987; Rao, 1990; Ramappa *et al.*, 2002; Nagaraja, *et al.*, 2007). The avoidable loss of grain yield and fodder yield in Gujarat was estimated to be 35.8 per cent and 43.7 per cent respectively (Prajapathi *et al.*, 2013). *P. grisae* also produces a phytotoxic metabolite, Pyrichalasin H, which have toxic effect on seed germination and seedling growth of blast resistant and susceptible finger millet varieties (Sanmathi Kumar *et al.*, 2006).

The Pathogen

Blast pathogen can infect around 50 grasses, including economically important crops like wheat, rice, barley and millets. Substantial work has been done with the rice blast pathosystem on pathogenic and genetic diversity, epidemiology and disease management through host-plant resistance. However, such studies are very limited with the finger millet-blast pathosystem. *Pyricularia* blast fungi are known to have coincided with the cultivated cereals in different parts around the world. Its sexual stage or teleomorph (*Magnaporthe grisea* Sacc.) is a haploid, filamentous, heterothallic ascomycetous fungus that parasitizes over 50 grasses. Species names were assigned to those pathogens according to the cultivated cereals from which they were isolated. Namely, pathogens of finger millet, foxtail millet, common millet, and rice have been designated as *P. setariae, P. panici-milliacei, P. penniceti* and *P. oryzae* respectively. Rossman *et al.* (1990) argued that all of these species should be referred to as *P. grisea,* which was originally assigned to the pathogen of crabgrass (*Digitaria sanguinalis*). Kato *et al.* (2000) suggested that the isolates from cereal crops form a single group with a common ancestor although they are pathogenic to taxonomically diverse plants. Despite the wide host range of the pathogen, *M. grisea* populations mainly exist as host-specific forms, capable of infecting a single host (Todman *et al.*, 1994; Viji *et al.*, 2000). Within a particular host organ specificity (leaf, neck and finger blast) most probably does not occur in this pathogen and cross infectivity might happen at the three plant growth stages (Puri and Kumar, 2012). Substantial work on biology and variability has been done in the rice-blast pathosystem, whereas such studies are very limited for the finger millet-blast pathosystem (Singh and Kumar 2010; Takan *et al.*, 2012). Genetic diversity and population structure of 72 *M. grisea* isolates collected from finger millet, foxtail millet, pearl millet and rice from major crop growing areas in India was studied using 24 SSR markers. Results suggested presence of high degree of genetic variability among isolates (Babu *et al.*, 2013).

Disease Development

Blast epidemiology is well studied in rice but less in finger millet. The pathogen is very sensitive to minute change of temperature and other environmental factors. The pathogen can infect the crop at all stages, from seedling to post flowering phase. Seedlings may be killed under epidemic condition. On leaf elliptical or diamond shaped lesions with grey, water-soaked centre are surrounded by chlorotic halo. The spots may coalesce to cover a large area of the leaf blade. During flowering stage the disease may appear on peduncle and/or on finger causing neck and finger blast respectively. Depending on the time and severity of infection the infected ears become completely chaffy or produce shriveled grains. The life cycle of the fungi in nature is very complicated. It survives in the crop residues and on other cereals which acts as collateral hosts. It can parasitize over 50 grasses. The pathogen is found in glumes, pericarp and endosperm. But embryo infection has not been observed. The initial inoculum most probably comes from weeds or some collateral hosts. The fungus spreads mainly by air borne conidia. A temperature of 25-30°C, RH of 90 per cent and above, cloudy days with intermittent rainfall, are favorable for the rapid spread of the disease. The disease prevails from the month of May to September, but it may extend even up to November, when the favorable conditions are prolonged. The maximum disease incidence has been observed in the crop sown in the month of August. The disease incidence increases with increasing N levels (Kumar and Yadav, 2012).

Screening for Resistance

Development of an efficient screening technique for evaluation of germplasm is critical to a successful breeding program. Though such technique is well established in rice, little work has been done on finger millet. At present mostly field screening techniques are practised to identify blast resistance under natural infection (Mantur and Madhukeshwara, 2001; Kumar *et al.*, 2006; Kumar and Kumar, 2009; Nagaraja *et al.*, 2010). Recently Babu et. al. (2012) worked on field screening technique involving growing of susceptible genotype at regular interval and artificial inoculation of plants at flowering. They claimed to have developed a field screening technique for neck and finger blast whereby finger millet germplasm can be effectively screened in the field conditions. For leaf blat disease rating scale of 0-5 has been used where, a score of 0= immune or highly resistant, 0.1-2.0= highly resistant, 2.1-10= resistant, 10.1-25= moderately resistant, 25.1- 50= susceptible and >50= highly susceptible. Severity of neck and finger blast have been measured in percentage using formula for Neck blast (per cent) = (Number of ears showing infection on peduncle or neck/Total number of ears in all the plants in two rows) X 100, and for Finger blast (per cent) = (Number of fingers infected in randomly selected five plants/total number of ears) X 100 (Nagaraja *et al.*, 2007; Kumar and Kumar, 2009).

Sources of Resistance

Almost all cultivars, both local and improved, presently under cultivation are susceptible to blast disease though the level of susceptibility reaction varies from cultivar to cultivar. Sources of resistance among Indian land races are limited and most of the accessions from Asian origin are susceptible to neck and finger blasts. Finger millet accessions especially from East Africa are reported to possess genes for blast resistance. Repository of germplasm has been screened for identification of resistance sources and their utilization in breeding programme. A cross-section of 1,941 accessions of finger millet germplasm was scored for different forms of blast disease and many sources of resistance for blast were identified. Twenty eight entries showed resistant reaction for both neck and finger blast, and only nine entries showed combined resistant reaction for leaf, neck and finger blast (Seetharam, 1983). Kumar and Kumar (2009) observed that finger millet genotypes viz; VL 234, SANJI 1, PRM 9802, VL 328, VL 333, ED 201-5A, VR 708, SANJI 1 and VL 324 were completely free from neck and finger blast disease under natural epiphytotic conditions in Himalayan hill station. Babu *et al.* (2012) reported 372 blast resistant accessions that belonged to one wild and four cultivated races of finger millet that originated from 19 countries indicating the wide geographical diversity among resistant accessions.

Management of Blast

Use of Resistant Varieties

Finger millet is predominantly grown by poor and marginal farmers who have little means of controlling diseases. Development of varieties with inbuilt genetic resistance is the best means of disease control. A number of stabilized selections from a blast resistance programme were tested for yield in different states (Gowda *et al.*, 1986). After the high yielding Indaf varieties and others were found susceptible to neck and finger blast, search for resistant varieties began in the early 1990's and as a result resistant varieties like GPU 28, VL 149, GPU 48 and L 5 were released in India. These varieties, however, had juvenile susceptibility to leaf blast (Nagaraja *et al.*, 2012). GKVK centre Bangalore, utilized IE 1012 as the resistant parent and PES 176, HPB 7-6, PR 202 and Indaf 8 as high yielding adopted parents. Gautami, GPU 28, BM 9-1, GPU 26, GPU 45, Chilika, VL 315 and GPU 48 are some of the blast tolerant/ resistant varieties which are presently cultivated in India (AICSMIP, 2013). Blat fungi is a highly variable pathogen. Therefore, durable resistance is more desired for increasing shelf life of a resistant cultivar. Other than conventional breeding hardly any approach has been practiced for development of varieties resistant to blast of finger millet. Recently Ignacimuthu and Ceasar (2012) developed leaf blast resistance finger millet transgenic using rice chitinase (*chi*11) gene.

Use of Chemicals and other Practices

Seed treatment, plant spacing and regulating the amount of nitrogenous fertilizer applied to the crop are important measures to minimize the occurrence of blast disease. Even early sowing of the crop may help escape blast incidence (Senthil *et al.*, 2012). In the absence of resistant varieties a vast number of chemicals were recommended to contain blast. Spraying of carbendazin or kitazin once at the seedling stage and twice at the earhead stage was claimed to be economical in reducing the disease incidence. Sprays of Carbendazim @ 0.1 per cent or Tricyclazole @ 0.05 per cent or Collimation of Mancozeb 63 per cent +Carbenadzim12 per cent was recommended by the AICSMIP (2006). However, spraying combined with seed treatments always gave better management of the disease. Dutta and Jha (1983) opined that seed treatment with effective fungicides will reduce the initial seed borne inoculum and eventually help in obtaining a healthy crop. Seed treatment with bioagent *Trichoderma harzianum* and one spray of *Pseudomonas fluorescens* @ 0.3 per cent at the time of flowering followed by second spray ten days later was found to control all three types of blasts (Nagaraja and Mantur, 2009). Kumar (2011) reported that seed treatment with *Pseudomonas fluorescens* Pf 2 (@ 0.6 per cent) + its two foliar sprays (@ 0.6 per cent) followed by two sprays of hinosan @ 0.1 per cent were significantly most effective in reducing blast disease and increasing yields. Nagaraja *et al.* (2012) observed that seed treatment to resistant variety with either Carbendazim @ 2 g/kg or *Pseudomonas fluorescens* 6gkg/ha were found to reduce blast disease incidence by two and a half times over control besides recording high mean yield of 25.67 and 24.98 q/ha as against 21.06 q in susceptible variety. They further suggested that instead of trying sprays of different chemicals on a susceptible variety, it is worthwhile to select resistant variety and give simple seed treatment with Carbendazim @ 2 g/kg or *Pf* @6 g/kg. Very recently Prajapathi *et al.* (2013) claimed that with the proper protection measures, 70.45, 67.72 and 62.97 per cent disease of leaf blast, neck blast and finger blast can be controlled, respectively. Seed treatment with bavistin (3g/kg of seed) + two spray of tricyclazole (0.6g/lit. of water) can protect the crop from blast. Botanical have also been tried for management of finger millet blast. Joshi *et al.* (2012) reported that crude aqueous extracts of *O. latifolia* significantly reduced leaf (43.7 per cent), neck (45.5 per cent) and finger blast (28.1 per cent) along with a significant yield increase (28.1 per cent) over control. Silica fertilizers *viz.*, calcium silicate ($CaSiO_3$) and sodium silicate ($Na_2SiO_3 . 9H_2O$) have been reported to reduce neck blast in finger millet (Krishnappa *et al.*, 2013).

Research Gaps and Future Priority

Pyricularia grisea, the causal fungi of finger millet blast, infects many plant hosts including some important cereal crops. The pathogen has been vastly studied in other pathosystems *e.g.*, rice, wheat etc. and huge amount of information relating to those pathosystem is available. Among those information that is general in nature and non-specific to a pathosystem can be used for understanding blast of finger millet. Moreover, rice and wheat being vastly grown high value crops the extent of research that can be justifiable to these crops may not be applicable for marginal crop like finger millet. Critical research gaps and future research priority in finger millet blast are described hereunder that might change with time.

1. Finger millet is low input crop and grown by poor and marginal farmers who have little means of controlling diseases. Under such situations use of pesticides for the control of pests, diseases and weeds is neither feasible nor practicable. Incorporation of inbuilt resistance in varieties is the best way of managing this disease. Therefore, development of varieties with durable resistance should continue to receive the highest priority.
2. In India most of the local as well as improved cultivars of finger millet carry susceptibility to blast. Though the level of resistance in present day cultivars are somewhat better, they need further improvement to avoid chances of future epidemic that may occur under changing climatic conditions. Germplasm of Asian origin is reported to be mostly susceptible to blast while that of African origin are good in blast resistance. Resistant genes available in local land races are preferred over exotic gene as the former reduces the chances of development of new races in pathogen. However, in finger millet there is little option for using locals as they are mostly susceptible. Therefore, identification of blast resistance in African germplasm and incorporation of the resistance to Indian cultivar through appropriate resistance breeding programme should be emphasized.
3. Blast finger millet appears at different stages of plant growth affecting different plant parts. The pathogen is a highly variable one in genetic make-

up and different biotypes of the blast fungus also could be present as in other crops. Information on this area is totally lacking in *Pyricularia grisea*-finger millet pathosystem, and the area of racial differentiation and virulence mapping in blast fungi is worth investigating.

4. Deployment of single dominant gene is a common practice for development of plant varieties resistant to pests and diseases. But such technique is often subjected to breakdown of resistance. Through conventional and molecular breeding many blast-resistant varieties have been developed in rice. Such works need to be initiated in finger millet in future. Breeding work should be focused on utilizing the broad spectrum of resistance genes and pyramiding genes and QTL. Molecular breeding for blast resistance is in its infancy in finger millet. The molecular biology tools and emerging new methods should be used to facilitate gene pyramiding.
5. Information is lacking in the field of epidemiology, disease forecasting, genetics of resistance mechanism of resistance and pathogenic variability in *Pyricularia grisea*-finger millet pathosystem. Regarding screening technique further standardization especially for green house screening techniques is required for increasing efficiency. Studies in these areas should be strengthened for better understanding of the disease, identification of resistance and efficient management of the blast disease in finger millet.

References

1. AICMIP (All India Coordinated Pearl Millets Improvement Project) 011-2014. Progress report. Mandore, Rajasthan, India: Indian Council of Agricultural Research.
2. AICSMIP (2006). Technology for increasing finger millet and other small millets production in India. Project Coordination Cell, All India Coordinated Small Millets Improvement Project, Gandhi Krishi Vigyan Kendra, Bangalore, pp. 1-41.
3. Anaso, C.E. and Anaso, A.B. (2010). Cost-effectiveness of seed dressing with a new formulation of metalaxyl (Apron Star 42 WS) for sustainable pearl millet production in northern Nigeria. *Arch. Phytopath. Pl. Prot*. 43: 154- 159.
4. Appadurai, R., Parambaramani, C., and Natarajan, U.S. 1975. Note on the inheritance of susceptibility of pearl millet to downy mildew. *Indian Journal of Agricultural Sciences* 45(4): 179-180.
5. Babu, T.K., Sharma, R., Upadhyaya, HD, Reddy, S.P.N., Deshpande, P., Senthilvel, S., Sarma, N.D.R.K. and Thakur, R.P. (2013). Evaluation of genetic diversity in *Magnaporthe grisea* populations adapted to finger millet using simple sequence repeats (SSRs) markers. *Physiol. Mol. Pl. Patholo*. 84: 10 -18.
6. Babu, T.K., Thakur, R.P., Reddy, P.N., Upadhyaya, H.D., Girish, A.G. and Sarma, N.D.R.K. (2012). Development of a field screening technique and identification of blast resistance in finger millet core collection. *Indian J. Pl. Prot*. 40: 45-51.

7. Bhat, S.S. 1973. Investigations on the biology and control of Sclel'Ospora gra711inicola on bajra. *Ph.D. Thesis*, University of Mysore, Mysore, Karnataka, India. 165 pp.

8. Bisht, I.S. (1987). Blast tolerance and yield loss in finger millet. *Indian J. Agri. Sci.* 50: 954-955.

9. Celia Chalam, V. (1996). Studies on the mechanisms of resistance in pearl millet genotypes to downy mildew disease. Acharya NG Ranga Agricultural University, Rajendranagar, Hyderabad, *Ph D Thesis*. pp. 132+xi.

10. Chandrashekhar, A., Sathyanarayana N., Deepak S., Amruthesh K.A, Shetty N.P. and Shetty H.S. (2007). Endophytic bacteria from different plant origin enhance growth and induce downy mildew resistance in pearl millet. *Asian J. Pl. Path.* 1: 1-11.

11. Dang, J.K., Thakur, D.P., and Grover, R.K. 1983. Control of pearl millet downy mildew caused by Sclerospora graminicoia with systemic fungicides in an artificially contaminated plot. *Annals of Applied Biology* 102: 99- 106.

12. Dangaria C.J., Dhedhi K.K., Raghavani K.L., Mungra K.D., Sorathia J.S., Atara S.D. and Bunsa B.D. (2013. A high yielding downy mildew resistant pearl millet hybrid GHB-732 for Gujarat, India. *Agric. Sci. Digest.*, 33 (3): 219 – 222.

13. Deepak S., Manjunath, G., Manjula, S., Niranjan-Raj, S., Geetha N. P., and Shetty H.S. (2008). Involvement of silicon in pearl millet resistance to downy mildew disease and its interplay with cell wall proline/hydroxyproline-rich glycoproteins. *Aus. Pl. Path.* 37: 498-504.

14. Deepak S., Shailasree S., Sujeeth, N., Kini, R.K., Shetty H.S. Mithöfer, A. (2007). Purification and characterization of proline/hydroxyproline-rich glycoprotein from pearl millet coleoptiles infected with downy mildew pathogen *Sclerospora graminicola*. *Phytochem*. 68: 298-305.

15. Deepak, S., Oros, G., Sathyanarayana, S.G., Shetty, N.P., Shetty, H.S. and Sashikanth, S. (2005). Antisporulant activity of leaf extracts of Indian plants against *Sclerospora graminicola* causing downy mildew disease of pearl millet. *Arch. Phytopath. Pl. Prot*. 38: 31-39.

16. Deswau, D.P., Govila, O.P. and Sheoran. R.K. (1998). Genetic analysis of downy mildew resistance in pearl millet. *Indian Phytopath*. 51 (3): 261-264.

17. Dutta, A.K. and Jha, D.K. (1983). Seed borne fungi of ragi [*Eleusine coracana* (L.) Gaertn.] in Chotanagpur and their control. *Pesticides* 17: 28.

18. Gill, K.S., Phul, P.S., Chahal, S.S., and Singh, N.B. 1978. Inheritance of resistance to downy mildew disease in pearl millet. *Cereal Research Communications* 6(1): 71-74.

19. Gowda, B.T.S., Seetharam, A., Viswanath S. and Sannegowda, S. (1986). Incorporating blast resistance to Indian elite finger millet cultivars from African cv. IE 1012. *Sabrao J*. 18: 119-120.

20. Gwary D.M., Bdliya B.S. and Bdliya J.A. (2009). Appraisal of Yield Losses in Pearl Millet Due to Downy Mildew Pathogen (Sclerospora graminicola) in Nigerian Sudan Savanna. *Arch. Phytopath. Pl. Prot*. 42: 1010-1019.

21. Hash C.T. and Witcombe J.R. (2001). Pearl millet molecular marker research. *International Sorghum and Millets Newsletter* 42: 8–15.

22. Hash C.T., Singh S.D., Thakur R.P. and Talukdar B.S. (1999). Breeding for disease resistance. pp. 337–379 In: *Pearl millet breeding* (Khairwal IS, Rai KN, Andrews DJ and Harinarayana G, eds.). New Delhi, India: Oxford and IBH.

23. Ignacimuthu, S. and Ceasar, S.A. (2012). Development of transgenic finger millet (*Eleusine coracana* (L.) Gaertn.) resistant to leaf blast disease *J. Biosci*. 37(1): 135–147.

24. International Crops Research Institute for the Semi-Arid Tropics, Patancheru 502 324, Andhra Pradesh, India.

25. Jones E.S., Breese W.A., Liu C.J., Singh S.D., Shaw D.S. and Witcombe J.R. (2002). Mapping quantitative trait loci for resistance to downy mildew in pearl millet: field and glasshouse screens detect the same QTL. *Crop Science* 42: 1316–1323.

26. Jones E.S., Liu C.J., Gale M.D, Hash C.T. and Witcombe J. R. (1995). Mapping quantitative trait loci for downy mildew resistance in pearl millet. *Theo. Appld. Genet*. 91: 448-456.

27. Joshi, D., Bhatt, J.C., Hooda, K.S. and Gupta, A. (2012). Management of finger millet (*Eleusine coracana*) blast under field conditions by plant extracts. *Indian J. Agric. Sci*. 82 (3): 284–287.

28. Kato, H., Masaaki, Y., Tomona, Y., Hiroyuki, K., Yutaka, I., Hitoshi, N., Yukio, T., Shigeyuki, M. and Naoki, M. (2000). Pathogenicity, mating ability and DNA restriction fragment length polymorphisms of *Pyricu/aria* populations isolated from Gramineae, Bambusideae and Zingiberaceae. *Plants J. Gen. Plant Pathol*. 66: 30-47.

29. Khairwal I.S., Rai K.N., Yadav O.P. and Bhatnagar S.K. (2004). Pearl millet cultivars. Mandor, Jodhpur, India: All India Coordinated Pearl Millet Improvement Project, Indian Council of Agricultural Research. 22 pp.

30. King, S.B., Werder, J., and Singh, S.D. 1989. Pathogenic variation in Sclerospora graminicola on pearl millet. *Phytopathology* 79: 1390. (Abstract).

31. Krishnappa, P., Ramaswamy, G.R., and Jyothi, V. (2013). Role of silica fertilizers on blast (*Pyricularia grisea* Sacc.) disease development in finger millet [*Eleusine coracana* (L.) Gaertn.]. *Pl. Dis. Res*. 28 (2): 207-208.

32. Kumar, B. (2011). Management of blast disease of finger millet (*Eleusine coracana*) in mid-western Himalayas. *Indian Phytopath*. 64 (2): 154-158.

33. Kumar, B. and Kumar, J. (2009). Evaluation of small millet genotypes against endemic diseases in mid-western Himalayas. *Indian Phytopath.* 62 (4): 518-521

34. Kumar, B. and Rashmi, Y. (2012). Influence of nitrogen fertilizer dose on blast disease of finger millet caused by *Pyricularia grisea*. *Indian Phytopath*. 65 (1): 52-55.

35. Kumar, R and Sagar. (2009). Effect of cytoplasm on downy mildew vulnerability in pearl millet. *Indian J. Genet. Pl. Breed*. 69:

36. Kumar, V.B.S., Kumar, T.B.A., Bhat, S.A. and Nagaraju, A. 2006. Screening of long duration finger millet (*Eleusine coracana* (L) Gaertn.) genotypes against neck and finger blast caused by *Pyricularia grisea* (Cke.) Sacc. *J. Pl. Prot. Environ*. 3: 136-139.

37. Latha, A.M., Rao K.V., Reddy, T.P., and Reddy V.D. (2006). Development of transgenic pearl millet (*Pennisetum glaucum* (L.) R. Br.) plants resistant to downy mildew. *Pl. Cell Report*. 25: 927-935.

38. Mani, M.T. and Hepziba, S.J. (2009). Biological management of pearl millet downy mildew caused by *Sclerospora graminicola*. *Arch. Pl. Prot*. 42: 129-135.

39. Mani, M.T. and Hepziba, S.J. (2009). Integrated disease management of pearl millet downy mildew caused by *Sclerospora graminicola*. *Arch. Phytopath*. Pl. Prot. 42: 136-141.

40. Manjunatha G, Niranjan Raj S, Geetha NP, Deepak S, Amruthesh KN, Shetty HS. (2009). Nitric oxide is involved in chitosan induced systemic resistance in pearl millet against downy mildew disease. *Pest Manag. Sci*. 65: 737-743.

41. Mantur, S.G. and Madhukeshwara, S.S. (2001). Evaluation of finger millet genotypes for resistance to blast. *Curr. Res*. 30: 191-192.

42. Mathur, R.L., and Dalela, G.G. (1971). Estimation of losses from greenear disease (*Sclerospora graminicola*) of bajra (*Pennisetum typhoides*) and grain smut (*Sphacelotheca sorghi*) of jowar (*Sorghum vulgare*) in Rajasthan. *Indian Phytopath*. 24: 101-104.

43. McRae, W. (1920). Detailed administration report of the government mycologist for the year 1919-20.

44. Mgonja, M.A., Lenne, J.M., Manyasa, E. and Sreenivasaprasad, S. (2007). Finger millet blast management in East Africa. Creating opportunities for improving production and utilization of finger millet. Proceedings of the First International Finger Millet Stakeholder Workshop, Projects R8030 and R8445 UK Department for International Development—Crop Protection Programme (International Crops Research Institute for the Semi-Arid Tropics) pp.196

45. Michelmore, R.W., Pawar, M.N., and Williams, R.J. 1982. Heterothallism in *Sclerospora graminicola*. *Phytopathology* 72: 1368-1372.

46. Mitter, J.H., and Tandon, R.N. (1930). A note on *Sclerospora graminicola* (Sacc.) Schroet. in Allahabad. *J. Indian* Bot. Soc. 9: 243.

47. Nagaraja, A. and Mantur, S.G. (2009). Management of finger millet blast through eco-friendly approaches. *Mysore J. Agric. Sci*. 43: 8-10.

48. Nagaraja, A., Gowda, J., Krishnappa, M. and Gowda K.T.K. (2008). GPU 28: A finger millet variety with durable blast resistance. *J. Mycopathol. Res*. 46: 109-111.

49. Nagaraja, A., Kumar, B., Raguchander, T., Hota, A.K,. Patro, T.S.S.K. Devaraje Gowda, P., Savita Ekka and Gowda M.V.C. (2012). Impact of disease management practices on finger millet blast and grain yield. *Indian Phytopath*. 65 (4): 356-359.

50. Nagaraja, A., Kumar, J., Jain, A.K., Narasimhudu, Y., Raghuchander, T., Kumar, B. and Gowda, H.B. (2007). *Compendium of Small Millets Diseases*. Project Coordination Cell, All India Coordinated Small Millets Improvement Project, UAS, GKVK Campus, Bangalore. p. 80.

51. Nagaraja, A., Reddy, Y.A.N., Reddy, B.A., Patro, T.S.S.K., Kumar, B., Kumar, J. and Gowda K.T.K. (2010). Reaction of finger millet recombinant inbred lines (RILs) to blast. *Crop Res.* 39: 120-122.

52. Panchbhai, S.D., Reddy, M.S., and Singh, S.D. 1991. A repeatable method of germination of oospores of Sclerospora.

53. Prajapathi, V.P. Sabalpara, A.N. and Pawar, D.M. (2013). Assessment of yield loss due to finger millet blast caused by *Pyricularia grisea* (Cooke) Sacc. *Trends in Biosci.* 6 (6): 876-878.

54. Puri, S., and Kumar, J. (2012). Characterization of leaf, neck and finger blast pathogen populations from *Eleusine coracana* and elucidating their interrelationships in *Eleusine-Magnaporthe* pathosystem. *Indian Phytopath*. 65 (2): 133-141.

55. Pushpavathi, B, Thakur, R.P. and Chandrashekara Rao, K. (2006). Fertility and mating type frequency in Indian isolates of *Sclerospora graminicola*, the downy mildew pathogen of pearl millet. *Pl Dis*. 90: 211-214.

56. Pushpavathi, B., Thakur R.P., and Chandrashekara Rao K. (2005). Inheritance of avirulence in *Sclerospora graminicola*, the pearl millet downy mildew pathogen. *Pl. Path. J.* 5: 54-59.

57. Ramappa, H.K., Ravishankar, C.R. and Prakash, P. (2002). Estimation of yield loss and management blast disease in finger millet (ragi). *Proc. Asian Cong. Mycol. Pl. Path*. Oct 1-4, 2002, Univ. of Mysore, pp. 195.

58. Rao, A.N.S. (1990). Estimates of losses in finger millet (*Eleusine coracana*) due to blast disease (*Pyricularia grisea*). *J. Agric. Sci*. 24: 57-60.

59. Rossman, A.Y., Howard, R.J. and Valent, B. (1990). *Pyricularia grisea,* the correct name for the rice blast disease fungus. *Mycologia* 82 : 509-512.

60. Roy, M.M. (2013). Project coordinator review. Annual workshop, All India Pearl Millet Improvement Project, JAU, Junagadh. p.11.

61. Safeeulla, K.M. 1976. Biology and control of the downy mildews of pearl millet, sorghum and finger millet. University of Mysore, Mysore, Karnataka, India. 304 pp.

62. Sanmathi, Kumar R.P., Shanthala, L., Anilkumar, T.B., and Sudharsana, L. (2006). Phytotoxins from Pyricularia grisea and Their Effect on Finger Millet. *J. Pl. Biochem. Biotechnol*. 15: 63-66.

63. Sastry J.G., Ramakrishna W., Sivaramakrishnan S., Thakur R.P., Gupta V.S., Ranjekar P.K. (1995). DNA Fingerprinting Detects Genetic Variability in the Pearl Millet Downy Mildew Pathogen (*Sclerospora graminicola*). *Theor. Appl. Genet.* 91: 856-861.

64. Seetharam, A. (1983). Identification of sources of resistance to ragi blast and its utilization in breeding. *Proc. of National Seminar on Breeding Crop Plants for Resistance to Pests and Diseases*: 152. Tamil Nadu Agricultural University, Coimbatore, India.

65. Senthil, R., Manikandan, R., Ramanathan,A. and Raguchander, T. (2012). Influence of dates of sowing on blast disease incidence in finger millet. *Madras Agric. J.*, 99 (1-3): 99-101.

66. Shailasree S., Kini KR., Deepak S., Kumudini B.S. and Shetty H.S. (2004). Accumulation of hydroxyproline-rich glycoproteins in pearl millet seedlings in response to *Sclerospora graminicola* infection. *Pl. Sci.* 167: 1227-1234.

67. Shankara Rao, R., Rao, K.e.S., Singh, S.D., Reddy, M.S., Rao, M.V.B., and Venkateswarlu, K. 1987. Fungicidal control of downy mildew of pearl millet. *Indian Journal of Plant Protection* 15: 146-151.

68. Sharma, R., Rao, V.P., Varshney, R.K., Prasanth, V.P. Kannan, S. and Thakur, R.P. (2010). Characterisation of pathogenic and molecular diversity in *Sclerospora graminicola*, the causal agent of pearl millet downy mildew. *Arch. Phytopath. Pl. Prot.* 43: 538-551.

69. Shetty, H.S., Mathur, S.B., and Neergaard, P. 1980. Sclerospora graminicola in pearl millet seeds and its transmission. *Transactions of the British Mycological Society* 74: 127-134.

70. Singh S.D., Wilson J.P., Navi S.S., Talukdar B.S., Hess D.E. and Reddy K.N. (1997). Screening techniques and sources of resistance to downy mildew and rust in pearl millet. Information Bulletin no. 48. Patancheru 502 324, Andhra Pradesh, India: International Crops Research Institute for the Semi-Arid Tropics. 36 pp.

71. Singh, F., Singh, R.K., Singh, RM., and Singh, RB. 1978. Genetic analysis of downy mildew (Sclerospora graminicola) resistance in pearl millet (*Pennisetum hyphoides* (Burm.) S. and H.). *Zeitschrift fur Pflanzenzuochtung* 81: 54- 59.

72. Singh, S.D. 1974. Studies on downy mildew disease (*Sclerospora graminicola* (Sacc.) Schroet. of bajra, *Pennisetum typhoides* (Burm. E) Stapf and c.B. Hubb.). PhD. thesis, Indian Agricultural Research Institute, New Delhi, India. 126 pp.

73. Singh, S.D., and Gopinath, R. 1985. A seedling inoculation technique for detecting downy mildew resistance in pearl millet. *Plant Disease* 69: 582-584.

74. Singh, S.D., and Gopinath, R. 1990. Effect of temperature and light on sporangia I germination and zoospore infectivity in Sclerospora graminicola on pearl millet. *Canadian Journal of Plant Pathology* 12: 25-30.

75. Singh, S.D., and King, S.B. 1988. Recovery resistance to downy mildew in pearl millet. *Plant Disease* 72: 425-428.

76. Singh, S.D., and Singh, G. 1987. Resistance to downy mildew in pearl millet hybrid NHB 3. *Indian Phytopathology* 40(2): 178-180.

77. Singh, S.D., and Williams, R.J. 1979. A study of the host range of Sclerospora graminicola. International Working Group on Graminaceous Downy Mildews (IWGGDM) *Newsletter* 1(2): 3.

78. Singh, S.D., King, S.B., and Weraer, J. (1993). Downy mildew disease of pearl millet. Information Bulletin no. 37. (In En. Summaries in Fr, Es.) Patancheru, A.P. 502 324, India: International Crops Research Institute for the Semi-Arid Tropics. 36 pp.

79. Singh, Y. and Kumar, J. (2010). Study of genomic fingerprints profile of *Magnaporthe grisea* from finger millet (*Eleusine Coracona*) by random amplified polymorphic DNA-polymerase chain reaction (RAPD-PCR). *African J. Biotechnol.* 9(46): 7798-7804.

80. Sivaramakrishnan, S., Thakur, R.P., Seetha Kannan and Rao, V.P. (2003). Pathogenic and genetic diversity among Indian isolates of *Sclerospora graminicoia* from pearl millet. *Indian Phytopath*. 56 (4): 392-397.

81. Sudisha, J., Ananda Kumar, S., Niranjana, S.R., Shetty, N.P. and Shetty H.S. (2009). Cloning and Development of pathotype-specific SCAR marker associated with *Sclerospora graminicola* isolates from pearl millet. *Aust. Pl. Pathol.* 38: 216-221.

82. Sudisha, J., Shivanna, R.K., Gnanaprakash, P.H., Shetty, H.S. (2010). Evaluation of plant growth-promoting rhizobacteria for their efficiency to promote growth and induce systemic resistance in pearl millet against downy mildew disease. *Arch. Phytopath. Pl. Prot.* 43: 368-378.

83. Sujeeth, N., Deepak, S., Shailasree S., Kini, R.K., Shetty H.S. and Hille, J. (2010). Hydroxyproline-rich glycoproteins accumulate in pearl millet after seed treatment with elicitors of defense responses against *Sclerospora graminicola*. *Physiolo. Mol. Pl. Patholo*. 74: 230-237.

84. Sundaram, N.V., Ramasastry, D.V., and Nayar, S.K. 1973. Note on the seed-borne infection of downy mildew (*Sclerospora graminicola* (Sacc.) Schroet.) of pearl-millet. *Indian Journal of Agricultural Sciences* 43: 215-217.

85. Takan, J.P., Chipili, J., Muthumeenakshi, S., Talbot, N.J., Manyasa, E.O., Bandyopadhyay, R. (2012). *Magnaporthe oryzae* populations adapted to finger millet and rice exhibit distinctive patterns of genetic diversity, sexuality and host interaction. *Mol. Biotechnol*. 50(2): 145 - 58.

86. Thakur R.P., Rai K.N., Khairwal I.S. and Mahala R.S. (2008). Strategy for downy mildew resistance breeding in pearl millet in India. J. *SAT Agric. Res*. 6: 1-11.

87. Thakur, D.P., and Kanwar, Z.S. 1978. Internal seed-borne infection and heat therapy in relation to downy mildew of *Pennisetum typhoides* Stapf. and Hubb. *Science and Culture* 43: 432- 434.

88. Thakur, R.P. (2003). An ICRISAT success story- Greenhouse screening to combat downy mildew in pearl millet.

89. Thakur, R.P. and Rao, V.P. (1997. Variation in virulence and aggressiveness among pathotypes of *Sclerospora graminicola* on pearl millet. *Indian Phytopath.* 50 (1): 41-47.

90. Todman, A.K., Pawar, D.R., Joshi, M.H. (1994). Host reactions to finger millet blast (*Pyricularia grisea* Sacc.). *Mysore J. Agric. Sci*. 28: 45 -46.

91. Upasana, R., Chaudhary, D.P. and Srivastava, M. (2010). Downy mildew and its effect on quality of pearl millet hybrids. *Crop Improvement* 37 (2): 206.

92. Venugopal, M.N., and Safeeulla, K.M. 1978. Chemical control of downy mildews of pearl millet, sorghum and maize. *Indian Journal of Agricultural Sciences* 48: 537-539.

93. Viji, G., Gnanamanickam, S.S., Levy, M. (2000). DNA polymorphisms of isolates of *Magnaporthe grisea* from India that is pathogenic to finger millet and rice. *Mycol. Res.* 104(2): 161 -167.

94. Viswanath, S. and Seetharam, A. (1986). Diseases of small millets and their management in India. Pages 237-253, In Seetharam, A., Riley, K.W., Harinarayana, G. (eds) small millets in global agriculture. *Proceedings of the first international small millets workshop*. Oxford and IBH Publishing Co. Pvt. Ltd. Bangalore, India. pp 413.

95. Werder, J., and BaU, S.L. 1992. Variability of pearl millet downy mildew (Sclerospora graminicola) and its implications for breeding for stable and durable resistance. *Tropical Pest Management* 38(1) 43-47.

96. Williams, R.J., Singh, S.D., and Pawar, M.N. 1981. An improved field screening technique for downy mildew resistance in pearl millet. *Plant Disease* 65: 239-241.

97. Yadav O. P., Sahai B., Bidinger F. R., Bhasker Raj A. G. (2005). Exotic elite populations as sources for diversification of landraces of pearl millet. *Indian J. Genet. Pl. Breed*. 65:

98. Yadav, O.P. (1996). Downy Mildew Incidence of Pearl Millet Hybrids with Different male-sterility Inducing Cytoplasms. *Theor. Appl. Genet*. 92: 278-280.

99. Zarafi A. (2007). An assessment of pearl millet genotypes for resistance to downy mildew (*Sclerospora graminicola*) under field conditions in Nigeria. *Agroecol. Sustain. Food Sys*. 29: 155-171.

100. Zarafi, A.B., Emechebe, A.M., Akpa, A.D and Alabi. O. (2005). Effect of Fertilizer Levels on Grain Yield, Incidence and Severity of Downy Mildew in Pearl Millet *Arch. Phytopathol. Pl. Prot*. 38: 11-17.

Millets: Ensuring Climate Resilience and Nutritional Security (2015)
Editors: Vilas A. Tonapi and J.V. Patil
Published by: DAYA PUBLISHING HOUSE, NEW DELHI

Pages 473–489

Chapter 17

Insect Pests of Millets and their Management

B. Subbarayudu and A Kalaisekar

ICAR-Indian Institute of Millets Research (IIMR)
Rajendranagar, Hyderabad – 500 030

In most developing countries, agriculture is the driving force for broad-based economic growth; and low agricultural productivity is a major cause of poverty, food insecurity, and malnutrition. Accelerated public investments are needed to facilitate agricultural growth through high-yielding varieties with resistance to biotic and abiotic stresses, environment-friendly production technologies, availability of reasonably priced inputs in time, dissemination of information, improved infrastructure and markets, and education in basic health care. The use of high-yielding varieties, irrigation, fertilizers, and pesticides has increased crop productivity five-fold in the past five decades. However, growth has been leveling off in the past two decades. Land and water resources are diminishing; there is no option but to increase crop productivity per unit area. There is a need to examine how science can be used to raise biological productivity without the associated ecological costs. Some productivity increase can be achieved through the application of modern biotechnology tools in integrated gene management, integrated pest management, and efficient post-harvest management. Biotechnology in agriculture and medicine can be a powerful tool to alleviate poverty and improve the livelihoods of the rural poor. Sorghum crop protection plays an important role under the glamour of production umbrella, which is useful in enhancement of productivity levels. Important insect pests of sorghum, their economic threshold levels and their hot spot locations are given in Tables 1 and 2 (Indira and Subbarayudu, 2005).

Table 1: Important Insect Pests of Sorghum and their Economic Threshold Levels (ETL)

Sl.No.	*Common Name*	*Scientific Name*	*Growth Stage*	*Research ETL*	*Economic ETL*
1.	Shoot fly	*Atherigona soccata* (Rondani)	7 DAE*	20 per cent DH** before thinning	30 per cent DH before thinning
2.	Stem borer	*Chilo partellus* (Swinhoe)	20 DAE	Leaf damage or >10 per cent DH	Leaf damage or >25 per cent DH
3.	Oriental army worm	*Mythimna separate* (Walker)	45 DAE	Leaf damage on 20 per cent plants	Leaf damage on 40 per cent plants before boot stage
4.	Mite	*Oligonychus indicus* (Hirst)	60 DAE	10 per cent plants infested	25 per cent plants infested
5.	Shoot bug	*Peregrinus maidis* (Ashmead)	Before 60 DAE	Damage or infestation in >20 per cent plants	Damage or infestation exceeds 30 per cent
6.	Corn leaf aphid	*Rhopalosiphum maidis* (Fitch)	60 DAE	>30 per cent plants infested at 28 DAE or >50 per cent plants infested at 50 DAE	>50 per cent aphid infestation at 50 DAE. Do not spray if predators are present.
7.	Ear head bug	*Calocoris angustatus* (Lethiery)	60 DAE	2 adults or 20 nymphs/panicle	5 adults or 50 nymphs/panicle
8.	Pod borer	*Helicoverpa armigera* (Hubner)	90 DAE	5 medium to large sized larvae/ panicle	No spraying
9.	Sorghum midge	*Contarinia sorghicola* (Coquillett)	60 DAE	2-3 midge flies/panicle	5-6 midge flies/panicle

* DAE: Days after Emergencel ** DH: Deadheart.

Table 2: Hot Spot Locations for the Important Insect Pests of Sorghum in India

Insect Pests	*Locations*	
	Kharif	*Rabi*
Shoot fly	Akola, Dharwad, Indore, Palem, Parbhani, Rahuri and Udaipur	Bijapur, Dharwad, Parbhani, Rahuri and Solapur
Stem borer	Delhi, Hisar, Palem, Surat and Udaipur	Parbhani
Midge	Dharwad, Rahuri, Surat and Kovilpatti	–
Head bug	Palem and Kovilpatti	–
Head worms	Dharwad, Indore and Surat	Parbhani and Dharwad
Sugarcane aphid	–	Bijapur, Parbhani and Rahuri
Shoot bug	Palem	Bijapur and Solapur

1. Shoot Fly: *Atherigona soccata* (Rondani) (Muscidae: Diptera)

Nature and Symptoms of Damage

The maggots bore into the stem of young plants of sorghum generally up to four weeks from germination and cause deadhearts. Maximum infestation is observed in the third and fourth week after sowing. The tillers coming out from the infested plants are also attacked in turn by the insect pest. The fly population exhibits considerable variations and normally very low in April to June, tends to increase in July and reaches the peak in August. September onwards the population gradually declines and remains at a moderate level till March. Its activity is influenced by extreme temperature and also by continuous rains. As a result of larval feeding, the central leaf wilts and later drives up, giving the typical deadheart symptom. The deadhearts can be easily pulled out and at the base emits a bad smell. The young whitish yellow maggot feeds only on the decaying tissue. Normally the damage occurs from one week to about one month after emergence. If the attack occurs a little later, plants may produce side tillers that may also be attacked (Padmaja *et al.*, 2010 and Kalaisekar *et al.*, 2013).

Integrated Management

Cultural Control

Cultural control involving manipulations of insect environment is the safest and sustainable component in the integrated pest management (IPM).

Deep Ploughing

Exposes the larval and pupal stages of the shoot fly present in the soil and stubbles.

Early Sowing

Shoot fly can be avoided by suitable adjustment of the planting time, so that the vulnerable stage of the crop does not coincide with its active period. In *Kharif* season sowing is done within 7 to 10 days of the onset of the monsoon and in *Rabi*, between last week of September to first week of October crop to escape from the incidence of the shoot fly is recommended.

High Seed Rate

In late sown crops, in both the seasons of *Kharif* and *Rabi* the shoot fly attack will be severe. In such conditions use of high seed rate *i.e.* 10-12 kg/ha and removal of affected and extra plants at thinning, four weeks after sowing should be adopted to cut down the cost of cultivation.

Intercropping

Adoption of the intercropping is always beneficial than sole crop not only in the reduction of the shoot fly damage but also in increasing the total gross returns. For *Kharif* season the intercropping, namely, sorghum + redgram; sorghum + green gram/ black gram; sorghum + groundnut and sorghum + sesamum and for *Rabi* season, sorghum + safflower were proved advantageous. Selecting the least susceptible cultivars with high yield potentials, namely, CSV 13, CSV 15, CSH 16, CSH 18, CSV 8R, CSV 14R, CSH 13R and CSH 15R is suggested.

Biological Control

Hiremath *et al.*, 1998 reported that release of *Trichogramma chilonis* @ 12.5 lakhs/ ha recorded the minimum percentage of shoot fly deadhearts.

Botanical Control

Spraying the crop with 5 per cent neem seed kernel extract (NSKE) found effective against certain insect pests in some crops. These insecticides are ecologically sound and having repellent action. Experimental findings of IPM in National Agricultural Technology Project also proved that the use of NSKE 5 per cent spray at 14 days after sowing reduced the shoot fly oviposition and deadhearts formation (Subbarayudu *et al.*, 2002).

Chemical Control

a. Seed Treatment

Seed treatment with imidacloprid @ 14ml/kg of seed was recorded the least number of shoot fly eggs and its deadhearts (Subbarayudu *et al.*, 2002, Subbarayudu and Indira, 2007). Seed soaking in insecticide solution (2ml of Endosulfan in one litre of water) for 8 hours before planting, gave good results. When planted late the pest can effectively be controlled by seed treatment with Furadan 50 SP @ 100g/kg seed. Under moderate levels of infestation, a mixture of 60 per cent treated and 40 per cent untreated seed could be used. Seed treatment with Carbofuran 50 WP @ 100 g/kg seed is also effective.

b. Soil and Foliar Application

On need basis, soil application of carbofuran 3 G granules @ 20 kg/ha in furrows, at the time of sowing recommended in heavy infestation areas. In case, soil granular application is not done, damage can be minimized by spraying the seedlings at 7 and 14 days after sowing with Endosulfan @ 2 ml/lt of water. Subbarayudu and Indira (2002) reported that using soil application of murate of potash @ 25 kg/ha, seed treatment with imidacloprid @ 14 ml/kg of seed and spray of 5 per cent neem seed kernel extract (NSKE) at 14 days after sowing recorded the low levels of shoot fly eggs and its deadhearts and also minimized the yield losses.

Management

The combination of carbofuran 3G soil application @ 20kg/ha high seed rate (10 kg/ha), thinning after 20 days after sowing and release of 5 lakhs of *Trichogramma chilonis* adults at 7,14 and 21 days after germination appeared to be a good practice for integrated pest management of shoot fly. Seed soaking for 8-10 hours either in *Aegal marmelos* (10 per cent) + endosulfan (0.07 per cent) or endosulfan (0.07 per cent), reduces the shoot fly incidence effectively and increased the seed yield. In *Rabi* season, soil application of carbofuran 3G @ 2g/metre row, seed soaking in Endosulfan 35 EC @ 0.07 per cent for 8-10 hours, two sprays of Endosulfan 35EC @ 0.05 per cent, seed treatment with carbosulfan 25 ST @ 5 per cent, two sprays of phoshamidon 85 EC @ 0.04 per cent and two sprays of cypermethrin 10 EC @ 0.02 per cent reduced shoot fly incidence and gave higher seed yield. A field trial conducted in *Rabi* 1989-90 to integrate sowing dates with different insecticides and bioagent for suppressing the sorghum shoot fly, indicated that the treatment combinations having early sowing with soil applications of Carbofuran 3G @ 2g/metre row and early sowing with the use of 40 per cent treated and 60 per cent untreated seed reduced shoot fly incidence effectively and produced highest grain yield.

2. Spotted Stem Borer: *Chilo partellus* (Swinhoe) (pyralidae: Lepidoptera)

Nature and Symptoms of Damage

The first indication that plant is infested in the appearance of small elongated windows in young whorl leaves where the larvae have eaten the upper surface of leaf but have left the lower surface intact as a transparent window. As the severity of attack increases, the plant may become very ragged in appearance. Subsequently larvae bore into the stem and shot holes appear in whorl leaves. Deadheart symptoms then develop in younger plants. After head emergence the stalk just below the panicle is often resulting in breaking of the panicle or it leads to complete or partial chaffy seeds. There is often extensive tunneling of the stem. The borer attacks all parts of the sorghum plant except roots.

Management

Cultural Control

Time of Sowing

Early sowing of the crop escapes from spotted stem borer incidence.

Crop Rotation

Crop rotation with non-host crops reduces pest infestation by interrupting the continuity of the food chain of oligophagous pests. Sorghum rotation with non-host crops, such as groundnut, castor and redgram in *Kharif* whereas with safflower in *Rabi* reduces the infestation (Kishore, 1996). Shyam Prasad *et al.* (2010) evaluated the indigenous sorghum [*Sorghum bicolor* (L.) Moench] germplasm based on reaction to stem borer, *Chilo partellus* (Swinhoe) under artificial infestation.

Tillage

Ploughing after harvest is a cultural practice known to destroy stubble, weeds and other alternate hosts of stem borers. Removing the available host material reduces the potential for carryover of the pest. A reduction in population of *Chilo partellus* and also its carryover to the next crop as a result of off-season tillage. It was suggested that this practice exposed the borers to their natural enemies and to adverse climatic factors such as high temperature and low humidity.

Plant Spacing

Singh (1986) found a positive correlation between the incidence of *C. partellus* and plant population per unit area in sorghum. This observation suggests that higher plant density probably helps in the dispersion of larval population in reducing infestation of *C. partellus* (Sheshu Reddy, 1985). Destruction of central shoots showing early pinhole damage (these contain a large number of young stem borer larvae, ready for dispersal to adjacent plants) has also been to be an effective practice.

Management of Crop Remnants

The over wintering of stem borer larvae in stubble left in the field or in the harvested stems stored for fodder. The carryover of the pest occurred in leftover stubble and stalks. Uprooting and burning of stubbles and chopping of stems prevent its carryover.

Field Sanitation

This practice involves the removal and destruction of crop residues, weeds and nearby wild host plants to eliminate a pest, by destroying its food and shelter. As with tillage, this practice is intended to reduce the ability of insects to carryover to the next season.

Water Management

Soil moisture influences crop damage by insect pests through its effect on plant vigour and growth. It has been shown that sorghum grown under drought stress suffers greater damage from *C. partellus* (Sharma, 1985). In rainfed agriculture, however, there is little scope for manipulating soil moisture except through moisture conservation.

Fertilizer Management

Fertilizer application is known to influence the susceptibility of crops to insects. Fertilizers enhance plant nutrition, which can influence the longevity and fecundity of insects, and the degree of damage. Infestations in grain sorghum of *C. partellus* have been shown to increase with enhanced levels of nitrogen applied to the crop.

Removal of Deadhearts

Removal and destruction of deadhearts in sorghum, if carried out by farmers over large areas, has been found to be a successful practice in East Africa. Field sanitation is practiced as a cultural practice to reduce stem borer infestations. Destruction of crop residues and other potential host carryover material has been widely recommended.

Mulching

Non-treated crop residues are often used to mulch the next sorghum crop in Uganda. Subsequent levels of stem borer infestation have been found to be far higher than normal for the area.

Intercropping

The principle of controlling insect pest populations by increasing the diversity of an agroecosystem. Intercropping has some potential as a cultural method to control stem borer infestation of sorghum. Growing sorghum in association with other crops has been shown to reduce *C. partellus* damage on sorghum, mungbean, urdbean, pigeonpea (Singh and Singh, 1974), cowpea. Intercropping of maize and sorghum without association with a cereal crop gives rise to higher incidence of *C. partellus.* also reported that the incidence of *C. partellus* in trials of sorghum and maize mono-crop and maize/sorghum di-crops was earlier than in trials intercropping these cereals with cowpea, which showed a significant delay in borer colonization.

Biological Control

Apanteles flaviceps and *Sturmiopsis inferens* are promising larval parasitoids. *Tetrastichus ayyari* is a pupal parasitoid, which exercises natural control. In a field trial conducted in 1984, inundative release of *Trichogramma chilonis* @ 0.35 up to 1 lakh/acre were on par with the chemical application in recording shorter length of the stem tunnel, less number of larvae/plant and higher grain yield. 10, 8 and 6 lakhs parasitoids/acre recorded 81.6, 74.5 and 70.3 per cent parasitism, respectively with less incidence of the pest. Similar trend was observed in the subsequent two years when four releases were made at 10, 20, 30 and 40 days after sowing.

Chemical Control

Quantities of insecticides can be reduced to an economic level by integrating their use with resistant varieties and cultural practices (Kishore and Jotwani, 1982). Endosulfan application can be integrated with stem borer resistant varieties to achieve marginal benefits under moderate levels of infestation (Kishore and Govil, 1982 and Kishore 1984). Stem borer of sorghum can be effectively controlled by leaf whorl placement of granular or dust applications of endosulfan, quinalphos, malathion and fenvalerate at 0.600 to 1.2 kg a.i./ha. This application targets the vulnerable stage of the pests, as the larvae move towards leaf whorls after hatching. The insecticide is less affected by rain and its quantity can be reduced through this application method. Hazards of pollution residues and effects on non-target organism can also be avoided. Whorl application of carbofuran 3 G @ 7.5 kg/ha at 25th and 35th days after emergence 2-3 granules per whorl will help in combating the pest. Spraying of endosulfan 35EC @ 2ml/lt of water, twice at an interval of 15 days from a month after sowing is an effective chemical measure. Effective control of the borer can be achieved by application of any of the following insecticides into the whorl *i.e.* endosulfan 4G/4D, malathion 10 D or furadan 3 G @ 8-12 kg/hg at 20 and 35 day after emergence. The treatment should only be given after ascertaining the infestation levels as evidenced by leaf injury symptoms.

3. Pink Borer: *Sesamia inferens* Walker (Noctuidae: Ledpidoptera)

Nature and Symptoms of Damage

The larva bores into the stem and kills the central shoot, causing deadhearts. One larva may attack several plants in its lifetime. The symptoms are similar to that of *Chilo partellus*.

Management

Spraying the crop with endosulfan 35 EC @ 1 litre per hectare at every 20 days interval after germination of the crop is effective in controlling the pest.

4. Oriental Armyworm: *Mythimna separata* (Walker) (Noctuidae: Lepidoptera)

Nature and Symptoms of Damage

The larvae feed on the leaves, leaving only the midrib uneaten. The immature panicles are also damaged. Armyworm attacks are sporadic but, if heavy, an entire crop may be lost.

Management

Cultural control : Plough and burn the stubbles immediately after harvest of the crop. Remove alternate host plants and used hosts to prevent breeding during off-season. Discourage application of high doses of nitrogenous fertilizers. Light traps with mercury vapour lamps attract moths.

Biological Control

Direct dropping of nematode suspension in plant whorls was observed to be superior to spraying of nematode suspension. Dropping of 4 to 6 ml suspension was comparable to larval mortality obtained by application of carbofuran granules. Spray nuclear polyhedral virus (NPV) @ 250 larval equivalents (LE)/ha during evening hours with 10g boric acid powder as adjuvant.

Chemical Control

Monochrotophos bait @ 50 kg/ha is effective in attracting and killing both adults and larvae. Bait must be spread thinly at the base of the plant. (Mix 250 ml monochrotophos 36 SL, 4 kg Jaggery and sufficient water with 5 kg rice bran in large container or bread flat form, bring into crumb structure, ferment for 24 hours). Spray endosulfan 35 EC @ 2ml/lt on 25 and 45 days crop. Apply carbofuran 3G or endosulfan 4G @ 8.0 kg/ha in leaf whorls of 35-40 days crops. Use of promising larval parasitoids, namely, *Apantelis ruficrus, Campoletis chloridae* and *Exorista xanthaspis* can check the incidence. Application of 7.5 kg/ha carbofuran 3G granules or spraying of 0.07 per cent endosulfan 35EC is recommended practices. The fermented poison bait @ 50 kg/ha (50 kg rice bran + 4 kg jaggery + 250 ml monochrotophos + 6 to 8 litres of water and fermented for 48 hours), was found to be most effective in killing 98 per cent larvae and 100 per cent moths present in the sorghum field. This fermented poison bait was instrumental in killing all the moths of armyworm, earhead caterpillar, groundnut defoliator, cutworm, pink borer, castor semilooper, groundnut semilooper, blue butterfly and millipedes present in the sorghum environment thus paving a way out for working out the effective strategies against noxious lepidopterous pests of many field crops.

5. Corn Leaf Aphid: *Rhopalosiphum maidis* (Fitch) (Aphididae: Homoptera)

Nature and Symptoms of Damage

The aphid is most commonly found deep in the whorl of the middle leaf, but also on the underside of leaves, on stems, or in panicles. The young and adults suck the plant juice. This frequently causes yellowish mottling of the leaves. They may cause some marginal leaf necrosis. The aphid produces an abundance of honeydew on which molds grow. In panicles, honeydew may hinder harvesting. The aphid also transmits maize dwarf mosaic virus.

Management

Botanical Control

The *Pongamia pinnata* (seed and leaf, hot water extract of 60 to 85 per cent) and *Vitex nigundo* (leaf extract of 60 to 80 per cent) were found promising against aphid control and were on par with endosulfan 35 EC spray.

Biological Control

Predators namely, ladybird beetles (Coccinellidae) and horse flies (Syrphidae) devour aphids. When infestation is severe, it has to be controlled by spraying endosulfan @ 2ml/lt of water. The higher dosage of 2 lakhs/ha was found significantly superior in reducing the aphid population.

Chemical Control

Dimethoate 35 EC @ (0.05 per cent), chlorpyn 20 EC @ 0.05 per cent were found promising against aphid. Spraying of metasystox 35 EC @ 1 lit/ha in 500 lit of water effectively control aphids.

6. Shoot Bug: *Peregrinus maidis* (Ashmead) (Delphacidae : Homoptera)

Nature and Symptoms of Damage

On sorghum, nymphs and adults suck sap causing the plants to appear unhealthy and yellow. Plant growth becomes stunted. In severe cases of infestation the top leaves start dieing first but leaf death gradually extends to older leaves and some times, death of the whole plant occurs. High infestation at the boot stage may twist the top leaves and prevent the emergence of panicles. Honeydew excretion favours the growth of sooty mold (Subbarayudu, 2002 and Singh and Seetharama, 2008).

Management

Cultural Control

Removal of alternate hosts namely, *Pennisetum typhoides, Sorghum halepanse, Setaria italica, Echinocloa colon* and *Paspalum scrobiculatum* in India. Namba and Higa in 1970 found that removal of *Brachiaria mutica, coix lacryma-jobi, Digitaria decumbens, Pennisetum purpureum* and *Saccharum officinaram* in Hawaii was found to be beneficial.

Biological Control

It has been reported that *Coccinella septumpunctata* and *Menochilus sexmaculatus* are promising naturally occurring bioagents to suppress the pest.

Chemical Control

Application of carbofuran 3G in leaf whorl controls the insect. Dusting of malathion 5 per cent D or quinalphos 1.5 per cent D or endosulfan 4 per cent D or phosalone 5 per cent D @ 25 kg/ha could be followed. Spraying endosulfan 35 EC @ 2 ml/lit over crop canopy is also effective. Application of endosulfan 4 G @ 8 kg/ha in the whorls can effectively check the incidence of the pest.

7. Earhead Bug: *Calocoris angustatus* (Lethiery) (Miridae: Heteroptera)

Nature and Symptoms of Damage

Both nymphs and adults infest the panicles as soon as they emerge from the boot leaf and suck sap from the developing grain. Consequently, grain attacked in an early stage of development is shriveled, reducing crop yield. Older grain shows distinct feeding punctures that reduce the grain quality.

Management

Cultural Control

Early planting reduces the chances of flowering period coinciding with the peak activity of head bugs so as to minimize the bug damage. Grow resistant varieties, *viz.*, SPV-462 and SB-1079 in *Kharif* (Subbarayudu *et al.*, 2002).

Chemical Control

Protect crop at milky stage with application of malathion 5 per cent D or endosulfan 4 per cent D @ 25 kg/ha. Application of endosulfan 4D @ 20kg/ha to panicle once at pre bloom, followed again at milk stage can effectively check the infestation.

8. Sorghum Midge: *Stenodiplosis sorghicola* (Coquillett) (Cecidomyiidae: Diptera)

Nature and Symptoms of Damage

Larvae feeding on the ovary, preventing normal grain development and resulting in a "blasted panicle" cause damage to sorghum.

Management

Cultural Control

Destroy crop residues systematically since hibernating larvae are mainly in the old sorghum heads during winter or dry season. Timely sowing of sorghum will help the crop to escape from the infestation. Enmass sowing with uniform maturity period will escape the pest incidence. Staggered planting results in build up of midge population. In the endemic areas of Karnataka sowing before the end of June helps in escaping the incidence, while August sown crops suffer heavily. Resistant variety, DSV-3 (ICSV – 745), which has been released in Karnataka has less than 10 per cent damage. Effective midge control requires the successful integration of several cultural practices that adversely affect its population density and potential to cause crop damage. Early planting in *Kharif* avoids the peak midge activity during the flowering period. Planting of uniform flowering varieties and hybrids and sowing of early flowering types under delayed sowing, so as to synchronies in a given regions avoids and minimizes the risk of heavy population build up. Prevention of carryover of diapausing larvae, by destruction of chaffy panicles and earhead residues, as well as removal of Johnson grass and grassy types, which serve as collateral hosts are highly beneficial.

Biological Control

Tetrastichus diplosidus and *Eupelmus popa* are good natural enemies. Release and recovery studies with reference to *Tetrastichus* and *Orius* were made in 1984 varying populations of midge. Twentyfive parasitoids of *Tetrastichus* per earhead were sufficient to cause maximum parasitization. Four adults of *Orius* were able to kill 36 ovipositing females (Hiremath *et al.*, 1998).

Chemical Control

Dust malathion 5 per cent D or endosulfan 4 per cent D @ 25kg/ha at 90 per cent earhead emergence stage. Two to three applications of 0.03 per cent per cent endosulfan 35 EC @ 700 litre per hectare at intervals of seven days commencing at 90 percent head emergence afford good protection. Chemical control measures are needed under high levels of midge incidence applications on the earheads at 90 per cent emergence (period to flower opening). The treatments should only be given after inspecting the incidence of midge on flowering panicles during early morning or late in the evening.

9. Bollworm: *Helicoverpa armigera* Hubner (Noctuidae: Lepidoptera)

Nature and Symptoms of Damage

The larva infests the sorghum head where it feeds on the developing grain. It also feeds on the tender whorl leaves.

Management

Biological Control

Trichogramma achaeae and *T. chilonis* (egg parasitoid), *Campoletis chloridae and Bracon kirkapatriki* (larval parasitoids) are promising natural enemies which can be released @ 5, 7.5 and 10 lakhs/ha respectively at 70 days after sowing. Spray with nematode suspension three times from grain formation stage at 10 days interval gave higher mortality of larvae.

Chemical Control

Application of malathion 5 per cent dust @ 20kg/ha prevents the damage due to caterpillars. Spraying of monochrotophos 36EC or endosulfan 35EC (@ 1 lit/ha in 500 lt. of water) effectively check the incidence of earhead worms.

10. Mite: *Olygonychus indicus* Hirst. (Tetranychidae: Acarina)

Nature and Symptoms of Damage

Spider mites suck sap from the underside of leaves, beginning along the midrib of the lower functional leaves. The infested leaf areas are pale yellow initially and later become reddish on the topside of the leaves. The entire leaf may turn brown. As mite numbers increase on the lower, the infestation spreads upwards through the plant. The underside of heavily infested leaves has a dense deposit of fine webbing. In cases of severe infestations the mites may invade and web sorghum panicles. Killed plants may subsequently lodge. Population density increases generally after sorghum head emergence.

Management

Spraying of dicofol @ 0.5ml/lit of water control the mite populations. Spraying of monochrotophos 35 EC also has been found most effective in reducing the mite population.

Insect Pests of Pearl Millet (Bajra) and their Management

The insects noticed on sorghum and finger millet infest this crop also. They include, *Amsacta albistirga, A. moorei, Estigmene lactinea, Sesamia inferens, Chilo partellus, Orthacris simulans, Myllocerus discolour, M viridanus, Eublemma cilicula, Nezara viridula, Dolycoris indicus,* etc. However, the shoot fly, *Atherigona* approximate and the ear midge, *Geromyia penniseti* are important.

Shoot Fly: *Athrigona approximate* (Muscidae: Diptera)

This is a serious pest of bajra (Cumbu) in Tamil Nadu during the cold weather season. it causes 'dead hearts' in young plants and in the bearing stage due to feeding by maggots the set is affected considerably resulting in chaffy thin panicles. The percentage of incidence has been noticed to be 29.9 to 58.6 percent and the grain loss from 23.3 to 47.4 per cent. In the southern district of Tamil Nadu the grubs and adults of the beetle, *Lema downesi* B. (Galerucidae) scrape the green matter of leaves in rainfed crop causing whitening and ultimate drying. This gives a brunt up appearance. In Mandapam block (Ramanathapuram district) of Tamil Nadu the roots of Rainfed crop are damaged by the grubs of the beetle *Arthrodeis* sp. (Tenebrionidae) causing yellowing and gradual death of plants.

Insect Pest of Ragi and their Management

Pink Borer: *Sesamia inference* Wlk. (Noctuiade: Lepidoptera)

It is the important pest of finger millet in India. The larvae hatch out from the eggs inside the leaf sheath. The larva becomes full grown in 25 to 54 days and is pink, brown, smooth and cylindrical with a reddish brown head. A larva may migrate and attack a number of tillers and in a tiller sometimes up to 5 larvae can be met with. It pupates inside the stem and emerges as adult in 8-12 days. The control measures are similar to that of sorghum stem borer (Gadiyappanavar and Channabasavanna 1973)

White Borer: *Saluria inficita* W. (Phycitidae : Lepidoptera)

A specific pest of finger millet in south india and rarely infests tenai and rice. The larva is cream white with yellow head and is usually found at the basal portion of tillers close to soil level. The adult is a medium sized, Dark brown moth with a pale white band along the margin of fore wings and white hind wings. Control is similar to that of the pink borer.

Cutworm: *Spodoptera exigua* (Rb.) (Noctuidae: Lepidoptera)

The insect is widely distributed in India and defoliates the plants especially in the nursery. Apart from finger millet it also infests crops such as onion, brinjal, cotton, cowpea, chillies, safflower, daincha, castor etc. and feed on leaves. The brownish moth with white hind wings lays clusters of eggs on the lower portions of young plants. The larvae hide in the soil during daytime and feed on leaves at night and become full grown in 10 to 16 days. The brownish green larvae with wavy lines on, dorsal surface and yellow stripe laterally, pupate in the soil in earthen cocoons and emerge as adults in 7 to 11 days. Spraying methyl parathion 0.025 percent controls the pest.

Black Hairy Caterpiller: *Estigmene lactinea* G. (Arctiidae: Lepidoptera)

The larva with black head and hairs all over its body feeds on leaves. Its other host plants include *bajra*, sorghum, pulse crops, etc. The adult is large sized creamy white moth with characteristic crimson markings on the head, thorax and,abdomen-

Spraying methyl parathion @ 0.05 per cent or toxaphene for the control of,grown up larvae was found useful.

Flea Beetle: *Chaetocnema pusaensis* M. (Alticidae: Coleoptera)

The flea beetle is dark blue with enlarged hind femur. The adults' bite holes in the leaves of young plants, affect their vigour both in the nursery and in the young transplanted crop. Spraying methyl parathion @ 0.05 per cent for the control of grown up larvae was found useful.

Other Leaf Feeding Insects

Occasionally the beetles of *Monolepta signata* (Galerucidae), *M. viridanus* Fb., nymphs and adults of the wingless *Orthacris Simulans* Bol. (Acrididae) feed on foliage.

Sucking Insects

The plant lice that attack the crop are (i) *Schizaphis graminum* found on the under face of leaves, (ii) *Rhapalosiphum maidis* infesting the central whorls of leaves, and fiil *Hysteroneura setariae* sucking up the sap from leaves and ears. They are controlled by application of methyl demeton @ 0.05 per cent or dimethoate @ 003 per cent. The mealy bug, *Hererococcus rehi* is sometimes found within the leaf sheaths of tillers.

Insects Infesting Ears

The larvae of *Cryptoblabes angusripenllella, Eublemmma silicula* Hb., *Sitotroga cerealella* and *Stathmopoda theoris* Meyr. (Heliodinidae) feed on the grains in ears. The pentatomid bugs, *Dolycons indicus* and *Nezara viridllla,* the rice bug, *Leptocorisa acuta* (Thunb.) and the mirid, *Calocoris angustatus* Leth. suck up the sap from developing grains.

Root Aphid: *Tetraneura nigriahdominalis* (Sasaki)

The pink globular small aphids occur in large numbers and suck up the sap from the roots causing gradual wilting and drying of plants in patches. The black ant, *Camponotus compressus* is seen attending on the aphids for their honeydew excretion and the presence of the ants at the base of tillers is a sure indication of root aphid infested plants.

White Grub: *Holotrichia consanguinea* Blanch (Melolonthidae: Coleoptera)

In a few places in Tamil Nadu during July to September the roots of finger-millet plants are severely damaged by the grubs of the beetle resulting in death of large number of plants and ultimate loss in grain yield. Soil application of carbofuran 3G granules @ 20 kg/ha effectively check the incidence of grubs.

Conclusion

Millets are the low cost crops cultivated in the semiarid tropics. Use of chemicals to manage their insect pests is not cost effective. Though, breeding for insect resistance is the most suitable, economical approach, sources of durable resistance are not available. Alternative approaches for insect management should aim at seed treatment with other components of integrated pest management. This is of great potential interest since these strategies are eco-friendly and stimulate the plants natural defense mechanism and thus limit the further spread. It is hoped that the deployment of emerging novel strategies can successfully bridge the gap between food supply and demand in this century. Adoption of integrated pest management practices for the pests of sorghum and millets significantly reduces the pressure of pesticides load. It is also evidenced from many reports that integration of all the methods in a compatible manner resulted in low damage due to insect pests and also resulted in higher yields. Therefore integrated pest management strategies definitely useful to the farmers, by virtue of its economic viability, ecofriendly and sociological acceptability. In future, integrated pest management practices should be given in a capsule mariner to the farmers for the management of insect pests.

Integrate IPM components and validate their effectiveness for insects of millets management

- ✰ Evaluate beneficial microorganisms such as *Bacillus thuringiensis,* HaNPV, *Metarhizium anisopliae, Beauveria bassiana,* and natural plant products
- ✰ Evaluate IPM modules for management of millet pests
- ✰ Develop technologies for mass production, storage and utilization of microorganisms pathogenic to insect pests
- ✰ Study low-cost agronomic practices for integrated pest management, and develop strategies to manage pests
- ✰ Work with national research and extension agencies and NGOs to learn how farmers view pest
- ✰ Problems, what control options they prefer, and how these can be applied on-farm. This will accelerate technology adoption and adaptation
- ✰ Develop integrated pest modules that serve multiple goals: for example, crop diversification, introduction of potential cash crops, control methods that also increase soil fertility (eg, organic amendments, legume intercrops or rotations).

References

1. Gadiyappanavar, R.D. and Channabasavanna, G.P. 1973. *Mysore Journal of Agricultural Sciences.* **7 (3)** : 436-444.
2. Hiremath, I.G. Bhuti, S.C., Balikai, R.A. and Nandihalli, B.S.1998. *Sorghum Research in North Karnataka.* Manohar Printing Press, Dharwad, pp.46-78.
3. Indira, S. and Subbarayudu, B. 2005. Status and prospects of Integrated Pest Management in Selected Crops : Minor Millets : Text book chapter in the book *"Principles and Application of Integrated Pest Management"* Published by NCIPM, New Delhi and CBS, New Delhi Volume **2** pp. 127-162.
4. Indira S, Rao S V, Ravi Kumar S, Rao B.D, Aruna, C, Ratnavati, C V, Subbarayudu B, Kanna Babu N and Umakanth, A.V. 2005. Jonna Pantaku Asinchu Cheda Purugulu : Samagra Sasya Rakshna Paddatulu (Insect Pests of Sorghum and

Their Integrated Management : Text Book) : *Sorghum Production Technology*: National Research Centre For Sorghum (28th December, 2005). pp. 1-55.

5. Jotwani, M.G. 1980. Present status of field pests of sorghum and millets in India. *Tropical Pest Management*, **26**: 128-151.
6. Kalaisekar, A., Patil, J.V., Shyamprasad, G., Bhagwat, V.R., Padmaja, P.G., Subbarayudu, B., Srinivas Babu and Zeenat Rahman. 2013. Time-lapse tracing of biological events in an endophytic schizophoran fly, *Atherigona soccata* Rondani (Diptera : Muscidae). *Current Science*. **105** (5).
7. Kishore, P.1996.Anew approach to develop shoot fly and stem borer resistant cultivars for insect pest pest management in sorghum through crossing of intermediate in resistance derivatives. *Journal of Entomological Research* : **20 (2)** : 173-175.
8. Mahadevan, N.R. and Chelliah, S. 1986. Influence of intercropping legumes with sorghum on the infestation of the stem borer *Chilo Partellus* (Swinhoe) in Tamil Nadu, India. *Tropical Pest Management*, **32** (2): 162-163.
9. Padmaja, P.G., Madhusudana, R. and Seetharama, N. 2010. Sorghum shoot fly. Directorate of Sorghum Research, Rajendranagar, Hyderabad 500 030, Andhra Pradesh, India. ISBN 81-89335-29-4. 82 pp.
10. Seshu Reddy, K.V. 1985. Integrated approach to the control of sorghum stem borers. *In Proceedings of the International Sorghum Entomology Workshop*, 15-21 July, 1984. College Station, TEXAS, USA. International Crop Research Institute for Semiarid Tropics, Patancheru, Andhra Pradesh -502 324, India, pp.205-215.
11. Sharma, H.C. 1985. Strategies for pest control in sorghum, India. *Tropical Pest Management*, **31 :** 167-185
12. Sharma, H.C. 2006. Integrated pest management research at ICRISAT: present status and future priorities. Patancheru 502 324, Andhra Pradesh, India: International Crops Research Institute for the Semi-Arid Tropics. 48 pp.
13. Shyam Prasad, G; Elangovan, M. Bhagwat, V.R Kalaisekar, A; and Subbarayudu, B. 2010. Assortment of indigenous sorghum [*Sorghum bicolor* (L.) Moench] germplasm based on reaction to stem borer, *Chilo partellus* (Swinhoe) under artificial infestation. *Indian Journal of Plant Protection*. **38 (2)** : 126-130.
14. Singh, S.P. 1986. Screening of forage genotypes for resistance to shoot fly, *Atherigona soccata* (Rondani) and stem borer, *Chilo partellus* (Swinhoe) and estimation of avoidable losses. *Ph.D. Thesis, Haryana Agricultural University, Hissar, Haryana, India*, pp.135.
15. Singh, B. U. and Seetharama,.N. 2008. Host plant interactions of the corn planthopper, *Perigrinus maidis* (Ashmead) in maize and sorghum agroceosystems. *Arthropod- Plant interactions* **2** : 163-196.
16. Subbarayudu, B., Rao, M.H. and Rana, B.S. 2002. Effect of abiotic factors on the incidence of Head bug, *Calocoris angustatus* Lethiery in sorghum. *Journal of Research ANGR Agricultural University* **30** (1) : 8-10.

17. Subbarayudu, B., Indira, S. and Rana, B.S. 2001. Effect of Integrated Pest Management modules on the incidence of Shoot fly, *Atherigona soccata* Rondani in sorghum. *Journal of Research ANGR Agricultural University*, Hyderabad, India **30** (2) : 22-29.
18. Subbarayudu, B.2002. Shoot Bug Incidence on Sorghum in Southern India. International Sorghum and Millets Newsletter. International crops Research Institute for the Semi-Arid Tropics, Patancheru, Hyderabad, India. **43** : 19-20.
19. Subbarayudu,B. and S. Indira 2002. Integrated Pest Management for the shoot fly, *Atherigona soccata* (Rondani) in Sorghum. *Technical Bulletin* published by National Research Centre For Sorghum, Rajendranagar, Hyderabad- 500 030, India, pp. 1-16.
20. Subbarayudu, B. and Indira, S.2007. Integrated Pest Management for the shoot fly, *Atheriona soccata* (Rondani) in sorghum (*Sorghum bicolor*) in Andhra Pradesh. *Indian Journal of Agricultural Sciences*. 77 (1) : 51-54.

Part IV

Molecular Approaches

Millets: Ensuring Climate Resilience and Nutritional Security (2015) *Pages* **493–515**
Editors: **Vilas A. Tonapi and J.V. Patil**
Published by: **DAYA PUBLISHING HOUSE, NEW DELHI**

Chapter 18

Molecular Marker-Aided Approaches for Improvement of Millets

R. Madhusudhana, D. Balakrishna and B. Venkatesh Bhat

ICAR-Indian Institute of Millets Research (IIMR) Rajendranagar, Hyderabad – 500 030

For centuries, sorghum and millets have been important staples in the semi-arid tropics of Asia and Africa. These crops are still the principal sources of energy, protein, vitamins and minerals for millions of the poorest people in these regions; these "coarse" cereals which in fact are "nutritious" cereals. They perform well in harsh environments where other crops grow or yield poorly. The growers in developing countries are a multitude of smallholder farmers including India. They are grown in traditional agricultural systems with limited water resources and usually without application of any fertilizers or other purchased inputs. Sorghum or great millet (*Sorghum bicolor* (L.) Moench), pearl millet (*Pennisetum glaucum* (L.) R. Br.) and finger millet (*Eleusine coracana* (L.) Gaertn.) are the three major millet crops cultivated in larger areas of India.

The progress of plant breeding in India has been phenomenal during the last five decades. The successful adoption of improved varieties along with improved cultivation practices helped to rapidly increase food grain production, from 50 million tonnes to more than 200 million tones during the above period. However, much of this increase (155 mt out of 200 mt) has come from two major cereals, rice and wheat; the contribution of other crops is quite low in spite of the fact that high yielding widely adapted cultivars were also released for the nutritious cereals. The high yielding varieties have made more impact in high external input areas that are mostly irrigated. In low external input areas (dryland), the impact of improved cultivars on crop productivity has been limited. Sorghum and millets fall in the latter category.

Uncontrolled population growth leading to conversion of agricultural land for other purposes, urbanization and government policies have pushed sorghum and millet cultivation into drier, more marginal lands. However, with demographic and agro-climatic transitions and the brighter prospects for alternate and new end-uses for sorghum and millets using new technology can perhaps help realize the importance of these crops for national economy where livelihood in dryland, nutrition, fodder and resource conservation (mainly water) are factors of immediate importance.

Genetic improvement of millets has progressed over the last several decades with impressive gains in productivity and adaptability of high-yielding cultivars. However, the traditional approaches have been slow in addressing the major loss causing or yield-destabilizing traits such as susceptibility to insects and diseases and drought stress. This scenario was mainly due to the complex nature of these traits which are greatly influenced by environment and the absence of reliable genetic clues to manipulate them using the tools and know-how available with the erstwhile breeding programmes.

Sorghum is a major food and fodder crop for dryland farmers in India and elsewhere. One of the major reasons for lower yields in this crop is due to the problem of insect pests such as stem borer and shoot fly, and diseases such as grain mold and ergot, drought. Similarly, in pearl millet, the important losses of yields are mainly due to fungal diseases such as downy mildew, ergot and rust. One pearl millet hybrid, HB3, which was adopted widely in India during 1960s was devastated by downy mildew and had to be withdrawn from cultivation. Blast caused by the fungus *Magnoporthe grisea* is the most important disease of finger millet. In India, the disease appears regularly every year causing annual yield losses as high as 50 per cent in disease endemic areas and in favourable seasons. Classical breeding for blast resistance has not been rewarding, apparently due to the polygenic control of host resistance, significant environmental influence and the high level of variability present within the pathogen.

The advent of molecular technologies for the mapping and manipulating of complex genomes now permit wholly new approaches to the improvement of millets. While the conventional breeding has much to contribute to meet the growing demand, it is also essential to take full advantage of biotechnology to enhance the efficiency of the crop improvement programs. Combining marker-assisted selection methods with conventional breeding schemes can increase the overall selection gain and, therefore, the efficiency of a breeding programme.

Marker-assisted selection (MAS) is based upon the localization and estimation of the effects of QTL in segregating populations. MAS involves three main steps (i) mapping the genetic variation for a targeted trait, (ii) QTL mapping to identify and locate the markers that vary with the trait on the linkage map, and finally (iii) use of markers in breeding program to track the accumulation of desired segments of the genome into desired cultivars. For MAS to succeed, we require critical information on component traits, accurate phenotyping, the identification of candidate genes and quantitative trait loci (QTL), the relationship between QTL and genes, the contribution of individual QTL to the phenotype, and their variability across different locations and different crop seasons and even under different genetic backgrounds. Several molecular markers are available in millets all of which are essentially based on DNA polymorphism.

Sorghum

Genetic Linkage Maps of Sorghum

Construction of a linkage map is the most fundamental step required for a detailed genetic study and marker-assisted breeding approach. Sorghum genome mapping based on DNA markers began in the early 1990s, and since then several genetic maps of sorghum have been constructed (Table 1). Initially the genetic maps of sorghum were based largely on DNA probes previously mapped in maize genome (Hulbert *et al.,* 1990; Pereira *et al.,* 1994). Chittenden *et al.* (1994) used sorghum genomic DNA probes to prepare a complete map of 10 linkage groups. Tao *et al.* (1998) published a map using maize and sugarcane DNA and cDNA probes, SSRs and cereal anchor probes.

The maps with common markers across sorghum populations allow for comparison and facilitate the choice of markers that are evenly distributed across the genome, or specific to certain genomic regions. Using a common RIL population, Subudhi and Nguyen (2000) aligned the 10 linkage groups of all major sorghum RFLP maps based on sorghum probes used in earlier maps besides using maize probes and cereal anchor probes. Haussmann *et al.* (2002) aligned the markers mapped in their populations with other published maps by employing previously mapped markers as anchors. By comparing the map positions of the common grass markers in all sorghum maps reported, Menz *et al.* (2002) determined that these reference markers were essentially collinear in all published maps.

Identification of Markers for Important Economic Traits in Sorghum

Molecular markers and quantitative trait loci (QTL) mapping offer plant breeders a more efficient approach for working with quantitative traits, whether or not specific genes or gene products are known (Tanksley *et al., 1989*). A number of molecular markers have been tagged to oligogenic and polygenic-controlled traits of importance in sorghum (Tables 2 and 3). While the markers for the oligogenic traits can be almost readily used in MAS if the populations contain them, markers for most of the polygenic traits controlled by QTL are in the process of validation by testing against different genetic backgrounds and diverse environments. Molecular markers linked to disease resistance, tolerance to environmental stress, leaf phenotypic traits, seed and panicle characteristics, grain quality, yield components, and plant status (height, flowering, maturity, tillering) have been identified. Depending on their relative effects and position, many QTL could be targeted for MAS.

Table 1: Overview of Published Sorghum Genetic Linkage Maps

Reference	*Size and Type of Mapping Population*	*Parent Lines*	*Mapped Markers*	*Linkage Groups*	*Genome Length [cM]*[a]
Hulbert *et al.*, 1990	55 F_2 plants	Shanqui Red (koaliang from China) M91051 (zera zera from East Africa)	37 RFLP maize probes or cloned genes	8	283 R
Binelli *et al.*, 1992	149 F_2 plants	IS 18729 (caudatum-bicolor from Texas) IS 24756 (durra-caudatum from Nigeria)	21 RFLP maize probes	5	440 U
Whitkus *et al.*, 1992	81 F_2 plants	IS2482C (*S. bicolor* ssp. *bicolor*) IS18809 (*S. bicolor* ssp. *arundinaceum*)	85 RFLP maize probes 7 Isoenzymes	13	949 H
Berhan *et al.*, 1993	55 F_2 plants	Same parents as Hulbert *et al.*, 1990	96 RFLP maize probes or cloned genes	15	709 R
Chittenden *et al.*, 1994	56 F_2 plants	BTx 623 (*S. bicolor*) Unnamed accession of *S. propinquum*	256 RFLP sorghum probes 20 RFLP probes of maize, rice or oat	10	1445 U
Pereira *et al.*, 1994	78 F_2 plants	CK 60 (*S. bicolor* ssp. *bicolor*) PI229828 (*S. bicolor* ssp. *drummondii*)	191 RFLP maize probes 10 RFLP sorghum probes	10	1530 U
Xu *et al.*, 1994	50 F_2 plants	BTx 623 (zera zera × kafir) IS 3620C (guinea line)	179 RFLP sorghum probes 11 RFLP maize probes	14	1789 K
Ragab *et al.*, 1994	93 $F_{2:3}$ families	BSC 35BTx 631	38 RFLP sorghum probes 33 RFLP maize probes	15	633 H
Lin *et al.*, 1995	370 F_2 plants	Same parents as Chittenden *et al.*, 1994	78 RFLP sorghum probes 124 RFLP loci mapped by Chittenden *et al.*, 1994	11	935 K
Tuinstra *et al.*, 1996, 1997	98 RIL ($F_{5:7\text{-}8}$)	TX 7078B35 (durra sorghum from Ethiopia)	150 RAPD20 RFLP probes from maize or sorghum	17	ca. 1580 R
Dufour *et al.*, 1997	110 RIL (F_5)	IS 2807 (caudatum from Zimbabwe) 379 (guinea from South Africa)	126 RFLP maize probes 19 RFLP sugarcane probes 4 cloned genes, 2 morphological markers	13	977 H

Contd...

Table 1–*Contd...*

Reference	*Size and Type of Mapping Population*	*Parent Lines*	*Mapped Markers*	*Linkage Groups*	*Genome Length [cM][a]*
	91 RIL (F_5)	IS 2807 (caudatum from Zimbabwe) 249 (guinea from Burkina Faso)	115 RFLP maize probes 8 RFLP sugarcane probes 4 cloned genes, 1 morphological marker	12	878 H
	Composite map of the two populations		164 RFLP maize probes 19 RFLP sugarcane probes 3 cloned genes, 2 morphological markers	13	1095 H
Taramino *et al.*, 1997	68 F_2 plants	Same parents as Pereira *et al.*, 1994	7 SSR added to map of Pereira *et al.*, 1994	10	1575 U
Ming *et al.*, 1998	56 F_2 plants	Same parents as Chittenden *et al.*, 1994	328 RFLP probes of sorghum and other cereals	10	1750 K
Tao *et al.*, 1998	120 RIL (F_5)	QL39QL41 (both elite lines from Australia)	68 RFLP sorghum probes 87 RFLP of maize, sugarcane, rice, oat, barley 8 SSR, 3 morphological traits	21	1400 U
Boivin *et al.*, 1999	Composite map with populations as Dufour *et al.*, 1997		343 RFLP and morphological markers	11	1352 H
	110 RIL (F_5)	First population of Dufour *et al.*, 1997	128 RFLP sorghum, rice, oat or barley probes 176 AFLP151 RFLP loci mapped by Dufour *et al.*, 1997	11	1899 H
Crasta *et al.*, 1999	96 RIL ($F_{6:7}$)	B35 (durra from Ethiopia)Tx 430	142 RFLP sorghum, maize, rice or wheat clones	14	1602 K
Peng *et al.*, 1999	137 RIL ($F_{6\text{-}8}$)	Same parents as Xu *et al.*, 1994	323 RFLP from sorghum, maize, rice, oat, barley	10	1347 K
Bennetzen *et al.*, 2000	Framework map derived from comparison of the maps of Kong *et al.*, 2000, Peng *et al.*, 1999, Pereira *et al.*, 1994, and Berhan *et al.*, 1993.		154 RFLP34 SSR, 10 morphological markers	10	1450 U

Contd...

Table 1–*Contd...*

Reference	*Size and Type of Mapping Population*	*Parent Lines*	*Mapped Markers*	*Linkage Groups*	*Genome Length [cM][a]*
Bowers *et al.*, 2000	65 F_2 plants	Same parents as Chittenden *et al.*, 1994	2399 Loci based on 1925 RFLP probes	10	1200 U
Kong *et al.*, 2000	137 RIL (F_{6-8})	Same parents as Xu *et al.*, 1994	33 SSR111 RFLP probes from Peng *et al.*, 1999	10	1287 K
Tao *et al.*, 2000	152 RIL (F_5)	Same parents as Tao *et al.*, 1998	101 RFLP17 SSR166 markers mapped by Tao *et al.*, 1998	14	1871 U
Xu *et al.*, 2000	98 RIL (F_7)	B 35 (durra from Ethiopia) Tx 7000 (elite public line used in USA)	162 RFLP sorghum and maize probes, cloned genes or sequenced DNA probes	10	837 H
Bhattramakki *et al.*, 2000	137 RIL (F_{6-8})	Same parents as Xu *et al.*, 1994	116 SSR 354 RFLP or SSR markers mapped by Peng *et al.*, 1999 or Kong *et al.*, 2000	10	1406 K
Klein *et al.*, 2001b	125 RIL (F_5)	RTx 430 (elite line from USA) Sureño	44 SSR 85 AFLP, 1 morphological trait	10	970 K
Haussmann *et al.*, 2002	225 RIL ($F_{3:5}$)	IS 9830 (caudatum line from Sudan) E 36-1 (Ethiopian guinea-caudatum)	125 AFLP 45 SSR, 14 RFLP, 3 RAPD	10	1265 H
	226 RIL ($F_{3:5}$)	N 13 (Indian durra) E 36-1 (Ethiopian guinea-caudatum)	158 AFLP54 SSR, 16 RFLP	12	1410 H
	Composite map of 2 populations		339 AFLP, SSR, RFLP and RAPD markers	11	1424 H
Menz *et al.*, 2002	137 RIL (F_{6-8})	BTx623 x IS 3620C	203 cDNA, 2454 AFLP, 136 SSR markers	10	1713 K

[a] H, K: Map distances estimated using the mapping functions of Haldane (1919) and Kosambi (1944), respectively; U: Mapping function not specified; R: Recombination frequency (per cent).

Table 2: Molecular Markers Associated with Monogenic and Oligogenic Traits in Sorghum

Molecular Marker	*Probe/primer*	*Associated Trait*	*Reference*
RFLP	pSBTXS560 and pSbTXS 1294	Resistance to head smut	Oh *et al.* (1994)
RAPD	OPG 5	Resistance to head smut	Oh *et al.* (1994)
RAPD	OPL 20a and OPL20b	Plant height	Zhang *et al.* (1994)
RFLP	PSB 766 and PSB 195	Shattering	Paterson *et al.* (1995)
RFLP	TXS 713	Reaction to organophosphate insecticides	Toure *et al.* (1997); Tao *et al.* (1998)
RFLP	SSCIR 203	Awn	Tao *et al.* (1998)
RFLP	TXS 636	Mesocarp thickness	Tao *et al.* (1998)
RFLP	CSU6 and UMC 34	Juicy midrib	Xu *et al.* (2000)
RFLP	UMC 44	Red coleoptile	Xu *et al.* (2000)
RFLP	TXS 584	Red pericarp	Xu *et al.* (2000)
RAPD	OPD 12	Resistance to leaf blight	Boora *et al.* (1999)
AFLP	*Xtxa2582*	Pollen fertility restoration	Klein *et al.* (2001a)
SSR	*Xtxp18* and *Xtxp250*	Pollen fertility restoration	Klein *et al.* (2001a)
RAPD	OPI 16 and OPD 12	Resistance to Anthracnose	Panday *et al.* (2002)
STS/CAPS	LW7, LW8 and LW9	Pollen fertility restoration	Wen *et al.* (2002)

Table 3: Summary of Quantitative Trait Loci Studies in Sorghum

Trait	*Number of QTL*	*Phenotypic Variation Explained (per cent)*	*Effects of Individual QTLs (Range, in per cent)*	*Number of Linkage Groups*	*Reference*
Plant Height	6	71.0		5	Lin *et al.* (1995)
	1	54.8		1	Lin *et al.* (1995)
	4	63.4	9.2-28.7	4; orthologous to maize	Pereira and Lee (1995)
	2	67.3			Rami *et al.* (1998)
	2	29.6-35.0	11.8-17.8	2	Klein *et al.* (2001b)
	5	52.0-65.8	7.6-40.1	4	Hart *et al.* (2001)
Tallest Basal tiller height	2	29.9-30.5	10.2-19.7	2	Hart *et al.* (2001)
Tiller Number	4	23.7			Paterson *et al.* (1995)
	3	57.0-70.8	23.0-39.0	2	Hart *et al.* (2001)
Tillers with heads	4	48.9-86.3	9.7-48.7	3	Hart *et al.* (2001)
Photoperiod	8-10	–	13.6-18.0	3	Chantereau *et al.* (2001)
Flowering	3	85.7			Lin *et al.* (1995)
	2	8.5-10.94	5-10	2	Kebede *et al.* (2001)
Maturity	2	57.0	17-40	2	Crasta *et al.* (1999)
	3	8.2-33.1	8.2-22.0		Hart *et al.* (2001)
Pre-flowering stress tolerance	6	14.0-43.0			Tuinstra *et al.* (1996)
	4	25.7	11.9-22.2	3	Kebede *et al.*(2001)
Stay Green (Post-flowering stress	6		2 Major	6	Tuinstra *et al.* (1997)
tolerance)	7	63.0	3 major QTL- 42.0 4 minor QTL- 25.0	6	Crasta *et al.*(1999)
	2-3	–	10.3-15.3	3	Tao *et al.* (2000)

Contd...

Table 3–*Contd...*

Trait	*Number of QTL*	*Phenotypic Variation Explained (per cent)*	*Effects of Individual QTLs (Range, in per cent)*	*Number of Linkage Groups*	*Reference*
	4	30-46		3	Xu *et al.* (2000)
	4	30-54	9.1-29.2	3; Confirm Xu *et al.* (2000)	Subudhi *et al.* (2000)
	9	15.5-26.1	10.2-15.5	7; 3 QTL confirm earlier studies	Kebede *et al.*(2001)
	5-8	31-42	5-26	8; 3 QTL consistent across genotypes and years	Haussmann *et al.* (2003)
Chlorophyll content	3		21-32	2; all coincide with 3 of 4 stay-green QTL	Xu *et al.* (2000)
Leaf angle	3	35.8-64.5	7.4-45.3	3	Hart *et al.* (2001)
Lodging tolerance	3	17.8	14.6-19.1	3	Kebede *et al.*(2001)
Striga resistance	2-8	27-94	20-76	8	Haussmann *et al.* (2003)
Molds during seed germination (per cent)	3	33.8			Rami *et al.* (1998)
Molds after harvest	1	28.3			Rami *et al.* (1998)
Grain mold	2	20.5-37	10.0-23.6	2	Klein *et al.* (2001b)
Anthracnose	1	55.5	55.5	1	Klein *et al.* (2001b)
Zonal leaf spot	1	32.1	32.1	1	Klein *et al.* (2001b)
Bacterial leaf stripe	1	32.8	32.8	1	Klein *et al.* (2001b)
Oval leaf spot	1			1	Klein *et al.* (2001b)
Rust resistance	4		6.8-42.6	4	Tao *et al.* (1998)
Shootfly resistance	8		34-46.5	4	Sajjanar (2002)

Contd...

Table 3–*Contd...*

Trait	*Number of QTL*	*Phenotypic Variation Explained (per cent)*	*Effects of Individual QTLs (Range, in per cent)*	*Number of Linkage Groups*	*Reference*
Midge resistance - antixenosis	2	27.0	12.0-15.0	2	Tao *et al.* (2003)
Midge resistance -antibiosis	1	34.5	34.5	1	Tao *et al.* (2003)
Green bug resistance and tolerance	9	53-79	5.6-38.4	7	Agrama *et al.* (2002)
Pre-harvest sprouting resistance	2	83	53	2	Lijavetzsky *et al.* (2000)
Panicle length	6	70.0	9-25		Pereira *et al.*(1995)
	4	54.1			Rami *et al.* (1998)
	3	20.6-31.6	8.6-20.6	3	Hart *et al.* (2001)
Panicle compactness	3	40.3			Rami *et al.* (1998)
Panicle width	7	47.9-85.9	9.5-32.9	5	Hart *et al.* (2001)
Seed branch length	5	69.0	8-37		Pereira *et al.*(1995)
Length of sterile portion of seed branch	2	34.0	17 each		Pereira *et al.*(1995)
Peduncle length	2	32-44	10.9-32.7	2	Klein *et al.* (2001b)
Peduncle diameter	6	43.0	7-10		Pereira *et al.*(1995)
Awn length	1	6.7	6.7	1	Hart *et al.* (2001)
Number of seed branches/panicle	3	30.0	9-14		Pereira *et al.*(1995)
Number of kernels/panicle	1	12.3		1	Rami *et al.* (1998)
Kernel weight/panicle	1	31.3		1	Rami *et al.* (1998)
Seed size	9	51.7	5.3-11.9	8	Paterson *et al.* (1995)
Seed number	4	19.1	4.2-6.8	4	Paterson *et al.* (1995)
100-seed weight	3	28.0	9-16		Pereira *et al.* (1995)
1000-kernel weight	1	35.2		1	Rami *et al.* (1998)

Contd...

Table 3–*Contd...*

Trait	*Number of QTL*	*Phenotypic Variation Explained (per cent)*	*Effects of Individual QTLs (Range, in per cent)*	*Number of Linkage Groups*	*Reference*
Threshing percentage	1	15.3		1	Rami *et al.* (1998)
Dehulling yield (per cent)	3	43.0			Rami *et al.* (1998)
Kernel flouriness	2	57.1			Rami *et al.* (1998)
Kernel friability	1	13.7		1	Rami *et al.* (1998)
Kernel hardness	4	44.8			Rami *et al.* (1998)
Amylose content (per cent)	2	38.9			Rami *et al.* (1998)
Protein content(per cent)	2	26.0			Rami *et al.* (1998)
Lipid content (per cent)	1	14.7		1	Rami *et al.* (1998)
Germination rate	2	43.0			Rami *et al.* (1998)
Rhizomatousness	3	21.8			Paterson *et al.* (1995)
Ratooning ability	6	29.9			Paterson *et al.* (1995)
Grain mill hardness	2	29.7	14.4-15.3	2	Klein *et al.* (2001b)

Pearl Millet

Markers and Genetic Linkage Maps of Pearl Millet

The first molecular marker-based genetic linkage map of pearl millet was built with restriction fragment–length polymorphism (RFLPs), the marker system of choice in the early 1990's (Liu *et al.*, 1994). This map has served as the base for subsequent pearl millet marker-based studies. Genetic maps in pearl millet were constructed principally using 4 mapping populations (Table 4). A total of 597 loci were mapped over the four crosses. The crosses with LDG-1-B-10 × ICMP 85410 and 81B x ICMP 451 were mapped most extensively with RFLP markers while SSR markers were mainly mapped in the latter two crosses and in the 81B x ICMP 451 population. In total, 13 loci were mapped in all four crosses, 37 in three of the crosses, 66 in two crosses and 302 loci were mapped in only one cross.

Table 4: Overview of Pearl Millet Genetic Linkage Maps

Reference	*Size and Type of Mapping Population*	*Parent Lines*	*Mapped Markers*	*Linkage Groups*	*Genome Length [cM]*[a]
Liu *et al.*, 1994 Allouis *et al.*, 2001	133 F_2 plants	LDG-1-B-10 × ICMP 85410	242	7	473
Hash *et al.* (unpublished)	157 F_2 plants	81B x ICMP 451	198	7	442
Yadav *et al.*, 2004	149 F_2 plants	ICMB 841 x 863B	89	7	296
	131 F_2 plants	PT 732B x P1449-2	68	7	394
Gale *et al.*, 2001. Consensus map		Different mapping populations	418	7	

An interesting feature of the genetic maps of pearl millet is the presence of large gaps in the distal chromosome regions. It has been reasoned that the large gaps in the distal regions of the genetic maps indeed represent regions of high recombination, rather than being caused by a lack of markers (Qi *et al.*, 2004). Two of the pearl millet linkage (LGs 3 and 5) groups are still extremely short of around 30-40 cM. Using comparative maps with rice as a guide, it would appear that these linkage groups are relatively complete (Devos *et al.*, 2000). Two putative duplications have been detected in pearl millet involving the distal regions of LGs 1 and 4 and have been found to be orthologous to the duplication between the short arms of rice chromosomes 11 and 12 (Devos *et al.*, 2000).

With the development of polymerase chain reaction (PCR) that allow the rapid and inexpensive amplification of small quantities of DNA precisely targeted to know regions, the microsatellite markers (simple sequence repeats-SSRs) were developed (Allouis *et al.*, 2001; Qi *et al.*, 2001, Budak *et al.*, 2003; Qi *et al.*, 2004) (Table 5). Some 111 SSR markers are now available in pearl millet as DNA primers or as DNA sequences. The pearl millet programme aims to develop another 100 SSR markers (Gale *et al.*, 2001), and also the use of new marker system based on SNPs (Bertin *et al.*, 2005).

Table 5: Markers Development in Pearl Millet

Sl.No.	Marker Type	No.	Reference
1	RFLPs	500	
2	SSRs	25	Allouis *et al.*, 2001
2	SSRs	24	Qi *et al.*, 2001
3	SSRs	18	Budak *et al.*, 2003
4	SSRs	44	Qi *et al.*, 2004
4	SNP		Bertin *et al.*, 2005

Table 6: Summary of Quantitative Trait Loci Studies in Pearl Millet

Trait	Number of QTL	Phenotypic Variation Explained (per cent)	Effects of individual QTLs (Range, in per cent)	Linkage Groups
Downy mildew*	9	22-65.5	4.1-40.0	1, 2, 4, 6, 7
	2	16-27	10-60	1, 2
Flowering time$	9	62.9-72.7	12.4-41.7	2, 3, 4, 6
	3			4, 5, 6
Panicle number$	6	23.8-58.9	10-33.1	2, 4, 6, 7
	4			2, 4, 6
100-grain mass$	4	29.6-40.8	11.3-23.1	2, 6, 7
Panicle grain number$	9	48.5-53.7	9.0-49.6	1, 2, 6, 7
Grain yield$	2	9.3-10.8	9.3-10.8	1, 2
	3			2, 4, 6
Stover yield$	8	34.3-56.0	11.4-39.8	2, 3, 4, 6
	3			2, 4, 6
Biomass$	6	8.4-43.1	8.4-23.1	2,4, 6
	3			1, 2, 6, 7
Harvest index$	6	11.3-49.4	11.3-26.7	2, 4, 6, 7
	4			
100-grain mass	3	11.5-21.5	11.5-16.1	2, 6
Panicle number	1	15.3	15.3	2
Panicle grain number	1	10.3	10.3	1
Grain yield	3	13.3-23.0	13.3-23.0	1, 2
Stover yield	3	10.6-28.1	10.6-13.2	2, 4, 6
Biomass	2	10.7-16.2	10.7-16.2	2, 6
Harvest index	3	10.1-18.0	10.1-18.0	2
Domestication syndrome#				2, 5, 6, 7

Source: *Jones *et al.*, 1995 and 2002; $Yadav *et al.*, 2002 and 2003; #Poncet *et al.*, 2000 and 2002.

Progress in the Identification of Molecular Markers in Pearl Millet

Compared to sorghum, the work on mapping QTLs in pearl millet is less. Only few studies are available linking molecular markers with agronomically important traits (Table 6). Among the target traits identified for improvement of pearl millet are breeding for durable resistance to downy mildew and enhanced drought tolerance.

Genetic Control of Domestication Traits

Quantitative trait loci (QTLs) controlling the morphological differences between pearl millet (*Pennisetum glaucum* ssp. *glaucum*) and its wild ancestor (*Pennisetum glaucum ssp. monodii,* form *mollissimum*) indicated that the most critical adaptive changes resulting from the domestication process involved the spikelet structure: non-shedding seeds with reduced bracts and bristles and long involucral pedicel (Poncet *et al.,* 2000). Major differences also concerned characters describing the plant architecture, phenology and spike sizes. Many morphological differences could be attributed to the effect of a small number of loci concentrated on four linkage groups (2, 5, 6 and 7) with relatively large effects. The loss of shedding ability, due to the absence of a functional abscission layer, is controlled by a single locus on linkage group 6 (al6). Genetic control of the other spikelet traits involved factors with large effects which are located in the region of linkage group 6 close to al6 and to an esterase gene, Esterase-E. Moreover, QTLs with large effects on plant and spike morphology traits such as plant height, number of spikes and weight of the spike were also mapped on linkage groups 6 and 7. This strong linkage of factors in the domestication syndrome may be involved in the maintenance of the phenotypic identity of wild and cultivated populations.

Finger Millet

Markers and Genetic Linkage Maps in Finger Millet

In finger millet, there are no published genetic linkage maps available at present. The work on development of mapping populations and molecular markers is going on at University of Agricultural Sciences (UAS), Bangalore, India in collaboration with University of Georgia, USA. Five mapping populations have been developed (Table 7) targeting mapping of drought and blast disease, the most important traits in

Table 7: Available Mapping Population in Finger Millet (Shailaja *et al.,* 2003, under publication)

Reference	*Size and Type of Mapping Population*	*Parent Lines*	*Linkage Groups*
Shailaja *et al.,* 2003, under publication	220 RILs (F_{12})	Indaf5 x IE1012	–
	850 RILs(F_{12})	IE1012 x INDAF5	–
	165 RILs(F_{12})	IE-2885 x INDAF-9	Progressing
	152 RILs(F_{12})	IE 2885 x IE2912	–
	174RILs(F_{12})	GPU 28 x INDAF 9	A skeleton map is being developed

addition to other agronomically important characters (Shailaja *et al.*, 2003. under publication). In all populations the parental polymorphism was tested with RAPD, SSRs and candidate genes. Similarly, there are no published data on the molecular markers developed and utilized in finger millet. However, the work at UAS, Bangalore and University of Georgia, USA has led to the development of 315 SSR markers, 20 candidate genes, 15 novel marker types (Shailaja *et al.*, 2003, unpublished data).

Comparative Genome Mapping

Comparative genome mapping (CGM) of major crops have shown extensive synteny (conservation of genetic linkages) and co-linearity (conservation of gene order). This facilitates cross-referencing genetic information across groups of diverse species such as major cereals like rice, barley, wheat, maize, and sorghum, and millets (Paterson *et al.*, 1995; Devos and Gale, 1997). CGM helps in easy isolation of genes based on map-positions in other related species and also provides alternate allelic forms for transgenic breeding. This will be of increasing importance, as we want to decipher the action of QTL and their interactions for practical crop improvement. In this regard, the dense annotated maps and genome sequence in rice will play a crucial role in our research approach to other cereals like sorghum and millets. Considering the syntenic relationships among different cereals, it is desirable to use a set of the same or similar anchor probes in different crops so that opportunities to examine the relevance of selected genome regions in one crop can be readily tested in others. Comparative mapping has shown that two stay green QTLs identified in sorghum corresponded to stay green QTL regions in maize (Kebede *et al.*, 2001). Four QTL clusters controlling plant height and one QTL controlling flowering in sugarcane corresponded closely to four of the six plant-height QTLs (Ming *et al.*, 1998) and one of three flowering QTLs of sorghum respectively (Dufour *et al.*, 1997).

A comparative analysis of the foxtail millet genome, pearl millet, and also rice revealed a simple relationship between the chromosomes. Despite the close taxonomic relationship of pearl millet and foxtail millet, their genomes were highly rearranged. The larger pearl millet genome appears to have undergone many rearrangements relative to foxtail millet and rice although gene orders have remained conserved within each of the translocated segments (Devos *et al.*, 2000). Most of these arrangements are likely to be specific to pearl millet. The comparative data further demonstrated the presence of a major duplication between regions of linkage group 1 and 4 in pearl millet which were also found in foxtail millet and rice. The integrated maps can help in range of applications including, identification of candidate genes, gene prediction, fine-mapping, and elucidation of metabolic pathways.

Gene Pyramiding through MAS

Initial applications of MAS in sorghum may be on a selective basis and may even be alternated with empirical selection across segregating generations as in barley (Han *et al.*, 1997). With the current rate of progress in gene cloning, it will be possible to isolate QTL across crops and test these with suitable systems, and also examine allelic diversity for specific gene sequences relevant for the improvement of millets. In the long run, the identification of QTL for economically important complex traits will enable us to re-evaluate germplasm for positive and most effective alleles, and pyramid these alleles. The materials and the data developed will help us to better understand physiological and developmental phenomena in millets. Once the markers underlying the tolerance to stresses and improved agronomic characters are identified, they can be pyramided into the commercial or locally accepted cultivars. In *Rabi* sorghums, traits such as drought tolerance, resistance to charcoal rot and insects can be pyramided. Similarly, in pearl millet, resistance to downy mildew and drought tolerance at field level are desirable in the locally adapted varieties.

Practicality of MAS

With respect to any developing country program, even the convention research itself is quite resource intensive. Application of MAS would require far more resources not only for genotyping, but also for phenotyping as well. Fortunately, the cumulative experience in biotechnology will be more readily transferable than from the conventional breeding. Undoubtedly, new approaches for crop improvement are needed for complex stresses, as the progress with the conventional approach has been very slow, and mostly incidental. Therefore, we need to develop a suitable strategy to deal with this. Following are the some of the considerations that deserve serious attention.

- ☆ Choose well-characterized target sites, plant traits (or operational traits), and at least a few different genetic backgrounds for comparison. To make MAS a reality, one could consider incorporating several QTL simultaneously. As a first attempt, the best subset of QTL for different traits as suggested for maize (Ribaut *et al.*, 1997), could be used.
- ☆ For the production F_1 hybrids QTL mapping results obtained with different testers may be examined in the near future. (However, results with maize indicate that the consistency of QTL results across testers is disappointingly low; Lubberstedt *et al.*, 1997).
- ☆ Club related objectives to set practical milestones and goals. For example, terminal drought resistance, biomass and grain yield, stover quality and lodging resistance can all be targeted together such that gain for any one trait will have practical impact on sorghum productivity in the target region and beyond.
- ☆ Set realistic targets for improvement, and may even tone down expectations, especially because the crop management options in the target regions, especially with reference to Low external input agriculture. An improved downy mildew resistant version of the popular early-maturing pearl millet hybrid HHB 67, has been developed through marker-assisted selection and was released by the Haryana State Varietal Release Committee as HHB67-2.
- ☆ Test-hypotheses using well-characterized target sites, fine-tune new selections to target environment before large-scale adaptation.
- ☆ Develop or participate in consortium and networks to share responsibilities, resources, information and collective thinking to make the needed impact.

- ✰ Open mind to accept other approaches and ideas, and timely move to absorb new and complimentary technologies.

For MAS to succeed in our Indian programmes, we must ensure following

1. Team work, linking skills in various disciplines and in different test locations for accurate phenotype data
2. Rapid genotyping – use of automated machines
3. Skills for database development, data warehousing, and to rapidly analyze and interpret data; sharing information rapidly

At NRCS, we are attempting to build a strong consortium for using MAS for millets. With the imminent publication of sorghum sequence data, more exciting possibilities exist. In pearl millet and finger millet, more SSR markers need to be developed. In the long run, comparative mapping is a feasible proposition to identify usable markers in these crops. Further, use of wild species to introgress alleles for those traits for which the variability is limiting in the cultivated species will be essential.

Outlook

The molecular tools developed will be very useful to sorghum and millet breeders in developing better varieties for grain and fodder purposes. Efforts are needed to enhance the number of common markers in different sorghum and millets maps, to allow for a more comprehensive comparison and integration. The search should continue for new, highly polymorphic SSRs that would map to regions in the sorghum genome which are currently poorly covered with SSRs. Physical mapping is also expected to help comparative genetic analysis for sorghum improvement using MAS (Klein *et al.*, 2000). All the molecular markers essentially constitute indirect approaches towards assessing DNA sequence differences: single nucleotide polymorphisms (SNPs) and insertions/deletions (indels). A direct analysis of sequence difference between many individuals at a large number of loci has now become practical (Antoni, 2002). Dramatic advances in sequencing technology have resulted in the determination of near-complete DNA sequences of the model plants, a*Rabi*dopsis and rice. The next important objective is to determine sequence diversity of genic and regulatory regions in these and other species. This would allow the understanding of the relationship between phenotypic diversity and genetic diversity. The development and applications of SNP genetic markers in rice, maize and other crop plants are contributing towards the understanding of the organization of genetic diversity in plants. Such an understanding is expected to increase the means and effectiveness of allele tagging and gene-pyramiding approaches for markers aided breeding of sorghum and millets.

Another important issue is to develop mapping populations using genetically diverse parents for the important and novel traits. For example, in sorghum sugar accumulation in culm is becoming an important trait with the need for biofuel and other alternate uses. Traits associated with quality of the grain and stover are also to be considered in our future efforts. A strong demonstrated linkage between traits in relation to crop performance under the target environment and the molecular marker is a crucial step for effective MAS. Only such robust markers can be used for selection of the desired traits in breeding, gene pyramiding, and screening germplasm. Suitable breeding schemes combined with efficient QTL detection methods, such as composite interval mapping, could provide a good tool for creating new productive millet cultivars with appropriate characteristics. Depending on their relative effects and position, markers for oligogenic and polygenic traits can be used as targets for MAS and provide an opportunity for accelerating the efficiency of breeding programmes for these millet crops.

Finally, the interest of developed countries and private sector in millets is only marginal. Therefore, India should take lead in these crops and serve the needy dryland farmers and consumers of millets in India and the rest of developing world.

References

1. Agrama, H.A.; Widle, G.E.; Reese, J.C.; Campbell, L.R.; Tuinstra, M.R. (2002). Genetic mapping of QTLs associated with greenbug resistance and tolerance in *Sorghum bicolor Theor. Appl. Genet.* 104: 1373-1378.

2. Allouis S, Qi X, Lindup S, Gale MD, Devos KM (2001). Construction of a BAC library of pearl millet, *Pennisetum glaucum*. *Theor. Appl. Genet.*, 102: 1200–1205.

3. Antoni, R.J. (2002). Novel genetic mapping tools in plants: SNPs and LD-based approaches. *Plant Science* 162: 329-334.

4. Bennetzen, J.L.; Subramanian, V.; Xu, J.; Salimath, S.S.; Subramanian, S.; Bhattramakki, D.; Hart, G.E. (2000). A framework genetic etic map of *Sorghum*. Final Abstracts Guide, Plant and Animal Genome VIII Conference, Jan. 9-12, 2000, San Diego, CA, http://www.intl-pag.org/pag/8/abstracts/pag8712.html.

5. Berhan AM, Hulbert SH, Butler LG, Bennetzen JL (1993). Structure and evolution of the genomes of *Sorghum bicolor* and Zea mays. *Theor. Appl. Genet.* 86: 598-604.

6. Bertin I, Zhu JH, Gale MD (2005). SSCP-SNP in pearl millet—a new marker system for comparative genetics. *Theor. Appl. Genet.*, 110: 1467–1472.

7. Bhattramakki D, Dong J, Chhabra AK, Hart GE (2000). An integrated SSR and RFLP linkage map of *Sorghum bicolor* (L.). Moench. *Genome* 43: 988-1002.

8. Binelli GL, Gianfranceschi L, Pe ME, Taramino G, Busso C, Stenhouse J, Ottaviano E (1992). Similarity of maize and sorghum genomes as revealed by maize RFLP probes. *Theor. Appl. Genet*. 84: 10-16.

9. Boora KS, Frederiksen R, Magill C (1999). A molecular marker that segregates with sorghum leaf blight resistance in one cross is maternally inherited in another. *Mol. Gen. Genet*. 261: 317-322.

10. Boivin K, Deu M, Rami JF, Trouche G, Hamon P (1999). Towards a saturated sorghum map using RFLP and AFLP markers. *Theor. Appl. Genet*. 98: 320-328.

11. Bowers JE, Schertz KF, Abbey C, Anderson S, Chang C, Chittenden LM, Draye X, Hoppe AH, Jessup R, Lennington J, Lin Y, Liu S, Ming R, Qiang D, Reischmann

K, Skinner N, Wang Y, Paterson AH (2000). A high-density 2399-locus genetic map of Sorghum. Final Abstracts Guide, Plant and Animal Genome VIII Conference, Jan. 9-12, 2000, San Diego, CA, http://www.intl-pag.org/pag/8/abstracts/pag8712.html.

12. Budak H, Pedraza F, Cregan PB, Baenziger PS, Dweikat I (2003). Development and utilization of SSRs to estimate the degree of genetic relationships in a collection of pearl millet germplasm. *Crop Sci.,* 43: 2284–2290.

13. Chantereau, J.; Trouche, G.; Rami, J.F.; Deu, M.; Barro, C.; Grivet, L. (2001). RFLP mapping of QTLs for photoperiod response in tropical sorghum. *Euphytica* 120: 183-194.

14. Chittenden, L.M.; Schertz, K.F.; Lin, Y.R.; Wing, R.A.; Paterson, A.H. (1994). A detailed RFLP map of *Sorghum bicolor ´ S. propinquum*, suitable for high-density mapping, suggests ancestral duplication of *Sorghum* chromosomes or chromosomal segments. *Theor. Appl. Genet.* 87: 925-933.

15. Crasta, O.R.; Xu, W.W.; Rosenow, D.T.; Mullet, J.; Nguyen, H.T. (1999). Mapping of post-flowering drought resisatnce traits in grain sorghum: association between QTLs influencing premature senescence and maturity. *Mol Gen Genet.,* 262: 579-588.

16. Devos, K.M.; Gale, M.D. (1997). Comparative genetics in the grasses *Plant Mol. Biol.,* 35: 3-15.

17. Devos KM, Pittaway TS, Reynolds A, Gale MD (2000). Comparative mapping reveals a complex relationship between the pearl millet genome and those of foxtail millet and rice. *Theor. Appl. Genet.,* 100: 190–198.

18. Dufour, P.; Deu, M.; Grivet, L.; D'Hont, A.; Paulet, F.; Bouet, A.; Lanaud, C.; Glaszmann, J.C.; Hamon, P. (1997). Construction of a composite sorghum genome map and comparison with sugarcane, a related complex polyploid. *Theor. Appl. Genet.,* 94: 409-418.

19. Gale M.D., Devos K.M., Zhu J. H., Allouis S., Couchman M.S., Liu H., Pittaway T.S., Qi X Q., Kolensnikova-Allen and Hash C.T. 2001. New Molecular marker technologies for pearl millet improvement. *Intl. Sorghum Millets Newslett.*42: 16-22.

20. Han, F.; Romagosa, I.; Ullrich, S.E.; Jones, B.; Hayes, P.M.; Wesenberg, D.M. (1997). Molecular marker assisted selection for malting quality traits in barley. *Molecular Breeding,* 3: 427-437.

21. Hart, G.E.; Schertz, K.F.; Peng, Y.; Syed, N.H. (2001). Genetic mapping of *Sorghum bicolor* (L.). Moench QTLs that control variation in tillering and other morphological characters. *Theor. Appl. Genet.,* 103: 1232-1242.

22. Haussmann, B.I.G.; Hess, D.E.; Seetharama, N.; Welz, H.G.; Geiger, H.H. (2002). Construction of a combined sorghum linkage map from two recombinant inbred populations using AFLP, SSR, RFLP, and RAPD markers, and comparison with other sorghum maps. *Theor. Appl. Genet.,* 105: 629-637.

23. Haussmann, B.I.G.; Mahalakshmi, V.; Reddy, B.V.S.; Seetharama, N.; Hash, C.T.; Geiger, H.H. (2003). QTL mapping of stay-green in two sorghum recombinant inbred populations. *Theor. appl. Genet*,. 106: 133-142.

24. Hulbert, S.H.; Richter, T.E.; Axtell, J.D.; Bennetzen, J.L. (1990). Genetic mapping and characterization of sorghum and related crops by means of maize DNA probes. *Proc Natl Acad Sci* (USA), 87: 4251-4255.

25. Jones ES, Liu CJ, Gale MD, Hash CT, Witcombe JR (1995). Mapping quantitative trait loci for downy mildew resistance in pearl millet. *Theor. Appl. Genet.*, 91: 448–456.

26. Jones ES, Breese WA, Liu CJ, Singh SD, Shaw DS, Witcombe JR (2002). Mapping quantitative trait loci for resistance to downy mildew in pearl millet: Field and glasshouse screens detect the same QTL. *Crop Sci.*, 42: 1316–1323.

27. Kebede, H.; Subudhi, P.K.; Rosenow, D.T.; Nguyen, H.T. (2001). Quantitative trait loci influencing drought tolerance in grain sorghum (*Sorghum bicolor* L. Moench). *Theor. Appl. Genet.*, 103: 266-276.

28. Klein, P.E.; Klein, R.R.; Morishige, D.T.; Childs, K.L.; Cartinhour, S.W.; Mullet, J.E. (2000). Construction of an integrated genetic and physical map of sorghum. Final Abstracts Guide, Plant and Animal Genome VIII Conference, Jan 9-12, 2000, San Diego, California, http://www.intl-pag.org/pag/8/abstracts/pag8102.html.

29. Klein, R.R.; Klein, P.E.; Chhabra, A.K.; Dong, J.; Pammi, S.; Childs, K.L.; Mullet, J.E.; Rooney, W.L.; Schertz, K.F. (2001a). Molecular mapping of the *rf1* gene for pollen fertility restoration in sorghum (*Sorghum bicolor* L.). *Theor. Appl. Genet.*, 102: 1206-1212.

30. Klein, R.R.; Rodriguez-Herrera, R.; Schlueter, J.A.; Klein, P.E.; Yu, Z.H.; Rooney, W.L. (2001b). Identification of genomic regions that affect grain-mould incidence and other traits of agronomic importance in sorghum. *Theor. Appl. Genet.*, 102: 307-319.

31. Kong, L.; Dong, J.; Hart, G.E. (2000). Characteristics, linkage map positions, and allelic differentiation of *Sorghum bicolor* (L.). Moench DNA simple-sequence repeats (SSRs). *Theor. Appl. Genet.*, 101: 438-448.

32. Lijavetzsky, D.; Martínez, M.C.; Fernando, C.; Hopp, H.E. (2000). *QTL analysis and mapping of pre-harvest sprouting resistance in Sorghum Euphytica*, 112 : 125-135.

33. Lin, Y.R.; Schertz, K.F.; Paterson, A.H. (1995). Comparative analysis of QTLs affecting plant height and maturity across the Poaceae, in reference to an inter-specific sorghum population. *Genetics*, 141: 391-411.

34. Liu C.J, Witcombe J.R, Pittaway T.S, Nash M, Busso C.S, Hash C.T, Gale M.D. (1994). An RFLP-based genetic map of pearl millet (Pennisetum glaucum). *Theor. Appl. Genet.*, 89: 481–487.

35. Luebberstedt, T.; Melchinger, A.E.; Schon, C.C.; Utz, H.; Klein, D. (1997). QTL mapping in testcrosses of European flint lines of maize.1. Comparison of different testers for forage yield traits *Crop Sci.*, 37: 921-931.

36. Menz, M.A.; Klein, R.R.; Mullet, J.E.; Obert, J.A.; Unruh, N.C.; Klein, P.E. (2002). A high-density genetic map of *Sorghum bicolor* (L.). Moench based on 2926 AFLP, RFLP and SSR markers. *Plant Mol Biol.*, 48: 483-499.

37. Ming, R,; Liu, S.C.; Lin, Y.R.; da Silva, J.; Wilson, W.; Braga, D.; van Deynze, A.; Wenslaff, T.F.; Wu, K.K.; Moore, P,H,; Burnquist, W.; Sorrells, M.E.; Irvine, J.E.; Paterson, A.H. (1998). Detailed alignment of Saccharum and sorghum chromosomes: comparative organization of closely related diploid and polyploid genomes. *Genetics*, 150: 1663-1682.

38. Oh, B.J.; Frederiksen, R.A.; Magill, C.W. (1994). Identification of molecular markers linked to head smut resistance gene (Shs). in sorghum by RFLP and RAPD analyses. *Phytopathol.*, 84: 830-833.

39. Panday, S.; Sindhu, A.; Boora, K.S. (2002). RAPD based DNA markers linked to anthracnose disease resistance in *Sorghum bicolor* (L.). Moench. *Indian J. Exp. Biol.*, 40: 206-211.

40. *Paterson, A.H.; Lin, Y.R.; Li, Z.; Schertz, K.F.; Doebley, J.F.; Pinson, S.R.M.; Liu, S.C.; Stansel, J.W.; Irvine, J.E. (1995).* Convergent Domestication of Cereal Crops by Independent Mutations at Corresponding Genetic Loci. *Science, 269: 1714-1718.*

41. Peng, Y.; Schertz, K.F.; Cartinhour, S.; Hart, G.E. (1999). Comparative genome mapping of *Sorghum bicolor* (L.). Moench using an RFLP map constructed in a population of recombinant inbred lines. *Plant Breeding*, 118: 225-235.

42. Pereira, M.G.; Lee, M.; Bramel-Cox, P.; Woodman, W.; Doebley, J.; Whitkus, R. (1994). Construction of an RFLP map in sorghum and comparative mapping in maize. *Genome*, 37: 236-243.

43. Pereira, M.G.; Lee, M. (1995). Identification of genomic regions affecting plant height in sorghum and maize. *Theor. Appl. Genet.*, 90: 380-388.

44. Poncet V, Lamy F, Devos KM, Gale MD, Sarr A, Robert T (2000). Genetic control of domestication traits in pearl millet (*Pennisetum glaucum* L., Poaceae). *Theor. Appl. Genet.*, 100: 147–159.

45. Poncet V, Martel E, Allouis S, Devos KM, Lamy F, Sarr A, Robert T (2002). Comparative analysis of QTLs affecting domestication traits between two domesticated × wild pearl millet (*Pennisetum glaucum* L., Poaceae). crosses. *Theor. Appl. Genet.*, 104: 965–975.

46. Qi X, Lindup S, Pittaway TS, Allouis S, Gale MD, Devos KM (2001). Development of simple sequence repeat markers from bacterial artificial chromosomes without subcloning. *BioTechniques*, 31: 355–361.

47. Qi X, Pittaway TS, Lindup S, Liu H, Waterman E, Padi FK, Hash CT, Zhu J, Gale MD, Devos KM (2004). An integrated genetic map and a new set of simple sequence repeat markers for pearl millet, *Pennisetum glaucum*. *Theor. Appl. Genet.*, 109: 1485–1493.

48. Ragab, R.A.; Dronavalli, S.; Saghai-Maroof, M.A.; Yu, Y.G. (1994). Construction of a sorghum RFLP linkage map using sorghum and maize DNA probes. *Genome*, 37: 590-594.

49. Rami, J.F.; Dufour, P.; Trouche, G.; Fliedel, G.; Mestres, C.; Davrieux, F.; Blanchard, P.; Hamon, P. (1998). Quantitative trait loci for grain quality, productivity, morphological and agronomical traits in sorghum (*Sorghum bicolor* L. Moench). *Theor. Appl. Genet.*, 97: 605-616.

50. Ribaut, J.M.; Jiang, C.; Gonzalez-de-Leon, D.; Edmeades, G.O.; Hoisington, D.A. (1997). Identification of quantitative trait loci under drought conditions in tropical maize. 2. Yield components and marker-assisted selection strategies. *Theor. Appl. Genet.*, 94: 887-896.

51. Sajjanar, G.M. (2002). Genetic analysis and molecular mapping of components of resistance to shoot fly (*Atherigona soccata* Rond.). in sorghum [*Sorghum bicolor* (L.). Moench.]. Unpublished Thesis, submitted to University of Agricultural Sciences, Dharwad, 580 005. India.

52. Subudhi, P.K.; Nguyen, H.T. (2000). Linkage group alignment of sorghum RFLP maps using a RIL mapping population. *Genome* 43: 240-249.

53. Tanksley, S.D.; Young, N.D.; Paterson, A.H.; Bonierbale, M.W. (1989). RFLP mapping in plant breeding: new tools for old sciences. *Bio/technology*, 7: 257-264.

54. Tao, Y.Z.; Jordan, D.R.; Henzell, R.G.; McIntyre, C.L. (1998). Construction of a genetic map in a sorghum RIL population using probes from different sources and its comparison with other sorghum maps. *Aust. J. Agric. Res.*, 49: 729-736.

55. Tao, Y.Z.; Henzell, R.G.; Jordan, D.R.; Butler, D.G.; Kelly, A.M.; McIntyre, C.L. (2000). Identification of genomic regions associated with stay green in sorghum by testing RILs in multiple environments. *Theor. Appl. Genet.*,100: 1225-1232.

56. Tao, Y.Z.; Hardy, A.; Drenth, J.; Henzell, R.G.; Franzmann, B.A.; Jordan, D.R.; Butler, D.G.; McIntyre, C.L. (2003). Identifications of two different mechanisms for sorghum midge resistance through QTL mapping. *Theor. Appl. Genet.*, 107: 116-122.

57. Taramino, G.; Tarchini, R.; Ferrario, S.; Lee, M.; Pe, M.E. (1997). Characterization and mapping of simple sequence repeats (SSRs). in *Sorghum bicolor*. *Theor Appl. Genet.*, 95: 66-72.

58. Toure, A.; Xu, W.; Rosenow, D.T.; Peterson, G.C.; Nguyen, H.T. (1997). Inheritance of insecticide phytotoxicity in sorghum. Plant and Animal Genome V Conference, San Diego, California. pp. 204.

59. Tuinstra, M.R,; Grote, E.M.; Goldsbrough, P.B.; Ejeta, G. (1996). Identification of quantitative trait loci associated with pre-flowering drought tolerance in sorghum. *Crop Sci.*, 36; 1337-1344.

60. Tuinstra, M.R.; Grote, E.M., Goldsbrough PB, Ejeta G (1997). Genetic analysis of post-flowering drought tolerance and components of grain development in *Sorghum bicolor* (L.). Moench. *Mol Breed* 3: 439-448.

61. Wen, L.; Tang, H.V.; Chen, W.; Chang, R.; Pring, D.R.; Klein, P.E.; Childs, K.L.; Klein, R.R. (2002). Development and mapping of AFLP markers linked to the sorghum fertility restorer gene *rf4*. *Theor. Appl. Genet.*, 104: 577-585.

62. Whitkus, R.; Doebley, J.; Lee, M. (1992). Comparative genome mapping of sorghum and maize. *Genetics*, 132, 1119-1130.

63. Xu, G.W.; Magill, C.W.; Schertz, K.F.; Hart, G.E. (1994). A RFLP linkage map of *Sorghum bicolor* (L.). Moench. *Theor. Appl. Genet.*, 89: 139-145.

64. Xu, W.; Subudhi, P.K.; Castra, O.R.; Rosenow, D.T.; Mullet, J.E.; Nguyen, H.T. (2000). Molecular mapping of QTLs conferring stay green on grain sorghum (*Sorghum bicolor* L. Moench). *Genome*, 43: 461-469.

65. Yadav RS, Hash CT, Bidinger FR, Cavan GP, Howarth CJ (2002). Quantitative trait loci associated with traits determining grain and stover yield in pearl millet under terminal drought-stress conditions. *Theor. Appl. Genet.*, 104: 67–83.

66. Yadav RS, Bidinger FR, Hash CT, Yadav YP, Yadav OP, Bhatnagar SK, Howarth CJ (2003). Mapping and characterization of QTL × E interactions for traits determining grain and stover yield in pearl millet. *Theor. Appl. Genet.*, 106: 512–520.

67. Yadav RS, Hash CT, Bidinger FR, Devos KM, Howarth CJ (2004). Genomic regions associated with grain yield and aspects of post-flowering drought tolerance in pearl millet across stress environments and tester background. *Euphytica*, 136: 265–277.

68. Zhang, X.; Pammi, S.; Miller, F.R.; Hart, G.E.; Schertz, K.F.; Mullet, J.E. (1994). Identification of RAPD markers associated with a plant height gene (Dw2). in sorghum. *Intl. Sorghum Millets Newslett.* 35: 92.

Part V

Nutritional and Health Benefits

Millets: Ensuring Climate Resilience and Nutritional Security (2015) *Pages* **519–537**
Editors: **Vilas A. Tonapi and J.V. Patil**
Published by: **DAYA PUBLISHING HOUSE, NEW DELHI**

Chapter 19

Nutritional and Health Benefits of Millets

B. Dayakar Rao and C.V. Ratnavathi
ICAR-Indian Institute of Millets Research (IIMR)
Rajendranagar, Hyderabad – 500 030

After almost 67 years of Independence, malnutrition continues to plague India. Even while vast segments of resource-poor people suffer from undernutrition, particularly micronutrient deficiencies (hidden hunger), there is a growing incidence of obesity and chronic diseases like diabetes, cardiovascular diseases, cancer etc. Both the ends of this grim spectrum are at least partly due to changing food habits, loss of millets from the diet being one of them. Millets offer unique advantage for health being rich in micronutrients, particularly minerals and B vitamins as well as nutraceuticals. These phytochemicals have been shown to mitigate above mentioned chronic diseases. Millets can also withstand environmental stress being resistant to drought and warming. Thus, these are grains for the future and "harbingers for evergreen revolution". Currently, they sustain the livelihood of over 60 per cent of small and marginal farmers (NAAS, 2013).

Millets are comparable to major cereals with respect to their nutritional features and are good sources of carbohydrates, micronutrients and phytochemicals with nutraceutical properties. They contain 7-12 per cent protein, 2-5 per cent fat, 65-75 per cent carbohydrates and 15-20 per cent dietary fibre. Among them, pearl millet contains considerably high proportion of proteins (12-16 per cent) as well as lipids (4-6 per cent) whereas; finger millet contains lower levels of protein (6-8 per cent) and fat (1.5-2 per cent). The essential amino acid profiles of the millet proteins is better than maize. The niacin content in pearl millet is higher than all other cereals whereas, finger millet proteins are unique because of the sulphur rich amino acid contents.

Similar to cereal proteins, the millet proteins are poor sources of lysine, but they complement well with lysine-rich vegetable (leguminous) and animal proteins and form nutritionally balanced composites of high biological value. Small millets are more nutritious compared to fine cereals. Finger millet is the richest source of calcium (300-350 mg/100 g) and other small millets are good source of phosphorous and iron.

Table 1: Nutrient Composition of Millets Compared to Fine Cereals (per 100 g)

Food Gain	*Carbo-hydrates (g)*	*Protein (g)*	*Fat (g)*	*Energy (KCal)*	*Crude Fibre (g)*	*Mineral Matter (g)*	*Ca (mg)*	*P (mg)*	*Fe (mg)*
Finger millet	72.0	7.3	1.3	328	3.6	2.7	344	283	3.9
Kodo millet	65.9	8.3	1.4	309	9.0	2.6	27	188	0.5
Proso millet	70.4	12.5	1.1	341	2.2	1.9	14	206	0.8
Foxtail millet	60.9	12.3	4.3	331	8.0	3.3	31	290	2.8
Little millet	67.0	7.7	4.7	341	7.6	1.5	17	220	9.3
Barnyard millet	65.5	6.2	2.2	307	9.8	4.4	20	280	5.0
Sorghum	72.6	10.4	1.9	349	1.6	1.6	25	222	4.1
Bajra	67.5	11.6	5.0	361	1.2	2.3	42	296	8.0
Wheat (whole)	71.2	11.8	1.5	346	1.2	1.5	41	306	5.3
Rice (raw, milled)	78.2	6.8	0.5	345	0.2	0.6	10	160	0.7

Source: Nutritive Value of Indian Foods, NIN, 2007.

Table 2: Vitamin Profile of Millets and Major Cereals

Millet	*Thiamin (mg)*	*Niacin (mg)*	*Ribo-flavin*	*Vit A (Carotene) (mg/100g)*	*Vit B6 (mg/100g)*	*Folic Acid (mg/100g)*	*Vit B5 (mg/100g)*	*Vit E (mg/100g)*
Foxtail	0.59	3.2	0.11	32	–	15.0	0.82	31.0
Proso	0.41	4.5	0.28	0	–	–	1.2	–
Finger	0.42	1.1	0.19	42	–	18.3	–	22.0
Little	0.3	3.2	0.09	0	–	9.0	–	–
Barnyard	0.33	4.2	0.1	0	–	–	–	–
Kodo	0.15	2.0	0.09	0	–	23.1	–	–
Sorghum	0.38	4.3	0.15	47	0.21	20.0	1.25	12.0
Bajra	0.38	2.8	0.21	132	–	45.5	1.09	19.0
Rice	0.41	4.3	0.04	0	–	8.0	–	–
Wheat	0.41	5.1	0.1	64	0.57	36.6	–	–

Source: Nutritive Value of Indian Foods, NIN, 2007.

MILLET in your Meals, http://www.sahajasamrudha.org/

Table 3: Essential Amino Acid Profile of Millets Compared to Fine Cereals (mg/g of N)

Millet	*Arginine*	*Histidine*	*Lysine*	*Tryptophan*	*Phenyl Alanine*	*Tyrosine*	*Methionine*	*Cystine*	*Threonine*	*Leucine*	*Isoleucine*	*Valine*
Foxtail	220	130	140	60	420	–	180	100	190	1040	480	430
Proso	290	110	190	50	310	–	160	–	150	760	410	410
Finger	300	130	220	100	310	220	210	140	240	690	400	480
Little	250	120	110	60	330	–	180	90	190	760	370	350
Barnyard	270	120	150	50	430	–	180	110	200	650	360	410
Sorghum	240	160	150	70	300	180	100	90	210	880	270	340
Bajra	300	140	190	110	290	200	150	110	140	750	260	330
Rice	480	130	230	80	280	290	150	90	230	500	300	380
Wheat	290	130	170	70	280	180	90	140	180	410	220	280

Source: Nutritive Value of Indian Foods, NIN, 2007.

Millet grain contains about 65 per cent carbohydrate, a high proportion of which is in the form of non starchy polysaccharides and dietary fibre which help in prevention of constipation, lowering of blood cholesterol and slow release of glucose to the blood stream during digestion. Lower incidence of cardiovascular diseases, duodenal ulcer and hyperglycemia (diabetes) are reported among regular millet consumers. Millet grains are also rich in important vitamins *viz.*, Thiamine, riboflavin, folin and niacin. Millets are comparable to rice and wheat or rich in some of the minerals as well as fatty acids. Millets vary largely in composition of carbohydrates as proportion of amylose and amylopectin content vary from 16-28 per cent and 72-84 per cent, respectively. The exact nutritional profile is given in Tabless 1–6.

Table 4: Micro-nutrient Profile of Millets Compared to Fine Cereals (mg/100g)

Cereals/Millets	*Mg*	*Na*	*K*	*Cu*	*Mn*	*Mb*	*Zn*	*Cr*	*Su*	*Cl*
Foxtail	81	4.6	250	1.40	0.60	0.070	2.4	0.030	171	37
Proso	153	8.2	113	1.60	0.60	–	1.4	0.020	157	19
Finger	137	11.0	408	0.47	5.49	0.102	2.3	0.028	160	44
Little	133	8.1	129	1.00	0.68	0.016	3.7	0.180	149	13
Barnyard	82	–	–	0.60	0.96	–	3	0.090	–	–
Kodo	147	4.6	144	1.60	1.10	–	0.7	0.020	136	11
Sorghum	171	7.3	131	0.46	0.78	0.039	1.6	0.008	54	44
Bajra	137	10.9	307	1.06	1.15	0.069	3.1	0.023	147	39
Rice	90	–	–	0.14	0.59	0.058	1.4	0.004	–	–
Wheat	138	17.1	284	0.68	2.29	0.051	2.7	0.012	128	47

Source: Nutritive Value of Indian Foods, NIN, 2007.

MILLET in your Meals, http://www.sahajasamrudha.org/

Table 5. Fatty Acid Composition of Millets Compared to Fine Cereals

Cereals/Millets	*Palmitic*	*Palmoleic*	*Stearic*	*Oleic*	*Linoleic*	*Linolenic*
Foxtail	6.40	–	6.30	13.0	66.50	–
Proso	–	10.80	–	53.80	34.90	–
Finger	–	–	–	–	–	–
Little	–	–	–	–	–	–
Sorghum	14.0	–	2.10	31.0	49.0	2.70
Bajra	20.85	–	–	25.40	46.0	4.10
Rice	15.0	–	1.90	42.50	39.10	1.10
Wheat	24.50	0.80	1.00	11.50	56.30	3.70

Source: Nutritive Value of Indian Foods, NIN, 2007.

MILLET in your Meals, http://www.sahajasamrudha.org/

Table 6: Amylose and Amylopectin Content of Millets

Cereal Grain	*Amylose (per cent)*	*Amylopectin (per cent)*
Proso millet	28.2	71.8
Foxtail millet	17.5	82.5
Kodo millet	24.0	76.0
Finger millet	16.0	84.0
Sorghum	24.0	76.0
Bajra	21.1	78.9
Short Grain Rice	12-19	88-81
Wheat	25.0	75.0

Source: *MILLET in your Meals*, http://www.sahajasamrudha.org/

Nutritional Profile of Sorghum

Grain sorghum has certain properties which makes it suitable to be consumed by population suffering from chronic disorders. Each sorghum nutrient has specific nutritional significance, which is suggestible to prevent and control life style diseases and disorders.

Protein content and composition of sorghum varies from cultivar to cultivar due to agronomic conditions (water availability, soil fertility, temperatures and environmental conditions during grain development) and genotype. Sorghum proteins are chiefly located in the endosperm (80 per cent), germ (16 per cent) and pericarp (3 per cent) (*Taylor and Schussler, 1986*). Kafirins or prolamins and glutelins comprise the major fractions in sorghum (*Warsi and Wright, 1973*) and located in endosperm while germ is rich in albumins and globulins.

The sorghum carbohydrate content is composed of starch, soluble sugar and fiber. Carbohydrates in sorghum are classified into non-structural (sugars, starch and fructosans) and structural (cellulose, hemicelluloses and pectin substances).

Sorghum contains slow digestible starch (SDS) in good amounts, which has functional property, prolongs digestion and absorption of carbohydrates in intestine. This SDS is favorable for dietary management and also for metabolic disorders such as diabetes and hyperlipidemia (*Asp, 1994; Wursch, 1997*). Sorghum also contains good amount of dietary fiber that is 9.7-14.3g and plays the role of bulking agent, binding agent of cholesterol, increases transit time and retards carbohydrates absorption (*Narasinga Rao,2003*), which has a significant positive effect on preventing and managing the diseases like constipation, irritable bowel syndrome and obesity.

Lipid content of Sorghum grain is minor and located in the scutellar area of the germ. Sorghum lipid has potential to lowerthe cholesterol. A study was done on effect of cholesterol absorption and plasma non-HDL cholesterol concentration in hamsters fed with grain sorghum lipid extract. In this study, hamsters were fed with grain sorghum lipid extract (GSL) comprising different proportions of the diet and compare with control. Liver cholesterol ester concentration was also significantly reduced in hamsters fed with GSL. GSL diet lowers non-HDL cholesterol, at least in part, by inhibiting cholesterol absorption; GSL extract consist plant sterols that reduce cholesterol absorption efficiency significantlyand policosano isthat inhibit endogenous cholesterol synthesis. Research findings further indicate that grain sorghum contains components that could be used as food ingredients or dietary supplements to manage cholesterol levels in humans (Carr *et al, 2005*).

Sorghum is a good source of minerals and vitamins and mostly located in aleurone layer and germ. Sorghum is an important source of B vitamins except for vitamin B12 (Gazzaz *et al.*, 1989).Yellow colored sorghum grain is rich in beta carotene, leutin and zeaxanthin. However, it varies due to environment and genes.It is poor in vitamin – C, but can be synthesized with soaking and germination process. Detectable amounts of other vitamins E, K and D are found to be present in sorghum grain. Decortications and degermination reduces water and fat soluble vitamins since the micro nutrients are located primarily in aleurone and germ portions of caryopsis. Grain sorghum is good source of potassium and adequate source of Mg, Fe, Zn and Cu but a poor source of calcium and sodium.

Sorghum is excellent source of polyphenols, flavonoids and condensed tannins which are antioxidant potent in nature. The phenolic acids such as hydroxyl benzoic acids (galliac, vanillic, syringic, p- hydroxyl benzoic and protocatechic acids), hydroxycinnamic acids (coumaric, caffeic, ferulic and sinapinic acids) are located in sorghum pericarp, endosperm and aleurone layer (McDonough *et al.*, 1986). The anthocyanins are the major class of flavonoids present in sorghum. Sorghum anthocyanins do not contain the hydroxyl group in the 3-position of the C-ring and called as 3-deoxyanthocyanins. These anthocyanins are responsible for stability of grain at higher pH and natural color. Sorghum 3-deoxyanthocyanins are apigeninidin (yellow), luteolinidin (orange) (*Awika et al.,2004*). Sorghum pericarp contains highest amounts of 3-deoxyanthocyanins and flavonoids such as luteoforol and apiforol. These flavan-4-ols play an important role in mould resistance. Other identified flavonoids in sorghum grains are flavones (apigenin and luteolin), Flavanones (riodictyol and eriodictyol 5-glucoside), flavonol (kaempferol and taxifolin). Condensed tannins, also known as proanthocyanidins or procyanidins, are high-molecular weight polyphenols which are also present in sorghum (*Dykes and Rooney, 2006*)

Health Benefits of Sorghum

- ☆ Gluten free food - Recommended for gluten intolerant and celiac patients.
- ☆ Relatively low glycemic index and low glycemic load food – Reduces the risk of Diabetes *mellitus*
- ☆ Invisible fat – lowers Low-density lipoprotein (LDL) and Improves High-density lipoprotein (HDL) cholesterol
- ☆ Low calorie food - Beneficial for obesity, diabetes and CVD
- ☆ Rich source of antioxidants and polyphenols - Reduces oxidative stress and carcinogenic properties
- ☆ Bile acid and steroid binding properties - Beneficial for obesity and CVD

- ☆ Good Source of Nitrilosides-salicylates - Fights against Arthritis and Rheumatism
- ☆ Rich in dietary fiber- Beneficial for obesity, diabetes, CVD and inflammatory bowel disease and acts as a detoxifying agent
- ☆ Rich in magnesium - Subdues depression

Nutritional Facts of Sorghum

Celiac Disease

Celiac disease (CD) is one of the most common genetic diseases, by which genetically predisposed people suffer a reaction to gluten proteins found in wheat and other cereals. This disease is caused by an adverse reaction of the immune system to gluten and it may lead to severe abdominal pains. Sorghum can be a healthy diet for those who are ailing from celiac disease as it is gluten free. Sorghum products could not modify the level of anti-transglutaminase antibodies after prolonged consumption (Carolina *et al, 2007*).

Obesity

Obesity is an emerging problem in India and it is positively associated with several chronic diseases including diabetes and CVD. Empirical evidences suggest that intake of highdietary fibre decreases the incidence of obesity (*Alfieri et al, 1995; Burkitt and Trowel, 1975*). Foods rich in dietary fibre improve the large bowel function and slows the digestion and absorption process, thereby reducing the risk of chronic diseases (*Ali et al.,1982; Schneeman and Tietyen, 1994*). Sorghum is rich in dietary fibre and has unique chemical and physical characteristics (bulk to the diet, viscosity, water holding and absorption capacity) which determine the subsequent physiological behavior. It aids to the hunger satisfaction, increases satiety and thereby reducing the risk of development of obesity.

Diabetes

Diabetes Mellitus- complex metabolic diseases a major health concern in many countries. The prevalence of diabetes type 2 is on the rise across the globe at an alarming rate, especially in India where the prevalence of diabetes has reached 14.3 per cent. Prevention of type 2 diabetes will require measures to promote healthy dietary pattern and life style which includes balanced diet and physical activity. People who eat three or more servings of whole grains in a day, especially from high-fiber cereals, are less likely to develop insulin resistance and the metabolic syndrome, common precursors of both T2D and Coronary Heart Disease (CHD) (*Mckeownet al., 2004*). Diets rich in whole grain foods tend to decrease LDL cholesterol, triglycerides, blood pressure, and increase HDL cholesterol (*Anderson, 2003*). Sorghum is rich in dietary fibre and low glycemic index, which could help in prevention and control of T2D in Indians. The fibre, magnesium, vitamin –E, phenolic compounds and tannins present in foods reduces the risk of diabetes as they slower the sudden increase of blood glucose and insulin levels (*Montonen et al, 2003*).

National Institute of Nutrition (ICMR) in 2010 assessed Glycemic Index (GI) of sorghum based foods in collaboration with the Indian Institute of Millets Research, Hyderabad under National Agricultural Innovation Project (NAIP). The results revealed that sorghum based foods have low GI and reduces the postprandial blood glucose level, glycosilated hemoglobin. Another study also points to the fact that blood glucose level of non-obese patients with non-insulin-dependent diabetes mellitus (NIDDM), who consumed sorghum bran papadi, showed considerable reduction.(Shinde, *2004*).

Coronary Heart Disease

Empirical evidences (Slavin, *2003;* Slavin, *2004*) suggest that regular consumption of whole grains reduces the risk of CVD (Anderson et al, 1999). A long term prospective study by Heidemann *et al*, 2008 showed that regular consumption of balanced diet which includes whole grains, vegetable, fruit, fish and poultry diet reduce the risk of CVD and total mortality. In 2004, Harvard professionals (*Jensen et al, 2004*) analyzed the diet records of 27,000 men aged between 40-75 years for a period of 14 years and found intake of whole grains 40g/per day reduces the risk of CHD by 20 per cent.

Cancer

Anti-carcinogenic properties of sorghum have been well documented. Vivo and *in vitro* studies have shown consumption of sorghum has positive health impacts on cancer. The polyphenols and tannins present in sorghum have anti mutagenic and carcinogenic properties (Grimmer *et al, 1992*) and can act against human melanoma cells, as well as positive melanogenic activity (Gomez-Cordovez *et al, 2001*). Parbhoo *et al* (*1995*) reported that in rat liver procyanidin extracts may induce cytochrome P-450, a protein that is capable of converting certain promutagens to mutagenic derivatives. Epidemiological data from Sachxi Province, China and in different parts of the world (*Van Rensburg, 1981*) showed that Incidence of oesophageal cancer was low with sorghum consumption. In each country, the authors studied 21 communities over a period of 6 years and found consumption of sorghum showed lower mortality from oesophageal cancer than wheat and corn. It can be concluded that anti-carcinogenic compounds present in sorghum lowered risk of oesophageal cancer.

Oxidative Stress

Free radicals are atoms or molecules with an unpaired number of electrons and can be formed with interaction of other molecules. These free radicals are responsible for oxidative stress related pathogenesis of various diseases such as Alzheimer's disease, myocardial infarction, atherosclerosis, Parkinson disease, auto-immune diseases etc. Antioxidants play an important role in preventing oxidation process, thereby reducing cellular damage. Sorghum has potential antioxidant capacity which can act against reactive oxidative species.

Nutritional Profile of Pearl Millet

Pearl millet (*Pennisetum glaucum*) is the most widely grown type of millet. It has been grown in Africa and the Indian subcontinent since prehistoric times. In different pearl millet genotypes the starch content of the grain varied from 62.8 to 70.5 percent, soluble sugar from 1.2 to 2.6 percent and amylose from 21.9 to 28.8 percent (*Jambunathan and Subramanian, 1988*). Lower values for starch (56.3 to 63.7 percent) and amylose (18.3 to 24.6 percent) have been found in some high-yielding Indian pearl millet varieties (Singh and Popli, 1973). Subramanian and Jambunathan found that the predominant component of total soluble sugar (2.16 to 2.78 percent) was sucrose (66 percent), followed by raffinose (28 percent). Other sugars detected in measurable amounts were stachyose, glucose and fructose. The proportion of sucrose in total sugar was lower in pearl millet than in sorghum.

Pearl millet, like sorghum, is generally 9 to 13 percent protein, but large variations in protein content, from 6 to 21 percent, have been observed (Serna-Saldivar, *et al.*, 1991). Lysine is the first limiting amino acid of pearl millet protein. A significant inverse correlation has been reported between the level of protein in the grain and the Lysine content of the protein (Deosthale *et al.*, 1971). In high-protein varieties of pearl millet with protein content ranging from 14.4 to 27.1 percent, significant inverse correlations have also been observed between protein and threonine, methionine and tryptophan. The essential amino acid profile shows more lysine, threonine, methionine and cystine in pearl millet protein than in proteins of sorghum and other millets. Its tryptophan content is also higher.

Differences in lipid extraction procedures as well as genetic variability were shown to contribute to differences in the fatty acid content of pearl millet (Jellum and Powell, 1971). The principal fatty acids in both free and bound were found to be linoleic, oleic and palmitic acids. Distinct differences in fatty acid composition were noted in the neutral lipid, phospholipid and glycolipid fractions (Osagie and Kates, 1984). Neutral lipid was highest in linoleic acid and lowest in palmitic acid; phospholipid was lowest in oleic acid and highest in palmitic acid; and glycolipid was highest in linolenic acid.

The total dietary fibre in pearl millet (20.4 percent) and finger millet (18.6 percent) was higher than that in sorghum (14.2 percent), wheat (17.2 percent) and rice (8.3 percent) also found that the total dietary fibre content of pearl millet was 17 percent (Kamath and Belavady, 1980).

Health Benefits of Pearl Millet

- ✰ Millets are a great source of starch, making it a high-energy food. It is also an excellent source of protein and fiber. It is said that the amino acids in the pearl millet are more easily digestible than the ones found in wheat.
- ✰ Due to essential nutrients such as methionine (an amino acid), B complex vitamins (niacin, thiamin, and riboflavin), folic acid, lecithin, potassium, magnesium, manganese and zinc, millets are very effective in several roles. Niacin reduces cholesterol while magnesium is essential for maintaining good heart health, as it lowers blood pressure and reduces the risk of heart attacks.
- ✰ Pearl millet is a rich source of phosphorus, which plays an important part in the structure of body cells. Phosphorus, found in pearl millets, is a

significant component of several necessary compounds including adenosine triphosphate (ATP). This element is also a crucial component of nucleic acids, which are the building blocks of the genetic code. Phosphorus is a constituent of lipid-containing structures such as cell membranes and nervous system structures.

- ☆ Recent studies have proven that regular consumption of pearl millets help in preventing gallstones in women. They contain insoluble fibres which not only speed up intestinal transit time but also reduce the secretion of bile acids. Pearl millets are known to increase insulin sensitivity and lower the level of triglycerides.
- ☆ Regular intake of millets provides protection against breast cancer in pre-menopausal women. Apart from that, it has also shown a considerable reduction in the occurrence of wheezing and asthma in children.
- ☆ Millets contain an essential phytonutrient, lignin, which is very beneficial for the human body. With the help of natural flora, lignans get converted to mammalian lignans and they fight against hormone-dependent cancers and reduce the risk of cardiac arrests.
- ☆ Consumption of pearl millets helps in minimizing the risk of type 2 diabetes. Being a good source of magnesium, millets act as a cofactor in a number of enzymatic reactions.

Nutritional Facts of Pearl Millet

Due to its rich composition of minerals and proteins, Pearl Millet has many health benefits. Pearl Millet has the highest protein content. It contains many essential minerals like magnesium, phosphorus, zinc etc. It contains essential amino acids and vitamins also which contribute to its therapeutic properties.

- ☆ **Beneficial in treating stomach ulcers**: Pearl millet is recommended for curing stomach ulcers. The most common cause for stomach ulcers is excess acidity in the stomach after food intake. Pearl millet is one of the very few foods that turns the stomach alkaline and prevents formation of stomach ulcers or reduces the effect of ulcers.
- ☆ **Beneficial for Heart health:** The lignin and phytonutrients in millet act as strong antioxidants thus preventing heart related diseases. This is why, pearl millet is considered good for heart health. High amounts of magnesium present in pearl millet have been shown to control blood pressure and relieve heart stress.
- ☆ **Beneficial due to high amount of magnesium** : Pearl millet contains high concentration of magnesium which helps reduce severity of respiratory problems for asthma patients and is also effective in reducing migraine attacks
- ☆ **Helps in bone growth development and repair:** Pearl millet has a large amount of phosphorus. Phosphorus is very essential for bone growth and development as well as for development of ATP which is the energy currency of our body.

- ☆ **Reduces cancer risk:** All millets are known to reduce the risk of cancer occurrence and pearl millet is no exception. Though scientists are not sure how this is, they believe it has something to do with the high amount of magnesium and the compound phytate.
- ☆ **Helps in weight loss:** The biggest challenge faced by people trying to lose weight is controlling their food intake. Pearl millet can aid the process of weight loss as it is high in fibre content. Owing to its fibre content it takes longer for the grain to move from the stomach to the intestines. This way, pearl millet satiates hunger for a long period of time and thus helps in lowering the overall consumption of food.
- ☆ **Beneficial for diabetes:** Pearl millet is very effective for controlling diabetes. Because of its high fibre content, it digests slowly and releases glucose into the blood at a slower rate as compared to other foods. This effectively helps in maintaining the blood sugar level constant in diabetes patients for a long period of time.
- ☆ **Beneficial for Celiac Disease:** Celiac disease is a condition in which a person cannot tolerate even a small amount of gluten in his/her diet. Unfortunately, most of the common grains like rice, wheat, etch have gluten present in them. Millets are the only type of grains which do not have any gluten present. Thus this is suitable for people with celiac disease.
- ☆ **Reduces Cholesterol:** It is common knowledge that Pearl Millet is suggested for people suffering from high cholesterol levels. Pearl millet contains a type of phyto chemical called phytic acid which is believed to increase cholesterol metabolism and stabilise the levels of cholesterol in the body.
- ☆ **Contains all the essential amino acids:** Amino acids are essential for smooth functioning of our body. Pearl millet is one of the few foods which have all the essential amino acids required in it. Unfortunately, most of these amino acids are lost in the process of cooking because these amino acids cannot stand high temperatures. Thus it is better to consume in a low cooked form so as to preserve as many of these amino acids as possible.
- ☆ **Beneficial in Preventing Gall stones:** The high fibre content in pearl millet is also known to reduce the risk of gall stone occurrence. The insoluble fibre content in pearl millet reduces the production of excessive bile in our system. Excessive amount of bile secretion in our intestine often leads to aggravate the condition of gall stones.
- ☆ **Anti allergic properties**: Pearl millet is a treasure trove of beneficial properties. The grain is very digestible as such and has a very low probability of causing allergic reactions. Due to its hypo allergic property, it can be safely included in the diets of infants, lactating mothers, elderly and convalescents.

Nutritional Profile of Finger Millet

Finger millet (*Eleusine coracana*) is an annual plant widely grown as a cereal in the arid areas of Africa and Asia. It remains one of the main ingredients of the staple diet in Karnataka. Nutritionally, finger millet is good source of nutrients especially of calcium, other minerals and fibre. Total carbohydrate content of finger millet has been reported to be in the range of 72 to 79.5 per cent (*Bhatt et al., 2003*). The carbohydrates include starch as the main constituent being 59.4 to 70.2 per cent (*Mittal, 2002*). Finger millet starch granules exhibit polygonal rhombic shape. About 80 to 85 per cent of the finger millet starch is amylopectin and remaining 15 to 20 per cent is amylose. The non-starch polysaccharide accounts for 20 to 30 per cent of the total carbohydrates in finger millets. It contains around 1.5 per cent reducing sugar and 0.03 per cent non-reducing sugar (*Bhatt et al., 2003*). The white varieties have higher protein content than the brown varieties of the finger millet. Finger millet contains 44.7 per cent essential amino acids (*Mbithi et al., 2000*) of the total amino acids, which is higher than the 33.9 per cent essential amino acids. Since *ragi* does not contain gluten, it is a wonderful grain alternative for people who are gluten-sensitive.

The total dietary fibre (TDF), insoluble dietary fibre (IDF), and soluble dietary fibre (SDF) con tent in finger millet was found to be 12, 11 and 2 per cent, respectively (*Ramulu and Rao 1997*). It was reported that 18.6 per cent dietary fibre and 3.6 per cent crude fibre in finger millet (*Kamath and Belavady, 1980*). The health benefits associated with high fibre foods are delayed nutrient absorption, increased faecal bulk, lowering of blood lipids, prevention of colon cancer, barrier to digestion, mobility of intestinal contents, increased faecal transit time and fermentability characteristics (*Tharanathan and Mahadevamma, 2003*). Finger millet is a good source of energy for weight watchers, as it contains the amino acid tryptophan that reduces the appetite.

Total antioxidant capacity of finger, little, foxtail and proso millets were found to be higher and their total carotenoids content varied from 78–366 mg/100 g in the millet varieties (*Mathanghi and Sudha, 2012*). Finger millet has the higher amount of calcium (344mg) and potassium (408mg). Calcium helps in keeping your bones and teeth healthy. It has higher dietary fiber, minerals, and sulphur containing amino acids compared to white rice, the current major staple in India (*Shobana et al., 2013*). However, the finger millet also contains phytates (0.48 per cent), polyphenols, tannins (0.61 per cent), trypsin inhibitory factors, and dietary fiber, which were once considered as "anti nutrients" due to their metal chelating and enzyme inhibition activities (*Thompson, 1993*) but nowadays they are termed as neutraceuticals. It is now established that phytates, polyphenols and tannins can contribute to antioxidant activity of the millet foods, which is an important factor in health, aging and metabolic diseases. It functions as an amazing common relaxant and serves to battle fidgetiness, a sleeping disorder and discouragement. Additionally assists in treating headache cerebral pains. Antioxidant effects of finger millet on the dermal wound healing process in diabetes induced rats with oxidative stress-mediated modulation of inflammation were studied by *Rajasekaran et al.* (2004). They reported that the role of finger millet feeding on skin antioxidant status, nerve growth factor (NGF) production and wound healing parameters in healing the impaired early diabetic rats.

Health Benefits of Finger Millet (Mathanghi and Sudha, 2012)

- ☆ Finger millet also is known to have several potential health benefits. Some of the health benefits are attributed to its polyphenol contents. The phenolic acid content of brown finger millet 96 per cent higher compared to white variety

- ☆ It contains more lysine, threonine, and valine than other millets. In addition, black finger millet contains 8.71 mg/g dry weight fatty acid and 8.47 g/g dry weight protein
- ☆ Finger millet is reported to have antiulcerative properties and finger millet diets lowered blood glucose and cholesterol in diabetic rat models. (36 per cent reduction in blood glucose levels)
- ☆ Finger millet seed coat matter which is a rich source of dietary fibre and phenolic compounds were found to exhibit blood glucose and cholesterol lowering, nephroprotective and anticataractogenic properties in streptozotocin induced diabetic rat models
- ☆ Supplementing infants with the germinated finger millet-based food showed a general improvement on hemoglobin status
- ☆ Lower serum cholesterol and triacyl glycerol levels (43 per cent and 62 per cent, respectively) compared to diabetic controls
- ☆ Finger millet extracts were also reported to possess free radical scavenging, anti-protein glycation, anti-cataractogenic and antimicrobial properties in 'in vitro'. Inhibition of snake venom phospholipases by finger millet phenolics in 'in vitro'
- ☆ High reducing power for seed coat polyphenol extract compared to the polyphenol extract from finger millet whole flour. Higher antioxidant activity for finger millet seed coat polyphenol extract (86 per cent) compared to polyphenol extract from finger millet whole flour (27 per cent).
- ☆ Offered protection against mucosal ulceration, epithelialization, increased synthesis of collagen, activation of fibroblasts, and mast cells

Nutritional Facts of Finger Millet

- ☆ **Finger Millet/Ragi for losing weight**: Ragi contains an amino acid called Tryptophan which lowers appetite and helps in keeping weight in control. Ragi gets digested at a slower rate thus keeps one away from intaking excessive calories. Also, fibers present in ragi give a feeling of fulness thus controls excessive food consumption.
- ☆ **Finger Millet/Ragi for bone health**: Ragi is rich in Calcium which helps in strengthening bones. It is an excellent source of natural calcium for growing children and aging people. Ragi consumption helps in development of bones in growing children and in maintenance of bone health in adults. Ragi keeps diseases such as osteoporosis at bay and could reduce risk of fracture.
- ☆ **Finger Millet/Ragi for diabetes**: Finger Millet's phytochemicals help in slowing digestion process. This helps in controlling blood sugar level in condition of diabetes. In a study conducted in 2000, it was found that Finger Millet based diet helps diabetics as it contains higher fiber than rice

and wheat. Also, the study found that diet based on whole finger millet has lower glycemic response *i.e.* lower ability to increase blood sugar level. This is due to presence of factors in ragi flour which lower digestibility and absorption of starch.

- **Finger Millet/*Ragi* for lowering blood Cholesterol**: Finger Millet contains amino acids Lecithin and Methionine which help in bringing down cholesterol level by eliminating excess fat from Liver. Finger Millet also contains Threonine amino acid which hinders fat formation in the liver, which brings cholesterol level of the body down.
- **Finger Millet/*Ragi* for Anaemia**: *Ragi* is a very good source of natural Iron. Ragi consumption helps in condition of Anaemia.
- **Finger Millet/Ragi for Relaxation**: *Ragi* consumption helps in relaxing body naturally. It is beneficial in conditions of anxiety, depression and insomnia. *Ragi* is also useful for migraines.
- **Finger Millet/*Ragi* for Protein/Amino Acids**: *Ragi* is rich in Amino Acids which are vital in normal functioning of body and are essential for repairing body tissues. Finger Millet contains Tryptophan, Threonine, Valine, Isoleucine and Methionoine amino acids. Isoleucine helps in muscle repair, blood formation, contributes to bone formation and improves skin health. Valine is essential amino acid which facilitates metabolism, helps in muscle coordination and repair of body tissues. It helps in balancing nitrogen in the body. Another essential amino acid, not found in most cereals, is Methionine which is useful in various body processes, helps in eliminating fat from the body, and is main provider of sulfur in body. Sulfur is essential for production of Glutathione - body's natural antioxidant.
- **Finger Millet for other health conditions**: If consumed regularly, *Ragi* could help in keeping malnutrition, degenerative diseases and premature aging at bay. Green *Ragi* is recommended for conditions of blood pressure, liver disorders, asthma and heart weakness. Green *Ragi* is also recommended to lactating mothers in condition of lack of milk production.
- ***Ragi*** is an extremely nutritious cereal and is very beneficial for maintaining a good health. However, its high intake could increase quantity oxalic acid in the body. Therefore, it is not advised to patients having kidney stones (Urinary Calculi). *Ragi* could be enjoyed in different forms and preparations. *Ragi Roti, Ragi Dosa, Ragi* Porridge, *Ragi Upma, Ragi* Cakes, *Ragi* Biscuits are few popular dishes of *Ragi.*

Nutritional Profile Foxtail Millet

Foxtail millet (*Setaria italica* L.) is also known as Italian millet. It is one of the world's oldest cultivated crops. In the northern area of China it has been widely used as a nourishing gruel or soup for pregnant and nursing women and has been applied to food therapy. It has been recorded that millet has many nutritious and medical functions (*Prashant et al., 2005*). Foxtail yellow seeded cultivars, medicinally used as astringent, digestive, emollient and stomachic. It is also used in the treatment of dyspepsia, poor digestion and food stagnancy in abdomen (*Yeung, 1985*). White seeds are refrigerant and used in the treatment of cholera and fever while the green seeds are diuretic and strengthening to virility (*Duke and Ayensu, 1985*).

This millet contains 12.3 per cent crude protein and 3.3 per cent minerals (*Vithal and Machewad, 2006*). Foxtail and Barnyard millet showed comparable amounts of crude protein which was highest among all the millets studied. Foxtail millet protein characterization showed that its protein concentrate is a potential functional food ingredient and the essential amino acid pattern suggests possible use as a supplementary protein source to most cereals because it is rich in lysine (*Ravindran 1992*). The work of Liang et al. (2010) presented the general properties of foxtail millet oil and its fatty acid profile. It is apparent that millet oil could be a good source of natural oil rich in linoleic acid and tocopherols (*Liang et al., 2010*).

The foxtail varieties revealed to have significant antioxidant activity by (DPPH) method (*Devi et al., 2011*). Foxtail Millet is calculated to be 364Cal per 100 grams. It is rich sourse of minerals like copper and iron. Unlike rice, foxtail millet releases glucose steadily without affecting the metabolism of the body. The incidence of diabetes is rare among the population which consumes foxtail millet diet. Scientists studied its health benefits in diabetic rats, and concluded that the millet produced a "significant fall (70 per cent) in blood glucose" while having no such effect in normal rats. Diabetic rats fed millet also showed significantly lower levels of triglycerides, and total/ LDL/VLDLcholesterol, while exhibiting an increase in HDL cholesterol.

Health Benefits of Foxtail millet

- ☆ Helps control Blood sugar levels when consumed on regular basis. It showed lowered triglyceride levels, LDL/VLDL Cholestrol and increase in HDL Cholestrol.
- ☆ It is known for its Low Glycemic index- gradual increase in blood sugar after food intake when compared to rice.
- ☆ Ideal food for people suffering from Diabetes and Gastric problems.
- ☆ Reduces risk of Heart Attack.
- ☆ Helps in the development of Body Tissue and Energy Metabolism.
- ☆ Rich in Anti-oxidants.
- ☆ Foxtail Millet help to control blood sugar and cholesterol

Nutritional Profile of Proso Millet

Proso millet (*Panicum miliaceum* L.) is the best alternative crop for diversifying and intensifying winter wheat-based dryland production systems. Proso Millet is calculated to be 356Cal per 100 grams. The protein content is similar to that of wheat, but it contains no gluten and by itself is not suitable for yeast-leavened bread. The protein content was found to be (11.6 per cent of dry matter) and was significantly richer in essential amino acids (leucine, isoleucine, and methionine) than wheat protein (Kalinova and Moudry 2006). It is rich in vitamins and minerals such as copper and magnesium. Proso millet also improved glycemic responses and plasma levels (*Park et al., 2008*). In addition, proso millet protein concentrate has protective effects against D-galactosamin-induced liver injury in rats (*Ito et al., 2008*). *Choi et al.* (*2005*) and *Park et al.* (*2008*) concluded that proso millet protein could be a potential therapeutic intervention in type 2diabetes. *Devi et al.* (*2011*) review the nature of polyphenols and dietary fiber of finger millet and their role with respect to the health benefits associated with millet. The composition of free and bound lipids in proso millet (Panicum miliaceum) flours and brans were analyzed and found that, in the free lipids, hydrocarbons, sterol esters, triacylglycerols, diacylglycerols, and free fatty acids were present. The predominant fatty acids in the free lipids were linoleic, oleic, and palmitic acids, though, in the bound lipids, monogalactosyl diacylglycerols, digalactosyl diacylglycerols, phosphatidylethanolamine, phosphatidyl serine, and phosphatidyl choline were tentatively identified (*Bagdi et al., 2011*).. reported the total The total lipid content in the foxtail, proso, and finger millets ranged from 5.2 to 11.0 per cent (dry basis), while it ranges from 5.1 to 8.3 per cent in the little, kodo, and barnyard millets (*Saldivar, 2003*). It is apparent that millet oil could be a good source of natural oil rich in linoleic acid and tocopherols (*Liang et al., 2010*). The main polyphenols in cereals are phenolic acids and tannins, whilst flavonoids are present in small quantities; they act as antioxidant and play many roles in the body immune system defence (*Devi et al., 2011*). Whole cereal grains are considered a rich source of fibre. However, foods from grains have marked differences in the amount and type of dietary fibre (*Shukla and Srivastava, 2011*). The dietary fibre content in cereal-based food varies greatly, depending on the extent of milling. Finger millet shows relatively higher than other cereals carbohydrate (72 per cent) comprises of starch as the main constituent and the non-starchy polysaccharides which amounts to 15–20 per cent of the seed matter as an unavailable carbohydrate dietary fibre content and complements which are the health benefits of the millet (*Devi et al., 2011*). Millets are good sources of magnesium and phosphorus. Magnesium has the ability to help reduce the effects of migraine and heart attacks, while, phosphorus is an essential component of adenosine triphosphate (ATP) a precursor to energy in the body (Devi *et al.*, 2011). Proso millet is rich source of B vitamins, especially vitamin-B6 and folic acid.

References

Basavaraj G, Parthasarathy Rao P, Bhagavatula S and Ahmed W. 2010. Availability and utilization of pearl millet in India. *Journal of SAT Agricultural Research* 8.

DAC. 2010. Status Paper on Millets. Directorate of Millets Development, Ministry of Agriculture, Department of Agriculture and Cooperation, Jaipur.

Dayakar Rao B, Reddy S and Seetharama N. 2007. Reorientation of investment in R&D of millets for food security: The case of sorghum in India. *Journal of Agricultural Situation in India*. 64(7): 303-305.

GOI. 2010-11. *Economic survey*. New Delhi, India: Ministry of Finance.

Gulati A and Kelly T. 2000. *Trade Liberalization and Indian agriculture*. Oxford University Press, New Delhi, India.

NRCS. 2009. *Vision 2025 Document - Perspective Plan*. Rajendranagar, India: National Research Centre for Sorghum.

Parthasarathy Rao P, Basavaraj G, Bhagavatula S and Ahmed W. 2010. An analysis of availability and utilization of sorghum grain in India. *Journal of SAT Agricultural Research* 8.

Parthasarathy Rao P, Birthal PS, Belum Reddy VS, Rai KN and Ramesh S. 2006. Diagnostics of Sorghum and Pearl Millet Grains-based Nutrition in India, International Sorghum and Millets Newsletter.

Bagdi, A., G. Balázs, J. Schmidt, M. Szatmári, R. Schoenlechner, E. Berghofer and S. Tö mö skö zia. 2011. Protein characterization and nutrient composition of Hungarian proso millet varieties and the effect of decortication. *Acta Alimentaria* 40:128–141.

Bhatt A, Singh V, Shrotria PK, Baskheti DC (2003). Coarse Grains of Uttaranchal: Ensuring sustainable Food and Nutritional Security. *Indian Farmer's Digest*. pp. 34-38

Choi, Y. Y., K. Osada, Y. Ito, T. Nagasawa, M. R. Choi and N. Nishizawa. 2005. Effect of dietary protein of Korean foxtail millet on plasma adiponectin, HDL-cholesterol, and insulin levels in genetically type 2 diabetic mice. *Biosc. Biotechnol. Biochem*. 69:31–37.

Deosthale, Y.G., Visweswara Rao, K., Nagarajan, V. and Pant, K.C. 1971. Varietal differences in protein and amino acids of grain bajra (*Pennisetum typhoides*). *Indian J. Nutr. Diet*. 8: 301-308.

Devi, P. B., R. Vijayabharathi, S. Sathyabama, N. G. Malleshi and V. B. Priyadarisini. 2011. Health benefits of finger millet (*Eleusine coracana* L.) polyphenols and dietary fiber: a review. *J. Food Sci. Technol*. DOI: 10.1007/s13197-011-0584-9

Duke, J.A. and E.S. Ayensu, 1985. *Medicinal Plants of China*. Publications, Inc, ISBN 0-917256-20-4.

Ito, K., H. Ozasa, Y. Noda, S. Arii and S. Horikawa. 2008. Effects of free radical scavenger on acute liver injury induced by Dgalactosamine and lipopolysaccharide in rats. *Hepatol. Res*. 38:194–201.

Jambunathan, R. and Subramanian, V. 1988. Grain quality and utilization of sorghum and pearl millet. In Biotechnology in tropical crop improvement. *Proceedings of the International Biotechnology Workshop*, Patancheru, Inde, 12-15 janvier 1987, p. 133139. Patancheru, ICRISAT)

Jellum, M.D. and Powell, J.B. 1971. Fatty acid composition of oil from pearl millet seed. *Agron. J*., 63: 29-33

Kamath M V, Belavady B(1980). Unavailable carbohydrates of commonly consumed Indian foods.*J. Sci. Food Agric*.,31: 194-202.

Kalinova J, Moudry J. 2006. Content and quality of protein in proso millet (*Panicum miliaceum* L.) varieties. *Plant Foods Hum Nutr* 61: 45–9.

Liang, S., G. Yang and Y. Ma. 2010. Chemical characteristics and fatty acid profile of foxtail millet bran oil. *J. Am. Oil Chem. Soc*. 87: 63–67.

Mathanghi SK and K. Sudha, (2012) Functional and phytochemical properties of finger millet (*Elusine coracana*L.) for health, international journal of pharmaceutical, chemical and biological sciences, *IJPCBS* 2(4), 431-438.

Mbithi-Mwikya S, Ooghe W, Van Camp J, Nagundi D, Huyghebaert A (2000). Amino acid profile after sprouting, Autoclaving and lactic acid fermentation of finger millet (*Elusine coracana*) and kidney beans (*Phaseolus vulgaris* L.). *J. Agric. Food Chem.*, 48 (8): 3081-3085.

MILLET in your Meals, Sahaja Samrudha Organic producer Company Ltd. No 569, Sajjan Rao Road, Near Trishal lodge V.V.Puram, Bangalore-560004. (http://www.sahajasamrudha.org).

Mittal M (2002). Development of finger millet and barnyard millet based convenience mixes for food products and their evaluation for nutritional quality, storage stability and acceptability. Thesis. *Ph.D.* G.B. Pant Univ. of Agric and Technol. Pantnagar. p. 260.

Nutritive Value of Indian Foods (NVIF) By C. Gopalan, BV Raman Sastri and SC Balasubramaniam, Revised and Updated (1989) by BS Narasinga Rao, YG Deosthala and KC Pant (Reprinted 2007, 2011).

Osagie, A.U. and Kates, M. 1984. Lipid composition of millet (*Pennisetum americanum*) seeds. *Lipids*, 19: 958-965.

Park, K. O., Y. Ito, T. Nagasawa, M. R. Choi and N. Nishizawa. 2008. Effects of dietary Korean proso-millet protein on plasma adiponectin, HDL cholesterol, insulin levels, and gene expression in obese type 2 diabetic mice. *Biosc. Biotechnol. Biochem.* 72(11): 2918–2925..

Prashant SH, Namakkal SR, Chandra TS 2005. Effects of the antioxidant properties of millet species on oxidative stress and glycemic status in alloxan-induced rats. *Nutr Res.* 25: 1109–1120.

Rajasekaran NS, Hegde PS, Chandra TS. 2004. Effects of the antioxidant properties of millet species on oxidative stress and glycemic status in alloxan-induced rats. *Nutr Res.*, 25: 1109–1120.

Ramulu P,Udayaekhara Rao P(1997). Effect of processing on dietary fibre content of cereals and pulses. *Plant Foods Hum. Nutr.*, 50: 249-257.

Ravindran G (1992) Seed protein of millets: amino acid composition, proteinase inhibitors and *in vitro* protein digestibility. *Food Chem* 44: 13–17.

Saldivar, S. 2003. Cereals: dietary importance. In: B. Caballero, L. Trugo and P. Finglas (Eds.). pp. 1027–1033. *Encyclopedia of Food Sciences and Nutrition*, Reino Unido: Academic Press, Agosto, London.

Serna-Saldivar, S.D., McDonough, C.M. and Rooney, L.W. 1991. The millets. In K.J. Lorenz and K. Kulp, Eds. *Handbook of cereal science and technology* p. 271- 300. New York, Marcel Dekker.

Shobana S, Krishna Swami K, Sudha V, Malleshi NG, Anjana RM, Palaniappan L, Mohan V. 2013. Finger millet (Ragi, Eleu*sine coracana* L.): a review of its nutritional properties, processing, and plausible health benefits. *Adv Food Nutr Res.* 69: 1-39. doi: 10.1016/B978-0-12-410540-9.00001-6.

Singh, R. and Popli, S. 1973. Amylose content and amylolytic studies on high-yielding varieties of bajra (*Pennisetum typhoides*). *J. Food Sci. Technol.* 10: 31-33.

Shukla, K. and S. Srivastava. 2011. Evaluation of finger millet incorporated noodles for nutritive value and glycemic index. *J. Food Sci. Technol.* DOI: 10.1007/s13197-011-0530-x.

Tharanathan RN, Mahadevamma, S. 2003. Grain legumes" a boon to human nutrition. *Trends Food Sci Technol* 14: 507–518.

Thompson LU (1993) Potential health benefits and problems associated with antinutrients in foods. *Food Res Int.*, 26: 131–149.

Vithal, D.S. and G.M. Machewad, (2006). Processing of foxtail millet for improved nutrient availability. *J. Food Proces. Preser.*, 30: 269-279.

Yeung, H.C., 1985. *Handbook of Chinese Herbs and Formulas*, Institute of Chinese, Los Angeles.

FAO: http://faostat3.fao.org/download/FB/FBS/E

Rai KN, Gowda CLL, Reddy BVS and Sehgal S. 2008. Adaptation and Potential Uses of Sorghum and Pearl Millet in Alternative and Health Foods. *Comprehensive Reviews in Food Science and Food Safety*—Vol. 7, 2008.

NAAS 2013. Role of Millets in Nutritional Security of India. Policy Paper No. 66, National Academy of Agricultural Sciences, New Delhi: 16 p.

Part VI
Utilization and Commercialization

Millets: Ensuring Climate Resilience and Nutritional Security (2015)
Editors: **Vilas A. Tonapi and J.V. Patil**
Published by: **DAYA PUBLISHING HOUSE, NEW DELHI**

Pages **541–549**

Chapter 20

Utilization Pattern of Millets

B. Dayakar Rao

ICAR-Indian Institute of Millets Research (IIMR)
Rajendranagar, Hyderabad – 500 030

The role of sorghum and millets as food staples at the aggregate level has declined since 1980. In Asia, sorghum for food use contributed 5.7 kg capita^{-1} yr^{-1} and millets 5.2 kg capita^{-1} yr^{-1} in 1980. In 2011, it declined to 1.9 kg capata^{-1}yr^{-1} for sorghum and 2.9 kg capata^{-1}yr^{-1} for millets. During the same period the per capita availability of all cereals in Asia also decreased marginally from 164 kg capita^{-1}yr^{-1} to 156.4 kg capita^{-1}yr^{-1}. Per capita availability of sorghum and millets declined in all countries including China, India and Pakistan. In India, the per capita availability of sorghum reduced by more than half and for millets by less than one-third.

Table 1: Millets Consumption across Countries

Country/Region	*Total Cereals*			*Sorghum*			*Millets*		
	1980	*2000*	*2011*	*1980*	*2000*	*2011*	*1980*	*2000*	*2011*
Asia									
Bangladesh	168	179	190.4	0.1	0	0	0.4	0.4	0.1
China	186	183	151.5	4.5	0.9	0.7	4.4	0.9	0.5
India	140	159	152.1	13.4	6.6	5.1	12.2	9	8.9
Myanmar	192	215	137	0	0	1.6	1.9	2.9	0.4
Nepal	164	194	186.6	0	0	0	7.2	10.7	9.8
Pakistan	144	149	142.3	2.3	1.3	0.7	1.1	0.6	0.9

Contd...

Table 1–*Contd...*

Country/Region	*Total Cereals*			*Sorghum*			*Millets*		
	1980	*2000*	*2011*	*1980*	*2000*	*2011*	*1980*	*2000*	*2011*
Saudi A*Rabia*	145	154	158.3	12.4	9.7	3.9	0.7	0.5	0.6
Syrian Arab Republic	171	177	166.3	0	0	0	0.6	0	0.1
Thailand	153	126	133.4	0	0	0	0	0	
Yemen	148	164	162.8	64	18.3	16.6	7.8	3.3	3.4
Region									
Asia	164	171	156.4	5.7	2.3	1.9	5.2	3	2.9
South Asia	143	160	157	10.7	5.1	3.7	9.8	7.1	6.7
World	150	155	147.2	5.3	4	3.6	4.6	3.6	3.3

Source: FAO, http://faostat3.fao.org/download/FB/FBS/E.

Despite the low and declining per capita availability of sorghum and millets at the aggregate level their importance as food crops cannot be underestimated (Parthasarathy *et al.*, 2003). For example, in the Indian state of Gujarat, Rajasthan, Maharashtra and Karnataka, coarse cereals accounted for 30-50 per cent of the total cereal production and consumption in 2012-13.

Consumption Trends of Nutri-Cereals in India

According to Parthasarathy Rao et al. 2010 and Basavaraj et al. 2010 the annual per capita consumption of sorghum at the all-India level has declined sharply by 74 per cent (8.5 to 2 kg) in urban areas and by 81 per cent (19.2 to 3.5 kg) in rural areas of India between 1972–73 and 2009–10. Similarly, the pearl millet consumption both in rural and urban areas has fallen very steeply from 11.5 kg to 3.06 kg (by 73 per cent) in rural areas and from 4 kg to 1.08 kg (by 73 per cent) in urban areas. This is due to increase in per capita income, growing urbanization, changing tastes and preferences (Chand 2007) rendering sorghum and millet as inferior goods with low to negative income elasticity of demand and positive price elasticity.

Apart from decline in consumption, these crops are gradually disappearing in traditional areas due to access to irrigation and markets, which have enabled farmers to shift their area under cultivation to high value crops (Chandrakanth and Akarsha 2011). The food security strategy of the Government of India of supplying subsidized rice and wheat through the Public Distribution System (PDS) has been the major factor contributing to the decline in consumption of sorghum and other cereals like pearl millet and fi nger millet in rural areas and urban centers. Thus, the popularity of sorghum and pearl millet faded, resulting in negative growth in area, production practices and hence their cultivation became uneconomical. The PDS system in India is based on the wheat and rice model, which is less relevant in many areas and especially in the dryland farming areas, where millets, sorghum and pulses are traditionally the staple grains for household consumption (Dayakar Rao *et al.*, 2007). Despite the decline in per capita consumption, sorghum grain is an important staple for the low and middle income consumers in regions where they are grown. For example, in rural areas of central Maharashtra, per capita annual consumption of sorghum is around 60 kg, accounting for almost half (48 per cent) of the per capita consumption of all cereals. Similarly, among the major pearl millet producing regions, per capita consumption was highest (69 kg/year) in rural Rajasthan and in the dry areas of Gujarat (59 kg/year). In these two regions, pearl millet accounts for more than 50 per cent of cereal consumption, contributing about 20 to 40 per cent of the total energy and protein intake (Parthasarathy Rao *et al.*, 2009).

The main reasons for this decline in demand for direct consumption (human consumption) is due to rising incomes, urbanization and subsequent changes in consumer tastes and food habits. Apart from decreasing consumption demand other reasons that led to decrease in the area and production are due to lack of competitiveness vis-à-vis other competing crops such as pulses, oilseeds and commercial crops, low profitability etc.

Table 2: Trend in Direct Consumption of Cereals and Food Grains

Kg/person/year

Commodity	*1973-74*	*1983 -84*	*1993-94*	*2003-04*	**2013-14*
			Rural		
Rice	84.0	80.7	85.4	82.7	83.80
Wheat	42.8	54.3	53.5	51.7	51.13
Sorghum	19.0	12.5	9.7	5.7	3.9
Coarse cereals	56.8	45.1	24.1	16.5	13.2
Total cereals	183.6	180.1	163.0	151.6	148.59
Food grains			172.8	162.2	159.9
			Urban		
Rice	65.5	64.7	64.2	59.4	60.11
Wheat	52.6	58.6	57.4	56.8	53.42
Sorghum	11.0	6.0	4.9	3.3	2.5
Coarse cereals	19.7	14.1	7.7	6.0	5.4
Total cereals	137.7	137.5	129.3	122.4	119.37
Food grains			140.4	135.1	134.3

Source: Various NSSO reports, GOI, and Dayakar Rao *et al* (2014).

*: Estimated utilization.

Note: Projection assumes 7.1 per cent growth rate in economy: 5 per cent for rural and 12 per cent for urban population.

If the trend of sorghum and coarse cereals consumption continues in the current scenario, after the next two decades they will be completely vanished and replace by fine cereals such as rice and wheat. However, the prospects of sorghum and millets in direct food consumption tends to rise as interventions in its food processing and promotion has been successfully taken out by IIMR in collaboration with DAC and other millets research institutes.

Impending demand for sorghum stalks for meeting their domestic livestock fodder needs has become lifeline for continuation of sorghum farming in the dry-land regions despite low consumption demand for grain. Thus utilization of *Kharif* sorghum grain (by product) became a question in event of deterioration of sorghum grain quality (due to blackening (caught up in October rains)) making it unfit for human consumption. Further, the scale of sorghum farming is linked to the extent of domestic fodder requirements in *Kharif* season, thus resulting in meager marketed surpluses. Therefore, *Kharif* sorghum grain was evolved as a regular source of raw material for above mentioned alternate uses. However, due to smallness of marketed surplus volumes, supply chain management has become a major constraint and making it unsustainable. *Dayakar et al* reveals that decrease in consumption of sorghum was found to be proportional to increase in consumer expenditure. Increased income is accompanied by increased consumption of wheat and rice, as products made from these cereals are easy to prepare and have better keeping quality.

Similarly, Pearl millet consumption also declined by 67 per cent in urban areas and 59 per cent in rural areas during 1972-73 to 2004-05 (Basavaraj *et al* 2010). The food use of Pearl millet has declined sharply at the country level, but it is still an important crop in the major producing states. The increase in Pearl Millet production juxtaposed with its declining food use implies that its use in alternative uses has been increasing mainly from animal feed which includes mainly dairy, and to some extent in poultry, alcohol industry, starch industry, processed food industry and export demand. The potential demand for food processing, though at a early stage, presents encouraging prospects for value addition.

Another study conducted by Indian Institute of Millets Research (IIMR) on "Macro and Micro Level Changes in Consumption of Nutritious Cereals in India" (Dayakar et al 2004) also pointed that sorghum in all the States and country as a whole showed decline in sorghum consumption among urban and rural masses, whereas Pearl millet consumption in major producing States like Gujarat (129 per cent) and Rajasthan exhibited increased growth in urban areas, while in rural Gujarat, it has declined (-18.16 per cent) and the same has increased in rural Rajasthan (39.1 per cent). The magnitude of reduction in pearl millet consumption was highest in rural Tamil Nadu (-95.5 per cent) followed by rural Andhra Pradesh (-80.95 per cent). In Finger millet there is a slight decline of 9 per cent in consumption in urban Karnataka, whereas in Rural Karnataka the decline in consumption was four fold (36 per cent) than in urban consumption. The Andhra Pradesh rural consumption of finger millet also declined drastically (-67.27 per cent).

Causal Factors for the Decline in Consumption of Nutritious Cereals

1. Easy availability of fine cereals like rice and wheat at a cheaper price to that of nutritious cereals under Public Distribution System (PDS).
2. Rapid urbanization.
3. Nutritious cereals are socially less valued necessitated the decline in their consumption on rising of capita income.
4. Tedious and time consuming preparation of nutritious cereals foods.
5. Preference to cash payment by labourers rather than taking food grain in lie of work performed.
6. Change in food habit particularly preference to fast food by younger generation.

Measures for Increasing the Consumption/Demand of Nutritious Cereals

1. Creating public awareness about the nutritional superiority of millets.
2. Supply of nutritious cereals through PDS.
3. Value-addition in nutritious cereals.
4. Provisions for supply of millet foods under Mid -Day Meal/Other Food Supplement programmes being arranged by various States.
5. Utilization of millet foods under larger Scheme like Adolescent Girl Nutrition Scheme, under consideration of Ministry of Women and Child Welfare.

Utilization of Millets for Alternative Uses

Millets are traditionally used for preparation of fermented foods and beverages. Grains with high starch (65-72 per cent) and low protein (8-12 per cent) could be utili zed for production of alcohol. The utilization of *Kharif* sorghum grain as a raw material in various industries is increasing, given the limited prospects of rainy season (*Kharif*) sorghum for human consumption.

The important traditional/fast food and industrial usages of millets are given in the Table 3.

Most recently pearl millet is being used as a low cost substitute for maize in poultry and dairy feed, which is gaining momentum. In addition, there are other niches for pearl millet:

- ☆ Due to its rich protein content, the grain can be used for preparation of protein-rich foods.
- ☆ The bran is rich in oil; hence it can complement rice bran in oil extraction.
- ☆ Pearl millet can also be used for brewing in those favorable areas with marketed surplus where the factories are located.

Table 3: Important Food and Industrial Uses of Millets

Millets	*Food Products*	*Industrial products*
Sorghum	Roti, ugali, popped sorghum, malt food, snack/roasted mix grains.	Malting, high fructose syrup, starch, Jaggery, bakery, value added products for diabetics, poultry and animal feed.
Pearl millet	Roti, ugali, fermented food products, pizza, roasted mix grains.	Brewing/malting, starch, bakery, poultry and animal feed.
Finger millet	Roti, dumpling, popped millet, malt-food.	Malting/brewing, baby foods, bakery and food for diabetics.
Other millets	Roti, cooked cereals.	Value added food for devotees (Barnyard millet), Feed, value added food products for diabetics.

☆ The high fiber content in the grain can be exploited to prepare health foods. Diabetic food products can also be prepared.

The main industries currently using sorghum and millets in India are the poultry feed, animal feed, potable alcohol distilleries, etc. At present poultry feed sector is using approximately 2.0 million tonnes sorghum annually; animal feed sector uses about 0.60 million tonnes sorghum followed by alcohol distillers (about 0.49 million tonnes).

1. **Poultry feed/Animal Feed**: Poultry feed industry in the country is progressing well. Several independent feeding trials conducted on layer and broiler birds could conclusively demonstrate that sorghum and pearl millet as good as maize although, the latter is marginally superior in total energy value. Sorghum and millets based feed in layer birds has to be supplemented with carotene to ensure yellow colour to egg yolk. In the case of cattle feed, sorghum based feed is found satisfactory. Moulded *Kharif* grain is found as an acceptable grain component in the feed of goat and swine too. Sorghum is used when maize is in short supply and priced up to 20 per cent higher than sorghum.
2. **Starch:** Sorghum is rarely used by the starch industry in view of processing problem and starch yield as the starch extraction machineries are adapted to the bigger grain size of maize. While there is scope to improve starch recovery from sorghum by modified extraction, selective breeding and use of specifically suited machineries, starch production from the present sorghum cultivars be economically viable only if the grain price is 20 per cent lesser than that of maize. Sorghum grain is found more suitable for production of glucose and liquid glucose.
3. **Potable alcohol:** The sorghum grain offers an excellent ethanol, and feed stock with a high alcohol recovery of 422-448 litres per tonne. This provides scope for utilization of rain-damaged blackened and less market value sorghum grain for utilization as raw material for malting and production of quality potable alcohol. Currently more than 10 distilleries are using sorghum grain (mainly blackened) as raw material for potable alcohol.

Government of Maharashtra has announced incentives for those sorghum grain based distilleries which are established as alternative to sugarcane molasses. The setting of such industry in the potential areas would benefit the farmers in getting assured market with remunerative price through contract farming. Like sorghum, pearl millet is also being used for production of potable alcohol.

4. **Malt**: Another potential area could be the use of ragi and sorghum malt as an adjunct in brewing industry. Malted sorghum is used for brewing beer in Ghana and Nigeria. Sorghum's comparative advantage would be its low output price, especially in production regions of rainy sorghum in India over that of existing raw material.
5. **Bio-ethanol**: Sweet stalked and high energy sorghum has the potential to emerge as a major bio-energy crop for production of industrial alcohol, gasohol and even electricity. While the national trials on sweet sorghum had shown its potential productivity as 50-60 t stalk and 2-3 t grain/ha in 125-130 days, reports from China indicate a far high productivity to the tune of 90 t stalk and 6 t grain/ha from the crop in 140-150 days. This productivity potential places the crop at an advantage over sugarcane or molasses at their ruling prices for ethanol production. As a bio-energy crop sweet sorghum is more efficient than sugarcane because of its high productivity with relatively low levels of fertilisation and irrigation. Sweet sorghum, in this context, could be effectively exploited as an energy crop. IIMR and ICRISAT and many R&D organizations were actively involved in fine-tuning technologies related to sweet sorghum to project it as an alternative feed stock for bioethanol production. However, government policy support in declaring high ethanol pricing will determine its current utilization and future demand.
6. Production of jaggery and brown or colourless syrup or high fructose syrup are other opportunities from sweet sorghum. These diversifications could be done at village or farm level as small scale enterprises.

Limitations with Storage and Consumption of Millets

The awareness about the high nutritive values of millets is very low both among the producers and consumers. Besides, being small seeded and low priced commodity, the produce is not properly cleaned, graded and dried before they are brought to the markets, fetching low price to the farmers and poses storage problems. Besides, rains at the time of maturity/harvest also lead to poor quality of *Kharif* produce. The high fat, fibre, inert matters coupled with high moisture content/high humidity leads to infestation of insect/fungus during storage. Grains stored with high moisture content often develop Mycotoxins/Aflatoxin which is highly poisonous both for human and live stock. The shelf life of flour of millets is also low because of high fat content in the grains. The dark seed coat color of some millet such as ragi and bajra also possess problem in acceptance of food products of these commodities. Thus, enhancement of shelf life could be an important challenge to process grain to products and market them sustainability.

Government Policies that Determine Millets Production and Consumption in the Country

Minimum Support Price (MSP)

Sorghum, pearl millet and finger millet are covered under MSP while small millets are not covered so far. An analysis of MSP announced for millets vis-à-vis wheat and rice during 1977-78 to 2012-13 reveals an increase in MSP of millets at par with wheat and rice and of late even higher. However, MSP could not help in sustainability of millets because of Lack of procurement system for millets through public agencies and Lack of provision for supply of millets under PDS like wheat and rice. Millets are included in the new Food security bill, thus it is expected the millets are going to be procured, however, it needs to be seen how this is going to be operationalized. There is no MSP for small millets, some states have suggested for coverage of small millets under MSP and supply of millet grains under PDS for regional food security.

The Government of India has increased the Minimum Support Price (MSP) of bajra or pearl millet of Fair Average Quality (FAQ) from Rs. 980 per quintal (100 Kilograms) to Rs 1,175 per quintal of the crop year 2012-2013 and marketing season 2013-2014. This is an increase of Rs. 195 a quintal or 19.89 percent over last year's Minimum Support Price fixed for bajra. Similarly highest increase is witnessed in sorghum

Public Distribution System (PDS)

PDS provided a substantive captive market for rice and wheat produced in Green Revolution areas. PDS has had another important effect—on dietary habits of Indians. Effectively, PDS has made rice and wheat the staple diet across the country, replacing millets and other cereal

The GoI production policy for coarse grains/millets is similar to that for wheat and rice, but there is less focus on these crops. Also, the government's support for coarse grain via the procurement and PDS operation is very limited and ineffective compared to rice and wheat. Unlike wheat and rice, the government does not typically maintain a buffer stock of coarse grains to keep process in check. GoI also announce Minimum Support Price for coarse grains and there has been sharp increase in MSP.

Decentralized procurement, storage and distribution by elected bodies such as Panchayats or even state governments remains a dream despite the obvious economic benefits. If one were to assume that decentralized procurement will provide communities with the choice to procure produce of their liking, then the possibility of coarse grains and cereals finding a way back into the dietary habits of people will exist. Alas, the political support for this is lacking.

The obvious choice, then, is to hope that the markets can do the trick. Ask children what they would like in their mid-day meal and the chances are that noodles will figure in the top five choices.

The millets may be targeted in public funded welfare programmes such as Mid-day meals in those states where millets are traditionally grown and consumed, at least 2 times a week to begin with.

Development Programmes

MMA

Developmental programmes of millets are being supported through Centrally Sponsored Scheme of Macro Management in Agriculture (MMA). The Scheme provides flexibility to the States to develop and pursue activities on the basis of their location specific needs. The States are also free to include new/innovative interventions in their Work Plan provided; it is not covered under any other Scheme of the Central Government or is not part of any on going State Schemes. Schemes also have a provision for utilization of at least 33 per cent funds for small, marginal and women farmers, who have larger stake in millets cultivation. Assistance to the States/UTs is provided as 100 per cent Grant-in-aid. All States are eligible to receive assistance under MMA Scheme.

INSIMP

Initiative for Nutrition Security through Intensive Millets Promotion (INSIMP), a sub-scheme of Rashtriya Krishi Vikas Yojna (RKVY) which aims to increase production of millets especially Sorghum, Pearl millet, Finger millet and Small millets in the country by demonstrating post-harvest technologies in an integrated manner. This scheme is also expected to generate consumer demand for millet based food products through processing and value addition techniques. Government of India allotted Rs. 300 Crore in 2011-12 under RKVY for promotion of millets as Nutri-cereals and is being continued till date. It has components of support of on farm production, seed production, establishment of processing clusters, popularization and promotion of millets and marketing linkages. The INSIMP established 3 Centres of Excellence (COE) (a) on sorghum processing at IIMR, Hyderabad, (b) on pearl millet at CCSHAU, Hisar and (c) on Small millets processing at UAS, Bangalore. The major objectives of CoE includes refinement of technology, retrofitting of machinery and their demonstration and Providing entrepreneurship development and training on sorghum post-harvest technologies and value addition. In XII plan the activities of INSIMP are going to be merged with National Food Security Mission (NFSM).

Millets: Ensuring Climate Resilience and Nutritional Security (2015) *Pages* **551–574**
Editors: **Vilas A. Tonapi and J.V. Patil**
Published by: **DAYA PUBLISHING HOUSE, NEW DELHI**

Chapter 21

Creating Demand of Millets through Value Addition

N.G. Malleshi[1] and B. Dayakar Rao[2]

[1]*Former Head, Grain Science Technology, CFTRI, Mysore*
[2]*ICAR-Indian Institute of Millets Research (IIMR) Rajendranagar, Hyderabad – 500 030*

Traditional Practices and Contemporary Methods of Millets Processing

Sorghum, the major millets (pearl millet and finger millets) and also the minor millets (foxtail, little, kodo, proso and barnyard millets), the indigenous African cereals have become traditional crops of India and many of the Asian countries. However, Fonio (*Digitaria exilis*) and Tef (*Eragrostic tef*), also the indigenous minor cereals of Africa are still confined to Nigeria and Ethiopia respectively. These grains represent the major source of dietary energy and protein for more than a billion people in the semi-arid tropics including India. Sorghum and some of the millets are also grown in Australia, Europe and North and South America, mainly for ruminant and bird feed. Not only are sorghum and millets nutritionally important, but also they have unique functional characteristics and are potentially the sources of value added products, especially snacks and foods for the population at risk of physiological disorders such as diabetes, obesity and CVD. Hence, these minor cereals need to be suitably processed to upgrade their food value as well as consumer acceptability to provide them in the form that are readily acceptable to non-traditional millet consumers.

Like any other cereals, sorghum and millets are neither ready-to-eat nor ready-to-cook grains and need some kind of processing for human consumption. Most commonly followed processing methodologies for these are: milling including decortications or seed coat separation and size gradation in to semolina and flour, popping, malting, fermentation and cold extrusion. In recent years the contemporary food processing technologies such as extrusion cooking for RTE food and breakfast cereals, pasta and vermicelli-noodles and bakery products, malting and brewing, wet milling for starch preparation are employed for these grains.

In India and most other Asian and African countries, these grains form the staple for the traditional consumers and also to the people of lower economic strata. This is mainly due to the non-availability of ready-to-use or ready-to-cook products similar to rice and wheat to the consumers and also some form of stigma attached to them as 'coarse' or hardy cereals. Efforts to replace the term 'Coarse cereals' by 'Nutri-cereals' are yet to catch the imagination of the cereal consumers. Now a days, as a result of increased awareness about the nutritional and health benefits of these grains, many of the non-traditional consumers are also looking for suitable products for substituting rice and wheat in their regular diet. But still, the convenience food products from them are scanty which is the major hindrance for enlarging their food base. However, recent studies towards processing of sorghum and millets have demonstrated that, it is possible to provide them in convenience as well as consumer friendly forms for preparation of conventional and also value added and health oriented products.

Processing quality of any cereals is determined predominantly by the morphological features such as kernel shape, size, density, hardness, endosperm texture, nature of pigmented testa and pericarp. Sorghum and millets are not exception to these features; sorghum, pearl millet and finger millet are practically naked grains and resemble wheat and do not have distinct husk tissue covering kernels, except glumes. On the other hand, small millets (foxtail, little, kodo, proso and barnyard millets) resemble paddy and have husk encaping the edible matter or the endosperm and embryo. Hence, processing in general and milling in particular vary for sorghum, pearl and finger millets from the other minor millets.

Milling

In the case of SAM the term milling refers to preparation of flour as well as decortications or pearling. In sub-Saharan countries, the pestle and mortar as well as hand pounding are still the main tools used to grind sorghum and millets, even though, manually grinding or pounding grains is painful and time-consuming. However, of late mechanical grinders have made inroads. In the Indian subcontinent as well as other part of the world, sorghum, pearl millet and finger millet are pulverised to flour for food uses without any pretreatment except cleaning. For custom milling, stone, iron or emery disc mills are used where as for large scale milling hammer mills and also roller flour mills are also used. The meal is commonly sieved to separate out a small percentage of coarse seed coat or bran. Hence, the meal is a good source of dietary fibre, minerals and the vitamins derived from the bran but the keeping quality of such meals is low due to oil content from the bran. But to prepare flour with low bran contamination (refined flour), it is recommended that, the grains are moistened with 3-5 per cent addition water, conditioned for 10 -15 min, pulverized and sieved to separate out the coarse bran fraction. Moist conditioning minimises the bran fragmentation and facilitates its segregation to a larger extent.

Sorghum and pearl millet are also decorticated or debranned to obtain pearled grains, but for efficient operation, the seed coat should be moistened prior to pearling. For the purpose, rice huller or rice polishers could be used. However, finger millet because of its unique endosperm texture (soft texture to which the seed coat is rigidly attached) cannot be pearled/polished or decorticated in its native form. Hence, it is pulverized invariably and the flour is used for food preparation. But the recent developments with respect to finger millet processing have enabled its decorticate similar to any other cereals. On the other hand, foxtail, little, kodo, barnyard and proso millets could be dehusked and debranned similar to rice using rubbler roll or centrifugal sheller with suitable adaption or dehusked even in the disc mills by the expert operator. Decorticated grains could be cooked similar to rice or may be flaked or size graded to semolina as well as to flour for conventional food uses and also for value added food products.

Kernels, with vitreous endosperm or with higher proportion of hard endosperm dehull effectively and yield higher proportion of pearled head grains where as grains with soft or floury endosperm disintegrate during dehulling. But hardening the endosperm by hydrothermal treatment, the softer grains could also be pearled effectively. No doubt, milling to separate the seed coat or decortication reduces protein, dietary fibre, vitamins and mineral contents of the grains to some extent but this is compensated by better consumer acceptability, improved bioavailability of the nutrients and enhanced product making qualities.

The bran fraction from pearl millet and some of the small millets is very good source of dietary fibre and edible oil. Hence, it could serve as an extender to the rice bran for oil extraction. The deoiled millet bran may be used as source of dietary fibre in formulating high fibre foods as it contains negligible less of silica compared to deoiled rice bran.

As on date dehullers for sorghum and pearl millet have been developed which have not found application by the common millet consumers because of poor yield of pearled grains. They need substantial improvement.for their acceptance by the common rural population. But till date, no exclusive mill is available for the decortication of millets except lab models or prototypes developed by somé of the R&D Institutions and Agric-Universities. It may be worthwhile to mention here that, small millets milling is carried out on large scale in Nasik in Maharastra and also in Teni in TN and the milled millets are marketed under the trade name '*Bhagar*'. But these mills are highly capital intensive and the yield of the milled grains is hardly 50 per cent. Hence, there is need for development of a an integrated mill comprising of sheller for dehullung, separation of unhulled grains and debranning the dehulled grains in a single unit. Sorghum and peal millet are also milled in roller mills to prepare refined flour (suitable for bakery products), but the flour from these mills is not normally suitable for traditional food preparation such as roti or mudde, because of low damaged starch content.

In some of the African countries, milo or sorghum with red pigmented seed coat is common which is cultivated as high tannin-bird resistant sorghum. Because of high proportion of tannins the food products from these cultivars are not readily acceptable. Decortication of such sorghum is also difficult because of softer texture of their endosperm. Hence, to improve their food value, hydrothermal treatment may come handy. Since, tannins on partial hydrolysis serve as antioxidants by suitable processing methodology the products from these may serve as health food components.

Popping

Popping of SAM is one of the traditional and popular dry-heat (high temperature short time – HTST) processing methods followed to prepare ready-to-eat products. The popped grains contain pre-gelatinised starch and possess highly desirable aroma and crunchy texture. Since, popped SAM contain the bran layer, they are good source of dietary fibre and nutraceuticals. Normally, the expansion volume of the popped SAM ranges from 7 - 18 ml/g, highest being for sorghum and lowest for barnyard millet. The varieties with hard endosperm and medium thick pericarp exhibit superior popping qualities. During popping the lipases enzyme gets denatured and hence, shelf-life of popped products is better than other kinds of SAM products. This is highly advantageous with respect to pearl millet as processed pearl millet has very low shelf-life. HTST treatment ensures that popped foods are generally microbiologically safe. Popped grains serve as snacks after seasoning and also could be used for preparation of sweet meats such as *laddu* or *sattu* and *chikki etc.* Popped SAM could be blended with toasted or puffed legumes, oilseeds and jaggery or sugar to prepare delicious and nutritionally balanced convenience supplementary foods for growing children and lactating mothers. The possibility of using popped SAM exists very much in the state sponsored nutrition intervention programmes as substitute or supplement for toasted wheat which is in vogue. The snacks and supplementary foods from them will be nutritionally superior over similar products from rice and wheat.

Popping is practiced by subjecting the grains to HTST treatment using hot sand as a heat transfer media till date in India. The sand particles adhering the product affect the eating quality as well as the food value and hence, hot air popping which is a feasible proposition should be followed. A food machinery fabricator in Coimbatore manufactures hot air toasting machine which could be successfully used for the HTST treatment for popping SAM. Alternately, the gun-popping technology followed in East Asian should be popularized in India also.

Expanded Grains

Expanded products which resemble rice poori or *murmura* would be the new generation snacks from SAM. To prepare such a product from SAM, the grins should be parboiled, pearled, bumped and then subjected to HTST treatment. The recent R&D work at CFTRI Mysore has shown the feasibility of preparation of expanded grains from millets. In fact, such a product from pearl millet is in the Indian market. Expanded grains are novel and high value products and can find application as ingredients for snacks and crispy in confectioneries and also as thickener in soup mixes.

Flaking

Cereal flakes are of three kinds in India and their methods of preparation also differ from each other, namely, using edge runner, roller flaker and extrusion cooker and flaker. Flakes from SAM could be prepared following all these methods.

1. Conventional Method

SAM soaked in cold or hot (<70 C) water for hydrating to their equilibrium moisture content (EMC), subjected to HTST treatment using hot air as heat transfer media and flattened in the edge runner. During the process the seed coat (both husk and bran) get pulerised where as the endosperm gets flattened (DSR Hyderabad is actively producing such type of flakes from sorghum). These flakes find usage similar to rice flakes.

2. Roller Flaking

The pearled grains are hydrated to their EMC, steamed to gelatinize the starch and dried to about 20 per cent moisture level and then flattened by single point impact in a heavy duty roller flaker and then dried (sorghum flakes following this method are prepared by a food industry at Bijapur, Karnataka). This type flakes are more suitable to prepare snacks after deep oil frying.

3. Breakfast Cereals

The technology followed is similar to that followed for preparation of breakfast cereals from corn. It is an elaborate process but the product will have international acceptance. The flour from SAM will be extrusion cooked and the extrudate will be cut into small bits, and then flattened in the roller flaker, dried, blistered and then coated with sugar and flavouring agents. Normally there flakes are taken along with cold or hot milk as breakfast cereals (Breakfast cereals from finger millet are successfully prepared and marketed by two food industries from Bangaluru).

The process of flaking gelatinizes the starch and also inactivates the lipase. Hence, the flakes are RTE products and will normally have better shelf-life. This will be a boon to pearl millet as its products normally develop rancidity quickly. The flakes from SAM would be a new avenue for their extended utilization as value added products and consumer friendly snacks for the non-traditional millet consumers. SAM flakes could also be used as component of midday meals and other feeding programmes.

Currently, cereal flakes are used as ingredients of muesli. Flakes from roller flaking process and also breakfast cereals from SAM could go well along with oats and other cereals flakes in muesli and add to the nutritional quality of the products.

Malting and Brewing

Malting is one of the very early biotechnological processes adopted for cereal processing for food and brewing. Although, barley has the place of pride for malting, sorghum and finger millet malting is also practiced extensively. Finger millet malting is mostly followed in India for specialty food product formulations. Pearl millet has very limited scope for malting as the malt will have poor keeping quality; likewise, other minor millets are at disadvantage because of the low level hydrolytic enzymes in their malts. Malted finger millet being a good source of amylases and micro-nutrients is termed as "Amylase Rich Food" (ARF). The malt is a store house of active hydrolytic enzymes which hydrolyse the starch to sugars and maltodextrin during cooking the slurry. The cooked product will be of low bulk density and calorie dense food. This characteristic feature is gainfully utilised to prepare a wide range of specialty foods such as nutrient dense weaning and infant foods, medical foods and milk based beverages. Thus it could be stated that, "Finger Millet Malt is a New Ingredients in Food Industry". Incorporation of about 5 per cent finger malt mandatory in the energy food formulation being used in the state sponsored nutrition programmes to improve the texture and digestibility of the product.

Sorghum malting is carried out on industrial scale in South Africa, Zimbabwe and Nigeria and it serves as principle raw material for brewing. The commercial beers manufactured in Africa are the lactic acid fermented thick brew and also clarified beer similar to barley beer. In India also SAM are used as adjuncts in brewing to a limited extent by a select few brewing houses.

Pasta/Vermicelli/Noodles

Pasta and vermicelli/noodles are generally prepared from wheat because of the beneficial properties of gluten. But SAM being non-glutinous cereals, their starch has to act as a binder between the particles to prepare pasta or vermicelli-noodles. Hence, the flours or the fine semolina from SAM need special pretreatment to partially gelatinize the starch to extrude into strands. Very often some kinds of functional ingredients such as gums are also used to facilitate binding. However, efforts to prepare noodles from these grains have not been fruitful till date and the composite flour consisting of wheat and FAM are used for the purpose. Such products are marketed in Karnataka and Tamil Nadu. Such products exclusively from SAM will have high market potential especially in Europe and North America because of their non-glutinous nature.

Roti/Tortilla

Roties are highly popular traditional food products from SAM which are normally consumed fresh. But now a days, dry rotis are sold in some places in the country but such products are brittle and need to be softened prior to consumption. Moreover they are not suitable for warming prior to consumption. These factors limits their acceptance. Since, most of the consumers prefer soft and warm rotis, products like Mexican tortilla will find buyers in the country. It is needless to mention here that, ready-to-warm rotis from SAM will have very good market as the normal Indian consumer prefers chapathi or roti to bread. Alternately, pre-processed roti flours that enable mixing with cold water and rolling similar to wheat flour will offer convenience to the time starved consumers. Roti making machines suitable for house hold to hostels and catering institutions have been developed and are being used in the country, but qualities of the products form these are not up to the mark and hence, R&D work to improve the roti making machines deserves attention.

Bread and Bakery

It has been very well documented that, SAM flours singly are not suitable for preparation of bread and bakery products similar to wheat. But composite flours consisting of wheat blended with 20 - 30 per cent SAM could be used for preparation of such products without affecting the texture and taste. In fact, the products from the composite flour would be nutritionally superior to wheat based products due to the phytochemical contents of SAM. However, there is need for developing the processes for preparation of bread and bakery products from SAM similar to the Scandinavian rye bread. Obviously, such products would be gluten free and find widespread application as health foods.

Starch, Dextrins and Ethanol

SAM are basically starchy grains and have potential for industrial level production of starches for food and allied application. Technologically and economically, sorghum has the potential for production of such products. It is learnt that, a substantial quantity of sorghum is diverted for starch production in the country. Very often sorghum caught in rain at pre-harvest situation and also fungal infested due to faulty storage could be profitably for ethanol production. This may find application in the years to come, although, the present regulations of GOI do not permit for this. Despite this regulation, it is learnt that considerable quantity sorghum is diverted for this purpose.

Papad and such other Meal Adjuncts

Papad, sandige, murukku, chakkuli and such other products prepared normally at home or cottage industry level are important adjuncts in the Indian diets. *Papads* from finger millet are popular. Utlilisation of pearl and foxtail millets for deep fried adjuncts will be of highly desirable quality, probably because of their inherent lipid contents. Suitably SAM flours blended with legumes (moong bean or horse gram) could be sheeted and cut into products of required shape and size, and could be toasted or deep oil fried or blistered in hot air to prepare ready-to-eat multi-grain snack products. Low fat and high fibre snacks similar to taco-chips from SAM will find ready acceptance by one and all.

Parboiling or Hydrothermal Treatment

Parboiling or hydrothermal treatment hardens the endosperm texture and improves the decortication or milling and also the cooking characteristics of cereals. The process also increases retention of vitamins, minerals and some of the antioxidants of the cereals and thus upgrades their nutritional value. Parboiling of rice is carried out extensively whereas parboiling of wheat (bulgur) is done to a limited extent. The R&D work on parboiling of SAM is scanty even though the potential has been established. Parboiled grains on cooking form less sticky food and suit preparation of multi-grain food products similar to *Kashipilaf* marketed in US. Parboiled and milled millets could be utilized similar to rice for preparation of convenience foods such as *bilibelebath, puliogare* and ready-to-eat expanded cereals. Very recently, the parboiling technology has been successfully applied to finger millet for production of decorticated ragi (ragi rice), which is altogether a new product. Parboiled millets are also marketed by a food industry in the country.

Extrusion Cooking

Among the various contemporary cereal processing technologies, extrusion cooking is highly popular and are largely followed for corn and rice. SAM could also be extrusion cooked to prepare ready-to-eat (RTE) products successfully. The products will have crunchy texture and can be coated with traditional spice and condiments. Alternately, the grits could be mixed with spice and condiments prior to extrusion to obtain RTE snacks of desirable taste and palate. Extrusion cooked products being of ready-to-eat nature will have greater scope for use as weaning and supplementary foods and also as a component of edible films. Adapting these technologies, it is possible to prepare multigrain snacks or supplementary foods or health bars. Extrusion cooking of SAM has very high potential for production of pet foods, the demand for which is expanding in the country.

Feed Formulations

In most of the developed countries, SAM find extensive usage as feed components for bird, cattle and pigs. Normally to improve the biological value of the feed formulations, small millets are dehusked where as sorghum and pearl millet are steamed and bumped.

Sorghum and millets are excellent sources of carbohydrates, micronutrients and phytochemicals with nutraceutical properties. They contain 7-12 per cent protein, 2-5 per cent fat, 65-75 per cent available carbohydrates and 15-20 per cent dietary fibre. Niacin and pyridoxine contents in pearl millet are higher than all other cereals, whereas, finger millet proteins are unique in that they contain higher levels of sulphur rich amino acids. Impressively, finger millet also contains about 10 times higher calcium than the other cereals. Also, the millets are admirable sources of B-group vitamins. The higher levels of dietary fibre and micronutrient contents and complex nature of starch and also the phytochemicals with nutraceutical characteristics provide a place of pride for these cereals as health grains. Hence, the image of sorghum and millet as a poor man's food could be overcome by developing nutritionally and texturally improved products that are attractive and appealing even to the elite and affluent population. It may be noted that, the urban consumers want food products that deliver convenience with desirable taste, texture, color and also shelf-stability at affordable price. In view of this innovative methods of processing them to make available in more and more user friendly manner deserves attention.

The knowledge base on SAM grain quality, processing and value addition is now increasing with recent R&D work across the world. Special nutritional and technological features of these food grains has been adequately documented but still a lot has to be done for widening the scope of their food and allied utilization. There exists great potential for production and marketing of identity preserved ready-to-cook convenience products. The prominent food and allied products from SAM that will have consumer acceptance would be shelf-stable roti flour, extrusion cooked and expanded ready-to-eat snacks and supplementary foods, noodles, malt foods, *papad* and such other food adjuncts, flakes, quick cooking cereals, health foods, breakfast cereals. pet foods and bird feeds. Processing technologies may not be the major obstacle to successful production of value added products from SAM as there is substantial knowledge in that arena. However, consistent supply of good quality SAM for industrial level processing, suitable marketing strategy for the products and creating awareness about their health benefits need to be addressed. This aspect deserves greater attention as Government of India is contemplating supply of these grains under PDS under the food security programme.

In summary, the following aspects of sorghum and millets need immediate attention with respect to post-harvest processing and value addition;

1. Design and development of a versatile millet mill (suitable for custom milling; about 100 Kg/hr milling capacity) for pearling sorghum, pearl millet and finger millet and also a compact mill consisting of dehusker (rubber roll sheller), gravity separator for segregating unhusked grains form the dehusked (similar to paddy separator) and polisher for the dehusked grains.
2. Development of simple technologies for preparation of semolina and flours for various end uses suitable for preparation of traditional foods as well as value added products such as breakfast cereals, vermicelli noodles, pasta etc,
3. Development of a mini-malting unit integrating steeping, germination and kilning operations in a single unit,
4. Hot air based HTST unit for sand less popping,
5. R&D work on generating evidence based information on the phytochemicals with nutraceutical characteristics and authentication on the anti-diabetes, anti-inflammatory and hypo-cholesterolemic properties, by undertaking clinical and nutritional surveys. And development of functional foods for diabetes and obese populations based on SAM. Inclusion of these cereals in the midday meal and ICDS and such other mass feeding programmes,
6. Identifying factors for diminishing the consumption of sorghum and millets in the country and to initiate measures to improve their consumption,
7. Breeding to improve the levels of lysine and tryptophan and also screening the germplasms for specific end uses such as milling, popping, malting and vermicelli noodles etc.,
8. Setting up of a training-cum-demonstration centre for integrated processing of sorghum and millets,
9. Creating awareness about the nutritional and health benefits of sorghum and millets by print and audiovisual media,
10. Undertaking survey of sorghum and millets food and allied industries, for bringing out a directory and also to share knowledge base for modernization of the SAM processing industries.

Opportunity of Developing Value Added Millet Products

Over the years, there has been a drastic change in farming, production and consumption of sorghum and millets in rural and urban India. Since millets are termed as poor man's food, the effect is more visible in rural areas. Generally, the demand for such crops is higher where the farming is cost effective, highly marketable, long shelf life, readily processed and that creates employment through food processing. Previously, millets farmers had little interest in cultivation, as there was neither support from government policies (like no subsidies as rice, wheat and not included in PDS, etc.) nor incentive in terms of higher remuneration for its cultivation. Finding market for the cultivated grain is so difficult for the farmers making them divert to other cereals and cash crops. The grain keeping quality is also very low and hence farmers cannot store the grain for long season. *Rabi* season sorghum grain is of good quality but the yields are very low and the *Kharif* sorghum gives high yields but the quality is very poor which is not suitable for consumption. Primary and secondary processing of sorghum and millets is a difficult process that make rural and urban population shift to other cereals. These are few reasons for the decline in cultivating millets by the farmers. A gap analysis done by IIMR points out the inconveniences associated with the existing methods of food preparations which are time consuming and involve drudgery. This is primarily due lack of processing interventions in sorghum and millets zones. Also, the absence of processing machinery suitable to mill and process them compared to in rice, wheat and corn render millets cultivation non-remunerative. Thirdly, the nutritional merits of millets over other fine cereals are not exploited due to lack of awareness among the consumers. The shift in preference to other foods in urban population can be addressed by innovation in secondary processing methods. Under the NAIP sponsored project, Indian Institute of Millets Research under Value chain approach has come up with excellent ready-to-cook and ready-to-eat sorghum value added products which gained great demand in urban markets where there is growing concern for health related problems. Some of the products commercialized by IIMR includes sorghum rich multi grain atta, sorghum rawa (fine and coarse rawa for upma, dosa, idli etc.), extruded products (vermicelli and pasta), flaking (breakfast cereal), baked products (biscuits and cookies), instant mixes (dosa mix, idli mix, peda mix, etc). These technologies can now be replicated for other millets too. These products are now being successfully commercialized on a pilot scale in the city of Hyderabad. IIMR with millets promotion as mandate is aggressively promoting these technologies for replication and horizontal expansion activities targeting private sector either through startup entrepreneurs or even big players in the form of technology transfer driven process.

Need for Product Development

Food processing industry plays a vital role in the economic growth of the country. It helps to increase the production and income of farmers, reduce the post-harvest losses, generates employment and creates a market for agricultural products. Food processing is defined as a process of modifying or adding value to the agricultural products by various methods. The main objective of Product Development is to create food products with different characteristics that give additional benefits to the consumers. Products that are developed with innovative processing technologies improve the shelf life, quality, palatability and nutritional composition of foods.

In India, food processing industry has emerged as a key component in the Manufacturing sector. Previously, food processing was majorly confined to the food preservation, packaging and transportation. Emerging processing technologies made it easier to produce RTE and RTC foods. Current government policies as well as economic growth of the country have contributed to the growth of processing industries, development of various food products and expansion of the market. The demand for processed and convenience foods is increasing constantly because of greater domestic demand, raised income levels, urbanization, growing middle class and younger generation, life-style and food habits of the people, rising export opportunities and proximity to key destinations. Among the food processing industries, grain processing industry is the key component of the food sector which shares about 40 per cent of the market.

Over the past decade, millets are gaining more importance for human consumption due to its nutritional and health benefits. India is one of the primary or secondary centers of diversity for sorghum and other several millet crops. The cost of cultivation of sorghum is also less compared to other fine cereals. However, millets utilization was limited in processing sector and its consumption declined due to urbanization, previous government policies, lack of processing technologies and awareness on health and nutritional benefits, laborious and time-consuming process involved in preparation and non-availability of convenient foods. These resulted in the shift towards more convenient/processed products of fine cereals like rice and wheat from the Sorghum. In some parts of India, this trend has resulted in increasing malnourishment among low economic population.

Food and Nutritional security are the major challenging constraints among developing nations. In India, more than one third of the population has been estimated to be poor and, one half of it is malnourished. This conflicts the assurance of food security in a country like India. Rice and wheat alone cannot suffice meeting food and nutritional security in the wake of increasing rate of growth in population. Cereals and millets should offer a dependable source with their rich nutritional profile. However, efforts in processing millets are highly inadequate to make them more acceptable as convenient foods.

Millets have unique nutritional properties that promise the nutritional security of our nation. Labeling and flagging of these nutritional benefits in combination with processing interventions such as milling, baking, flaking and extrusion cooking is expected to raise the demand for its increased utilization. Further, processing interventions in millets will enable emergence of ready to eat and cook foods which would appeal to most of the working, and middle class groups.

Sorghum and millets can be milled to produce starch and grits (semolina) from which many Indian traditional recipes can be made. The methods of primary and secondary processing systems are not widely practiced and products made from them have limitations for wider and appropriate utilization of the millets. There is a need to find if millets preparations are appealing and could be consumed daily in place of other cereals. Appropriate processing technologies should be used to produce shelf stable and palatable convenient foods

PPP Opportunities in Developing and Marketing Millets Based RTC/RTE Products

First initiative on public private partnership (PPP) in a big way happened when IIMR linked MoU with ITC (ABD) in 2008 under NAIP. Many other linkages across the millets growing belts are following with Government's programme INSIMP (Initiative for Nutritional Security through Intensive Millets Promotion) is launched in 2011-12. Backward integration of Sorghum value chain was pilot tested by ITC (ABD) Ltd under IIMR led consortium on Millets Value Chain funded by NAIP/ ICAR (World Bank aided). It was enabled with its support under *e-choupal* platform where beneficiary farmers are market assured to ascertain their functionality, feasibility and sustainability of intensive sorghum cultivation. It is relatively a new phenomenon in India Millets scenario, which paved the way for Supply chain management and further strengthened the development and commercialization of sorghum products. Further, the product specific on farm production led to procurement, aggregation and linking further with other stakeholders in the value chain ultimately linking of producers to consumers. This sorghum value chain though existed it was in a disintegrated manner. Since the Fair Average Quality (FAQ) grains and their quantity were predetermined under the PPP, both the parties could rely on each other; it provided a better price to the farmers and sufficient quantity and desired quality of sorghum to the processors for smooth functioning of their business.

End product-specific cultivars developed/identified by IIMR, are technologically backstopped to all the beneficiary farmers in Maharashtra and AP states. Here, the ITC (ABD) ltd for the first time took up farm extension services in sorghum on pilot scale along with a buyback arrangement. The identity preserved grain was procured and bulked to overcome supply chain issues under low marketed surplus regime.

Developing sorghum products *vis-à-vis* determining their nutritional values, and simultaneously promoting and marketing them has enhanced the awareness and preference of this food in recent years. Rural *choupal haats* were arranged to create awareness on this new initiative. The value chain approach thus provided end to end solutions so that forward integration may also be enabled for processing and commercialization by the small and medium scale entrepreneurs. The impact of this successful PPP collaboration has enhanced productivity and higher incomes to the stakeholders. There was an increased allocation of better lands to sorghum cultivation which can be ascribed to the impact of the PPP initiative.

Other initiatives which is worth mentioning is MoU entered between IIMR and M/s Britannia Industries Ltd, Bangalore, a multinational company in food sector to undertake

Joint R&D on sorghum biscuits and bakery products for three years where the IIMR will be identifying cultivars suitable for these products and pilot test backward integration while Britannia will come up with some sorghum based products with their functional portfolio and rheological properties studied in depth by both the partners. This PPP is expected to lead to create demand for sorghum as the uptake is going to be in large volumes and the primary stakeholder may be benefitted in long run. Britannia Industries is expected to categorize the sorghum based products under their nutri-choice banner. Besides many other MoUs with potential food players in Hyderabad and from non-traditional cities such as Delhi NCR, Bangalore and Pune, M/s Ensigns Life Sciences Pvt. Ltd. has approached IIMR for export purpose. Negotiation has also been underway with M/s Parle for horizontal expansion.

Status of Millets Processing and Commercialization

Primary processing of millets is necessary to remove the inconveniences in millets preparation and consumption. Secondary processing has started with the increase in the awareness of millets' nutritional values by the consumers. With the implementation of INSIMP project initiated by Department of Agriculture and Cooperation (DAC), Government of India, there has been a mass pilot scale millets processing taking place across the millets growing states.

Few sorghum processing technologies have been developed in the country with IIMR playing the key role in developing them for promotion and commercializing them as health foods meant for all age groups and niche segments such as diabetic patients, old aged groups and school children. This knowledge has also been disseminate to the farmers through entrepreneurship development programs so that they are either aware of the various developments that have been taking place or learn to start their own processing at farm level.

Since there is no machinery in the market exclusively meant for sorghum processing, the available machineries that are meant for processing other fine cereals such as wheat and rice are procured and retrofitted for sorghum processing.

These processing technologies have gain wide momentum from the private firms and entrepreneurs owing to growing awareness of nutritional benefits of millets and preference of healthy and convenient foods among the consumers. The intervention also perceived as the best method by the policy makers to promote millets consumption and demand in the country.

Seasonality vs. Quality of Sorghum

Though the *Kharif* Sorghum cultivars produce higher grain and fodder yields, they are prone to mould due to excess or untimely rain at the time of harvest usually during October. This affects the grain quality and makes it unfit for human consumption. Thus the price of *Kharif* grain is lowered due to its poor quality and is increasingly diverted for non-food industrial uses almost all the years (*i.e.*, atleast thrice in every five years).

Current trend of *Kharif* Sorghum is mainly determined for the fodder requirement from the crop. Introduction of hybrids witnessed a major change in Indian Sorghum farming specifically in *Kharif* season.

Though the productivity is comparatively lower than the rain-bound *Kharif* grain, the *Rabi* Sorghum grain is far more superior in quality, due to its maturity in cold climate. However, of late some of the *Rabi* Sorghum grain is cultivated under protective irrigation yielding almost double the quantities. M 35-1 is ruling for a long period with excellent grain quality traits though yields are not so high. Recently released *Rabi* varieties, *Phule* series, CSV 14R and CSV 18 were also better received by farmers. In general, due to its better quality, the profit realized by *Rabi* grain is almost double to that of *Kharif* grain.

There has been an increase in use of machinery after the advent of technology through retrofitting process, especially for dehulling after post NAIP. Subsequent promotion from Government of India's INSIMP programme launched across the country, which has a component of post-harvest clusters supply to beneficiaries the use of machineries technologies increased (IIMR is technologically backstopping them). These prototypes are used by R&D institutes and small scale entrepreneurs. The Dehuller decorticate outer layer. Though it may lead to partial loss of minerals and vitamins, such grains can be used for direct food consumption; and amenable for product development and used for preparation of various RTC/RTE sorghum products which are convenient as well as with improved taste benefits.

The market price of de-hulled grains will be higher than the moulded grains. Due to changing scenario of sorghum emergence as a commercial crop, it has a good potential to be used for various industrial purposes. It has already established its ground with regard to use as cattle feed, poultry feed and potable alcohol. Thus, *Kharif* sorghum grain has primary utilization in poultry, alcohol and other industrial sectors and as secondary role as direct staple food.

Prospective Areas (Districts) where Backward Linkages as well as Production can be Taken Up

As per the latest data derived from the Directorate of Economics and Statistics, ten districts each in Maharashtra and Karnataka and five districts of Andhra Pradesh during *Rabi* season have high potential where sorghum food processing can be sustainably taken up through sustainable linkages with the farmers. Tables 1 and 2 reveals the three years average *i.e.*, 2007-08 to 2009-10 average in case of Karnataka and 2008-09 to 2010-11 average in case of Andhra Pradesh and Maharashtra.

Processing Interventions at IIMR

Processing methods have been initiated by Indian Institute of Millets Research (IIMR) led consortium under National Agricultural Innovation Project (NAIP). IIMR has developed and fine-tuned different processing technologies such as extrusion, flaking, milling, parboiling, baking, popping etc., to prepare good quality of sorghum processed products for better utilization of sorghum and provide choices to the consumers in the form of convenient foods throughout the year. Similarly interventions have been made by other R&D institutions and State Agriculture Universities (SAU) to attempt processing interventions in sorghum and millets as a means of value addition. However, the value addition approach for revival of demand for sorghum through interventions and flagging nutritional merits of sorghum is quite novel and successful pilot model.

Table 1: Major *Rabi* Sorghum Producing Districts of Andhra Pradesh, Maharashtra and Karnataka

Karnataka				*Maharashtra*				*Andhra Pradesh*			
District	*Area*	*Production*	*Yield*	*District*	*Area*	*Production*	*Yield*	*District*	*Area*	*Production*	*Yield*
Bagalkot	1.16	1.32	1.16	Ahmednagar	5.23	3.32	0.64	Adilabad	0.21	0.17	0.77
Belgaum	1.30	1.30	1.01	Beed	1.93	1.58	0.82	Guntur	0.04	0.27	6.33
Bellary	0.24	0.30	1.29	Jalna	1.38	1.24	0.90	Kurnool	0.56	1.10	1.90
Bijapur	2.06	1.96	0.96	Jalgaon	0.48	0.73	1.53	Mahabubnagar	0.10	0.09	0.87
Dharwad	0.37	0.25	0.70	Osmanabad	2.74	2.18	0.80	Medak	0.16	0.18	1.07
Gadag	0.60	0.47	0.79	Parbhani	1.68	1.19	0.70				
Gulbarga	2.74	2.61	0.96	Pune	4.34	3.07	0.71				
Haveri	0.28	0.25	0.89	Sangli	1.80	1.05	0.58				
Koppal	0.46	0.27	0.60	Satara	1.44	1.19	0.83				
Raichur	1.05	1.34	1.40	Solapur	6.62	3.96	0.60				

Source: Directorate of Economics and Statistics, DAC, MoA, Govt of India.

The *Kharif* production is comparatively lesser. Most of the districts produced less than a lakh tonnes since they are cultivated mostly for fodder purpose and not for food purpose.

Table 2: Major *Kharif* Sorghum Producing Districts of Andhra Pradesh, Maharashtra and Karnataka (Area: In lakh Hectare, Production: In lakh Tonnes, Yield: Tonnes/Hec)

Karnataka				*Maharashtra*				*Andhra Pradesh*			
District	*Area*	*Production*	*Yield*	*District*	*Area*	*Production*	*Yield*	*District*	*Area*	*Production*	*Yield*
Belgaum	0.30	0.48	1.64	Amravati	0.46	0.47	1.02	Adilabad	0.25	0.38	1.53
Bellary	0.30	0.34	1.18	Beed	0.34	0.30	0.90	Kurnool	0.08	0.23	2.95
Bidar	0.56	0.70	1.28	Buldhana	0.50	0.79	1.59	Mahaboobnagar	0.42	0.31	0.77
Chamarajanagar	0.19	0.12	0.64	Dhule	0.23	0.32	1.35	Medak	0.21	0.21	0.96
Davangere	0.20	0.42	2.18	Hingoli	0.39	0.44	1.12	Rangareddy	0.16	0.17	1.01
Haveri	0.11	0.15	1.46	Jalgaon	0.80	1.73	2.09				
Koppal	0.12	0.13	1.14	Latur	1.22	1.32	1.08				
				Mandurbar	0.28	0.44	1.60				
				Nanded	1.21	1.10	0.91				
				Osmanabad	0.62	0.37	0.61				
				Parbhani	0.62	0.68	1.10				
				Sangli	0.73	0.71	0.92				
				Satara	0.46	0.77	1.68				

Source: Directorate of Economics and Statistics, DAC, MoA, Govt. of India.

1. Destoning and Grading

Destoner and Grader machine at IIMR is a single unit, retrofitted from destoner machine and seed grader machine, which perform dual function of both separation of impurities from grains and grading. Removal of impurities from grain is a laborious process. With this machine sorghum grains can be destoned and graded.

Grains are fed through feed hopper with feed control gate. These dropped grains are then subjected to the aspirator fan that clears dust, straw, chaff etc. Dust and other particles are collected through separate openings. Destoned grains are passed through different layers of sieves. Through agitation process, as per the width and thickness of grains are graded though different layers of the sieves. Graded grains were collected through one outlet and broken grains through another outlet. 300kg/hr grains can be destoned and graded using this equipment.

2. Dehulling

Dehulling/decortication/debranning is done primarily to remove the pericarp/outer layer. Normally, pestle and motor or hand pounding are used to dehull the grains. It is not an efficient method of processing because, part of the husk remains in the grain and there is considerable loss of grain. Improperly dehulled grain cannot be grounded to fine quality flour.

The removal of fibrous layers from sorghum can be operated mechanically. The prototype of dehuller consists of grinding discs and acts as abrasive dehulling surfaces. Grain is dropped in the feeder, enters the dehulled chamber where the grinding stones decorticate the grain. Through the inspection door the grain can be checked for dehulling quality. Then the grain is released from discharging door which is at the bottom of the machine. Pearling of sorghum grain removes bitterness, increases palatability of the flour (Desikachar, 1982) and improves *roti* quality (Vimala *et al*, 1996).

Kernels, with vitreous endosperm or with higher proportion of endosperm are more suitable for dehulling and yield higher fractions of dehulled grains; whereas soft endosperm does not yield high and operational losses are more. But exposing soft endosperm kernels to steam for few minutes may harden the endosperm and makes it suitable for dehulling.

Indian sorghum genotypes are free from/or with very low tannin contents makes them more suitable for dehulling. But if Sorghum grain is caught up in rains at the time of harvest, it develops moulds and causes discoloration of pericarp of grain during *Kharif* season. The application of dehulling processing becomes important to remove moulds and phytochemicals on grain to improve the eating quality.

Now with the introduction of dehulling machines, it is possible to dehull the sorghum grains more efficiently and economically. However, there is a loss of nutrients in the process as a result of removal of pericarp (fiber, minerals) though the coarseness is removed. It is advisable to go for partial dehulling so as to minimize the loss of nutrients. Dehulling after parboiling helps in redistribution of micro nutrients into endosperm. Dehulling is a prescribed to obtain fineness to the end products such as bakery (biscuits, cake and bread).

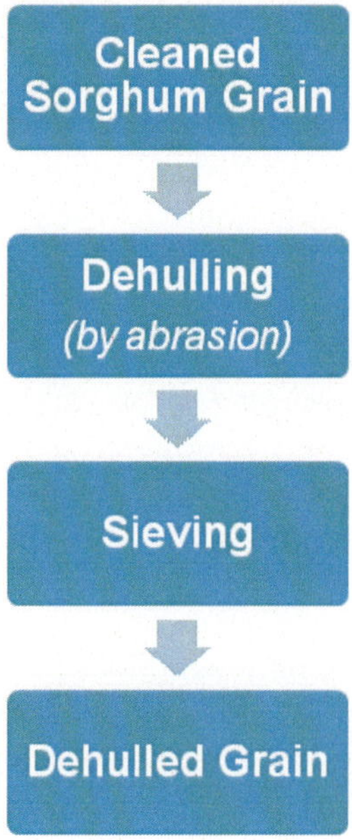

Figure 1: Process Flow Chart for Dehulling of Sorghum.

3. Milling

Milling is a process of separating the bran and germ from the starchy endosperm so that the endosperm can be ground into flour and *rawa*. Milling for food grains can be done by wet or dry methods.

Wet milling for corn is usually practiced extensively to isolate starch for food purposes. This practice is done by steeping the sorghum grain aqueous solution with or without antimicrobial agent, sulphur dioxide (Shandera *et al*, 1995). Dry milling involves cleaning and milling of the grain into the desired product. It is the separation of the anatomical parts (bran and germ) of the grain and reducing the particle size of endosperm.

Milling quality of sorghum can be determined by kernel shape, size, endosperm nature, hardness. Sorghum milling can be done by *chakki* (Emery stones)/Hammer/ Plate mill. For custom milling, emery or iron disc mills and at industrial level hammer and roller millers are practiced. Milling yield depends on type of cultivar, endosperm nature, type of mill and also pre-treatment (3-5 per cent moisten grain before processing).

4. Flaking

Ready-to-eat breakfast flakes are popular snacks made from cereals. Rice and corn flakes are produced and consumed throughout the world. Traditional process flaking involves pounding the paddy in a wooden mortar and pestle. *Poha* and *chuduwa* made from rice flakes is an important breakfast item in semi-urban and rural areas

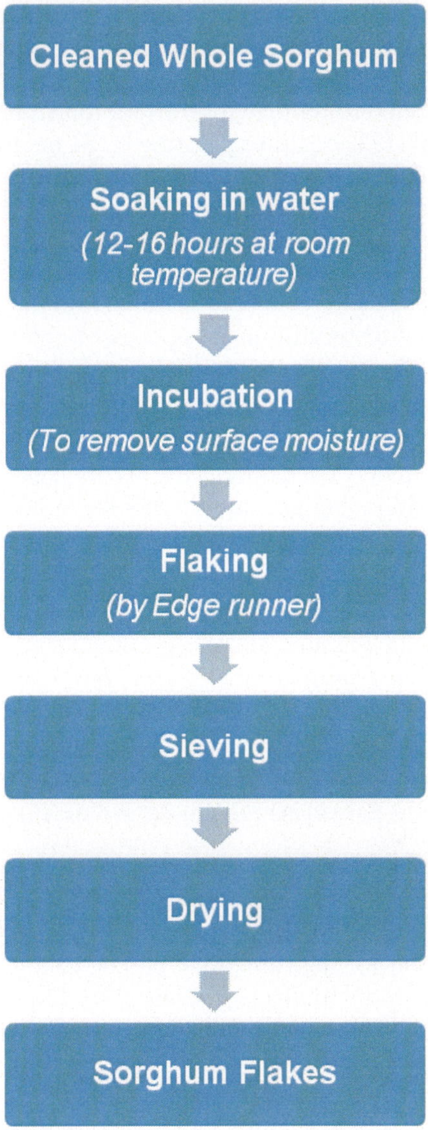

Figure 2: Process Flow Diagram of Sorghum Flakes using Edge Runner.

and middle class families of urban India. There is also a fairly large bulk market for flakes; therefore it is extensively used all over the country round the year. However, sorghum flaking is relatively new; this may trigger an overall demand for sorghum, especially in states like Maharashtra who are currently consuming rice flakes. Switching from rice flakes to jowar flakes depends on the availability and price of jowar flakes.

Flaking can be done for cereals and millets by conventional (edge runner), roller, flaker and extrusion cooking methods. IIMR has retrofitted the rice flaking machine for sorghum and optimized the conditions. Using paddy flaking equipment, that is, edge runner (flaking machine) along with roaster has made it possible to produce flakes from sorghum. In this method, sorghum grain was hydrated to equilibrium moisture content, incubated to remove the surface moisture and subjected to high temperature short time (HTST) treatment. During HTST treatment, starch granules of grain get gelatinized and these grains were fed to edge runner, where the gelatinized endosperm gets flattened and bran is eventually separated, pulverized and collected from another outlet. Sorghum flakes obtained from this technology resembles rice flakes and is texturally more appealing due to quick hydration characteristics and less chewy properties. The yield recovery fractions from this technology depend on cultivar type, nature of endosperm, and the optimized conditions and the resultant output is 50-65 per cent. The broken flake powder is a by-product which can be used for making value added sweets for increasing the overall profitability of flaking for commercial purposes.

Roller flaking is commonly used for making of corn flakes and is capital intensive compared to the conventional edge runner. This technology involves pearling of grains and hydrating, steamed to gelatinize the starch and then flattened between rollers. This type of flakes is texturally different from conventional method flakes, as they are more leathery and chewy and are more suitable for blistering or deep oil frying.

Flaking is done using extrusion technology and produces breakfast cereals which can be consumed with cold or hot milk. This technology is widely used for only corn. However, there has been some efforts to adopt it for sorghum flaking, the grain flour or *suji* will be cooked and extrude and sheared into small balls and flattened in roller flaker, dried, blistered and with or without sugar and flavour coating.

5. Baking (Biscuits Preparation)

Bakery products especially biscuits are becoming popular in developing countries as indicated by the fact that in the last decade the production has been more than doubled. The chief reason for the same may be fast changing food habits due to urbanization. In addition, industrialization, availability, low cost, ready to eat nature, high nutritional values are factors for the in increased use (Kamaliya, 2005).

Biscuits/cookies are ready to eat, convenient and low cost food products, containing nutrient principles of vital importance (Kulkarni, 1997). The trend in biscuits making is to improve their value by improving their functional and health portfolio such as high fiber biscuits.

Under NAIP, IIMR has standardized and optimized various types of biscuits from pure sorghum to composite flour with different flavors. In this method, fat and sugar blended in planetary mixer with a flat beater. Flour was mixed with baking powder and sieved for better mixing. And to cream mix, milk solids, flour mix and essence were added well mixed to make dough. Dough was sheeted using a rolling pin and sheared with specific moulds to make a desirable shape and then baked in an aluminum tray at 150°C for 25 minutes. After baking, biscuits were cooled and packed.

Sorghum/Millets Products

The processing interventions of ICAR-IIMR led consortium are aimed to create demand for millets products owing to declining trend in consumption of millets as staple food. Priority was given to sorghum so that once it is successful the same technology may be replicated for other millets.

Traditionally, consumption of sorghum is limited to *roti, kanji, mudde and annam.* Preparation of sorghum is cumbersome, time consuming and requires skill, as it does not contain a protein called gluten (as in wheat) which makes the dough elastic and helps in rolling. Using the processing technologies (milling, flaking, baking, blending and extrusion) developed by IIMR, sorghum based recipes can be prepared conveniently. Sorghum products developed by IIMR have lower glycemic index and glycemic load than wheat-based products and are nutritionally superior over refined wheat (*maida*) and rice based products available on shelf in the market. The product wise details are given below:

Jowar Rich Multigrain Flour

Jowar rich multigrain flour is a blend of sorghum and other cereals such as wheat, ragi, Pulses (black gram dal/soya) and fenugreek (methi). The blended flour meets the emerging nutritional needs of the people in the wake of preference for modern and healthy food habits. Different grains have varied advantages. Sorghum and other whole grains and pulses add minerals, dietary fiber and nutrients to *rotis* than normal rotis made from wheat/Jowar. Addition of wheat to the dough, makes it pliable, allows better shaping and retaining the original wheat roti taste. Sorghum rich multigrain flour can be stored for a period of 2 months at ambient temperature.

Jowar Flakes

Sorghum flakes have been developed by IIMR with the advent of technology, Edge runner (flaking machine) along with Roaster. Sorghum flakes are ready to eat foods, white in color and easily digestible. They contained higher protein, fibre, calcium, magnesium, iron, zinc and folic acid than rice flakes. They can be used for preparation of variety of products such as *upma or poha, porridge, chuduwa*, fried and seasoned mixtures etc. Sorghum flakes can be stored for 4 months at ambient temperature.

Jowar Rawa

Sorghum semolina locally called as rawa or suji can be made through milling technology. IIMR has developed three different types of rawa (coarse, medium and fine) for the preparation of various recipes. It is a ready-to-cook product which has nutritional significance especially in case of obese and diabetic patients. Sorghum rawa can be stored for a period of 3 months at ambient temperature. Availability of sorghum in suji form will help in overcoming drudgeries involved in suji based recipes.

Jowar Biscuits

Biscuits are ready to eat foods. IIMR has prepared pure sorghum biscuits with sorghum flour, trans free fat, sugar, naturally identical flavoring substances and baking powder. They are rich in protein, fibre, calcium, magnesium, iron, zinc and folic acid and less sugar than the market products. Pure sorghum biscuits can be stored for a period of 6 months.

Jowar Vermicelli and Pasta

Jowar vermicelli and pasta are ready-to-cook products, prepared through cold extrusion. The texture of products becomes hard and brittle after drying and can be cooked in few minutes. Jowar vermicelli and pasta are prepared from sorghum and wheat semolina in the ratio of 7:3 respectively. Blending with wheat semolina offers visco-elasticity to the dough as sorghum lacks gluten. Jowar vermicelli and pasta improves the digestibility, decreases the sudden rise of blood glucose levels, and has obtained nutritional benefits thereby increasing the consumption of sorghum. Sorghum vermicelli and pasta can be stored for 6 months at ambient temperature.

Way Forward and Future Plan

With the given importance of sorghum and millets in the human life, the necessity of reviving millets demand for food and fodder is substantial. To promote the millets as food product, value-addition is appropriate in each stage, which gives the optimal level of returns to consumer as well as producers/processors. Moreover, Consumer centric market economy must be supported by demand driven farm production, procurement, storage and handling systems. One dynamic approach to promoting commercialization of millets will be guidance to individuals or groups for setting up micro enterprise at rural level. In these efforts, the IIMR has also planned to set up technology incubation center with all the required millet processing facility for the dual purpose - both sustainability and upscaling this innovative approach towards commercialization of millets in the country. As far as demand is concerned 'consumption' parameter is an important determinant of millets economy. The value addition technology includes, winnowing, threshing, grading, standardizations, quality control, storage, processing, package, transportation, brand name, packing, selling of produce through suitable market channels. This value addition will also innovate the best alternative technology; helps in reduce post-harvesting losses and provide suitable remedial measures for the dryland farmers and to all the stakeholders.

One of the major causes of lack of processing and commercialization interventions in millets is due to lack of desired quantity and quality of raw material (sorghum/millets grain) due to low marketable surplus on the part of the farmers. However, the marketable surplus are determined by various factors, and unless otherwise if the remuneration received from the crop is comparable to the competing crops, they subsequently led to decline in such business, which so happens in case of sorghum and millets in the country.

The following points are derived from the millets value chain interventions, which are in fact innovative steps that are seen to be a sustainable approach through which the demand of millets can be created in the country.

1. Technological innovations:
    a. Identification of product specific varieties in millets

   b. Nutritional evaluation of millet foods and using data for labeling highlighting nutritional features
   c. Shelf –life enhancement through improved packaging to meet modern food habits by providing balanced nutrition
   d. Standardization of millet rich multi-grain flour and food products
   e. Fine-tuning of technologies in millet foods

2. Change from commodity-based, on-farm production to market-driven end-product specific farm production and market assurance to the millet farmers through buy-back facilitation.
3. Public-private participative (PPP) model of holistic farm-extension services on seed supply, fertilizers, plant protection chemicals, crop management and training.
4. Ready-to-eat and convenient foods to overcome cumbersome and time-consuming food preparation of millets.
5. To establish more linkages with line departments of government
6. To identify and standardize designer foods targeting mid-day meals of School children.
7. Branding millets as health foods through value-addition through nutritional evaluation and certification, and implementation of HACCP.
8. Innovative approaches for popularization such as road shows, wet sampling, and millet tableaux for Republic day or Independence Day parade, and at important public places and functions.
9. Coalition building with other research and private groups working on millets elsewhere in the world through exchange visits and conducting national seminar and workshops on millets and further global meet on millets.
10. Recognition and value-addition to land races nursed by farmers for long; possibly also securing IPR for local innovations.
11. "Model of support" to dry land farmers by linking with government schemes such as mid-day meal scheme in pilot scale in one district each of Andhra Pradesh, Maharashtra and Karnataka under INSIMP, so as to extract value, and make public welfare system more efficient and locally relevant educating on nutritious use of millets formulations.

Way Forward Plan of ICAR-IIMR

- Horizontal expansion- Upscaling of existing commercialized technologies and Replication
- Exploring more business tie-ups with Nestle, ITC foods, Bambino to engage in commercialization of targeted products in volumes
- Creating aggressive awareness strategies with communication plans such as advertisement and through celebrities

- ☆ To establish more linkages with line departments of government
- ☆ Mainstreaming millets in public funded welfare programmes
- ☆ Technology incubation for handholding at IIMR
- ☆ Establish mini commercial scale millet processing facility

Millets: Ensuring Climate Resilience and Nutritional Security (2015) *Pages* **575–592**
Editors: **Vilas A. Tonapi and J.V. Patil**
Published by: **DAYA PUBLISHING HOUSE, NEW DELHI**

Chapter 22

Commercialization of Millet Based Food Products

B. Dayakar Rao, K.N. Ganapathy and J.V. Patil

ICAR-Indian Institute of Millets Research (IIMR)
Rajendranagar, Hyderabad – 500 030

The market for health and wellness foods in India is estimated to be worth $7.5-10 billion by 2015 according to a study undertaken by the Tata Strategic Management group; and the demand for fortified foods, whole grain, high fibre, low cholesterol, natural and organic foods is expected will grow (Madhavan, 2013). However, the perception of consumers dubbing millets as poor man's crops has been a hindrance to their commercialization. Also, according to Nagarajan and Smale the minor millets namely, Finger millet (*Eleucine corocona*), foxtail millet (*Setaria italica*) and little millet (*Panicum milliaceum*) are considered 'minor' because of the lack of research investment they attract, and their limited commercial importance in terms of area, production and consumption. They are insignificant even in national and international marketing system (pearl millets have none) in the regions where they are traditionally grown (Rai *et al.*, 2008).

However, millets have high nutraceutical properties, besides, various nutritional studies suggested millets as better choices than fine cereals such as wheat and rice for the nutritional security of the rural and urban poor who have limited access to other sources of dietary components. In these regard, various attempt by various research institutes and organization for value addition in millets. However, they are hardly known by the consumers as they are not available on-shelf in the market. But for commercialization, the products need to be standardized oganoleptically with high functional properties added to them, and acceptable to the consumers.

In the Kolli Hills of Tamil Nadu, despite a consumption preference among the farming communities for minor millets, they have been cultivated on a subsistence scale, and marketing was considered the main constraint for scaling up the production. As a response, the M.S. Swaminathan Research Foundation (MSSRF) based in Chennai has led targeted conservation cum commercialization intervention programs over the last 7-9 years, to establish a robust local market for minor millets- within 100 km radius- by using promotion campaign, awareness, and introduced new recipes in order to add value locally. The MSSRF established market linkages at the local and regional levels. To do so, self-help groups and enterprises were formed to institutionalize market operations. The various groups have also been brought under a single system of federation that was linked to the Tribal Cooperative Marketing Development Federation of India Ltd. (TRIFED). Marketing mechanisms and particular economic linkages were negotiated and resulted in a Memorandum of Understanding. Other initiatives in this regard include branding and innovative packaging. They have also encouraged lots of entrepreneurs to take up these technologies in their business.

In regard to sorghum, efforts of ICAR-IIMR led intervention in public private partnerships mode under NAIP (2008-2014) have led to commercialization of some millet products which have been mentioned in the previous chapter, though scaling up still remains a challenge; however, they have proved the point of possibility of commercializing convenient millet foods on par with rice and wheat so as to offer choices to urban health conscious consumers. Similarly there are efforts from major private player such as Britannia industries Ltd through their nutria-choice segment and ITC's multigrain biscuits which appear to play a major role in millets in the days to come. In recent times Britannia Industries have entered a research MoU with IIMR on sorghum biscuits and bakery products which is again a land mark. Apart from big players there are many small scale players who produce extruded snacks from sorghum, and pearl millet in Mumbai, Rajasthan, Tamil Nadu and Andhra Pradesh. IIMR has entered more than 20 MoU's with private partners. However, advocacy and policy prescription would take the consumption to greater heights that would generate not only demand for millets but also raise standard of living of the poor dryland farmers growing millets.

Among the millets, finger millet based malt is more acceptable product both in traditional and non-traditional consuming areas. Some of these products have premium prices and are acceptable as up market products cater to high income groups. Gluten-free sorghum products are commercialized by an entrepreneur based in Kolkata catering to celiac patients. Processed small millets are commercialized in a big way by entrepreneurs from Nasik, Maharashtra and are being exported. Though substantial studies are not attempted on the bioavailability of minerals in millets the general awareness that is being created across the country and world over that millets are nutri-cereals and offer promise to health seeking population who desire to keep lifestyle diseases at bay.

Steps Towards Commercialization of Sorghum/Millets Products

Branding

A brand can be understood simply as a name termed for a product, service, commodity, etc. The most vital and important part of a brand is its name and logo that distinguishes one seller's product from those of others. In other word, a brand can easily be understood if you refer to a brand as a person.

Effective branding can result in higher sales of not only one product, but of other products associated with that brand. For example, if a customer loves sorghum biscuits and trusts the brand, he or she is more likely to try other sorghum products offered by the company - such as sorghum cookies and other sorghum ready-to-eat products. Brand is the personality that identifies a product, service or company (name, term, sign, symbol, or design, or combination of them) and how it relates to key constituencies: customers, staff, partners, investors etc.

eatrite TM

Eat Jowar - Stay Healthy

The IIMR brand *Eatrite* was launch for commercialization of sorghum product technologies in the twin cities of Secunderabad and Hyderabad on a pilot scale with the tagline 'eat jowar-stay healthy' and 'health meet taste'. The name itself refers its health benefits and nutritional superiority over alternate products available on-shelf in the market. IIMR being a government agency received a wide momentum from across the organization and individuals, and incidentally attracted lots of entrepreneurs from this initiative that many firms entered MoU with the institute to market or for technology licensing as they became confident of the demand for these health products.

Labeling

In labeling the most important part is to mark the product as per the standard classifications. For *e.g.* if the product is a vegetarian product, we have to mark it green, as is seen in Eatrite logo.

Barcode

These days the product can be uniquely marked with its details encrypted in a barcode. It really proves very helpful for maintaining stock inventory, faster billing at POSs, product search on internet using smart mobiles. Generally for encryption EAN-13 codes, code 126 formats etc are used.

Choice of Material

Depending upon the shelf life required, product type and the finish/look we have to choose the material. If a product is fragile and might get crushed easily, then its better we pack it in a paperboard box packing or plastic containers etc. You can meet packaging experts to get a perfect understanding of your product packing needs.

Packaging

Packaging is a science, art and technology of enclosing or protecting products for distribution, storage, sale and use. It also refers to the process of design, evaluation, and production of packages. Packaging depends upon the need and product. There are various packaging types like poly packaging, paperboard, foil, etc.

The product and the package are two different things. The product is the thing we want or need, but its packaging is what we will throw away after taking or making used of the product. However, packaging adds value to the product with respect to customer's perception. A well design packaging can complement or even enhance the product it was designed to carry. Good design isn't just about the product – it's about good packaging as well.

Packaging Layout

Always the brand name should be at the centre of the layout with maximum prominence. The product identity should get equal prominence so as to describe the customer what is inside the pack. It is necessary to always mention how to use or make the product. Also if you can mention the benefits of the product, then people are like to understand your product thoroughly. Other necessary requirements are credentials, contact and mandatory packing information should be mentioned behind the pack like Quantity, Net Weight, MRP, Expiry Date, Batch No., Barcode etc.

Promotion and Commercialization under *Eatrite* Brand

"Eatrite" is a brand promoted by the Indian Institute of Millets Research, in India. This brand has been initiated under the National Agricultural Innovation Project (NAIP) which is quite uncommon for an institutional brand promotion is unique.

Eatrite name is chosen to reflect the characteristics of carbohydrates present in sorghum which are complex in nature that are released slowly, regulating hunger which helps naturally not to over eat but eating right quantities. Eating sorghum is also right because it has high fibre, minerals, vitamins and anti oxidants. Eating sorghum is right because it has manifold health benefits which are enlisted earlier in the report.

The range of products under this brand includes: Sorghum Rich Multigrain Flour, Sorghum Semolina, Sorghum Pasta, Sorghum Vermicelli, Sorghum Flakes and Sorghum Roasted Flakes, and Sorghum Biscuits/Cookies. These products are tested for their nutrient profile and nutritional suitability by National Institute of Nutrition (NIN), Hyderabad and ascertained that they are recommended for those who are ailing from life style disorders such as diabetes and obesity, and are beneficial to all age groups.

Initially we started with aggressive promotion such as organizing Road shows and health awareness campaigns in public parks, malls, etc., wherein the product technologies were displayed and samples are distributed. Later with the increase in awareness by the denizens in Hyderabad, mass media started approaching IIMR, which further spread its popularity. With the introduction of IIMR brand *EATRITE*, with the message 'Eat Jowar, Stay Healthy' it has gain a great momentum from consumers especially in urban areas where people have started looking for health foods. The impact of popularization and increase of awareness among the masses stimulated entrepreneurs who started approaching IIMR for either exclusive marketing or manufacturing and marketing the product technologies in their own brand.

IIMR has now signed Memorandum of Understanding with more than 20 entrepreneurs of which, MoU with multinational company Britannia industries Ltd is a most prominent achievement. Others ToR include adopting processing technologies at pilot scale for commercialization, technology licensing on multigrain atta and biscuits and taking up Eatrite products for commercialization. Many more are in pipeline from cities like Delhi, Bangalore and Chennai. Under this, 5 formats of business plans are commercialized for Jowar products which evolved under their relative merit assessed in terms of farmer's share in the consumer rupee.

Branding and labeling with nutritional profile, recipes, mandatory information and packing was unique in sorghum as there were no products available in the market except multigrain atta. This has huge impact on consumer perception on product diversification and flagging of nutritional superiority. The acceptability among health conscious urban consumers is great so much so the response of entrepreneurs is overwhelming to produce and market this future alternate nutri-cereal and made MoU with IIMR.

Salient Features of Commercialization and Scaling Up

'Eatrite' brand name was registered and launched by IIMR with unique selling proposition 'Eat Jowar Stay Healthy' for commercialization of sorghum product technologies developed under NAIP- Millets Value Chain.

Successfully commercialized ***Eatrite* products** on pilot scale in Hyderabad through Heritage Retail Stores across the city and through unorganised retail sector.

For sustainability, IIMR has Signed MoU with 17 prospective entrepreneurs from Hyderabad, New Delhi and Bangalore with variable emphasis on production, promotion and marketing sorghum product technologies.

An MoU is signed with M/s Britannia, an MNC (first time in the history of ICAR) by IIMR to work jointly on R&D on bakery and biscuit products.

The commercialization is solidly backed with the popularization measures adopting various communication strategies.

Horizontal expansion is envisaged in Bangalore, Chennai, Pune and Delhi.

These convenient and healthy products are targeted primarily at urban population. The products are aggressively promoted through road-shows, nutritional campaigns, print and electronic media across the Hyderabad City on a pilot scale. Currently they are marketed by M/s Heritage Fresh retail chain stores and are explored for their market in unorganized market. The response has been very good and there is an increasing demand for their availability across the city and other major metros. Particularly, multigrain *atta*, pure *jowar* biscuits, flakes have good demand across all age groups.

Introduction of Eatrite Products in Retail Chain: A Pilot Testing in Hyderabad City

Eatrite products are now marketed through Heritage Fresh and ITC retail chain stores in Hyderabad. Simultaneously, outsourced event managers for popularization of sorghum products (360 degree communication, brand designing logo, etc with BTL and ATL strategies implemented) in urban markets and New age Media. Massive awareness is created on sorghum as health and nutria food through Road shows (100+) in public parks, malls, and institutes *etc* in Hyderabad and in exhibitions by imparting awareness of sorghum to across 40000 consumers through fabricated *Jowar Rath* in Pune, Bangalore, Jabalpur, Chennai, Coimbatore, New Delhi etc. This whole commercialization started with the soft launching of IIMR brand *'Eatrite'* at ICAR Krishi Bhavan and other 2 NAFED Outlets in New Delhi.

Marketing of Millet Products

Under aegis of National Agricultural Innovation Project (NAIP), IIMR has developed many post-harvest technologies for efficient and enhanced use of sorghum for the benefit of the consumers and various stakeholders making relevant to the society. The ultimate goal is benefit not only providing convenient and healthy options for enhanced sorghum consumption but to revive the sorghum crop cultivation through interventions such as diversified food processing and nutritional tagging for stabilization of farm incomes of our primary stakeholder, sorghum farmer. Three products have been identified by IIMR as most suitable processing technologies for any entrepreneurial activities. These are:

1. Jowar Based Multigrain Atta
2. Jowar Based Biscuits
3. Jowar Based Flakes

All of these technologies has been developed and standardized as a value addition in the Sorghum/Jowar to provide healthy foods at the affordable price to the customer.

Advantage of the Jowar (Sorghum) Based Products

- ☆ Complete family product- individual, family, young, aged, kids, health conscious and traditional.
- ☆ Healthy product- wholesome nutrition for all members of the family, no restriction for diabetic/old/ill members, energy powerhouse, perfect diet management.
- ☆ Perfectly substitutable- No variation in terms of preparation, appearance and taste of regular atta.
- ☆ Convenience- no extra effort required and time consuming.
- ☆ Addressing both health and taste-

- Available products either positioned "Health (cholesterol management and diabetic (Saffola), Gluten free, Diabetic) or "Taste" (Ashirvaad, Pillsbury, Annapurna etc.)
- Available biscuits are either positioned "Health (digestive, light, nutritious, diabetic) or "Taste" (Good day, Hide and Seek, Butter Bite, or loose (Nan-Khatai, fresh backed).

State-wise Potential Area/Markets for Launching Jowar Products

Table 1: State-wise Potential Markets

State	*Areas*	*Supply Options*	*Market Entry*	*Effort Required*
Andhra Pradesh	Vijayawada, Vizag, Karnool	Central manufacturing	Easy with low marketing cost	Consumer acceptance study with launching strategy
Karnataka	Bangalore, Dharwad, Gulbarga, Bidar, Hubli	Central manufacturing	Easy with low marketing cost	Consumer acceptance study with launching strategy
Rajasthan	Jaipur, Ajmer, Jodhpur, Pali	Outsourcing/ Tie-up	Easy, moderate marketing cost	Opportunity analysis, Market survey, Pre-launch consumer study
Gujarat	Bharuch, Vadodara	Outsourcing/ Tie-up	Easy, moderate marketing cost	Opportunity analysis, market survey, pre-launch consumer study
Tamil Nadu	Chennai, Coimbatore	Central manufacturing	Easy with low marketing cost	Opportunity analysis, market survey, pre-launch consumer study

Industry Size Estimation of Sorghum

A. Multigrain Atta Industry

- ☆ The market size of multigrain atta is very small compared to the normal atta market.
- ☆ It is estimated to be around 15-25 thousand tons per annum. (on an average 1-1.5 per cent of total packed atta industry of India)
- ☆ Mostly organized players dominate the segment and national brands are available in the market.
- ☆ Few regional players are also there in the market.
- ☆ Multigrain atta is sold at premium. (Ashirvaad, Pillsbury, Safola).
- ☆ Growth of this particular segment is estimated to be 3- 4 per cent annually.

B. Biscuits Industry

- ☆ Biscuit industry size is estimated to be around 25-30 lakh tonnes per annum (Indian Biscuit's Manufacturing Association (IBMA).
- ☆ About 55-65 per cent of the total volume is catered to by the organized players. Glucose biscuits constitute 45-50 per cent, followed by Marie to the tune of 10-15 per cent in total biscuits consumption basket.
- ☆ Share of cookies is estimated at around 3-4 per cent of the total market size.
- ☆ Growth of overall biscuit sector has been estimated at around 12-15 per cent per annum (Indian Biscuit's Manufacturing Association (IBMA).

C. Rice Flakes Industry

- ☆ Rice flake market size is estimated to be around 8-10 lakhs tons per annum. (source: Foretell Internal research)
- ☆ The unorganized players dominate the segment.
- ☆ No branded players exist in this segment.
- ☆ It is sold under private labels in most of the urban areas.
- ☆ The consumption growth is estimated to be 5-7 per cent annually.

S.W.O.T. Analysis of Jowar Food Products

The S.W.O.T Analysis of the jowar food products is essential for establishing its uniqueness amongst the available products in the market in same category. Few clear representations from the table make jowar as one of the healthy food products because of its nutritional content. However it has few weaknesses like its shelf life and lack of commercial business model but its opportunity gives and advantage and put its potential to emerge as one of healthy food products.

In food product segment like multigrain atta and biscuits, consumer are looking for healthy option. The awareness bringing the health consciousness as a most important factor while purchasing food item. The changing life style and with increase in number of diseases people are becoming health conscious and trying to have healthy foods as a part of their eating habit (Table 2).

Product wise S.W.O.T. Analysis

A. Jowar Multigrain Atta

The S.W.O.T analysis of multigrain atta in table shows that people have developed a liking towards the multigrain atta based on jowar. The health benefit of the atta *i.e.* its gluten free content making it suitable for diabetic patients. There are weakness also related to the product like smell and shelf life, however opportunity is very big as very few players are there in the market who are selling jowar based multigrain atta. Also, continuous search from the consumer for the better health food item also provide them and opportunity to tap the current multigrain atta market.

Table 2: SWOT of Jowar Food Products

Strength	*Weakness*
☆ Awareness of Jowar as healthy and nutritious food is favorable ☆ Appealing response for quality certification from NIN ☆ Strong research and development facility ☆ Only brand (IIMR's brand 'Eatrite') offering versatile product range based on Jowar in the market ☆ Easy adoptable in terms of appearance, preparation and taste when compared to traditional products ☆ Taste of the product is liked by the respondents People are willing to consider the product if "Eatrite" launched in the market	☆ Infant brand in the market ☆ Lack of awareness among the non-traditional respondents ☆ Lack of commercial business operation, sales/ distribution network and presence in the market ☆ Immature economy of scale ☆ Lack of communication and advertisement about the product ☆ Shelf life of the product is short (1 month) ☆ The typical Jowar smell has received disliking response from non-traditional consumers ☆ Jowar has difficulties in preparation/cooking for urban consumers Inferior appearance attribute compared with traditional product
Opportunity	*Threat*
☆ Growing consumer's eagerness to try new products with health benefit is high ☆ Specialty consumers (diabetic, Obese) are excited on favorable result ☆ Able to substitute traditional product giving variety and choice ☆ Health and Wellness food market size is around Rs. 10000 crores and expected to grow by CAGR 23% (TSMG) Increasing disposable incomes, diminishing culinary skills and the rising need for convenience would be the driving factors of Jowar products in traditional areas.	☆ Initial hesitation of the retailers to keep the products in non-traditional areas ☆ Consistency about the quality is doubtful in consumer's mind, as supplier is not a branded Fast Moving Consumer Goods (FMCG) supplier ☆ Sustainability in the short term is high due to marketing set-up cost ☆ Increasing raw material price and uncertainly in the production of Jowar ☆ Negative health related perceptions (non-edible varieties for some seasons, dental disease issues) ☆ The health product segment size is very small and many consumers fall in the 'Late Majority' category.

B. Jowar Biscuits

The S.W.O.T of the jowar biscuits in table represent the fact that currently either no or very few players have jowar based biscuits in their product portfolio. Also, with the increase in awareness about the content of normal biscuits, people are trying to have more balanced and nutritional biscuits in their normal day- to-day life. The opportunity for this segment is very high as it is still under very early stage.

The jowar based biscuit being new to market can capture the market size of the nutritious biscuits segment as its nutritional contents are high compare to other millets and cereal crops.

Table 3: SWOT of Jowar Multigrain Atta

Strength	*Weakness*
☆ The product is very much liked by most of the respondents who have tasted the product. ☆ More than 90% of the survey respondents expressed their willingness to purchase the product. ☆ Health benefit of Jowar atta is the major acceptance criteria in traditional areas. ☆ The respondents have felt no significant difference between Jowar atta and wheat atta. ☆ Jowar atta is an ideal product for diet plans and weight management. ☆ Suitable for all family members, and no restriction for diabetic, ill, kids and aged.	☆ Jowar multigrain atta is only product available in the market rich in (containing) Jowar. ☆ Existing players in the multigrain atta segment in the market are very few. ☆ The multigrain atta consumers are not brand loyal as they seek more health benefits from such types of atta. ☆ Increasing health consciousness, disposable incomes, diminishing culinary skills and the rising need for convenience could be driving factor of Jowar products in urban traditional Jowar consuming areas.
Opportunity	*Threat*
☆ Shelf-life- The shelf life of Jowar atta is 2 month. ☆ Smell- Typical Jowar smell in the atta during preparation is not liked by a few consumers (especially the new consumers). ☆ Swelling- swelling attribute of chapathi was not liked by the Delhi consumers. ☆ Stickiness and texture of Jowar atta is another issue to be addressed ☆ No existing business experience, supply chain is still in formation stage. ☆ No professional manufacturing, marketing and sales department ☆ Economy of scale, learning curve, expertise in commercial operation, and packaging are yet to be fine-tuned.	☆ Size of the multigrain atta industry in very small, and dominated by the national brands. ☆ Growth of this segment also very less (3-4 % yearly) ☆ Higher price of Jowar seed would in turn increase the price of Jowar multigrain atta. ☆ Cost of marketing and distribution is very high for the new product. ☆ Usage of Jowar as food product is restricted to the traditional consuming areas.

Financial Analysis

Production Plan

Table depicts the production plan of the Multigrain atta, Biscuits and Flakes. It is considered that plant will run for 8 hours on a daily basis. The monthly working day is assumed to be 25.

The financial analysis has been done for the 4 year, and the production of these products will be increased on yearly basis as depicted in the Table 5.

Table 4: SWOT of Jowar Biscuits

Strength	*Weakness*
✰ Jowar Biscuit is very much liked by the most of the respondents. ✰ More 90% of the survey respondents showed their willingness to purchase the product. ✰ Health benefits associated with Jowar biscuit is the major acceptance criteria in the traditional areas. ✰ Jowar biscuit is suitable for all family members and no restriction for diabetic, ill, kids and aged. ✰ It is an ideal product for diet plans and weight management. ✰ Taste-wise Jowar biscuits are not very much different from the existing cookies.	✰ Jowar biscuit is the only product available in the market rich in (containing) Jowar. ✰ Product can compete with other cookies targeted at the health conscious consumer segment. ✰ The health biscuit consumers are not brand loyal as they seek more health benefits from such type of cookies. Increasing health consciousness disposable incomes, diminishing culinary skills and the rising need for convenience could be driving factor of Jowar products in urban centric traditional areas.
Opportunity	*Threat*
✰ Shelf-life- The product shelf life is very short (2 month). ✰ Smell- Typical Jowar smell in the biscuit is not liked by some consumers. ✰ Crispiness- Coconut biscuits were less crispy while normal biscuits were hard while chewing. ✰ The Jowar biscuits gumminess (while chewing biscuits getting broken into particles and becoming ready to swallow) is less as compared to the normal cookies. ✰ Fullness of taste after eating and mouth feel is not liked by some consumers. ✰ No existing business experience, supply chain is still in formation stage. ✰ No professional manufacturing, marketing, sales department ✰ Lack of production and operational advantages like economy of scale, learning curve, expertise in commercial operation, packaging, etc.	✰ Share of "cookies" in the total biscuit market is very small. (5-7 % of total) ✰ Healthy-Cookies are niche product and dominated by the organized players or brands. ✰ Growth of this segment (Cookies) also very less (3-4 % yearly (CAGR 2008-13) as compared to 12-15 % for all). (Source- Data monitor) ✰ Higher price of Jowar seed would in turn increase the price of Jowar biscuit. ✰ Cost of marketing and distribution is very high for the new product. Usage of Jowar as food product is restricted to the traditional consuming areas.

Table 5: Daily, Monthly and Annual Production Quantity of the Above Three Products

Particulars	*Daily Production in tons*	*Month Production in tons*	*Annual Production in tons*
Multi-Grain Atta	1.2	30	360
Biscuits	0.36	9	108
Flakes	0.8	20	240

Table 6 packaging size has been identified keeping in mind the movement of the product from the shelf. Table indicates the pack size of the selected products

Table 6: Packs Size for Different Products

Particulars	*Pack Sizes*	*Total Annual SKU Quantity*
Multi-Grain Atta	1 kg	270000
	5 kg	18000
Biscuits	0.1 kg or 100 g	810000
	0.25 kg or 250 g	108000
Flakes	1 kg	180000
	5 kg	12000

Financial implication of putting up the plant for the production of these 3 products were worked out in table. It is advisable that multi-grain atta and biscuits plant should be considered together at the machineries are common up to certain extent in the early processing of raw jowar and other material and preparation of input for manufacturing of both products. Flakes require altogether different machines and set-up. However, if an entrepreneur (early start-up) is planning for these technologies, he should go for all the three technologies together(This will not only help him in reducing or sharing the cost in terms of putting up manpower but also provides variety to the consumer leading to enhancement of the sales).

The fixed cost in the below calculation has been put separately for all the products, however while calculating the IRR (Internal rate of return) and NPV (Net present value), we have taken the fixed cost of the biscuit and flakes together as the machineries will be same in case as of biscuits in case of multigrain *atta*.

The plan has assume that it should have internal storage in order to have greater control over the raw material to be used for the production. It is also advised that the production plant should be near to the area where sorghum and other inputs are easily available in order to have smooth supply of the input for the production.

Table 7: Financial for the First Year of Operations (All cost in Rs. Lakh)

Particulars	*Multi-Grain Atta*	*Biscuits*	*Flakes*	*Total*
Building and Plant including plant machineries and other legal and setup cost	82.8	77.8	99.8	166.5
Variable cost				
Cost of raw Material	68.8	52	96	**216.8**
Operational Cost which includes electricity, depreciation of the machineries, packaging of the products etc.	12.66	13.75	9.93	**36.34**
Marketing cost	18	18	18	**54**
Manpower cost	7.62	7.62	7.62	**22.86**

Contd...

Table 7–*Contd...*

Particulars	*Multi-Grain Atta*	*Biscuits*	*Flakes*	*Total*
Channel Margin (@ 25 per cent), transportation and logistics (@5 per cent) and taxes (@ 4 per cent)	39.9	33.72	48.94	**122.56**
Total Variable cost	146.39	125.09	180.49	452.56
Total Sales. It is assumed that 10 percent of the total SKU will be sales return and 10 of smaller SKU will be used in promotional activities. Therefore actual quantity for sales will be 80 per cent of the smaller SKU 90 per cent of the bigger SKU	148.5	131.22	195.3	475.02
Net profit before tax	2.1	6.34	14.8	23.24
Net Profit after tax (@30 per cent)	1.47	4.44	10.36	16.27

Based on the first year operations the projection for the next 3 years have been made in terms of total number of quantity to be produce on annual basis of each product.

Table 8 provides a suggestive estimated quantities of the three products that needs to be produced year on year basis in order to achieve the breakeven point and recover the initial investment in first 4 year of the operations.

Table 8: Annual Production Quantity (in ton) for Next Three Year

Particular	*2nd Year*	*3rd Year*	*4th Year*
Multi-Grain Atta	420	420	420
Biscuits	120	120	120
Flakes	300	360	420

Based on the production plan for all four years, IRR and NPV has been calculated in table which comes out to be 9 per cent and 48.4 per cent respectively.

Table 9: IRR and NPV of the Project

Targets			*Fixed Cost*	*Year*	*Revenue*	*Cost (Variable)*	*Revenue-Cost*
Multi-Grain Atta	*Biscuit*	*Flakes*	*166.5*				*–166.5*
1200	360	800		1	475.0	451.6	**23.3**
1400	400	1000		2	563.0	513.0	**50.0**
1400	400	1200		3	612.0	548.2	**63.8**
1400	400	1400		4	660.8	583.0	**77.8**
						IRR	**9 per cent**
						NPV (Rs lakhs)	**48.40**

It is also important to highlight that the quantities on which the first year calculation is based are close to the breakeven point of the variable cost. So, this means that quantities below this will lead to loss on the operational front. It was also identified during the financial calculation that single product or technologies based on sorghum will not be able to give you viable income. Therefore it is suggested that combination of products should be consider before starting the fresh venture. However, if entrepreneur is already in this business and he is diversifying his product mix can go for the single product or technology.

Marketing Plan

It is suggested that for the first year, entrepreneur has to go for an intensive marketing strategies, in order to establish their brand.

The Marketing plan for product will have both strategies *i.e.* above territory line (ATL) and Below territory line (BTL). A combination of mass campaign, road shows, paper insertion, cable TV advertisement and newspaper advertisement will be given in order to create awareness of the brand in front of consumer, leading to the fast movement of the product from the retailer shelf.

It is also suggested that these activities should be conducted on regular intervals *i.e.* month on month basis in order to grasp the attention of the consumers.

Apart from the above mentioned activities, entrepreneur should also have banners, posters, T-shirts, Caps etc. in order promote the brands in the market. Banners and posters should be put on the key locations which receives maximum footfall of the customers. Sales and marketing team should wear the T-shirts and caps to create a buzz in the market. T-shirts and caps should be also given during the mass campaign and road shows, so that brands reaches to the consumer household and create word of mouth publicity.

The bundling of the products can also be considered during the product promotion activities. In Table 10 few combinations has been suggested, which entrepreneur can consider for their bundling plan during the promotional activities-

Table 10: Suggestive Combination for Bundling of Three Products

Product Purchased	*Quantity*	*Free Gifts (Product) with Purchased Product*
Multigrain Atta	5 kg pack	100 gm pack of biscuit free
Flakes	1 kg pack	100 gm pack of biscuit free
Flakes	5 Kg Pack	250 gm pack of biscuit free
Multigrain atta+ Flakes	5 Kg of Atta+ 5 kg of Flakes	1 kg of Multi grain Atta pack free

The number of activities in terms of Below Territory Line (BTL) and Above Territory Line (ATL) should be take on month on month basis in the initial year are listed in Table 11. This is essential in the initial period or years in order to establish the brand in the market. Once people will start recognizing the brand, it will give the momentum to the sales of the products. Also, the tie-up chances with big retail chains to get their shelf space will increase.

Table 11: Brief Activity under BTL and ATL

Sl.No.	*Marketing Strategy*	*Activities per Month or on Quarter Basis*
1	Below Tertiary line	
1.1	Road Shows on the most happening places and the road on which footfall of customers are high	3/month
1.2	Mass Campaign with the help of umbrella and Standees on the key locations of the market	2/month
1.3	Paper Insert	3/month
2	Above Tertiary line	
2.1	News Paper Advertisement	1/quarter
2.2	Cable Television advertisement	1/month

Alternative Strategies for the Selling of the Products

The business plan also suggests an alternative marketing or sales plan. The entrepreneur or enterprise can also become a contract manufacture for the big players in the market. They can ensure the supply of the products for particular region with the timely supply commitment. This can also facilitate the development of the cluster based sorghum production center in the major sorghum producing state. This development will also lead to the increase in the production area and also provide a dryland based farmers to and opportunity to improve their livelihood.

It is proposed that if IIMR can take this initiative in tying the entrepreneurs or enterprise with the bigger players having greater experience and expertise in the marketing and selling of the products. The entrepreneurs can played a role in doing the backend manufacturing for those bigger players. The companies which will be buying the products can implement their quality standard for maintaining the quality of the products.

The Table 12 provides and better idea with the differentiation between the roles of the different stakeholders involved in the partnership.

Table 12: Suggestive Roles of the Different Players Involved in the Partnership

Particulars	*Entrepreneur or Enterprise*	*Indian Institute of Millets Research*	*Big Indian or Multi-national Company*
Procurement and production	Manage and concentrate on the production of quality sorghum base products	Can help entrepreneur with backward integration with Farmers and provide technology for production of sorghum based products	Provide technical guidance on setting up the plants and help in implementation and usages of the quality standards for production
Marketing and sales	Focus on the quality production with attractive packing	Technical help in highlighting the key Uniqueness of the products.	Through their established distribution network provide an necessary momentum for the sales

Risk Factors

Availability and Accessibility of the Jowar

The availability of jowar for the production of value added and nutritional product is a critical factor. The plant needs to have very strong backend integration in terms of procurement of the grain. It is advisable that plant should be established near the production area. Also, there should be integration with the farmers under the contract farming agreement in order to have smooth and continuous functioning of the plant. The procurement planning and mapping of the required quantity should be planned well ahead before the start of the season.

The price fluctuation in the sorghum is also one of the critical factor. If crop production goes down, price of sorghum grain hit the range of Rs. 4400/quintal. Therefore the backward integration in terms of the procurement will lead to have better control over the input cost for the production.

Entrepreneur Skills and Capabilities

This is also important to highlight this factor as these products are premium products in the market and have very less market share in terms of regular atta or biscuits available in the market. The entrepreneur should have very strong business acumen and should be able to lead the business from the front. He should have the understanding of the market channels and should be able to crack the right channel. He should also have understanding about the various ways through he can procure the sorghum while keeping his operations unaffected due to price variation.

Social Impact

The products identified have huge social impact on the society. Sorghum based products are more nutritional in comparison to the wheat or rice based products.

Gluten free content makes it right food for the health conscious people. It is also recommended to the people having diabetes.

The children can have healthy food which helps in balanced growth and development of their body. This can also lead to the increase in area under the cultivation and can give the dryland farmers a better price for the product which potential is still not recognized by the market. The farm family engaged in the sorghum production in dryland area can earn good income as the product made from sorghum are fetching premium price from the market.

This will also help in the development of sorghum based product industry and can provide employment to the rural youth by engaging them in the process of production, procurement, manufacturing of the products and its sales and marketing.

Based on analysis of data, the following emerges as some major conclusion:

1. First time entrepreneur or enterprises, needs to get two or more than two jowar based food products to start their business. This will ensure that business model be more economical viable. However, if existing entrepreneur or enterprises wants to diversify their existing business they can go for any

number of jowar based products as the processing of the raw material into the final product is same.

2. Data analyzed indicates minimum quantity of each food products, which needs to be manufactured on monthly or yearly basis. These quantity will ensure the recovery of investment at the end of the 4th year of the operations.
3. The identified key districts where the sorghum cultivation having potential may be explored for the setting up of the manufacturing plant. This will ease down the difficulty in raw material procurement for the manufacturing of the food products for the entrepreneurs and makes it more cost effective.
4. It is clear that Jowar based food products are relatively new to the market. Hence it is essential that entrepreneurs develop competent marketing skills for sales and marketing of these food products. These products also needs focused marketing activities in terms of ATL or BTL. This is essential to develop the brand in the market place leading to customer preferences towards purchase of these food products.
5. As an alternative strategy to establish sole entrepreneur mode tri-partite partnership between the entrepreneur, IIMR and a big player (who is already in business) is also suggested. The entrepreneur will have the role of manufacturing standard quality food products, IIMR can provide him with backward integration for the raw material procurement and forward integration by big players through development of supply chain for the food products. The entrepreneur can act as a contract manufacturer for the big companies. The company can be provide technical guidance to the entrepreneur for the manufacturing of the good quality food products and play the role of marketer in which they have develop their expertise.

Part VII
Seed Systems

Millets: Ensuring Climate Resilience and Nutritional Security (2015) *Pages* **595–606**
Editors: **Vilas A. Tonapi and J.V. Patil**
Published by: **DAYA PUBLISHING HOUSE, NEW DELHI**

Chapter 23

Sustaining Viability of Informal Millet Seed Systems

Ch. Ravinder Reddy[1] and Vilas A. Tonapi[2]

[1]*International Crops Research Institute for Semi-Arid Tropics, (ICRISAT), Patancheru, Telangana*
ICAR-Indian Institute of Millets Research (IIMR) Rajendranagar, Hyderabad – 500 030

The main purpose of alternative seed delivery system is to address the seed availability problems of smallholder farmers. Most of the community-based seed production models/schemes are initiated because farmers are concerned about the non-availability of quality seed at planting time. Many farmers don't have access to improved varieties; and wouldn't be able to afford them even if they were. So introduction of alternative seed systems models must impact farmers' access to seeds of improved varieties at affordable cost. The quality of seed produced by community-based system or farmer seed systems is guaranteed only by its seller or village seed committee, because they are not processed and are uncertified seed. The seed so produced is low priced, and available at farmers' doorsteps at the right time, and provides access to all farmer groups in the village.

The regulatory and legal framework of national seed rules and regulations in many countries hampers the development of informal seed systems. National seed regulations are mostly based on international standards, which are often incompatible or irrelevant to the realities of farmers' seed systems. The restrictions imposed by national seed authorities on free exchange and marketing of seed, especially compulsory variety registration and seed certification, as practiced by many developed and developing countries are constraints on the efficient functioning of the formal seed sector and on the development of alternative seed systems (Annette von Lossau 2000). On the other hand regulatory frameworks are crucial for the development of a national seed system (Tripp 2003).

The major source of seed for small-scale farmers comes from their own on-farm savings, seed exchange, borrowings and local traders. Nevertheless, farmer's community systems of seed supply are under pressure due to recurring natural calamities such as drought, crop failure, storage problems and poverty. In the drought situations farmers depend on subsidized seed supply by government agencies, which meets only 30–40 per cent seed requirement of smallholder farmers (Ravinder Reddy 2005). In order to strengthen the seed delivery system, interventions are required to strengthen informal seed supply systems, such as establishing village-based seed banks as alternative seed systems for seed security. The alternate village based seed delivery models that may enable sustainability of community seed systems in the dry land ecosystems have following objectives:

1. To improve seed availability and access to improved varieties of seed to small and resource-poor farmers
2. Build capacity of stakeholders at the community level to enhance sustainable supply of good quality seed, and timely supply at affordable price.

Alternate Village Based Seed Delivery Models

Model 1: Individual Farmer as Seed Bank

This model (Figure 1) can be developed as a local seed system for different crops. Most effectively this seed system will benefit farmers for crops that require high seed rate, which are bulky in nature, or crops that involve high transport and package costs, for example groundnut pod. This model involves training a couple of farmers in each village in seed production technology and supporting them by supplying breeder seed and technology backstopping.

The Pros and Cons of the Model

- ☆ This model can be tried even in remote areas where NGO are unwilling to take up operations.
- ☆ External finance is not required as all the costs are usually met by the farmer/seed producer.
- ☆ Effective and provides wider scope for dissemination and adoption of improved varieties through informal seed channels.
- ☆ Technical institutional services not justifiable for individual farmers.
- ☆ Farmers are still unwilling to save seed because of storage pests and other financial debts.
- ☆ Procurement of breeder seed is difficult at the farmer level once the project is completed.

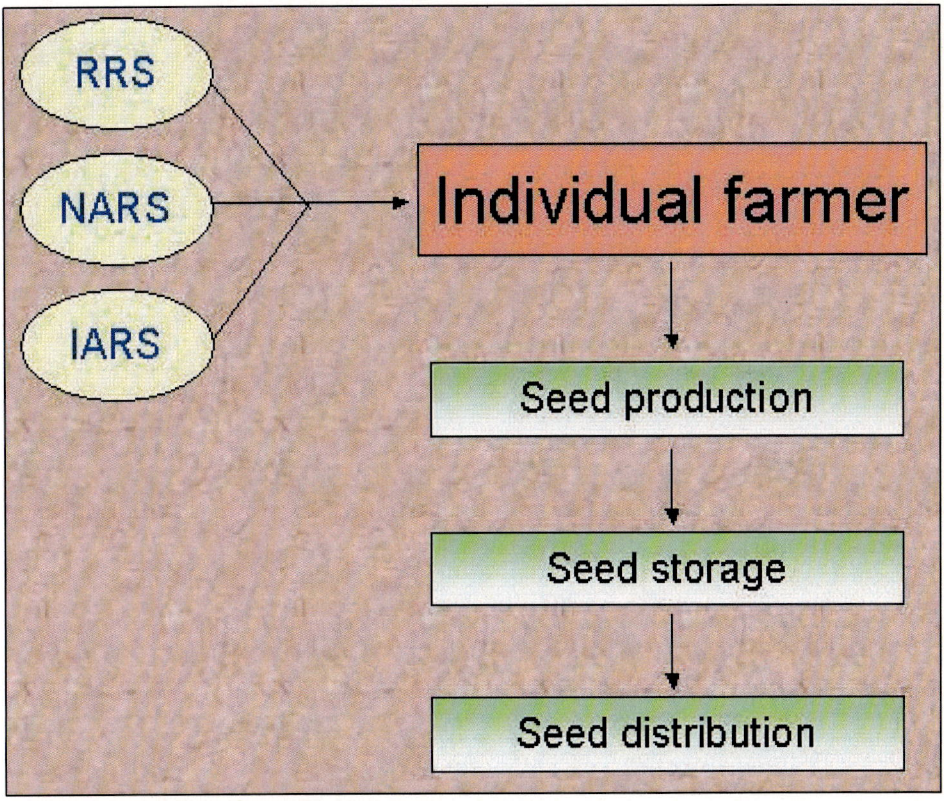

Figure 1: Model 1: Individual Farmer as Seed Bank.

- ☆ There is no control on the selling price of seed.
- ☆ There is no control on seed distribution to different communities in the village.
- ☆ Seed distribution is limited to select groups.

Model 2: Village Based Seed Banks

The concept of 'seed bank' (Figure 2), which advocates village self-sufficiency in production and distribution of quality seeds, is fast gaining ground. Seed villages or village seed banks operate with utmost transparency, mutual trust and social responsibility. Though this is not an entirely new concept, it is being promoted to reduce farmers' dependence on external inputs.

Advantages of Village Based Seed Banks

- ☆ Availability of improved varieties in sufficient quantity within the village
- ☆ Assured and timely supply of seed
- ☆ Decentralized seed production
- ☆ Availability of improved variety seed at a low price

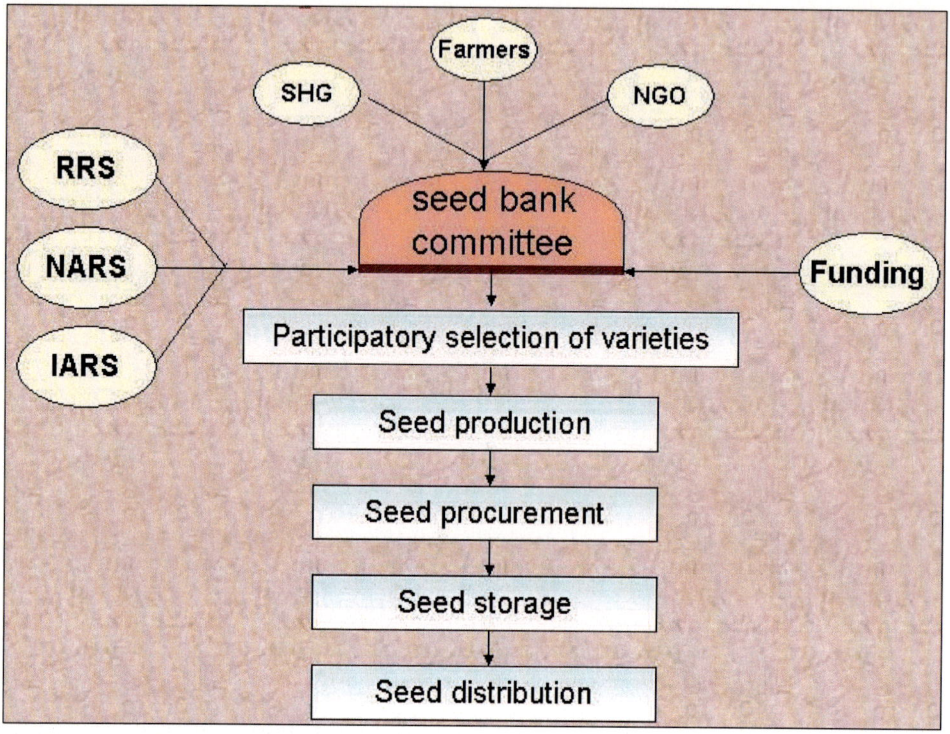

Figure 2: Model 2: Village Based Seed Banks.

- ☆ Improved seed delivery system to resource-poor farmers
- ☆ Reduced dependence on external seed sources and hence an effective measure to curb spurious seed trade
- ☆ Encourages village level trade and improves village economy
- ☆ Social responsibility of the seed production and delivery system
- ☆ A step ahead towards sustainable crop production
- ☆ Avoidance of diseases carried through seed (seed-borne pathogens) that have been produced and imported from different agro-ecoregions
- ☆ Scope for farmers' participatory varietal selection
- ☆ Availability of true-to-type varieties and healthy seed.

Constraints

- ☆ Reluctance of farmers when it comes to adopting quality seed production practices
- ☆ Additional investment for inputs in seed production
- ☆ Lack of buy-back assurance to farmers from SHGs/NGOs
- ☆ Proper seed storage facilities and management in villages

- ✩ Lack of funds with SHGs/NGOs for seed procurement, packaging, storage and transportation
- ✩ Fixing of a minimum support price for seed procurement
- ✩ Technical support for seed production and its monitoring
- ✩ Responsibility of quality control aspects and monitoring of seed production
- ✩ Regular availability, accesses, and procurement of breeder seed for seed production
- ✩ Willingness of farmers to participate in seed bank activities

Small-scale Seed Enterprises Models

Model 3: SHG-Mediated System

In this model, SHG is empowered to take up the task of seed production (Figure 3). Members, however, need to develop skills in planning and seed production techniques as well as support in terms of storage. Alternatively arrangements may be worked out with nearest market yard or state godowns to have the seed storage facility. The most critical aspect in this model is technical support and supply of breeder seed. Given the proper support, this model could provide significant benefits to farmers as it ensures the opportunity for all members and groups to share the profits of seed production. This model performs two tasks: meeting seed requirements of farmers as well as conserves a wide range of crop genetic diversity.

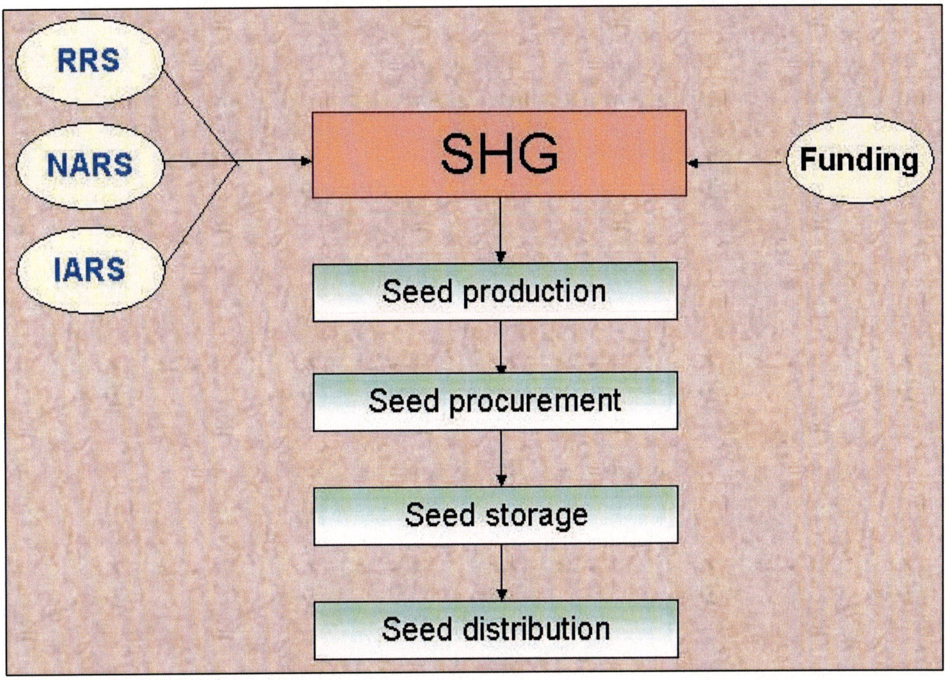

Figure 3: Model 3: SHG–Mediated System.

In most developing countries, the formal sector is far smaller than the informal seed sector. The later is the major source of planting material for smallholder farmers, contributing 80–90 per cent of the seed requirement of smallholder farmers (Monyo *et al.*, 2003; Ravinder Reddy 2004b). Strategies to improve quality of seed, accesses and availability of improved varieties, multiplication and dissemination, availability of seed on time at affordable prices to resource-poor smallholder farmers can bring about changes in the food security in developing countries. Support from state/ national governments, and international organizations or any other funding agencies should be targeted at improving the efficiency of these investments by strengthening technical capabilities of SHGs, NGOs, farmer cooperatives, CBO's, KVK's and schemes to improve or develop village based seed programmes through seed multiplication, quality control and marketing activities.

Pros and Cons

- ✰ Improved availability and access to improved varieties by all groups of farmers.
- ✰ Minimum overheads.
- ✰ Seed is stored in the village.
- ✰ Seed available at reasonable price and at the right time.
- ✰ Control on fixing procurement and selling price of seed.
- ✰ Priority for farmers' preferred varieties.
- ✰ Need for institutional support for technical backstopping and supply of breeder seed.
- ✰ Fund for procurement of seed.

Model 4: NGO-Mediated System

Non-Governmental Organizations (NGOs)

In this model (Figure 4), an NGO may be given the responsibility for a cluster of villages. These organizations select and engage farmers in seed production on a contract basis and preference is given to crops and varieties that are in demand in particular area. Basically NGOs are involved in mobilizing farmers/seed producers, planning seed multiplication, training, procuring, processing and marketing seed. Similar to other models, NGOs has to depend on other institutions for procuring foundation/basic seed stocks for multiplication.

Pros and Cons

- ✰ Operates in couple of villages.
- ✰ Seed production operations in couple of villages (3-5).
- ✰ Storage of seed within the village.
- ✰ Seed distribution within the operational areas.
- ✰ Selling price can be fixed by discussions with farmers.
- ✰ Improved seed availability and access for all groups of farmers.

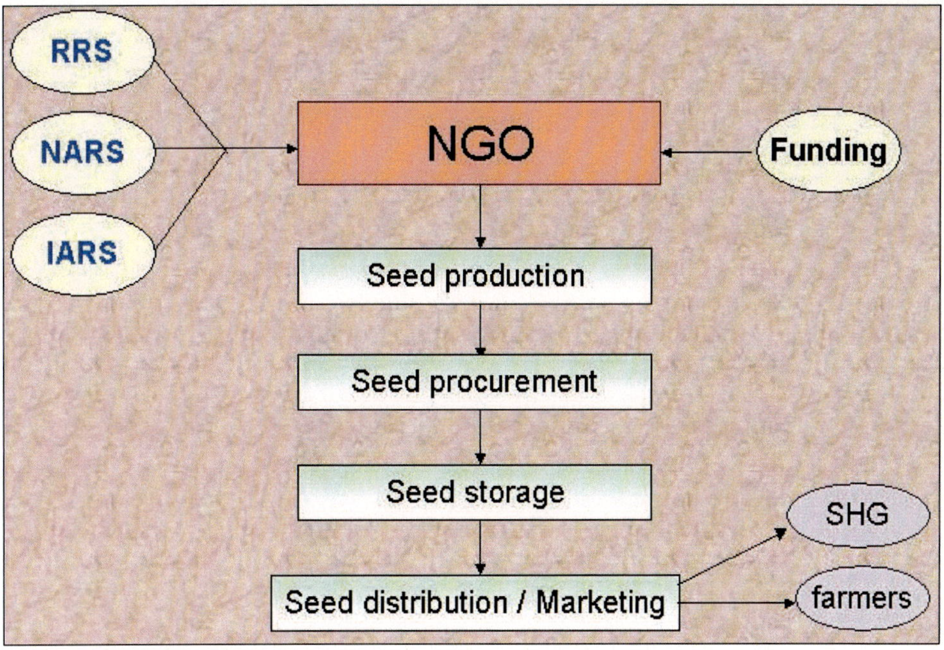

Figure 4: Model 4: NGO–Mediated System.

- ☆ Minimum overhead costs.
- ☆ Need institutional support for technical backstopping and supply of breeder seed.
- ☆ Fund required for seed procurement.

Model 5: KVK-Mediated System

In this model (Figure 5) KVK's are given the responsibility to implement the model.

Pros and Cons

- ☆ Targets, large operational area (many villages).
- ☆ Centralized production; needs large storage place (such as godowns)
- ☆ Comparatively greater overhead costs
- ☆ Involves more than one crop and variety in production
- ☆ Less scope for farmers' participation and their choice of varieties
- ☆ More inclined towards commercial seed trade
- ☆ No scope for involving farmers in fixing procurement and selling prices
- ☆ Model needs infrastructure
- ☆ Usually KVKs has fairly good technical capabilities
- ☆ Generally institution funds are available to some extent.

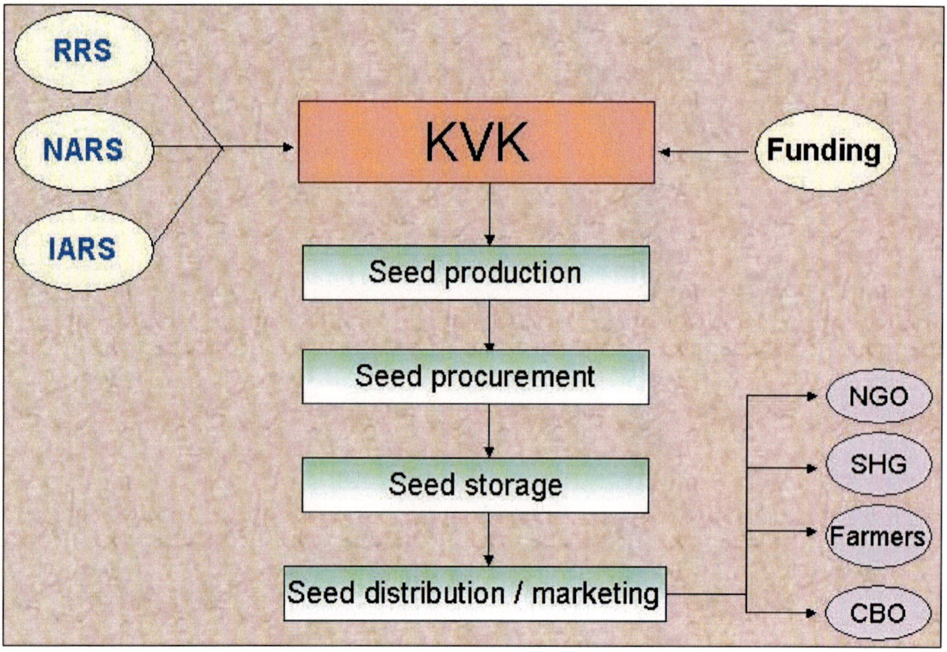

Figure 5: Model 5: KVK–Mediated System.

Each of five models tested can be harbingers of seed system sustainability upon implementation based on the environments and regions they operate. The comparative statement comparing all the five models across each component working towards seed system sustainability is given in Table 1.

Steps for Strengthening Community Seed Production, Seed Saving and Storage

1. Appropriate policies for seed production, and seed distribution, to help focus efforts of government supported initiatives on the varietal needs of resource poor farmers in particular, with attention, where necessary, on the needs of women farmers, and minor crops that are inadequately covered by the private sector.
2. Provide, and promote small-scale seed enterprises and strengthen linkages between gene banks, plant breeding organizations, seed producers, and small-scale seed production and distribution enterprises;
3. Strengthen seed quality control schemes for small scale enterprises and provide appropriate incentives, credit schemes, etc., to facilitate the emergence of seed enterprises, paying attention to the needs of the small farming sector, of women and of vulnerable or marginalized groups;
4. Provide support to and strengthen farmers' organizations in order that they can more effectively express demand for their seed requirements, paying particular attention to the needs of women and of vulnerable or marginalized groups;

Table 1: Comparitive Chart of different Seed Systems Models

Components	*Models*				
	Model 1	*Model 2*	*Model 3*	*Model 4*	*Model 5*
Organization/ community involved	Individual farmer	VBSB	SHG	NGO	KVK
Breeder seed source	Research institute or project scientist	Research institute or project scientist	Research institute or project scientist	Research institute or project scientist	Self or research institutions
Responsibility for transport of source seed	Research institute or project scientist	Research institute or project scientist	Research institute or project scientist	Research institute or project scientist	Self
Sourcing of other inputs	Farmer	Seed bank committee/ farmers	Farmers	Farmers/NGO	Farmers/KVK
Choice of crop/ variety	Farmers	Farmers	Farmers	NGO/farmer	KVK
Training in seed production	Project scientist (PS)	Farmers	Farmers	NGO	KVK
Seed production monitoring	Project scientist	PS, NGO, VSBC members	PS, SHG	PS, NGO.	PS, KVK
Seed quality assurance	Farmer	VSBC	SHG	NGO	KVK
Cleaning, packing, and transportation	Farmer	Farmers	Farmers	NGO	KVK
Marketing	Farmer	VSBC	SHG	NGO	KVK
Fixing procurement and selling price	Farmer	VSBC	SHG	NGO	KVK
Funding for seed production	Farmer	Farmers	Farmers	Farmers/NGO	Farmers/KVK

Contd...

Table 3.2–*Contd...*

Components	*Models*				
	Model 1	*Model 2*	*Model 3*	*Model 4*	*Model 5*
Funding for seed procurement	Farmer	VO/SBC	VO, self	VO, self, other org.	Self, other org.
Sustainability issues	Technical support, supply of breeder seed	Technical support, supply of breeder seed, funding, takes over of role once project completed Incentives for farmers for maintaining quality.	Incentives for farmers for maintaining quality, technical support, breeder seed supply, funding for seed procurement	Farmer produce fetch low price because there is no external quality control, certification. Supply of breeder seed, funding	Marketing, cost of seed, selection of varieties, Incentives for farmers for maintaining seed quality, certification

5. Provide training and infrastructural support to farmers in seed technology, in order to improve the physical and genetic quality of farmer saved seed.
6. Develop approaches to support small-scale, farmer level seed distribution, learning from the experiences of community and small-scale seed enterprises already underway in some countries.
7. Seed Quality of Farm Processed Seed can be as good as and often better than Certified Seed if farmers takes the first step by selecting the right variety, controlling purity with good rotation, and follow standard agronomic practices' to achieve disease and weed control. They can choose a Mobile Seed Processor that can offer the equipment, management and expertise to achieve the standard you require.

Conclusion

An effective means of improved seed distribution is farmer-to-farmer seed exchange. This may be primed to a limited extent by supplies of improved seed from public agencies, agricultural research stations and non-governmental organizations to farmers in easily accessible villages. However, such a system is very slow. To speed up the flow of adapted improved varieties to farmers, there is a need to form a network, formal and informal or integrated seed systems between community-based organizations and research institutes, public and private seed multiplication agencies, involved in various aspects of seed production. This network will identify bottlenecks in the seed production chain, and catalyze or instigate applied and adaptive research and policy changes that may be required to ensure rapid movement of new cultivars into local seed delivery system benefiting smallholder farmers small and resource poor farmers who need them. This approach will require continued interaction between the various institutions, policy makers and stakeholders.

References

Almekinders CJM, Louwaars NP and de Bruijn GH. 1994. Local seed systems and their importance for an improved seed supply in developing countries. *Euphytica*. 78:207-216.

Ball SS and Douglas JE. 1992. Designing successful farmer-managed seed systems. Winrock International Institute for Agricultural Development. *Development Studies Paper Series*. Winrock, Arlington, USA.

Cormwell E, Friis-Hansen E and Turner M. 1992. The seed sector in developing countries: A framework for performance analysis. ODI, London.

Dayakar Rao B, Tonapi VA, Hyma Jyothi S and Seetharama N. 2004. Supply and distribution of quality seed: Need for sustained approach. Paper presented at XVII Annual Conference on Agricultural Marketing, Organised by the Indian Society of Agricultural Marketing (ISAM), Acharya NG Ranga Agricultural University, Hyderabad, Andhra Pradesh, 5-7 February 2004. Abstract *in Indian Journal of Agricultural Marketing (Conference special)*. 17: 260.

Delouche, J.C. 1982. Seeds quality guidelines for the small farmer. In: Proceedings of the Conference on Improved Seed for the Small Farmer, 9-13 August, CIAT, Cali, Colombia.

Louwaars N. 1994. Seed supply systems in the tropics: International course on seed production and seed technology. Wageningen, The Netherlands: International Agricultural Center.

Monyo ES, Mgonja MA and Rohrbach DD. 2003. An analysis of seed systems development, with special reference to smallholder farmers in southern Africa: Issues and challenges. Paper presented at the workshop on Successful community based seed production strategies, co-organised by CIMMYT and ICRISAT, 3-6 August 2003, Harare, Zimbabwe.

Ravinder Reddy, Ch. 2004a. Alternative seed delivery systems – Village based seed banks. *In* ICRISAT-Asia in-house review held at ICRISAT, Patancheru, 24-26 November 2004.

Ravinder Reddy, Ch. 2004b. Seed systems in the semi-arid tropics of Andhra Pradesh. *In* Enhanced livelihoods of poor livestock keepers through increasing use of fodder, 2-3 September 2004.

Tripp R. and W. Van der Burg. 1997. "The conduct and reform of seed quality control," in R. Tripp (ed.) *New seed and old laws: regulatory reform and the diversification of national seed systems.* London: Intermediate Technology Publications. Uganda seed Project. 2000. *Annual report*. Kampala: Kawanda Agricultural Research Institute.

Tripp R. 2003. How to cultivate a commercial seed sector. Paper presented for symposium Sustainable Agriculture in the Sahel, 1-5 December 2003 Bamako, Mali.

Millets: Ensuring Climate Resilience and Nutritional Security (2015) *Pages* **607–626**
Editors: **Vilas A. Tonapi and J.V. Patil**
Published by: **DAYA PUBLISHING HOUSE, NEW DELHI**

Chapter 24

Developing Post Rainy Sorghum Seed System in India

Ch Ravinder Reddy[1], A. Ashok Kumar[1], Sharad Gadakh[2], Uttam Chavan[2], H.K. Kalpande[3] and Vilas A. Tonapi[4]

[1]*International Crops Research Institute for Semi-Arid Tropics, Patancheru*
[2]*Mahatma Phule Krishi Vidyapeeth, Rahuri, Maharashtra*
[3]*Marathawada Agricultural University, Parbhani, Maharashtra*
[4]*ICAR-Indian Institute of Millets Research (IIMR) Rajendranagar, Hyderabad – 500 030*

The crops grown under rainfed agriculture are described as farming practices that rely on rainfall for crop production and their seed systems describe, how farmers in these regions are sourcing seed for cultivating these crops. The objective of this chapter is to share information, experiences and some success stories of seed value chain models developed for production and supply of improved varieties to resource poor farmers to enhance production of rainfed crops. Majority of crop varieties grown under rainfed agriculture system are open pollinated varieties or self-pollinated crops especially food crops, cereals and legumes grown in semi-arid tropics of the globe. The importance of rainfed agriculture varies regionally but produces most food for poor communities in developing countries. In sub-Saharan Africa more than 95 per cent of the farmed land is rainfed, while the corresponding figure for Latin America is almost 90 per cent, for South Asia about 60 per cent, for East Asia 65 per cent and for the Near East and North Africa 75 per cent (FAOSTAT, 2005).

The problems in semi-arid tropical (SAT) regions are exacerbated by adverse biophysical growing conditions and the poor socioeconomic infrastructure. The uncertain climatic conditions or otherwise called as climate change effects, makes SAT farmers more vulnerable. The most essential input of crop production is seed and is the cheapest of all inputs in rainfed agriculture. A good quality seed of improved variety can enhance production by 20 per cent and with improved crop production practices can increase yield by 35-40 per cent. The major types of seed constraints in rainfed agriculture are 1) Seed insecurity due to frequent droughts, and natural disaster 2) Poverty and food insecurity leading to seed insecurity 3) Quality seed and new varieties and development of appropriate seed systems.

Availability and accessibility to seed of improved varieties in these areas is a big task. During drought years or and natural calamities subsidized seed or international relief programs meet the requirement of seed supply which nullifies farmers preference and forces to adopt the variety available. But it's a temporary relief for farmers, the year and again the problem of seed security repeats in SAT regions. The SAT is the home to 38 per cent of the developing countries' poor, 75 per cent of whom live in rural areas. Over 45 per cent of the world's hungry and more than 70 per cent of its malnourished children live in the SAT. The institutional mechanisms to multiply the farmers preferred varieties of crops grown in rainfed regions is poorly developed and private seed sector is not showing interest in such crops because of economic reasons. Public sector research and development organizations do develop varieties to enhance production and productivity in these regions but, public sector extension mechanisms are unable to disseminate the technologies available to farmers to meet their requirements fully.

Existing Seed Systems of Post Rainy Sorghum in India

Sorghum is grown both in rainy and post rainy (*Rabi*) seasons in India. The majority of *Rabi* sorghum grain and stover production is concentrated in districts across the states of Maharashtra, Karnataka and Andhra Pradesh (Trivedi, 2008; Rana *et al.*, 1999; Hosmani and Chittapur, 1997; Murty *et al.*, 2007; Pray and Nagarajan, 2009, (Figure 1). Sorghum seed system is very unique in the country with contrasting situations. Hybrids are the cultivar choice in rainy season sorghum and hybrid adoption by farmers is up to 95 per cent in states like Maharashtra though there are wide variations in adoption across the states in India. The public and private sector seed companies developed hybrids, rule the market and seed requirement is predominantly met by the vibrant formal seed system led by private sector seed companies and public sector seed agencies like National Seeds Corporation (NSC) and state seed development corporations in different states and Mahabeej in Maharashtra. In case of postrainy season, open-pollinated varieties are the cultivar choice because of stringent quality considerations and lack of appropriate hybrids and inadequate hybrid seed production and supply chain.

The post rainy sorghum crop accounts for 45 per cent of the total sorghum area under cultivation and 32 per cent of the total sorghum production in India (Sajjanar *et al.*, 2011). Although post rainy (*Rabi*) sorghum is highly valued due for its good grain quality, its yields are lower (750 kg/ha) compared to *Kharif* sorghum

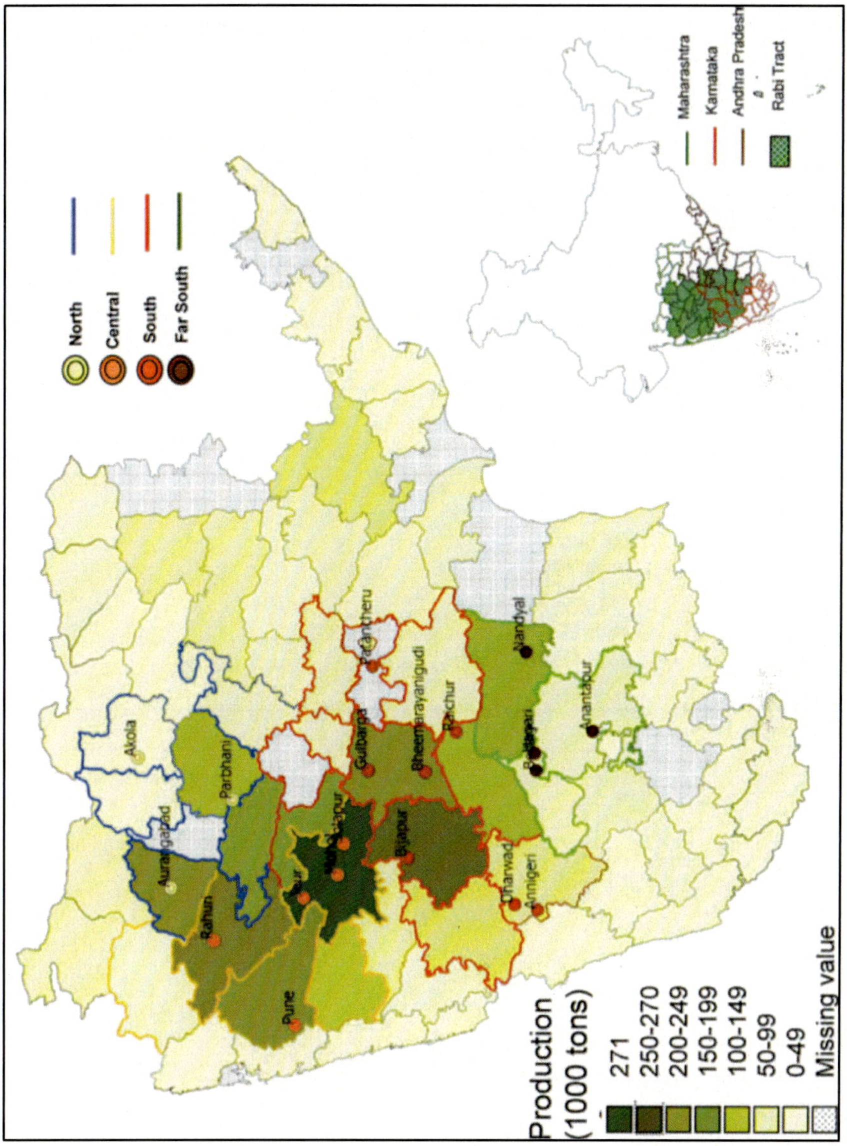

Figure 1: Map of the Sorghum Rabi Production Tract in India with Highlighted Four Major Production Zones – Central, Northern, Southern and Far South. The inset map showsthe location of the Indian states (Maharashtra, Karnataka and Andhra Pradesh) and the major rabi tract.(*Source*: J. Kholová *et al.*, Field Crops Research 141 (2013) 38–46)

(1100 kg/ha) (Anon, 2006).This low productivity rate of post rainy sorghum increase calls for a change in production strategy including breeding, targeting varieties for different soil depths and improved seed systems to make improved variety seed available to small scale farmers in India. The postrainy season sorghum crop was grown on 4.8 million ha (CMIE, 2007) in India (Table 1), Maharashtra has highest area of 3.2 m ha under sorghum, requires 32,000 tons of seed @ 10kg ha^{-1} seed rate. Formal sector is able to meet ~12 per cent of seed requirement and balance ~88 per cent seed supply is from informal sector, mostly from farmers own saved seed.

Table 1: Region-wise Area Under Post Rainy Season Sorghum Cultivation and Seed Sources in India (2011-2012)

Region/state	*Area Under Cultivation (lakh ha)*	*Present Varieties in Cultivation*	*Total Quantity of Seed Requirement (in tons)*	*Seed Supplied by Formal Seed Sector (in tons)*	*Seed Supplied by Informal Seed Sector (in tons)*
Maharashtra	32	M 35-1, Dagadi, Phule Vasudha and Parbhani Moti	32,000	4,000* (12.5 per cent)	28,000** (87.5 per cent)
Karnataka	13	M35-1, Muguti (5-4-1), Annigiri(A-1), DSV-4 and DSV-5	13,000	<10 per cent	>90 per cent
Andhra Pradesh	2	M35-1, Budda Mallelu, Udgir local, Saayi Jonnalu, Dagdi local, CSV216R	2,000	INA	INA
Gujarat	0.1	BP 53, Surat 1, GJ 108, Malvan, Solapuri, Gundari	100	INA	INA
Other states	1	INA	1000	INA	INA
Total	**48.1**		**48,100**		

* Formal sector-supply by private/corporations (Maharashtra state seed Development Corporation/ NSC)

** Informal sector-Farmers own saved seed, local markets, friends, relatives, govt. subsidised seed supply.

INA: Information not available.

High yielding and improved cultivar's seed availability is not a constraint in rainy season sorghum but it is a major issue in postrainy season sorghum in India, and majority of varieties are age old still ruling the major area under cultivation. Most notable local varieties popular among the farmers include M 35-1 (Maldandi), and Dagadi grown by 80-90 per cent of farmers in India. However, M 35-1, a landrace selection from Maldandi cultivated by the farmers is traditionally in these areas for several decades, was selected in 1938, nearly 75 years ago and is still dominating the postrainy season tracts (Maharashtra, Karnataka and Andhra Pradesh) in India (Belum Reddy *et al.*, 2012). Several improved varieties such as Phule Yashoda, Phule Anuradha, Phule Chitra, Phule Revathi, Parbhani Moti and Parbhani Jyothi developed by SAU have been released in the recent past by the All India Coordinated Sorghum Improvement Project (AICSIP), (Table 2).

Table 2: Sorghum Varieties Released by Agricultural Universities in Maharashtra

Sl.No.	*Variety*	*Year of Release*	*Variety Released by Agriculture University*
1.	M 35-1	1938	MPKV-Rahuri
2.	Parbhani Moti	2002	VNMKV- Parbhani
3.	Phule Vasudha	2007	MPKV-Rahuri
4.	Phule Anuradha	2008	MPKV-Rahuri
5.	Phule Revati	2010	MPKV-Rahuri
6	Phule Yashoda	2009	MPKV-Rahuri
7	Phule Suchitra	2012	MPKV-Rahuri
8.	Parbhani Jyoti	2005	VNMKV Parbhani

The private sector is not forthcoming for multiplying the open pollinated varieties (OPVs) of sorghum for various reasons:

1. There is no proprietary advantages in multiplying public domain varieties
2. In case of postrainy season adapted varieties (or hybrids), the seed produced in postrainy season has to be marketed in next postrainy season which means they need to wait for 8 months to market them and hence the returns on investment are realized late.
3. The margins for private seed companies are low in production and marketing of open pollinated varieties (OPVs) compared to hybrids market, and there are not many improved hybrids with all the farmers preferred traits in post rainy sorghum.
4. It is primarily 10-12 per cent of total seed requirement of Maharashtra state which is fulfilled by the public sector seed agencies and state agricultural universities partially catering to the needs of farmers for sorghum seed supply in postrainy season (Table 3).
5. Major portion (90 per cent) of farmer's seed source is met from farmers own -saved seed (Informal sector).

In India rainy season sorghum is cultivated in around 2.6 m ha which is predominantly grown with hybrids. Considering the high adoption rates of hybrids up to 95 per cent in Maharashtra state, reveals the strength of seed companies, genetic materials adoption in different agro-ecological zones with varied climate and soils. The scenario of hybrid cultivars shows wide variation among states (only 10 per cent adoption in Bihar and Odisha) in adoption of improved cultivars. Similarly, the adoption of Hybrids varies from 2-12 per cent in Eastern and Southern Africa (ESA) and Western and Central Africa (WCA) (Figure 2).

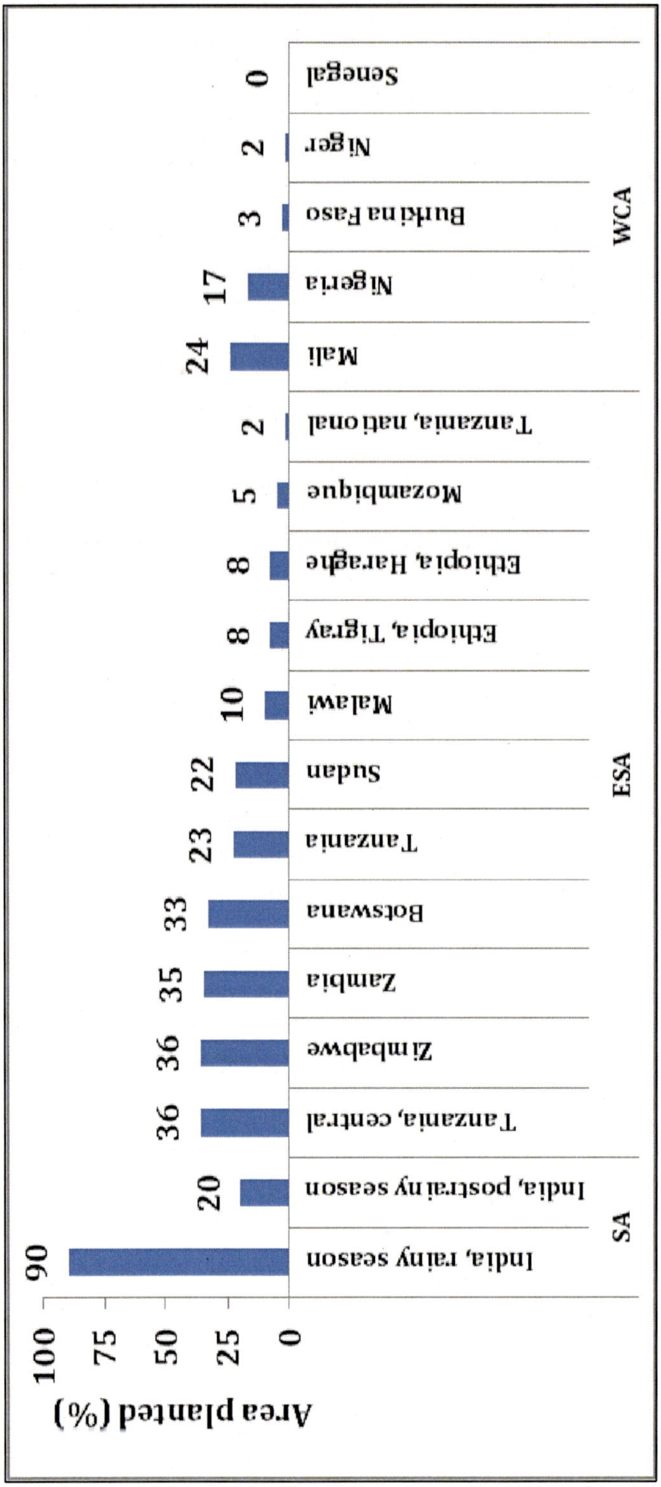

Figure 2: Adoption Levels of Improve Varieties in Sorghum Rainy and Post-rainy Seasons in India and other Countries.

Post-rainy sorghum grain is staple food of Maharashtra state and every farmer grows sorghum for his own food use and stover for livestock. Hence, sorghum is an important crop in crop –livestock cropping system which feeds humans and livestock. The seed required for post rainy sorghum is predominantly produced by the public sector (state seed development corporations and Agriculture Universities) organizations which meets 10-12 per cent of total seed requirement (32000 t) of the state and more over these organizations still produce 80 per cent of their total production that comprises of old varieties (Table 3). Baseline survey report (Pokarkar and Chavan *et al.,* 2014) reveals that 93-98 per cent of the seed sources are from farmers own saved seed and balance component is met by Public and Private sector and other informal seed sources (Table 4). Improved variety seeds are available at seed stores in the market and the takers are very few. The percentage of improved varieties seed sold is 0.6–12 per cent and old local variety constitutes 76 per cent of the total sale of seed (Table 5).

Table 3: Seed Production of Post Rainy Sorghum by State Seed Development Corporation of Maharashtra (Mahabeej) in 2012-13

Sl.No.	*Variety*	*Area under Seed Production (ha)*	*Seed Production (in tons)*
1.	M-35-1*(old local variety)	2336	2564
2.	Parbhani Jyoti	49.60	54
3.	Parbhani Moti	315.20	336
4.	Phule Anuradha	2.00	2
5.	Phule Chitra	30.80	20.6
6.	Phule Revati	77.20	64.7
7.	Phule Vasudha	92.40	76
8.	PKV Kranti	108.20	94.4
	Total	**3011.4**	**3211.7**

* Old variety

Research and development programs of state Agricultural universities (SAU) have developed improved varieties and are available in public domain for several years (Table 2). Due to poor infrastructure and institutional mechanisms to disseminate improved variety seed by state extension department, the yields of the post rainy sorghum is stagnant at 500-700 kg ha^{-1} for last couple of decades. To augment the seed production and dissemination of improved varieties, "seed consortium" model developed by ICRISAT (Figure 3) comprising various partner institutions like Department of Agriculture, Agriculture Universities, Seed certification agency, State and National seed development corporations, Private seed companies, NGOs, SHGs, KVKs which were brought on to one platform with basic objective to enhance availability of improved variety seed at right time and for right price to increase production and productivity of sorghum.

Table 4: Procurement of Seed by the Farmers from different Sources in Maharashtra State during 2013-14.

Seed Source	*Per cent Farmers*					
	Sholapur Dist.	*Pune Dist.*	*Ahmednagar Dist.*	*Beed Dist.*	*Jalna Dist.*	*Parbhani Dist.*
Own saved seed	93.33	98.33	98.33	93.33	95.00	92.5
Borrowed from others	0	0	26.67	1.66	0.00	0.00
Village Market	8.33	3.33	18.33	1.66	0.00	0.00
Local Market at Taluka level	3.33	30.00	15.00	0.00	5.00	5.00
Village Landlords	0	0	1.67	3.03	0.00	2.50
Private seed company*	8.33	3.33	13.33	6.67	8.033	1.25
Govt. Subsidized seed supply**	66.67	0	0	5.0	1.67	1.25
SAU*	0	0	18.33	0.00	0.00	10.00

*: Multiply and supply improved variety seed

**: Multiply and supply 80-90 per cent local variety seed and 10-12 per cent improved variety seed

Source: Baseline Survey Report of SAU, 2014.

Table 5: Post Rainy Sorghum Varieties Seed Sold by Seed Dealers in Maharashtra State during 2013-14

Variety	*Per cent of Total Seed Sale in Project Areas*	
	Marathwada Region (Eastern Maharashtra)	*Western Maharashtra*
M 35-1*(old local variety)	78.66	77.45
Parbhani Moti	11.93	–
Parbhani Jyoti	0.49	–
Phule Anuradha	0.18	0.17
Phule Chitra	0.38	–
Phule Revathi	0.77	–
Phule Vasudha	1.51	–
PKV Kranti	2.87	–
DJ 4005	0.05	–
Deccan Pearl	–	3.11
Kopargaon	–	1.86
Mahabeej	–	0.6
Suvarna	0.06	13.08
Swati	–	0.62
Vimal	3.1	3.11

* Old local variety and rest all improved released varieties

Interventions to Enhance Production

Postrainy sorghum is an important food and fodder crop of India is grown on residual soil moisture and gets affected by terminal moisture stress. Despite good efforts by Indian National agricultural Research services (NARS), the productivity remained low because of poor adoption of improved varieties and crop management technologies by farmers and above all the recurring droughts affecting the grain yields. Further, the market opportunities for grain and fodder are limited restricting it to a subsistence production system.

One of the reasons for the dismal performance of many development projects and programmes is that they are often designed to address problems of target populations as perceived by government officials and other outsiders rather than as perceived by the people themselves. The danger with this approach is that in many cases, considerable amount of resources are expended on trivial problems while priority problems of people are left unattended. The result is that as soon as external funding and assistance are withdrawn, the initiated development activities which are supposed to be continued by the people are not carried beyond the life of the project duration.

To enhance the on-farm productivity and to develop a sustainable integrated postrainy sorghum seed value chain, ICRISAT partnered with Government, Private, Public and people institutions; the project partners are Directorate of Sorghum Research, Hyderabad, Vasantrao Naik Marathwada Krishi Vidyapeeth (VNMKV) – Parbhani and Mahatma Phule Krishi Vidyapeeth (MPKV), Rahuri, Krishi Vignana Kendra's (KVKs), State seed development corporation(Mahabeej), National seeds corporation (NSC), Private seed companies, and Farmers associations (FAs), with an aim converge individual resources and knowledge to increase crop productivity and farmers' incomes by reducing the gap between on-station sorghum yields (2.5–3.0 t ha^{-1}) and farmers yields (0.75-0.58 t ha^{-1}) and by adding value to grain and stover and by providing market linkages.

Bill and Melinda Gates Foundation (BMGF) funded the project on "Harnessing the Opportunities for Productivity Enhancement" (HOPE) in Sorghum and Millets implemented in 11 countries including India during 2009-2013 (HOPE.icrisat.org/). Earlier efforts on postrainy sorghum improvement resulted in release of improved varieties with farmers preferred traits. In the absence of suitable seed system to deliver these varieties, farmers' still use old local varieties like Maldandi(M35-1), Dagadi, Zipri, Jut, Dudhmogra, Bondri, Cheetpuri, Shalu, Khadki, Manthi, and Dukrilocal. Concerted efforts by HOPE team in supplying the improved seed, technologies, knowledge, input and credit systems, value addition and market linkages made a remarkable change in the postrainy sorghum value chain in the project areas in Maharashtra, India. The project commenced in 2009, more than 300,000 farmers benefitted directly and indirectly through the project interventions.

Project locations consisting six clusters " Sanpuri (District: Parbhani), Limbaganesh (Dist. Beed) and Wakulni (Dist. Jalna) in Marathwada area (Eastern Maharashtra) and Hivare Bazar (Dist. Ahmednagar), Borkarwadi (Dist. Pune) and Aurad (Dist. South Sholapur) under Western Maharashtra region were selected for technology dissemination considering the large area under postrainy sorghum in these clusters. Seed of selected (farmers participatory varietal selection) and released varieties such as Parbhani Moti, Parbhani Jyoti, Phule Vasudha, Phule Chitra and Akola Kranti were distributed for cultivation in the project areas and farmers were trained to use improved crop management practices. The fallowing improved crop production technologies were demonstrated across all villages of the project areas.

1. *In-situ moisture conservation*: Opening up of furrows in preceding legume crop at every 3rd or 4th row for improving drainage, moisture and soil conservation which have good effect on the yield of succeeding post rainy season sorghum. Three intercultural operations from 2nd week to 8th week after sowing will help to reduce weeds, loosening the soil for root development; improve soil aeration and moisture conservation.
2. *Plant population geometry and spacing*: Plant population ranging from 1.25 to 1.35 lakhs ha^{-1} is optimum depending on receding soil moisture. Soil types play an important role in deciding on plant population. Wider row spacing of 45cm X 10cm and 10-12 cm depth of sowing must be adopted for better crop yields; a seed rate of 10kg ha^{-1} is recommended for optimum plant population. A seed drill developed for wider row spacing and deep placement of seed and fertilizer should be used for sowing under rainfed cultivation of sorghum.
3. *Seed treatment and fertilizer management*: Seed treatment with Thiamethoxam (70WS) @ 3 g kg^{-1} of seed is a must to manage the shoot fly an important pest on sorghum. Application of basal dose (before sowing) of 40kg Nitrogen and 20kg Phosphorus fertilizer is recommended based on soil analysis reports.

Table 6: Average Yields of Improved Varieties Demonstrated in Project Locations in Maharashtra (2009-2013)

Region	*Grain Yield* (qt. ha^{-1})*	*Fodder Yield (qt. ha^{-1})**	*Percent Increase Over Local Variety*		*No. of Farmers Participated*
			Grain	*Fodder*	
Marathwada region	9.61(6.8)	29.19(23)	41	24	17,143
Western Maharashtra	12.6(9.2)	29.0(22)	36	33	16,100

* One qt. =100kg

In the first five years of project implementation, the project directly covered 33,000 farmers and the impact of the interventions reached more than 300,000 farmers in Maharashtra state. The implementation of technologies led to significant increase in grain productivity by 39 per cent and stover productivity by 29 per cent (Table 6) in project villages. The early adoption results indicated that the HOPE interventions enhanced technology adoption rates, reduced the yield gaps (by 30 per cent), increased the productivity and gave higher returns to farmers (36-41 per cent). They also indicated that for every single farmer covered by HOPE project directly, 5-6 non-

HOPE farmers benefitted. Dissemination of technologies (improved variety seeds) through secondary channels like farmer to farmer, relatives, friends and gifts to their kith and kin spurred the production (http://www.cgiar.org/consortium-news/hope-leads-to-increased-sorghum-yields/).

With these interventions farmers got convinced and are showing more inclination for growing improved varieties. To mitigate the deficiency of seed production and supply, an innovative seed system model (Figure 1) developed and for its implementation and monitoring a 'Seed Consortium' was formed during 2013 involving Department of Agriculture, Indian NARS, public and private sector seed agencies, seed certification agency, KVKs, SHGs, NGOs, and farmers became the stakeholders of the consortium with respective roles and responsibilities in production and distribution of seed.

Implementation of Seed System Model

i) Main Objective

To improve farmers accessibility and availability of seed of improved varieties at affordable price at right time for enhancing yield, income and household seed and food security

Specific Objectives

1. Conduct regional survey to understand the existing seed system
2. Develop appropriate seed system model through participation of Private-Public-People -partnership to improve accessibility and availability of improved varieties seed to small-scale farmers.
3. Identifying the varieties suitable for different agroecological regions through farmers participatory selection.
4. Harnessing the farmers knowledge and preferences in improving the seed system and research.
5. Improve the cultivar replacement rate (CRR) and seed replacement rate (SRR) to enhance the production.

ii) Approach

The proposed conceptual and organizational approach, strategies and partners, the linkages and support from formal sector institutions were planned and developed a seed consortium model that includes private, public sector seed companies, State Agriculture Universities(SAU), Krishi Vignana Kendras (KVKs), Self-help groups(SHGs), Non-Government organizations (NGOs), and farmers.

iii) The Model

Based on the findings of baseline survey on existing seed systems of post rainy sorghum in Maharashtra, a model (seed consortium) was developed to multiply improved varieties and distribute to farmers was implemented (Figure 3).

In this model, we envisaged a decentralized seed production and centralized seed procurement and distribution in initial years but eventually shifted to

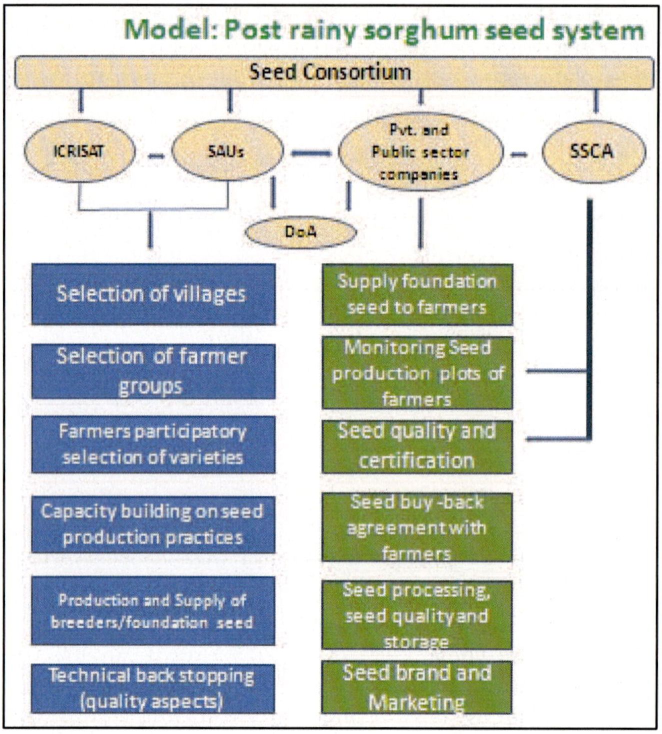

Figure 3: Seed Consortium Model.

decentralized seed production and distribution. The public sector research and development institutions (ICAR, SAUs and ICRISAT) and private sector seed companies engaged in partnership research to develop improved products and the private sector seed companies and public sector seed agencies played a critical role in seed production, procurement and dissemination of these improved products to farmers. This is akin to the rainy season hybrid seed production and distribution model in the country, which is one of the most successful examples in the developing world for having a strong seed system. Thus we have a sustainable commercial model to replicate to strengthen the postrainy sorghum value chain.

The seed consortium developed a work plan to produce 29,000 tons of improved varieties seed to meet the requirement of 2.9 m ha in Maharashtra in 4 years. The early adoption studies on improved varieties by the farmers conducted by ICRISAT, revealed that secondary dissemination of seed is very active in the region, each farmer on an average is sharing seed with 2 to 6 other farmers across the districts in the state. The trends of secondary dissemination of seed by the farmers will able to meet the seed requirements of farmers in Maharashtra, covering 3.2 m ha by the end of 4 year project duration.

iv) Roles and Responsibilities of Consortium Members

For establishing a seed consortium, ICRISAT and Partners worked for developing integrated postrainy sorghum value chain by harnessing the power of genetics, crop management, value addition and markets under the HOPE project. Based on the strength of achievements under HOPE project for postrainy sorghum productivity enhancement, a 'Seed Consortium' was formed during 2013 under the chairmanship of commissioner of Agriculture, Maharashtra state involving private and public sector partners to sustain HOPE interventions. Under the Consortium, an innovative seed system model developed and delineated responsibilities to partners for achieving seed production targets fixed during the meeting (Table 7).

Table 7: Seed Production Targets of the Project

Year	*Seed Production Area (Ha)*	*Seed Production (tons)**	*Expected Area covered with Improved Varieties (Lakh Ha)*
1st year (2013-14)	500	1000	1.0
2nd year (2014-15)	2000	4000	4.0
3rd year (2015-16)	4000	8000	8.0
4th year (2016-17)	8000	16000	16.0

*: Indicative seed yield 2.0 tons ha^{-1}

1. *State Agricultural Universities*: Two agricultural universities MPKV and VNMKV are the members of the consortium and have developed new varieties. The selected released varieties Phule Vasudha, Phule Chitra, Phule Revathi, Parbhani Mothi were multiplied on research farm and supplied Breeder/and Foundation seed to public sector seed company (Mahabeej) for production of certified seed. the cost of foundation seed production is borne by the project. The universities and Mahabeej scientists jointly selected villages and farmers for seed production. The Govt. of Maharashtra is encouraging farmers by paying an incentive of Rs. 500/- per quintal (100kg) of seed produced by the farmers under seed village scheme (Anon. 2009).
2. *Public sector seed company (Mahabeej)*: Mahabeej has agreed on the work plan (Table 7) and also agreed in principle to reduce gradually production of local variety M35-1 (Maldandi) to promote improved released varieties. They have entered into buy-back agreement with farmers with a prefixed minimum price of seed procurement and agreed to pay 20 per cent more over the grain price in the market at the time of procurement. The seed harvesting and transportation to processing plant is the responsibility of farmers and processing, grading, branding and marketing is Mahabeej responsibility.
3. *NGOs, KVKs*: These organizations agreed to promote farmers in growing seed in addition to village seed programme to meet the target area under seed production. However, the organizations has a programme of seed development which was merged with our programme and they are benefited by access to foundation seed supply, training programmes for farmers and other crop production incentives and market linkages through consortium.

4. *State seed certification agency*: Mahabeej has taken responsibility to register farmers name and area for seed certification. The main objective of seed certification agency is to monitor genetic purity of the variety and certify the quality and quantity of seed produced by the farmers. The expenses incurred for monitoring the seed crop and issuing the certificate for seed produced by the farmer was borne by the project.
5. *Department of Agriculture*: Agriculture Department was established in 1883. Work started with the aim of helping the rural community to achieve higher productivity in agriculture. Development of hybrid varieties of different crops since 1965-66 laid down the foundation of Green Revolution. Introduction of intensive agriculture, comprising of large scale use of improved seed, fertilizers, pesticides and available water helped to increase in agriculture production. Later on, considering the need for providing guidance to the farmers for proper and judicious use of these inputs, Training and Visit Scheme was launched. Valuable contribution of this scheme through effective implementation of programmes like Crop Demonstrations, Field Visits, Corner meetings, Workshops, Fairs, Exhibitions etc. aimed at transfer of technology from Agriculture Universities to farmers fields increase agricultural production.

Agriculture department considers farmer as the focal point and the whole department is organized in such a fashion that a single mechanism is working to facilitate the farmer for adoption of advanced technology and sustainable use of available resources. There are several government schemes for the farmer's producer group like National Food Security Mission, National Horticulture Mission, Seed production scheme, Exposure visit, ATMA project scheme. Again there are many schemes crop wise for all the farmers from Maharashtra.

v) Capacity Building

Seed producing farmers were trained on seed production technique, isolation distance, sowing practices, other agronomic practices to be followed for the sorghum crop. Training was organized during flower initiation stage of the seed crop. The seed growing farmers were trained to identify off types, rogues and its removal from the seed plots and to maintain the quality of seed production and other agronomic practices, plant protection measures and harvesting methods. Further training was organized after harvest and at the time of seed processing to impart knowledge on seed cleaning, seed grading, seed treating, seed storage, seed packaging aspects, and how to draw the representative seed sample, for seed testing/local seed testing method to assess the seed germination and quality.

Training programmes were conducted on-station (University) and on- farm (in the villages) by technical staff of universities jointly with seed certification officials. Mahabeej staff joined the programmes to announce their buy-back agreement and assurance of seed procurement to develop confidence levels in seed producers. During seed production period university technical staff visited the farmers' fields and wave technical advice to farmers on crop production. Most of these villages were earlier adopted under HOPE project. Hence, almost all farmers in the villages are well-versed with improved crop production technologies giving fillip to the seed production programme.

vi) Seed Production

The first seed consortium meeting was conducted at Pune in April 2013, under the chairman ship of commissioner of Agriculture. The members of department of agriculture, Vice chancellor, Director of Research, Advisor to Dryland Agriculture Mission, Seed certification agency director, General Manager- Mahabeej and Private seed company's representatives participated in the meeting.

There was consensus among the consortium partners to develop robust seed system for post rainy sorghum in Maharashtra and the commissioner of Agriculture extended all support under seed village program for the benefit of seed growing farmers (Anon, 2009).

Table 8: Seed Production during 2013

Partners	*Target Seed Production Area (ha)*	*Seed Production Area in ha*	*Quantity of Seed Produced in tonns*
MPKV (Eastern Maharashtra)*	250	243	263
VNMKV (Western Maharashtra)**	250	198(98)	166
Total	500	341	429

* MPKV area Average yield -10.18 q/ha

**VMMKV area average yield- 17 q/ha from 98 ha

During year 1 (2013-14) of project implementation we could grow seed crop on 341 ha and produce 429 tons of seed (Table 8). We could not achieve target area and yield for various administrative and adverse climatic conditions. Due to short time in implementing the project, MKV could not reach the targets due to unfavorable climatic conditions (heavy rains and high speed winds) during harvesting period. All the seed produced in both the regions was procured by the Mahabeej as per buy-back agreement signed with the farmers.

Table 9: Seed Production during 2014

Partners	*Target Seed Production Area (ha)*	*Certified Seed Production Area Sown (ha)*	*Quantity of Seed Produced (tons)**
MPKV (Eastern Maharashtra)	1000	2000	2400
VNMKV (Western Maharashtra)	1000	210	315
Total	2000	2210	2715

* Expected yield.

During second year (2014), the targets of area were doubled over year1 (Table 9). Foundation seed production area was damaged during year 1 in western Maharashtra so they did not have sufficient seed to meet the target area of 1000 ha in year 2. To reduce the deficit of other partner, eastern Maharashtra partners increased their seed production area by 2000 ha to meet the target area of the year.

Farmers in Maharashtra acquired post rainy sorghum seed through various modes (Table 4) in varying proportions depending up on the variety, rate of seed replacement, social networks, and market integration. Sorghum is grown in rainy season purely under rainfed conditions and in post rainy seasons it is grown on receding soil moisture condition. On the contrary the seed system operating in rainy season in Maharashtra is 95 per cent formal and acquisition varies from place to place, hybrids are main choice of farmers. Whereas, 93 per cent of seed sourced informally during post rainy season. This indicates that the seed sowing by farmers vary greatly during seasons and across ecoregions. The proportion of seed is sourced by the farmers varies with in the system and among the regions of the state (Table 4). Majority of farmers in Maharashtra save their own seed and use it for sourcing next year, this practice is likely to alter the sourcing the seed from other two mode (purchasing and sharing). Farmers living in the vicinity of SAU procured improved varieties seed from SAU sales counter and mostly they are big farmers and are aware of varieties and sources of seed and by virtue of their location and accessibility to seed source.

It is not uncommon that the innovative farmers use seed (cultivars) from formal and informal sectors (relatives, neighbours, own-saved seed) and grow all on same piece of land leading to contamination due to lack of isolation. Farmers select the variety and save the seed for next season sowings seed is shared with their friends and relatives. Varietal purity and identity thus become blurred. With this sort of farmers practice, the purity of the good old variety M 35-1(Maldandi) released during 1938 is questionable. Still farmers prefer this variety and covers 80 per cent share in post rainy sorghum cultivation.

The components of formal and informal seed systems of post rainy sorghum operating in India and the flow of the genetic material from one system to other is depicted in Figure 4.

The replacement of old varieties with new varieties is the major task for govt. extension department to make availability of new variety seed in time on regular basis to the farmers. Usually, the seed production by State Seed Development Corporation (largest seed producing agency of post rainy sorghum) produces improved varieties seed only 2 per cent of total seed production (Table 3), because there is no uptake of improved varieties seed by the farmers; the social reason for not preferring new varieties is that the people prefer the taste of the roti they make from old variety and animals like the fodder of old variety.

Cultivar replacement indicates how effectively seeds of new cultivars are adopted by the farmers and produced and supplied by the seed agencies. The factors which determine the rate of replacement are how government popularizes the cultivar, superiority of new cultivar to the existing ones which they intend to replace. The higher and quicker replacement depends on superiority of the cultivar base yield, price of seed, and seed quality of farmers saved seed. The resuls indicate that farmers

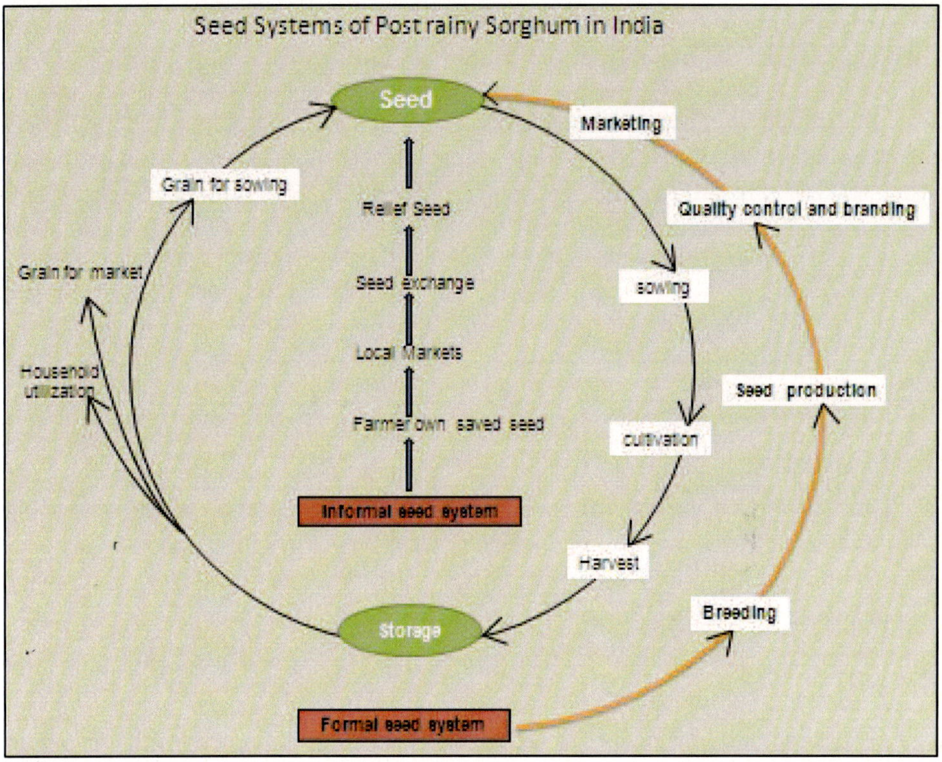

Figure 4: Flow of Genetic Material from Formal and Informal Systems in Post Rainy Sorghum in India.

did not adopt improved varieties released by AICSIP over a period of 7 years (Table 2) inspite of higher yields of grain and fodder (Table 6) for the reasons mentioned above. Assessing varietal or cultivar replacement rate (CRR) is not very easy and many indices of varietal replacement have been proposed (Brennan, J.P. and Byerlee, D.1989; Byerlee, D. and Heisey, P.W. 1990), but these indices can be obtained from statistics on breeder or certified seed production data and field surveys on adoption of new varieties (Witcombe *et al.*, 1998).

The project partners of state agricultural universities (MPKV and VNMKV) are committed to produce breeder and foundation seed (Table 10) and supply to Mahabeej for production of certified seed under seed village program of Govt. of India (Anon, 2009). With available foundation seed, Mahabeej will be able to take up seed production on 5000 ha in 2015, inturn will be able to produce 7500 t (average seed production 1.5 t/ha) of certified seed, which will be sufficient for sowing 7,50,000 ha area with improved varieties to cover 23 per cent of total area of post rainy sorghum area in Maharashtra.

Table 10: Seed Production Programme of Agricultural Universities-2014

Varieties	*Seed Production*		
	Nucleus (Kg)	*Breeder (Kg)*	*Foundation (Kg)*
Phule Revathi	50	1000	14,800
Phule Vasudha	50	1000	14,800
Phule Suchitra	50	1000	14,800
Phule Anuradha	25	500	4800
Parbhani Moti	10	11880	2430
Parbhani Jyoti	10	2160	1080
Total	**195**	**17540**	**52710**

Way Forward

Some important issues that could provide way forward for sustainable seed value chain to meet the demand of improved varieties of Post rainy sorghum in India are:

1. Varietal Denotification

i) A review of existing list of released and notified varieties do reveal that old varieties still find place in package of practices

ii) Continued production of seed of old varieties by state corporations is rather counter productive

iii) Denotifying old and obsolete varieties irrespective of whether they are from public or private sector to allow the seed multiplication of the new improved cultivars.

2. Cultivar Replacement Rate (CRR)

i) State must ensure production of Breeder/Foundation seed of rainfed crops and multiplication and replacement of seed to increase CRR progressively.

ii) CRR will happen through technology upgradation and extension work and govt. policies.

iii) For achieving the desired levels of CRR, adequate quantities of seed of improved varieties has to be produced and made available to farmers.

iv) Varietal replacement rate is a continuous process, the new varieties released from time to time should flow into seed chain to improve income and profitability.

3. Seed Mission

Developing and implementing rainfed agriculture seed mission -with a built in mechanism of supporting the cost of seed production for five years through Public–Private–Partnership with effective coordination and convergence mechanisms will help to increase productivity.

4. Advocacy

By increasing access to high yielding varieties/hybrids on priority basis to enhance adaptation rate will bridge the productivity gap and increase production.

5. Selection of Cultivars

Appointing a joint committee comprising of Indian council of agricultural Research (ICAR), State agriculture Universities (SAUs), public and private seed sector representatives and farmer groups to select rainfed crop varieties /hybrids suitable for different agro-ecological areas.

6. Seed Production

Promoting contract seed production by advance indenting of the seed of specific improved cultivars by both Public and Private Sector Seed Companies including KVKs, and community based organizations with technical support and capacity building for production of quality seed will strengthen the seed system.

7. Policy and Funding Support Framework

i) An enabling policy environment does help in production and dissemination of improved variety seed of rainfed crops.

ii) Provision of funds and support for seed multiplication and dissemination activities at least for 5 years.

iii) Strengthening extension services for creating awareness and demonstration of rainfed agricultural technologies.

References

Anon. 2009. Guidelines for Implementation of Seed village Scheme.http://seednet.gov.in/material/Guidelines per cent 20and per cent 20Schemes/Guidelines per cent 20for per cent 20seed per cent 20village per cent 20sceme.pdf. Joint Secretary (Seeds) DAC, Krishi Bhavan, New Delhi. Ministry of Agriculture and Cooperation.

Anon. 2006. AICSIP Co-ordinating Res. Rep. p. 1- 120.

Belum VS Reddy1, P Sanjana Reddy2, AR Sadananda3, Elango Dinakaran1, A Ashok Kumar1, SP Deshpande1, P Srinivasa Rao1, HC Sharma1, Rajan Sharma1, L Krishnamurthy1 and JV Patil 2012. Postrainy season sorghum: Constraints and breeding approaches. *SAT e Journal*: Vol 10:1-12.

Byerlee, D. and Heisey, P.W. 1990. Wheat varietal diversification over time and space as factors in yield gains and rust resistance in the Punjab. In: Accelerating the Transfer of wheat Breeding Gains to farmers: A study of the Dynamics of Varietal Replacement in Pakistan, P.W. Heisey (ed.). CIMMYT Research Report No.1 Mexico, D.F. Mexico.pp. 5-24.

Brennan, J.P. and Byerlee, D.1989. Measuring the rate of crop varietal replacement. Draft paper. Mexico, D.F.: CIMMYT.

Chavan and Ch Ravinder Reddy 2014. Baseline Scenario of post rainy season sorghum seed system in Marathwada region of Maharashtra state. Pages 21.

CMIE. 2007. Agriculture. Mumbai, India: Centre for Monitoring Indian Economy. 360 pp.

FAOSTAT2005.The FAO Statistical Yearbooks 2005/2006. http://faostat.fao.org/site/395/default.aspx

Hosmani, M.M., Chittapur, B.M., 1997. Sorghum Production Technology. University of Agricultural Sciences, Dharwad, India.

Jana Kholováa,, Greg McLeanb, Vincent Vadeza, Peter Craufurda, Graeme L. Hammerc 2013. Drought stress characterization of post-rainy season (*Rabi*) sorghum in India. *Field Crops Research* 141 (2013) 38–46.

Macklin, Michael 1992. Agricultural extension in India.World Bank technical paper, ISSN 0253-7494 ; no. 190) ISBN 0-8213-2291-5.

Murty, M.V.R., Piara Singh, Wani, S.P., Khairwal, I.S., Srinivas, K., 2007. Yield gap analysis of sorghum and pearl millet in India using simulation modeling. Global Theme on Agroecosystems Report no. 37. International Crops Research Institute for the Semi-Arid Tropics, Patancheru 502324, Andhra Pradesh, India, 82 pp.

Patil, JV, P. Sanjana Reddy, Prabhakar, A. V. Umakanth, Sunil Gomashe and K. N. Ganapathy 2014. History of post-rainy season sorghum research in India and strategies for breaking the yield plateau Indian Society of Genetics and Plant Breeding, *Indian J. Genet.*, 74(3): 271-285.

Pray, C.E., Nagarajan, L., 2009. Pearl millet and sorghum improvement in India. IFPRI discussion paper 00919. http://www.ifpri.org/sites/ default/files/publications/ifpridp00919.pdf

Pokarkar VG and Ch Ravinder Reddy 2014. Baseline Scenario of post rainy season sorghum seed systems in Western Maharashtra. Mahatma Phule Krishi Vidyapeeth, Rahuri, Maharastra, 20 p.

Rana, B.S., Rao, M.H., Indira, S., Singh, B.U., Appaji, C., Ionapi, V., 1999. Technology for Increasing Sorghum Production and Value Addition. National Research Centre for Sorghum (Indian Council of Agricultural Research), Rajendranagar, Hyderabad, India.

Sajjanar G. M., Biradar B. D. and Biradar S. S. 2011.Evaluation of crosses involving *Rabi* landraces of sorghum for productivity traits. *Karnataka J. Agric. Sci.*, 24: 227-229.

Trivedi, T.P., 2008. Handbook of Agriculture. Directorate of Information and Publications of Agriculture, Indian Council of Agricultural Research, New Delhi, India.

Witcombe JR, AJ Packwood, AGB Raj, and DS Virk.1998. The external and rate of adoption of modern cultivars in India. www.envirobase.info/PDF/RLPSRbookchap5.pdf.

Index

The Authors

Dr. J.V. Patil

Director,

ICAR-Indian Institute of Millets Research (IIMR)

Hyderabad – 500 030, India

Email: director.millets@icar.gov.in

Dr. Patil is the Director of Indian Institute of Millets Research (IIMR) Hyderabad. During his 28 years research career, he has made significant contributions towards crop improvement through the development of several varieties in sorghum (10), chickpea (5), safflower (1), mungbean (1), rajmabean (1) and pigeonpea (1). In sorghum, he introduced the concept of breeding for specific soil depths for rabi season. Concerted efforts towards this has led to the development of *Phule Vasudha* and *Phule Revathi* for heavy soils, *Phule Chitra* for medium soil and *Phule Anuradha* for shallow soils. He also conceptualized and developed several innovative technologies for sustainable production of *rabi* sorghum especially *in situ* moisture conservation strategies. He has also involved in development of sorghum varieties for specific end uses. He is elected fellow of National Academy of Agricultural Sciences -2012. He is a recipient of several international awards including CGIAR's Baudouin Team Award (2002) and Doreen Mashler Team Award (2002). For his contribution towards developing high yielding varieties in various crops he has received two state awards Baliraja's Late Annasaheb Shinde Smruti Krishi Sanshodhan Gauro Puraskar and Vasantrao Naik Krishi Puraskar awards. He is also recipient of Agricultural Literature award 2010 and 2012 from Govt. of Maharashtra.

Dr. Vilas A. Tonapi

Principal Scientist (Seed Science and Technology)

ICAR-Indian Institute of Millets Research (IIMR)

Hyderabad – 500 030, India

E-mail: vilastonapi@hotmail.com; tonapi@millets.res.in

Dr. Tonapi has joined Indian Institute of Millets Research (IIMR) as Scientist in 1989. He was the first Officer Incharge of Centre on Rabi Sorghum, Solapur. From July 2011 to July 2013 he also headed Division of Seed Science & Technology, IARI, New Delhi. He has visited United States of America, Mexico, Malaysia, Singapore and Australia. He did his post-doctoral research at the University of Queensland, Australia and CRC for Tropical Plant Pathology, Brisbane, Australia. His specialization includes seed industry and seed systems development, IPRs, Plant Variety Protection and seed vigour and viability. He has received several awards, and some of these are Seedsmen Association Medal and Award (2009) for service to seed industry, Consultancy to FAO (2008), Best Scientist award of Directorate of Sorghum Research (2006 & 2007), J.N.TATA Endowment award (2000-02), R.V. Swaminathan Gold Medal for highest OGPA & Best Thesis Award in Seed Technology at TNAU, Coimbatore (1988), ICAR JRF and SRF, National Merit Scholarship, Government of India, District Merit Scholarship, Government of Karnataka. He has 50 refereed research papers, 13 book chapters, 20 conference papers, 65 abstracts and 10 popular articles to his credit. He has authored 20 books, 10 technical bulletins, 10 filders, 10 posters, 23 reports, and edited 5 books.

Dr. A. Seetharam

Former Project Coordinator,

All India Small Millets Improvement Project,

UAS Bangalore, India

E-mail: annada.seetharama@gmail.com

Dr. A. Seetharam is a plant breeder of repute and well known for his contributions in small millets and sunflower. He assembled more than 11,000 collections of six small millets and made available the elite genetic stocks to researchers all over the country. He developed the first sunflower hybrids in the country. Dr. Seetharam is the chairman of the International Small Millets Steering Committee. He was the leader of the National Jai Vigyan Scienceand Technology Mission of Government of India on "Household food and Nutritional Security for Tribal and Hilly areas in the country from 1999-2003. He has guided 20 M.Sc and 22 Ph.D students and three of his Ph.D. students received Jawaharlal Nehru Award of ICAR for outstanding Ph.D. research. He has published more than 250 scientific research papers and several chapters in national and international books. He is the receipient of many awards including the coveted "Shri Om Prakash Bhasin Award", New Delhi, "Prof. R. D. Asana Endowment

Award" of ICAR, "Dr. Harbhajan Singh Memorial Award" of Indian Society of Plant Genetic Resources, New Delhi; Distinguished Alumni Award" of the University of Agricultural Sciences, Bangalore and "Out standing Teacher Appreciation Award" from the Karnataka Civil Defence Corps and Rotary, Bangalore. He is elected fellow of National academy of Agricultural Sciences, New Delhi, Honorary fellow of Indian Society of Oil Seeds Research, Fellow of Indian Society of Genetics and Plant Breeding, New Delhi and Fellow of Indian Society of Plant Genetic Resources, New Delhi.

Dr. Prabhakar

Principal Scientist (Genetics & Cytogenetics) &

Officer In-charge, Centre on Rabi Sorghum,

Solapur, India

E-mail: prabhakar@millets.res.in

Dr. Prabhakar started his scientific career at Central Marine Fisheries Research Institute, Cochin in May 1985, working on Marine Algae for a short period. In February 1987, he joined National Research Centre for Soybean (now IISR), Indore. He has worked on genetic improvement of soybean and released two soybean varieties, besides maintenance and evaluation of soybean germplasm and seed production aspects. In 1996, he joined Centre on Rabi Sorghum (IIMR) and at present, functioning as Principal Scientist and Officer In-charge. He has worked on breeding for high yielding rabi sorghum cultivars and released one rabi variety. He has developed diverse parental lines and new rabi hybrids. He has visited USA in 1993 for acquainting National Agricultural Research System of USDA and UK in 1999 for planning Participatory plant breeding with farmers in rabi sorghum in India under DFID. He has 3 international, 45 national journal research articles, 11 technical bulletins, 3 book chapters, and other publications in his publication basket. He has released 3 varieties and registered 3 genetic stocks and developed new diverse rabi sorghum parental lines (15 CMS lines, 25 B lines and 36 R lines and 20 test hybrids) to his credit.

Dr. C.V. Ratnavathi

Principal Scientist (Biochemistry),

ICAR-Indian Institute of Millets Research (IIMR)

Hyderabad – 500 030, India

E-mail: ratnavathi@millets.res.in

Dr. C.V. Ratnavathi joined Agricultural Research Service, ICAR in 1985 and started her career at CTRI, Rajahmundry. She joined sorghum family at IIMR, Hyderabad in 1989 and has research experience of more than 30 years. Dr. Ratnavathi has significant research contributions to focus sweet sorghum as a bio-energy crop for ethanol production and grain sorghum as a major health cereal. She developed many semi-processed and processed foods like flakes and pasta from sorghum. She published 30 research articles which include mostly international. She is recognized as a reviewer for biofuel projects by MNRE and for health foods by DBT. Dr. Ratnavathi has undertaken several projects such as NATP project on value addition, ICAR network project on food safety, NAIP project on millet foods, NFBSFARA project on sweet sorghum and DBT project on functional foods. She has vast experience in food safety research and her work was cited by CODEX committee. She is a resource person for many national and international training programmes and a recognized reviewer for many international journals. She has also participated in international research programmes at Australia and USA.

Dr. S.S. Rao

Principal Scientist

ICAR-Indian Institute of Millets Research (IIMR)

Hyderabad – 500 030, India

Email : ssrao@millets.res.in

During 1991 he joined the Indian Institute of Millets Research (IIMR), Hyderabad as physiologist. Presently he is working as Principal Scientist at IIMR. His has specialization in phenotyping kharif sorghum (grain, dual-purpose, sweet and forage sorghums) for climate change especially drought and heat tolerance) adaptation, mapping population development for MAS/QTL identification for drought and heat stress tolerance in dryland sorghum. He is also looking after AICSIP Physiology Coordination as Principal Investigator. He has several publications and recognitions to his credit.

Dr. B. Dayakar Rao

Principal Scientist (Agricultural Economics),

ICAR-Indian Institute of Millets Research (IIMR)

Hyderabad – 500 030, India

E-mail: dayakar@millets.res.in

Dr. Dayakar joined in Agricultural Research Services (ARS) in 1991 and started the research career in Indian Institute of Horticultural Research (IIHR), Bangalore and worked on Betel vine, a perennial crop economics. In 1993 he joined Indian Institute of Millets Research (IIMR), Hyderabad. He has taken key role in assessment of trends in sorghum production and consumption in India and identified policy measures to fill those gabs. Has developed a national database on millets that revealed the constraints of sorghum production and suggested most suitable technologies and cultivation practices zone wise to enhance productivity, higher profits and enhance food and nutritional security. Developed sustainable model of millets value chain through processing interventions, nutritional studies, promotion and marketing, industrial liaison of research centers, and entrepreneurship development; which was greatly applauded by the Ministry of Agriculture and the World Bank that has resulted to setting up of 200 processing clusters across the country backstopping by IIMR under INSIMP- a Rs. 300 crore Initiative by DAC, Ministry of Agriculture, GoI appointing the nominee as the Nodal Officer to mentor the stakeholders. At present, he holds the position as PI of NAIP-Millets Value Chain, DSR-Britannia Joint Project, and Nodal Officer of INSIMP (GoI programme), PI of Centre of Excellence on sorghum processing and value addition. His research interest is to developed sustainable and ideal model of value addition on Agricultural commodities by identifying aspects and prospects from farm level through post harvest management (processing and entrepreneurship development) and finally promotion and commercialization.

Dr. H.S. Talwar

ICAR-Indian Institute of Millets Research (IIMR)

Hyderabad – 500 030, India

E-mail: talwar@millets.res.in

Dr. H.S. Talwar started his professional carrier during 1981, after receiving M.Sc. from CCHAU, Hisar, when he joined pulse physiology unit of legume programme in ICRISAT. Later on, he obtained his Ph.D. from PAU Ludhiana in 1990. During his employment of >19 years with ICRISAT, he worked in the area of "Improving cold tolerance in Chickpea", and "identification of traits associated with drought tolerance in groundnut. The work done during his initial years of service on development of cold tolerant chickpeas has made impact in various countries including northern India and Australia. During 1990s, he was engaged, primarily in drought physiology in groundnut and contributed significantly in developing in-expansible to make selections for water-use efficiency. He was awarded a Visiting Research Fellowship for two years (1995-1997) by JIRCAS, Japan to work on the mechanism of heat tolerance in groundnut. His accomplishment includes a screening technique for heat tolerance in groundnut. Dr. Talwar joined ICAR in year 2000 as a Senior Scientist at CAZRI, RRS Bikaner and contributed significantly in elucidating the mechanism of stomatal regulation & its relationship with root traits in arid legumes. Since 2006, Dr. Talwar is with IIMR, where he is working primarily on post-flowering drought tolerance in post-rainy sorghum. Dr. Talwar participated in many national and international projects. Since 2008, he is coordinating an international collaborating project between ICRISAT-DSR-ILRI-UQ Australia on postflowering drought tolerance in post-rainy sorghum. In addition to this, he was involved in NICRA (National Initiative on Climate Resilient Agriculture) project through a sub-project funded under its competitive grant. During his fairly long carrier of >32 years, Dr. Talwar got exposure to both international and national research environments and published his research work in the form of 87 papers as journal articles, book chapters, popular / technical articles etc. in highly reputed international journals.

Dr. C. Aruna

Principal Scientist (Plant Breeding),

ICAR-Indian Institute of Millets Research (IIMR)

Hyderabad – 500 030, India

E-mail: aruna@millets.res.in

Dr. Aruna started her scientific career at Indian Institute of Millets Research (IIMR), Hyderabad in 1993. She is now working as a Principal Scientist (Plant Breeding) at IIMR. She has been working on sorghum improvement including grain and forage sorghums. In grain sorghum she is working on different aspects such as hybrid development, improvement of grain quality, and resistance for important biotic stresses. She is also working on forage sorghum improvement for yield and quality. She is acting as Principal Investigator forage sorghum breeding of AICSIP, nodal officer RFD and Co-nodal officer for DUS testing of Sorghum. She has released three high yielding kharif sorghum hybrids, CSH 16, CSH 27 and CSH 30. She has registered twelve high yielding B lines, five B lines with high brix, five B lines with improved stover quality, three high yielding R lines, one shoot fly resistant line with NBPGR. She has 14 international, 18 national journal research articles, and over 20 conference papers to her credit.

Dr. A.V. Umakanth

Principal Scientist (Plant Breeding),

ICAR-Indian Institute of Millets Research (IIMR)

Hyderabad – 500 030, India

E-mail: umakanth@millets.res.in

Dr. Umakanth started his scientific career in March 1997 at the Indian Institute of Millets Research (IIMR). Since then, he is working as a sorghum breeder involved in the genetic enhancement of sorghums and is currently the Principal Investigator for sweet sorghum breeding under the AICRP on Sorghum. He is the PI for two externally funded projects on sorghum for biofuels. He is actively involved in the genetic enhancement of sorghums for high grain and fodder yields through different approaches. He was associated in the development of an early maturing and high yielding sorghum hybrid CSH 23 and a dual-purpose sorghum variety CSV 27 in India. He has also done extensive work on brown midrib sorghums and has developed genetic stocks, utilized for fodder purposes and also as a feedstock for lignocellulosic biofuel production. He has 7 international, 40 national journal research articles and many popular articles in his publication basket.

Dr. Sujay Rakshit
Principal Scientist (Plant Breeding),
ICAR-Indian Institute of Millets Research (IIMR)
Hyderabad – 500 030, India
E-mail: rakshit@millets.res.in

Dr. Sujay Rakshit started his scientific career at Indian Institute of Pulses Research, Kanpur in 1996. In September 2000 he joined the Directorate of Maize Research, New Delhi as well as faculty member in the Division of Genetics, IARI. In November 2008 he joined the Indian Institute of Millets Research (IIMR) as Principal Scientist (Plant Breeding). He has worked on genomics and molecular breeding of chickpea, maize and rice. He was actively involved in development of DUS testing guidelines in maize. Presently he is involved in genomics-assisted sorghum improvement. He is also acting as Principal Investigator grain sorghum breeding of AICSIP, AICSIP Nodal Officer and Coordinator, AICRP-Sorghum Trials & Nurseries. He is recipient of the following recognitions: NE Borlaug Fellow (USDA) – 2010, JSPS Post-doctoral Fellowship for Foreign Researchers – 2003, BOYSCAST Fellowship – 2003, CSIR Scientist Travel Fellowship – 2002, Indian Science Congress Association Young Scientist Award – 1999, ICAR Jawaharlal Nehru Award – 1999, UNESCO Fellowship in Biotechnology – 1998. He has 15 international, 24 national journal research articles, 20 technical bulletins, 10 book chapters and over 25 conference papers in his credit.

Dr. J.S. Mishra
Principal Scientist, Agronomy,
Directorate of Sorghum Research,
Rajendra Nagar Hyderabad-500 030 (A.P.)
E-mail: jsmishra31@gmail.com

Dr J S Mishra has obtained his M. Sc. (Ag.) from NDUAT, Faizabad and Ph D (Agronomy) from JNKVV, Jabalpur. He started his career as Scientist in Agricultural Research Service (ARS) in 1991, and served for 18 years in the Directorate of Weed Science Research (DWSR), Jabalpur in various positions. He moved to the Directorate of Sorghum Research, Hyderabad in 2008 as Principal Scientist (Agronomy). He has made outstanding contributions in the field of Weed Management Research and Popularization of Sorghum Cultivation in Rice-fallows of Coastal Andhra Pradesh. His pioneering contributions are in the area of Biology of parasitic weed *Cuscuta campestis* and its management, conservation tillage and weed management in rice/soybean-based cropping systems, influence of tillage systems on weed seedbank dynamics, and identification and popularization of sorghum hybrids for rice-fallows. He has published 90 research papers, 7 reviews, 8 books, 22 book chapters, 25 technical/extension bulletins, 68 popular articles and 31 technical articles and reports.

He has also presented over 70 papers in national and international forum. Dr. Mishra is a recipient of PS Deshmukh Young Agronomist Award (1999) for Outstanding Contributions in Agronomy Research and Best Scientist Award (2007) of DWSR, Jabalpur. He is also the Fellow of Indian Society of Weed Science (2007) and Indian Society of Agronomy (2010).

Dr. B. Venkatesh Bhat
Principal Scientist (Genetics and Cytogenetics)
ICAR-Indian Institute of Millets Research (IIMR)
Hyderabad – 500 030, India
E-mail: bhatv@millets.res.in

Dr. Venkatesh Bhat obtained his doctoral degree in Genetics & Plant Breeding from UAS, Bangalore. He was at the Indian Grassland and Fodder Research Institute (IGFRI) Jhansi from 1995 till 2005 before joining the Indian Institute of Millets Research (IIMR) Hyderabad. Dr. Bhat has been doing research on genetic improvement of forage crop at IGFRI, Jhansi. He is working on genetic improvement of forage sorghum research at IIMR. He has concentrated on utilization of sweet sorghum and sudan grass lines for forage sorghum. He is also developing new male sterile lines with high biomass potential for use in forage and sweet sorghum hybrid development. He has demonstrated that by using anti-sense technology, the HCN content in sorghum forage can be substantially reduced. Dr. Bhat has also been researching on the phenomena of apomixis in plants and its utilization for sorghum improvement. He is currently developing sorghum plants with RNAi mediated somatic embryogenesis receptor kinase levels to induce elements of apomixis in sorghum. Dr. Bhat has had several research projects from DBT and NFBSRA. He has published many research articles in reputed Indian and foreign journals.

Dr. M. Elangovan
Principal Scientist (Genetic Resources),
ICAR-Indian Institute of Millets Research (IIMR)
Hyderabad – 500 030, India
E-mail: elangovan@millets.res.in

He started his career as Scientist in ICAR at CPCRI and posted at Regional Station Vittal (1997-98) and positioned as Scientist In-Charge at International Coconut Genebank for South-Asia, Seed Farm, Kidu (1998-1999) in the remote area and moved to Head Quarters and worked under AICRPP (1999-2000). Then he joined at IIMR in the year 2000. During the last 15 years, he is involved in Research, Training, Management and Coordination and activities in Sorghum Genetic Resources

Management, apart from projects in addition to DUS Testing, Plant Variety Protection, Mega seed project and IPR Management. He has visited Putra University Malaysia, Kuala Lampur during 1999, Wageningen International, The Netherlands during 2006 and National Institute of Agricultural Botany (NIAB), Cambridge University, United Kingdom during 2010. He has 3 international, 33 national journal research articles, 18 full paper articles, 16 books, 9 book chapter, 15 course materials, 29 popular articles, 6 pamphlets, 6 manuals, 28 exploration reports, 74 conference papers in his publication basket.

Dr. R. Madhusudhana

Principal Scientist (Plant Breeding)

ICAR-Indian Institute of Millets Research (IIMR)

Hyderabad – 500 030, India

E-mail: madhu@millets.res.in

Dr. R. Madhusudhana did his masters from the UAS, Dharwad (1993) and doctoral studies at IARI, New Delhi (1996) in Plant Breeding and Genetics. In 1996, he joined this institute (IIMR), and he is working as a sorghum breeder for the development of new hybrids and varieties. He was involved in the development and release of one sorghum hybrid (CSH 23), and one sorghum variety (CSV 20). He went to UK to pursue his post-doctoral work on molecular markers for sorghum improvement under DST fellowship. Later on his work is focused on the use of molecular markers for sorghum improvement. Presently, he is working on mapping of *Rf* genes in sorghum, and on MAS for shoot fly resistance, guiding students for their doctoral studies. He was PI for several DBT funded projects, and published his works in many international (15) and national (25) journal articles.

Dr. K. Hariprasanna

Senior Scientist (Plant Breeding)

ICAR-Indian Institute of Millets Research (IIMR)

Hyderabad – 500 030, India

E-mail: hari@millets.res.in

After completing M. Sc. And Ph. D. in Genetics (1996-2002) from IARI, New Delhi he joined as Research Associate under NATP-MM on Plant Biodiversity at NBPGR, New Delhi. He joined ARS of ICAR and started his career as Scientist (Plant Breeding) at IIGR, Junagadh in 2003. He joined Indian Institute of Millets Research (IIMR), as Senior Scientist (Plant Breeding) during 2008. His research areas include breeding for micronutrient enrichment in sorghum, genetic augmentation of sorghum parental lines and hybrid development, and identification of molecular markers related to combining ability and heterosis. Since August 2012 he is acting as Nodal officer for DUS testing in sorghum. He is a life member of the Indian Society of Genetics and Plant Breeding, and Society for Millets Research. He is working as referee for a number of papers from Indian and International journals, and actively involved in technical editing of Annual Report of the Directorate.

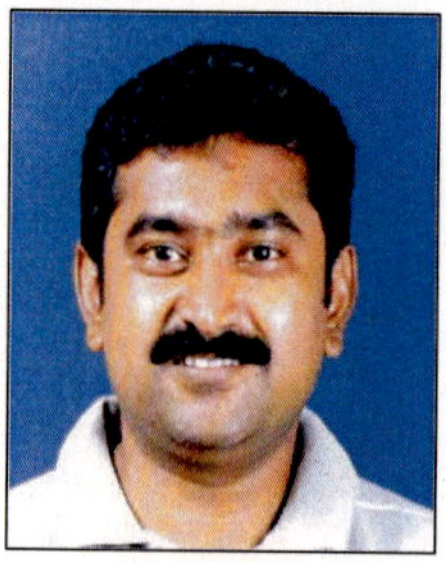

Dr. D. Balakrishna

Senior Scientist (Plant Biotechnology)

ICAR-Indian Institute of Millets Research (IIMR)

Hyderabad – 500 030, India

E-mail: balakrishna@millets.res.in

He did his masters from the University of Hyderabad, Hyderabad (Plant Science) and pursued his doctoral studies at JNTU, Hyderabad (Biotechnology). He joined Indian Institute of Millets Research (IIMR) in 1998. Since then he is working in the area of sorghum biotechnology research on tissue culture, genetic transformation and marker assisted selection programs. He developed *Bt* transgenic sorghum for stem borer resistance and evaluated under confined field conditions (BRL-1), approved by RCGM/GEAC, DBT, GOI for the first time. He also developed sorghum transgenics for salinity tolerance. He worked as a visiting scientist at ICRISAT during 2002 in Asian Development Bank (ADB) sponsored programme. He was PI for 2 and Co-PI for 3 externally funded projects, guided students for their post-graduation and doctoral studies and published 7 international journal articles.

Dr. K.N. Ganapathy

Scientist (Plant Breeding),

ICAR-Indian Institute of Millets Research (IIMR)

Hyderabad – 500 030, India

E-mail: ganapathy@millets.res.in

Dr. Ganapathy joined Indian Institute of Millets Research (IIMR), Hyderabad as Scientist (Plant Breeding) in 2009. He is involved in development of sorghum lines with improved grain quality traits especially protein digestibility by using both conventional breeding methods and genomic tools such as association mapping. He has identified promising lines for high semolina recovery, protein digestibility and more recently he is involved in identification of end product specific lines under the DSR-Britannia Project. Recently he is actively involved in development of single cut forage lines with improved forage yield and quality. He has been the Co-Nodal officer for Breeder seed and ICAR Mega Seed projects during 2012-13. He is recipient of Thiru Sivasailam award, Gold medal, TNAU merit scholarship during M.Sc.

Programme and Thiru S. Subramania Pillai Eightieth Birthday Prize award during B.Sc. programme. He has recipient of Barwale Travel grant fellowship-2006. He was also selected as Group Leader for international training on application of molecular markers and gene gun use for crop improvement. Presently he is acting as Editorial board member for Electronic Journal of Plant Breeding and reviewer for Euphytica, Australian journal of crop science, Crop and Pasture science journals. He has published 10 research articles in reputed international journals like Genetic resources and crop evolution, Euphytica, Australian Journal of Crop Science etc. and 9 publications in national research journals.

Dr. Sunil S. Gomashe

Scientist (Plant Breeding),

ICAR-Indian Institute of Millets Research (IIMR)

Hyderabad – 500 030, India

E-mail: sunil@millets.res.in

Dr. Sunil Gomashe completed his M. Sc. in Cytogenetics & Plant Breeding (2001-03) from Mahatma Phule Krishi Vidyapeeth (MPKV), Rahuri and Ph.D. in Genetics & Plant Breeding (2004-07) from Marathwada Agricultural university (MAU), Parbhani. He joined as a scientist (Plant Breeding) at Indian Institute of Millets Research, Hyderabad in 2009. At present he is involved in development of MS and R lines for earliness and photoperiod insensitivity, germplasm evaluation for Value for Cultivation and Use (VCU). He has been awarded with Dr. Panjabrao Deshmukh Memorial Gold Medal, ASPEE Gold Medal, Late. Renurao alias Annasaheb Saikhedkar Gold Medal, College of Agriculture Akola Silver Jubilee Commemoration silver medal, Shri. Govind Krishna Vaidya Silver Medal, He has published 06 international, 06 national journal research articles, and several popular articles.

Dr. I.K. Das

Principal Scientist (Plant Pathology)

ICAR-Indian Institute of Millets Research (IIMR)

Hyderabad – 500 030, India

E-mail: das@millets.res.in

Dr. Das Joined IARI, New Delhi for doing M.Sc. in Mycology & Plant Pathology (1992). From the same institute he completed PhD in Plant Pathology in 1997 and joined ICAR as ARS scientist in IIMR and posted at its sub-station at Solapur. He worked there on rabi sorghum diseases for almost eight years and moved to Headquarter during end of 2004. He has been Associate Editor for Indian Phytopathology, Reviewer for many National and International journals, recognised as guide for student in Osmania University, Nodal Officer PIMS-ICAR, Secretary

IRC, Chairman/ member of various selection committees member. He handled many Research Projects including Institutional, externally funded and ICAR-ICRISAT projects. Recently he has been appointed as Principal Investigator, Pathology for AICRP on sorghum. His major work is on grain mold management exploiting host resistance. He has published 15 research papers in International and many in national journals.

Dr. S. Ravi Kumar
Sr. Scientist (Agronomy)
ICAR-Indian Institute of Millets Research (IIMR)
Hyderabad – 500 030, India
ravikumar@millets.res.in

He is working at IIMR as agronomist since 1984. He is specialized in validation of sorghum developmental model and crop modeling, He has worked specifically on Modelling environmental effects on phenology and canopy development of diverse sorghum genotypes. He has several publications and recognitions to his credit.

Dr. B. Subbarayudu
Senior Scientist (Agrl. Entomology)
ICAR-Indian Institute of Millets Research (IIMR)
Hyderabad – 500 030, India

He is working at IIMR since 1997. He is specialized in Influence of abiotic factors on shoot bug population and plant characters offering resistance to the shoot bug, Development of crop management options for key production constraints in target sorghum production ecologies and Promotion of sorghum cultivation in rice-fallows in non- traditional areas. He has several publications and recognitions to his credit.

Dr. A. Kalaisekar
Senior Scientist (Agrl. Entomology)
ICAR-Indian Institute of Millets Research (IIMR)
Hyderabad – 500 030, India
Email : kalaisekar@millets.res.in

Dr. Kalaisekar is senior entomologist working on systematics and population biology of shoot flies since 2008 at IIMR. He has several publications and recognitions to his credit.

Dr. N.G. Malleshi
Former Head,
Dept. of Grain Science and Technology,
Central Food Technology Research Institute,
Mysore, India

He obtained his M.Sc. from Karnataka University, Dharwad and Ph.D from Dept. of Food Science and Technology from university of Mysore. He has 35 years of experience at Central Food Technological Research Institute (CFTRI) Mysore, in the areas of food Science and Technology of Cereals. Post retirement he served as consultants to FAO, USAID and UNIODO and adviser to Madras diabetes research foundation, Chennai. He has designed and developed machinery for cereals, millets and legumes. He is the Fellow of the Association of the Food Scientists and Technologists (India), Visiting Scientists Fellowships for undertaking advanced research on Cereal Science and Technology at Carlsberg Research Centre, Copenhagen, Denmark; Kansas State University, Manhattan, KS, USA; Oxford Brookes University, Oxford, UK; and at National Research Institute of Brewing, Higashi-Hiroshima, Japan. He has published about 95 articles in international and national journal, 16 books and proceedings, 21 scientific/popular articles and more than 19 national and international patents filed. He is member of a large number of professional bodies, expert committ ee of ICAR, NRDC, DST, mid-day meal programme and Editorial boards of a good number of scientific journals and also Reviewer of several International journals on Food Science and Technology.

Dr. Ch. Ravinder Reddy
Senior Scientist (Technology Exchange)
ICRISAT, Patancheru, Telangana, India.
Email : c.reddy@cgiar.org

At present, he is coordinating agriculture developmental projects funded by National and international organizations. The main activities are Planning, implementation, monitoring and evaluation. Programme Coordination with partner organisation, (private, public, national, international) in different countries; thematic and technical back stopping, providing technical, strategic advice to the partners, promote principles of innovation, liaising and support to the core team, planning training and capacity building activities to the key project actors, liaising with the donor organisations and supervising bodies. He is specialized in seed system development. He has many publications and awards to his credit.

Dr. A. Ashok Kumar

Senior Scientist (Sorghum Breeder)

ICRISAT, Patancheru, Telangana, India.

Email : a.ashokkumar@cgiar.org

Improving post rainy season sorghums for yield and adaptation and biofortification are major areas of his interest besides the resistance improvement for shoot fly and grain mold, heat tolerance among others. He has published more than 60 research papers in reputed journals and developed 2 varieties and >30 hybrid parents which are widely used by NARS, seed industry and farmers. He authored 15 book chapters, edited 4 books and guided 3 students. He received young wheat scientist award, best research center award under All India Coordinated system; research mobilizer award from ICRISAT and special recognition award from DG, ICRISAT for his contribution to ICRISAT Crop Adaptation to Climate Change. Most recently he along with the team was awarded the outstanding partnership award for the post rainy sorghum research and dissemination work in South Asia. He has many publications and awards to his credit.

Dr. Uttam D. Chavan

Professor and Senior Cereal Food Technologist

Sorghum Improvement Project

Mahatma Phule Krishi Vidyapeeth,

Rahuri, Maharashtra

Email : uttamch2002@yahoo.co.in

Most of his work is focused on nutritional quality of food crops and developing new processing technologies for cereals, pulses, oilseeds and fruits and vegetables. His innovations in developing several technologies for preparation of several value added sorghum products and for preservation of sorghum hurda and pops will have significant impact in enhancing the small farmer's incomes from growing sorghum. As a CO-PI of the project has conducted several capacity building activities for small farmers and entrepreneurs in sorghum cultivation and business development under HOPE and CRP seed system project funded by ICRISAT. He has many publications and awards to his credit.

Dr. Sharad R. Gadakh

Professor of Agriculture Botany and Senior Sorghum Breeder

Mahatma Phule Krishi Vidyapeeth, Rahuri, Maharashtra

Email : rahuri@sorghum.res.in

He is in-charge of All India Sorghum Improvement Project, MPKV Rahuri and Coordinator of Rabi Sorghum of Maharashtra State. He is also PI of HOPE Project and CRP seed systems project funded by ICRISAT, dealing R&D and Extension of research of Rabi Sorghum in Maharashtra State. His specialization is plant physiology and has worked on abiotic stresses extensively. He has many publications and awards to his credit.

Dr. H.V. Kalpande

Officer In-Charge & Sorghum Breeder

Sorghum Research Station

Vasantrao Naik Marathwada Krishi Vidyapeeth

Parbhani (M.S.) India

Email : parbhani@sorghum.res.in

Dr. Hirakant V. Kalpande, completed his Ph. D. in Genetics & Plant Breeding (1995 - 99) from Dr. Panjabrao Deshmukh Krishi Vidyapeeth (Dr. PDKV), Akola. He started his career as Plant Breeder in Mahabeej, Akola (1999- 2001). He joined as Assistant Professor during May 2001 and was selected as Associate Professor in February 2007 at Dept. of Agril. Botany, College of Agriculture, Vasantrao Naik Marathwada Krishi Vidyapeeth (VNMKV), Parbhani. He joined as Officer In-Charge and Sorghum Breeder, Sorghum Research Station, VNMKV, Parbhani in July 2013. At present he is involved in monitoring AICSIP activities as an Officer In-Charge for Sorghum breeding research and trials. He is also actively involved in ICRISAT collaborative research projects like CFC, HOPE, CRP and Harvest plus. He has several publications and recognition to his credit.